Lecture Notes in Artificial Intelligence   10938

Subseries of Lecture Notes in Computer Science

More information about this series at http://www.springer.com/series/1244

Dinh Phung · Vincent S. Tseng
Geoffrey I. Webb · Bao Ho
Mohadeseh Ganji · Lida Rashidi (Eds.)

Advances in Knowledge Discovery and Data Mining

22nd Pacific-Asia Conference, PAKDD 2018
Melbourne, VIC, Australia, June 3–6, 2018
Proceedings, Part II

Springer

Editors
Dinh Phung
Deakin University
Geelong, VIC
Australia

Vincent S. Tseng
National Chiao Tung University
Hsinchu City
Taiwan

Geoffrey I. Webb ⓘ
Monash University
Clayton, VIC
Australia

Bao Ho
Japan Advanced Institute
  of Science and Technology
Nomi, Ishikawa
Japan

Mohadeseh Ganji
University of Melbourne
Melbourne, VIC
Australia

Lida Rashidi
University of Melbourne
Melbourne, VIC
Australia

ISSN 0302-9743   ISSN 1611-3349   (electronic)
Lecture Notes in Artificial Intelligence
ISBN 978-3-319-93036-7   ISBN 978-3-319-93037-4   (eBook)
https://doi.org/10.1007/978-3-319-93037-4

Library of Congress Control Number: 2018944425

LNCS Sublibrary: SL7 – Artificial Intelligence

Printed on acid-free paper

This Springer imprint is published by the registered company Springer International Publishing AG part of Springer Nature
The registered company address is: Gewerbestrasse 11, 6330 Cham, Switzerland

PC Chairs' Preface

With its 22nd edition in 2018, the Pacific-Asia Conference on Knowledge Discovery and Data Mining is the second oldest conference and a leading venue in the area of knowledge discovery and data mining (KDD). It provides a prestigious international forum for researchers and industry practitioners to share their new ideas, original and latest research results, and practical development experiences from all KDD-related areas, including data mining, data warehousing, machine learning, artificial intelligence, deep learning, databases, statistics, knowledge engineering, visualization, and decision-making systems.

This year, we received 592 valid submissions, which is the highest number of submissions in the past 10 years. The diversity and reputation of PAKDD were also evident from the various regions from which submissions came, with over 25 different countries, noticeably from North America and Europe. Our goal was to continue to ensure a rigorous reviewing process with each paper assigned to one Senior Program Committee (SPC) member and at least three Technical Program Committee (TPC) members, resulting in an ideal minimum number of reviews of four for each paper. Owing to the unusually large number of submissions this year, we had to increase almost doubling the number of committee members, resulting in 72 SPC members and 330 TPC members. Each valid submission was reviewed by three PC members and meta-reviewed by one SPC member who also led the discussion. This required a total of approximately 2,000 reviews. The program co-chairs then considered recommendations from the SPCs, the submission, and the reviews to make the final decision. Borderline papers were discussed intensively before final decisions were made. In some cases, additional reviews were also requested.

In the end, 164 out of 592 papers were accepted, resulting in an acceptance rate of 27.9%. Among them, 58 papers were selected for long presentation and 107 papers were selected for regular presentation. This year, we introduced a new track in Deep Learning for Knowledge Discovery and Data Mining. This track was particularly popular (70 submissions); however, in the end, the number of papers accepted as the primary category for this track was moderate (six accepted papers), standing at 8.8%. The conference program contained 32 sessions in total. Long presentations were allocated 25 minutes and regular presentations 15 mins. These two types of papers, however, are not distinguished in the proceedings.

We would like to sincerely thank all SPC members, TPC members, and external reviewers for their time, effort, dedication, and services to PAKDD 2018.

April 2018

Dinh Phung
Vincent S. Tseng

General Chairs' Preface

Welcome to the proceedings of the 22nd Pacific-Asia Conference on Knowledge Discovery and Data Mining (PAKDD). This conference has a reputable tradition in bringing researchers, academia, developers, practitioners, and industry together with a focus on the Pacific-Asian regions. This year, PAKDD was held in the wonderful city of Melbourne, Australia, during June 3–6, 2018.

The single most important element of PAKDD is the technical contributions and submissions in the area of KDD. We were very pleased with the number of submissions received this year, which was well close to 600, showing a significant boost in the number of submissions and the popularity of this conference. We sincerely thank the many authors from around the world who submitted their work to the PAKDD 2018 technical program as well as its data competition and satellite workshops. In addition, PAKDD 2018 featured three high-profile keynote speakers: Professor Kate Smith-Miles, Australian Laureate Fellow from Melbourne University; Dr. Rajeev Rastogi, Director of Machine Learning at Amazon; and Professor Bing Liu from the University of Illinois at Chicago. The conference featured three tutorials and five satellite workshops in addition to a data competition sponsored by the Fourth Paradigm Inc. and ChaLean.

We would like to express our gratitude to the contribution of the SPC, TPC, and external reviewers, led by the program co-chairs, Dinh Phung and Vincent Tseng. We would like to thank the workshop co-chairs, Benjamin Fung and Can Wang; the tutorial co-chairs, Wray Buntine and Jeffrey Xu Yu; the competition co-chairs, Wei-Wei Tu and Hugo Jair Escalante; the local arrangements co-chairs, Gang Li and Wei-Luo; the publication co-chairs, Mohadeseh Ganji and Lida Rashidi; the Web and content co-chairs, Trung Le, Uyen Pham, and Khanh Nguyen; the publicity co-chairs, De-Chuan Zhan, Kozo Ohara, Kyuseok Shim, and Jeremiah Deng; and the award co-chairs, James Bailey, Bart Goethals, and Jinyan Li.

We are grateful to our sponsors: Deakin University as the host institution and gold sponsor; Monash University as the gold sponsor, University of Melbourne, Trusting Social, and the Asian Office of Aerospace Research and Development/Air Force Office of Scientific Research as silver sponsors, Springer as the publication sponsor, and the Fourth Paradigm, CodaLab and ChaLearn as the data competition sponsors.

April 2017

Tu-Bao Ho
Geoffrey I. Webb

Organization

Organizing Committee

General Co-chairs

Geoffrey I. Webb	Monash University, Australia
Bao Ho	Japan Advanced Institute of Science and Technology, Japan

Program Committee Co-chairs

Dinh Phung	Deakin University, Australia
Vincent Tseng	National Chiao Tung University, Taiwan

Tutorial Co-chairs

Wray Buntine	Monash University, Australia
Jeffrey Xu Yu	Chinese University of Hong Kong, Hong Kong, SAR China

Workshop Co-chairs

Benjamin Fung	McGill University, Canada
Can Wang	Griffith University, Australia

Data Competition Co-chairs

Wei-Wei Tu	Fourth Paradigm Inc., China
Hugo Jair Escalante	INAOE Mexico, ChaLearn, USA

Publicity Co-chairs

De-Chuan Zhan	Nanjing University, China
Kozo Ohara	Aoyama Gakuin University, Japan
Kyuseok Shim	Seoul National University, South Korea
Jeremiah Deng	University of Otago, New Zealand

Publication Co-chairs

Mohadeseh Ganji	University of Melbourne, Australia
Lida Rashidi	University of Melbourne, Australia

Local Arrangements Co-chairs

Gang Li	Deakin University, Australia
Wei Luo	Deakin University, Australia

Web and Content Co-chairs

Trung Le   Deakin University, Australia
Uyen Pham   Vietnam National University, Vietnam

Award Co-chairs

James Bailey   University of Melbourne, Australia
Bart Goethals   University of Antwerp, Belgium
Jinyan Li   University of Technology Sydney, Australia

Steering Committee

Co-chairs

Ee-Peng Lim   Singapore Management University, Singapore
Takashi Washio   Institute of Scientific and Industrial Research,
                               Osaka University, Japan

Treasurer

Longbing Cao   Advanced Analytics Institute, University
                               of Technology, Sydney, Australia

Members

Ee-Peng Lim   Singapore Management University, Singapore
                               (member since 2006, co-chair 2015–2017)
P. Krishna Reddy   International Institute of Information Technology,
                               Hyderabad (IIIT-H), India (member since 2010)
Joshua Z. Huang   Shenzhen Institutes of Advanced Technology, Chinese
                               Academy of Sciences, China (member since 2011)
Longbing Cao   Advanced Analytics Institute, University of
                               Technology, Sydney (member since 2013)
Jian Pei   Simon Fraser University, Canada (member since 2013)
Myra Spiliopoulou   Otto von Guericke University Magdeburg,
                               Germany (member since 2013)
Vincent S. Tseng   National Chiao Tung University, Taiwan
                               (member since 2014)
Tru Hoang Cao   Ho Chi Minh City University of Technology,
                               Vietnam (member since 2015)
Gill Dobbie   University of Auckland, New Zealand
                               (member since 2016)
Kyuseok Shim   Seoul National University, South Korea

Life Members

Hiroshi Motoda   AFOSR/AOARD and Osaka University, Japan
                          (member since 1997, co-chair 2001–2003, chair
                          2004–2006, life member since 2006)
Rao Kotagiri   University of Melbourne, Australia (member since
                          1997, co-chair 2006–2008, chair 2009–2011,
                          life member since 2007, treasury Co-sign since
                          2006)
Huan Liu   Arizona State University, USA (member since 1998,
                          treasurer 1998–2000, life member since 2012)
Ning Zhong   Maebashi Institute of Technology, Japan
                          (member since 1999, life member since 2008)
Masaru Kitsuregawa   Tokyo University, Japan (member since 2000,
                          life member since 2008)
David Cheung   University of Hong Kong, SAR China (member since
                          2001, treasurer 2005–2006, chair 2006–2008,
                          life member since 2009)
Graham Williams   Australian National University, Australia (member
                          since 2001, treasurer since 2006, co-chair 2009–
                          2011, chair 2012–2014, life member since 2009)
Ming-Syan Chen   National Taiwan University, Taiwan, ROC
                          (member since 2002, life member since 2010)
Kyu-Young Whang   Korea Advanced Institute of Science and Technology,
                          South Korea (member since 2003, life member since
                          2011)
Chengqi Zhang   University of Technology Sydney, Australia (member
                          since 2004, life member since 2012)
Tu Bao Ho   Japan Advanced Institute of Science and Technology,
                          Japan (member since 2005, co-chair 2012–2014,
                          chair 2015–2017, life member since 2013)
Zhi-Hua Zhou   Nanjing University, China (member since 2007,
                          life member since 2015)
Jaideep Srivastava   University of Minnesota, USA (member since 2006,
                          life member since 2015)
Takashi Washio   Institute of Scientific and Industrial Research, Osaka
                          University (member since 2008, life member since
                          2016)
Thanaruk Theeramunkong   Thammasat University, Thailand (member since 2009)

Past Members

Hongjun Lu   Hong Kong University of Science and Technology,
                          Hong Kong, SAR China (member 1997–2005)
Arbee L. P. Chen   National Chengchi University, Taiwan, ROC (member
                          2002–2009)
Takao Terano   Tokyo Insitute of Technology, Japan
                          (member 2000–2009)

Senior Program Committee

Albert Bifet	Universite Paris-Saclay, France
Andrzej Skowron	University of Warsaw, Poland
Benjamin C. M. Fung	McGill University, Canada
Byung Suk Lee	University of Vermont, USA
Chandan Reddy	Virginia Tech, USA
Chuan Shi	Beijing University of Posts and Telecommunications, China
Dat Tran	University of Canberra, Australia
Dinh Phung	Deakin University, Australia
Eibe Frank	University of Waikato, New Zealand
Feida Zhu	Singapore Management University, Singapore
Gang Li	Deakin University, Australia
Geoff Holmes	University of Waikato, New Zealand
George Karypis	University of Minnesota, USA
Guozhu Dong	Wright State University, USA
Hanghang Tong	City University of New York, USA
Hu Xia	Texas A&M University, USA
Hui Xiong	Rutgers University, USA
Jae-Gil Lee	KAIST, South Korea
James Bailey	University of Melbourne, Australia
Jeffrey Xu Yu	Chinese University of Hong Kong, Hong Kong, SAR China
Jia Wu	Macquarie University, Australia
Jian Pei	Simon Fraser University, Canada
Jianyong Wang	Tsinghua University, China
Jiliang Tang	Michigan State University, USA
Jiuyong Li	University of South Australia, Australia
Joshua Huang	Shenzhen Institutes of Advanced Technology, Chinese Academy of Sciences, China
Kai Ming Ting	Federation University, Australia
Kamalakar Karlapalem	International Institute of Information Technology, Hyderabad, India
Krishna Reddy P.	International Institute of Information Technology, Hyderabad, India
Kyuseok Shim	Seoul National University, South Korea
Latifur Khan	University of Texas at Dallas, USA
Longbing Cao	University of Technology Sydney, Australia
Masashi Sugiyama	University of Tokyo, Japan
Michael Berthold	University of Konstanz, Germany
Ming Li	Nanjing University, China
Min-Ling Zhang	Southeast University, China
Nikos Mamoulis	University of Ioannina, Greece
Niloy Ganguly	IIT, Kharagpur, India
Nitin Agarwal	University of Arkansas at Little Rock, USA

Olivier DeVel	DST, Australia
Osmar Goethals	University of Antwerp, Belgium
Patrick Gallinari	LIP6, Université Pierre et Marie Curie, France
Paul Montague	DST, Australia
Peter Christen	Australian National University, Australia
R. K. Agarwal	Jawaharlal Nehru University, India
Rajeev Raman	University of Leicester, UK
Reza Haffari	Monash University, Australia
Sang-Wook Kim	Hanyang University, South Korea
Seungwon Hwang	Yonsei University, South Korea
Shengjun Huang	Nanjing University of Aeronautics and Astronautics, China
Takashi Washio	Institute of Scientific and Industrial Research, Osaka University, Japan
Trung Le	Deakin University, Australia
Truyen Tran	Deakin University, Australia
Tu Nguyen	Deakin University, Australia
U. Kang	Seoul National University, South Korea
Vincenzo Piuri	Università degli Studi di Milano, Italy
Wee Keong Ng	Nanyang Technological University, Singapore
Wei Wang	University of California, Los Angeles, USA
Weidong Cai	University of Sydney, Australia
Wen-Chih Peng	National Chiao Tung University, Taiwan
Xiangjun Dong	Qilu University of Technology, China
Xiaofang Zhou	University of Queensland, Australia
Xiaohua Hu	Drexel University, USA
Xindong Wu	University of Vermont, USA
Xing Xie	Microsoft Research Asia, China
Xintao Wu	University of Arkansas, USA
Xuan Vinh Nguyen	University of Melbourne, Australia
Xuan-Hong Dang	IBM T. J. Watson Research Center, USA
Yan Wang	Macquarie University, Australia
Yanchun Zhang	Victoria University, Australia
Yu Zheng	Microsoft Research Asia, China
Yue Xu	Queensland University of Technology, Australia
Zhao Zhang	Soochow University, China

Program Committee

Adriel Cheng	Defence Science and Technology Group, Australia
Aijun An	York University, Canada
Aixin Sun	Nanyang Technological University, Singapore
Akihiro Inokuchi	Kwansei Gakuin University, Japan
Angelo Genovese	Università degli Studi di Milano, Italy
Anne Denton	North Dakota State University, USA

Arnaud Giacometti	François Rabelais University, France
Arnaud Soulet	François Rabelais University, France
Arthur Zimek	University of Southern Denmark, Denmark
Athanasios Nikolakopoulos	University of Minnesota, USA
Bay Vo	Ho Chi Minh City University of Technology, Vietnam
Bettina Berendt	Katholieke Universiteit Leuven, Belgium
Bin Liu	IBM T. J. Watson Research Center, USA
Bing Xue	Victoria University of Wellington, New Zealand
Bo Jin	Dalian University of Technology, China
Bolin Ding	Microsoft Research, USA
Brendon Woodford	University of Otago, New Zealand
Bruno Cremilleux	Université de Caen Normandie, France
Bum-Soo Kim	Korea University, South Korea
Canh Hao Nguyen	Kyoto University, Japan
Carson Leung	University of Manitoba, Canada
Chao Lan	University of Wyoming, USA
Chao Qian	University of Science and Technology of China, China
Chedy Raissi	Inria, France
Chen Chen	Nankai University, China
Chengzhang Zhu	University of Technology Sydney, Australia
Chenping Hou	National University of Defence Technology, China
Chia Hui Chang	National Central University, Taiwan
Choochart Haruechaiyasak	National Electronics and Computer Technology Centre, NECTEC, Thailand
Chuan Shi	Beijing University of Posts and Telecommunications, China
Chulyun Kim	Sookmyung Women's University, South Korea
Chun-Hao Chen	Tamkang University, Taiwan
Dao-Qing Dai	Sun Yat-Sen University, China
Dat Tran	University of Canberra, Australia
David Anastasiu	San José State University, USA
David Taniar	Monash University, Australia
David Tse Jung Huang	University of Auckland, New Zealand
De-Chuan Zhan	Nanjing University, China
Defu Lian	University of Electronic Science and Technology of China, China
Dejing Dou	University of Oregon, USA
De-Nian Yang	Academia Sinica, Taiwan
Dhaval Patel	IBM T. J. Watson Research Center, USA
Dinh Quoc Tran	University of North Carolina at Chapel Hill, USA
Divyesh Jadav	IBM Research, USA
Dragan Gamberger	Rudjer Boskovic Institute, Croatia
Du Zhang	California State University, USA
Duc Dung Nguyen	Institute of Information Technology, Vietnam
Elham Naghizade	University of Melbourne, Australia
Enhong Chen	University of Science and Technology of China, China

Jing Zhang	Nanjing University of Science and Technology, China
Jingrui He	IBM Research, USA
Jingwei Xu	Nanjing University, China
Jingyuan Yang	Rutgers University, USA
Joao Vinagre	LIAAD – INESC Tec, Porto, Portugal
Johannes Bloemer	University of Paderborn, Germany
Jörg Wicker	University of Auckland, New Zealand
Joyce Jiyoung Whang	Sungkyunkwan University, South Korea
Jun Gao	Peking University, China
Jun Luo	Lenovo, Hong Kong, SAR China
Junbin Gao	University of Sydney, Australia
Jundong Li	Arizona State University, USA
Jungeun Kim	KAIST, South Korea
Jun-Ki Min	Korea University of Technology and Education, South Korea
Junping Zhang	Fudan University, China
K. Selçuk Candan	Arizona State University, USA
Keith Chan	Hong Kong Polytechnic University, Hong Kong, SAR China
Kevin Bouchard	Université du Quebec a Chicoutimi, Canada
Khoat Than	Hanoi University of Science and Technology, Vietnam
Ki Yong Lee	Sookmyung Women's University, South Korea
Ki-Hoon Lee	Kwangwoon University, South Korea
Kitsana Waiyamai	Kasetsart University, Thailand
Kok-Keong Ong	La Trobe University, Australia
Kouzou Ohara	Aoyama Gakuin University, Japan
Krisztian Buza	University of Bonn, Germany
Kui Yu	University of South Australia, Australia
Kun-Ta Chuang	National Cheng Kung University, Taiwan
Kyoung-Sook Kim	Artificial Intelligence Research Centre, South Korea
Latifur Khan	University of Texas, USA
Le Wu	Hefei University of Technology, China
Lei Gu	Nanjing University of Post and Telecommunications, China
Leong Hou U	University of Macau, SAR China
Liang Hu	Jilin University, China
Liang Hu	University of Technology Sydney, Australia
Liang Wu	Arizona State University, USA
Lida Rashidi	University of Melbourne, Australia
Lijie Wen	Tsinghua University, China
Lin Liu	University of South Australia, Australia
Lin Wu	University of Queensland, Australia
Ling Chen	University of Technology Sydney, Australia
Lizhen Wang	Yunnan University, China
Long Yuan	University of New South Wales, Australia
Lu Zhang	University of Arkansas, USA

Luiza Antonie	University of Guelph, Canada
Maciej Grzenda	Warsaw University of Technology, Poland
Mahito Sugiyama	National Institute of Informatics, Japan
Mahsa Salehi	Monash University, Australia
Makoto Kato	Kyoto University, Japan
Marco Maggini	University of Siena, Italy
Marzena Kryszkiewicz	Warsaw University of Technology, Poland
Md Zahidul Islam	Charles Sturt University, Australia
Meng Chang Chen	Academia Sinica, Taiwan
Meng Jiang	University of Illinois, USA
Miao Xu	RIKEN, Japan
Michael E. Houle	National Institute of Informatics, Japan
Michael Hahsler	Southern Methodist University, USA
Ming Li	Nanjing University, China
Ming Tang	Chinese Academy of Sciences, China
Ming Yin	Microsoft Research and Purdue University, USA
Mingbo Zhao	Donghua University, China
Min-Ling Zhang	Southeast University, China
Miyuki Nakano	Advanced Institute of Industrial Technology, Japan
Mohadeseh Ganji	University of Melbourne, Australia
Mohit Sharma	Walmart Labs, USA
Mostafa Haghir Chehreghani	Telecom Paristech, France
Motoki Shiga	GIFU University, Japan
Muhammad Aamir Cheema	Monash University, Australia
Murat Kantarcioglu	University of Texas at Dallas, USA
Nam Huynh	Japan Advanced Institute of Science and Technology, Japan
Nayyar Zaidi	Monash University, Australia
Ngoc-Thanh Nguyen	Wroclaw University of Technology, Poland
Nguyen Le Minh	Japan Advanced Institute of Science and Technology, Japan
Noseong Park	University of North Carolina at Charlotte, USA
P Sastry	IISc, India
P. Krishna Reddy	International Institute of Information Technology Hyderabad, India
Pabitra Mitra	Indian Institute of Technology Kharagpur, India
Panagiotis Papapetrou	Stockholm University, Sweden
Patricia Riddle	University of Auckland, New Zealand
Peixiang Zhao	Florida State University, USA
Pengpeng Zhao	Soochow University, China
Philippe Fournier-Viger	Harbin Institute of Technology, China
Philippe Lenca	IMT Atlantique, France
Qi Liu	University of Science and Technology of China, China
Qiang Tang	Luxembourg institute of Science and Technology, Luxembourg

Qing Wang	Australian National University, Australia
Qingshan Liu	Nanjing University of Information Science and Technology, China
Ranga Vatsavai	North Carolina State University, USA
Raymond Chi-Wing Wong	Hong Kong University of Science and Technology, Hong Kong, SAR China
Reza Zafarani	Syracuse University, USA
Rong-Hua Li	Shenzhen University, China
Rui Camacho	University of Porto, Portugal
Rui Chen	Samsung Research America, USA
Sael Lee	SUNY, South Korea
Sangkeun Lee	Korea University, South Korea
Sanjay Jain	National University of Singapore, Singapore
Santu Rana	Deakin University, Australia
Sarah Erfani	University of Melbourne, Australia
Satoshi Hara	Osaka University, Japan
Satoshi Oyama	Hokkaido University, Japan
Shanika Karunasekera	University of Melbourne, Australia
Sheng Li	Adobe Research, USA
Shirui Pan	University of Technology Sydney, Australia
Shiyu Yang	University of New South Wales, Australia
Shoji Hirano	Shimane University, Japan
Shoujin Wang	University of Technology Sydney, Australia
Shu Wu	NLPR, China
Shu-Ching Chen	Florida International University, USA
Shuhan Yuan	University of Arkansas, USA
Shuigeng Zhou	Fudan University, China
Sibo Wang	University of Queensland, Australia
Silvia Chiusano	Polytechnic University of Turin, Italy
Simon James	Deakin University, Australia
Songcan Chen	Nanjing University of Aeronautics and Astronautics, China
Songlei Jian	University of Technology Sydney, Australia
Steven Ding	McGill University, Canada
Suhang Wang	Arizona State University, USA
Sunhwan Lee	IBM Research, USA
Sunil Gupta	Deakin University, Australia
Tadashi Nomoto	National Institute of Japanese Literature, Japan
Takehiro Yamamoto	Kyoto University, Japan
Takehisa Yairi	University of Tokyo, Japan
Tanmoy Chakraborty	University of Maryland, College Park, USA
Teng Zhang	Nanjing University, China
Tetsuya Yoshida	Nara Women's University, Japan
Thanh Nguyen	Deakin University, Australia
Thin Nguyen	Deakin University, Australia
Tho Quan	John Von Neumann Institute, Vietnam

Tong Xu	University of Science and Technology of China, China
Toshihiro Kamishima	National Institute of Advanced Industrial Science and Technology, Japan
Trong Dinh Thac Do	University of Technology, Sydney, Australia
Tru Cao	Ho Chi Minh City University of Technology, Vietnam
Tuan-Anh Hoang	Leibniz University of Hanover, Germany
Tzung-Pei Hong	National University of Kaohsiung, Taiwan
Vien Ngo	Queen's University Belfast, UK
Viet Huynh	Deakin University, Australia
Vincenzo Piuri	University of Milan, Italy
Vineeth Mohan	Arizona State University, USA
Vladimir Estivill-Castro	Griffith University, Australia
Wai Lam	Chinese University of Hong Kong, Hong Kong, SAR China
Wang-Chien Lee	Pennsylvania State University, USA
Wei Ding	University of Massachusetts Boston, USA
Wei Kang	University of South Australia, Australia
Wei Liu	UTS, Australia, Australia
Wei Luo	Deakin University, Australia
Wei Shen	Nankai University, China
Wei Wang	University of New South Wales, Australia
Wei Zhang	ECNU, China
Weiqing Wang	University of Queensland, Australia
Wenjie Zhang	University of New South Wales, Australia
Wilfred Ng	HKUST, China
Woong-Kee Loh	Gacheon University, South Korea
Xian Wu	Microsoft Research Asia, China
Xiangfu Meng	Liaoning Technical University, China
Xiangjun Dong	Qilu University of Technology, China
Xiangliang Zhang	King Abdullah University of Science and Technology, Saudi Arabia
Xiangmin Zhou	RMIT University, Australia
Xiangnan He	National University of Singapore, Singapore
Xiangnan Kong	Worcestor Polytechnic Institute, USA
Xiaodong Yue	Shanghai University, China, China
Xiaofeng Meng	Renmin University of China, China
Xiaohui (Daniel) Tao	University of Southern Queensland, Australia
Xiaoying Gao	Victoria University of Wellington, New Zealand
Xin Huang	Hong Kong Baptist University, Hong Kong, SAR China
Xin Wang	University of Calgary, Canada
Xingquan Zhu	Florida Atlantic University, USA
Xintao Wu	University of Arkansas, USA
Xiuzhen Zhang	RMIT University, Australia
Xuan Vinh Nguyen	University of Melbourne, Australia

Xuan-Hieu Phan	University of Engineering and Technology – VNUHN, Vietnam
Xuan-Hong Dang	UC Santa Barbara, USA
Xue Li	University of Queensland, Australia
Xuelong Li	Chinese Academy of Science, China
Xuhui Fan	University of Technology Sydney, Australia
Yaliang Li	University at Buffalo, USA
Yanchang Zhao	CSIRO, Australia
Yang Gao	Nanjing University, China
Yang Song	University of Sydney, Australia
Yang Wang	University of New South Wales, Australia
Yang Yu	Nanjing University, China
Yang-Sae Moon	Kangwon National University, South Korea
Yanjie Fu	Missouri University of Science and Technology, USA
Yao Zhou	Arizona State University, USA
Yasuhiko Morimoto	Hiroshima University, Japan
Yasuo Tabei	RIKEN Centre for Advanced Intelligent Project, Japan
Yating Zhang	RIKEN AIP Centre/NAIST, Japan
Yidong Li	Beijing Jiaotong University, China
Yi-Dong Shen	Chinese Academy of Sciences, China
Yifeng Zeng	Teesside University, UK
Yim-ming Cheung	Hong Kong Baptist University, Hong Kong, SAR China
Ying Zhang	University of New South Wales, Australia
Yi-Ping Phoebe Chen	La Trobe University, Australia
Yi-Shin Chen	National Tsing Hua University, Taiwan
Yong Guan	Iowa State University, USA
Yong Zheng	Illinois Institute of Technology, USA
Yongkai Wu	University of Arkansas, USA
Yuan Yao	Nanjing University, China
Yuanyuan Zhu	Wuhan University, China
Yücel Saygın	Sabancı University, Turkey
Yue-Shi Lee	Ming Chuan University, Taiwan
Yu-Feng Li	Nanjing University, China
Yun Sing Koh	University of Auckland, New Zealand
Yuni Xia	Indiana University – Purdue University Indianapolis (IUPUI), USA
Yuqing Sun	Shandong University, China
Zhangyang Wang	Texas A&M University, USA
Zhaohong Deng	Jiangnan University, China
Zheng Liu	Nanjing University of Posts and Telecommunications, China
Zhenhui (Jessie Li)	Pennsylvania State University, USA
Zhiyuan Chen	University of Maryland Baltimore County, USA
Zhongfei Zhang	Binghamton University, USA
Zhou Zhao	Zhejiang University, China

Zhu Xiaofeng	Guangxi Normal University, China
Zijun Yao	Rutgers University, USA
Zili Zhang	Deakin University, Australia
Josh Jia-Ching Ying	Feng Chia University, Taiwan
Ja-Hwung Su	Cheng Shiu University, Taiwan
Chun-Hao Chen	Tamkang University, Taiwan
Chih-Ya Shen	National Tsing Hua University, Taiwan
Chih-Hua Tai	National Taipei University, Taiwan
Chien-Liang Liu	National Chiao Tung University, Taiwan
Ming-Feng Tsai	National Chengchi University, Taiwan
Hon-Han Shuai	National Chiao Tung University, Taiwan
Hoang Trong Nghia	MIT, USA
Bo Dao	Deakin University, Australia
Dang Nguyen	Deakin University, Australia
Binh Nguyen	Deakin University, Australia

Sponsors

Contents – Part II

Spatial-Temporal, Time-Series and Stream Mining

Graphical Models, Latent Variables and Statistical Methods

Probabilistic Topic and Role Model for Information Diffusion in Social Network

Hengpeng Xu[1], Jinmao Wei[1(✉)], Zhenglu Yang[1(✉)], Jianhua Ruan[2], and Jun Wang[3]

[1] College of Computer and Control Engineering, Nankai University,
Tianjin 300071, China
xuhengpeng@mail.nankai.edu.cn, {weijm,yangzl}@nankai.edu.cn
[2] Department of Computer Science, University of Texas at San Antonio,
San Antonio, TX 78249, USA
jianhua.ruan@utsa.edu
[3] College of Mathematics and Statistics Science, Ludong University,
Yantai 264025, Shandong, China
junwang@mail.nankai.edu.cn

Abstract. Information diffusion, which addresses the issue of how a piece of information spreads and reaches individuals in or between networks, has attracted considerable research attention due to its widespread applications, such as viral marketing and rumor control. However, the process of information diffusion is complex and its underlying mechanism remains unclear. An important reason is that social influence takes many forms and each form may be determined by various factors. One of the major challenges is how to capture all the crucial factors of a social network such as users' interests (which can be represented as topics), users' attributes (which can be summarized as roles), and users' reposting behaviors in a unified manner to model the information diffusion process. To address the problem, we propose the joint information diffusion model (TRM) that integrates user topical interest extraction, role recognition, and information diffusion modeling into a unified framework. TRM seamlessly unifies the user topic role extraction, role recognition, and modeling of information diffusion, and then translates the calculations of individual level influence to the role-topic pairwise influence, which can provide a coarse-grained diffusion representation. Extensive experiments on two real-world datasets validate the effectiveness of our approach under various evaluation indices, which performs superior than the state-of-the-art models by a large margin.

Keywords: User topic · User role · Information diffusion
Social network

D. Phung et al. (Eds.): PAKDD 2018, LNAI 10938, pp. 3–15, 2018.
https://doi.org/10.1007/978-3-319-93037-4_1

1   Introduction

Information diffusion focuses on how a piece of information (knowledge) spreads and reaches individuals in or between networks [13,15]. The process of the information diffusion is crucial for spreading technological innovations [11], word of mouth effects in marketing [10], and opinion formulations [14]. In reality, the information diffusion process is complex, as is the influence of one user on another. Central to information diffusion is the estimation of influence strength. The strength of social influence depends on many factors, such as the characteristics and positions of individuals in the network, the impacts of the message content and temporal effects. Furthermore, not only the social influence reflects the changes of user behaviors, the user behaviors can also reflect the social influence in turn. Meanwhile, different types of social ties have essentially different influence on social actions, since users may have different attributes. Hence, how to make full use of these factors to effectively quantify the influence strength of individuals is the key problem in modeling information diffusion, especially, for the repost prediction. Namely, how should we model the information diffusion process so that the model can capture the intrinsic relations between all these elements, such as individual attributes, users' topical interests, and actions?

Nowdays, the modeling information diffusion problem has attracted much interest from researchers, and extensive efforts have been made in this field [14, 17]. From the perspective of structure of network availability, the information diffusion model can be classified into network structure-based methods and non-structure based methods. Non-structure based approaches are limited by the fact that they ignore the topology of the network and only forecast the evolution of the rate at which information globally diffuses [1]. In network structure-based methods, two representative models are used, namely, independent cascade (IC) model [6] and linear threshold model (LT) model [8]. These models assume that the network structure determines the flow of information and focuses on the structure of the process. As these two models require a diffusion probability between every two users, thus they have high computational complexity. To overcome this problem, [14,17] introduce the topic model such as LDA [3] to make users with the same topic distribution share the same behavior pattern. Xu [14] assumed that the user posting behavior is mainly influenced by three factors: breaking news, posts from social friends and user's intrinsic interests. Furthermore, topic-aware diffusion models assumed that either the topics associated with the diffusion process is specified in advance or independent with the user structural attributes [5].

Although significant progress has been made, the results of existing work are not satisfactory because of the following limitations:

1. Most information diffusion models used only portions of the available social network information. For example, Zhang [16] only considered the network structure information into consideration but ignored the differences between the users themselves, such as user's preferences or interests.

2. Highly volatile user behaviors usually cause difficulty in accurately uncovering diffusion patterns for the approaches between individuals.

3. The underlying mechanism of information diffusion remains unclear. One important reason is that social influence takes many forms and each form may be determined by different factors.

Consequently, several interesting questions emerge: Is there any dynamics or mutual influence between the three factors, user interests, social roles, and users repost behaviors? To what extent do they influence the information diffusion process? If, for example, a famous artificial intelligent expert and a normal political-science major student both retweet the same two messages, one about AlphaGO and the other one about the president Trump, will the followers of each user retweet the two different messages equally? Specifically, will the two users have the same strength of influence on their common followers? Finally, will the repost actions of followers affects their followees' post behaviors? This paper offers a new perspective.

Topics and social roles are both hidden. Pipeline approaches to extract these two factors in sequence fail to capture their interdependence. Although in recent years an array of techniques [4,15,17] have been developed for jointly leveraging these two critical factors, these techniques fall short of properly modeling the correlations between them. Besides the task of simultaneously extracting topics and social roles, we are even required to accurately characterize the role-aware topic-level information diffusion process with temporal factors.

To address the aforementioned issues, we introduce a novel TRM model that integrates user structural attributes, text of information, user repost action to uncover and explore temporal diffusion. The joint information diffusion model seamlessly unifies the extractions of user topic, role recognitions, and modeling of information diffusion. In TRM, we model topics and roles in a unified latent framework, and extract role-aware topic level influence dynamics. Furthermore, we group the users based on their structural properties and reposted information, and translate the calculations of individual level influence to the role-topic pairwise influence, which can provide a coarse-grained diffusion representation. These effective technologies facilitate our TRM model to accurately characterize the role-aware topic-level information diffusion process, and better predicts and analyzes the diffusion.

To summarize, we make the following contributions:

- We propose to integrate user structural attributes, user interests, and user repost behaviors into a unified probabilistic generative framework, which extracts the role-aware topic level influence dynamics. Concretely, we systematically study on the building joint models to explore mutual influence for user topics and roles in the process of information diffusion, and translate the calculations of individual level influence to the role-topic pairwise influence, which can provide a coarse-grained diffusion representation. It brings up a new perspective to the information diffusion process. To the best of our knowledge, such a new angle has not been studied previously.
- We introduce a latent model to uncover the hidden topics and roles as well as capture the information diffusion, which can model the process of information

diffusion better than other models. We further devise a Gibbs sampler to estimate the parameters.

- An effective diffusion prediction approach is developed which leverages the information diffusion patterns with user topics and social roles. We conduct extensive experiments to validate the proposed model over several baselines by employing two large real-world network as experimental datasets. Experimental result demonstrates that the proposed model performs much better than the state-of-the-art methods.

2 Related Work

Information Diffusion. There are two representative information diffusion models, i.e., IC model [6] and LT model [8]. Both types of models have the computational problem of selecting the set of initial users that are more likely to influence the largest number of users in the social network [12], and also have the over-fitting problem resulting from their large number of unknown parameters to learn. TRM addresses these two problems by allowing users with the same social role and user topical interests to share the same diffusion patterns, thereby significantly reducing the number of parameters.

Topic-Aware Influence. Although most of the preceding studies have utilized the network structural and timing information to model the information diffusion process, a different line of work has considered analyzing the available textural information and using the latent topics of the messages as the user's interests [10,13]. Topics are the collections of user's interests to post a message and provide the intentions for user engagement in social networks [14,16]. In [13], the authors proposed a mixture latent topic model to predict the user's reposting behaviors. Most of the topic-aware information diffusion models consider the topic of the user or the twitter, but neglect the user's structural attributes. In contrast to what these models do, the diffusion process emphasized in this study not only considers how the topical interests may influence such a process but also considers the different roles of users. In particular, the social role and user topical interest distributions of each user are not only determined by her structural attributes and the contents of the reposted messages respectively, but also by her diffusion behaviors.

Some state-of-the-art works have endeavored to combine topic model and information diffusion process, e.g., [15], as presented in the submitted paper. However, these studies usually analyze partial social network information in local observation views, i.e., neglecting the actual effects of role-topic pairs on the information diffusion process. In contrast, our TRM model pays attention to the role-aware topic level diffusion analysis, which emphasizes the interplays between user role-topic pairs and their influence on information diffusion. The advantage of our model is experimentally demonstrated in the evaluation section, i.e., Table 2, by comparing with the state-of-the-art ones.

Table 1. Notations

Symbol	Description
R, K, H, W	Number of social roles, topics, attributes and unique words in the dataset
T	The largest timestamp in a given diffusion model
N_d	The number of words in the dth messages
e_{iuv}^t	A variable denoting whether user v reposts the message i posted by user u at time t
k_d	Topic associated with post d
ϕ_v	Multinomial distribution over topics specific to user v
ψ_k	Multinomial distribution over words specific to topic k
θ_v	Multinomial distribution over roles specific to user v
λ_r	geometric distribution over Δt associate with role r
ρ_{rk}	Bernoulli distribution over decision to repost a message associate with topic k and role r
u_{rh}	Mean of h-th attribute specific to role r
δ_{rh}	Standard deviation of h-th attribute specific to role r

3   TRM Model

3.1   Formulation

Let $G = (V,E,X)$, where V is the set of all the users and $E \subseteq V \times V$ is a set of relationships between users. Each factor $e_{ij} = <v_i, v_j> \in E$ represents user v_i follows v_j, in other words, v_i is the follower of v_j and v_j is the followee of v_i in turn. Each user v_i has H-dimensional attribute vector x_i, where H is the number of all attributes. Each factor $x_{vh} \subset X$ denotes the h-th attribute of user v. We can define the user's attributes such as PageRank score [9], in degree and network constraint score [4], based on the structure of the social network. For each user $v \in V$, we use $N(v) = \{u | u \in V, e_{uv} \in E\}$ to denote the set of followees of v. For a message, whether a user activates her followers may also depend on the role she plays and the intention she chooses. The notations used in this paper are listed in Table 1.

To model the intuition that a user may have different interest topics and take different roles in the information diffusion process, we associate each with an interest topic and social role distribution respectively:

Definition 1 *Topic Distribution.* In social networks, a user is often interested in multiple topics. Formally, each user is associated with a vector $\phi_v \in V^K$, where K is the number of topics ($\sum_k \phi_{vk} = 1$).

Definition 2 *Role.* Each user may play multiple different roles, denoted as $r = [1, 2, ..., R]$. Each role has a set of parameters for the distribution that the attributes conform to. Here we use Gaussian distribution. If a user plays role r, its $h - th$ attribute conforms to $(u_{rh}, \delta_{rh}^{-1})$.

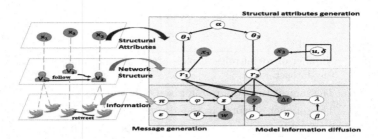

Fig. 1. Illustration of TRM: the left part depicts the input of TRM (i.e., user's structural attributes, network structure, and information on the network). The right part is the graphical model representation of TRM. The components with different color correspond to that of three processes in Sect. 3.2.

Definition 3 *Role Distribution.* Each user has a multinomial distribution over roles, which is denoted as θ. θ_v denotes the probability for user v to play role r, and is subject to $\sum_r \theta_{vr} = 1$.

Definition 4 *Topic-Role Pair.* Whether a user v would repost a message posted by her followee u depends on the role that user u plays and the topic she chooses. We use ρ to denote the distribution of topic-role pairs over reposting actions. In the information diffusion process, the actions of reposting messages only contains two cases, so we can use a Bernoulli distribution to model the distribution of topic-role pairs over actions. In other words, ρ_{rk} denotes the influence strength that a user plays role r and chooses the topic k to successfully activate one of her followers for a message.

3.2   Model Description

Based on the preceding definitions, the proposed TRM model is explained. Our goal is to devise a probabilistic generative model for extracting the user topical interests, learning user social roles, and modeling information diffusion simultaneously. Figure 1 illustrates the model. We use the content of user's reposted messages to determine her topic distribution and use the user's attributes to determine her role distribution, which are all used as priors to guide the sampling for the user's actions. Overall, the TRM model we proposed consists of three parts: the user's messages generation, the user's attributes generation, and modeling the information diffusion process.

User's Messages Generation. Here, we associate a single hidden variable with each message to indicate its topic due to the limitations in the number of characters in a single message. The generative process is described in Algorithm 1.

Social Attributes Generation. Each user may play several roles in different information diffusion processes and is subject to a certain distribution over attributes, denoted by θ_v. Each user has a random mixture of roles and can be denoted by $v = (x_1...x_h...x_H)$. The generative process of the value of attribute h for user v in a social network is described in Algorithm 2.

Model Information Diffusion Process. We introduce topic-role parameters ρ_{rk} which denotes the probability that one user plays role r and successfully activates another user specific to topic k, and a per-role parameters λ_r, which denotes the probability that cause a one-timestamp delay in information diffusion. At anytime, user v will become active and at least one of her followees activate her successfully. We use Independent Cascade Model as diffusion function in TRM model. Specifically, we first generate the influence strength and diffusion delay corresponding to ρ_{rk} and λ_r, respectively. Consider a message i posted by user u at time t, u will have only one chance to activate her follower v. The generative process of information diffusion is described in Algorithm 3

Algorithm 1. Message generation process

1: **begin**
2: **for** each message i posted by user v **do**
3:     sample the user distribution over topics, $\phi_v | \pi \sim Dir(\pi)$;
4:     sample topic indicator, $k|\phi_{vk} \sim Mul(\phi_v)$;
5:     sample topic distribution over words, $\psi_k | \varepsilon \sim Dir(\varepsilon)$;
6:     **for** each word w_{dn} in post d posted by user v, $n = 1, 2, 3, , , N_d$ **do**
7:         sample word, $w_{dn} | \psi_k \sim Mul(\psi_k)$;
8:     **end for;**
9: **end for;**
10: **end;**

Algorithm 2. User attributes generation

1: **begin**
2: **for** each user v **do**
3:     sample the user distribution over roles, $\theta_v | \alpha \sim Dir(\alpha)$;
4:     sample role indicator, $r|\theta_{vr} \sim Mul(\theta_v)$;
5:     **for** each attribute x_{rh}, $h = 1, 2, 3, , , H$ **do**
6:         sample attribute x_{rh} of user v, $x_{rh} | (u_{rh}, \delta_{rh}^{-1}) \sim N(u_{rh}, \delta_{rh}^{-1})$;
7:     **end for;**
8: **end for;**
9: **end;**

Algorithm 3. Modeling information diffusion process

1: **begin**
2: **for** each action of repost. For instance, user u posted a message i at time t, and user v who followees user u, reposted the message i at time $t' = t + \Delta t + 1$. **do**
3:     sample topic indicator, $k|\phi_{uk} \sim Mul(\phi_u)$;
4:     sample role indicator, $r|\theta_{ur} \sim Mul(\theta_u)$;
5:     sample the role-topic pair influence strength, $\rho_{rk} | \beta \sim Beta(\beta)$;
6:     sample the temporal influence of role r, $\lambda_r | \eta \sim Beta(\eta)$;
7:     sample the Δt, $\Delta t | \lambda_{rt} \sim Geo(\lambda_r)$;
8:     take a coin, sample y, $e_{iuv}^t | \rho_{rk} \sim Bern(\rho_{rk})$;
9: **end for;**
10: **end;**

3.3 Model Learning

Learning the model aims to find a configuration for the parameters $\{\theta, \phi, \rho, \lambda\}$ to maximize the log-likelihood objective function. The posterior probability of k_d, which denotes that the latent topic k for the post d of user u to activate her follower is calculated by:

$$p(k_d = k | k_{-ud}) = \frac{n_{-uk} + \pi}{\sum\limits_{k}(n_{-uk} + \pi)} \times \prod_{n=1}^{N_d}(\frac{n_{w_{dn}}^k + \beta}{\sum\limits_{w}(n_w^k + \beta)}), \qquad (1)$$

where the counter n_{uk} denotes the number of times topic k is sampled with user u, w_{dn} is the n-th word in post d, and n_w^k denotes the number of times word w is assigned to topic k. The subscript $-uk$ on the counters indicates exclusion of the current observation (resp. the message d posted by user u) from the counts. According to [2], we adopt:

$$u_{rh} = \frac{\tau_0\tau_1 + n_{rh}\overline{x}_{rh}}{\tau_1 + n_{rh}}, \quad \delta_{rh} = \frac{2\tau_2 + n_{rh}}{2\tau_3 + n_{rh}s_{rh} + \frac{\tau_1 n_{rh}(\overline{x}_{rh} - \tau_0)^2}{\tau_1 + n_{rh}}}. \qquad (2)$$

Similarly, after Gibbs sampling, parameters $\{\theta, \phi, \rho, \lambda\}$ can be estimated by:

$$\theta_{ur} = \frac{n_{ur} + \alpha}{\sum\limits_{r}(n_{ur} + \alpha)}, \quad \phi_{uk} = \frac{n_{uk} + \pi}{\sum\limits_{k}(n_{uk} + \pi)},$$

$$\rho_{rk} = \frac{n_{erk(e=1)} + \beta_1}{n_{0rk} + \beta_0 + n_{1rk} + \beta_1}, \quad \lambda_r = \frac{n_r + \eta_1}{s_r + \eta_0 + \eta_1}, \qquad (3)$$

where $\overline{r}, \overline{k}, \overline{\Delta t},$ and $\overline{e}$ respectively represent a new observation of r, k, Δt, and e. In terms of time consumption, the computation cost for the core iteration unit, i.e., sampling the user role, user topic, time delay, and the binary variable, is constant across different posts. Therefore, the time complexity for Gibbs sampling in TRM is linear w.r.t the value of $O(N * K * R)$, which is much smaller than the IC model whose complexity is linear of $O(N * N)$, since the value of $K * R$ is far less than the value of user number N.

4 Experiments

4.1 Experimental Setup

Datasets. We evaluate the effectiveness of the proposed model on two real-world datasets belonging to two different social networks:

- **Weibo** is a dataset from Sina Weibo, the largest microblogging service in China. The Weibo data we used in our experiment is from [11] with 66,348 users and 13,487,120 repost actions. We select the original posts that were reposted by more than 6 users, and use the remaining 129,560 original posts for experiments. For a given tweet from a user, we would like to predict who will repost the tweet.

Table 2. Performance comparison on two datasets evaluated by P@N and MAP.

Weibo					CND			
Method	P@10	P@50	P@100	MAP	P@10	P@50	P@100	MAP
Count	0.007	0.006	0.006	0.013	0.089	0.029	0.017	0.127
LDA	0.112	0.039	0.020	0.085	**0.153**	**0.049**	**0.030**	0.198
MUPB	0.405	0.137	0.079	0.415	0.122	0.038	0.022	0.307
Rain	0.407	0.146	0.083	0.427	0.121	0.038	0.021	0.299
TRM	**0.429**	**0.156**	**0.088**	**0.458**	0.143	0.043	0.024	**0.345**

- **Citation Network Database (CND)** is extracted from DBLP, ACM, MicroSoft Academic Graph, and other sources [17]. We select the original paper that was cited by more than 6 users, and use the remaining 67,414 original papers for experiments. For a given paper, we would like to predict which author will cite this paper next.

Since the retweet or cite action prediction is much similar to a ranking problem, we prefer the precision at top ranked results as the evaluations of our proposed model. Specifically, given a message or a paper i produced by user v, we calculate the reposting or citation probability of each of $v's$ followers, and we use P@10(precision of top-10 predictions), P@50, P@100, and Mean Average Precision(MAP) to evaluate the ranking prediction results for each message and aggregate the results for all messages together.

Baselines. We compare TRM with several representative methods for user's prediction Count, LDA [3], MUPB [14] and Rain [15].

- **Count:** Here, the probability of a user reposting a message is in direct proportion to the number of followees who have reposted message i.
- **LDA:** In LDA [3], the probability of a user reposting a message based on the topic distributions of message i and v's all reposted posts in the past.
- **MUPB:** Although MUPB [14] can help extract the topics of users to a certain extent, it fails to capture the real motivation of users to publish content, because user behavior can easily be affected by the structure of the network, other than user interest.
- **Rain:** Rain [15] predicts whether user $v's$ will repost message i based on the user $v's$ role distribution and the information diffusion attributes corresponding to each role.
- **TRM:** This is the proposed method.

Following some state-of-the-art works [3,15,17], we fix the hyper-parameters in the experiments for both TRM and the baselines for a fair comparison. We set the model parameters as $R = 10$, $K = 10$, $\alpha = 0.1$, $\pi = 0.1$, $\varepsilon = 0.1$, $\beta = (1,1)$, and $\eta = (1,1)$.

4.2   Experimental Results

Better Performance. The performance comparison of the two datasets evaluated by P@10, P@50, P@100 and MAP is illustrated in Table 2. We can discover that the TRM model clearly outperforms Count, LDA, MUPB, and Rain on nearly all metrics in Weibo (+0.076 $\sim$ 0.445 improvement in terms of average MAP) and CND (+0.049 $\sim$ 0.218 improvement in terms of average MAP). Due to the lack of supervised information, Count performs worst on both datasets, whereas the Count and LDA model outperform themselves better in CND than in Weibo. Since an author usually only focuses on one or two fields of study, the users in CND database usually cite the papers with the similar topics. The Count and LDA model all ignore the user's social structural attributes. Prediction of the user's reposting action based on LDA only depends on the user's history reposts, and ignores the situation where a user needs or topic distributions may change over time. TRM also outperforms MUPB and Rain on all metrics. Although MUPB and Rain also considers user topics and social roles, respectively, they still ignore the correlations and mutual influence between topics and roles.

Effect of Mutual Influence. We also examine the nature and the effectiveness of the associated latent factors on the mutual influence between user topics and social roles, and Fig. 2 demonstrates their feasibility in our modeling. Note that if we do not incorporate the latent role or topic factor, our TRM model becomes the traditional MUPB or Rain approach, respectively. This condition shows that the latent factors consistently enhance the precision (evaluated by P@10,P@50 and MAP) for the repost prediction. For example, the latent topic and role factor respectively improve the MAP by 4.8% (from 29.5% to 34.3%), and improve the MAP by 3.5% (from 41.5% to 45.8%) for the repost prediction. These results also illustrate that the user's topics and social roles are very crucial to modelling the information diffusion.

Social Role Analysis. The learned parameters ρ represent the influence strength of a user for different topics and roles. The method also learns u_{rh}, which denotes the mean value of social attribute k for role r, and ρ_{rk} denotes the topic-role pair activation probability of topic k for role r. Thus, we can use the $P(r) = \sum_k \rho_{rk}$ to denotes the influence strength of role r. Figure 3 shows the correlations between a role's social attribute and its influence strength. We discover that the correlation follows a logarithm function. We try different forms

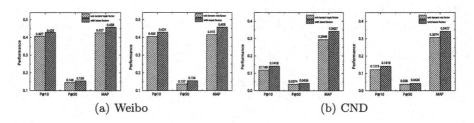

(a) Weibo   (b) CND

Fig. 2. The contribution of topic and role on repost prediction.

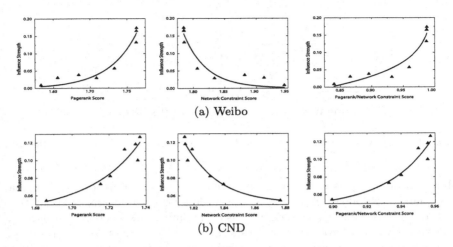

(a) Weibo

(b) CND

Fig. 3. Role analysis on two datasets: the correlation of influence strength with *Pagerank, Network Constraint and Pagerank/Network Constraint Score.*

of functions to fit the remaining data points and select the logarithm function of R^2. Similar to Fig. 3(a), (b) shows that people who have a higher PageRank score or smaller network constraint score will have a stronger influence in information diffusion than ordinary people. The reason is that people with a higher PageRank score tend to have a larger number of followers, and their posted messages are more likely to be reposted. People with a smaller network constrain score tend to be a structural hole spanner that connects two or more communities, and their posted messages are more easily be propagated to different social network communities.

Correlation Between User Topic and Social Role. Inspired by the work of [4, 9, 15, 17], we classify users into three groups according to their structural properties, i.e., network constrain score and Pagerank score. For example, we assume that the user with small network constrain score tends to be a structural hole spanner that connects two or more communities. Similarly, the user with high PageRank score may be an opinion leader. Furthermore, the learned parameter ϕ represents the topic distribution for different users. Inspired by [7], we compute the entropy of user's topical distributions to measure how much topical a user's interests or topics are. For a user v, the entropy of her topical interests distribution is computed as follows: $Hp(v) = \sum_k \phi_{vk} \log(\phi_{vk})$. To further analyze the correlations between users' topics and roles, we continue to calculate the average entropy for each role as follows: $Hp(r) = 1/|N_r| \sum_{v \in A_r} Hp(v)$, where A_r denotes the set of users to be assigned to role r, and N_r denotes the number of users in A_r. The higher the entropy, the less topical the role is. Figure 4 demonstrates the correlations between user topics and roles. The higher the entropy, the less topical the role is. Thus, the most topical would be a user that is interested in only a single topic, whereas the least topical would be a user that is

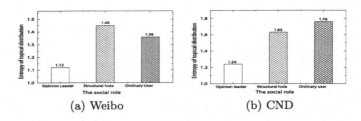

(a) Weibo   (b) CND

Fig. 4. The average entropy of topical distributions for each role on two datasets.

interested in all topics with equally preferences. The phenomenon in Fig. 4 may be explained as follows: most of opinion leaders post the messages about their areas of expertise and they tend to focus on regional and specialized topics, the structural spanners have broad interests than opinion leaders because they usually focus on more general topics that tend to propagate from one community to another more easily, and the ordinary users have more broader interests because they behave more randomly. In Fig. 4(b), the structural spanners have a higher entropy value than ordinary users. The phenomenon may be explained as follows: when a person has published many papers in different regions, she may become more open-minded and tend to accept new ideas from others.

5 Conclusions

In this paper, We propose to integrate user structural attributes, user interests, and user repost behaviors into a unified probabilistic generative framework, which extracts the role-aware topic level influence dynamics. Experimental result demonstrates that the proposed model performs much better than the state-of-the-art methods.

Acknowledgements. This work was supported in part by the National Natural Science Foundation of China under Grant No. 61772288, U1636116 and 11431006, and the Research Fund for International Young Scientists under Grant No. 61650110510 and 61750110530.

References

1. Aslay, C., Barbieri, N., Bonchi, F., Baeza-Yates, R.A.: Online topic-aware influence maximization queries. In: Proceedings of the 17th EDBT, pp. 295–306 (2014)
2. Bernardo, J.M., Smith, A.: Bayesian Theory. Wiley, New York (2009)
3. Blei, D.M., Ng, A.Y., Jordan, M.I.: Latent Dirichlet allocation. J. Mach. Learn. Res. **3**, 993–1022 (2003)
4. Burt, R.S.: Structural Holes: The Social Structure of Competition. Harvard University Press, Cambridge (2009)
5. Chen, S., Fan, J., Li, G., Feng, J., Tan, K.L., Tang, J.: Online topic-aware influence maximization. PVLDB **8**(6), 666–677 (2015)

6. Goldenberg, J., Libai, B., Muller, E.: Talk of the network: a complex systems look at the underlying process of word-of-mouth. Mark. Lett. **12**(3), 211–223 (2001)
7. Grabowicz, P.A., Ganguly, N., Gummadi, K.P.: Distinguishing between topical and non-topical information diffusion mechanisms in social media. In: Proceedings of the 10th ICWSM, pp. 151–160 (2016)
8. Granovetter, M.: Threshold models of collective behavior. Am. J. Sociol. **83**(6), 1420–1443 (1978)
9. Page, L., Brin, S., Motwani, R., Winograd, T.: The PageRank citation ranking: bringing order to the web. Technical report (1999)
10. Tang, J., Sun, J., Wang, C., Yang, Z.: Social influence analysis in large-scale networks. In: Proceedings of the 15th ACM SIGKDD, pp. 807–816 (2009)
11. Tu, C., Liu, Z., Sun, M.: Prism: profession identification in social media with personal information and community structure. In: Proceedings of the 4th SMP, pp. 15–27 (2015)
12. Wu, S., Hofman, J.M., Mason, W.A., Watts, D.J.: Who says what to whom on Twitter. In: Proceedings of the 20th WWW, pp. 705–714 (2011)
13. Xiong, F., Liu, Y., Zhang, Z., Zhu, J., Zhang, Y.: An information diffusion model based on retweeting mechanism for online social media. Phys. Lett. A **376**(30), 2103–2108 (2012)
14. Xu, Z., Zhang, Y., Wu, Y., Yang, Q.: Modeling user posting behavior on social media. In: Proceedings of the 35th SIGIR, pp. 545–554 (2012)
15. Yang, Y., Tang, J., Leung, C.W.-k., Sun, Y., Chen, Q., Li, J., Yang, Q.: Rain: social role-aware information diffusion. In: Proceedings of the 29th AAAI, pp. 367–373 (2015)
16. Zhang, J., Liu, B., Tang, J., Chen, T., Li, J.: Social influence locality for modeling retweeting behaviors. In: Proceedings of the 23rd IJCAI, pp. 2761–2767 (2013)
17. Zhang, J., Tang, J., Zhuang, H., Leung, C.W.K., Li, J.: Role-aware conformity influence modeling and analysis in social networks. In: Proceedings of the 28th AAAI, pp. 958–965 (2014)

Topic-Sensitive Influential Paper Discovery in Citation Network

Xin Huang, Chang-an Chen, Changhuan Peng, Xudong Wu, Luoyi Fu,
and Xinbing Wang[✉]

Shanghai Jiao Tong University, Shanghai, China
hxin18@gmail.com, {chen-chang-an,598095762,xudongwu,
yiluofu,xwang8}@sjtu.edu.cn

Abstract. Discovering important papers in different academic topics is known as topic-sensitive influential paper discovery. Previous works mainly find the influential papers based on the structure of citation networks but neglect the text information, while the text of documents gives a more precise description of topics. In our paper, we creatively combine both topics of text and the influence of topics over citation networks to discover influential articles. The observation on three standard citation networks shows that the existence of citations between papers is related to the topic of citing papers and the importance of cited papers. Based on this finding, we introduce two parameters to describe the topic distribution and the importance of a document. We then propose MTID, a scalable generative model, which generates the network with these two parameters. The experiment confirms superiority of MTID over other topic-based methods, in terms of at least 50% better citation prediction in recall, precision and mean reciprocal rank. In discovering influential articles in different topics, MTID not only identifies papers with high citations, but also succeeds in discovering other important papers, including papers about standard datasets and the rising stars.

Keywords: Citation network · Generative model
Academic recommendation

1 Introduction

In academic research, the prior arts are essential for the future works. One bottleneck in research is that as the amount of available scientific literatures on the Internet becomes larger, it would be increasingly difficult for researchers to identify the masterpiece among numerous papers. This problem becomes even more complicated given the fact that important papers are only influential in one or several domains of knowledge. As a result, how to effectively identify the milestone papers in different academic topics is a crucial task in data mining.

The goal of finding important papers in different academic topics is to discover documents which are of great significance in a specific topic. In the

© Springer International Publishing AG, part of Springer Nature 2018
D. Phung et al. (Eds.): PAKDD 2018, LNAI 10938, pp. 16–28, 2018.
https://doi.org/10.1007/978-3-319-93037-4_2

researches of citation network, most works try to use the network to discover the interaction [3,8] and the evolution of topics [5,15] in the collection of documents, while little attention is paid to finding influencers in different academic topics. Among limited number of works, which indeed focus on influential paper discovery in citation network, Wang et al. [15] adopt Latent Dirichlet Allocation (LDA) [1] to generate citation networks. They view the reference of a paper as a "bag of citation" and learn the topic-document distribution from the citation network. This distribution describes the importance of documents in a particular topic. Lu et al. [6] extend the method by taking into account additional factors that influence the importance of papers, such as authorship and published venues. The model proposed by Lu et al. [6] could discover the important papers for different topics, authors and venues.

In spite of [6,15] mentioned above, finding influential nodes in citation networks remains an open problem. One direction is to add topics of textual information into influencer detection. The existing works only take into consideration the network structure and ignore the text. Although [6,15] use "topic" in the description of their methods, the topic defined in [6,15] is actually a cluster of documents. He et al. [4] describe this kind of topic as "DocTopic", which could be simultaneously related to distinct "WordTopics", i.e. topics extracted from the text. Thus, the topics described in [6,15] are too general but imprecise. The other direction is to determine the important factors that affect papers to cite. As for this direction, we conduct an observation on three standard citation networks. The result shows that whether one paper cites another depends on the topic of the citing paper and the importance of the cited paper.

In this paper, we study the problem of influential paper discovery in citation networks. One feature that distinguishes our method from other related works is that we solve this problem by covering both directions mentioned in the previous discussion. During this process, a challenge is to precisely describe the factors that affect papers to cite. To accomplish this, we introduce two parameters to represent the topic distribution and the importance of a document. Based on these two parameters, we generate citation networks. While we defer a more detailed description of our methods in Sect. 4, we would like to point out that our method succeeds in adding the topic of papers into the generation of citation networks. During this process, the importance of papers is learnt from the data. The topic of node could be obtained by topic modeling, e.g. LDA, or any other methods that transform a document into a topic vector. Another advantage is that our learning schema could be separated into a set of independent convex optimization problems. This propriety indicates that our model is scalable and easy to initiate. The following three aspects are our core contributions.

- We conduct an empirical observation based on three standard datasets: AAN, DBLP and ACM. There are two fundamental conclusions. One reveals that papers with similar topic distribution are likely to cite similar papers as references. The other states that papers with high citations are likely to be selected as the papers to refer.

- We propose a new, robust model: Model for Topic-sensitive Influential paper Discovery (MTID), which is parallelizable and compatible to all methods representing documents with topic vectors. MTID is inspired by our observation and models the citation network with two parameters of papers: (1) An importance parameter, M, that captures the importance of cited papers (2) A topic parameter, N, which describes the topic distribution of citing papers.
- We evaluate our model on citation prediction and influential paper discovery. The first part proves that our model outperforms other topic-based citation prediction methods with an improvement over 50% in recall, precision and mean reciprocal rank. In the second part, we not only effectively identify the papers with high citations, but also discovered other important papers such as papers about standard datasets and the rising stars for an academic topic.

The rest of this paper is organized as follows: Sect. 2 summarizes the related work. Section 3 presents our empirical observations about how topics of documents influence their citations. Section 4 introduces the MTID along with the learning method. Then, we report the experimental results and the model applications in Sect. 5. Finally, the conclusion is presented in Sect. 6.

2   Related Work

In recent years, with increasing number of digital libraries such as ACM Digital Library[1] and DBLP[2] come into use. There is a growing interest on the analysis of citation networks in the research community.

As the citation network is a kind of network with rich textual information, one important direction in the study of citation network is to use the network structure in understanding the topic of text. In this direction, some researchers extend the classical topic model to joint versions in order to model both text and citation for documents. These works succeed in enhancing the quality of topics [12] and reflecting the interaction among topics [3,8]. Others make use of the characteristic that papers could only cite papers published earlier to study the evolution of topics in academic fields [5,15].

Another important direction is to detect the influential papers in the citation network. To assess the importance of papers, ranking algorithms such as PageRank and its variants are applied [11,14]. These methods, however, detect the general influential papers in citation networks and ignore the topic context of documents. In fact, as a document only contains information in one or several knowledge domains, the influential papers vary in different topics. As a result, discovering influencers in different topics is of great value in the citation network.

While most works for topic-sensitive influential node discovery in networks aim at identifying important users in social networks [9,16], little attention is paid to citation networks. Among limited amount of related works, Wang et al. [15] use topic model to generate citation networks. They introduce the notion

---

[1] https://dl.acm.org/.
[2] http://dblp.uni-trier.de/.

of "bag of citation" and consider a topic as a mixture of documents. Then, they learn the topic-document distribution from the citation network. The distribution describes the importance of documents in a specific topic. Lu et al. [6] extend the method by considering additional factors that influence the importance of papers, such as authorship and published venues. The model proposed by Lu et al. [6] discovers the important papers for different topics, authors and venues.

In our model, different from [6,15], we use the topics extracted from textual information. In this way, we can make full use of the rich information in text. Another difference lies in the generation of networks. In [6,15], the reference for a document is determined by sampling cited documents according to the topic-document distribution. In this case, the same document could appear more than once in the reference. Our model, however, overcomes this problem by modeling the probability of whether one paper cites another.

3   Empirical Observation

One important direction of discovering important papers in the citation network is to figure out how papers cite other papers. In this section, we adopt empirical observations on the academic network to discover the factors that affect papers to connect with each other by citations. In general, we mainly cope with two important questions. How topics of a document affect the way it cites? Which kind of documents are frequently cited?

We observe three academic datasets: AAN, DBLP and ACM, the detailed description is shown in Sect. 5.1. For each paper, we extract the topic from the text with LDA [1]. According to the topic diversity of papers in datasets, we set the number of topics to 10 for AAN and 100 for DBLP and ACM.

First, we analyze how topics affect the way documents cite. To do this, for papers published last year in each dataset, we select two papers with more than 10 references and calculate the cosine similarity of their topic distributions. The higher the value is, the more similar the articles are. We repeat this process among all paper pairs of the last-year publication in each dataset. Then, we divide the pairs into 8 different parts according to the topic similarity. Finally we analyze the relation between the size of the overlapping references and their topic similarity within the document pairs. Figure 1(a), (b) and (c) respectively show the result of dataset AAN, DBLP, and ACM. For DBLP and ACM, due

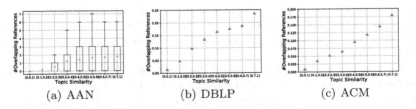

|(a) AAN|(b) DBLP|(c) ACM|

Fig. 1. The relation between topic similarity and overlapping references

to the large topic diversity, only a tiny portion of paper pairs have overlapping references. Thus, we only plot the average number of overlapping references for these two datasets. In these figures, the number of overlapping references grows when the similarity of document pairs increases. The result shows that papers with similar topic distribution are likely to cite similar papers.

Second, we analyze, in the citation network, what documents are likely to be cited. For early publications, e.g. papers published in the first three years, we construct two sets based on a timestamp, e.g. the penultimate year. The first set contains the citations before this timestamp, the second set includes citations after this timestamp. For example, for an early publication in ACM dataset, the first set contains citations before 2007 and the second set includes citations in or after 2007. We then compare the size of these two sets for each early publication.

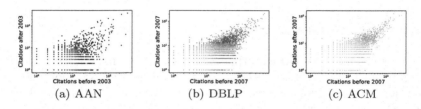

(a) AAN   (b) DBLP   (c) ACM

Fig. 2. The relation between exiting citations and incremental citations

Figure 2 shows the relation of the size of two citation sets. It shows that the number of citations in later years is positively correlated to that in early years. In other words, papers with high citations are likely to be selected as the papers to refer. As the number of citations is positively related to the importance of a paper, papers with higher importance are more frequently to be cited.

4   Proposed Model

4.1   Generation of Citation Network

Based on previous observations, we can conclude that whether papers are linked in citation networks is related to the topic of the citing paper and the importance of the cited paper. The former is the intrinsic characteristic of the article and the latter depends on the network structure.

In order to precisely represent these two factors that affect papers to cite, we introduce two new parameters, $\mathbf{N}$ and $\mathbf{M}$, to represent respectively the topic distribution and the importance of a document.

$\mathbf{N}$ is the parameter which represents the topical distribution of citing papers. It can be labelled manually or extracted from the text. N_u is a column vector which describes the topic representation of document u and N_{ui} represents how likely the topic i could describe the document u. In topic modeling, N_u is the document-topic distribution in the document u and N_{ui} is the proportion of topic i in this document.

$\mathbf{M}$ is the parameter which represents the importance of papers in different topics. It describes how likely a paper would be cited. M_v is a column vector which represents the importance of document v in different topics. $M_{vj} = 0$ indicates that the document v is not important in topic j. The larger this parameter is, the more important the paper is in the topic.

We then present MTID (Model for Topic-sensitive Influential paper Discovery), a probabilistic model which generates the citation network and models the importance of papers in different topics simultaneously. In the generating process, MTID follows the idea presented in [17] and models the citation network with Poisson distribution.

Suppose that in a citation network, a non-negative random variable X_{uv} represents the latent connection strength for the pair of papers (u, v). We define that paper u cites paper v if and only if $X_{uv} > 0$. Now we consider the case of a single topic. We define X_{uv}^i as the random variable of latent connection strength in topic i for the pair of papers (u, v), this random variable follows the Poisson distribution with the parameter $N_{ui} \cdot M_{vi}$. Then the total connection strength X_{uv} is the sum of X_{uv}^i, with the additivity of the Poisson random variable:

$$X_{uv} = \sum_{i=1}^{K} X_{uv}^i \qquad X_{uv}^i \sim Poission(N_{ui} \cdot M_{vi})$$

The total connection strength X_{uv} follows the Poisson distribution with the parameter $\sum_{i=1}^{K} N_{ui} \cdot M_{vi}$, where K denotes the number of topics. The probability $P(u \rightarrow v)$ is defined as follows:

$$P(X_{uv} > 0) = 1 - P(X_{uv} = 0) = 1 - \exp\left(-\sum_{i=1}^{K} N_{ui} \cdot M_{vi}\right) = 1 - \exp\left(-M_v^\mathsf{T} N_u\right)$$

Finally, MTID learns the importance matrix $\mathbf{M}$ and maximizes the log likelihood of the observed network G. The problem could be formalized as follows:

$$\hat{\mathbf{M}} = \arg\max(L(\mathbf{M})) \tag{1}$$

where the nonnegative matrix $\mathbf{M} \in \mathbb{R}^{K \times N}$ and K, N denote the number of topics and nodes, respectively. The log likelihood can be written as below:

$$L(\mathbf{M}) = \sum_v \left\{ \sum_{u \in R_v} \log(1 - \exp(-M_v^\mathsf{T} N_u)) - \sum_{u \notin R_v, u \in C_v} M_v^\mathsf{T} N_u \right\} \tag{2}$$

R_v is a set of papers that cite paper v and C_v is a set of papers that are published later than paper v.

4.2   Parameter Learning

We solve the optimization problem defined in Eq. 1 through block coordinate gradient ascent. At each iteration, we update the importance vector M_v for each paper v with M_u for all other papers $u \neq v$ fixed. To update the importance parameter M_v for paper v, we solve the following subproblem:

$$\hat{M}_v = \arg\max(L(M_v)) \tag{3}$$

where $L(M_v)$ is the part of $L(M)$ defined in Eq. 2 that involves M_v, i.e.,

$$L(M_v) = \sum_{u \in R_v} \log(1 - \exp(-M_v^\mathsf{T} N_u)) - \sum_{u \notin R_v, v \in C_v} M_v^\mathsf{T} N_u \tag{4}$$

Noticing M_v is a non-negative vector, this subproblem can be further solved by projected gradient ascent.

$$M_{v_{new}} \leftarrow \max(0, M_{v_{old}} + \alpha_{M_v}(\nabla L(M_v))$$

α_{M_v} is the step size computed by backtracking line search [2], and the gradient is:

$$\frac{dL(M_v)}{dM_v} = \sum_{u \in R_v} N_u \frac{\exp(-M_v^\mathsf{T} N_u)}{1 - \exp(-M_v^\mathsf{T} N_u)} - \sum_{u \notin R_v, u \in C_v} N_u \tag{5}$$

During the iterations, only the calculation of the first term in Eq. 5 is required and the second term is a constant given a paper v. This constant can be computed in $O(In_{degree}(v))$ time according to Eq. 6 and cached during the training process.

$$\sum_{u \notin R_v, v \in C_v} N_u = \sum_{u \in C_v} N_u - \sum_{u \in R_v} N_u \tag{6}$$

Thus, the computation of Eq. 5 requires $O(In_{degree}(v))$ time. As the real-world citation networks are extremely sparse ($In_{degree}(v) \ll N$), we can update M_v for each iteration in near-constant time.

4.3   Initialization and Parallelization

One advantage of our model is that the optimization problem shown in Eq. 4 is concave. In this case, parameters will converge to the same result with different initializations, thus we could randomly initiate the vector M_v for each paper v. Another advantage is that our approach also allows for parallelization, which further increases the scalability of MTID. When updating M_v for each paper v, we observe that each subproblem is separable. That is, updating the value of M_v for a specific node v does not affect the updates of M_u for all other nodes u. Consequencely, parallelization does not affect the final result of the model. The implementation is available in https://github.com/hxin18/mtid.

5   Experiment

5.1   Dataset

We evaluate our model with three citation networks, AAN, DBLP and ACM. We use LDA [1] to extract topics from the text of documents. We note that, at the same time, topics extracted by other topic modeling methods are also compatible with our model and tend to have similar results.

ACL Anthology Network. ACL Anthology Network (AAN) [10] is a dataset which includes papers about *Natural Language Processing*. AAN includes 19,435 papers published from 1980 to 2013 with full text and reference. For the text, we remove invalid tokens and stop words. As AAN only contains papers in one scientific field, topics detected in AAN are more specific and some of them are quite similar. As a result, we set the number of topics to 10.

DataBase Systems and Logic Programming. DataBase systems and Logic Programming (DBLP) [13] is a dataset on computer science journals and proceedings. From the dataset, we extract 298,840 papers published from 1996 to 2007 with abstract and reference to build the training set. For the textual information, we remove the invalid tokens and stop words. DBLP contains papers in all sub-fields of computer science. As a result, we set the number of topics to 100.

Association for Computing Machinery. Association for Computing Machinery (ACM) [13] is an online dataset on computer science journals and proceedings. From the dataset, we extract 413,373 papers published from 2003 to 2007 with abstract and reference to build the training set. For the text, we remove the invalid tokens and stop words. ACM contains papers in all sub-fields of computer science, as a result, we set the number of topics to 100.

5.2   Citation Prediction

In this section, we evaluate MTID by predicting citations for new documents. The whole dataset is divided into a training set and a test set. For the test set of each dataset, we include the late publications with at least 10 references. We fit our model with the training set and predict citations for the papers in the test set.

Procedure. For a new query document with topic distribution N_{new}, the MTID recommends citations among existing papers based on the importance parameter $\mathbf{M}$. In details, we compute the citation strength of each existing paper to the query document, then we rank the papers according to the strength and recommend them based on the ranking. The strength is defined as $S_d = M_d^{\mathsf{T}} N_{new}$.

Baselines. We utilize random selection and three topic-based citation prediction methods as baselines:

- **Random:** For each query in test set, we randomly recommend papers to cite. The result of this method is the average of 10 measures.
- **TopicSim:** TopicSim is to compare the topic similarity between queries and the cited papers. For each query, it returns the papers with the most similar topic distribution. The topic distribution of documents is measured by LDA.
- **Link-PLSA-LDA:** Link-PLSA-LDA [8] is a mixed membership method that models both text and citation. In the citation prediction, the cited papers are ranked in terms of the conditional probability of citations associated with the topic distribution of query.
- **Topic PageRank:** This method considers not only the topic similarity between queries and the cited papers, but also the importance of cited papers in the network. For a query, cited papers are ranked in terms of the multiplication of the weight of cited documents in PageRank and the similarity between cited documents and queries.

Metric. We adopt Precision and Recall at number N (P@N and R@N) as the evaluation metrics for citation prediction. R@M is defined as the percentage of correct references that appear in the top-N prediction. P@N is used to quantify whether correct references are ranked top for the query. A higher recall and precision indicate a better result. Furthermore, it is important that ground-truth references appear earlier in the prediction. Therefore, we adopt Mean Reciprocal Rank (MRR) as a metric. The MRR is defined as $\frac{1}{|S_{test}|}\sum_{d\in S_{test}}\frac{1}{rank(d)}$, where S_{test} denotes the test set and $rank(d)$ denotes the rank of first correct citation for query d.

Result. Table 1 shows the result of citation prediction for three datasets. Topic-based methods significantly outperform random selection. Among all topic-based models, TopicSim performs the worst because it only exploits information in text. For other three methods that consider both text and citation, MTID significantly outperforms other two methods. We can also notice that performance on AAN is much better than other two datasets. It is because that DBLP and ACM are large networks with wider range of topics. This makes the prediction more difficult. The result proves the effectiveness of MTID in citation prediction.

5.3 Finding Influential Papers

In this section, we adopt MTID in discovering influential papers of different topics in the citation network. To do this, for each topic, we rank the papers according to the importance in this topic, which can be reflected by the parameter **M** in our model. Tables 2, 3 and 4 display five most important papers of three topics selected in each dataset, the keywords of topic are displayed in the left of the table. For each topic i, we rank the papers according to the value of M_i. In general, the importance

Table 1. Result of citation prediction

Dataset	Model	P@10	P@20	R@20	R@50	MRR
AAN	Random	0.000967	0.000896	0.001381	0.003419	0.006369
	TopicSim	0.001872	0.002433	0.003669	0.014069	0.014534
	Link-PLSA-LDA	0.016341	0.012858	0.018686	0.034708	0.059297
	Topic PageRank	0.037406	0.041362	0.062698	0.109224	0.095652
	MTID	**0.141056**	**0.101980**	**0.150841**	**0.22537**	**0.309743**
DBLP	Random	0.000030	0.000034	0.000067	0.000148	0.000371
	TopicSim	0.002197	0.001551	0.002880	0.005391	0.008760
	Link-PLSA-LDA	0.013687	0.010525	0.025611	0.034949	0.058961
	Topic PageRank	0.013621	0.010598	0.019777	0.037148	0.056319
	MTID	**0.040889**	**0.032202**	**0.058351**	**0.101983**	**0.136039**
ACM	Random	0.000014	0.000025	0.000050	0.000091	0.000225
	TopicSim	0.000978	0.000889	0.001958	0.004279	0.005619
	Link-PLSA-LDA	0.014373	0.011093	0.022353	0.039949	0.053244
	Topic PageRank	0.008274	0.006572	0.013942	0.025045	0.039397
	MTID	**0.022046**	**0.017035**	**0.035423**	**0.060776**	**0.085698**

Table 2. Important papers for Topic 5 of AAN

Topic	M	Paper title	Year	#Citation
Model feature data training set use learning using word result	0.123106	A Maximum Entropy Approach To Natural Language Processing	1996	390
	0.117956	Discriminative Training Methods For Hidden Markov Models: Theory And Experiments With Perceptron Algorithms	2002	351
	0.098413	Word Representations: A Simple and General Method for Semi-Supervised Learning	2010	133
	0.081476	Building A Large Annotated Corpus Of English: The Penn Treebank	1993	1008
	0.081431	A Maximum Entropy Model For Part-Of-Speech Tagging	1996	281

of papers in the citation network is positively related to the number of citations. However, there are some exceptions in our result.

In Table 2, most papers describe *Machine Learning* for *Natural Language Processing*, while the fourth important paper *Building A Large Annotated Corpus Of English: The Penn Treebank* is about parsing and contains little information about the *Machine Learning*. However, it is considered as an influential paper in *Machine Learning* and cited by papers in this topic. The reason is that serves as

Table 3. Important papers for Topic 1 of DBLP

Topic	M	Paper title	Year	#Citation
Network networks sensor wireless routing detection nodes mobile protocol performance	0.436223	Directed diffusion: a scalable and robust communication paradigm for sensor networks	2000	450
	0.417814	Ad-hoc On-Demand Distance Vector Routing	1999	456
	0.407508	System Architecture Directions for Networked Sensors	2000	351
	0.365696	Chord: A scalable peer-to-peer lookup service for internet applications	2001	703
	0.353928	GPSR: greedy perimeter stateless routing for wireless networks	2000	361

a standard dataset for papers in *Machine Learning*. For example, [7] uses "gold standard" to describe *The Penn Treebank*. In this case, *The Penn Treebank* plays a role as an important dataset in the field of *Machine Learning*.

Table 3 presents the important papers in the topic of *Wireless Sensor Network*, the first three papers focus exactly on this field. However, *Chord: A scalable peer-to-peer lookup service for internet applications*, which ranks the fourth in this topic, is about *Content Distributed Network*. The reason is that *Content Distributed Network* and *Wireless Sensor Network* are two highly correlated topics. Methods utilized by papers in *Content Distributed Network* are frequently referenced by papers in *Wireless Sensor Network*. In this case, important papers in *Content Distributed Network*, such as *Chord*, are also considered as important references in the topic of *Wireless Sensor Network*.

Table 4. Important papers for Topic 38 of ACM

Topic	M	Paper title	Year	#Citation
Data query database queries mining search databases efficient processing time	0.149107	Aurora: a new model and architecture for data stream management	2003	114
	0.134329	Compressed full-text indexes	2007	20
	0.114798	Issues in data stream management	2003	80
	0.099354	Load shedding in a data stream manager	2003	72
	0.096368	What's hot and what's not: tracking most frequent items dynamically	2003	68

In Table 4, paper *Compressed full-text indexes* ranks the second with only 20 citations. Recalling that the ACM dataset contains papers published from year 2003 to 2007, we can know that paper *Compressed full-text indexes* gains 20 citations in less than one year. In the academic network, papers like *Compressed full-text indexes* are considered as rising stars. Thus, the paper *Compressed full-text indexes* should be recognized as an important paper in *Data Management*.

The examples above prove that MTID not only recommends papers with high citations but also discovers important references such as the papers about standard datasets and raising stars. This propriety improves the performance of our model in academic recommendation.

6   Conclusion

In this paper, we study the problem of topic-sensitive influential paper discovery in citation networks. We study how papers cite other papers by observing three standard citation networks and find that the citations are related to the topic of citing papers and the importance of cited papers. Based on the observations, we bring in two parameters to represent the topic and the importance of documents. Combining these two parameters, we propose MTID, a generative model to generate citation networks and learn the importance of papers in different topics from the data. Extensive experiments show that MTID significantly outperforms other topic-based methods in citation prediction. Furthermore, we demonstrate that MTID not only identifies papers with high citations, but also succeeds in discovering other important papers in different topics, including papers about standard datasets and the rising stars.

References

1. Blei, D.M., Ng, A.Y., Jordan, M.I.: Latent Dirichlet allocation. J. Mach. Learn. Res. **3**, 993–1022 (2003)
2. Boyd, S., Vandenberghe, L.: Convex Optimization. Cambridge University Press, Cambridge (2004)
3. Chang, J., Blei, D.M.: Relational topic models for document networks. In: International Conference on Artificial Intelligence and Statistics, pp. 81–88 (2009)
4. He, J., Huang, Y., Liu, C., Shen, J., Jia, Y., Wang, X.: Text network exploration via heterogeneous web of topics. In: 2016 IEEE 16th International Conference on Data Mining Workshops (ICDMW), pp. 99–106. IEEE (2016)
5. He, Q., Chen, B., Pei, J., Qiu, B., Mitra, P., Giles, L.: Detecting topic evolution in scientific literature: how can citations help? In: Proceedings of the 18th ACM Conference on Information and Knowledge Management, pp. 957–966. ACM (2009)
6. Lu, Z., Mamoulis, N., Cheung, D.W.: A collective topic model for milestone paper discovery. In: Proceedings of the 37th International ACM SIGIR Conference on Research & Development in Information Retrieval, pp. 1019–1022. ACM (2014)
7. Madhyastha, P.S., Carreras Pérez, X., Quattoni, A.: Learning task-specific bilexical embeddings. In: Proceedings of the 25th International Conference on Computational Linguistics, pp. 161–171 (2014)

8. Nallapati, R.M., Ahmed, A., Xing, E.P., Cohen, W.W.: Joint latent topic models for text and citations. In: Proceedings of the 14th ACM SIGKDD International Conference on Knowledge Discovery and Data Mining, pp. 542–550. ACM (2008)
9. Pal, A., Counts, S.: Identifying topical authorities in microblogs. In: Proceedings of the Fourth ACM International Conference on Web Search and Data Mining, pp. 45–54. ACM (2011)
10. Radev, D.R., Muthukrishnan, P., Qazvinian, V.: The ACL anthology network corpus. In: Proceedings of the 2009 Workshop on Text and Citation Analysis for Scholarly Digital Libraries, pp. 54–61. Association for Computational Linguistics (2009)
11. Sayyadi, H., Getoor, L.: Futurerank: ranking scientific articles by predicting their future pagerank. In: Proceedings of the 2009 SIAM International Conference on Data Mining, pp. 533–544. SIAM (2009)
12. Sun, Y., Han, J., Gao, J., Yu, Y.: iTopicmodel: information network-integrated topic modeling. In: Ninth IEEE International Conference on Data Mining, ICDM 2009, pp. 493–502. IEEE (2009)
13. Tang, J., Zhang, J., Yao, L., Li, J., Zhang, L., Su, Z.: Arnetminer: extraction and mining of academic social networks. In: KDD 2008, pp. 990–998 (2008)
14. Walker, D., Xie, H., Yan, K.K., Maslov, S.: Ranking scientific publications using a model of network traffic. J. Stat. Mech. Theory Exp. **2007**(06), P06010 (2007)
15. Wang, X., Zhai, C., Roth, D.: Understanding evolution of research themes: a probabilistic generative model for citations. In: Proceedings of the 19th ACM SIGKDD International Conference on Knowledge Discovery and Data Mining, pp. 1115–1123. ACM (2013)
16. Weng, J., Lim, E.P., Jiang, J., He, Q.: Twitterrank: finding topic-sensitive influential twitterers. In: Proceedings of the Third ACM International Conference on Web Search and Data Mining, pp. 261–270. ACM (2010)
17. Yang, J., Leskovec, J.: Overlapping community detection at scale: a nonnegative matrix factorization approach. In: Proceedings of the Sixth ACM International Conference on Web Search and Data Mining, pp. 587–596. ACM (2013)

Course-Specific Markovian Models for Grade Prediction

Qian Hu[✉] and Huzefa Rangwala

Department of Computer Science, George Mason University, Fairfax, VA, USA
qhu3@gmu.edu, rangwala@cs.gmu.edu

Abstract. Over the past 15 years, the average six-year graduation rates for colleges and universities across the Unites States have remained stable at around 60%. This vehemently impacts society in terms of workforce development, national productivity and economic activity. Educational early-warning systems have been identified as an important approach to tackle this problem. The key to these systems are accurate grade prediction algorithms. In this paper we propose application of markovian models for the problem of predicting next-term student performance. Traditional approaches predict student's grade in a course by using a subset of prior courses and content features. However, these models ignore the dynamic evolution of student's knowledge states, which is a strong influence on student's learning and performance. We developed course-specific Hidden Markov Models and Hidden Semi-markov Models for the problem of next-term grade prediction. Our experimental results on datasets from a large public university show that the proposed approaches outperform prior state-of-the-art methods. We show by a case study the application of these methods for early identification of at-risk students.

1 Introduction

Over the past decade higher education institutions have been facing many challenges related to low graduation rates and high number of drop outs. According to the National Center for Education Statistics, the six-year graduation rate for first-time full-time undergraduate students in the United States is 59% [1]. The longer graduation time leads to increased cost for students. One of the main reasons students need longer time to graduate is the lack of sufficient and timely advisement [2].

Educational technologies in the form of degree planners and early warning systems help students stay on track and graduate on time. Degree planners are decision guidance systems that identify a personalized course plan that ensures students' timely graduation. Early-warning systems inform university officials of at-risk students so that these students can be reached out for advising, training and mentorship. The foundations of these educational technologies are based on accurate prediction of students' future academic performance. Grade prediction can help students plan for their success in courses by identifying course/topics that they may be deficient in [3].

© Springer International Publishing AG, part of Springer Nature 2018
D. Phung et al. (Eds.): PAKDD 2018, LNAI 10938, pp. 29–41, 2018.
https://doi.org/10.1007/978-3-319-93037-4_3

Several approaches have been developed in the past few years to tackle the problem of next-term grade prediction [4]. In particular course-specific approaches predicting a student's grade in a course by using the grades on a subset of courses taken prior to the target course [5,6] have shown promising results. Given the sequential aspect of academic programs; where a chain of courses build fundamental concepts and lead to training (education) of students; these models assume that a subset of related prior courses can provide the necessary knowledge for future courses. Course-specific models are based on regression or matrix factorization. One of the limitations of these course-specific models is that they ignore the temporal dynamics associated with the evolution of a student's knowledge state. The concept of knowledge state is proposed in mathematical psychology literature for assessment of a student's mastery of knowledge. Assessments uncover the particular state of a student and are used for predicting student's future performance and abilities. Latent factor models are useful for modeling students' knowledge state evolution [7].

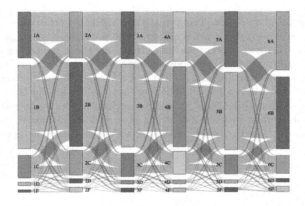

Fig. 1. Change in student Term GPA for the first six semesters. The digit of the text label denotes the term and the letter denotes the GPA. E.g., 3B implies term 3 and GPA of B (3.0)

To model the student learning behavior and predict student's performance we propose the Hidden Markov Model (HMM) and Hidden Semi-Markov Model (HSMM). In these models, students' knowledge states are modeled as hidden states. For HMM, the sojourn time is the number of steps spent in one state before transitioning to another state and is geometrically distributed. However, its performance degrades when the data exhibits long-term temporal dependency [8] as in the case of student knowledge state. For example, a student with strong capability is likely to be a high performing student in the next several semesters, instead of suddenly transitioning to a hidden state indicative of a low performing student. Figure 1 shows this property. This figure illustrates the dynamics in students' term GPA across all majors at George Mason University for the first

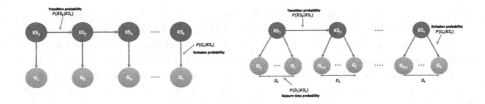

(a) Graphical Model of the HMM. KS_i represents knowledge state, G_i is the grade in a course.

(b) Graphical Model of the HSMM. KS_i represents knowledge state, G_i is the grade in a course. D_i is the sojourn time of state KS_i.

Fig. 2. Graphical model of HMM vs. HSMM

six semesters (excluding Summer terms). We only present full letter grades (i.e., A, B, C, D, F) for this figure. The width of the flow from one semester to another shows the number of grades and illustrates that given a student with a particular GPA in one semester, the GPA in the next semester will probably remain at the same level or off by one grade point with a high likelihood. If we consider more refined grade points (i.e., the full letter grades plus A+, A−, B+, B−, ..., C−), the statistics of the grade data shows that 24.3% of students have the same GPA from one semester to another, 66.84% and 84.33% of students have their next-term GPA within one and two ticks of their current-term GPA, respectively. Ticks measure the deviations from the true letter grade and is explained in Sect. 5. Thus it is very likely for a student to have similar GPA for the next term, which shows that a student's performance does not change frequently or abruptly.

To capture this long-term dependency property of students' knowledge evolution, we propose HSMM for grade prediction. Compared to HMM, the sojourn time of the knowledge state in HSMM is modeled explicitly. Each hidden state in a HMM emits one observation while in HSMM each hidden state emits a sequence of observations. The number of observations, i.e., the duration d, produced in a hidden state is determined by the sojourn time distribution of that state. Figure 2a and b shows the difference between HMM and HSMM, respectively. In this work, the distribution of sojourn time is assumed to be nonparametric and learned from data. Our experimental results show that the proposed methods improve the prediction performance up to 42.96% and 30.34% in terms of MAE and RMSE compared to baselines, respectively.

2   Related Work

2.1   Academic Performance Prediction

Students' performance prediction is often treated as a classification or regression problem. Classical classification algorithms such as decision tree, neural network, support vector machine and Naive Bayes have been applied [9]. To predict a student's performance, these approaches extract various features from different databases available at a higher education institution. Exemplar features include student's final grades, cumulative grades, high school background and features related to courses and instructors [4]. Inspired by recommender systems, popular algorithms such as collaborative filtering [10] and matrix factorization [4,11] have been proposed for grade prediction. These models use a "one size fits all" strategy. However, due to different characteristics of students, these approaches are limited in modeling students' learning behaviors and academic performance.

To overcome this issue, Polyzou et al. proposed models specific to students and courses [5]. Morsy et al. [12] represents course-knowledge spaces as latent vectors and proposed cumulative knowledge-based regression models for next-term grade prediction. Elbadrawy et al. proposed personalized multi-regression models to predict students' performance [13]. Elbadrawy et al. [14] proposed domain-aware algorithms for grade prediction and course recommendation.

2.2 Markovian Models for Educational Data

Hidden Markov Models (HMM) were first introduced and extensively studied by Baum et. al. [15,16]. The application of HMM to education domain was first proposed by Corbett et al. [17] to model the acquisition of procedural knowledge in intelligent tutoring system as Bayesian Knowledge Tracing (BKT). Extending BKT, several models have been proposed such as the individualized BKT [18] to improve the prediction performance.

HMM have also been applied for predicting dropouts in Massively Open Online Courses (MOOCs). Balakrishnan et al. [19] used a novel Input-Output HMM to predict student retention. Geigle et al. proposed a two-layer Hidden Markov Model to model student's behavior in MOOCs and found that the features extracted from the two-layer Hidden Markov Model correlated with educational outcomes [20]. The observation layer of the two-layer HMM is used to model the sequence of students' interactions with the learning management systems [20].

Hidden Semi-Markov Model (HSMM) was first proposed in the area of speech recognition [21] and applied to areas including computer vision, genomics and financial time series [22]. To the best of our knowledge, HSMM models have not been applied to educational datasets and for the problem of next-term grade prediction.

3   Problem Formulation and Notations

Assume that we have records of n students and m courses; comprising a $n \times m$ sparse grade matrix $\mathbf{G}$. Entry $G_{s,c}$ in $\mathbf{G}$ represents the grade of student s in course c. In addition we have the time stamp information for each grade $G_{s,c}$. Besides the grade matrix $\mathbf{G}$, we have information associated with the student (e.g., academic level, previous GPA and major) and course offering (e.g., discipline, course level and difficulty) that can be combined to extract features.

Table 1. Notations

Symbol	Description
O	Observation
KS	Knowledge State
π	Initial state distribution
D	The maximum number of duration of the hidden states
$g_{s,c}$	The true grade of student s in course c
$\hat{g}_{s,c}$	The predicted grade of student s in course c
a_{ij}	The probability of transition from state i to state j
$b_j(G_t)$	The probability of observing G_t at state j
$d_j(u)$	The probability of staying at state j for u steps

We denote the feature vector as $\mathbf{x}$ of p dimensions. As a convention, bold upper-case letters are used to represent matrices (e.g., $\mathbf{X}$) and bold lowercase letters represents vectors (e.g., $\mathbf{x}$) (Table 1).

In this work, given the collection of students' historical grades data our objective is to train a machine learning algorithm to model students' knowledge evolution and predict their grades in future courses.

4 Methods

4.1 Hidden Markov Model (HMM)

Model Description. HMM seeks to capture the dynamic evolution of student's knowledge state. Student's knowledge state is modeled as the latent (hidden) states in HMM. The grades of a student are modeled as the observations. Compared to existing discriminative models, one of the key advantages of the HMM approach is that it introduces stochasticity/uncertainty. For example, a student with high capability has the chance to get a low grade by slipping [17], which is hard to model using discriminative models.

In HMM, the evolution of student's knowledge state is modeled as a Markov chain and has the assumption that the next state only depends on current state. The transition distribution of the model determines the evolution of students' knowledge state, as shown in Eq. 1.

$$a_{ij} = P(\text{KS}_{t+1} = j | \text{KS}_t = i) \tag{1}$$

The emission distribution determines a student's performance, given his knowledge state, given by Eq. 2.

$$b_j(\text{G}_t) = P(\text{G}_t | \text{KS}_t = j) \tag{2}$$

where G_t is the student's grade at time t.

A student's knowledge state cannot be observed; only the grades are observable. The space of the knowledge states and the observations are discrete.

To use HMM for modeling student's knowledge state evolution and predict performance in next course, two related questions need to be answered:

– Given an observation sequence and a model, what is the likelihood of the observation sequence? This question can be solved by using forward algorithm [16].
– Given a set of sequences, in our case they are the sequence of students' grades, how do we infer the parameter set of the model? This can be done by using the classical EM algorithm [16].

Grade Prediction. To predict the grade $\hat{g}_{s,c}$ for student s in a future course c, we first extract the grades of student s in a series of courses $c_1, c_2, ..., c_T$ taken prior to course c and form them as a sequence $G_s = g_{s,c_1}, g_{s,c_2}, ..., g_{s,c_T}$. Assume that there are N possible grades student s could get in course c, in our case, the possible grade x is in (4, 4, 3.67, 3.33, 3, 2.67, 2.33, 2, 1.67, 1, 0). Then we have the following:

$$\hat{g}_{s,c} = \max_x P(x|g_{s,c_1}, g_{s,c_2}, ..., g_{s,c_T})$$

$$= \max_x \frac{P(g_{s,c_1}, g_{s,c_2}, ..., g_{s,c_T}, x)}{P(g_{s,c_1}, g_{s,c_2}, ..., g_{s,c_T})} \tag{3}$$

$$\propto \max_x P(g_{s,c_1}, g_{s,c_2}, ..., g_{s,c_T}, x)$$

The grade $\hat{g}_{s,c}$ is predicted using maximum likelihood.

4.2   Hidden Semi-Markov Model (HSMM)

Model Description. The HMM model proposed above assumes that a single knowledge state emits grade distributions for one course only. Further, the number of time steps spent in a given state (i.e., sojourn time) in a HMM model has geometric distribution as show by Eq. 4 [23].

$$d_i(u) = P(S_{t+u+1} \neq i, S_{t+u} = i, S_{t+u-1} = i, ...,$$
$$S_{t+2} = i|S_{t+1} = i, S_t \neq i) \tag{4}$$
$$= a_{ii}^{u-1}(1 - a_{ii})$$

where $d_i(u)$ is the probability of staying at state i for u steps.

However, a student's knowledge state has long-term temporal dependency. It is demonstrated that a student with strong academic capability is unlikely to become low performing in a short time.

To better model student's knowledge state evolution, we propose Hidden Semi-Markov Model (HSMM) shown in Fig. 2b. For HSMM, the underlying process is assumed to be a semi-Markov chain. Each state can emit variable number of observations. In other words, each knowledge state is responsible for performance in multiple courses. The sojourn time of HSMM is explicitly modeled and different hidden states have different sojourn time distribution. For modeling student's knowledge state evolution, the sojourn time distribution for a given knowledge state j is defined as following.

$$d_j(u) = P(\mathrm{KS}_{t+u+1} \neq j, \mathrm{KS}_{t+u-v} = j, v = 0, ..., u - 2|\mathrm{KS}_{t+1} = j, \mathrm{KS}_t \neq j) \quad (5)$$

which is assumed to be nonparametric in this work (i.e. categorical distribution). The state transition distribution of the semi-Markov chain determines the evolution of knowledge state; and is show in Eq. 6.

$$a_{ij} = P(\mathrm{KS}_{t+1} = j|\mathrm{KS}_{t+1} \neq i, \mathrm{KS}_t = i) \quad (6)$$

where $j \neq i$, $\sum_{j \neq i} a_{ij} = 1$ and $a_{ii} = 0$.

The emission distribution of HSMM determines a student's performance, given their knowledge state. Similar to HMM, for student's knowledge state modeling and grade prediction we need to compute the likelihood of a sequence and infer the parameters of HSMM which can be done by using forward and EM algorithms, respectively. The prediction of a student's grade in a future course by using HSMM is the same as HMM shown in Eq. 3.

4.3 Baseline Methods

Bias Only (BO). The Bias Only method only takes into consideration student's, course's and global bias. The predicted grade is estimated using Eq. (7).

$$\hat{g}_{s,c} = b_0 + b_s + b_c \quad (7)$$

where b_0, b_s and b_c are the global bias, student bias and course bias respectively.

Matrix Factorization (MF). The use of MF for grade prediction is based on the assumption that the students' and courses' knowledge space can be jointly represented in low-dimensional latent feature space [5]. The grade is estimated as:

$$\hat{g}_{s,c} = b_0 + b_s + b_c + \boldsymbol{p}_s^T \boldsymbol{q}_c \quad (8)$$

where $\boldsymbol{p}_s$, $\boldsymbol{q}_c$ are the latent vectors representing student s and course c, respectively. We also applied course-specific matrix factorization (CS$_{\mathrm{MF}}$) for grade prediction, which utilizes a course-specific subset of data to estimate a matrix factorization model [6].

Course-Specific Regression with Prior Courses (CSR$_{PC}$). CSR$_{\mathrm{PC}}$ predicts the grade of a student s in a future course c as a sparse linear combination of grades in the courses taken prior to course c [6].

Course-Specific Regression with Content Features. CSR$_{\mathrm{CF}}$ predicts the performance of a student in a course using content features such as academic level, difficulty level and instructor information [24].

Course-Specific Hybrid Model (CSR_{HY}). This model is obtained by combining the content feature vector and prior course vector [24].

5   Experiments

5.1   Dataset Description and Preprocessing

We evaluate the proposed methods on a dataset from George Mason University, the largest public university in Virginia and enrolled about 36000 students in Fall 2017. We extracted student and course related data from the largest five undergraduate majors in terms of student enrollment. These included: (I) Computer Science (CS), (II) Electrical and Computer Engineering (ECE), (III) Biology (BIOL), (IV) Psychology (PSYC) and (V) Civil Engineering (CEIE).

We used data from the period of Fall 2009 to Spring 2016. Using the University catalog [25] we selected student records for courses that are required by the major program and electives offered by the department offering the major. We also removed courses that did not result in a grade score (in between A–F) but were only pass/fail courses. If a course was taken more than once by a student, only the last grade was kept. For a course, if the number of students who had taken the course was smaller than the number of the prior courses of this course we removed this course from training and test sets. If the number of test instances of a course was smaller than 5, we removed it.

To simulate the real-world scenario of predicting the next-term grades for students we use the data extracted from the latest term as testing data and all the data from terms prior to the latest term as the training set. The training data was split into 80/20, of which 80% was training data and 20% was validation data for choosing the hyperparameters associated with the model. After selection of hyperparameters, the model was retrained on the entire training set before final evaluation on the last term (test set).

5.2   Evaluation Metrics

The performance of the methods were assessed by three different evaluation metrics: (i) mean absolute error (MAE), (ii) root mean squared error (RMSE) and (iii) tick error. MAE and RMSE are computed by pooling together all the grades across all the courses.

To gain deeper insights regarding the performance of the methods for course selection and degree planning, we report an application-specific metric called tick errors [5]. Tick error measures the deviation of the predicted grades from the true grades. The performance of a model is assessed based on how many ticks away the predicted grades are from the actual grades. The grading system has 11 letter grades (A+, A, A−, B+, B, B−, C+, C, C−, D, F) which correspond to (4, 4, 3.67, 3.33, 3, 2.67, 2.33, 2, 1.67, 1, 0). To compute the tick error for a predicted grade, the real value prediction outputs are first converted to the closest letter grades.

Table 2. Comparative performance of different models using MAE and RMSE. (↓ is better)

Method	MAE					RMSE				
	CS	ECE	BIOL	PSYC	CEIE	CS	ECE	BIOL	PSYC	CEIE
BO	0.7257	0.6902	0.5411	0.5951	0.5863	0.9824	0.9413	0.7784	0.7831	0.7734
MF	0.7184	0.6790	0.5420	0.6099	0.5796	0.9715	0.9557	0.7495	0.7977	0.7517
CS_{MF}	0.7151	0.6666	0.5365	0.5673	0.5733	0.9576	0.9623	0.7762	0.7545	0.7756
CSR_{PC}	0.6805	0.6739	0.5372	0.4933	0.601	0.9288	0.9699	0.7943	0.7348	0.8058
CSR_{CF}	0.7183	0.6775	0.4769	0.4743	0.6091	0.9539	0.9680	0.7205	0.6732	0.7941
CSR_{HY}	0.6693	0.6630	0.5057	0.4859	0.5839	0.9200	0.9542	0.7679	0.7283	0.7701
CS_{HMM}	0.601	0.4532	0.4634	0.3362	**0.3632**	0.9202	0.7638	0.7806	0.6665	**0.6463**
CS_{HSMM}	**0.555**	**0.3782**	**0.4231**	**0.3023**	0.3676	**0.8307**	**0.6647**	**0.7038**	**0.5313**	0.6613

6   Results and Discussion

6.1   Comparative Performance

Table 2 shows the average MAE and RMSE for the different methods across the five majors on the test set. The results show that the CS_{HSMM} model achieves the best performance on all the majors. The CS_{HMM} outperforms previously developed Course-Specific regression and factorization models in terms of MAE and RMSE. The proposed course-specific Markovian models are able to take into account the temporal dynamics associated with the evolution of student's knowledge states in comparison to prior course-specific approaches. The CS_{HSMM} model outperforms CS_{HMM} on almost all the majors and has similar performance on CEIE. The students' knowledge state which is modeled by CS_{HMM} and CS_{HSMM} as hidden states tends to stay in the same state for some time instead of changing constantly. Rather than a geometric distribution, the ideal duration distribution should have lower probabilities on longer or shorter durations but higher probabilities on medium durations. By modeling the transition of hidden states as semi-Markov model rather than Markov model, CS_{HSMM} achieves better performance than CS_{HMM}. The exception on CEIE is because of the flexibility in the particular degree program (i.e., there are many electives for students to choose).

To have better insights into what kind of errors different methods make, we evaluate the approaches using tick error metrics. Table 3 presents the results of the best performed traditional course-specific methods (i.e., CSR_{PC} and CSR_{HY}) and the proposed Markovian methods with respect to tick errors. For exact prediction (i.e., 0 tick error) and one tick error, the CS_{HMM} and CS_{HSMM} have the best performance. For two tick errors, the CS_{HMM} and CS_{HSMM} win for most of the majors, while traditional course-specific model CSR_{HY} show better performance in CS majors. The traditional course-specific models are poorer than CS_{HMM} and CS_{HSMM}, as they ignores students' knowledge evolution dynamics. The reason that CSR_{CF} and CSR_{HY} show better performance in some cases

Table 3. Comparative performance of different models using tick error (↑ is better)

#Ticks	Method	CS	ECE	BIOL	PSYC	CEIE
Zero tick	CSR_{PC}	19.57	20.77	28.84	34.08	27.17
	CSR_{CF}	13.44	16.39	28.03	27.39	29.13
	CS_{HMM}	**37.78**	45.36	43.13	**54.24**	**50.39**
	CS_{HSMM}	37.18	**46.99**	**46.49**	52.2	49.78
One tick	CSR_{PC}	48.22	55.19	62.80	61.15	52.76
	CSR_{HY}	49.80	55.19	67.38	61.78	53.15
	CS_{HMM}	54.08	68.85	66.58	79.32	**72.44**
	CS_{HSMM}	**57.85**	**78.14**	68.65	**79.66**	72.29
Two ticks	CSR_{PC}	74.31	73.22	81.40	79.62	69.69
	CSR_{HY}	**75.10**	74.32	82.75	78.66	69.29
	CS_{HMM}	70.97	82.51	83.29	85.76	86.22
	CS_{HSMM}	72.37	**87.98**	**84.32**	**88.81**	**87.01**

is that they incorporate content features which are informative for student's performance prediction and are not included within the Markovian approaches proposed here.

6.2   Case Study: At-Risk Students

An important application of grade prediction is to develop an early-warning system that is able to identify students at-risk of failing the courses that they plan to enroll in. We define at-risk students as those whose grade for a course is below 2.0. To assess the capability of the methods on catching at-risk students we treat the prediction as a classification problem. The experimental procedures are similar to grade prediction as discussed in Sect. 5.1 except that the predicted grade over 2.0 are treated as pass and below 2.0 as fail. We compare the best performed traditional course-specific methods CSR_{PC} and CSR_{HY} with models proposed here. The evaluation metrics are chosen as accuracy and F-1 score. Given the imbalanced nature of the dataset, F-1 score is a suitable classification metric. From Table 4, we see that the proposed CS_{HMM} and CS_{HSMM} outperform all the baseline methods. In most cases, the CS_{HSMM} outperforms the CS_{HMM} models. For the Psychology major (has the lowest proportion of at-risk students as shown by numbers in the table notes), some of the existing methods are not able to identify any of the at-risk students and their F-1 score is zero.

Table 4. Predictive power at identifying at-risk students (↑ is better)

Method	CS		ECE		BIOL		PSYC		CEIE	
	Acc	F-1	Acc	F-1	Acc	F-1	Acc	F-1	Acc	F-1
CSR_{PC}	0.8202	0.5561	0.7869	0.4507	0.8437	0.5397	0.9172	0	0.7143	0.5217
CSR_{HY}	0.8063	0.5333	0.7923	0.4412	0.8491	0.5625	0.9204	0	0.7532	0.6122
CS_{HMM}	0.8231	0.6276	0.8634	0.7126	**0.9027**	**0.7831**	**0.9492**	0.4828	**0.9004**	**0.6462**
CS_{HSMM}	**0.8549**	**0.7092**	**0.8962**	**0.7711**	0.8784	0.7594	0.9458	**0.5294**	**0.9004**	**0.6462**

The percentage of at-risk students for each major is CS (24.40%), ECE (19.14%), BIOL (18.79%), PSYC (9.80%), CEIE (12.97%).

7   Conclusions

In this paper, we propose Course-Specific Hidden Markov Model and Hidden Semi-Markov Model for student's next-term grade prediction. The proposed Markovian models are able to capture the temporal dynamic characteristics of students' knowledge state evolution. The limitation of HMM is that its hidden state duration is inherently geometrically distributed. To better model student's knowledge state evolution, we use Hidden Semi-Markov Model for grade prediction to model the distribution of state duration explicitly.

We conducted extensive experiments and compared the proposed Markovian models with other state-of-the-art grade prediction algorithms. The experimental results showed that the proposed models achieved better grade prediction performance than the baselines. One important application of grade prediction is early-warning systems. We evaluated the performance of the proposed methods for identifying at-risk students. For this task, our proposed methods achieved the best performance in comparison to other state-of-the-art methods.

Acknowledgement. This research work was supported by the National Science Foundation grant #1447489. The experiments were run on ARGO, a research computing cluster provided by the Office of Research Computing at George Mason University, VA. (URL: http://orc.gmu.edu).

References

1. Aud, S., Wilkinson-Flicker, S., Kristapovich, P., Rathbun, A., Wang, X., Zhang, J.: The condition of education 2013. nces 2013–037. National Center for Education Statistics (2013)
2. Anschuetz, N.: Breaking the 4-year myth: Why students are taking longer to graduate? (2015)
3. Holman, C., Aguilar, S.J., Levick, A., Stern, J., Plummer, B., Fishman, B.: Planning for success: how students use a grade prediction tool to win their classes. In: Proceedings of the Fifth International Conference on Learning Analytics and Knowledge, pp. 260–264. ACM (2015)
4. Sweeney, M., Rangwala, H., Lester, J., Johri, A.: Next-term student performance prediction: A recommender systems approach (2016). arXiv preprint arXiv:1604.01840

5. Polyzou, A., Karypis, G.: Grade prediction with course and student specific models. In: Bailey, J., Khan, L., Washio, T., Dobbie, G., Huang, J.Z., Wang, R. (eds.) PAKDD 2016. LNCS (LNAI), vol. 9651, pp. 89–101. Springer, Cham (2016). https://doi.org/10.1007/978-3-319-31753-3_8
6. Polyzou, A., Karypis, G.: Grade prediction with models specific to students and courses. Int. J. Data Sci. Anal. **2**(3–4), 159–171 (2016)
7. Schrepp, M.: About the connection between knowledge structures and latent class models. Methodology **1**(3), 93–103 (2005)
8. Duong, T.V., Bui, H.H., Phung, D.Q., Venkatesh, S.: Activity recognition and abnormality detection with the switching hidden semi-markov model. In: 2005 IEEE Computer Society Conference on Computer Vision and Pattern Recognition, CVPR 2005, vol. 1, pp. 838–845. IEEE (2005)
9. Romero, C., Ventura, S., Pechenizkiy, M., Baker, R.S.: Handbook of Educational Data Mining. CRC Press, Boca Raton (2010)
10. Bydžovská, H.: Are collaborative filtering methods suitable for student performance prediction? In: Pereira, F., Machado, P., Costa, E., Cardoso, A. (eds.) EPIA 2015. LNCS (LNAI), vol. 9273, pp. 425–430. Springer, Cham (2015). https://doi.org/10.1007/978-3-319-23485-4_42
11. Sweeney, M., Lester, J., Rangwala, H.: Next-term student grade prediction. In: 2015 IEEE International Conference on Big Data (Big Data), pp. 970–975. IEEE (2015)
12. Morsy, S., Karypis, G.: Cumulative knowledge-based regression models for next-term grade prediction. In: Proceedings of the 2017 SIAM International Conference on Data Mining, pp. 552–560. SIAM (2017)
13. Elbadrawy, A., Studham, R.S., Karypis, G.: Collaborative multi-regression models for predicting students' performance in course activities. In: Proceedings of the Fifth International Conference on Learning Analytics and Knowledge, pp. 103–107. ACM (2015)
14. Elbadrawy, A., Karypis, G.: Domain-aware grade prediction and top-n course recommendation. In: Proceedings of the 10th ACM Conference on Recommender Systems, pp. 183–190. ACM (2016)
15. Baum, L.E., Petrie, T., Soules, G., Weiss, N.: A maximization technique occurring in the statistical analysis of probabilistic functions of markov chains. Ann. Math. Stat. **41**(1), 164–171 (1970)
16. Rabiner, L.R.: A tutorial on hidden markov models and selected applications in speech recognition. Proc. IEEE **77**(2), 257–286 (1989)
17. Corbett, A.T., Anderson, J.R.: Knowledge tracing: modeling the acquisition of procedural knowledge. User Model. User-Adap. Interact. **4**(4), 253–278 (1994)
18. Yudelson, M.V., Koedinger, K.R., Gordon, G.J.: Individualized bayesian knowledge tracing models. In: Lane, H.C., Yacef, K., Mostow, J., Pavlik, P. (eds.) AIED 2013. LNCS (LNAI), vol. 7926, pp. 171–180. Springer, Heidelberg (2013). https://doi.org/10.1007/978-3-642-39112-5_18
19. Balakrishnan, G., Coetzee, D.: Predicting student retention in massive open online courses using hidden markov models. Electrical Engineering and Computer Sciences University of California at Berkeley (2013)
20. Geigle, C., Zhai, C.: Modeling student behavior with two-layer hidden markov models. JEDM-J. Educ. Data Min. **9**(1), 1–24 (2017)
21. Ferguson, J.D.: Variable duration models for speech. In: Proceedings of the Symposium on the Applications of Hidden Markov Models to Text and Speech, pp. 143–179 (1980)

22. Bulla, J.: Application of hidden markov models and hidden semi-markov models to financial time series (2006)
23. Guédon, Y.: Estimating hidden semi-markov chains from discrete sequences. J. Comput. Graph. Stat. **12**(3), 604–639 (2003)
24. Hu, Q., Rangwala, H.: Enriching course-specific regression models with content features for grade prediction (2017)
25. GMU: George mason university catalog (2017)

A Temporal Topic Model for Noisy Mediums

Rob Churchill$^{(\boxtimes)}$, Lisa Singh, and Christo Kirov

Georgetown University, Washington, D.C., USA
{rjc111,lisa.singh}@georgetown.edu, ckirov@gmail.com

Abstract. Social media and online news content are increasing rapidly. The goal of this work is to identify the topics associated with this content and understand the changing dynamics of these topics over time. We propose Topic Flow Model (TFM), a graph theoretic temporal topic model that identifies topics as they emerge, and tracks them through time as they persist, diminish, and re-emerge. TFM identifies topic words by capturing the changing relationship strength of words over time, and offers solutions for dealing with flood words, i.e., domain specific words that pollute topics. An extensive empirical analysis of TFM on Twitter data, newspaper articles, and synthetic data shows that the topic accuracy and SNR of meaningful topic words are better than the existing state.

1 Introduction

Enormous amounts of content are being generated on social media, e.g., over 500 million tweets [9] and over 420 million status updates on Facebook [12] are posted daily. One can use topic models to generate meaningful topics from these text streams. Unfortunately, current topic model algorithms have a number of weaknesses. First, the topics generated are often bogged down by noise words or impacted by noise-generating bots, e.g., approximately 15% of Twitter accounts are bots [16]. Second, domain specific corpora often have domain specific words that are not discriminative of topics, but appear across all different topics. For example, current topic models that generate topics about the extremist group ISIS will generate topics that all rank the term ISIS (or a variant) amongst the top terms for the topic. While obviously an important term, including it in specific topics about ISIS does not improve the quality of the topics and may make those topics look too similar. Finally, in some domains, the topics themselves change so quickly that current methods have difficulty keeping an up-to-date sketch of the active topics through time.

To begin addressing some of these limitations, we propose the Topic Flow Model (TFM) for monitoring the ebb and flow of topics in noisy text streams, e.g., Twitter, blogs, and online news. TFM identifies groups of content-rich topic words from a semantic graph by finding meaningful subgraphs of words that represent topics, and using relationship strength and frequency of words to determine their importance to different topics through time. This approach also effectively identifies and "drains" flood words, making the top ranked words for the

© Springer International Publishing AG, part of Springer Nature 2018
D. Phung et al. (Eds.): PAKDD 2018, LNAI 10938, pp. 42–53, 2018.
https://doi.org/10.1007/978-3-319-93037-4_4

topic more discriminative. **The contributions of this paper are as follows:** (1) we introduce TFM, a temporal topic modeling algorithm that identifies topics as they emerge; (2) we introduce the concept of flood words and offer solutions to gracefully deal with them, and (3) we conduct an empirical analysis of TFM on Twitter data, newspaper data, and synthetic data, showing its effectiveness for identifying and monitoring changing topic dynamics under different conditions.

2   Related Work

Many topic modeling algorithms have been proposed in the last two decades. The most popular algorithms use probabilistic generative models. Latent Dirichelet Allocation (LDA) [5] and its many variants [2,3,11,14,15,17] belong to this group of models, which rely on the assumption that documents are generated following a known distribution of terms. LDA finds the parameters of the topic/term distribution that maximizes the likelihood of the documents in the data set. These models have been successful for longer text documents written by a smaller number of authors that have a fixed vocabulary, i.e., new words (hashtags) are not being created continually. Another direction for research considers methods that have been used in the dimensionality reduction and clustering literature [10,13,18]. For example, Yan et al. perform topic modeling by applying non-negative matrix factorization to a term correlation matrix [18], an approach that works better on short documents than generative models [18]. A third direction of research uses a semantic graph to identify topics [1,7]. Topic Segmentation [1], for instance, uses an undirected term co-occurence graph and the Louvain modularity algorithm [6] to find topics in a data set. Cataldi et al.'s Emerging Topic Detection (ETD) [7] employs a directed term correlation graph, and uses a double depth-first search to find emerging topics in a temporal topic modeling setting. Finally, some research focuses on post-processing the output of topic models to make them more meaningful [4,8].

Our work is closest in spirit to Cataldi et al. [7] since we also employ a directed semantic graph and use that graph to identify topics. Our work differs from their work in the following ways; (1) we track all topics through time, as they emerge, persist, and diminish (Cataldi et al. focus on emerging topics), (2) because we are interested in topics through time, we do not regenerate topics at every time step, but instead use the topic knowledge from the previous time step to help determine the changes to existing topics and identify new ones, (3) we employ a new metric, the Energy-Nutrition ratio, to identify the most important terms in the semantic graph and avoid using flood words to build our topics, and (4) we employ a more efficient graph traversal procedure (a constant depth BFS) that identifies more accurate topic terms.

3   Background and Definitions

A document is an ordered list of terms $(w_1, ... w_k)$, where k is the length of the document. A *topic* T is a set of words believed to describe a theme or subject.

In our models, we will output M topics from a data set of D documents. The set of documents are partitioned into τ time periods. We therefore take as input D_t documents for each $t \in \tau$, and output M_t topics at time t.

Not all words in a document are useful for topic generation. Stop words are obvious words that are frequent, but content-poor. Noise words and spam both detract from topic quality, polluting the topics. Another type of word that pollutes topics is a *flood word*. A flood word is an important domain-specific word occurring so frequently that it is relevant to nearly every topic. For example, suppose we are interested in a data set about the recent presidential election. It would not be surprising if every topic contained Trump and/or Clinton in the top words related to the topic. While clearly relevant, we define these words to be flood words since they are domain-relevant and frequent, but do not add value to potential topics. Therefore, it is imperative to deal with flood words so that they are not the dominant words in every topic generated by our model. To help us keep track of the changing dynamics of words in topics, we now define nutrition, energy, and Energy-Nutrition Ratio.

The *nutrition* of a term is an indicator of how popular a term is in the document collection. More formally, $nutrition(w) = (1-c) + c * tf(w)/tf(w_i^*)$ where w_i^* is the most frequent term in document d_i, $tf(w)$ is the term-frequency of the input word w in d_i, and c is some constant between zero and one. $nutrition(w)_t$ is the sum of these nutritions over D_t in time period t. We then normalize nutrition by $|D_t|$ to account for change in data set size over time.

The *energy* of a term considers the change in nutrition of that term over time. More formally, $energy(w) = \Sigma_{i=1}^{s}(nutrition(w)_t^2 - nutrition(w)_{t-i}^2) \times \frac{1}{i}$ where s is the number of previous time periods before t.

Because energy is a sum of squared differences, it is biased toward higher nutrition terms. A high nutrition term that sees a small change over time intervals might still have a high energy compared to a low nutrition term that has a bigger change relative to its original nutrition. We account for this relative change with the Energy-Nutrition Ratio: $ENR(w) = \frac{energy(w)}{nutrition(w)}$ A term with high nutrition that sees a small change will see a relatively low change in its ENR, whereas a term with low nutrition that sees a large change will see a big change in ENR. We can compare a term's current ENR and previous ENR to decide whether the rate of growth is accelerating, constant, or decelerating. Because energy is a polynomial function of nutrition, it grows and shrinks faster than nutrition. This growth, or lack thereof, is captured in ENR.

Problem Statement: Formally, given τ time intervals, and D_t documents for each $t \in \tau$, find the set of topics M_t and flood words $flood_t$ for each t.

4   Topic Flow Model

4.1   TFM Overview

We now provide an overview of our proposed approach, Topic Flow Model (TFM). The high level algorithm can be found in Algorithm 1. The input to

Algorithm 1. Topic Flow Model (TFM)

```
 1: Input: D_t for each t ∈ (1, ...τ)
 2:    α, β, γ, δ, θ
 3: Output: M_t, flood_t, and L_t for each t ∈ (1, ...τ)
 4: repeat
 5:    nutrition_t = compute_nutrition(D_t)
 6:    energy_t = compute_energy(α, nutrition_t, nutrition_{t-1}, D_t)
 7:    ENR_t = compute_ENR(nutrition_t, energy_t)
 8:    emerging = select_emerging_terms(α, β, γ, nutrition_t, energy_t, ENR_t, D_t)
 9:    flood_t = identify_flood_words(α, nutrition_t)
10:    C = compute_term_correlations(D_t)
11:    G = create_term_correlation_graph(δ, energy_t, C)
12:    M_t = {}
13:    for term ∈ emerging do
14:       new_topic = discover_topic(G, term, θ)
15:       M_t = M_t + new_topic
16:    end for
17:    for T_i ∈ M_{t-1} do
18:       persistent_topics = identify_persistent_topic(G, leaders(T_i), θ)
19:       M_t = M_t + persistent_topics
20:    end for
21:    M_t = merge_topics(M_t)
22:    L_t = identify_leaders(M_t)
23: until t = τ
24: return M, flood, L
```

the algorithm is the set of documents for each time period and a set of tuning parameters that will be described later in this section. For each time interval, our algorithm begins by using nutrition, energy, and the ratio between the two (ENR) to identify emerging terms, terms that have become important in the current time period (line 8). It then creates a directed term-correlation graph (line 11) and identifies the topics from the previous time window that persist in the current time window (line 18). It does this using a double Breadth-First Search (BFS) on the graph. Once all topics have been identified, topics are compared and merged if sufficiently similar (line 21). Leader nodes are chosen for each topic based on their centrality scores within their topic (line 22). TFM outputs a set of topics, topic leaders, and flood words for each time window. The remainder of this section describes the main parts of the algorithm in more detail - identifying emerging terms, building and using the semantic graph, and determining and merging topics. Throughout this section, we will make use of the example presented in Fig. 1. In the example, there are three time periods, each containing three documents. The semantic graph, document frequency of each word, and each word's energy are shown for each time period.

4.2   Identifying Important Terms

One of the keys to generating good topics is identifying important terms. We use nutrition, energy, and ENR to aid in this process. After computing these values for each term, we can select the terms that fit within the thresholds of each criterion. We define three separate tuning parameters, one for each value. Flood words, by definition, have the highest nutrition. To avoid adding them to our topics, we set an upper bound for nutrition (α). For energy, we do not want to

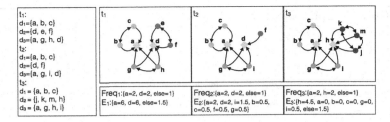

Fig. 1. Example of TFM graphs over time. Orange diamond nodes represent flood words. Topics are color-coded. Red, blue, and green topics emerge in t_1, and persist until t_2 or t_3. The Purple topic emerges in t_3. (Color figure online)

consider any words with exceedingly low energy values, so we set a lower bound on energy (β). In Fig. 2, we show how these initial two thresholds cut swathes of terms from the list of potential emerging terms. Within the set of remaining terms with high energy and high nutrition, we set an ENR threshold (γ) to weed out terms whose growth is low compared to the previous time window.

For our running example, assume $s = 2$, $\alpha = 1$, $\beta = 0$, and $\gamma = 1$. In Fig. 1, we see that terms a, and d have the highest frequency in t_1 and t_2, and terms a and h have the highest frequency in t_3. Looking at nutrition, energy, and ENR, we see that a, d, and h have too high nutrition values and are identified as flood words in different time periods instead of qualifying as emerging terms. At time t_2, the nodes returning from time t_1 have lower energy levels than i, leading to

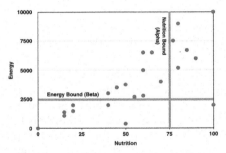

Fig. 2. The effect of α and β thresholds on term selection

i being the only emerging term identified in t_2. As we will show, emerging terms are important because we start traversing the graph for topics from nodes representing emerging terms.

4.3 Semantic Graph Construction

In order to determine topics, we construct a directed term-correlation graph. Any word that is not a stop word can be a node in the graph. For edges, we compute asymmetric term correlations [7], and then selectively add edges to our semantic graph based on these correlations. The term correlation $c_{k,z}^t$ of two terms k, z at time t is:

$$c_{k,z}^t = log\left(\frac{n_{k,z}/(n_k - n_{k,z})}{(n_z - n_{k,z})/(|D_t| - n_z - n_k + n_{k,z})}\right) \cdot \left|\frac{n_{k,z}}{n_k} - \frac{n_z - n_{k,z}}{|D_t| - n_k}\right| \quad (1)$$

where: $n_{k,z}$ is the number of documents in D_t that contain both k and z; n_z is the number of documents in D_t that contain z; n_k is the number of documents in D_t that contain k. The *term-correlation* of two terms is the correlation of the first term to the second term at time t, considering both co-occurrence and individual occurrence of each term. An edge (u, v) in the graph will have a weight equal to the correlation of u to v. In order for an edge to be added to the graph, its weight must be greater than the median term correlation plus δ times the standard deviation of correlations, where δ is a tuning parameter used to control the number of edges in the graph. A higher value of δ will result in a smaller, less connected graph, whereas a zero value will result in a maximally connected graph. Many flood words have high-correlation incoming edges, but because they are connected to so many other terms, their outgoing correlations are well below the median value, preventing these edges from being added to the graph.

In our example, we see the connections between nodes that occur together in a document. As we can see in time period t_1, node a has incoming edges with nodes b and c, because they co-occur in document d_1, and it also has incoming edges from nodes g, and h, which co-occur with node a in document d_3. Note that there are no outgoing edges from a or d because their correlations to other terms are below the threshold for adding an edge in our example.

4.4   Finding Topics

Using the graph G, we build a set of topics for our current time interval. We define two types of terms to focus on topic discovery: leader terms, and origin terms. A leader term is a term that represents a specific topic, chosen by centrality score. An origin term corresponds to a node from which we can start a topic search in G. As we will describe below, an origin term can be an emerging term or an existing topic leader. There are three main steps for finding topics: identifying persistent topics, discovering emerging topics, and merging similar topics.

Emerging Topics. Starting at an origin, we perform a breadth-first search (BFS) up to the depth limit θ to find other potential terms in the topic. We run a second breadth-first search from any node found during the forward pass, looking specifically for the origin. Since the graph is directed, we are not guaranteed to find a path back to the origin term from every node. If we do find a path during the backward pass, we assume the term is strongly connected to the origin, and include it in the associated topic. Flood words have exceptionally low correlations to other terms, and so even when included in the graph, it is unlikely that they will be included in a cycle from an origin node.

Persistent Topics. To identify topics that have persisted from the previous time period to the current window, we run our double BFS algorithm using the topic's existing leader as the origin, and compare the returned topic terms to the existing set. We compute a *topic distance*: $td_{t_1,t_2} = \frac{min(|t_1 \setminus t_2|, |t_2 \setminus t_1|)}{|t_1 \cap t_2|}$. A smaller distance implies that the topics are more similar. A distance of zero means that one topic is a subset of the other. If the 'new' topic is sufficiently similar to its old self, we keep only the new version. Returning to our example, suppose $\theta = 1$.

From each emerging term, we do a search to a maximum depth of one, where the depth at the root is 0. The resulting topics are $\{b, c\}$, $\{b, c\}$, $\{e, f\}$, $\{e, f\}$, $\{g, h\}$, and $\{g, h\}$, with leader sets $\{b\}$, $\{f\}$, and $\{g\}$, respectively. These are the topics we want to see since they do not include flood words. Duplicates occur because the emerging terms had directed edges in both directions.

Merging Similar Topics. Once we have found a set of topics for our time period, we must decide whether any are similar enough to merge into one. We compare the shared terms of each pair of topics in the set using the distance defined above. If the two topics share enough terms, we merge them by taking the union of their terms, and choosing new leaders. Returning to our example, we find duplicates of each topic because there are multiple emerging terms identified for each emerging topic. Using our method of merging similar topics, we will compare the topic membership of different topics and merge the duplicate topics. Our final result in t_1 is: $\{b, c\}$, $\{e, f\}$, $\{g, h\}$. In t_2 and t_3, topics $\{b, c\}$ and $\{g, h\}$ persist. In t_3, topic $\{j, k, m\}$ emerges.

5   Empirical Evaluation

This section presents an empirical analysis of TFM and other state of the art methods on a Twitter data set, a newspaper data set, and synthetic data sets.

5.1   Data Sets

For our synthetic data sets, we generate topics from a set of words, assigning each a normally distributed random probability of appearing in a document with that topic. By assigning different probabilities to different words, we are simulating the nature of tweets containing few content-rich, important words mixed in with many less useful words.[1] The synthetic data sets we generated each contain 200 vocabulary items, 500 documents, and seven topics over seven time periods. Our synthetic data sets contain varying levels of flood words, 0%, 1%, 5%, 10%, and 15% of the total number of words in the vocabulary.

The Twitter data set is a daily random sample of 5,000 tweets about Donald Trump from August and September 2016. We have a total of 280,000 tweets split into weekly time periods. The newspaper data set consists of news articles about Trump and Clinton from the Washington Post. The newspaper data set contains 14,269 articles and spans the same time frame as the Twitter data set.

For our data sets, we evaluate the accuracy and quality of topics using recall and Signal to Noise Ratio (SNR). The SNR is the ratio of terms in the approximated topic that belong to the true topic to the terms in the approximated topic that do not belong in the true topic. Let T_{noise} be the set of noise words in topic T, and T_{signal} be the set of signal words in topic T, then $SNR = \frac{T_{signal}}{T_{noise}}$.

---

[1] Another way to simulate this is to sample from a Zipfian distribution. Our data generator allows for distribution changes. For these experiments, we create a mixture that is noisier and harder to generate topics from than a Zipfian sample.

5.2   Synthetic Data Evaluation

Using our synthetic data sets, we evaluate four methods: TFM, Cataldi et al. [7], LDA [5], and Topic Segmentation (TS) [1]. For the static topic model algorithms, we rerun them in each time period to generate topics. Our settings for TFM are: $\alpha = 4, \beta = 6, \gamma = 1, \delta = 1.5, \theta = 2$. For both LDA and TS, we need to specify the number of topics. We show the results from the best performing number of topics – LDA $= 7$ and TS $= 10$. The results are shown in Table 1. The first column shows the fraction of flood words in the data set.

Table 1. Evaluation of synthetic data sets varying the % of flood words

Flood word %	Metric	LDA	Cataldi	TFM	TS
0%	Precision	0.11	0.00	0.67	0.04
	Recall	0.37	0.00	0.52	0.20
	SNR	0.95	0.05	4.10	0.39
1%	Precision	0.09	0.33	0.80	0.00
	Recall	0.30	0.33	0.37	0.00
	SNR	1.19	0.51	5.47	0.21
5%	Precision	0.23	0.78	1.00	0.48
	Recall	0.25	0.63	0.83	0.62
	SNR	1.66	0.05	5.52	1.38
10%	Precision	0.11	0.00	0.60	0.00
	Recall	0.27	0.00	0.40	0.00
	SNR	0.50	0.05	2.48	0.20
15%	Precision	0.03	0.10	0.54	0.00
	Recall	0.10	0.05	0.42	0.00
	SNR	0.48	0.22	2.26	0.12

In general, TFM performs significantly better across all three metrics at all the different fractions of flood words. It is also interesting to point out that its precision remains high even when the fraction of flood words in the data set is high. This is because it has been designed to avoid generating topics around flood words. In contrast, LDA performs best in terms of recall when the fraction of flood words is low or nonexistent, and best in terms of precision at 5%. The other three methods perform best when the fraction of flood words is set to 5%.

5.3   Twitter and Newspaper Evaluation

Researchers at Gallup worked with our research team to semi-manually create a set of popular topics for the presidential election campaign in 2016. For each week we have topics that persist, diminish and emerge. Because that initial topic set was generated without considering tweets from Twitter, we augmented the Gallup topics with appropriate hashtags and other topic words used on Twitter. The average number of topics being discussed each week is 5.5.

Accuracy & SNR. For this empirical evaluation, we compare TFM, Cataldi et al. (Cataldi) [7], LDA [5], HDP [15], Topic Segmentation (TS) [1], Topics over Time (ToT) [17], and Naïve Graph Properties on the Trump Twitter data set. The Naïve Graph Properties method attempted to discover topics using graph invariants, including degree, betweenness centrality, and eigencentrality. We show only the best graph invariant results in this analysis. We set the number of topics for LDA to 12 and 24 and for TS to 10 and 20. Using these parameter settings for LDA and TS led to fewer noise words than other settings.

We present our findings in two forms: the number of ground truth topics identified in each time interval in Fig. 3(a) (x-axis = time period, y-axis = number of ground truth topics identified), and the average signal to noise ratio of identified topics in Fig. 3(b) (x-axis = time period, y-axis = SNR). In order for a ground truth topic to be considered accurately identified by a model, the model must output a topic with an SNR of at least 0.5 in reference to that ground truth topic.

We see that the best performing algorithms are TFM and Cataldi, followed by different variants of LDA. LDA does identify a significant number of topics. TS identified one ground truth topic at time zero, but failed to identify any others, while HDP and Naïve Graph Properties identified no ground truth topics. TS's failure to identify more topics stemmed from the large number of noise words that it picked up in comparison to the number of ground truth words when generating topics. HDP's failure to identify ground truth topics seems to be a result of overfitting of the topics. In each of its topic sets, every topic contained almost the exact same terms. Notice that for both TFM and Cataldi, a warm-up period is needed - neither perform well in the first time period. In terms of SNR, TFM has the highest SNR of all the methods. Cataldi and LDA are comparable to TFM in time periods in which they find the same number of topics.

We tested the best performers, TFM, Cataldi, and LDA on the Washington Post data set, using the same ground truth topics used for the Trump Twitter data set since we were interested in topics related to Trump. For TFM, we used the following parameter values: $\alpha = 10, \beta = 2, \gamma = 2, \delta = 1.5, \theta = 2$. We present our accuracy results in Fig. 3(c). TFM outperformed Cataldi and LDA, identifying nearly every topic in every time interval except for the first. In Fig. 3(d), we see that it has a high SNR across time windows. Cataldi identified one or more topics in every time window except for the first, while both LDA options find one topic in the third time window. The SNR of TFM was consistent across all time windows. In the last time window, Cataldi had a higher average SNR.

Finally, due to space limitations, we cannot present a sensitively analysis for all the different parameters. However, we pause to mention that small variations of α, β, and γ do not impact the identified important terms significantly. Large differences, on the other hand, do (as we will show in the next subsection). In terms of the semantic graph, keeping θ low (around 2) reduces noise in the discovered topics and improves efficiency. δ impacts the number of edges in G. We have found that while the number of edges decreases significantly when δ increases, the accuracy of selected topics does not decrease for our data sets.

5.4   TFM Flood Words Evaluation

We now present two cases that demonstrate the graceful handling of flood words.

Flood Word Retainment. The α parameter controls the removal of flood words prior to generating G. In this experiment, we consider the case of setting $\alpha = \infty$, removing no flood words. When doing this on the Trump tweets data set, we find that TFM only identifies two topics as emerging, `hillary` and `#debatenight`. In both cases, no other words are identified as significant terms in those topics. This suggests that the two terms were so much more frequent than any other term that no other term could be reasonably assumed to be emerging relative to these terms. When α is smaller, these two terms are correctly labeled as flood words and over two dozen unique topics are identified. The debate topic is still identified even though the associated hashtag is a flood word. We pause

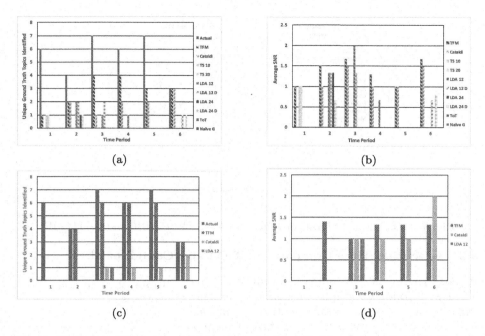

Fig. 3. (a) Trump Twitter: ground truth topics identified by each method (b) Trump Twitter: SNR of each method (c) Newspaper: ground truth topics identified by each method (d) Newspaper: SNR of each method

to point out that it is typical for some flood words to be in the graph, but the most extreme to not be. For the Trump tweet data set, the degree of the average flood word in G is 935 and its average correlation is 0.29, while the degree for the average topic word is 24 and the average correlation is 1.18. This highlights the importance of not focusing topic discovery on flood words.

TFM vs. Cataldi Emerging Terms. For this case study, we compare the emerging terms of TFM and Cataldi on the Trump tweet data set for two different weeks, September 4th and September 11th. Figure 4 shows the TFM emerging terms, the TFM flood words, and the Cataldi emerging terms as a Venn diagram. This figure highlights a few interesting findings. First, the emerging terms of Cataldi are all flood words returned by TFM, except for one word, `tax`. Second, most of the flood words identified by TFM are very general domain words, e.g. `president`, `campaign`, `policy`, `people`. These words are content-poor within the domain because they cross a large number of more meaningful, content-rich topics. There are a few terms that are more content-rich, `Putin` being the most notable. In these cases, either the flood word was an emerging term in the previous time window, so the topic has already been discovered, or the term rose so rapidly that its presence in G would cause it to be the center of a topic that is really a cluster of smaller topics merged into one.

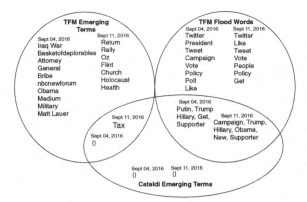

Fig. 4. Distribution of terms in Trump data set, 09/04/2016 and 09/11/2016

5.5   Execution Time Comparison: TFM and Cataldi et al.

In this experiment, we compare execution times of TFM & Cataldi on graphs of similar size. These experiments were run on an Ubuntu 16.04 machine with a 4.00 GHz processor and 16 GB of memory. Table 2 shows the results. We set the edge inclusion parameters to levels such that the number of edges in the respective graphs are similar. We tested algorithms on a graph size of 350,000 edges, 190,000

Table 2. Execution time comparison

# Edges	TFM	Cataldi
350,000	43 s	5 days
190,000	39 s	2 h
110,000	37 s	28 s

edges, and 110,000 edges. The specific numbers were chosen because the former was the approximate number of edges seen using TFM's optimal parameter settings, the latter was the optimal for Cataldi, and the middle gave reasonable, albeit worse, results for both models, with respect to topic quality. Table 2 shows that TFM's execution time is significantly smaller as the number of edges increases. This occurs because the DFS used in Cataldi requires traversal of a larger number of paths than the constant depth search used by TFM.

6   Conclusion

In this paper, we introduce the Topic Flow Model, and demonstrate its abilities to not only identify emerging topics, but to track those topics through time. We introduce the notion of flood words, and demonstrate how their graceful handling is integral to identifying concise topics in noisy data such as tweets, and even in less noisy data. We compare Topic Flow Model to state of the art topic modeling algorithms and show that it identifies topics more accurately with less noise than other methods. In future work, we plan to design extensions of this model for other types of text, understand the impact of pre-processing on topic model algorithms, and develop methods for reducing noise in topics.

Acknowledgements. This work was supported by the Massive Data Institute (MDI) at Georgetown University.

References

1. de Arruda, H.F., da Fontoura Costa, L., Amancio, D.R.: Topic segmentation via community detection in complex networks. CoRR (2015). http://arxiv.org/abs/1512.01384
2. Bhadury, A., Chen, J., Zhu, J., Liu, S.: Scaling up dynamic topic models. In: WWW. SIAM (2016)
3. Blei, D.M., Lafferty, J.D.: Dynamic topic models. In: ICML. IEEE (2006)
4. Blei, D.M., Lafferty, J.D.: Visualizing topics with multi-word expressions. arXiv e-prints (2009)
5. Blei, D.M., Ng, A.Y., Jordan, M.I.: Latent Dirichlet allocation. J. Mach. Learn. Res. **3**, 993–1022 (2003)
6. Blondel, V.D., Guillaume, J.-L., Lambiotte, R., Lefebvre, E.: Fast unfolding of communities in large networks. J. Stat. Mech. Theor. Exp. **2008**(10), P10008 (2008)
7. Cataldi, M., Di Caro, L., Schifanella, C.: Emerging topic detection on Twitter based on temporal and social terms evaluation. In: MDM-KDD. ACM (2010)
8. Chang, J., Boyd-Graber, J., Wang, C., Gerrish, S., Blei, D.M.: Reading tea leaves: how humans interpret topic models. In: NIPS. AAAI (2009)
9. InternetLiveStats: Twitter usage statistics. http://www.internetlivestats.com/twitter-statistics/. Accessed 05 May 2017
10. Kasiviswanathan, S.P., Melville, P., Banerjee, A., Sindhwani, V.: Emerging topic detection using dictionary learning. In: CIKM. ACM (2011)
11. Lafferty, J.D., Blei, D.M.: Correlated topic models. In: NIPS, pp. 147–154. AAAI (2006)
12. Noyes, D.: The top 20 valuable Facebook statistics - updated May 2017. https://zephoria.com/top-15-valuable-facebook-statistics/. Accessed 05 May 2017
13. Shahnaz, F., Berry, M.W., Pauca, V., Plemmons, R.J.: Document clustering using nonnegative matrix factorization. Inf. Process. Manage. **42**, 373–386 (2006)
14. Sleeman, J., Halem, M., Finin, T., Cane, M., et al.: Modeling the evolution of climate change assessment research using dynamic topic models and cross-domain divergence maps. In: Symposium on AI for Social Good. AAAI (2016)
15. Teh, Y.W., Jordan, M.I., Beal, M.J., Blei, D.M.: Hierarchical Dirichlet processes. J. Am. Stat. Assoc. **101**(476), 1566–1581 (2006)
16. Varol, O., Ferrara, E., Davis, C.A., Menczer, F., Flammini, A.: Online human-bot interactions: detection, estimation, and characterization. In: ICWSM. AAAI (2017)
17. Wang, X., McCallum, A.: Topics over time: a non-Markov continuous-time model of topical trends. In: KDD. ACM (2006)
18. Yan, X., Guo, J., Liu, S., Cheng, X., Wang, Y.: Learning topics in short texts by non-negative matrix factorization on term correlation matrix. In: SDM. SIAM (2013)

A CRF-Based Stacking Model with Meta-features for Named Entity Recognition

Shifeng Liu[(✉)], Yifang Sun, Wei Wang, and Xiaoling Zhou

The University of New South Wales, Sydney, Australia
shifeng.liu@unsw.edu.au, {yifangs,weiw,xiaolingz}@cse.unsw.edu.au

Abstract. Named Entity Recognition (NER) is a challenging task in Natural Language Processing. Recently, machine learning based methods are widely used for the NER task and outperform traditional handcrafted rule based methods. As an alternative way to handle the NER task, stacking, which combines a set of classifiers into one classifier, has not been well explored for the NER task. In this paper, we propose a stacking model for the NER task. We extend the original stacking model from both model and feature aspects. We use Conditional Random Fields as the level-1 classifier, and we also apply meta-features from global aspect and local aspect of the level-0 classifiers and tokens in our model. In the experiments, our model achieves the state-of-the-art performance on the CoNLL 2003 Shared task.

Keywords: Named Entity Recognition · Stacking
Feature engineering

1 Introduction

Named Entity Recognition (NER) is a fundamental stage in Natural Language Processing (NLP). In the sentence "Trump dined at the Trump National Hotel.", NER aims to identify "Trump" as a person and "Trump National Hotel" as an organization. The identified entities can be then used in downstream applications (e.g., information extraction systems) and other NLP tasks (e.g., named entity disambiguation and relation extraction).

Early NER systems are based on handcrafted rules. The rules are defined by human experts, which makes them labour consuming to develop and, more importantly, impossible to cover all the cases. Recently, machine learning techniques become more popular and effective in solving the NER problem. Supervised learning is one of the most widely used learning approaches for the NER task, which takes a set of human labelled documents as the training data,

---

This work was partially supported by D2DCRC DC25002 and DC25003, and ARC DP 180103411.

and predicts the named entities in the given documents (i.e., test data). Linear classifiers, such as Hidden Markov Model [8] and Conditional Random Fields (CRF) [7,10], are proved to be effective. However, the performance of these models is highly associated with carefully designed features, whose effectiveness is usually restricted within a specific domain. In recent years, an enormous amount of research effort has been put into neural network based methods [4,11,12,16], and these methods have achieved the state-of-the-art performance. Most neural network based methods simply use word embedding vectors as input [14,15]. However, to train the word embedding vectors, it requires a huge amount of unlabelled data in order to achieve high performance. These neural network based models are hard to adjust due to the non-trivial hyperparameter tuning process. The training stage is also much more time-consuming than linear classifiers, even with high performance GPUs.

Stacking [2,25] is an alternative way to improve the accuracy of a machine learning task. As a two-phrase method, the first step of stacking is to train a set of the level-0 classifiers (i.e., the base classifiers) using different models on the training dataset. In the second step, the level-1 classifier (e.g., a linear regression classifier) is trained on the training dataset with the predicted results from the level-0 classifiers as features. As such, the level-1 classifier is expected to achieve better performance than the level-0 classifiers.

Although traditional stacking method has shown its effectiveness on many machine learning tasks, there are still some potential improvements that can be explored, especially for the NER task. In this paper, we propose a CRF based stacking model with carefully designed meta-features to solve the NER problem. Comparing with the traditional stacking model, our model has two major differences. Firstly, we use CRF instead of linear regression as the level-1 classifier. The idea is inspired by the fact that CRF has shown its advantage as a sequential model in the NER task, while linear regression only works on independent instances. Secondly, we use a mix of meta-features and local features to improve the accuracy of the stacking model. Previous works either simply use the predicted results from the level-0 classifiers, or only extract features from the surface of tokens [7,10,18] (i.e., local features). We observe that the stacking model can also benefit from the non-local information. For example, the distribution of a token on different named entity types acts like the prior knowledge when we make the prediction. Moreover, our proposed model shows its robustness even when the performance of the level-0 classifiers is not good. This would be very helpful when users want to involve commercial NER systems (which usually cannot be tuned or re-trained).

The contributions of this paper are as follows:

- We proposed the CRF based stacking model named as SMEF, which took meta-features into consideration.
- We proposed a set of meta-features and local features and integrated these features in the stacking model to achieve better performance. We presented the details of the model and features in Sect. 3.

– We conducted extensive performance evaluation against the state-of-the-art NER systems. The proposed model outperforms all of them and achieves the overall F_1 score of 92.38% in the CoNLL 2003 Shared task. Moreover, the experiment results also show the effectiveness of each type of features and the robustness of the proposed model. Section 4 reports the experiment results and analyses.

2 Related Work

Named Entity Recognition is a research topic with a long history. Most recent approaches to NER have focused on CRF model and neural network models. CRF is a sequence model which could be used to predict sequences of labels based on the handcrafted features [7,9,10,18]. Neural network takes word embedding vectors as input features and learns a dense score vector for each possible named entity types [4,11,12,16]. Moreover, CRF model can be used as the output layer in neural network based models. Some of the most recent works [11,12,16] have shown the effectiveness of this combination.

Stacking has been proposed for many years [2,25] as a way to combine multiple classifiers (the level-0 classifiers) into one model (the level-1 classifier) in order to achieve better accuracy. While the level-0 classifier can be any machine learning model, the level-1 classifier is usually linear regression [25], stacking trees and ridge regression [2]. FWLS [21] uses a linear combination of meta-features to formulate the weight of the level-0 classifiers in the level-1 classifier, and achieves good performance in the Netflix Prize competition. [17,27] add meta-features extracted from the level-0 classifiers to the level-1 classifier to improve the stacking performance.

Stacking has been applied to NLP tasks such as Part-of-Speech Tagging [3,22] and NER [6,23,24,26]. Tsukamoto et al. [23] and Wu et al. [26] apply an extension of AdaBoost to learn the level-1 classifier. They take the sequence label information into consideration, and use handcrafted features from tokens. [6,24] use CRF model as the level-1 classifier to solve biomedical NER tasks. They also make use of the handcrafted local features.

3 Model

Named Entity Recognition takes a token sequence $\mathbf{x}_i = (x_{i1}, \ldots, x_{is})$ as input, and predicts a corresponding label sequence $\mathbf{y}_i = (y_{i1}, \ldots, y_{is})$, where $\mathbf{x}_i$ is taken from the i-th sentence in the dataset $\mathcal{X}$ and s is the length of the token sequence. As a named entity could span several tokens, we do not directly use the named entity types as labels. Instead, we apply chunking encoding for these labels. There are two popular encoding methods: BIO (i.e., **B**egin, **I**nside, and **O**utside token of a named entity) and BIOES (i.e., **B**egin, **I**nside, **O**utside, **E**nd, and **S**ingle token of a named entity). Table 1 shows an example.

As there are various named entity types, we form the NER task as a multiclass classification problem. As such, for each x_{ij}, a classifier (either a level-0

Table 1. Example of chunking encoding

	x_{i1}	x_{i2}	x_{i3}	x_{i4}	x_{i5}	x_{i6}	x_{i7}	x_{i8}
Token	Trump	dined	at	the	Trump	National	Hotel	.

	y_{i1}	y_{i2}	y_{i3}	y_{i4}	y_{i5}	y_{i6}	y_{i7}	y_{i8}
BIO	B-PER	O	O	O	B-ORG	I-ORG	I-ORG	O
BIOES	S-PER	O	O	O	B-ORG	I-ORG	E-ORG	O

classifier or the level-1 classifier) returns a vector $\mathbf{c}(x_{ij})$ with m dimensions, where m is the number of classes. More specifically, in the NER task, m is the number of predefined named entity types. Generally, each dimension in $\mathbf{c}(x_{ij})$ is a binary value, i.e., 1 for the predicted class and 0 for the rest classes. Real numbers can be used in some classifiers to represent the probability or the confidence score of each class.

In this section, we will firstly give more detail about the stacking method [2, 25]. Then we will introduce our model along with different meta-feature.

3.1 Stacking

Stacking is a two-phrase method. All the level-0 classifiers are trained on dataset $\mathcal{X}$. We denote the prediction of token sequence $\mathbf{x}_i$ by the k-th level-0 classifier as $\mathbf{c}_k(\mathbf{x}_i) = (\mathbf{c}_k(x_{i1}), \ldots, \mathbf{c}_k(x_{ij}), \ldots, \mathbf{c}_k(x_{is}))$, where $\mathbf{c}_k(x_{ij})$ is a m dimensional one-hot encoding vector. In this paper, all vectors are assumed to be column vectors unless noted. With predicted results from the level-0 classifiers, we generate a new dataset $\mathcal{Z}$, which consists of $\{(\mathbf{z}_i, \mathbf{y}_i), i = 1, \ldots, N\}$, where N is the size of dataset $\mathcal{X}$, $\mathbf{y}_i$ is the ground truth label sequence corresponding to the token sequence $\mathbf{x}_i$, and $\mathbf{z}_i$ is a vector sequence corresponding to $\mathbf{x}_i$. Each $\mathbf{z}_{ij}$ in $\mathbf{z}_i$ consists of $(x_{ij}, \mathbf{c}_1(x_{ij}), \ldots, \mathbf{c}_L(x_{ij}))$, where L is the number of the level-0 classifiers.

In [2], the loss function of stacking method is $\mathcal{L} = \sum_{i,j}(y_{ij} - \sum_k w_k c_k(x_{ij}))^2$, where w_k is the weight of $c_k(x_{ij})$, and both y_{ij} and $c_k(x_{ij})$ belong to $\mathcal{R}$.

However, this loss function is designed for regression problem thus can not be directly applied for a multi-class classification problem. Moreover, it does not consider the label dependency between sequential tokens.

In order to resolve the above issues, in this paper, we propose to use the Conditional Random Fields (CRF) model as the level-1 classifier instead of a linear regression model. The loss function of CRF is $\mathcal{L} = -\sum_i \log(p(\mathbf{y}_i \mid \mathbf{z}_i))$. Given each vector sequence $\mathbf{z}_i$, the probability of the corresponding label sequence $\mathbf{y}_i$ can be formulated as

$$p(\mathbf{y}_i \mid \mathbf{z}_i) = \frac{\exp(\sum_{j=0}^{s} U(\mathbf{z}_{ij}, y_{ij}) + \sum_{j=0}^{s-1} T_{y_{ij}, y_{i(j+1)}})}{\sum_{y' \in \mathbf{Y}_{\mathbf{z}_i}} \exp(\sum_{j=0}^{s} U(\mathbf{z}_{ij}, y'_{ij}) + \sum_{j=0}^{s-1} T_{y'_{ij}, y'_{i(j+1)}})}, \quad (1)$$

where $T_{y_{ij},y_{i(j+1)}}$ is the transition score between y_{ij} and $y_{i(j+1)}$, $\mathbf{Y_{z_i}}$ is the set of all the possible label sequences, and $U(\mathbf{z}_{ij}, y_{ij})$ is the unary potential score of $\mathbf{z}_{ij}$ with the corresponding label y_{ij}. More specifically, $U(\mathbf{z}_{ij}, y_{ij})$ is defined as

$$U(\mathbf{z}_{ij}, y_{ij}) = \sum_{win} \sum_{k} \mathbf{w}_{k,win,y_{ij}}^{\top} \mathbf{c}_k(x_{i(j+win)}) + bias_{y_{ij}}, \tag{2}$$

where $\mathbf{w}_{k,win,y_{ij}}$ is the weight vector of the predicted result from the level-0 classifier c_k for token $x_{i(j+win)}$, and $bias_{y_{ij}}$ is the learned bias corresponding to label y_{ij}. win denotes the token offset in the context window. In this paper, we set two as the size of the context window, i.e., $win \in [0, \pm 1, \pm 2]$.

3.2   Stacking with Meta-features

We have observed that the characteristic of the level-0 classifiers is useful for the level-1 classifier. For example, if a classifier consistently recognizes "Jordan" as a location, we should not give such classifier too much trust when dealing with "Jordan" as "Jordan" could also be a person's name.

In this subsection, we propose several meta-features based on the statistic information of dataset $\mathcal{Z}$. More specifically, for each level-0 classifier, we will extract meta-features from its prediction results on dataset $\mathcal{X}$. Note that the values of meta-features vary from different datasets such as training set, development set, and testing set.

The unary potential part of the model is therefore modified as

$$U(\mathbf{z_{ij}}, y_{ij}) = \sum_{win} \sum_{k} \mathbf{u}^{\top} \mathbf{w}_{meta,k,win,y_{ij}}^{\top} F_{meta}(x_{i(j+win)}, c_k) \mathbf{c}_k^{\top}(x_{i(j+win)}) \mathbf{u} \\ + bias_{y_{ij}}, \tag{3}$$

where $F_{meta}(x_{ij}, c_k)$ returns a vector of meta-features extracted from the level-0 classifier c_k for token x_{ij}, $\mathbf{w}_{meta,k,win,y_{ij}} \in \mathcal{R}^{|F_{meta}| \times m}$ is the weight matrix, and $\mathbf{u}$ is an all-ones vector with m dimension. Note that since there is only one nonzero element in $\mathbf{c}_k(x_{ij})$, only one column of weights in $\mathbf{w}_{meta,k,win,y_{ij}}$ are activated.

In our proposed model, $F_{meta}(x_{ij}, c_k)$ consists of four meta-features: constant, token label prior, token majority label, and token label entropy.

Constant. A constant 1 is used to maintain the predicted label of token x_{ij} from c_k. This feature helps us improve the model without losing the original information. Note that if we only apply this feature, then Eq. 3 falls back to Eq. 2, which is the standard CRF model.

Token Label Prior. Each token has a prior probability of being a named entity type $type_t$. For example, if token x_{ij} appears 11 times in the dataset and 9 of them are predicted as a person by classifier c_1, then we can approximate its token label prior of being a person as $\frac{9}{11}$. For each $type_t$, we use $F_{meta,prior,type_t}(x_{ij}, c_k, type_t)$ to denote the token label prior.

Token Majority Label. We define $F_{meta,major}(x_{ij}, c_k)$ to denote whether $\mathbf{c}_k(x_{ij})$ is consistent with the majority label of token x_{ij} for classifier c_k. For example, assume the surface of token x_{ij} is "Jordan" and the majority label of "Jordan" under the prediction of c_1 is person. Then we set $F_{meta,major}(x_{ij}, c_1) = 1$ for $\mathbf{c}_k(x_{ij} = $ "Jordan") indicates person and $F_{meta,major}(x_{ij}, c_1) = 0$ otherwise. We use $F_{meta,major}(x_{ij}, c_k)$ to enhance the impact of $\mathbf{c}_k(x_{ij})$.

Token Label Entropy. The token majority label feature may not be effective when the majority prediction is not distinguishable. Therefore, we apply the entropy of named entity types for token x_{ij}, denoted as $F_{meta,entropy}(x_{ij}, c_k)$, to further improve the performance of our model.

3.3 Stacking with Joint Meta-Features

We also observe that for a given token, the predicted labels over all the level-0 classifiers are helpful for the final decision. For example, if four out of five level-0 classifiers recognize token x_{ij} as a person, then most likely it is a person. However, this can not be captured by meta-features as they consider information from the level-0 classifiers independently.

We propose a set of meta-features which consider the joint information from the predicted labels of the given token from each level-0 classifier, and name them joint meta-features.

Similarly, we add the joint meta-features into the unary potential part and change it to

$$
\begin{aligned}
U(\mathbf{z_{ij}}, y_{ij}) = &\sum_{win} \sum_k \mathbf{u}^\top \mathbf{w}_{meta,k,win,y_{ij}}^\top F_{meta}(x_{i(j+win)}, c_k) \mathbf{c}_k^\top (x_{i(j+win)}) \mathbf{u} \\
&+ \sum_{win} \mathbf{w}_{joint,win,y_{ij}}^\top F_{joint}(\mathbf{c}_1(x_{i(j+win)}), \dots, \mathbf{c}_L(x_{i(j+win)})) + bias_{y_{ij}},
\end{aligned} \tag{4}
$$

where $F_{joint}(\mathbf{c}_1(x_{ij}), \dots, \mathbf{c}_L(x_{ij}))$ is a vector of the joint meta-features extracted from the predicted labels of the level-0 classifiers, and $\mathbf{w}_{joint,win,y_{ij}}$ is the weight vector.

We propose the following two joint meta-features: context prior, and joint label.

Context Prior. Under different local contexts, the probability of a token being $type_t$ should be different. For example, in sentence "Jordan is in Asia.", "Jordan" is more likely to be a location. We use the portion of the number of labels predicted as $type_t$ labels over the number of the level-0 classifiers to approximate the local context prior, denoted as $F_{joint,prior,type_t}(\mathbf{c}_1(x_{ij}), \dots, \mathbf{c}_L(x_{ij}))$. For example, the context prior of "Jordan" being predicted as a location is $\frac{2}{3}$ when two of three level-0 classifiers predict "Jordan" as a location.

Joint Label. Following the sequence of the level-0 classifiers i.e., $c_1, \dots, c_k$, we connect their predicted labels as the joint label (i.e., PER-LOC-PER). We denote this joint label feature as $F_{joint,joint}(\mathbf{c}_1(x_{ij}), \dots, \mathbf{c}_L(x_{ij}))$. It reserves all the information in the predicted labels rather than only keeps the majority label.

3.4   Stacking with Local Embedding Features

In addition, we also apply the local embedding features to enhance our model. Note that even if the level-0 classifiers have applied these local embedding features, there is no duplicated usage of local embedding features as we use a different model.

In order to apply the local embedding features, the unary potential part is modified to

$$
\begin{aligned}
U(\mathbf{z_{ij}}, y_{ij}) = & \sum_{win} \sum_{k} \mathbf{u}^\top \mathbf{w}_{meta,k,win,y_{ij}}^\top F_{meta}(x_{i(j+win)}, c_k) \mathbf{c}_k^\top (x_{i(j+win)}) \mathbf{u} \\
& + \sum_{win} \mathbf{w}_{joint,win,y_{ij}}^\top F_{joint}(\mathbf{c}_1(x_{i(j+win)}), \dots, \mathbf{c}_L(x_{i(j+win)})) \quad (5) \\
& + \sum_{win} \mathbf{w}_{local,win}^\top F_{local}(x_{i(j+win)}) + bias_{y_{ij}},
\end{aligned}
$$

where $F_{local}(x_{ij})$ extracts the local embedding features from token x_{ij} and $\mathbf{w}_{local,win}$ is the weight vector.

Following [9], we cluster word embeddings [15] by using the batch k-means clustering algorithm [20] with different numbers of clusters. For token x_{ij} and the number of clusters, we use the clustering id as one local embedding feature in our model. The number of clusters is set as 500, 1000, 1500, 2500 and 3000.

4   Experiment

We evaluate our model on two public NER benchmarks: the CoNLL 2003 Shared task [19] and the ACE 2005 dataset. Our model achieves the state-of-the-art performance on both benchmarks. In this section, we will firstly give the details about the datasets and the evaluation metrics. Then we will describe our training process. After that, we will show the overall performance of our model, and feature effectiveness results. In the end, we will show the robustness of our model when using existing low-performance classifiers as the level-0 classifiers.

4.1   Dataset and Evaluation

The CoNLL 2003 Shared Task. (CoNLL03) consists of news articles from the Reuters RCV corpus. There are four predefined named entity types: PER (Person), LOC (Location), ORG (Organization), and MISC (Miscellaneous). It includes standard tokenized training, development and test sets. We use the English data of the shared task. The details of the dataset can be found in Table 2.

The ACE 2005 Dataset. (ACE05) consists of articles from diverse sources including Broadcast News, Broadcast Conversations, Newswire, Weblog, Usenet, and Conversational Telephone Speech. Seven named entity types are predefined

Table 2. Statistics of The CoNLL 2003 Shared task [19]

	Articles	Sentences	Tokens	LOC	MISC	ORG	PER
Train	946	14,987	203,621	7,140	3,438	6,321	6,600
Dev.	216	3,466	51,362	1,837	922	1,341	1,842
Test	231	3,684	46,435	1,668	702	1,661	1,617

in the dataset, including FAC (Facility), GPE (Geo-Political Entity), LOC (Location), ORG (Organization), PER (Person), VEH (Vehicle), and WEA (Weapon). As we only have the full training set, we split the dataset into training (56%), development (24%), and test (20%) sets following [1]. Texts in the dataset are tokenized using the spaCy tokenizer. We work on the English data of the ACE 2005 dataset. The details of the ACE05 dataset can be found in Table 3.

Table 3. Statistics of The ACE 2005 dataset

	Articles	Sentences	Tokens	FAC	GPE	LOC	ORG	PER	VEH	WEA
Train	337	12,965	164,539	130	3,175	161	1,546	4,506	66	16
Dev.	145	5,142	73,411	81	1,565	83	713	1,804	20	11
Test	117	4,348	60,186	52	1,393	70	674	1,518	23	7

We evaluate the performance of different models by comparing the predicted results on the test set using *Precision, Recall*, and F_1 score. The predicted results can be classified into true positives (TP), true negatives (TN), false positives (FP) and false negatives (FN), according to the ground truth. *Precision* (P) is defined as $\frac{TP}{TP+FP}$, *Recall* is defined as $\frac{TP}{TP+FN}$, and F_1 score is the harmonic average of P and R. Following the previous work [4, 11, 16, 18], named entities are evaluated in phrase level.

4.2 Training

We use the BIOES chunking encoding as we find it bringing slightly better performance than the BIO encoding. This is also consistent with the observation in the previous work [18]. We pre-process the text by lowercasing all the tokens and replacing all the digits with 0 following [4].

We use the following NER classifiers as our level-0 classifiers:

- spaCy[1] is based on CNN with Glove word embedding vectors [15] as input;
- CoreNLP [13] is based on CRF with handcrafted features;
- UIUC [18] is based on handcrafted features and a set of gazetteers by using a regularized averaged perceptron;

---

[1] https://spacy.io/.

- MITIE[2] makes use of structural SVM and word embedding vectors;
- NeuroNER [16] has a BLSTM-CRF architecture with Glove word embedding vectors [15] as input.

Following the standard procedure of stacking [25], we generate the training set (i.e. $\mathcal{Z}_{train}$) for the level-1 classifier as follows. Firstly, we separate the original training set (i.e. $\mathcal{X}_{train}$) evenly into five parts (i.e. $\mathcal{X}_{train}^1, \dots, \mathcal{X}_{train}^5$). Then we train each level-0 classifier with four of them (e.g. $\mathcal{X}_{train}^1, \dots, \mathcal{X}_{train}^4$) and get the predicted results on the left part (e.g. $\mathcal{Z}_{train}^5$). These predicted results (i.e. $\mathcal{Z}_{train}^1, \dots, \mathcal{Z}_{train}^5$) are combined together to form $\mathcal{Z}_{train}$. For the development set and test set, we get the predicted results of the level-0 classifiers trained on the original training set (i.e. $\mathcal{X}_{train}$).

We use binary values in $\mathbf{c}(\mathbf{x}_i)$ as described in Sect. 3. For those level-0 classifiers which provide scores for their predicted named entities, we use the score to replace the binary values.

In all the experiments, our model is optimized using stochastic gradient descent with l_2 regularization. We train the proposed model with different C (the coefficient for l_2 regularization) and select the one with the best performance on the development set as the final C of the model. Following [4,16], we train our model ten times and report the average value for each metric. In addition, we also report the standard deviation to show the robustness of model.

4.3 Overall Results

In Table 4, we compare the performance of our model with the following state-of-the-art models:

- NeuroNER (2017) [5] has a neural network architecture of BLSTM-CRF with Glove word embedding vectors [15] as input;
- UIUC (2009) [18] is based on handcrafted features and a set of gazetteers by using a regularized averaged perceptron;
- Lample et al. (2016) [11] combines BLSTM and CRF with input word embedding vectors trained on different corpora from Glove [15];
- Ma and Hovy (2016) [12] is a neural network architecture with combination of BLSTM, CNN and CRF;
- Chiu and Nichols (2016) [4] is a hybrid of BLSTM and CNNs along with a set of gazetteers;
- TagLM (2017) [16] combines GRUs and CRF as well as an external bidirectional neural language model trained on a one billion token corpus.

For the first four models, we show the best results of the reported results in the original papers and the results from our experiments. For the last two models, we list the reported results as we do not have the corresponding models and source codes. In the CoNLL03, our model achieves the best average F_1 score of 92.38%. It shows a significant improvement compared with the previous best result of 91.93% ± 0.19% from TagLM [16]. Moreover, our model is more stable than previous methods.

---

[2] https://github.com/mit-nlp/MITIE.

Table 4. Test set performance comparison in The CoNLL 2003 Shared task

Model	$P \pm std$	$R \pm std$	$F_1 \pm std$
NeuroNER (2017) [5]	90.54%	90.78%	90.66%
UIUC (2009) [18]	91.20%	90.50%	90.80%
Lample et al. (2016) [11]	-	-	90.94%
Ma and Hovy (2016) [12]	91.35%	91.06%	91.21%
Chiu and Nichols (2016) [4]	91.39%	91.85%	91.62% ± 0.33%
TagLM (2017) [16]	-	-	91.93% ± 0.19%
SMEF	**92.95% ± 0.08%**	91.83% ± 0.04%	**92.38% ± 0.03%**

4.4 Effectiveness of Our Model and Meta-features

Table 5 reports the performance of the level-0 classifiers and SMEF in the CoNLL03 and the ACE05 datasets. The best level-0 classifier in the CoNLL03 scores 90.66% F_1. Our model has an increase of 1.72% in F_1 score compared with it. In the ACE05, our model increases the F_1 score by 2.43% compared with the best level-0 classifier.

In SMEF, we use three types of features including meta-features, joint meta-features, and local embedding features. Table 6 shows the results with different types of features on both datasets.

In order to justify our choice of using CRF as the level-1 classifier, we also implement a stacking model with logistic regression classifier as the level-1 classifier (i.e., LR in Table 6). Our model achieves better performance thanks to the sequence inference ability of CRF.

We show that all types of features are effective. Generally speaking, one can always achieve better F_1 score by applying more features. When using all the features, our model achieves the best performance in both datasets.

Meta-features are the most effective. All the models with meta-features consistently outperform those without meta-features. The model with only meta-features (i.e., meta in Table 6) shows a decent result which surpasses the previous best result of 91.93% from TagLM [16] in the CoNLL03. According to our analysis, the most effective meta-feature is the **token label prior**. The other two

Table 5. Test set performance of the level-0 classifiers and SMEF

Dataset	CoNLL03			ACE05		
Model	$P \pm std$	$R \pm std$	$F_1 \pm std$	$P \pm std$	$R \pm std$	$F_1 \pm std$
spaCy	82.62%	80.44%	81.51%	84.30%	74.55%	79.13%
CoreNLP [13]	87.41%	79.32%	83.17%	75.52%	33.34%	46.26%
UIUC [18]	90.10%	80.42%	84.98%	85.06%	84.56%	84.81%
MITIE	88.72%	86.88%	87.79%	80.55%	77.58%	79.03%
NeuroNER [5]	90.54%	90.78%	90.66%	85.80%	**87.82%**	86.80%
SMEF	**92.95% ± 0.08%**	**91.83% ± 0.04%**	**92.38% ± 0.03%**	**91.01% ± 0.18%**	87.52% ± 0.19%	**89.23% ± 0.07%**

Table 6. Effectiveness of meta-features

Dataset	CoNLL03			ACE05		
Model	$P \pm std$	$R \pm std$	$F_1 \pm std$	$P \pm std$	$R \pm std$	$F_1 \pm std$
LR	91.71% ± 0.00%	90.63% ± 0.00%	91.17% ± 0.00%	89.29% ± 0.00%	86.54% ± 0.00%	87.89% ± 0.00%
base_CRF	92.39% ± 0.08%	90.77% ± 0.04%	91.57% ± 0.04%	90.52% ± 0.37%	86.83% ± 0.41%	88.63% ± 0.13%
joint	92.53% ± 0.08%	90.79% ± 0.05%	91.65% ± 0.02%	90.14% ± 0.64%	87.10% ± 0.44%	88.59% ± 0.20%
local	92.47% ± 0.10%	90.95% ± 0.07%	91.75% ± 0.04%	90.80% ± 0.23%	86.92% ± 0.20%	88.82% ± 0.05%
joint + local	92.58% ± 0.08%	91.03% ± 0.04%	91.80% ± 0.05%	90.84% ± 0.32%	86.80% ± 0.21%	88.77% ± 0.13%
meta	92.73% ± 0.08%	91.49% ± 0.07%	92.11% ± 0.05%	90.77% ± 0.21%	87.36% ± 0.38%	89.03% ± 0.10%
meta + joint	92.69% ± 0.04%	91.54% ± 0.06%	92.11% ± 0.03%	90.65% ± 0.48%	87.49% ± 0.38%	89.04% ± 0.06%
meta + local	92.82% ± 0.10%	91.77% ± 0.05%	92.29% ± 0.05%	**91.11%** ± 0.31%	87.40% ± 0.23%	89.22% ± 0.08%
all	**92.95%** ± 0.08%	**91.83%** ± 0.04%	**92.38%** ± 0.03%	91.01% ± 0.18%	**87.52%** ± 0.19%	**89.23%** ± 0.07%

meta-features also show their effectiveness especially for ORG and MISC in the CoNLL03, and PER and ORG in the ACE05.

4.5 Our Model with the Existing Level-0 Classifiers

Some NER systems, especially commercial NER systems, cannot be tuned or re-trained on a specified dataset. Thus they may not be able to present satisfactory results. Our model offers a solution to deal with new data by only using these existing low-performance classifiers as the level-0 classifiers.

Since these existing classifiers are not trained on the specified dataset, our model is essentially an ensemble model without changing the loss function of CRF. These existing classifiers are usually trained on different datasets, even with different predefined named entity types, which could also be different from the specified dataset. For example, CoreNLP, MITIE, and NeuroNER are trained on the CoNLL03 and have four named entity types; spaCy and UIUC are trained on OntoNotes 5.0 with eighteen named entity types.

Table 7 shows the performance in the ACE05 with the existing level-0 classifiers. Our model achieves F_1 score of 88.87%, which is much better than any of the existing level-0 classifiers. It also outperforms the best trained classifier NeuroNER (whose F_1 score is 86.80%). Comparing with the SMEF model

Table 7. Performance in The ACE 2005 dataset using the existing level-0 classifiers

	Model	$P \pm std$	$R \pm std$	$F_1 \pm std$
Existing	CoreNLP [13]	37.16%	18.62%	24.81%
	NeuroNER [5]	43.40%	41.80%	42.58%
	MITIE	47.07%	40.86%	43.75%
	spaCy	44.40%	53.12%	48.37%
	UIUC [18]	44.61%	52.42%	48.20%
Our model	SMEF	**90.73%** ± 0.54%	**87.10%** ± 0.41%	**88.87%** ± 0.08%

with the trained level-0 classifiers, the one with the existing level-0 classifiers is just slightly worse (e.g., 0.46% lower F_1 score). Moreover, SMEF takes less than 5 min to train the model on the ACE05, which is much faster than a neural network (e.g., NeuroNER needs more than 2 h on the same dataset).

There are mainly three reasons for this. The first reason is that SMEF makes use of consistent and correlated named entity types between the level-0 classifiers and the ACE05. The second reason is that the level-0 classifiers provide prior distributions on corresponding named entity types for each token even though they have different predefined named entity types. Thus, meta-features would be effective in this scenario. The last reason is that local embedding features provide additional information beyond the named entity labels predicted by the level-0 classifiers.

5 Conclusion

In this paper, we propose a new stacking method with CRF model and meta-features for the NER task. These meta-features extract non-local information over the dataset for each level-0 classifier, and local information of the level-0 classifiers and tokens. Our approach, SMEF, achieves the state-of-the-art performance on the benchmark CoNLL 2003 Shared task. Besides, even with existing low-performance classifiers as the level-0 classifiers, our model can still achieve robust performance on the evaluated dataset.

References

1. Bansal, M., Klein, D.: Coreference semantics from web features. In: ACL (1), pp. 389–398. The Association for Computer Linguistics (2012)
2. Breiman, L.: Stacked regressions. Mach. Learn. **24**(1), 49–64 (1996)
3. Chen, H., Zhang, Y., Liu, Q.: Neural network for heterogeneous annotations. In: EMNLP, pp. 731–741. The Association for Computational Linguistics (2016)
4. Chiu, J.P.C., Nichols, E.: Named entity recognition with bidirectional LSTM-CNNs. TACL **4**, 357–370 (2016)
5. Dernoncourt, F., Lee, J.Y., Szolovits, P.: NeuroNER: an easy-to-use program for named-entity recognition based on neural networks. In: EMNLP (System Demonstrations), pp. 97–102. Association for Computational Linguistics (2017)
6. Ekbal, A., Saha, S.: Stacked ensemble coupled with feature selection for biomedical entity extraction. Knowl.-Based Syst. **46**, 22–32 (2013)
7. Finkel, J.R., Grenager, T., Manning, C.D.: Incorporating non-local information into information extraction systems by Gibbs sampling. In: ACL (2005)
8. Florian, R., Ittycheriah, A., Jing, H., Zhang, T.: Named entity recognition through classifier combination. In: CoNLL, pp. 168–171. ACL (2003)
9. Guo, J., Che, W., Wang, H., Liu, T.: Revisiting embedding features for simple semi-supervised learning. In: EMNLP, pp. 110–120. ACL (2014)
10. Krishnan, V., Manning, C.D.: An effective two-stage model for exploiting non-local dependencies in named entity recognition. In: ACL (2006)
11. Lample, G., Ballesteros, M., Subramanian, S., Kawakami, K., Dyer, C.: Neural architectures for named entity recognition. In: HLT-NAACL (2016)

12. Ma, X., Hovy, E.H.: End-to-end sequence labeling via bi-directional LSTM-CNNs-CRF. In: ACL (1). The Association for Computer Linguistics (2016)
13. Manning, C.D., Surdeanu, M., Bauer, J., Finkel, J.R., Bethard, S., McClosky, D.: The Stanford CoreNLP natural language processing toolkit. In: ACL (System Demonstrations), pp. 55–60. The Association for Computer Linguistics (2014)
14. Mikolov, T., Sutskever, I., Chen, K., Corrado, G.S., Dean, J.: Distributed representations of words and phrases and their compositionality. In: NIPS (2013)
15. Pennington, J., Socher, R., Manning, C.D.: Glove: global vectors for word representation. In: EMNLP, pp. 1532–1543. ACL (2014)
16. Peters, M.E., Ammar, W., Bhagavatula, C., Power, R.: Semi-supervised sequence tagging with bidirectional language models. In: ACL (1) (2017)
17. Rajani, N.F., Mooney, R.J.: Stacking with auxiliary features. In: IJCAI (2017)
18. Ratinov, L., Roth, D.: Design challenges and misconceptions in named entity recognition. In: CoNLL, pp. 147–155. ACL (2009)
19. Sang, E.F.T.K., Meulder, F.D.: Introduction to the CoNLL-2003 shared task: language-independent named entity recognition (2003)
20. Sculley, D.: Combined regression and ranking. In: KDD, pp. 979–988. ACM (2010)
21. Sill, J., Takács, G., Mackey, L.W., Lin, D.: Feature-weighted linear stacking. CoRR abs/0911.0460 (2009)
22. Sun, W., Peng, X., Wan, X.: Capturing long-distance dependencies in sequence models: a case study of Chinese part-of-speech tagging. In: IJCNLP, pp. 180–188. Asian Federation of Natural Language Processing/ACL (2013)
23. Tsukamoto, K., Mitsuishi, Y., Sassano, M.: Learning with multiple stacking for named entity recognition. In: CoNLL. ACL (2002)
24. Wang, H., Zhao, T., Tan, H., Zhang, S.: Biomedical named entity recognition based on classifiers ensemble. IJCSA 5(2), 1–11 (2008)
25. Wolpert, D.H.: Stacked generalization. Neural Netw. 5(2), 241–259 (1992)
26. Wu, D., Ngai, G., Carpuat, M.: A stacked, voted, stacked model for named entity recognition. In: CoNLL, pp. 200–203. ACL (2003)
27. Zenko, B., Dzeroski, S.: Stacking with an extended set of meta-level attributes and MLR. In: Elomaa, T., Mannila, H., Toivonen, H. (eds.) ECML 2002. LNCS (LNAI), vol. 2430, pp. 493–504. Springer, Heidelberg (2002). https://doi.org/10.1007/3-540-36755-1_41

Adding Missing Words to Regular Expressions

Thomas Rebele[1]([✉]), Katerina Tzompanaki[2], and Fabian M. Suchanek[1]

[1] Télécom ParisTech, Paris, France
thomas.rebele@gmail.com, suchanek@enst.fr
[2] ETIS lab/ENSEA/Cergy-Pontoise University/CNRS, Cergy-Pontoise, France
atzompan@u-cergy.fr

Abstract. Regular expressions (regexes) are patterns that are used in many applications to extract words or tokens from text. However, even hand-crafted regexes may fail to match all the intended words. In this paper, we propose a novel way to generalize a given regex so that it matches also a set of missing (previously non-matched) words. Our method finds an approximate match between the missing words and the regex, and adds disjunctions for the unmatched parts appropriately. We show that this method can not just improve the precision and recall of the regex, but also generate much shorter regexes than baselines and competitors on various datasets.

1 Introduction

Regular expressions (regexes) find applications in many fields: in information extraction, DNA structure descriptions, document classification, spell-checking, spam email identification, deep packet inspection, or in general for obtaining compact representations of string languages. To create regexes, several approaches learn them automatically from example words [2–5]. These approaches take as input a set of positive and negative example words, and output a regular expression. However, in many cases, the regexes are hand-crafted. For example, projects like DBpedia [10], and YAGO [19] all rely (also) on manually crafted regexes. These regexes have been developed by human experts over the years. They form a central part of a delicate ecosystem, and most likely contain domain knowledge that goes beyond the information contained in the training sets.

In some cases, a regex does not match a word that it is supposed to match. Take for example the following (simple) regex for phone numbers: $\backslash d\{10\}$. After running the regex over a text, the user may find that she missed the phone number 01 43 54 65 21. An easy way to repair the regex would be to add this number in a disjunction, as in $\backslash d\{10\}$ | 01 43 54 65 21. Obviously, this would be a too specific solution, and any new missing words would have to be added in the same way. A more flexible repair would *split* the repetition in the original regex and *inject* the alternatives, as in $(\backslash d\{2\}\backslash s?)\{4\}\backslash d\{2\}$.

D. Phung et al. (Eds.): PAKDD 2018, LNAI 10938, pp. 67–79, 2018.
https://doi.org/10.1007/978-3-319-93037-4_6

In this paper, we propose an algorithm that achieves such generalizations automatically. More precisely, given a regex and a small set of missing words, we show how the regex can be modified so that it matches the missing words, while maintaining its assumed intention. This is a challenging endeavor, for several reasons. First, the new word has to be mapped onto the regex, and there are generally several ways to do this. Take, e.g., the regex `<h1>.*</h1>` and the word `<h1 id=a>ABC</h1>`. It is obvious to a human that the id has to go into the first tag. However, a standard mapping algorithm could just as well map the entire string `id=a>ABC` onto the part `.*`. This would yield `<h1>?.*</h1>` as a repair – which is clearly not intended. Second, there is a huge search space of possible ways to repair the regex. In the example, `<h1(>.*| id=a>ABC)</h1>` is certainly a possible repair – but again clearly not the intended one. Finally, the repair itself is non-trivial. Take, e.g., the regex `(abc|def)*` and the word `abcabXefabc`. To repair this regex, one has to find out that the word is indeed a sequence of `abc` and `def`, except for two iterations. In the first iteration, the last character of `abc` is missing. In the second iteration, the first character of `def` has to be replaced by an `X`. Thus, the repair requires descending into the disjunction, removing part of the left disjunct and part of the right one, before inserting the `X` into one of them, yielding `(abc?|[dX]ef)*` as one possible repair.

Existing approaches typically require a large number of positive examples as input in order to repair or learn a regex from scratch. This means that the user has to come up with a large number of cases where the regex does not work as intended – a task that requires time, effort, and in some cases continuous user interaction (see Sect. 2 for examples). We want to relieve the user from this effort. Our approach requires not more that 10 non-matching words to produce meaningful generalizations. The contributions of this paper are as follows:

- we provide an algorithm to generalize a given regex, using string-to-regex matching techniques and adding unmatched substrings to the regex;
- we show how such repairs can be performed even with a small set of examples;
- we run extensive and comparative experiments on standard datasets, which show that our approach can improve the performance of the original regex in terms of recall and precision.

This paper is structured as follows. Section 2 starts with a survey of related work. Section 3 introduces preliminaries, and Sect. 4 presents our algorithm. Section 5 shows our experiments, before Sect. 6 concludes.

2   Related Work

In this paper we consider repairing regexes that fail to match a set of words provided by the user. We discuss work relevant to our problem along three axes: (1) matching regexes to strings, (2) automatic generation of regexes from examples, and (3) transformation of an existing regex based on examples.

Regex Matching. Many algorithms (e.g., [6]) aim to match a regex efficiently on a text. Another class of algorithms deals with matching the input regex

to a given input word – even though the regex does not match the string
entirely [8, 14]. Other algorithms for approximate regex matching [15, 20] optimize for efficiency. We build on the seminal algorithm of [14]. Different from all
these approaches, our work aims not just to match, but also to repair the regex.

Regex Learning. Several approaches allow learning a regex automatically from
examples. One approach [5] uses rules to infer regexes from positive examples
for entity identifiers. Other work [2–4] uses genetic programming techniques to
derive the best regex for identifying given substrings in a given set of strings.
The work presented in [16] follows a learning approach to derive regexes for
spam email campaign identification. In the slightly different context of combining
various input strings to construct a new one, the work of [7] proposes a language
to synthesize programs, given input-output examples. In the same spirit, the
authors of [9] proposed an interactive framework in which users can highlight
example subparts of text documents for data extraction purposes.

All of these works take as input a set of positive and negative examples, for
which they construct a regex from scratch. In our setting, in contrast, we want
to repair a given regex. Furthermore, we have only very few positive examples.

Regex Transformation. There are several approaches that aim to improve a
given regex. One line of work [11] takes as input a set of positive and negative
examples as well as an initial regex to be improved. As in our setting, the goal
is to maximize the F-measure of the regex. The proposed approach makes the
regex stricter, so that it matches less words. Our goal is different: We aim to
relax the initial regex, so that it covers words that it did not match before.

Similar to us, the work of [13] attempts to relax an initial regex. The approach
requires the user in the loop, though, while our method is autonomous. Only two
works [1, 17] can relax a given regex automatically. However, as we will see in our
experiments, both works produce very long regexes (usually over 100 characters).
Our approach, in contrast, produces much shorter expressions – at comparable
or even better precision.

3   Preliminaries

We assume that the reader is familiar with the basics of regexes. We write $L(r)$
for the language of a regex, and $T(r)$ for the syntax tree of a regex. Figure 1
shows an example of a regex with its syntax tree, along with a *matching* of a
string to the regex. We further define a matching as follows:

Definition (Matching). *Given a string s and a regex r, a* matching *is a partial
function m from $\{1, ..., |s|\}$ to leaf nodes of r's syntax tree $T(r)$, denoted $m|_{1,...,|s|}$,
such that one of the following applies:*

- *r is a character or character class and $\exists i : s_i \in L(r)$*
- *$r = pq$ and $\exists i$ s.t. $m|_{1,...,i}$ is a matching for p and $m|_{i+1,...,|s|}$ is a matching
  for q*
- *$r = p|q$ and m is a matching for p or for q*

Syntax tree $T(r)$

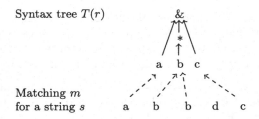

Matching m
for a string s

Fig. 1. The syntax tree and a matching for the regex `ab*c`

- $r = p*$ and $\exists i_1, \ldots, i_j : i_1{=}1 \wedge i_1 < \cdots < i_j \wedge i_j{=}|s|+1 \wedge \forall k \in \{1, \ldots, j-1\} :$
  $m|_{i_k, \ldots, (i_{k+1})-1}$ *is a matching for p*

A matching is *maximal*, if there is no other matching that is defined on more positions of the string. Figure 1 shows a maximal matching for the regex `ab*c` and the (non-matching) string `abbdc`. Maximal matchings can be computed with the algorithm proposed in [14]. The problem that we address is the following:

Problem Statement. *Given a regular expression r, a set S of positive examples, and a set E^- of negative examples, s.t. $|S| \ll |E^-|$, find a "good" regular expression r' s.t. $L(r) \subseteq L(r')$, $S \subseteq L(r')$, and $|L(r') \cap E^-|$ small.*

In other words, we want to generalize the regex so that it matches all strings it matched before, plus the new positive ones. For example, given a regex $r = $ `[0-9]+` and a string $s = $ "12 34 56", a possible regex to find is $r' = $ `([0-9]?)+`. This regex matches all strings that r matched, and it also matches s.

Now there are obviously trivial solutions to this problem. One of them is to propose $r = $ `.*`. This solution matches s. However, it will most likely not capture the intention of the original regex, because it will match arbitrary strings. Therefore, one input to the problem is a set of negative examples E^-. The regex shall be generalized, but only so much that it does not match many words from E^-. The rationale for having a small set S and a large set E^- is that it is not easy to provide a large set of positive examples: these are the words that the hand-crafted regex does not (but should) match, and they are usually few. In contrast, it is somewhat easier to provide a set of negative examples. It suffices to provide a document that does not contain the target words. All strings in that document can make up E^-, as we show in our experiments.

Another trivial solution to our problem is $r' = r|s$. However, this solution will not capture the intention of the regex either. In the example, the regex $r' = $ `[0-9]+|12 34 56` will match s, but it will not match any other sequence of numbers and spaces. Hence, the goal is to *generalize* the input regex appropriately, i.e., to find a "good" regex that neither over-specializes nor over-generalizes.

4   Repairing Regular Expressions

Given a regex r, a set of strings S, and a set of negative examples E^-, our goal is to extend r so that it matches the strings in S. Table 1 shows an example with only a single string $s \in S$ and no negative examples. Our algorithm is shown in Algorithm 1. It takes as input a regex r, a set of strings S, a set of negative examples E^-, and a threshold α. The threshold α indicates how many negative examples the repaired regex is allowed to match. Higher values for α allow a more aggressive repair, which matches more negative examples. $\alpha = 1$ makes the algorithm more conservative. In that case, the method will try to match at most as many negative examples as the original regex did. The algorithm proceeds in 4 steps, which we will now discuss in detail.

Table 1. Example regex reparation

Original regex r: `(\d{3}-){2}\d{4}`
String $s \in S$: `(234) 235-5678`
Repaired regex: `\(?\d{3}(-|\) )\d{3}-\d{4}`

Algorithm 1. Repair regex

INPUT: regex r, set of strings S, negative examples E^-, threshold $\alpha \geq 1$
OUTPUT: modified regex r
  1: $M \leftarrow \bigcup_{s \in S} \mathit{findMatchings}(r, s)$
  2: $\mathit{gaps} \leftarrow \bigcup_{m \in M} \mathit{findGaps}(m)$
  3: $\mathit{findGapOverlaps}(r, \mathit{gaps})$
  4: $\mathit{addMissingParts}(r, S, \mathit{gaps}, E^-, \alpha)$

Step 1: Finding the Matchings. For each word $s \in S$, our algorithm finds the maximal matchings (see again Sect. 3). We use Myers' algorithm [14] for this purpose. The maximal matchings are collected in a set M.

Step 2: Finding the Gaps. The matching tells us which parts of the regex match the string. To fix the regex, we are interested in the parts that do *not* match the string. For this purpose, we introduce a data structure for the gaps in the string (i.e., for the parts of the string that are not mapped to the regex). Formally, a gap g for string s in a matching m is a tuple of the following:

- $g.start$: The index in the string where the gap starts (possibly 0).
- $g.end$: The index in the string where the gap ends (possibly $|s| + 1$).
- $g.span$: The substring between $g.start$ and $g.end$ (excluding both).
- $g.m$: The matching m, which we store in the gap tuple for later access.
- $g.parts$: An (initially empty) set that stores sequences of concatenation child nodes. The sequences are disjoint, and partition the regex. Each $p \in g.parts$ is a possibility to inject $g.span$ into the regex as $(p\,|\,g.span)$.

Example (Finding the gaps): When matching the word (234) 235–5678 onto the regex \d\d\d-\d\d\d-\d\d\d\d, we encounter two gaps: gap_1 embraces the substring "(" with $gap_1.start = 0$, $gap_1.end = 2$. gap_2 embraces the substring ")" with $gap_2.start = 4$, $gap_2.end = 7$.

For each matching m, we find all gaps g that have no matching character in between (i.e., $\nexists k : g.start < k < g.end \land m(k)$ is defined), and where at least one of the following holds

- there is a character in the string between $g.start$ and $g.end$,
  i.e., $g.start < g.end - 1$
- there is a gap in the regex between $m(g.start)$ and $m(g.end)$,
  i.e., $m(g.start) \notin previous(m(g.end))$.

This set of gaps is returned by the method *findGaps* in Algorithm 1, Line 2.

Step 3: Finding Gap Overlaps. Gaps can overlap. Take for example the regex $r = 01234567$ and the missing words 0x567 and 012y7. One possible repair is 0(12)?(34|x|y)?(56)?7. We can find this repair only if we consider the overlap between the gaps. In this example, we have two gaps: one with span 1234 and one with span 3456. We have to partition the concatenation for the first gap into 12 and 34, and for the second gap into 34 and 56.

This is what Algorithm 2 does. It takes as input the regex r and the set of gaps *gaps*. It walks through the regex recursively, and treats each node of the regex. We split quantifiers $r\{min, max\}$ with $max < 100$ into $r\{\ldots\}r\{\ldots\}$, if the gap occurs between iterations. For other quantifiers, Kleene stars, and disjunctions, we descend recursively into the regex tree (Line 5).

For concatenation nodes, we determine all gaps that have their start point or their end point inside the concatenation (Line 7). Then, we determine the partitioning boundaries (Line 8; $l \in r$ means that regex r has a leaf node l). We consider each gap g (Line 9). We find whether the start point or the end point of any other gap falls inside g. This concerns only the boundaries between s and e (Lines 10–11). We partition the concatenation subsequence $c_s \ldots c_{e-1}$ by cutting at the boundaries (Line 12). Finally, the method is called recursively on the children of the concatenation that contain the start point or the end point of any gap (Lines 13–14).

Step 4: Adding Missing Parts. The previous step has given us, for each gap, a set of possible partitionings. In our example of the regex 01234567, the word 012y7, and the gap 3456, we have obtained the partitioning 34|56. This means that both 34 and 56 have to become optional in the regex, and that we can insert the substring y as an alternative to either of them: 012(34|y)(56)?7 or 012(34)?(56|y)7. Algorithm 3 will take this decision based on which solution performs better on the set E^- of negative examples. It may also happen that none of these solutions is permissible, because they all match too many negative examples. In that case, the algorithm will just add the word as a disjunct to the original regex, as in 01234567|012y7. To make these decisions, the algorithm

Algorithm 2. Find Gap Overlaps

INPUT: regex r, gaps $gaps$
OUTPUT: modified regex r, modified gaps $gaps$

1: **if** r is quantifier $q\{min, max\}$ with $max < 100$ **then**
2:     $r \leftarrow q\{\ldots\}q\{\ldots\}$ with appropriate ranges
3:     $findGapOverlaps(r, gaps)$
4: **else if** r is disjunction or Kleene star or quantifier **then**
5:     **for** child c of r **do** $findGapOverlaps(c, gaps)$
6: **else if** r is concatenation $c_1 \ldots c_n$ **then**
7:     $gaps' \leftarrow \{g : g \in gaps \land g.m(g.start) \in r \lor g.m(g.end) \in r\}$
8:     $idx \leftarrow \{i + 1 : g \in gaps' \land g.m(g.start) \in c_i\} \cup \{i : g \in gaps' \land g.m(g.end) \in c_i\}$
9:     **for** $g \in gaps'$ **do**
10:        $s \leftarrow$ i+1 if $g.m(g.start) \in c_i$, else 1
11:        $e \leftarrow$ i   if $g.m(g.end) \in c_i$, else $n + 1$
12:        $g.parts \leftarrow g.parts \cup \{c_i \ldots c_{j-1} : i, j \in idx \land s \leq i < j \leq e \land$
                                $(\nexists k : k \in idx \land i < k < j)\}$
13:    **for** $c_i \in \{c_i : \exists g \in gaps'. g.m(g.start) \in c_i \lor g.m(g.end) \in c_i\}$ **do**
14:        $findGapOverlaps(c_i, gaps)$

---

will compare the number of negative examples matched by the repaired regex with the number of negative examples matched by the original regex. The ratio of these two should be bounded by the threshold α.

Algorithm 3 takes as input a regex r, a set of gaps $gaps$ with partitionings, negative examples E^-, and a threshold α. The algorithm first makes a copy of the original regex (Line 1) and treats each gap (Line 2). For each gap, it considers all parts (Line 3). In the example, we will consider the part 34 and the part 56 of the gap 3456. The algorithm transforms the part into a disjunction of the part and the span of the gap. In the example, the part 34 is transformed into (34|y) (Lines 4). If the number of matched negative examples does not exceed the number of negative examples matched by the original regex times α (Line 5), the algorithm chooses this repair, and stops exploring the other parts of the gap (Lines 6–7). In Line 8, the algorithm collects all positive examples that are still not matched. The changes that were made for these words are undone (Line 9). Line 10 generalizes these words into one or several regexes. The generalization is adapted from [1]. First, we assign a group key to every word. The key is obtained by substituting substrings consisting only of digits with a (single) \d, lower or upper case characters with a [a-z] or [A-Z], and remaining characters with character class \W or \w. Finally we obtain the group regex r by adding $\{min, max\}$ after every character class, such that r matches all strings in that group. The algorithm then checks if the regex obtained this way is good enough (Lines 12–13). If this is the case, the regex is added as a disjunction (Line 13). Otherwise the words that contributed to that group are added disjunctively (Line 16). Table 1 shows how our method repairs the example regex.

Algorithm 3. Add Missing Parts

INPUT: regex r, set of strings S, gaps $gaps$, negative examples E^-, threshold $\alpha \geq 1$
OUTPUT: modified regex r
1: $org \leftarrow r$
2: **for** g $\in$ gaps **do**
3:    **for** part p $= (c_i \cdots c_j) \in g.parts$ **do**
4:       $r' \leftarrow r$ with $c_i \cdots c_j$ replaced by $(c_i \cdots c_j \,|\, g.span)$,
          and all other parts $c_x \ldots c_y$ in $g.parts$ made optional with $(c_x \ldots c_y \,|\,)$
5:       **if** $|E^- \cap L(r')| \leq \alpha \cdot |E^- \cap L(org)|$ **then**
6:          $r \leftarrow r'$
7:          **break**
8: $S' \leftarrow S \setminus L(r)$
9: undo all changes for $s \in S'$ not required by other repairs
10: $G \leftarrow$ generalize words in S'
11: **for** g $\in$ G **do**
12:    **if** $|E^- \cap L(r|g)| \leq \alpha \cdot |E^- \cap L(org)|$ **then**
13:       $r \leftarrow r|g$
14:    **else**
15:       **for** $s \in L(g) \cap S'$ **do**
16:          $r \leftarrow r|s$

Time Complexity. Let N' be the length of the input regex in expanded form (i.e., where quantifiers $r\{...\}$ have been replaced by copies of r). Let M be the sum of the lengths of the missing words. Let t be the runtime of applying a regex of length $O(N')$ to the negative examples. We show in our technical report [18] that our algorithm runs in $O(N'Mt)$.

5   Experiments

5.1   Setup

Measures. To evaluate our algorithm, we follow related work in the area [1,13, 17] and use a gold standard of positive example strings, $E^+ \supset S$. With this, the *precision* of a regex r is the fraction of positive examples matched among all examples matched:

$$prec(r) = \frac{|E^+ \cap L(r)|}{|L(r) \cap (E^+ \cup E^-)|}$$

The *recall* of r is the fraction of positive examples matched:

$$rec(r) = \frac{|E^+ \cap L(r)|}{|E^+|}$$

As usual, the *F1 measure* is the harmonic mean of these two measures.

Competitors. We compare our approach to both other methods [1,17] that can generalize given regexes (see Sect. 2). For [1], the code was not available upon request. We therefore had to re-implemented the approach. We think that our implementation is fair, because it achieves a higher F1-value (87% and 84%, as opposed to 84% and 82%) when run on the same datasets as in [1] (s.b.) with a full E^+ as input.

Datasets. We use 3 datasets from related work. The *Relie dataset*[1] [11] includes four tasks (phone numbers, course numbers, software names, and URLs). Each task comes with a set of documents. Each document consists of a span of words, and 100 characters of context to the left and to the right. Each span is annotated as a positive or a negative example, thus making up our sets E^+ and E^-, respectively. We manually cleaned the dataset by fixing obvious annotation errors, e.g., where a word is marked as a positive and a negative example in the same task. In total, the dataset contains 90807 documents.

The *Enron-Random dataset*[2] [12] contains a set of emails of Enron staff. The work of [1] uses it to extract phone numbers and dates. Unfortunately, there is no gold standard available for these tasks, and the authors of [1] could not provide one. Therefore, we manually annotated phone numbers and dates on 200 randomly selected files, which gives us E^+. As in [1], any string that is matched on these 200 documents and that is not a positive example will be considered a negative example. In this way, we obtain a large number of negative examples.

The *YAGO-Dataset* consists of Wikipedia infobox attributes, where the dates and numbers that were used to build YAGO [19] have been annotated as positive examples. This dataset is used in [17]. As in *Enron-Random*, all strings that are not annotated as positive examples count as negative examples.

We thus have 8 tasks: 4 for the Relie dataset, 2 for the Enron dataset, and 2 for the YAGO dataset. Each task comes with positive examples E^+ and negative examples E^-. Our algorithm needs as input an initial regex that shall be repaired. For the Enron and YAGO tasks, we used the initial regexes given in [1,17]. For Relie, we asked our colleagues to produce regexes by hand. For this purpose, we provided them with 10 randomly chosen examples from E^+ for each task, and asked them to write a regex. This gives us 5 initial regexes for each Relie task. Table 2 summarizes our datasets.

Runs. Our algorithm does not take as input the entire set of positive examples E^+, but a small subset S of positive examples. To simulate a real setting for our algorithm, we randomly select S from E^+. We average our results over 10 different random draws of S. For each draw, we use each initial regex that we have at our disposal, and average our results over these. Thus, we run our algorithm 50 times for each Relie task, 10 times for each Enron-Random task, and 150 times for each YAGO task, and we average the obtained numbers over these. Our competitors are not designed to work on a small set of positive examples. Therefore, we provide them with additional positive examples obtained from running the original regex on the input dataset. Our method does not need this

---

[1] http://dbgroup.eecs.umich.edu/regexLearning/.

[2] http://www.cs.cmu.edu/~einat/datasets.html.

Table 2. Statistics of the datasets

Task	Documents	Avg. size	$|E^+|$	Regexes
ReLie/phone [11]	41896	211	2657	5
ReLie/course [11]	569	210	314	5
ReLie/software [11]	44413	185	2307	5
ReLie/urls [11]	3929	176	735	5
Enron/phone [1]	225	1452	145	1
Enron/date [1]	225	1452	392	1
YAGO/dates [17]	100000	25	109824	15
YAGO/numbers [17]	100000	57	131149	15

step. All algorithms are implemented in Java 8. The experiments were run on an Intel Xeon with 2.70 GHz and 250 GB memory.

5.2   Experimental Evaluation and Results

F1-measure. Table 3 shows the F1-measure on all datasets for different algorithms: the original regex, the disjunction-baseline (which consists just of a disjunction of the original regex with the 10 positive words), the star-baseline (which is just `.*`), the method from [1], the method from [17], and our method with different values for α. The table shows the improvement of the $F1$ measure w.r.t. the dis-baseline, in percentage points. For example, for *Relie/phone* and $\alpha = 1.2$, our algorithm achieves an F1 value of $81.6\% + 2.3\% = 83.9\%$. We can see that, across almost all tasks and settings, our algorithm outperforms the original regex

Table 3. F1 measure for different values of the parameter α, improvement over the dis-baseline in percentage points. Bold numbers indicate the maximum F1 measure within each row. ▲ (and △) indicates significant improvement relative to the dis-baseline for a significance level of 0.01 (and 0.05).

Task	Original	Baseline		Competitors		Our approach		
		dis	star	[1]	[17]	$\alpha = 1.0$	$\alpha = 1.1$	$\alpha = 1.20$
ReLie/phone	81.6	82.1	12.3	−.5 ▼	+.3	+.8	+1.6 △	**+2.3** ▲
ReLie/course	45.8	46.0	48.4	**+6.4** ▲	+.2 ▲	+.5 ▲	+1.3 ▲	+1.4 ▲
ReLie/software	9.2	12.4	9.9	+.1 ▲	−.0	+.7 ▲	+3.9	**+4.6**
ReLie/urls	55.2	56.0	31.5	−30.3 ▼	+.3	+2.9	**+4.2** ▲	**+4.2** ▲
Enron/phone	61.7	61.7	.1	−7.7	+5.3 △	**+21.0** ▲	**+21.0** ▲	**+21.0** ▲
Enron/date	72.3	72.4	.0	−49.6 ▼	−.0	**+.6**	**+.6**	+.4
YAGO/number	40.1	40.1	31.0	−11.5 ▼	+1.8 ▲	**+3.4**	+2.2	+2.4
YAGO/date	70.1	70.1	34.3	−26.3 ▼	+.3 ▲	**+6.9** ▲	+6.7 ▲	+6.6 ▲

Table 4. Length of the repaired regexes (# of characters). Bold numbers indicate the shortest ones (without the original).

Task	Original	Baseline		Competitors		Our approach		
		dis	star	[1]	[17]	$\alpha = 1.0$	$\alpha = 1.1$	$\alpha = 1.20$
ReLie/phonenum	41.6	230.3	2.0	94.6	221.2	**46.8**	50.0	53.1
ReLie/coursenum	22.2	203.2	2.0	280.4	181.1	**33.7**	48.2	52.5
ReLie/softwarename	43.2	168.2	2.0	594.7	168.2	**54.4**	67.3	67.8
ReLie/urls	52.4	630.3	2.0	5826.1	570.2	**70.6**	74.7	74.7
Enron/phone	17.0	199.1	2.0	243.8	164.8	**41.4**	**41.4**	**41.4**
Enron/date	17.0	170.6	2.0	581.0	170.6	**34.2**	**34.2**	35.6
YAGO/number	65.4	223.2	2.0	19471.0	207.9	**119.4**	120.5	120.7
YAGO/date	191.4	336.2	2.0	4337.0	313.6	203.6	203.6	**203.5**

as well as the dis-baseline. We verified the significance of the F1 measures with a micro sign test [21]. Detailed results on recall and precision can be found in our technical report [18].

If α is small, the algorithm is conservative, and tends towards the dis-baseline. If α is larger, the algorithm performs repairs even if this generates more negative examples. As we can see, the impact of α is marginal. We take this as an indication that our method is robust to the choice of the parameter.

Regex Length. Table 4 shows the average length of the generated regexes (in number of characters). For our approach, the length depends on α: If the value is large, the algorithm will tend to integrate the words into the original regex. Then, the words are no longer subject to the generalization mechanism. Still, the impact of α is marginal: No matter the value, our algorithm produces regexes that are up to 8 times shorter than the dis-baseline, and nearly always at least twice as short as either competitor – at comparable or better precision and recall.

Runtime. For the ReLie and Enron dataset, all approaches take a time in the order of seconds for repairing one regex. Due to the much larger E^+, runtimes differ for the YAGO dataset: fastest system is [17] with 12 s on average, followed by our approach with 84 s. The runtime of our reimplementation of [1] lies in the order of minutes, as we did not optimize for runtime efficiency.

Example. Table 5 shows an example of a repaired regex in the Relie/phonenum task. Our algorithm successfully identifies the non-matched characters : and > at the beginning of a phone number. It introduces them as options at the beginning of the original regex, leaving the rest of the regex intact. The dis-baseline, in contrast, would add all words in a large disjunction. Our solution is more general and more syntactically similar to the original regex.

Table 5. Example of a scenario for the Relie/phonenum task

Original regex: ({0,1} \d{3}){0,1}(-|\.|) \d{3}(-|\.|) d{4}
Missing words: :734-763-2200 >317.569.8903 >443.436.0787 >512.289.1407 >734-615-9673 >734-647-8027 >734-763-5664 >734.647.3256 >773.339.3223
Repaired regex: ((|:|>)?\d{3})?(-|\.|)\d{3}(-|\.|)\d{4}

6   Conclusion

In this paper, we have proposed an algorithm that can add missing words to a given regular expression. With only a small set of positive examples, our method generalizes the input regex, while maintaining its structure. In this way, our approach improves the precision and recall of the original regex.

We have evaluated our method on various datasets, and we have shown that with few positive examples, we can improve the F1 measure on the ground truth. This is a remarkable result, because it shows that regexes can be generalized based on very small training data. What is more, our approach produces regexes that are significantly shorter than the baseline and competitors. This shows that our method generalizes the regexes in a meaningful and non-trivial way.

Both the source code of our approach and the experimental results are available online at https://thomasrebele.org/projects/regex-repair. For future work, we aim to shorten the produced regexes further, by generalizing the components into character classes.

Acknowledgments. This research was partially supported by Labex DigiCosme (project ANR-11-LABEX-0045-DIGICOSME) operated by ANR as part of the program "Investissement d'Avenir" Idex Paris-Saclay (ANR-11-IDEX-0003-02).

References

1. Babbar, R., Singh, N.: Clustering based approach to learning regular expressions over large alphabet for noisy unstructured text. In: Workshop on Analytics for Noisy Unstructured Text Data (2010)
2. Bartoli, A., Davanzo, G., Lorenzo, A.D., Mauri, M., Medvet, E., Sorio, E.: Automatic generation of regular expressions from examples with genetic programming. In: GECCO (2012)
3. Bartoli, A., Davanzo, G., Lorenzo, A.D., Medvet, E., Sorio, E.: Automatic synthesis of regular expressions from examples. IEEE Comput. **47**(12), 72–80 (2014)
4. Bartoli, A., De Lorenzo, A., Medvet, E., Tarlao, F.: On the automatic construction of regular expressions from examples. In: GECCO (2016)
5. Brauer, F., Rieger, R., Mocan, A., Barczynski, W.M.: Enabling information extraction by inference of regular expressions from sample entities. In: CIKM (2011)
6. Ficara, D., Giordano, S., Procissi, G., Vitucci, F., Antichi, G., Di Pietro, A.: An improved DFA for fast regular expression matching. SIGCOMM Comput. Commun. Rev. **38**(5), 29–40 (2008). https://doi.org/10.1145/1452335.1452339

7. Gulwani, S.: Automating string processing in spreadsheets using input-output examples. In: SIGPLAN Notices, vol. 46 (2011)
8. Knight, J.R., Myers, E.W.: Approximate regular expression pattern matching with concave gap penalties. Algorithmica **14**(1), 85–121 (1995)
9. Le, V., Gulwani, S.: FlashExtract: a framework for data extraction by examples. In: PLDI (2014)
10. Lehmann, J., et al.: DBpedia - a large-scale, multilingual knowledge base extracted from wikipedia. Seman. Web J. **6**(2), 167–195 (2015)
11. Li, Y., Krishnamurthy, R., Raghavan, S., Vaithyanathan, S., Jagadish, H.V.: Regular expression learning for information extraction. In: EMNLP (2008)
12. Minkov, E., Wang, R.C., Cohen, W.W.: Extracting personal names from email: applying named entity recognition to informal text. In: EMNLP (2005)
13. Murthy, K., Padmanabhan, D., Deshpande, P.M.: Improving recall of regular expressions for information extraction. In: Wang, X.S., Cruz, I., Delis, A., Huang, G. (eds.) WISE 2012. LNCS, vol. 7651, pp. 455–467. Springer, Heidelberg (2012). https://doi.org/10.1007/978-3-642-35063-4_33
14. Myers, E.W., Miller, W.: Approximate matching of regular expressions. Bull. Math. Biol. **51**(1), 5–37 (1989)
15. Navarro, G.: Approximate regular expression searching with arbitrary integer weights. Nord. J. Comput. **11**(4), 356–373 (2004)
16. Prasse, P., Sawade, C., Landwehr, N., Scheffer, T.: Learning to identify concise regular expressions that describe email campaigns. J. Mach. Learn. Res. **16**(1), 3687–3720 (2015)
17. Rebele, T., Tzompanaki, K., Suchanek, F.: Visualizing the addition of missing words to regular expressions. In: ISWC (2017)
18. Rebele, T., Tzompanaki, K., Suchanek, F.: Technical report: adding missing words to regular expressions. Technical report, Telecom ParisTech (2018)
19. Suchanek, F.M., Kasneci, G., Weikum, G.: Yago: a core of semantic knowledge. In: WWW (2007)
20. Wu, S., Manber, U., Myers, E.: A subquadratic algorithm for approximate regular expression matching. J. Algorithms **19**(3), 346–360 (1995)
21. Yang, Y., Liu, X.: A re-examination of text categorization methods. In: SIGIR (1999)

Marrying Community Discovery and Role Analysis in Social Media via Topic Modeling

Gianni Costa and Riccardo Ortale[✉]

ICAR-CNR, Via P. Bucci 41c, 87036 Rende, (CS), Italy
{costa,ortale}@icar.cnr.it

Abstract. We explore the adoption of topic modeling to inform the seamless integration of community discovery and role analysis. For this purpose, we develop a new Bayesian probabilistic generative model of social media, according to which the observation of social links and textual contents is governed by novel and intuitive relationships among latent content topics, communities and roles. Variational inference under the devised model allows for exploratory, descriptive and predictive tasks, including the detection and interpretation of overlapping communities, roles and topics as well as the prediction of missing links. Extensive tests on real-world social media reveal a superior accuracy of the proposed model in comparison to state-of-the-art competitors, which substantiates the rationality of the motivating intuition. The experimental results are also insightfully inspected from a qualitative viewpoint.

1 Introduction

The integration of community discovery and role analysis has been investigated to accurately explain network topology [3,4,6]. However, such efforts disregard the textual content of node interactions. A study of how roles influence content generation within networks of correspondents exchanging emails is developed in [9], without jointly considering the underlying community structure.

In principle, looking at the textual content of node interactions in the simultaneous analysis of communities and roles improves the accuracy of each task in isolation as well as their mutuality and interpenetration. This is in turn expected to be beneficial for a higher accuracy of the two tasks in tandem. Intuitively, awareness of text topics is helpful to refine (or, also, capture) affiliations of nodes to communities and roles, even with few or no interactions with other nodes. This is especially interesting with regard to communities, that can be participated by nodes because of either their social connections or, more simply, their interest in shared topics (even in the absence of social connections).

Jointly modeling community discovery, role analysis and topic modeling [2] is, to the best of our knowledge, an unexplored research line, that involves dealing with several issues. Firstly, links between nodes have to be explained in terms of their affiliations to the foresaid communities and roles. Secondly, the traditional

D. Phung et al. (Eds.): PAKDD 2018, LNAI 10938, pp. 80–91, 2018.
https://doi.org/10.1007/978-3-319-93037-4_7

notion of role has to be generalized, so that to influence node behavior not only in social networking but also in message wording. Thirdly, communities, roles and topics are unobserved network properties with no ascertained connections.

In this paper, we present a new model-based machine-learning approach to the unsupervised, joint analysis of communities and roles in social media, where nodes can interact with one another either in a purely social manner, i.e., by simply establishing connections (such as, e.g., in the case of friending, following, (dis)trusting) or, also, by using their ties in order to coauthor documents (e.g., project proposals, deliverables, technical reports, articles) as well as share text messages (e.g., news, comments, reviews, opinions, questions/answers, status updates, tweets). The devised approach consists in performing posterior inference in a latent-factor Bayesian generative model of the targeted social-media, referred to as NOODLES (*overlappiNg cOmmunities and rOles from noDe Links and messagES*), in which topic modeling is used to inform the seamless integration of community discovery and role analysis.

Extensive experiments on real-world social media reveal that NOODLES overcomes state-of-the-art competitors in community detection and link prediction, which confirms that the topic-aware integration of community discovery and role analysis is beneficial to improved accuracy. A qualitative evaluation of NOODLES is also provided, in order to demonstrate its behavior in practical applications.

2   Background

Notation and preliminaries are introduced below.

2.1   Observed Social-Media Properties: Topology and Messages

As with any complex system, social media can be viewed from the network-centric perspective as a graph $\mathcal{G} = \{N, A\}$, where $N = \{1, \ldots N\}$ is a set of nodes (numbered 1 through N) and $A \subseteq N \times N$ is a set of links.

Nodes correspond to entities (e.g., individuals, organizations), that interact in the network along with authoring (and exchanging) text messages.

Links represent directed interactions between nodes and are summarized into a binary adjacency matrix L, whose generic entry $L_{n,n'}$ is 1 iff a link is observed from node n to node n' (i.e., iff $\langle n, n' \rangle \in A$) and 0 otherwise.

The set of text messages[1] authored by node n is d_n. Let V be a vocabulary of V word tokens. Any message $d \in d_n$ is some suitable collection of words from the vocabulary V, i.e., $d \subseteq V$. Messages are summarized into the binary array F, whose arbitrary entry $F_{w,d}^{(n)}$ is 1 iff word $w \in V$ is chosen by node n in authoring message $d \in d_n$ and 0 otherwise.

In practical applications, matrices L and F are, usually, sparse.

---

[1] Notice that, in the case of collaboration networks, the term message refers to the corresponding type of coauthored content, such as, e.g., project proposals, deliverables and publications. In particular, one data set in Sect. 6 is chosen from the scientific collaboration domain and, in such a context, message is a synonym of publication.

2.2   Unobserved Social-Media Properties: Topics, Affiliations, Roles and Communities

Let $T = \{1, \ldots, T\}$ be a set of T topics. The text messages authored by nodes deal with one or more topics from T. Basically, the generic topic $t \in T$ is some assignment of relevance to the individual words of the vocabulary V, that ultimately identifies a semantically coherent theme. The relevance of each word $w \in V$ to every topic $t \in T$ is captured through the corresponding affiliation $\pi_{w,t}$. Notation $\boldsymbol{\Pi} \triangleq \{\pi_{w,t} | w \in V, t \in T\}$ concisely indicates the relevance of all words to the individual topics.

Communities are structures of $\mathcal{G}$, that can be formalized as a set $\boldsymbol{C} \triangleq \{C_1, \ldots, C_K\}$ of K overlapping groups of nodes (i.e., such that $C_k \subseteq \boldsymbol{N}$ with $k = 1, \ldots, K$). Any node n can participate in all communities, although with a different involvement. In particular, affiliation $\vartheta_{n,k}$ denotes the degree to which n takes part in the arbitrary community $C_k \in \boldsymbol{C}$. Overall, node affiliations to communities are compactly denoted as $\boldsymbol{\Theta} \triangleq \{\vartheta_{n,k} | n \in \boldsymbol{N}, C_k \in \boldsymbol{C}\}$. Within every community C_k (with $k = 1, \ldots, K$), nodes have social connections and/or a shared focus on some corresponding subset of topics. A same topic is differently relevant to the individual communities. Affiliation $\eta_{t,k}$ represents the relevance of topic t in the context of community C_k. Topic relevance to communities is succinctly indicated as $\boldsymbol{H} \triangleq \{\eta_{t,k} | t \in \boldsymbol{T}, C_k \in \boldsymbol{C}\}$.

Roles form a set $\boldsymbol{R} \triangleq \{R_1, \ldots, R_H\}$ of H behavioral classes. As discussed in Sect. 3, the arbitrary role R_h (with $h = 1, \ldots, H$) influences node behavior both in networking and authoring through, respectively, its inherently-characteristic interactions with the roles of the neighboring nodes and the intrinsic bias for specific topics. Formally, $\boldsymbol{E}_h \triangleq \{\epsilon_{h,h'} | R_{h'} \in \boldsymbol{R}\}$ is the behavioral pattern of role R_h, with affiliation $\epsilon_{h,h'}$ being the strength of interaction from R_h to any other role $R_{h'}$ (with $h' = 1, \ldots, H$). Notably, all nodes playing the same role behave identically with the roles of their neighbors, in compliance with mixed-membership block-modeling. As a whole, the behavioral patterns of roles are represented as $\boldsymbol{E} \triangleq \cup_{h=1}^{H} \boldsymbol{E}_h$. Moreover, $\boldsymbol{\Gamma}_h \triangleq \{\gamma_{t,h} | t \in \boldsymbol{T}, R_h \in \boldsymbol{R}\}$ is the bias of role R_h for the individual topics, with affiliation $\gamma_{t,h}$ being the appropriateness of topic t to role R_h. Together, the topic biases of all roles are formalized as $\boldsymbol{\Gamma} \triangleq \cup_{h=1}^{H} \boldsymbol{\Gamma}_h$. The generic node n can play all roles, though with a different attitude. Affiliation $\sigma_{n,h}$ is the extent to which n is suitable to play role $R_h \in \boldsymbol{R}$. Collectively, node affiliations to roles are denoted as $\boldsymbol{\Sigma} \triangleq \{\sigma_{n,h} | n \in \boldsymbol{N}, R_h \in \boldsymbol{R}\}$.

2.3   Problem Statement

Let $\mathcal{G}$ be an input network. We are interested in inferring the unobserved properties $\boldsymbol{\Theta}$, $\boldsymbol{\Sigma}$, $\boldsymbol{\Pi}$, $\boldsymbol{H}$, $\boldsymbol{\Gamma}$ and $\boldsymbol{E}$ from the observation of its topology $\boldsymbol{L}$ and the node messages $\boldsymbol{F}$. We aim to perform the following tasks:

– the unsupervised exploratory analysis of $\mathcal{G}$, i.e., the discovery of the latent organization of $\mathcal{G}$ in underlying communities, roles and topics (along with their respective relationships);

- the prediction of missing links;
- the interpretation of topics, communities and roles.

We deal with the unobserved network properties through posterior inference under a Bayesian probabilistic generative model of $\mathcal{G}$. Basically, such a model postulates the reliance of the observed properties on the unobserved ones. Variational inference stretches back to the latter by reversion.

Hereinafter, all elements of Θ, Σ, Π, H, Γ and E are viewed as random variables, being unknown and not directly measurable. Our network model is introduced in Sect. 3. Variational inference is developed in Sect. 4.

3   NOODLES

NOODLES (*overlappiNg cOmmunities and rOles from noDe Links and messagES*) is a network model, whose innovative peculiarity relies in the exploitation of topic modeling to inform the integration of community discovery and role analysis. This is achieved as an effort to explain the observation of node interactions L and text messages F in a targeted network $\mathcal{G}$, from a Bayesian probabilistic perspective. Essentially, NOODLES assumes that both are the result of a generative process, that is influenced by the latent network properties introduced in Sect. 2.2. Such a generative process operates on the basis of conditional (in)dependencies between the observations (i.e., L and F) and the latent network properties (i.e., Θ, Σ, Π, H, Γ and E). These conditional (in)dependencies are formally shown in the directed graphical representation of Fig. 1, where the observations are distinguished from the latent network properties, being reported in the form of shaded and unshaded random variables, respectively. The generative process under NOODLES accomplishes the realization of the random variables of Fig. 1 according to the algorithmic description reported in Fig. 2 (in which, at steps I., II., III. and IV., α_ψ and β_ψ with $\psi \in \{\vartheta, \sigma, \pi, \gamma, \eta, \epsilon\}$ are, respectively, shape and rate hyperparameters).

Basically, at steps I. through VI., the latent affiliations Θ, Σ, Π, H, Γ and E are individually sampled from respective Gamma priors. These enforce nonnegativity, which improves the interpretability of NOODLES, by favoring sparseness in its representation. Also, it ensures that the overall strength of affiliation for each node, role, topic or word does not equal 1, which avoids the typical inconvenient of mixed-membership modeling. Consequently, a very strong affiliation of any node, role, topic or word does not imply a corresponding drop in the overall strength of all other affiliations of that node, role, topic or word.

At step V., the presence/absence of directed links is sampled from a Poisson distribution, that is placed over L as the link data likelihood. In particular, the establishment of a directed link from a node n to a node n' is governed by the Poisson-specific rate $\delta_{n,n'}$, that captures the interaction from n to n' in terms of their affiliations to common communities, respective roles and corresponding behavioral patterns. More precisely,

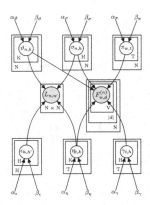

Fig. 1. Graphical representation of NOODLES in plate notation

$$\delta_{n,n'} \triangleq \sum_{k=1}^{K} \sum_{h,h'=1}^{H} \vartheta_{n,k} \sigma_{n,h} \epsilon_{h,h'} \sigma_{n',h'} \vartheta_{n',k} \tag{1}$$

Equation 1 is a generalization of pluralistic homophily [10], in which the latter is combined with link direction as well as role analysis.

Notice that, though being a model of directed networks, NOODLES can straightforwardly handle undirected networks.

At step VI., the presence/absence of word w in document d authored by node n is sampled from a Poisson distribution, that is placed over $\boldsymbol{F}$ as the text data likelihood. Specifically, the inclusion of w among the words of d on the part of n is governed through the below rate

$$\delta_{w,d}^{(n)} \triangleq \sum_{k=1}^{K} \sum_{h=1}^{H} \sum_{t=1}^{T} \vartheta_{n,k} \sigma_{n,h} \eta_{t,k} \gamma_{t,h} \pi_{w,t} \tag{2}$$

$\delta_{w,d}^{(n)}$ captures the appropriateness of word w to document d across all latent topics as well as their relevance to the communities and roles, which node n is affiliated to.

Remarkably, Poisson distributions at steps V. and VI. are well suited to deal with binary data and, also, beneficial for faster inference on sparse networks [7].

4   Posterior Inference

NOODLES assumes that the observation of links and messages, within a targeted network $\mathcal{G}$, is influenced by its unobserved properties according to the generative process of Fig. 2. In this section, we design a variational algorithm, with which to perform posterior probabilistic inference under NOODLES, for the purpose of quantifying the unobserved properties $\boldsymbol{\Theta}$, $\boldsymbol{\Sigma}$, $\boldsymbol{\Pi}$, $\boldsymbol{H}$, $\boldsymbol{\Gamma}$ and $\boldsymbol{E}$ given the observations $\boldsymbol{L}$ and $\boldsymbol{F}$. In principle, this is accomplished by computing the

I. For each word $w \in \boldsymbol{V}$
  - For each topic $t \in \boldsymbol{T}$
    - draw the relevance of word w to topic t, i.e., $\pi_{w,t} \sim \mathsf{Gamma}(\pi_{w,t} | \alpha_\pi, \beta_\pi)$.

II. For each node $n \in \boldsymbol{N}$
  - For each community $C_k \in \boldsymbol{C}$
    - draw the degree $\vartheta_{n,k}$ to which n is involved in community C_k, i.e., $\vartheta_{n,k} \sim \mathsf{Gamma}(\vartheta_{n,k} | \alpha_\vartheta, \beta_\vartheta)$.
  - For each role $R_h \in \boldsymbol{R}$
    - draw the extent $\sigma_{n,h}$ to which n is suitable to play role R_h, i.e., $\sigma_{n,h} \sim \mathsf{Gamma}(\sigma_{n,h} | \alpha_\sigma, \beta_\sigma)$.

III. For each topic $t \in \boldsymbol{T}$
  - For each community $C_k \in \boldsymbol{C}$
    - draw the relevance of topic t to community C_k, i.e., $\eta_{t,k} \sim \mathsf{Gamma}(\eta_{t,k} | \alpha_\eta, \beta_\eta)$.
  - For each role $R_h \in \boldsymbol{R}$
    - draw the relevance of topic t to role R_h, i.e., $\gamma_{t,h} \sim \mathsf{Gamma}(\gamma_{t,h} | \alpha_\gamma, \beta_\gamma)$.

IV. For each role $R_h \in \boldsymbol{R}$
  - For each role $R_{h'} \in \boldsymbol{R}$
    - draw the strength of interaction from role R_h to role $R_{h'}$, i.e., $\epsilon_{h,h'} \sim \mathsf{Gamma}(\epsilon_{h,h'} | \alpha_\epsilon, \beta_\epsilon)$.

V. For each node $n \in \boldsymbol{N}$
  - For each node $n' \in \boldsymbol{N}$
    - draw the presence/absence of a link $L_{n,n'}$ from n to n', i.e., $L_{n,n'} \sim \mathsf{Poisson}(L_{n,n'} | \delta_{n,n'})$, where $\delta_{n,n'}$ is the rate defined by Eq. 1.

VI. For each node $n \in \boldsymbol{N}$
  - For each text message $d \in \boldsymbol{d}_n$
  - For each word $w \in \boldsymbol{V}$
    - draw whether or not word w is chosen by node n in authoring text message d, i.e., $F_{w,d}^{(n)} \sim$ $\mathsf{Poisson}(F_{w,d}^{(n)} | \delta_{w,d}^{(n)})$, where $\delta_{w,d}^{(n)}$ is the rate defined by Eq. 2.

Fig. 2. The generative process under NOODLES

posterior distribution $\Pr(\boldsymbol{\Theta}, \boldsymbol{\Sigma}, \boldsymbol{\Pi}, \boldsymbol{H}, \boldsymbol{\Gamma}, \boldsymbol{E} | \boldsymbol{L}, \boldsymbol{F}, \boldsymbol{\zeta})$, where $\boldsymbol{\zeta} = \{\alpha_\vartheta, \beta_\vartheta, \alpha_\sigma, \beta_\sigma,$ $\alpha_\pi, \beta_\pi, \alpha_\eta, \beta_\eta, \alpha_\gamma, \beta_\gamma, \alpha_\epsilon, \beta_\epsilon\}$ is the set of all hyperparameters. However, exact posterior inference is generally intractable under Bayesian models of practical relevance, because of the complexity of the true posterior distribution. We obviate by employing mean-field variational inference under NOODLES to compute an analytical approximation of $\Pr(\boldsymbol{\Theta}, \boldsymbol{\Sigma}, \boldsymbol{\Pi}, \boldsymbol{H}, \boldsymbol{\Gamma}, \boldsymbol{E} | \boldsymbol{L}, \boldsymbol{F}, \boldsymbol{\zeta})$. Variational inference tends to be faster and more easily scalable on large networks than MCMC sampling [1].

We proceed by enriching NOODLES through the addition of auxiliary latent variables, which ultimately simplifies the algorithmic developments. Specifically, due to the additivity of Poisson random variables, each $L_{n,n'}$ can be redefined as $L_{n,n'} \triangleq \sum_{k=1}^{K} \sum_{h,h'=1}^{H} z_{n,n'}^{(k,h,h')}$, where $z_{n,n'}^{(k,h,h')} \sim Poisson(\vartheta_{n,k}\sigma_{n,h}\epsilon_{h,h'}$ $\sigma_{n',h'}\vartheta_{n',k})$. Here, the auxiliary random variable $z_{n,n'}^{(k,h,h')}$ captures the contribution of the interaction from node n to node n' inside community C_k, when R_h and $R_{h'}$ are their respective roles. Likewise, each $F_{w,d}^{(n)}$ can be redefined as $F_{w,d}^{(n)} \triangleq$ $\sum_{k=1}^{K} \sum_{h=1}^{H} \sum_{t=1}^{T} s_{w,d}^{(n,k,h,t)}$, with $s_{w,d}^{(n,k,h,t)} \sim Poisson(\vartheta_{n,k}\sigma_{n,h}\eta_{t,k}\gamma_{t,h}\pi_{w,t})$ being a contribution, that captures whether word w is used in document d by node n, when n plays role R_h within community C_k and t is the topic for w. Let $\boldsymbol{Z} = \{z_{n,n'} | n, n' \in \boldsymbol{V}\}$ and $\boldsymbol{S} = \{s_{w,d}^{(n)} | n \in \boldsymbol{V}, d \in \boldsymbol{d}_n, w \in \boldsymbol{V}\}$ be the sets of auxiliary variables added to NOODLES. Formally, the mean-field family of approximate distributions over $\boldsymbol{\Theta}, \boldsymbol{\Sigma}, \boldsymbol{\Pi}, \boldsymbol{H}, \boldsymbol{\Gamma}, \boldsymbol{E}, \boldsymbol{Z}$, and $\boldsymbol{S}$ is

$$\rho_{n,k}^{(shp)} = \alpha_\vartheta + \left[\sum_{n' \in N, R_h, R_{h'} \in R} L_{n,n'} \tau_{n,n'}^{(k,h,h')} + L_{n',n} \tau_{n',n}^{(k,h,h')} \right] + \left[\sum_{d \in d_n, w \in V, R_h \in R, t \in T} F_{w,d}^{(n)} \mu_{w,d}^{(n,k,h,t)} \right]$$

$$\lambda_{n,h}^{(shp)} = \alpha_\sigma + \left[\sum_{n' \in N, C_k \in C, R_{h'} \in R} L_{n,n'} \tau_{n,n'}^{(k,h,h')} + L_{n',n} \tau_{n',n}^{(k,h',h)} \right] + \left[\sum_{d \in d_n, w \in V, C_k \in C, t \in T} F_{w,d}^{(n)} \mu_{w,d}^{(n,k,h,t)} \right]$$

$$\kappa_{h,h'}^{(shp)} = \alpha_\epsilon + \left[\sum_{n,n' \in N, C_k \in C} L_{n,n'} \tau_{n,n'}^{(k,h,h')} \right]; \quad \xi_{w,t}^{(shp)} = \alpha_\pi + \left[\sum_{n \in N, d \in d_n, C_k \in C, R_h \in R} F_{w,d}^{(n)} \mu_{w,d}^{(n,k,h,t)} \right]$$

$$\omega_{t,k}^{(shp)} = \alpha_\eta + \left[\sum_{n \in N, d \in d_n, w \in V, R_h \in R} F_{w,d}^{(n)} \mu_{w,d}^{(n,k,h,t)} \right]; \quad o_{t,h}^{(shp)} = \alpha_\gamma + \left[\sum_{n \in N, d \in d_n, w \in V, C_k \in C} F_{w,d}^{(n)} \mu_{w,d}^{(n,k,h,t)} \right]$$

Fig. 3. Variational updates for Gamma-shape parameters

$$q(\Theta, \Sigma, \Pi, H, \Gamma, E, Z, S | \nu) = \prod_{n \in N, C_k \in C} q(\vartheta_{n,k} | \rho_{n,k}) \prod_{n \in N, R_h \in R} q(\sigma_{n,h} | \lambda_{n,h}) \cdot \prod_{w \in V, t \in T} q(\pi_{w,t} | \xi_{w,t}) \cdot$$
$$\prod_{t \in T, C_k \in C} q(\eta_{t,k} | \omega_{t,k}) \cdot \prod_{t \in T, R_h \in R} q(\gamma_{t,h} | o_{t,h}) \prod_{R_h, R_{h'} \in R} q(\epsilon_{h,h'} | \kappa_{h,h'}) \cdot$$
$$\prod_{n,n' \in N} q(z_{n,n'} | \tau_{n,n'}) \cdot \prod_{n \in N, d \in d_n, w \in V} q(s_{w,d}^{(n)} | \mu_{w,d}^{(n)}) \tag{3}$$

with $\nu \triangleq \{\rho_{n,k}, \lambda_{n,h}, \xi_{w,t}, \omega_{t,k}, o_{t,h}, \kappa_{h,h'} \tau_{n,n'}, \mu_{w,d}^{(n)} | n, n' \in N, C_k \in C, R_h, R_h' \in R, w \in V\}$ being the set of all variational parameters, that individually condition the respective factors on the right hand side of Eq. 3.

The approximate posterior distribution is obtained from Eq. 3 by fitting the variational parameters ν. Therein, we emphasize that due to the addition of the auxiliary random variables, NOODLES is a *conditionally conjugate* model. For such a class of models, the variational parameters can be fitted through a simple coordinate-ascent variational Algorithm [1].

Essentially, each variational parameter is iteratively optimized, while all others are held fixed. The updates used for optimization comply with the nature of the individual variational parameters, that can be understood by accounting for the functional forms of the factors on the right hand side of Eq. 3. More precisely, factors $q(z_{n,n'} | \tau_{n,n'})$ and $q(s_{w,d}^{(n)} | \mu_{w,d}^{(n)})$ are multinomial distributions, while all other factors are Gamma distributions.[2] Therefore, the generic $\tau_{n,n'}$ and $\mu_{w,d}^{(n)}$ are variational multinomial parameters. These are respectively fitted through the updates in Fig. 5, where $\Psi[\cdot]$ is the digamma function (i.e., the first derivative of *log* Γ function). Instead, $\rho_{n,k}, \lambda_{n,h}, \xi_{w,t}, \omega_{t,k}, o_{t,h}$ and $\kappa_{h,h'}$ are pairs of variational Gamma parameters, individually consisting of a shape (denoted by a superscript (shp)) and a rate (denoted by a superscript $(rate)$). The shape parameters are fitted through the updates in Fig. 3, whereas the rate parameters are fitted by means of the updates in Fig. 4.

---

[2] The derivation of both the functional forms of the factors on the right hand side of Eq. 3 and the updates of the respective variational parameters is omitted for brevity.

$$\rho_{n,k}^{(rate)} = \left[\sum_{n'\in N, R_h, R_{h'}\in R} \frac{\rho_{n',k}^{(shp)}}{\rho_{n',k}^{(rate)}} \frac{\kappa_{h,h'}^{(shp)}}{\kappa_{h,h'}^{(rate)}} \left(\frac{\lambda_{n,h}^{(shp)}}{\lambda_{n,h}^{(rate)}} \frac{\lambda_{n',h'}^{(shp)}}{\lambda_{n',h'}^{(rate)}} + \frac{\lambda_{n',h}^{(shp)}}{\lambda_{n',h}^{(rate)}} \frac{\lambda_{n,h'}^{(shp)}}{\lambda_{n,h'}^{(rate)}}\right)\right] +$$
$$\left[\sum_{d\in d_n, w\in V, R_h\in R, t\in T} \frac{\lambda_{n,h}^{(shp)}}{\lambda_{n,h}^{(rate)}} \frac{\omega_{t,k}^{(shp)}}{\omega_{t,k}^{(rate)}} \frac{o_{t,h}^{(shp)}}{o_{t,h}^{(rate)}} \frac{\xi_{w,t}^{(shp)}}{\xi_{w,t}^{(rate)}}\right] + \frac{\alpha_\vartheta}{\beta_\vartheta}$$

$$\lambda_{n,h}^{(rate)} = \left[\sum_{n'\in N, C_k\in C, R_{h'}\in R} \frac{\rho_{n,k}^{(shp)}}{\rho_{n,k}^{(rate)}} \frac{\rho_{n',k}^{(shp)}}{\rho_{n',k}^{(rate)}} \left(\frac{\kappa_{h,h'}^{(shp)}}{\kappa_{h,h'}^{(rate)}} \frac{\lambda_{n',h}^{(shp)}}{\lambda_{n',h}^{(rate)}} + \frac{\lambda_{n',h'}^{(shp)}}{\lambda_{n',h'}^{(rate)}} \frac{\kappa_{h',h}^{(shp)}}{\kappa_{h',h}^{(rate)}}\right)\right] +$$
$$\left[\sum_{d\in d_n, w\in V, C_k\in C, t\in T} \frac{\rho_{n,k}^{(shp)}}{\rho_{n,k}^{(rate)}} \frac{\omega_{t,k}^{(shp)}}{\omega_{t,k}^{(rate)}} \frac{o_{t,h}^{(shp)}}{o_{t,h}^{(rate)}} \frac{\xi_{w,t}^{(shp)}}{\xi_{w,t}^{(rate)}}\right] + \frac{\alpha_\sigma}{\beta_\sigma}$$

$$\kappa_{h,h'}^{(rate)} = \left[\sum_{n,n'\in N, C_k\in C} \frac{\rho_{n,k}^{(shp)}}{\rho_{n,k}^{(rate)}} \frac{\lambda_{n,h}^{(shp)}}{\lambda_{n,h}^{(rate)}} \frac{\lambda_{n',h'}^{(shp)}}{\lambda_{n',h'}^{(rate)}} \frac{\rho_{n',k}^{(shp)}}{\rho_{n',k}^{(rate)}}\right] + \frac{\alpha_\epsilon}{\beta_\epsilon}$$

$$\xi_{w,t}^{(rate)} = \left[\sum_{n\in N, d\in d_n, C_k\in C, R_h\in R} \frac{\omega_{t,k}^{(shp)}}{\omega_{t,k}^{(rate)}} \frac{o_{t,h}^{(shp)}}{o_{t,h}^{(rate)}} \frac{\lambda_{n,h}^{(shp)}}{\lambda_{n,h}^{(rate)}} \frac{\rho_{n,k}^{(shp)}}{\rho_{n,k}^{(rate)}}\right] + \frac{\alpha_\pi}{\beta_\pi}$$

$$\omega_{t,k}^{(rate)} = \left[\sum_{n\in N, d\in d_n, w\in V, R_h\in R} \frac{o_{t,h}^{(shp)}}{o_{t,h}^{(rate)}} \frac{\lambda_{n,h}^{(shp)}}{\lambda_{n,h}^{(rate)}} \frac{\rho_{n,k}^{(shp)}}{\rho_{n,k}^{(rate)}} \frac{\xi_{w,t}^{(shp)}}{\xi_{w,t}^{(rate)}}\right] + \frac{\alpha_\eta}{\beta_\eta}$$

$$o_{t,h}^{(rate)} = \left[\sum_{n\in N, d\in d_n, w\in V, C_k\in R} \frac{\lambda_{n,h}^{(shp)}}{\lambda_{n,h}^{(rate)}} \frac{\rho_{n,k}^{(shp)}}{\rho_{n,k}^{(rate)}} \frac{\xi_{w,t}^{(shp)}}{\xi_{w,t}^{(rate)}} \frac{\omega_{t,k}^{(shp)}}{\omega_{t,k}^{(rate)}}\right] + \frac{\alpha_\gamma}{\beta_\gamma}$$

Fig. 4. Variational updates for Gamma-rate parameters

$$\tau_{n,n'}^{(k,h,h')} \propto e^{\Psi[\rho_{n,k}^{(shp)}]-\log\rho_{n,k}^{(rate)}+\Psi[\lambda_{n,h}^{(shp)}]-\log\lambda_{n,h}^{(rate)}+\Psi[\kappa_{h,h'}^{(shp)}]-\log\kappa_{h,h'}^{(rate)}+\Psi[\rho_{n',k}^{(shp)}]-\log\rho_{n',k}^{(rate)}+\Psi[\lambda_{n',h'}^{(shp)}]-\log\lambda_{n',h'}^{(rate)}}$$

$$\mu_{w,d}^{(n,k,h,t)} \propto e^{\Psi[\rho_{n,k}^{(shp)}]-\log\rho_{n,k}^{(rate)}+\Psi[\lambda_{n,h}^{(shp)}]-\log\lambda_{n,h}^{(rate)}+\Psi[\omega_{t,k}^{(shp)}]-\log\omega_{t,k}^{(rate)}+\Psi[o_{t,h}^{(shp)}]-\log o_{t,h}^{(rate)}+\Psi[\xi_{w,t}^{(shp)}]-\log\xi_{w,t}^{(rate)}}$$

Fig. 5. Variational updates for multinomial parameters

5 Tasks

Posterior inference allows for a variety of exploratory, predictive, and descriptive tasks, that involve posterior expectations of respective random variables.

5.1 Exploratory Network Analysis

The organization of $\mathcal{G}$ into overlapping communities and roles is revealed by node affiliations. The degree to which node n is involved in community C_k is $\vartheta_{n,k}^* \triangleq E[\vartheta_{n,k}]$. The extent to which node n is suitable to role R_h is $\sigma_{n,h}^* \triangleq E[\sigma_{n,h}]$.

The social functions fulfilled by roles are unveiled in terms of their mutual interactions. The strength of interaction from R_h to $R_{h'}$ is $\epsilon_{h,h'}^* \triangleq E[\epsilon_{h,h'}]$.

5.2 Predictive Analysis

The formation of missing links $\langle n, n'\rangle \notin A$ is forecast by associating them with a respective score $s_{n,n'}$, that is used for ranking their future establishment. Specifically, $s_{n,n'} \triangleq E[\delta_{n,n'}]$, where $\delta_{n,n'}$ is defined by Eq. 1.

5.3 Descriptive Analysis

NOODLES allows for the interpretation of topics, communities and roles.

Essentially, topics involve words, that are relevant to corresponding themes. The relevance of word w to topic t is $\pi_{a,t}^* \triangleq E[\pi_{a,t}]$. A characterization of topic t is any suitable subset of topmost relevant words.

Communities and roles can be characterized in terms of especially pertinent topics. Therein, the pertinence of topic t to community C_k and role R_h is, respectively, $\eta_{t,k}^* \triangleq E[\eta_{t,k}]$ and $\gamma_{t,h}^* \triangleq E[\gamma_{t,h}]$.

6 Experimental Evaluation

In this section, we conduct an empirical assessment of NOODLES on real-world social media, that is both quantitative and qualitative.

6.1 Data Sets and Competitors

We carry out experiments on two benchmark data sets, i.e., *DBLP* and *Enron*.

DBLP (http://dblp.uni-trier.de/xml/) is a bibliographic archive of scientific publications in the domain of computer science. In particular, we focused on 33 conferences, i.e., AAAI, AAMAS, CVPR, ECCV, ECML, PKDD, EDBT, EURO-PAR, I3D, ICCP, ICCV, ICDE, ICDM, ICDT, ICML, IJCAI, INFOCOM, IPDPS, MOBICOM, NIPS, PACT, PAKDD, PODS, PPoPP, SIGCOMM, SIGGRAPH EMERGING TECHNOLOGIES, KDD, SIGMETRICS, SIGMOD CONFERENCE, UAI, UMAP, VLDB, WSDM. The papers in the 2010 proceedings of such conferences were gathered to form a collaboration network, in which nodes correspond to authors, links to co-authorships and publication titles to the textual content authored by the corresponding nodes. The resulting networks consists of 8,875 nodes and 17,122 edges.

Enron (http://www.cs.cmu.edu/~enron/) is a set of emails generated by the employees of the *Enron* Corporation. We considered the communication network shaped by 18,233 emails exchanged among the 148 different employees.

Both vocabularies were pruned through stemming and stop word removal. Word stems were further distilled on *Enron* by their respective TFIDF values.

On *DBLP*, the input parameters of NOODLES, i.e., K, H and T are set to 30, 3 and 6. On *Enron*, K is set to 8, H is fixed to 4 and T is set to 20.

NOODLES is compared both on *DBLP* and *Enron* against BHLFM [4], BH-CRM [3] and LDA-G [8].

6.2 Quantitative Evaluation

We comparatively test NOODLES in community discovery and link prediction.

The performance of all competitors in community discovery is assessed in terms of compactness [11], which is suitable to networks without any ground truth about their actual community structure and roles. Essentially, compactness is defined as the average of the shortest distances between community members.

We compute compactness in compliance with [11], i.e., by considering the top 10 members with highest probabilistic involvement in each community.

Figure 6(a) reports the compactness of the communities uncovered by the individual competitors in *DBLP* and *Enron*. The empirical evidence reveals that NOODLES uncovers the most compact communities in both data sets.

Link prediction allows for comparing the predictive accuracy of the different competitors, i.e., the degree to which such competitors reliably predict the presence or absence of links between nodes. The predictive performance of all competitors is evaluated through Monte-Carlo cross validation. In particular, 5 experiments of the predictive performance of each competitor are conducted both on *DBLP* and *Enron*. Each experiment consists of two steps. Preliminarily, the input network is split into a training set and a held-out test set. The latter is formed by randomly sampling the whole input network to pick an equal number of present and absent links, whose sum amounts to 15% of the overall number of links in the whole input network. Subsequently, all links in the held-out test set are predicted by the distinct competing models inferred from the training set. The details on link prediction in NOODLES are provided in Sect. 5.2, whereas those on LDA-G, BH-CRM and BHLFM appear in the respective references.

Figure 6(b) shows the average AUC values achieved by all competitors. NOODLES delivers a superior link-prediction performance on *DBLP* and *Enron*.

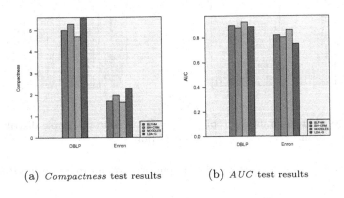

(a) *Compactness* test results   (b) *AUC* test results

Fig. 6. Performance comparison

The empirical evidence in Fig. 6 reveals a consistently higher accuracy of NOODLES both in community discovery and link prediction on the chosen data sets. This substantiates the rationality of using topic modeling for a tighter and more accurate integration of community discovery and role analysis.

6.3   Qualitative Evaluation

Here, we demonstrate the behavior of NOODLES on real-word social media, through a case study developed from the inspection of its results on *DBLP*.

Table 1. Summarization and interpretation of the topics found by NOODLES in *DBLP*

Topic 1 (Networks)	Topic 2 (Data mining)	Topic 3 (AI)	Topic 4 (Distributed and parallel computing)	Topic 5 (Databases)	Topic 6 (Graphics)
network	data	learn	multi	queri	model
wireless	mine	reinforc	agent	optim	imag
sensor	pattern	kernel	system	process	segment
mobil	cluster	activ	plan	databas	detect
schedul	topic	metric	task	search	face

Table 1 reports the 6 uncovered topics. Each topic is summarized by its top-5 most relevant (stemmed) words, whose clarity, specificity and coherence enable the intuitive interpretations in brackets as fields of research in computer science.

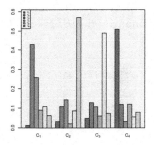

(a) Within-community topic distribution

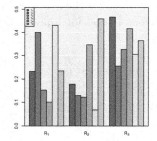

(b) Within-role topic distribution

Fig. 7. Explanation of selected communities (a) and roles (b) in *DBLP*

Figure 7 provides an explanation of a selection of the detected communities and roles via topic pertinence. Communities are mainly focused on respective subsets of topics. Roles can be interpreted as expertise in corresponding research fields.

Finally, Table 2 illustrates a summary of the communities of Fig. 7(a), that includes their top-5 members with strongest involvement according to our *DBLP* data set. It is evident that each community includes members, who are actively involved in the underlying topics reported in Fig. 7(a).

Table 2. Summarization of author involvement into the *DBLP* communities of Fig. 7(a)

Community 1	Community 2	Community 3	Community 4
Jiawei Han	Luc J. Van Gool	Jeffrey D. Ullman	David A. Maltz
Charu C. Aggarwal	Josef Kittler	Jennifer Widom	Sudipta Sengupta
Philip S. Yu	Krystian Mikolajczyk	Philip S. Yu	Aditya Akella
Christos Faloutsos	Bastian Leibe	Yannis E. Ioannidis	Jitendra Padhye
Ravi Kumar	Jurgen Gall	Hector Garcia-Molina	Balaji Prabhakar

7 Conclusions

We presented NOODLES, an innovative generative model of the relationships among content topics, overlapping communities and roles in social media.

It is interesting to consider the attributes of nodes along with their textual contents in the joint analysis of communities and roles in social media. Another research line aims to advance social recommendation [5], by also accounting for user communities, roles and reviews through suitable adaptations of NOODLES.

References

1. Blei, D., Kucukelbir, A., McAuliffe, J.: Variational inference: a review for statisticians. J. Am. Stat. Assoc. **112**(518), 859–877 (2017)
2. Blei, D.: Probabilistic topic models. Commun. ACM **55**(4), 77–84 (2012)
3. Costa, G., Ortale, R.: A Bayesian hierarchical approach for exploratory analysis of communities and roles in social networks. In: Proceedings of IEEE/ACM ASONAM, pp. 194–201 (2012)
4. Costa, G., Ortale, R.: A unified generative Bayesian model for community discovery and role assignment based upon latent interaction factors. In: Proceedings of IEEE/ACM ASONAM, pp. 93–100 (2014)
5. Costa, G., Ortale, R.: Model-based collaborative personalized recommendation on signed social rating networks. ACM Trans. Internet Technol. **16**(3), 20:1–20:21 (2016)
6. Costa, G., Ortale, R.: Mining overlapping communities and inner role assignments through Bayesian mixed-membership models of networks with context-dependent interactions. ACM Trans. Knowl. Discov. Data **12**(2), 18:1–18:32 (2018)
7. Gopalan, P., Hofman, J., Blei, D.: Scalable recommendation with hierarchical poisson factorization. In: Proceedings of UAI, pp. 326–335 (2015)
8. Henderson, K., Rad, T.E.: Applying latent dirichlet allocation to group discovery in large graphs. In: Proceedings of ACM SAC, pp. 1456–1461 (2009)
9. McCallum, A., Wang, X., Corrada-Emmanuel, A.: Topic and role discovery in social networks with experiments on enron and academic email. J. Artif. Intell. Res. **30**(1), 249–272 (2007)
10. Yang, J., Leskovec, J.: Overlapping community detection at scale: a nonnegative matrix factorization approach. In: Proceedings of ACM WSDM, pp. 587–596 (2013)
11. Zhang, H., Qiu, B., Giles, C., Foley, H., Yen, J.: An LDA-based community structure discovery approach for large-scale social networks. In: IEEE Intelligence and Security Informatics, pp. 200–207 (2007)

Text Generation Based on Generative Adversarial Nets with Latent Variables

Heng Wang[1], Zengchang Qin[1(✉)], and Tao Wan[2(✉)]

[1] Intelligent Computing and Machine Learning Lab, School of ASEE,
Beihang University, Beijing 100191, China
zcqin@buaa.edu.cn
[2] School of Biological Science and Medical Engineering,
Beijing Advanced Innovation Centre for Biomedical Engineering,
Beihang University, Beijing 100191, China
taowan@buaa.edu.cn

Abstract. In this paper, we propose a model using generative adversarial net (GAN) to generate realistic text. Instead of using standard GAN, we combine variational autoencoder (VAE) with generative adversarial net. The use of high-level latent random variables is helpful to learn the data distribution and solve the problem that generative adversarial net always emits the similar data. We propose the VGAN model where the generative model is composed of recurrent neural network and VAE. The discriminative model is a convolutional neural network. We train the model via policy gradient. We apply the proposed model to the task of text generation and compare it to other recent neural network based models, such as recurrent neural network language model and SeqGAN. We evaluate the performance of the model by calculating negative log-likelihood and the BLEU score. We conduct experiments on three benchmark datasets, and results show that our model outperforms other previous models.

Keywords: Generative adversarial net · Variational autoencoder
VGAN · Text generation

1 Introduction

Automatic text generation is important in natural language processing and artificial intelligence. For example, text generation can help us write comments, weather reports and even poems. It is also essential to machine translation, text summarization, question answering and dialogue system [18]. One popular approach for text generation is by modeling sequence via recurrent neural network (RNN) [18]. However, recurrent neural network language model (RNNLM) suffers from two major drawbacks when used to generate text. First, RNN based model is always trained through maximum likelihood approach, which suffers from exposure bias [1]. Second, the loss function used to train the model is at word level but the performance is typically evaluated at sentence level.

© Springer International Publishing AG, part of Springer Nature 2018
D. Phung et al. (Eds.): PAKDD 2018, LNAI 10938, pp. 92–103, 2018.
https://doi.org/10.1007/978-3-319-93037-4_8

There are some research on using generative adversarial net (GAN) to solve the problems. For example, Yu *et al.* [24] applies GAN to discrete sequence generation by directly optimizing the discrete discriminator's rewards. Li *et al.* [14] applies GAN to open-domain dialogue generation and generates higher-quality responses. Instead of directly optimizing the GAN objective, Che *et al.* [2] derives a novel and low-variance objective using the discriminator's output that follows corresponds to the log-likelihood. Lamb *et al.* [13] propose to provide D with the intermediate hidden vectors of G rather than its sequence outputs, which makes the model differentiable and achieves promising results in tasks like language modeling and handwriting generation. The gumbel-softmax technique is another approach to train GANs with discrete variables [11]. GAN based on recurrent neural networks with gumbel-softmax output distributions is differentiable. In GAN, a discriminative net D is learned to distinguish whether a given data instance is real or not, and a generative net G is learned to confuse D by generating highly realistic data. GAN has achieved a great success in computer vision tasks [4], such as image style transfer [5], super resolution and imagine generation [19]. Unlike image data, text generation is inherently discrete, which makes the gradient from the discriminator difficult to back-propagate to the generator [7]. Reinforcement learning is always used to optimize the model when GAN is applied to the task of text generation [24].

Although GAN can generate realistic texts, even poems, there is an obvious disadvantage that GAN always emits similar data [21]. For text generation, GAN usually uses recurrent neural network as the generator. Recurrent neural network mainly contains two parts: the state transition and a mapping from the state to the output, and two parts are entirely deterministic. This could be insufficient to learn the distribution of highly-structured data, such as text [3]. In order to learn generative models of sequences, we propose to use high-level latent random variables to model the observed variation. We combine recurrent neural network with variational autoencoder (VAE) [10] as generator G.

In this paper, we propose a generative model, called VGAN, by combining VAE and generative adversarial net to better learn the data distribution and generate various realistic text.

2 Preliminary

2.1 LSTM Architecture

A recurrent neural network is a class of artificial neural network where connections between units form a directed cycle [18]. This allows it to exhibit dynamic temporal behavior. Long short-term memory (LSTM) is an improved version of recurrent neural network considering long-term dependency in order to overcome the vanishing gradient problem. It has been successfully applied in many tasks, including text generation and speech recognition [6]. LSTM has a architecture consisting of a set of recurrently connected subnets, known as memory blocks. Each block contains memory cells and three gate units, including the input, output and forget gates. The gate units allow the network to learn when to forget previous information and when to update the memory cells given new information.

Given a vocabulary V and an embedding matrix $W \in R^{q \times |V|}$ whose columns correspond to vectors; $|V|$ and q denote the size of vocabulary and the dimension of the token vector, respectively. The embedding matrix W can be initialized randomly or pretrained. Let x denote a token with index k, $e(x) \in R^{|V| \times 1}$ is a vector with zero in all positions except $R_k^{|V| \times 1} = 1$. Given an input sequence $X_s = (x_1, x_2, \cdots, x_T)$, we compute the output sequence of LSTM $Y_s = (y_1, y_2, \cdots, y_T)$. When the input sequence passes through the embedding layer, each token is represented by a vector: $v_i = W \cdot e(x_i) \in R^{q \times 1}$. The relation between inputs, memory cells and outputs are defined by the following equations:

$$i^{(t)} = \sigma(W_{ix}v^{(t)} + W_{ih}h^{(t-1)} + W_{ic}c^{(t-1)}) \tag{1}$$

$$f^{(t)} = \sigma(W_{fx}v^{(t)} + W_{fh}h^{(t-1)} + W_{fc}c^{(t-1)}) \tag{2}$$

$$c^{(t)} = f^{(t)} \odot c^{(t-1)} + i^{(t)} \odot tanh(W_{cx}v^{(t)} + W_{ch}h^{(t-1)}) \tag{3}$$

$$o^{(t)} = \sigma(W_{ox}v^{(t)} + W_{oh}h^{(t-1)} + W_{oc}c^{(t-1)}) \tag{4}$$

$$h^{(t)} = o_{(t)} \odot tanh(c^t) \tag{5}$$

where $i^{(t)} \in R^{l \times 1}$, $f^{(t)} \in R^{l \times 1}$, $o^{(t)} \in R^{l \times 1}$ and $h^{(t)} \in R^{l \times 1}$ represent the input gate, forget gate, output gate, memory cell activation vector and the recurrent hidden state at time step t; l is the dimension of LSTM hidden units, σ and $tanh$ are the logistic sigmoid function and hyperbolic tangent function, respectively. $\odot$ represents element-wise multiplication [6].

2.2 Variational Autoencoder

An autoencoder (AE) is an unsupervised learning neural network with the target values to be equal to the inputs. Typically, AE is mainly used to learn a representation for the input data, and extracts features and reduces dimensionality [10]. Recently, autoencoder has been widely used to be a generative model of image and text. The variational autoencoder is an improved version based on the standard autoencoder. For variational autoencoder, there is a hypothesis that data is generated by a directed model and some latent variables are introduced to capture the variations in the observed variables. The directed model $p(x) = \int p(x|z)p(z)\,dz$ is optimized by using a variational upper bound:

$$
\begin{aligned}
-\log p(x) &= -\log \int p(x|z)p(z)\,dz \\
&\leq -KL(q(z|x)||p(z)) + E_{z \sim q(z|x)}[\log p(x|z)]
\end{aligned}
\tag{6}
$$

where $p(z)$ is a prior distribution over the latent random variable z; The prior distribution is unknown, and we generally assume it to be a normal distribution. $p(x|z)$ denotes a map from the latent variables z to the observed variables x, and given z it produces a distribution over the possible corresponding values of x. $q(z|x)$ is a variational approximation of the true posterior distribution. let $KL(q(z|x)||p(z))$ denote the Kullback-Leibler divergence between $q(z|x)$ and $p(z)$; The introduction of latent variables makes it intractable to optimize the model directly. We minimize the upper bound of the negative log-likelihood to optimize VAE. The training algorithm we use is Stochastic Gradient Variational Bayes (SGVB) proposed in [10].

2.3   Generative Adversarial Nets

For generative adversarial nets, there is a two-sided zero-sum game between a generator and a discriminator. The training objective of the discriminative model is to determine whether the data is from the fake data generated by the generative model or the real training data. For the generative model, its objective is to generate realistic data, which is similar to the true training data and the discriminative model can't distinguish. For the standard generative adversarial networks, we train the discriminative model D to maximize the probability of giving the correct labels to both the samples from the generative model and training examples. We simultaneously train the generative model G to minimize the estimated probability of being true by the discriminator. The objective function is:

$$\min_{G} \max_{D} V(D, G) = E_{x \sim p_{data}(x)}[\log D(x)] + E_{z \sim p_z(z)}[\log(1 - D(G(z)))] \quad (7)$$

where z denotes the noises and $G(z)$ denotes the data generated by the generator. $D(x)$ denotes the probability that x is from the training data with emprial distribution $p_{data}(x)$.

3   Model Description

3.1   The Generative Model of VGAN

The proposed generative model contains a VAE at every time step. For the standard VAE, its prior distribution is usually a given standard normal distribution. Unlike the standard VAE, the current prior distribution depends on the hidden state h_{t-1} at the previous moment, and adding the hidden state as an input is helpful to alleviate the long term dependency of sequential data. It also takes consideration of the temporal structure of the sequential data [3,22]. The model is described in Fig. 1. The prior distribution $z_t = p_1(z_t|h_{t-1})$ is:

$$z_t \sim N(\mu_{0,t}, \sigma_{0,t}^2), \quad where \quad [\mu_{0,t}, \sigma_{0,t}^2] = \varphi^{prior}(h_{t-1}) \quad (8)$$

where $\mu_{0,t}$, $\sigma_{0,t}^2$ are the mean and variance of the prior Gaussian distribution, respectively. The posterior distribution depends on the current hidden state. For the approximate posterior distribution, it depends on the current state h_t and the current input x_t: $z_t' = q_1(z_t|x_t, h_t)$.

$$z_t' \sim N(\mu_{1,t}, \sigma_{1,t}^2), \quad where \quad [\mu_{1,t}, \sigma_{1,t}^2] = \varphi^{posterior}(x_t, h_t) \quad (9)$$

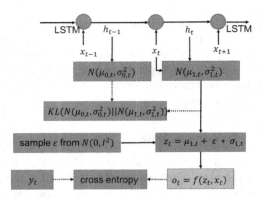

Fig. 1. The structure of the generator G_θ. The generator is composed of LSTM and VAE. x_t denotes the input at time step t; h_t denotes the LSTM hidden state; $N(\mu_{0,t}, \sigma_{0,t}^2)$ denotes the prior distribution; $N(\mu_{1,t}, \sigma_{1,t}^2)$ denotes the approximate posterior distribution; y_t denotes the target output at time step t, which is the same as x_{t+1}. o_t denote the estimated result. The dotted line denotes the optimization process in the pre-training stage.

where $\mu_{1,t}$, $\sigma_{1,t}^2$ are the mean and variance of the approximate posterior Gaussian distribution, respectively. φ^{prior} and $\varphi^{posterior}$ can be any highly flexible functions, for example, a neural network.

The derivation of the training criterion is done via stochastic gradient variational Bayes. We achieve the goal of minimizing the negative log-likelihood in the pre-training stage by minimizing $L(x_{1:T})$:

$$
\begin{aligned}
L(x_{1:T}) = -\log p(x_{1:T}) = -\log \int_{z_{1:T}} \frac{q_1(z_{1:T}|x_{1:T}, h_{1:T})}{q_1(z_{1:T}|x_{1:T}, h_{1:T})} \prod_{t=0}^{T-1} p(x_{t+1}|x_{1:t}, z_{1:t})\, dz_{1:T} \\
\leq -KL(q_1(z_{1:T}|x_{1:T}, h_{1:T})||p_1(z_{1:T}|x_{1:T-1}, h_{1:T-1})) \\
+E_{z_{1:T} \sim q_1(z_{1:T}|x_{1:T}, h_{1:T})} \left[\sum_{t=0}^{T-1} \log p(x_{t+1}|x_{1:t}, z_{1:t}) \right]
\end{aligned}
$$
(10)

where p_1 and q_1 represent the prior distribution and the approximate posterior distribution.

If we directly use the stochastic gradient descent algorithm to optimize the model, there will be a problem that some parameters of the VAE are not derivable. In order to solve the problem, we introduce the "reparameterization trick" [10]. For example, if we want to get the samples from the distribution $N(\mu_{1,t}, \sigma_{1,t}^2)$, we will sample from a standard normal distribution $\epsilon \sim N(0, I^2)$ and get the samples z_t' via $z_t' = \mu_{1,t} + \sigma_{1,t}\epsilon$. Before the adversarial training, we need to pre-train the generative model via SGVB. For example, given the input $X_s = (\mathbf{S}, i, like, it)$, and the target output is $Y_s = (i, like, it, \mathbf{E})$, where $\mathbf{S}$ and $\mathbf{E}$ are the start token and the end token of a sentence, respectively.

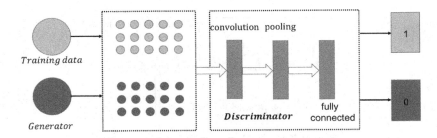

Fig. 2. The illustration of the discriminator. The discriminator D_ϕ is trained by using the true training data and the fake data generated by generator. The discriminator contains the convolution layer, the max-pooling layer and the fully connected layer.

3.2 Adversarial Training of VGAN

The parameters of CNN are less than that of RNN and are easier to optimize. CNN also has recently shown great success in text classification [8,9] and so we choose the CNN as our discriminator instead of RNN in this paper.

Let $v_i \in R^k$ denote a k-dimension vector corresponding to the i-th word in the sentence. $v_{1:n} = v_1 \oplus v_2 \oplus \cdots \oplus v_n$ denote a sentence of length n, where $\oplus$ is the concatenation operator and $v_{1:n} \in R^{n \times k}$ is a matrix. Then a filter $w_1 \in R^{h \times k}$ is applied to a window of h words to produce a new feature. For example, $c_i = f(w_1 \cdot v_{i:i+h-1} + b)$ where c_i is a feature generated by convolution operation. f denotes a nonlinear function such as the hyperbolic tangent or sigmoid; b is a bias term. When the filter is applied to a sentence $v_{1:n}$, a feature map $\mathbf{c} = [c_1, c_2, \cdots, c_{n-h+1}]$ is generated. We can use a variety of convolution kernels to obtain a variety of feature maps. We apply a max-over-time pooling operation to the features to get the maximum value $c' = \max\{\mathbf{c}\}$. Finally, all features are used as input to a fully connected layer for classification. After the generator is pre-trained, we use the generator to generate the negative examples. The negative samples and the true training data are combined as the input of the discriminator. The training process of discriminator is showed in Fig. 2. In the adversarial training, the generator G_θ is optimized via policy gradient, which is a reinforcement learning algorithm [20,23]. The training process is showed in Fig. 3. G_θ can be viewed as an *agent*, which interacts with the environment. The parameters θ of this *agent* defines a policy, which determines the process of generating the sequences.

Given a start token $\mathbf{S}$ as the input, G_θ samples a token from the generating distribution. And the sampled word is as the input at the next time. A whole sentence is generated word by word until an end token $\mathbf{E}$ has been generated or the maximum length is reached. For example, given a start token $\mathbf{S}$ as the input, the sequence $Y_{1:T} = (y_1, y_2, \cdots, y_T)$ is generated by G_θ. In time step t, the state s_t is the current produced tokens $(y_1, y_2, \cdots, y_{t-1})$ and the action a_t is to select the next token in the vocabulary. After taking an action a_t, the agent updates the state $(y_1, y_2, \cdots, y_t)$. If the agent reaches the end, the whole sequence has been generated and the reward will be assigned. During the training, we choose

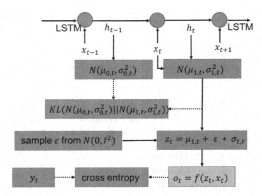

Fig. 3. The training process of the generator via policy gradient. The dotted line denotes sampling a token from the output distribution. The sampled token is as the input at the next time. **START** denotes a start token.

the next token according to the current policy and the current state. But there is a problem that we can observe the reward after a whole sequence. When G_θ generates the sequence, we actually care about the expected accumulative reward from start to end and not only the end reward. At every time step, we consider not only the reward brought by the generated sequence, but also the future reward.

In order to evaluate the reward for the intermediate state, Monte Carlo search has been employed to sample the remaining unknown tokens. In result, for the finished sequences, we can directly get the rewards by inputting them to the discriminator D_ϕ. For the unfinished sequences, we first use the Monte Carlo search to get estimated rewards. To reduce the variance and get more accurate assessment of the action value, we employ the Monte Carlo search for many times and get the average rewards. The objective of the generator model G_θ is to generate a sequence from the start token to maximize its expected end reward:

$$\max_\theta J(\theta) = E[R_T|s_0] = \sum_{y_1 \in V} G_\theta(y_1|s_0) \cdot Q_{G_\theta}^{D_\phi}(s_0, y_1) \tag{11}$$

where R_T denotes the end reward after a whole sequence is generated; $Q_{G_\theta}^{D_\phi}(s_i, y_i)$ is the action-value function of a sequence, the expected accumulative reward starting from state s_i, taking action $a = y_i$; $G_\theta(y_i|s_i)$ denotes the generator chooses the action $a = y_i$ when the state $s_i = (y_1, y_2, \cdots, y_{i-1})$ according to the policy. The gradient of the objective function $J(\theta)$ can be derived [24] as:

$$\nabla_\theta J(\theta) = E_{Y_{1:t-1} \sim G_\theta} \left[\sum_{y_t \in V} \nabla_\theta G_\theta(y_t|Y_{1:t-1}) \cdot Q_{G_\theta}^{D_\phi}(Y_{1:t-1}, y_t) \right] \tag{12}$$

we then update the parameters of G_θ as:

$$\theta \leftarrow \theta + \alpha \cdot \nabla_\theta J(\theta) \tag{13}$$

where α is the learning rate. After training the generator by using policy gradient, we will use the negative samples generated by the updated generator to re-train the discriminator D_ϕ as follows:

$$\min_\phi -E_{Y \sim p_{data}}[\log D_\phi(Y)] - E_{Y \sim G_\theta}[\log(1 - D_\phi(Y))] \tag{14}$$

The pseudo-code of the complete training process is shown in Algorithm 1. In this paper, we propose the VGAN model by combining VAE and GAN for modeling highly-structured sequences. The VAE can model complex multimodal distributions, which will help GAN to learn structured data distribution.

Algorithm 1. Generative Adversarial Network with Latent Variable

---

Require: generator: G_θ;
    discriminator: D_ϕ;
    True training dataset: X;
1 : Pre-train G_θ on X by Eq. (10);
2 : Generate negative samples Y by using G_θ;
3 : Pre-train D_ϕ on X and Y by minimizing the cross entropy;
4 : **Repeat:**
5 :   **for** $1 \sim$ m **do**
6 :   generate the data Y_1 by using G_θ;
7 :   use D_ϕ to get the reward of Y_1;
8 :   update the parameters θ of G_θ by Eq. (13);
9 :   **end for**
10:   **for** $1 \sim$ n **do**
11:     use G_θ to generate negative samples Y_2;
12:     update the parameters ϕ of D_ϕ on X and Y_2 by Eq. (14);
13:   **end for**
14: **Until: VGAN** converges.

---

4   Experimental Studies

In our experiments, given a start token as input, we hope to generate many complete sentences. We train the proposed model on three datasets: Taobao Reviews, Amazon Food Reviews and PTB dataset [17]. Taobao dataset is crawled on taobao.com. The sentence numbers of Taobao dataset, Amazon dataset are 400 K and 300 K, respectively. We split the datasets into 90/10 for training and test. For PTB dataset is relatively small, the sentence numbers of training data and test data are 42,068 and 3,370.

4.1 Training Details

In our paper, we compare our model to two other neural models RNNLM [18] and SeqGANs [24]. For these two models, we use the random initialized word embeddings, and they are trained at the level of word. We use 300 for the dimension of LSTM hidden units. The size of latent variables is 60. For Taobao Reviews and PTB dataset, the sizes of vocabulary are both 5 K. For Amazon Food Reviews, the size is 20 K. The maximum lengths of Taobao Reviews, PTB dataset and Amazon Food Reviews are 20, 30 and 30, respectively. We drop the LSTM hidden state with the dropout rate 0.5. All models were trained with the Adam optimization algorithm with the learning rate 0.001. First, we pre-train the generator and the discriminator. We pre-train the generator by minimizing the upper bound of the negative log-likelihood. We use the pre-trained generator to generate the negative data. The negative data and the true training data are combined to be the input of the discriminator. Then, we train the generator and discriminator iteratively. Given that the generator has more parameters and is more difficult to train than the discriminator, we perform three optimization steps for the discriminator for every five steps for the generator. The process is repeated until a given number of epochs is reached.

Table 1. BLEU-2 score on three benchmark datasets. The best results are highlighted.

Numbers of generated sentence	200	400	600	800	1000
RNNLM (Taobao)	0.965	0.967	0.967	0.967	0.967
SeqGAN (Taobao)	0.968	0.970	0.970	0.968	0.968
VGAN-pre (Taobao)	0.968	0.968	0.967	0.968	0.968
VGAN (Taobao)	**0.969**	**0.972**	**0.970**	**0.969**	**0.969**
RNNLM (Amazon)	0.831	0.842	0.845	0.846	0.848
SeqGAN (Amazon)	0.846	0.851	0.852	0.853	0.856
VGAN-pre (Amazon)	0.842	0.849	0.854	0.849	0.848
VGAN (Amazon)	**0.876**	**0.874**	**0.866**	**0.868**	**0.868**
RNNLM (PTB)	0.658	0.650	0.654	0.655	0.662
SeqGAN (PTB)	0.712	0.705	0.701	0.702	0.681
VGAN-pre (PTB)	0.680	0.690	0.694	0.695	0.671
VGAN (PTB)	**0.715**	**0.709**	**0.714**	**0.715**	**0.695**

4.2 Results and Evaluation

In this paper, we use the BLEU score and negative log-likelihood as the evaluation metrics. BLEU score is used to measure the similarity degree between the generated texts and the human-created texts. We use the whole test data as the references when calculating the BLEU score via nature language toolkit

(NLTK). For negative log-likelihood, we calculate the value by inputting the test data. Table 2 shows the NLL values of the test data. VGAN-pre is the pretrained model of VGAN.

$$NLL = -E \left[\sum_{t=1}^{T} \log G_\theta(y_t|Y_{1:t-1}) \right] \qquad (15)$$

where NLL denotes the negative log-likelihood;

Table 2. The comparison results (NLL) of VGAN to other models.

Dataset	Taobao dataset	Amazon dataset	PTB dataset
RNNLM [18]	219	483	502
SeqGAN [24]	212	467	490
VGAN-pre	205	435	465
VGAN	**191**	**408**	**423**

Table 1 shows the experimental results of BLEU-2 score, and numbers in Table 1 denote the numbers of the sentence generated. We calculate the average BLEU-2 score between the generated sentences and the test data. The descent processes of NLL values during the adversarial training are showed in the Fig. 4. From the experiment results, we can know that the training convergence is not very good. Because the training objective of GAN is inherently unstable and we optimize the model via reinforcement learning, which has high variance, and they lead to conditions that the training convergence is difficult. Here, we give some examples generated by the proposed model. Due to the page limit, only some of generated comments from Amazon Food Reviews are shown in Table 3 and more results will be available online in the final version of the paper. From Tables 1 and 2, we can see the significant advantage of VGAN over RNNLM and SeqGAN in both metrics. The results in the Fig. 4 indicate that applying adversarial training strategies to generator can breakthrough the limitation of generator and improve the effect.

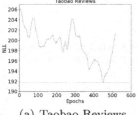 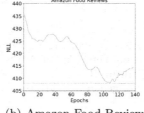 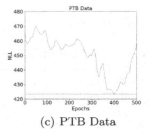

(a) Taobao Reviews   (b) Amazon Food Reviews   (c) PTB Data

Fig. 4. (a) NLL values of Taobao Reviews. (b) NLL values of Amazon Food Reviews. (c) NLL values of PTB Data.

Table 3. Generated examples from Amazon Food Reviews using different models.

RNNLM	1. This is a great product, if you liked canned jerky but this is probably okay because your taste is great too 2. We'll left never eating a bit of more bars. Will definitely buy again 3. But my friends and i love it mixes and they are hard in colored tap up; 4. We love this product, and our purchase was fast too 5. The soup is quite very good, its flax flavoring for the taste ... it is pronounced and yummy
SeqGAN	1. How good it is that i'll really cost again, this was my favorite 2. Service was super fast and good timely manner on each order 3. The risk is very important, but truly the best so, you'll probably love it! 4. Each bag isn't lower and use 5. I found that the seller was good to my own from amazon bags practically very frozen
VGAN	1. You just ate in first, but that is the best thing 2. But that did give me a much healthier and healthy 3. The tea powder in need is based on the label and is very well as well 4. Chips ahoy cookies are very hard. This is just not very tasty. Not what they say? 5. This is very nice and a little sweet. The red's very fresh

5   Conclusions

In this paper, we proposed the VGAN model to generate realistic text based on classical GAN model. The generative model of VGAN combines generative adversarial nets with variational autoencoder and can be applied to the sequences of discrete tokens. In the process of training, we employ policy gradient to effectively train the generative model. Our results show that VGAN outperforms two strong baseline models for text generation and behaves well on three benchmark datasets. In the future, we plan to use deep deterministic policy gradient [15] to train the generator better. In the addition, we will choose other models as the discriminator such as recurrent convolutional neural network [12] and recurrent neural network [16].

References

1. Bengio, S., Vinyals, O., Jaitly, N., Shazeer, N.: Scheduled sampling for sequence prediction with recurrent neural networks. In: NIPS, pp. 1171–1179 (2015)
2. Che, T., Li, Y., Zhang, R., Hjelm, R.D., Li, W., Song, Y., Bengio, Y.: Maximum-likelihood augmented discrete generative adversarial networks. arXiv:1702.07983 (2017)
3. Chung, J., Kastner, K., Dinh, L., Goel, K., Courville, A.C., Bengio, Y.: A recurrent latent variable model for sequential data. In: NIPS, pp. 2980–2988 (2015)
4. Denton, E.L., Chintala, S., Szlam, A., Fergus, R.D.: Deep generative image models using a Laplacian pyramid of adversarial networks. In: NIPS, pp. 1486–1494 (2015)

5. Gatys, L.A., Ecker, A.S., Bethge, M.: A neural algorithm of artistic style. arXiv: 1508.06576 (2015)
6. Hochreiter, S., Schmidhuber, J.: Long short-term memory. Neural Comput. **9**(8), 1735–1780 (1997)
7. Huszar, F.: How (not) to train your generative model: scheduled sampling, likelihood, adversary? arXiv:1511.05101 (2015)
8. Kalchbrenner, N., Grefenstette, E., Blunsom, P.: A convolutional neural network for modelling sentences. In: ACL, pp. 655–665 (2014)
9. Kim, Y.: Convolutional neural networks for sentence classification. In: EMNLP, pp. 1746–1751 (2014)
10. Kingma, D.P., Welling, M.: Auto-encoding variational Bayes. arXiv: 1312.6144 (2014)
11. Kusner, M.J., Hernndezlobato, J.M.: GANS for sequences of discrete elements with the Gumbel-softmax distribution (2016)
12. Lai, S., Xu, L., Liu, K., Zhao, J.: Recurrent convolutional neural networks for text classification. In: AAAI, pp. 2267–2273 (2015)
13. Lamb, A., Goyal, A., Zhang, Y., Zhang, S., Courville, A.C., Bengio, Y.: Professor forcing: a new algorithm for training recurrent networks. In: Neural Information Processing Systems, pp. 4601–4609 (2016)
14. Li, J., Monroe, W., Shi, T., Jean, S., Ritter, A., Jurafsky, D.: Adversarial learning for neural dialogue generation. arXiv:1701.06547 (2017)
15. Lillicrap, T.P., Hunt, J.J., Pritzel, A., Heess, N., Erez, T., Tassa, Y., Silver, D., Wierstra, D.P.: Continuous control with deep reinforcement learning. arXiv:1509.02971 (2016)
16. Liu, P., Qiu, X., Huang, X.: Recurrent neural network for text classification with multi-task learning. In: IJCAI, pp. 2873–2879 (2016)
17. Marcus, M.P., Marcinkiewicz, M.A., Santorini, B.: Building a large annotated corpus of English: the Penn Treebank. Comput. Linguist. **19**(2), 313–330 (1993)
18. Mikolov, T., Karafiat, M., Burget, L., Cernock, J., Khudanpur, S.: Recurrent neural network based language model. In: Interspeech, pp. 1045–1048 (2010)
19. Radford, A., Metz, L., Chintala, S.: Unsupervised representation learning with deep convolutional generative adversarial networks. arXiv: 1511.06434 (2016)
20. Ranzato, M., Chopra, S., Auli, M., Zaremba, W.: Sequence level training with recurrent neural networks. arXiv: 1511.06732 (2016)
21. Salimans, T., Goodfellow, I.J., Zaremba, W., Cheung, V., Radford, A., Chen, X.: Improved techniques for training GANs. In: NIPS, pp. 2234–2242 (2016)
22. Serban, I.V., Sordoni, A., Lowe, R., Charlin, L., Pineau, J., Courville, A.C., Bengio, Y.: A hierarchical latent variable encoder-decoder model for generating dialogues, vol. 1, arXiv: 1605.06069 (2016)
23. Sutton, R.S., Mcallester, D.A., Singh, S.P., Mansour, Y.: Policy gradient methods for reinforcement learning with function approximation. In: NIPS, pp. 1057–1063 (2000)
24. Yu, L., Zhang, W., Wang, J., Yu, Y.: Seqgan: sequence generative adversarial nets with policy gradient. arXiv:1609.05473 (2016)

GEMINIO: Finding Duplicates
in a Question Haystack

Sandya Mannarswamy[1]([✉]) and Saravanan Chidambaram[2]

[1] Conduent Labs India, Bangalore, KA, India
sandyasm@gmail.com
[2] Hewlett Packard Enterprise (HPE), Bangalore, KA, India
saravanan.chidambaram@hpe.com

Abstract. Effective reuse of existing crowdsourced intelligence present in Community Question Answering (CQA) forums requires efficient approaches for the problem of Duplicate Question Detection (DQD). Approaches which use standalone encoded representations for each of the questions in the question pair fail to use the cross question interactions between the two questions which impacts their performance adversely. In this paper, we propose two new schemes for DQD task. Our first approach leverages semantic relations and our second approach utilizes fine grained word level interactions across the two question sentences. We achieve test accuracy of 75.7% and 77.8% with our first and second approaches respectively on a publicly available DQD data set, demonstrating that cross question analysis information can help aid DQD task performance.

Keywords: Natural language processing · Neural Networks
Duplicate detection

1 Introduction

Community Question Answering (CQA) services are rich repositories of crowd sourced wisdom, sharing expert knowledge available 24×7, without boundaries. Since many users have similar informational needs, many of the new questions arising in these forums typically have already been asked and answered. Hence identifying duplicate questions is an essential requirement for ideal user experience in CQA forums. Providing similar/related questions still requires the user to browse through all the similar Q&A. Instead, identifying a pre-existing already answered duplicate question will result in the actual answer being displayed immediately to the user.

Identifying duplicate questions is a challenging task. Two questions with identical information needs can have widely different surface forms. For instance, one user may pose a question as 'does drinking coffee increase blood pressure?' whereas another may pose the question as 'has there been any known link between caffeine intake and hypertension?' These two questions while meeting the same information need, have poor lexico-syntactic similarity. On the other hand, consider the question pair:
Q1:   'Can you suggest some good multi-cuisine restaurants in Paris?'
Q2:   'Can you recommend some good multi-cuisine restaurants in London?'

© Springer International Publishing AG, part of Springer Nature 2018
D. Phung et al. (Eds.): PAKDD 2018, LNAI 10938, pp. 104–114, 2018.
https://doi.org/10.1007/978-3-319-93037-4_9

This pair while having high coarse grained semantic similarity are not duplicate questions since they satisfy totally different information needs. We focus on the problem of Duplicate Question Detection (DQD) in this paper.

There has been considerable work in the areas of paraphrase identification [1], textual entailment recognition [2], natural language inference [3–12] and semantic text similarity detection [9] which are closely related to the DQD problem. There has also been considerable work in similar question retrieval in CQA forums [13–23]. Most of the earlier work has based its foundation on some form of coarse grained semantic similarity measurement by creating a standalone representation of each sentence and then comparing these representations for semantic similarity [3]. While being conceptually simple, it does not leverage cross sentence interactions. Building upon these earlier approaches, we base our approaches intuitively on how humans would approach the task of duplicate question detection. We hypothesize that instead of comparing the two question sentence representations in isolation, humans typically tend to do cross sentence level checking for matches/mismatches. Hence we propose two new neural network based approaches for DQD task in CQA forums.

Our first scheme is based on using relation level matching at cross-question level to aid duplicate question detection. We call this scheme as REL-DQD (Relation aided Duplicate Question Detection). Our second scheme BraidNet-DQD is based on the notion that humans tend to focus on cross word level interaction for corresponding semantic similarity. We show that our two schemes achieve test accuracy of 75.7% and 77.8% respectively on a partition of the QUORA dataset [24].

The rest of the paper is organized as follows: In Sect. 2, we describe our two approaches. In Sect. 3, we discuss our experimental results. In Sect. 4, we provide a brief overview of related work and conclude in Sect. 5.

2   Description

In this section, we describe our two proposed approaches for this problem namely,

- Relation aided Duplicate Question Detection (REL-DQD)
- Braid Network for Duplicate Question Detection (BraidNet-DQD)

While these are independent approaches, exploring two totally different design choice points among the wide spectrum of techniques and implementation choices available for duplicate question detection, they follow the general architecture shown in Fig. 1. As in many earlier works, each question is processed and a fixed length individual sentence representation is built. However in addition to the individual sentence analysis in isolation, the model also makes use of cross sentence analysis.

2.1   Relation Aided Duplicate Question Detection (REL-DQD)

The overall block diagram of REL-DQD is shown in Fig. 2. It consists of the standard components of word embedding module, individual question representation module and the standard Multi-Layer Perceptron (MLP) Classifier module, which can be found

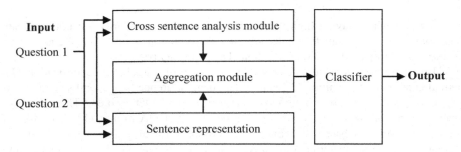

Fig. 1. General block diagram for our DQD approaches

in any typical Natural Language Sentence Matching pipeline. The additional components shown in bold in Fig. 2 are unique to our approach. These include the relation extractor module, relation aggregator module and relation conditioning module.

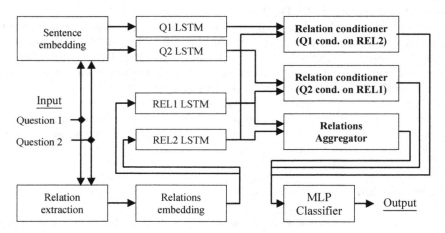

Fig. 2. REL-DQD block diagram

The key idea is to extract the semantic relations from each of the question sentences, and use the encoded relations as an additional aid in final sentence representations. Relation extractor module extracts the semantic relations from each question. The relations are encoded in relation encoder and the output of the relation encoders are aggregated in the relation aggregator module. The relation encoder outputs are also fed to the relation conditioning module. This builds the cross sentence relation conditioned representation of the each of the question sentences. The outputs of the relation conditioning modules along with the relation aggregator module output are fed to the MLP classifier module. We next briefly describe the individual blocks of our approach.

The relation extractor module takes each question in the question pair and extracts the semantic relations from it. Relation extractor is built using a combination of off-the-shelf of semantic relation extractor as well as hand coded relation extractor from the dependency information extracted from the sentence. The semantic relations are expressed as a relation triple <subject, relation, object>. We used the Stanford Open Information Extractor which is part of the Stanford CoreNLP toolkit [25] for extracting semantic relations. Since Stanford OpenIE extractor [26] is able to extract valid relations only in 77% of our dataset, we also created a hand coded relation extractor using rules on the dependency parse trees. The hand crafted relation extractor uses the extracted subject, object nodes in the dependency parse tree and the dependency path between the nodes as the relation. The task of the embedding layer is to generate the sentence matrix given a textual sentence. In the embedding layer, the word embedding sentence matrix is built for the two questions in the question pair using word embedding [29]. For each of the relations extracted also, the sentence embedding matrix is built and output from the embedding layer.

As is common practice in several NLP tasks, we use Recurrent Neural Networks (RNNs) with Long Short-Term Memory (LSTM) [27] units as our encoders for the basic sentence representation. Each question sentence is then encoded using a LSTM (referred to as Q1 LSTM and Q2 LSTM in Fig. 2). Each relation is also encoded using LSTM (denoted as REL1 LSTM and REL2 LSTM in Fig. 2). The key advantage of LSTM is that it contains memory cells which can store information for a long period of time and hence does not suffer from the vanishing gradient problem. We denote the encoded Question representations as H_{Q1} and H_{Q2}. Similarly the encoded relation representations are denoted as H_{R1} and H_{R2}.

Relation Aggregator module takes the encoded relation representations as input and outputs the aggregated relation as concatenated weighted output of [H_{R1}, H_{R2}. Diff (H_{R1}, H_{R2}), Prod (H_{R1}, H_{R2})] where Diff is the element wise difference and Prod is the element-wise multiplication. Thus the output of Relation Aggregator is denoted by

$$H_{RA} = W_{RA}[H_{R1}, H_{R2}, \text{Diff}(H_{R1}, H_{R2}), \text{Prod}(H_{R1}, H_{R2})] \qquad (1)$$

The relation encodings are also fed to the relation conditioning module to build cross sentence conditioned representation of the question. Let Y_{Q1} be the output matrix produced from Q1-LSTM and Y_{Q2} be the output matrix produced from Q2-LSTM encoders. The output of the relation conditioning module is generated as follows:

$$M_1 = \tanh\left(W_q Y_{Q1} + W_r H_{R2}\right) \qquad (2)$$

$$\alpha = \text{softmax}\left(W^T M_1\right) \qquad (3)$$

$$O_1 = \alpha Y_{Q1} \qquad (4)$$

The above equations are for generating cross sentence relation conditioned representation for Question Q1 using cross sentence relation R2. The equations for Question Q2 in the question pair would be similar with subscript 1 replaced by subscript 2 and vice versa. The relation conditioning module output and the relation aggregator module output are then concatenated and fed to the MLP classifier. Hence the input the MLP classifier would be [O_1, O_2, H_{RA}]. The MLP classifier is a standard multi-layer fully connected feed forward network, with the last layer being a softmax layer with 2 units. Next we describe our BraidNet-DQD scheme.

2.2 BraidNet Duplication Question Detection Scheme (BraidNet-DQD)

Our second scheme BraidNet attempts to capture cross sentence word level interactions by braiding the representation of the words from each of the two questions in pair and hence the name BraidNet. The overall block diagram of the BraidNet is shown in Fig. 3. As in the case of REL-DQD, we use Recurrent Neural Networks with Long Short Term Memory Units (LSTM) [27] to obtain the sequential representations.

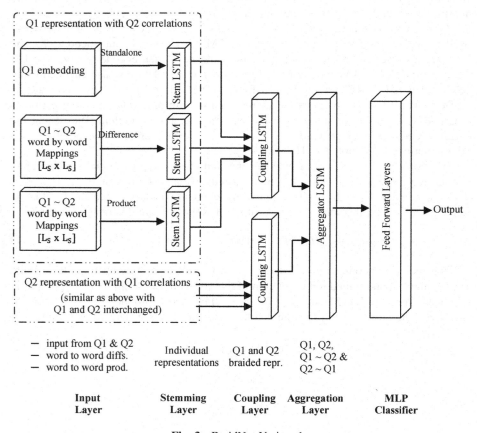

Fig. 3. BraidNet Variant-1

As mentioned before, the key idea is to enhance the standalone question sentence representations by representing the cross sentence word interactions. In our current approach, cross sentence word level interactions are obtained by taking the difference and Hadamard product of each the word pairs from the two questions in the question pair. Given a question pair <Q1, Q2> with question sequence of length Ls words, the first stage of BraidNet creates word interaction matrix of size (Ls × Ls), one for Hadamard product (element wise multiplication), and one for vector element wise difference of each word embedding pair from the two questions. The first column of the (Ls x Ls) word interaction matrix captures the interaction of word 1 of Q1 w.r.to all the words of Q2, second column captures the interaction of word2 of Q1 with respect to all the words of Q2 and so on. In Fig. 3, we use the notation (Q1 ~ Q2) to indicate Q1 and Q2 cross sentence word pair interactions.

Next stage of BraidNet consists of a set of LSTM units for encoding the word interaction matrices in addition to the standalone Q1 sentence embedding. We call these LSTM encoders as Stem LSTMs as it is the starting point of braiding the inter sequence interaction representation. There is a mirror image of the stem LSTM stage which takes the word interaction matrices (with Q1 and Q2 roles swapped) and the standalone Q2 sentence embedding, shown in the bottom left half of the Fig. 3.

Next stage houses the coupling LSTMs. Coupling LSTMs combine the outputs from the stem LSTM stage. Each of the stem LSTM unit encodes the word interaction level product matrix, word interaction difference matrix and the standalone question embedding for question Q1 and outputs the corresponding encoded representation. These three outputs are combined and encoded by the coupling LSTM stage. As in the case of stem LSTM stage, there is a mirror image coupling LSTM which performs the same task for the question Q2 taking inputs from the mirrored stem LSTM stage. The outputs of the coupling LSTMs are then aggregated in the aggregator LSTM, and the output of the aggregator LSTM is fed to the Multi-Layer Perceptron Classifier.

We also studied a second variant of the BraidNet scheme (named as BraidNet Variant-2), which is shown in Fig. 4. Instead of constructing and feeding the word interaction level matrices (namely element wise product and element wise difference) to first stage Stem LSTMs, we encode each word pair sequence through stem LSTM directly to encode word pair sequences. This results in (Ls x Ls) stem LSTMs. The outputs of the stem LSTMs are combined into Ls length sequences and fed to Coupling stage LSTMs (there are Ls coupling LSTMs) as shown in Fig. 4. The outputs of coupling LSTMs are combined along with standalone Q1, Q2 representations in the aggregator LSTM. The output of the aggregator LSTM is fed to the MLP classifier.

3 Experiments

3.1 Dataset

We used a partition of the publicly available Quora Duplicate Question Pairs data set [24]. Our dataset partition consists of 130000 question pairs selected randomly from the original Quora data set (this was done due to the resource constraints on our hardware infrastructure). The distribution of positive and negative examples in our partition was 54% and 46% respectively. We do not use any hand crafted features or magical features. We used a training data size of 100000 question pairs, with test set of 20,000 and 10000 question dev set for selecting hyper-parameters.

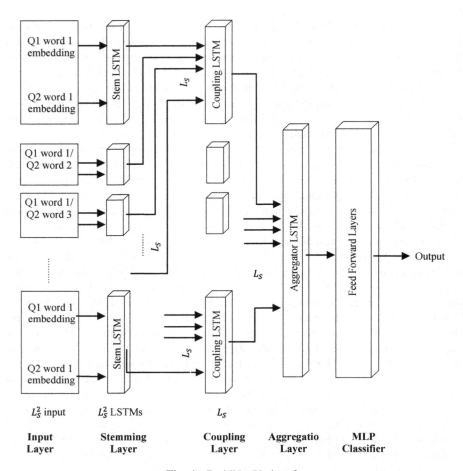

Fig. 4. BraidNet Variant-2

3.2 Experimental Setup

We have implemented our models using the TensorFlow library [28]. For both REL-DQD and BraidNet-DQD, the model training objective is cross-entropy loss. We use an Adam Optimizer with a learning rate of 0.001 and batch size of 32 for both our schemes. Dropouts are applied in-between the feed forward layers in the MLP classifier with drop-out rates randomly chosen in the range between (0.1, 0, 6). We used standard pre-trained word embeddings [29] which were not modified during training of the duplicate question detection task. Out of vocabulary words were handled by randomly initialized embeddings generated by sampling values uniformly from −0.05 and 0.05. The question sequences were pruned to a sequence of length 20, selected as part of Hyper-parameter tuning. We used the standard L2 regularization with L2 regularization strength of 0.001. In REL-DQD scheme, the hidden size for all LSTMs was 64 and the MLP classifier consists of 3 fully connected layers (256, 256, 256) dimensions followed by softmax. For BraidNet Variant-1, we used LSTM hidden size as 64, with MLP classifier dimensions of (256, 256, 2). For Variant-2, stem LSTM hidden size was 2, coupling LSTM hidden size was 20 and MLP classifier dimensions were (256, 128, 64, and 2).

3.3   Experimental Results

We report the results for our two approaches in Table 1. In order to put our numbers in perspective, we also report four other models from earlier work for comparison. These models are respectively referred to as Bowman LSTM fixed length representation [3], Attention LSTM [5], Attention word by word [5], and Bilateral Multi-Perspective Matching (BMPM) [9]. We implemented these models for comparison and their performance numbers are obtained by running these models on our selected dataset partition. We find that among our approaches, BraidNet Variant-1 achieves the best performance of 77.8%, followed by REL-DQD scheme which achieves a test accuracy of 75.7%. BraidNet Variant-2 achieves a test accuracy of 70.2%. In comparison to earlier models, we find that BraidNet Variant-1 performs better than Bowman LSTM, Attention LSTM and Attention word by word models and achieves results close to BMPM model [9], which is the current state of the art in sentence pair modelling task. Though our performance numbers are lower than that of state of the art BMPM model, our approach of using cross sentence information is complementary to BMPM model and can be combined with it. We plan to explore this in future work.

Table 1. Experimental results

Model	Test accuracy
Bowman LSTM [3]	65.7%
Attention LSTM [5]	67.1%
Word by word attention LSTM [5]	69.4%
BraidNet Variant-2	**70.2%**
REL-DQD	**75.7%**
BraidNet Variant-1	**77.8%**
BMPM [9]	79.1%

As we mentioned before, approaches using standalone representations of the questions perform poorly because cross sentence interactions are not taken into account. Attention LSTM [5] which implements a coarse grained form of cross sentence attention and word by word attention LSTM model [5] improve further over Bowman LSTM model [3]. We find that conditioning input questions on cross sentence semantic relations definitely improves performance as seen by REL-DQD achieving 75.7%.

We hypothesize that BraidNet Variant-1 achieves better performance by effectively capturing cross sentence word level interactions. We note that though semantic relations help to condition fixed length representations, they are still coarse grained, compared to fine grained word level cross sentence interaction modeling. This can explain why BraidNet_Variant-1 performs better than REL-DQD. When we compare BraidNet Variant-1 and Variant-2 representations, we find that we encode the cross sentence word level interactions into fixed length representations at a much earlier stage in Variant-2 and this is probably why Variant-2 performs poorly compared to Variant-1.

For REL-DQD, we also studied the effect of using only the off-the-shelf relation extractor vs a combination of off-the-shelf extractor and hand-coded relation extractor. We find that using only the Stanford OpenIE extractor [26] reduced the test accuracy by 1.1%. The results reported in Table 1 for REL-DQD is with using the off the shelf extractor along with the dependency path based hand coded relation extractor. This indicates that improving the relation extraction with better relation extractor tools should help further in improving the performance of REL-DQD, and we plan to explore other open information extractor tools for improving the relation extraction.

4   Related Work

We discuss the related work along the two dimensions of (i) Similar Question Retrieval and (ii) Natural Language Sentence Matching.

4.1   Similar Question Retrieval

Similar question retrieval is closely related to duplicate detection, but has more degrees of freedom since it needs to output a ranked list of related/similar questions instead of labelling a given question pair. Classical information retrieval model based techniques such as BM25 [13] and Language Modeling for information Retrieval (LMIR) [14] have been proposed in the past. Prior approaches have attempted to use word or phrase level translation models from Question-Answer pairs in parallel corpora of same language [15]. Topic model based approaches are based on learning latent topics from question-answer pairs and compute the similarity in the latent topic space [16, 17]. Recently, many neural network architectures have also been proposed [18–20].

4.2   Natural Language Sentence Matching

Semantic textual similarity, Natural Language Inference, Textual Entailment Recognition, Paraphrase Identification fall under the broad umbrella of Natural Language Sentence Matching (NLSM). Some of the earlier work on NLSM focused on building classifiers using hand crafted features such as N-gram overlap, word reordering and syntactic structural alignment [2]. The release of Stanford Natural Language Inference Corpus [3], a sufficiently large annotated data set for Natural Language Inference, saw many neural network models being employed for the task of Natural Language Inference [3–12]. Initial work was based on Siamese architecture wherein the each of the two sentences were encoded into fixed length sentence vectors and the entailment/paraphrase relation between sentences were solely decided by comparing the fixed length sentence vectors [3]. Building separate independent representations for each sentence does not capture the cross sentence interaction features between the two sentences. Later attention models were introduced which are used to capture the word level alignment across the sentences [5, 9]. Our two approaches build upon these earlier works adding in cross sentence relational features as well as word level interactions across the sentences.

5   Conclusion and Future Work

In this paper, we highlight the importance of duplicate question detection task and propose two new schemes for improving performance for DQD problem. We demonstrate by our experiments that using cross sentence analysis information such as semantic relations, word level interactions across sentences to fine-tune the standalone individual question representations helps to improve performance in duplicate question detection. As part of future work, we propose to investigate cross sentence relations using distant supervision in terms of external information sources.

References

1. Socher, R., Huang, E.H., Pennin, J., Manning, C.D., Ng, A.Y.: Dynamic pooling and unfolding recursive autoencoders for paraphrase detection. In: Advances in Neural Information Processing Systems, pp. 801–809 (2011)
2. Dagan, I., Roth, D., Sammons, M., Zanzotto, F.M.: Recognizing Textual Entailment: Models and Applications. Morgan & Claypool Publishers, July 2013. https://sites.google.com/site/excitementproject/
3. Bowman, S.R., Angeli, G., Potts, C., Manning, C.D.: A large annotated corpus for learning natural language inference. In: Proceedings of the 2015 Conference on Empirical Methods in Natural Language Processing (EMNLP) (2015)
4. Chen, Q., Zhu, X., Ling, Z., Jiang, H.: Enhancing and combining sequential and tree LSTM for natural language inference. arXiv preprint arXiv:1609.06038 (2016)
5. Rocktäschel, T., Grefenstette, E., Hermann, K.M., Kočisky, T., Blunsom, P.: Reasoning about entailment with neural attention. arXiv preprint arXiv:1509.06664 (2015)
6. Wang, S., Jiang, J.: Learning natural language inference with LSTM. arXiv preprint arXiv:1512.08849 (2015)
7. He, H., Lin, J.Q.: Pairwise word interaction modeling with deep neural networks for semantic similarity measurement. In: HLT-NAACL (2016)
8. Liu, P., Qiu, X., Huang, X.: Modelling interaction of sentence pair with coupled-LSTMs. arXiv preprint arXiv:1605.05573 (2016)
9. Wang, Z., Hamza, W., Florian, R.: Bilateral multi-perspective matching for natural language sentences. In: Proceedings of the Twenty-Sixth International Joint Conference on Artificial Intelligence, pp. 4144–4150 (2017)
10. Wang, S., Jiang, J.: A compare-aggregate model for matching text sequences. arXiv preprint arXiv:1611.01747 (2016)
11. Liu, Y., Sun, C., Lin, L., Wang, X.: Learning natural language inference using bidirectional LSTM model and inner-attention. arXiv preprint arXiv:1605.09090 (2016)
12. Mou, L., Men, R., Li, G., Xu, Y., Zhang, L., Yan, R., Jin, Z.: Natural language inference by tree-based convolution and heuristic matching. arXiv preprint arXiv:1512.08422 (2015)

13. Robertson, S.E., Walker, S., Jones, S., Hancock-Beaulieu, M.M., Gatford, M.: Okapi at TREC-3, pp. 109–126 (1996)
14. Zhai, C., Lafferty, J.: A study of smoothing methods for language models applied to information retrieval. ACM Trans. Inf. Syst. **22**(2), 179–214 (2004)
15. Zhou, G., Cai, L., Zhao, J., Liu, K.: Phrase-based translation model for question retrieval in community question answer archives. In: Proceedings of the 49th Annual Meeting of the Association for Computational Linguistics, PA, USA, pp. 653–662 (2011)
16. Cai, L., Zhou, G., Liu, K., Zhao, J.: Learning the latent topics for question retrieval in community QA. In: Proceedings of 5th International Joint Conference on Natural Language Processing (2011)
17. Ji, Z., Xu, F., Wang, B., He, B.: Question-answer topic model for question retrieval in community question answering. In: Proceedings of the 21st ACM International Conference on Information and Knowledge Management. ACM, New York (2012)
18. Jeon, J., Bruce Croft, W., Lee, J.H.: Finding similar questions in large question and answer archives. In: Proceedings of the 14th ACM International Conference on Information and Knowledge Management (CIKM 2005). ACM, New York (2005)
19. Das, A., Yenala, H., Chinnakotla, M.K., Shrivastava, M.: Together we stand: siamese networks for similar question retrieval. In: Proceedings of the 54th Annual Meeting of the Association for Computational Linguistics, ACL 2016 (2016)
20. Qiu, X., Huang, X.: Convolutional neural tensor network architecture for community-based question answering. In: Proceedings of the 24th International Conference on Artificial Intelligence (IJCAI 2015), pp. 1305–1311 (2015)
21. Zhang, K., Wu, W., Wu, H., Li, Z., Zhou, M.: Question retrieval with high quality answers in community question answering. In: Proceedings of the 23rd ACM International Conference on Information and Knowledge Management (CIKM 2014). ACM, New York (2014)
22. Zhou, G., Zhou, Y., He, T., Wu, W.: Learning semantic representation with neural networks for community question answering retrieval. Knowl. Based Syst. **93**(C), 75–834 (2016)
23. Xue, X., Jeon, J., Bruce Croft, W.: Retrieval models for question and answer archives. In: Proceedings of the 31st Annual International ACM SIGIR Conference on Research and Development in Information Retrieval. ACM, New York (2008)
24. https://data.quora.com/First-Quora-Dataset-Release-Question-Pairs
25. Manning, C.D., Surdeanu, M., Bauer, J., Finkel, J., Bethard, S.J., McClosky, D.: The stanford CoreNLP natural language processing toolkit. In: Association for Computational Linguistics (ACL) System Demonstrations (2014)
26. Angeli, G., Premkumar, M.J., Manning, C.D.: Leveraging linguistic structure for open domain information extraction. In: Proceedings of the Association of Computational Linguistics (ACL) (2015)
27. Hochreiter, S., Schmidhuber, J.: Long short-term memory. Neural Comput. **9**(8), 1735–1780 (1997)
28. www.tensorflow.org
29. Mikolov, T., Sutskever, I., Chen, K., Corrado, G.S., Dean, J.: Distributed representations of words and phrases and their compositionality. In: Advances in Neural Information Processing Systems, pp. 3111–3119 (2013)

Fast Converging Multi-armed Bandit Optimization Using Probabilistic Graphical Model

Chen Zhao[1](✉), Kohei Watanabe[2], Bin Yang[3], and Yu Hirate[3]

[1] University of Tsukuba, Tsukuba, Japan
s1630190@u.tsukuba.ac.jp
[2] University of Tokyo, Tokyo, Japan
watanabe@mi.t.u-tokyo.ac.jp
[3] Rakuten Institute of Technology, Tokyo, Japan
yangbin64@gmail.com, yu.hirate@rakuten.com

Abstract. This paper designs a strategic model used to optimize click-though rates (CTR) for profitable recommendation systems. Approximating a function from samples as a vital step of data prediction is desirable when ground truth is not directly accessible. While interpolation algorithms such as regression and non-kernel SVMs are prevalent in modern machine learning, they are, however, in many cases not proper options for fitting arbitrary functions with no closed-form expression. The major contribution of this paper consists of a semi-parametric graphical model complying with properties of the Gaussian Markov random field (GMRF) to approximate general functions that can be multivariate. Based upon model inference, this paper further investigates several policies commonly used in Bayesian optimization to solve the multi-armed bandit model (MAB) problem. The primary objective is to locate global optimum of an unknown function. In case of recommendation, the proposed algorithm leads to maximum user clicks from rescheduled recommendation policy while maintaining the lowest possible cost. Comparative experiments are conducted among a set of policies. Empirical evaluation suggests that Thompson sampling is the most suitable policy for the proposed algorithm.

Keywords: Multi-armed bandit · Markov random field
Bayesian network · Recommendation system · Machine learning

1 Introduction

This paper addresses the problem of quickly locating global maxima of an unknown stochastic function. A typical scenario where this problem draws attention is a newly built recommender system, which at cold start has no prior knowledge about upcoming popularity of its candidate items. In such a case, it is a demanding task that a recommendation engine quickly learns an accurate

© Springer International Publishing AG, part of Springer Nature 2018
D. Phung et al. (Eds.): PAKDD 2018, LNAI 10938, pp. 115–127, 2018.
https://doi.org/10.1007/978-3-319-93037-4_10

model discriminating between user favored items and unpopular ones. Click-through rate (CTR) as common metric on item popularity can be interpreted as probability that an item gets valid user response. When CTRs are treated as discrete function outputs, the global maxima is of our best interest for optimal decisions. So a predictive model is beneficial if it learns only a few samples and estimates function values since sampling CTR statistics online is expensive due to unbounded advertising fee. This paper sets up scenario that a recommender system learns optimal strategy to maximize expected reward under budget constraint assuming pay-per-click profit. Budge refers to the number of items a recommender is allowed to push, each standing for one-time user visit, or *"impression"*. Solution consists of two components. The first component, as major contribution of this paper, is a semi-parametric graphical model provably with fundamental property of a typical Gaussian Markov random field (GMRF) [3]. It is a predictive model in which CTRs of candidate items can be treated as indexable discrete random variables whose values are subject to a function of their indices. This function is called *environment*. This paper also assumes that practical environment is noisy and produces stochastic outputs. A graphical model has a runtime complexity constrained by its node count, compared to non-parametric models such as Gaussian process regression whose model size grows with the amount of training data. Experiments show that it maintains effective inference even in the case of data sparsity resulting from high dimensionality. The second solution component is decision making policies. To select the best random variable based on model inference, the multi-armed bandit (MAB) concept is used to seriously weigh exploitation-exploration trade-off [6,9,14,18] for optimal reward. We set up three meta-policies: acquisition functions, epsilon greedy strategy and Thompson sampling. Empirical outcomes show that Thompson sampling [15] is the best policy to work with the proposed graphical model.

The rest of the paper is structured as follows. First the GMRF model is defined in detail along with its inference process. Second MAB decision making policies used in the experiments are introduced. Then the complete algorithm is described along with experiment outcomes on theoretical environment. Next comes the adapted algorithm for real CTR test benches. Experiments on test bench data are also discussed before the paper comes to conclusion.

2   Model Definition

2.1   General Function Approximation

Function approximation in general cases can be defined as follows. Given a set of data samples $\{(i, r) \mid r = f(i)\}$ where f is ground truth representing the environment, we attempt to learn from data samples a model $\hat{f}$ so that for any $i \in \mathbb{R}^d$, $\hat{f}(i) \approx f(i)$. In practice i is a discrete index $i \in D^d$ where D^d is the index space in d dimensions. The set of samples in this paper is defined as $\mathbb{S}$.

$$\mathbb{S} = \{(i, r_i) \mid r_i = f(i), i \in D^d\} \tag{1}$$

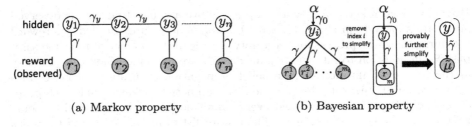

(a) Markov property   (b) Bayesian property

Fig. 1. Graphical model assumptions

The goal of optimization is to learn from $\mathbb{S}$ an estimated truth function $\hat{f}(i)$ so that $\text{argmax}_i \, \hat{f}(i) = \text{argmax}_i \, f(i)$, i.e., $\text{argmax}_i \, \hat{y}_i = \text{argmax}_i \, r_i$. It is also necessary to be aware that sampling (i, r_i) from environment incurs extra cost and $|\mathbb{S}|$ is thereby to be minimized.

2.2   Graphical Model Representation

This section defines the proposed graphical model, its probability interpretation and inference methods.

Markov Property. The first assumption is Markov property among nodes y_i in the Markov random field with Fig. 1(a) as an example, where y_i stands for a *hidden node* or *target node* that infers probability distribution of $f(i)$ in (1). For problems in this paper, we appreciate local Markov property provided by a Markov random field.

A hidden node probability conditioned on all its neighbors is independent from any non-adjacent node, or $y_v \perp\!\!\!\perp y_{u \notin J_v} | \{y_{u' \in J_v}\}$. In other words, belief of a hidden node does not propagate towards non-adjacent nodes given all its neighbors. The joint distribution is computed in clique factorization where a clique is defined as a fully connected subgraph. In Fig. 1(a) one clique only consists of two adjacent nodes. The clique joint probability is defined using a Gaussian function $p(y_i, y_{i+1} | \gamma_y) = \exp[-\frac{\gamma_y}{2}(y_{i+1} - y_i)^2]$. Here γ_y constrains the bonding strength between neighbors. Joint density over the field can be a product from all the cliques with $\boldsymbol{y}$ standing for the target node list.

$$p(\boldsymbol{y}|\gamma_y) = \prod_{i=1}^{n-1} \exp(-\frac{\gamma_y}{2}(y_{i+1} - y_i)^2) = \prod_{i,j \in E} \exp(-\frac{\gamma_y}{2}(y_i - y_j)^2) \qquad (2)$$

In general cases, graph nodes y_i are not necessarily linearly connected as in the particular example of Fig. 1(a). A more comprehensive expression of Markov joint density is expressed as (2) in factors of connected node pairs with E being the edge set.

Bayesian Property. Markov assumption above assumes that all the observed nodes are known. In practice values of observed nodes r_i are sampled as a reward list during learning process and certainty of y_i is under impact of multiple samples. So in addition to inter-node joint density, we introduce Bayesian property as the second assumption to take belief propagation from observed nodes into consideration. Figure 1(b) denotes the Bayesian network that models such property. Under Bayesian assumption, every target node y_i is conditioned on some prior α with bonding strength γ_0. The reward list of y_i is represented by nodes r_i separately indexed from 1 to m_i. Observed nodes r_i are conditionally independent given their hidden node y_i. Similar to the Markov random field we assign bonding coefficient γ between an observed node r_i and its target node y_i. Figure 1(b) also provides plate notation of the Bayesian network when there are n target nodes and m observations under a target node.

Factorization property of a Bayesian network says the joint distribution of a target node y_i and its children $r_j^{(i)}$ are defined below.

$$p(\boldsymbol{r}^{(i)}, y_i | \gamma, \gamma_0, \alpha) = p(y_i | \gamma_0, \alpha) \prod_{j=1}^{m_i} p(r_j^{(i)}, y_i | \gamma, y_i) \tag{3}$$

In (3), $\boldsymbol{r}^{(i)}$ is the reward list of y_i and $r_j^{(i)}$ is the jth observation. Computing the product of likelihood over observations using (3) gets expensive as sample size increases. Since we are more interested in the distribution of observations $\boldsymbol{r}^{(i)}$ than individual reward samples $r_j^{(i)}$, we approximate every $r_j^{(i)}$ into the mean of $\boldsymbol{r}^{(i)}$ so as to eliminate child indices of y_i.

$$p(\boldsymbol{r}^{(i)}, y_i | \gamma, \gamma_0, \alpha) = p(y_i | \gamma_0, \alpha)[p(\mu_i, y_i | \gamma, y_i)]^{m_i} \tag{4}$$

$$= \exp[-\frac{1}{2}\gamma_0(y_i - \alpha)^2] \exp[-\frac{m_i}{2}\gamma(y_i - \mu_i)^2] \tag{5}$$

$$\text{where } \mu_i = (\sum_{j=1}^{m_i} r_j^{(i)})/m_i \tag{6}$$

Similar to (2), joint density between connected nodes are modeled with Gaussian functions subject to bonding strength with (5). We further work on (5) by expanding all the terms in the exponential part.

$$p(\boldsymbol{r}^{(i)}, y_i | \gamma, \gamma_0, \alpha) = \exp\left\{-\frac{1}{2}[m_i\gamma(y_i - \mu_i)^2 + \gamma_0(y_i - \alpha)^2]\right\} \tag{7}$$

$$= \exp\left\{-\frac{1}{2}[(m_i\gamma + \gamma_0)y_i^2 - (2\gamma m_i\mu_i + 2\gamma_0\alpha)y_i + \gamma_0 y_i^2 + m_i\gamma\mu_i^2 + \gamma_0\alpha^2]\right\} \tag{8}$$

$$= \exp\left\{-\frac{1}{2}(m_i\gamma + \gamma_0)[y_i^2 - \frac{2\gamma m_i\mu_i + 2\gamma_0\alpha}{m_i\gamma + \gamma_0}y_i + C]\right\} \tag{9}$$

Expression (9) indicates that the constant C can easily be scaled so that the exponential term can be rewritten into a perfect square with respect to y_i.

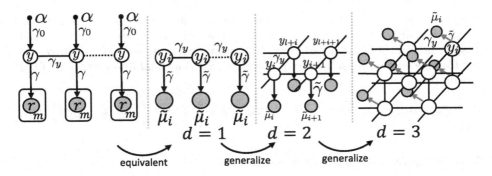

Fig. 2. Final graphical model - Gaussian Markov Random Field

$$p(\boldsymbol{r}^{(i)}, y_i | \gamma, \gamma_0, \alpha) = \exp\left\{ -\frac{1}{2}\tilde{\gamma}_i(y_i - \tilde{\mu}_i)^2 + C' \right\} \tag{10}$$

$$= \exp\{C'\} \exp\left\{ -\frac{1}{2}\tilde{\gamma}_i(y_i - \tilde{\mu}_i)^2 \right\} \tag{11}$$

$$\text{where } \tilde{\gamma}_i = m_i\gamma + \gamma_0 \text{ and } \tilde{\mu}_i = \frac{\gamma m_i \mu_i + \gamma_0 \alpha}{m_i\gamma + \gamma_0} \tag{12}$$

Expression (12) proves that the complete Bayesian network can be remodeled with only one edge parameter and interpolated sample mean $\tilde{\mu}_i$, whose values are μ_i rescaled by γ, γ_0 and α. The finally simplified Bayesian network is also displayed in Fig. 1(b).

Final Representation. The final graphical model as Fig. 2 is a consolidated graph containing both the simplified Bayesian component in Fig. 1(b) and the Markov random field in Fig. 1(a). This graph is used to infer distinct distributions of every target node y_i based on interpolated means $\tilde{\mu}_i$ calculated from samples on y_i, meanwhile modeling covariance among y_i as Gaussian kernels. This is equivalent to saying that random vector $\boldsymbol{y} = [y_1, y_2, \ldots, y_n]$ is in multivariable Gaussian distribution. Therefore node set $\{y_i\}$ constitutes a Gaussian Markov Random Field (GMRF). Total joint probability $p(\boldsymbol{y}, \boldsymbol{r} | \alpha, \gamma_y, \gamma_0, \gamma)$ of the final graph is computed as product of Bayesian probability in (10) for every target node and Markov joint density counting every edge using (2).

$$p(\boldsymbol{y}, \boldsymbol{r} | \alpha, \gamma_y, \gamma_0, \gamma) = p(\boldsymbol{y} | \gamma_y) \prod_i \left[p(\boldsymbol{r}^{(i)}, y_i | \gamma, \gamma_0, \alpha) \right] \tag{13}$$

To make (13) explicit, we now replace both probability parts with (11) and (2) ignoring the constant factor.

$$p(\boldsymbol{y}, \boldsymbol{r} | \alpha, \gamma_y, \gamma_0, \gamma) \propto \left\{ \prod_i \exp\left[-\frac{1}{2}\tilde{\gamma}_i(y_i - \tilde{\mu}_i)^2 \right] \right\} \left\{ \prod_{i,j \in E} \exp\left[-\frac{\gamma_y}{2}(y_i - y_j)^2 \right] \right\} \tag{14}$$

In (14) i, j refers to an edge connecting y_i, y_j given the edge set E and indices $i, j \in D^d$ can be d-dimensional as indicated in (1). A typical GMRF node y_i

has an increasing count of neighbors as dimension d goes up as is shown in Fig. 2. For instance there are 4 neighbors non-edge nodes in a 2-dimensional graph and 6 in 3-dimensional case. To continue working on (14), we take the log likelihood of joint probability $p(\boldsymbol{y}, \boldsymbol{r} | \alpha, \gamma_y, \gamma_0, \gamma)$ as final interpretation of the graph likelihood.

$$\ln p(\boldsymbol{y}, \boldsymbol{r} | \alpha, \gamma_y, \gamma_0, \gamma) \propto \sum_i \left\{ -\frac{1}{2} \tilde{\gamma}_i (y_i - \tilde{\mu}_i)^2 \right\} + \sum_{i,j \in E} \left\{ -\frac{\gamma_y}{2} (y_i - y_j)^2 \right\} \quad (15)$$

2.3 Graphical Model Inference

This section discusses, on top of the proposed graphical model, how inference is conducted. Now that (15) gives a compact representation of model likelihood, inference is now equivalent to maximizing this likelihood with optimal y_i values $\hat{y}_i$, or a target vector $\hat{\boldsymbol{y}}$, subject to currently available reward lists $\boldsymbol{r}$. Hyperparameters α, γ_y, γ_0, γ are initialized with dimension specific values to be stated in later experiments. Next we show that a closed-form solution $\hat{\boldsymbol{y}}$ exists for maximizing the model likelihood. Let $E(\boldsymbol{y}) \propto -2 \ln p(\boldsymbol{y}, \boldsymbol{r} | \alpha, \gamma_y, \gamma_0, \gamma)$ dropping any constant term.

$$E(\boldsymbol{y}) = \sum_i \left\{ \tilde{\gamma}_i (y_i - \tilde{\mu}_i)^2 \right\} + \sum_{i,j \in E} \left\{ \gamma_y (y_i - y_j)^2 \right\} \quad (16)$$

The optimal $\hat{\boldsymbol{y}}$ minimizes $E(\boldsymbol{y})$, which can be efficiently computed using matrix multiplication. Given a graph of n hidden nodes, let $\boldsymbol{B}$ be an $n \times n$ identity matrix multiplied by $\tilde{\boldsymbol{\gamma}}$ so that $\boldsymbol{B}_{ii} = \tilde{\gamma}_i$. Let $\boldsymbol{K}$ be an $n \times n$ adjacency matrix in which k_{ij} and k_{ji} is γ_y if node y_i and y_j are adjacent otherwise 0. In case of high dimensions $(d \geq 2)$, graph node indices are flattened before matrix construction so that target nodes are indexable with a 1-dimensional array of size n. In this way, $k_{ij} = k_{ji}$ so $\boldsymbol{K}$ is a symmetric matrix whose diagonal terms are set to 0. Furthermore, we define $\boldsymbol{k}$ as a length-n vector consisting of row/column sums of $\boldsymbol{K}$. So $k_x = \sum_j K_{xj} = \sum_i K_{ix}$. Alternatively, k_x can be treated as the neighbor count of the xth node. Define matrix $\boldsymbol{A} = \boldsymbol{B} - \boldsymbol{K} + \gamma_y diag(\boldsymbol{k})$ where $diag(\boldsymbol{k})$ is the diagonal matrix with k_x as its xth diagonal element.

$$\boldsymbol{A} = \boldsymbol{B} - \boldsymbol{K} + \gamma_y diag(\boldsymbol{k}) = \begin{pmatrix} \gamma_y k_1 + \tilde{\gamma}_1 & -\gamma_y & \cdots -\gamma_y & \cdots \\ -\gamma_y & \gamma_y k_2 + \tilde{\gamma}_2 & \cdots & \cdots & -\gamma_y \\ \vdots & \vdots & \ddots & \\ -\gamma_y & & \vdots & & \ddots \\ & -\gamma_y & & & \ddots \\ & & \cdots & & \gamma_y k_n + \tilde{\gamma}_n \end{pmatrix} \quad (17)$$

Having matrix A and B explicitly defined, we exploit a provable fact that $E(\boldsymbol{y})$ is equivalent to product of the following nested matrices.

$$E(\boldsymbol{y}) = (\boldsymbol{y}^\top \tilde{\boldsymbol{\mu}}^\top) \begin{pmatrix} \boldsymbol{A} & -\boldsymbol{B} \\ -\boldsymbol{B} & \boldsymbol{B} \end{pmatrix} \begin{pmatrix} \boldsymbol{y} \\ \tilde{\boldsymbol{\mu}} \end{pmatrix} = \boldsymbol{y}^\top \boldsymbol{A} \boldsymbol{y} - \tilde{\boldsymbol{\mu}}^\top \boldsymbol{B} \boldsymbol{y} - \boldsymbol{y}^\top \boldsymbol{B} \tilde{\boldsymbol{\mu}} + \tilde{\boldsymbol{\mu}}^\top \boldsymbol{B} \tilde{\boldsymbol{\mu}} \quad (18)$$

Recall that $\boldsymbol{y}$ is the target node list and $\tilde{\boldsymbol{\mu}}$ is the interpolated sample means of target nodes. $\left(\boldsymbol{y}^{\top}\tilde{\boldsymbol{\mu}}^{\top}\right)$ stands for horizontal concatenation of vectors $\boldsymbol{y}^{\top}$ and $\tilde{\boldsymbol{\mu}}^{\top}$ and similarly $\boldsymbol{y}$ and $\tilde{\boldsymbol{\mu}}$ can be vertically concatenated as well. Hence Formula (18) produces a product of three matrices of sizes $1 \times 2n$, $2n \times 2n$ and $2n \times 1$.

The optimal $\hat{\boldsymbol{y}}$ is $\boldsymbol{y}$ that minimizes $E(\boldsymbol{y})$ and equivalently maximizes the model probability. There exists a provable closed-form solution of $\hat{\boldsymbol{y}}$ defined with matrix notation below.

$$\hat{\boldsymbol{y}} = \underset{y}{\mathrm{argmax}} \left(\log p(\boldsymbol{r}, \boldsymbol{y}|\gamma_y, \gamma, \gamma_0, \alpha)\right) = \boldsymbol{A}^{-1}\boldsymbol{B}\tilde{\boldsymbol{\mu}} \text{ and } \boldsymbol{\sigma}_{\hat{y}}^2 = diag(\boldsymbol{A}^{-1}) \quad (19)$$

$\boldsymbol{\sigma}_{\hat{y}}^2$ is the posterior variance of $\hat{\boldsymbol{y}}$ taken from diagonal terms of $\boldsymbol{A}^{-1}$. Given $\boldsymbol{y} = \boldsymbol{A}^{-1}\boldsymbol{B}\tilde{\boldsymbol{\mu}}$, $E(\boldsymbol{y}) = E_{min} = \tilde{\boldsymbol{\mu}}^{\top}(\boldsymbol{B} - \boldsymbol{B}\boldsymbol{A}^{-1}\boldsymbol{B})\tilde{\boldsymbol{\mu}}$. Let $\Lambda = \boldsymbol{B} - \boldsymbol{B}\boldsymbol{A}^{-1}\boldsymbol{B}$. Given reward list $_r$ model probability distribution is expected to be

$$p(\boldsymbol{r}, \boldsymbol{y}|\gamma, \gamma_y, \gamma_0, \alpha) \propto \exp\left(-\frac{1}{2}\tilde{\boldsymbol{\mu}}^{\top}\Lambda\tilde{\boldsymbol{\mu}}\right) \quad (20)$$

Formula (20) indicates that graphical models in Fig. 2 conforms to GMRF property with multivariate Gaussian distribution such that $p(\boldsymbol{y}) \sim \mathcal{N}\left(\tilde{\boldsymbol{\mu}}, \Lambda^{-1}\right)$. Λ serves as the precision matrix.

Lastly, computing $\boldsymbol{A}$ defined in (17) requires sums of rows in $\boldsymbol{K}$. This leads to context overhead every time the model gets updated. We found that approximating $\boldsymbol{A}$ with $\boldsymbol{A}'$ greatly improves computational efficiency in practice.

$$\boldsymbol{A}' = \boldsymbol{B} - \boldsymbol{K} + 2d\gamma_y\boldsymbol{I}_n = \begin{pmatrix} 2d\gamma_y + \tilde{\gamma}_1 & -\gamma_y & \cdots & -\gamma_y & \cdots \\ -\gamma_y & 2d\gamma_y + \tilde{\gamma}_2 & \cdots & \cdots & -\gamma_y \\ \vdots & \vdots & \ddots & & \\ -\gamma_y & \vdots & & \ddots & \\ & -\gamma_y & & & \ddots \\ & \cdots & & & 2d\gamma_y + \tilde{\gamma}_n \end{pmatrix} \quad (21)$$

$\boldsymbol{A}'$ replaces neighbor counts k_x on the diagonal with constant $2d$. This in effect avoids counting neighbors by assuming that every d-dimensional GMRF node has $2d$ neighbors, an assumption true for all except edge nodes. Therefore we call approximation with $\boldsymbol{A}'$ an *edge normalization* method because edge nodes are treated as if they were non-edge nodes.

To summarize, the following closed-form solution is adopted as model prediction for experiments in this paper.

$$\hat{\boldsymbol{y}} = \underset{y}{\mathrm{argmax}} \left(\log p(\boldsymbol{r}, \boldsymbol{y}|\gamma_y, \gamma, \gamma_0, \alpha)\right) = \boldsymbol{A}'^{-1}\boldsymbol{B}\tilde{\boldsymbol{\mu}} \quad (22)$$

$$\boldsymbol{\sigma}_{\hat{y}}^2 = diag(\boldsymbol{A}'^{-1}) \quad (23)$$

$\hat{\boldsymbol{y}}$ is model prediction with uncertainty measured by the variance list $\boldsymbol{\sigma}_{\hat{y}}^2$.

2.4   Review on the Multi-armed Bandit Problem

The multi-armed bandit (MAB) is a widely studied [1,2,8] decision making problem associated with environment reward. Given a list of discrete random variables called "arms" $\{Y_i\}$ whose values represent sample rewards. In general, $\{Y_i\}$ are not of identical distribution, which leads to the problem nature how to select the arm of maximum expected reward. Given T rounds of attempts, an ideal policy collects maximized reward from the best arm. The loss due to failure in collecting optimal reward is measured in regret. MAB is formally defined as follows.

Given a random variable list $\{Y_i\}$ and μ_i as mean of Y_i, a policy decides the index $\pi(t)$ of the chosen arm at the tth step and $r_i^{(t)}$ is the observed reward sampled from the ith arm at step t. Under policy $\pi(t)$, the total regret after T rounds of observation is defined as $R_T = T\mu^* - \sum_{t=1}^{T} r_{\pi(t)}^{(t)}$ and $\mu^* = \max_k \mu_k$. So μ^* can only be approximated in practice. This paper models Y_i as $f(i)$ defined in (1). We assume $\{Y_i\} \sim \mathcal{N}(\tilde{\mu}, \Lambda^{-1})$ which the proposed graphical model conforms to. Graph node $\boldsymbol{y_i}$ corresponds to probability inference on Y_i.

3   Graphical Model Learning

3.1   Decision Making Policy

Policy produces π, i.e., a decision index $\pi^{(t)}$ at time t representing the potential location of optimum in the environment, given model prediction $\hat{\boldsymbol{y}}^{(t)}, \boldsymbol{\sigma}_{\hat{y}}^{(t)}$, whose closed-form solution is given as (22) and (23). This paper covers three suites of decision making policy as discussed below.

Acquisition Function. Constructing an acquisition function (ACQ) is a common approach in Bayesian optimization to determine the next optimal index to sample from. Commonly used acquisition functions include probability of improvement (PI), expected improvement (EI) and upper confidence bound (GP-UCB) [4,12]. We let $a(\hat{\boldsymbol{y}}, \boldsymbol{\sigma}_{\hat{y}})$ be the generic stereotype of an acquisition function that performs element-wise operation on input vectors and returns a vector of the same size so that policy produces $\pi \leftarrow \operatorname{argmax}_i a(\hat{\boldsymbol{y}}, \boldsymbol{\sigma}_{\hat{y}})$. Below is parameter setting of acquisition functions used in this paper.

$$a_{PI}(\hat{\boldsymbol{y}}, \boldsymbol{\sigma}_{\hat{y}}) = \Phi\left(\frac{\hat{\boldsymbol{y}} - r_{best} - \xi}{\boldsymbol{\sigma}_{\hat{y}}}\right)$$

where r_{best} is current best (largest) observation across all the nodes and $\xi = 0.01$ as constant bias. Φ is Gaussian cumulative density.

$$a_{EI}(\hat{\boldsymbol{y}}, \boldsymbol{\sigma}_{\hat{y}}) = z\Phi(z) + \boldsymbol{\sigma}_{\hat{y}}\phi(z)$$

where $z = \hat{\boldsymbol{y}} - r_{best} - \xi$ and ϕ is the Gaussian probability density.

$$a_{GP-UCB}(\hat{\boldsymbol{y}}, \boldsymbol{\sigma}_{\hat{y}}) = \hat{\boldsymbol{y}} + \sqrt{\beta(x)}\boldsymbol{\sigma}_{\hat{y}}$$

where x is total number of current observations and $\beta(x) = 2log(dx^2\pi^2/6\delta)$ [4]. d is dimension of node indices and $\delta = 0.9$. GP-UCB is considered a common option for Gaussian process instead of GMRF. This paper tentatively takes it for comparative experiments only.

Epsilon Greedy. Epsilon greedy (EPS) is a classical and naive exploration-exploitation trade-off heuristic that allocates a probability $1 - \epsilon$ for exploitation behavior and ϵ for else. $\epsilon = 0.2$ in this paper.

$$\pi \leftarrow \underset{i}{\text{argmax}}\, \hat{\boldsymbol{y}} \text{ with probability } 1 - \epsilon \text{ otherwise uniformly random } i \in D^d$$

Thompson Sampling. Thompson Sampling (TS) strictly adheres to model prediction and from every node y_i takes a random sample based on its posterior distribution $\mathcal{N}(\hat{y}_i, \sigma_i)$.

$$\pi \leftarrow \underset{i}{\text{argmax}}\, \{y_i | y_i \sim \mathcal{N}(\hat{y}_i, \sigma_i)\}$$

3.2   Experiments - Learning Synthetic Data

Combining the GMRF model and decision making policy, we develop Algorithm 1 as an online learning algorithm that in each iteration samples an environment reward at the best index as determined by some given policy. Sampled rewards serve as training data that helps update node inference using solution (22) and (23). An optimal policy is one that minimizes cumulative regret over T iterations. To find out which policy described in Sect. 3.1 eventually leads to minimal regret, we apply Algorithm 1 to synthetic functions $f(i)$ where $i \in D^d$. Three sets of experiments are carried out corresponding to 1, 2 and 3-dimensional indices. In each set of experiments a discrete interval $X = [-5, 5]$ with uniform

---

Algorithm 1. GMRF Learning

1: Initialize GMRF hyperparameters γ, γ_y, γ_0, α
2: Randomly initialize prior distribution $\hat{\boldsymbol{y}}$, $\boldsymbol{\sigma}_{\hat{y}} \leftarrow \mathcal{N}(0, \sigma^2)$
3: Initialize average accumulated regret $\bar{R}_T \leftarrow 0$ ▷ evaluation only
4: **for** t in iteration 1...T **do**
5:     $\pi^{(t)} \leftarrow$ from policy   ▷ ACQ/EPS/TS
6:     Sample observation at node of index $\pi^{(t)}$ as $r_{\pi^{(t)}}(t) \leftarrow f(\pi^{(t)})$
7:     Update node probability $\hat{\boldsymbol{y}}$, $\boldsymbol{\sigma}_{\hat{y}} \leftarrow p(\mathbf{r}, \mathbf{y}|\gamma, \gamma_y, \gamma_0, \alpha)$
8:     Calculate current regret $R_t \leftarrow max_i f(i) - r_{\pi^{(t)}}(t)$ ▷ evaluation only
9:     Update $\bar{R}_T \leftarrow \bar{R}_T \frac{t-1}{t} + R_t \frac{1}{t}$ ▷ evaluation only
10: **end for**

Algorithm 2. GMRF Learning for CTR Optimization

1: Initialize GMRF hyperparameters γ, γ_y, γ_0, α
2: Randomly initialize prior distribution $\hat{y}$, $\sigma_{\hat{y}} \leftarrow \mathcal{N}(0, \sigma^2)$
3: Initialize uniform impression weights $w_i^{(0)} \leftarrow 1/n$
4: Initialize average accumulated regret $\bar{R}_T \leftarrow 0$ ▷ evaluation only
5: **for** t in iteration 1...T **do**
6:     $\{w_i^{(t)}\} \leftarrow$ from modified policy
7:     Collect #clicks $\{C_i^{(t)} | C_i^{(t)} = \chi(\widetilde{f}(i), N w_i^{(t)})\}$ from the test bench χ ▷ online
8:     $\{r_i^{(t)}\} \leftarrow$ calculate CTRs $C_i^{(t)}/N w_i^{(t)}$
9:     **for** i in index 1...n **do**   ▷ completely offline
10:        Sample observation at node index i as $r_i^{(t)}$
11:        Update node probability $\hat{y}$, $\sigma_{\hat{y}} \leftarrow p(\mathbf{r}, \mathbf{y} | \gamma, \gamma_y, \gamma_0, \alpha)$
12:    **end for**
13:    Calculate current regret $R_t \leftarrow N max_i \widetilde{f}(i) - \sum_i C_i^{(t)}$ ▷ evaluation only
14:    Update $\bar{R}_T \leftarrow \bar{R}_T \frac{t-1}{t} + R_t \frac{1}{t}$ ▷ evaluation only
15: **end for**

---

increments is picked so that $D^1 = X$, $D^2 = X \times X$ and $D^3 = X \times X \times X$ and increments are respectively set as 0.01, 0.3 and 1.0 to allow for roughly the same size of D^d ($\|D^d\| \approx 1000$) for all cases of d. Synthetic functions f in each case are all configured as sums of three Gaussian functions to accommodate three separate maxima. Three Gaussian kernels of following variances/covariances are used, with $[0.0225, 2.25, 0.5]$ for $d = 1$; $[0.025, 2.25, 0.25]$ for $d = 2$; and $[0.05, 1, 0.25]$ for $d = 3$. Off-diagonal covariances are all 0 for $d = 2$ and $d = 3$ cases.

For all designated experiments, Algorithm 1 starts with hyperparameters initialization: $\alpha = 0.001$, $\gamma_y = 0.001$, $\gamma = 0.02d$, $\alpha = 0.02\gamma$ with d depending on index dimension. Experiments run $T = 1175$ iterations for $d = 1, 2$ and $T = 575$ iterations for $d = 3$. T in each case was manually tested and selected as the lowest for regret to converge. Within each iteration, policy determines the sampling index $\pi^{(t)}$ at which it explores a new reward sample. Environment throughout this paper carries a Gaussian noise $\sim \mathcal{N}(0, 0.025)$. Regret of current iteration complies with typical MAB notion and is defined as loss from global maximum $max_i f(i)$. Evaluation is given in the form of average regret across t steps. Policy that achieves the lowest average regret across T iterations, or $\bar{R}_T$ is considered optimal. Figure 3(a), (b) and (c) demonstrate average experiment regrets across 30 trials for distinct d. They reveal that Thompson sampling wins final $\bar{R}_T$ for all three cases. Some policies fail to locate the global maximum by getting stuck at local minima (as caused by larger covariances), such as acquisition function PI for all three cases and EI for $d = 3$.

4   Test Bench Experiments - Policy Variation

Minor policy modification is needed to apply Algorithm 1 to real CTR test bench, which simulates N total impressions. $N = 100,000$ is used in test bench experiments (Algorithm 2) in analogy to budget of 100,000 ads. We intend the

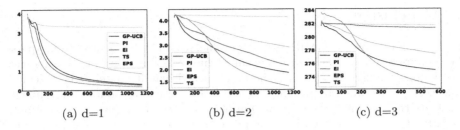

(a) d=1   (b) d=2   (c) d=3

Fig. 3. Experiments on synthetic functions

algorithm to deliver C_i ads on item y_i given n item candidates, or n graph nodes. So a weight list $\{w_i\}$ (Line 6) is desired from policy instead of a single decision index π. In case of acquisition functions, $\{w_i\}$ is a one-hot vector set at π so that all N impressions are devoted to item y_π at each step. For epsilon greedy policy, $w_\pi = 1 - \epsilon$ and $w_{i \neq \pi} = \epsilon/(n-1)$. For Thompson sampling, weights are computed as probability that an index is optimal, $w_i = p(y_i > y_{i \neq i}) = \prod_{j \neq i} p(y_i > y_j)$, where y_i is sampled from posterior so $y_i \sim \mathcal{N}(\hat{y}_i, \sigma_i)$ and $p(y_i > y_j)$ is directly available through the Gaussian cumulative distribution function of node y_j.

The CTR test bench χ simulates clicks on y_i as C_i given probability $\tilde{f}(i)$ and assigned impression Nw_i (Line 7). Probability $\tilde{f}$ is the 1-0 max-min scaled f over its discrete domain, where f is an environment function discussed in Sect. 3.2. C_i is simulated as binomial counts $B(Nw_i, \tilde{f}(i))$. CTR experiments are initialized with the same hyperparameter values in Algorithm 1. To increase sample sparsity, we adjust interval increments in X so that $\|D^d\| \approx 100$ instead of 1,000 in former experiments. The major advantage of Algorithm 2 lies within Line 9 − 12, where observation on every node is sampled instead of a single index in Algorithm 1. It is practically feasible as an offline process without extra cost because collecting user clicks is the only operation that acquires data from environment (Line 7). Figure 4 lists average missing clicks as regrets under different policies. Averaged evaluation across 30 trials showcases that $T = 100$ iterations suffice for regret to converge. Acquisition functions are prone to huge regret from devoting all impressions to wrong indices in higher dimensions, where early-stage prediction error is much more likely to occur. Thompson sampling stands out as optimal policy in practice.

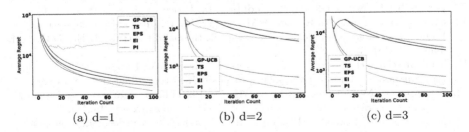

(a) d=1   (b) d=2   (c) d=3

Fig. 4. Regret evaluation on CTR simulator

5 Conclusion

This paper proposes a GMRF based graphical model that learns from sparse data samples and predicts distribution of function values at discrete indices. Predictions are used to optimize the multi-armed bandit model at best achievable cost under Thompson sampling as decision making policy. Experiments illustrate that the designated algorithm helps reduce online cost in case of data sparsity. Therefore our solution is applicable to profit improvement for recommendation engines as well as similar scenarios where seeking for optimum in unknown environment is desired.

References

1. Honda, J., Takemura, A.: An asymptotically optimal policy for finite support models in the multiarmed bandit problem. Mach. Learn. **85**(3), 361–391 (2011)
2. Maes, F., Wehenkel, L., Ernst, D.: Learning to play K-armed bandit problems. In: Proceedings of ICAART (2012)
3. Rue, H., Held, L.: Gaussian Markov Random Fields: Theory and Applications. Monographs on Statistics and Applied Probability. GMR:1051482. Chapman & Hall/CRC, Boca Raton (2005)
4. Srinivas, N., Krause, A., Kakade, S., Seeger, M.: Gaussian process optimization in the bandit setting: no regret and experimental design. In: Proceedings of ICML (2010)
5. Agrawal, S., Goyal, N.: Thompson sampling for contextual bandits with linear payoffs. In: Proceedings of ICML (2013)
6. Sui, Y., Gotovos, A., Burdick, J., Krause, A.: Safe exploration for optimization with Gaussian processes. In: Proceedings of ICML (2015)
7. Djolonga, J., Krause, A., Cevher, V.: High-dimensional Gaussian process bandits. In: Proceedings of NIPS (2013)
8. Wang, Z., Zhou, B., Jegelka, S.: Optimization as estimation with Gaussian processes in bandit settings. In: Proceedings of AISTATS (2016)
9. Desautels, T., Krause, A., Burdick, J.W.: Parallelizing exploration-exploitation tradeoffs in Gaussian process bandit optimization. J. Mach. Learn. Res. **15**, 3873–3923 (2014)
10. Wu, Y., György, A., Szepesvári, C.: Online learning with Gaussian payoffs and side observations. In: Proceedings of NIPS (2015)
11. Li, L., Chu, W., Langford, J., Schapire, R.E.: A contextual-bandit approach to personalized news article recommendation. In: Proceedings of WWW (2010)
12. Srinivas, N., Krause, A., Kakade, S.M., Seeger, M.W.: Information-theoretic regret bounds for Gaussian process optimization in the bandit setting. IEEE Trans. Inf. Theory **58**(5), 3250–3265 (2012)
13. Vanchinathan, H.P., Nikolic, I., De Bona, F., Krause, A.: Explore-exploit in top-N recommender systems via Gaussian processes. In: Proceedings of RecSys (2014)
14. Schreiter, J., Nguyen-Tuong, D., Eberts, M., Bischoff, B., Markert, H., Toussaint, M.: Safe exploration for active learning with Gaussian processes. In: Bifet, A., May, M., Zadrozny, B., Gavalda, R., Pedreschi, D., Bonchi, F., Cardoso, J., Spiliopoulou, M. (eds.) ECML PKDD 2015. LNCS (LNAI), vol. 9286, pp. 133–149. Springer, Cham (2015). https://doi.org/10.1007/978-3-319-23461-8_9

15. Chapelle, O., Li, L.: An empirical evaluation of Thompson sampling. In: Proceedings of NIPS (2011)
16. Zeng, C., Wang, Q., Mokhtari, S., Li, T.: Online context-aware recommendation with time varying multi-armed bandit. In: Proceedings of KDD (2016)
17. Nguyen, T.V., Karatzoglou, A., Baltrunas, L.: Gaussian process factorization machines for context-aware recommendations. In: Proceedings of SIGIR (2014)
18. Bubeck, S., Munos, R., Stoltz, G.: Pure exploration in multi-armed bandits problems. In: Gavaldà, R., Lugosi, G., Zeugmann, T., Zilles, S. (eds.) ALT 2009. LNCS (LNAI), vol. 5809, pp. 23–37. Springer, Heidelberg (2009). https://doi.org/10.1007/978-3-642-04414-4_7

Leveraging Label Category Relationships in Multi-class Crowdsourcing

Yuan Jin[1(✉)], Lan Du[2], Ye Zhu[1], and Mark Carman[2]

[1] School of Information Technology, Deakin University,
Burwood, Melbourne, VIC 3125, Australia
`yuan.jin@deakin.edu.au`, `ye.zhu@ieee.org`
[2] Faculty of Information Technology, Monash University,
Melbourne, VIC 3168, Australia
{`lan.du,mark.carman`}`@monash.edu`

Abstract. Current quality control methods for crowdsourcing largely account for variations in worker responses to items by interactions between *item difficulty* and *worker expertise*. Few have taken into account the *semantic relationships* that can exist between the response label categories. When the number of the label categories is large, these relationships are naturally indicative of how crowd-workers respond to items, with expert workers tending to respond with more semantically related categories to the categories of true labels, and with difficult items tending to see greater spread in the responded labels. Based on these observations, we propose a new statistical model which contains a *latent* real-valued matrix for capturing the *relatedness* of response categories alongside variables for worker expertise, item difficulty and item true labels. The model can be easily extended to incorporate prior knowledge about the semantic relationships between response labels in the form of a *hierarchy* over them. Experiments show that compared with numerous state-of-the-art baselines, our model (both with and without the prior knowledge) yields superior true label prediction performance on four new crowdsourcing datasets featuring large sets of label categories.

1 Introduction

Crowdsourcing is a process in which a *human intelligence* task is solved collectively by a large number of online workers who get paid to independently solve parts of the task that commonly overlap. In recent years, the process has been used by machine learning communities to cheaply collect large quantities of labelled training data, thanks to the development of online service providers, such as Amazon Mechanical Turk[1] and CrowdFlower[2]. While crowdsourcing has shown cost-effectiveness and scalability, it also produces noisy and biased labelled data as its online workforce is much less accurate than in-house experts.

---

[1] https://www.mturk.com/.
[2] https://www.crowdflower.com/.

© Springer International Publishing AG, part of Springer Nature 2018
D. Phung et al. (Eds.): PAKDD 2018, LNAI 10938, pp. 128–140, 2018.
https://doi.org/10.1007/978-3-319-93037-4_11

Furthermore, many crowdsourcing tasks in practice involve large numbers of unlabelled items under limited budgets, which often results in small numbers of responses collected for each item. Aggregating such small numbers of (oftentimes conflicting) labels using majority vote to infer the true label of each item can be unreliable.

To overcome the above issue, labels must be aggregated in such a way that the influence of "high-quality" responses should outweigh that of those "low-quality" responses for better estimating the true labels. This process is generally known as the Quality Control for Crowdsourcing (QCC). The QCC methods, largely based on statistical modeling, consider *expertise* of workers to govern the quality of labels they provide to items with greater expertise indicating higher quality of the labels [1–4]. Furthermore, some of the QCC methods also consider *difficulty* of items which counteracts worker expertise to undermine the quality of the labels [5–9]. All these methods have achieved overall superior performance over the majority vote. However, assuming individual crowdsourcing tasks contain small numbers of uncorrelated label categories, these methods inevitably ignores the impact of the *relationships* between response label categories on the quality of workers' responses to items. In practice, it is not unusual that crowdsourcing tasks can involve labeling data across label categories correlated to one another in terms of *large structural semantic relationships*. A typical example is the classification of objects in images for building the database of ImageNet[3] whose large number of label categories are related through the semantic relationships specified by WordNet[4]. Other examples include the classification of Webpages for the Open Directory Project, called "DMOZ"[5], and for DBpedia[6], whose large numbers of categories are connected through semantic relationships maintained by respective online volunteer communities. This paper focuses on leveraging semantic relationships between label categories for improving QCC performance in crowdsourcing problems especially involving *highly multi-class* labels. The semantic relationships are inherent in such problems and the conventional inference about them is based on human knowledge and reasoning which features prominently in crowdsourcing. Conversely, knowing the semantic relationships should contribute to accurate inference about how responses are formed in highly mult-class crowdsourcing.

When semantic relationships between categories exist in crowdsourcing, crowd-workers with greater expertise tend to respond to the same item with categories more related to the true label for the item. Moreover, the difficult items tend to see more variety in the responses (more distinct categories present) than simpler items. To be more specific, consider a simple measure of relatedness between two label categories k and k' shown below:

$$\text{Relatedness}(k, k') = \frac{1}{|\text{shortest\_path}(k, k')| + 1} \qquad (1)$$

---

[3] http://www.image-net.org/.

[4] https://wordnet.princeton.edu/.

[5] http://www.dmoz.org/.

[6] http://wiki.dbpedia.org/.

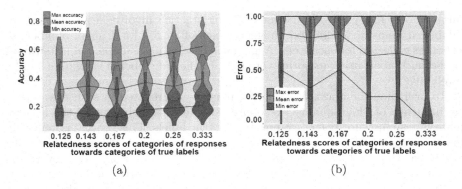

Fig. 1. (a) Worker response accuracy versus category relatedness. (b) Item difficulty (in terms of response error) versus category relatedness. (Color figure online)

where $|\text{shortest\_path}(\cdot, \cdot)|$ is the length of the shortest path between any pair of categories in some known semantic structure (e.g. a graph). Using this relatedness measure, Fig. 1 shows the relationship between the *relatedness* of the response category to the true label, and three *summary statistics* (namely the maximum, mean and minimum values) for the *response accuracy* of workers and the *item difficulty* (in terms of response error). The crowdsourcing task involved in this case is identifying breeds of dogs in images from ImageNet [10]. Every coloured "violin" area in each sub-figure of Fig. 1 represents the distribution of a particular summary statistics about either the response accuracy of workers or the response errors on items given the true labels. The medians of the areas with the same colours (i.e. the same summary statistics across different relatedness scores) are connected by straight lines in each sub-figure. We observe from Fig. 1 that:

- According to Fig. 1a, more related categories (with higher relatedness scores) to item true labels tend to be chosen more often as responses by workers with higher response accuracy;
- According to Fig. 1b, less related categories (with lower relatedness scores) to item true labels tend to be given more often as responses to more difficult items (i.e. ones with larger response errors in Fig. 1b).

In this paper, we leverage the above observed relationship between category relatedness and worker accuracy/item difficulty for improving the quality control of crowdsourced labels. This is done by encoding the correlations between categories into the conditional probability of a worker giving a label to an item given its true label. Such an encoding can help refine the estimation about the correctness of crowdsourced labels (which is modeled using those conditional probabilities in most QCC methods). The encoding is based on a *latent symmetric* relatedness matrix where each off-diagonal entry is a real-valued score representing how related categories are to one another. In this case, each category (as a true label/correct response) is associated with a continuous scale accommodating the latent relatedness scores of all the other categories as possible worker responses. We also model expertise of workers and difficulty of items on the same scale.

According to Fig. 1a, a worker with greater expertise and a category more related to the true label should have the estimated values for their respective variables reside further down the positive infinite end of the scale once learned from response data. Likewise from Fig. 1b, an item with greater difficulty and a less related category should have the estimated values situated towards the opposite end of the scale. The interactions between these variables on the scale are captured and transformed into the aforementioned conditional probabilities through an *ordered logit* model where the difference between item difficulty and worker expertise serves as the *response-specific slope*, and the off-diagonal terms in the same row of the latent relatedness matrix (corresponding to a latent true label) serve as the *intercepts* specific to different categories other than the true label. The off-diagonal terms in the matrix share a Normal prior, which can make use of prior knowledge (trees extracted from Wordnet and DMOZ) to better calibrate the estimates for the terms. The contributions of this paper are:

- A novel statistical model that leverages correlations/relationships between label categories for improving quality control of crowdsourced labels.
- The proposed model directly infers the latent relationships between label categories from crowdsourced labels.
- A priori knowledge of relationships between labels (in terms of a semantic hierarchy over concepts) is elegantly incorporated into the proposed model by modifying the prior over the latent relatedness variables.

2   Related Work

Two papers have considered leveraging relationships between label categories [11,12] for improving quality control of crowdsourced labels. In [12], a model called **SEEK** was proposed in which the conditional probability of any possible response category a worker can give to an item given its true label category is output from a *soft-max* function. The function takes in the observed relatedness scores of all the response categories to the true label of the item along with the difficulty of the item and the expertise of the worker. Inside the function, the difference between the difficulty and the expertise is multiplied by the relatedness score of every response category before the results are normalized to form the corresponding conditional probabilities. Since the difference value is the same for all response categories, the conditional probabilities are thus only proportional to the relatedness scores. The larger a score is, the higher the conditional probability of the corresponding response given the true label. In comparison, our model allows the conditional probabilities to be proportional to the joint interaction between the difference and the relatedness scores. In [12], each relatedness score between a pair of categories can vary from 0 to 1. It is 1 when the two categories are the same. It is between 0 and 1 only when one of the categories is a hypernym of the other. Otherwise, the score is always 0. Clearly,

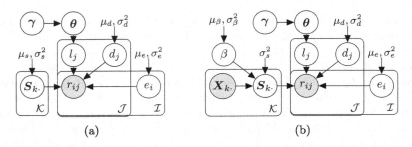

Fig. 2. The DELRA model with and without encoding observed knowledge matrix X specifying relationships between categories are shown in Fig. 2a and b.

this way of pre-computing the relatedness scores between categories constrains the quality control performance of SEEK in crowdsourcing tasks where most of the categories are not hypernyms. In [11], a model called **DASM** is proposed which share the same idea as SEEK except that the relatedness scores are pre-computed as the inverse of the Euclidean distances between categories in terms of their observed features. Both of these models rely on the availability of the external knowledge about the category relatedness, while our model is able to infer such relatedness directly from responses.

3   Problem Formulation and Proposed Model

Given a large but finite set of categories $\mathcal{K}$ and a set of items $\mathcal{J}$, a set of workers $\mathcal{I}$ have provided a set of responses $\mathcal{R}$ to $\mathcal{J}$. An item $j \in \mathcal{J}$ has one unknown true label $l_j = k$, where $l_j \in \mathcal{L}$, the set of corresponding true labels of individual items in $\mathcal{J}$, and k is a particular category in $\mathcal{K}$. For the set of categories $\mathcal{K}$, there exists a tree structure organizing them in terms of their semantic relationships. The relationships are quantified into an observed real-valued relatedness matrix $X \in \mathbb{R}^{|\mathcal{K}| \times |\mathcal{K}|}$. Each off-diagonal entry $x_{kk'}$ expresses how related a category k' (as a response to an item) is to another category k as the true label of that item. It is calculated by Eq. (1). Based on these inputs, our model should output a corresponding set of prediction $\hat{\mathcal{L}}$ for the latent item true labels $\mathcal{L}$ such that the overall difference between the former and the latter sets across their corresponding elements is as small as possible.

In this paper, we propose the Difficulty-Expertise-Label-Relationship-Aware (**DELRA**) model, characterized by a latent relatedness matrix $S \in \mathbb{R}^{|\mathcal{K}| \times |\mathcal{K}|}$. The matrix specifies how related a category k' (as a response to an item) is to another category k (as the true label of that item) in crowdsourcing. We assume S is symmetric so that $s_{kk'} = s_{k'k}$ where $s_{kk'}, s_{k'k} \in S$. This assumption is reasonable as if crowd-workers perceive category k' to be related to category k overall to a certain degree, they should also perceive the relatedness of category k to category k' to the same degree. Based on the assumption, the DELRA model is shown in Fig. 2a, and has the following generative process:

1. Draw true label category proportions $\boldsymbol{\theta} \sim \text{Dir}(\boldsymbol{\gamma})$;
2. For each pair of categories (k,k') where $k \neq k'$:
   (a) if $k' > k$ then draw relatedness $s_{kk'} \sim \mathcal{N}(\mu_s, \sigma_s^2)$;
       else $s_{kk'} \leftarrow s_{k'k}$[7];
3. For each item $i \in \mathcal{J}$:
   (a) Draw its true label $l_j \sim \text{Cat}(\boldsymbol{\theta})$;
   (b) Draw its difficulty $d_j \sim \mathcal{N}(\mu_d, \sigma_d)$;
4. For each worker $i \in \mathcal{I}$:
   (a) Draw her expertise $e_i \sim \mathcal{N}(\mu_e, \sigma_e^2)$;
5. For each worker-item pair (i, j):
   (a) Draw response $r_{ij} \sim \text{Cat}(\boldsymbol{\pi}_{ijl_j})$ where $\boldsymbol{\pi}_{ijl_j}$ is a $|\mathcal{K}|$-dimensional vector with each element $\pi_{ijl_jk} = P(r_{ij} = k|l_j)$ specified as the difference between consecutive sigmoid functions as follows:

$$\pi_{ijl_jk} = \delta_{ijl_jk} - \max_{k':\delta < \delta_{ijl_jk}} \delta_{ijl_jk'} \quad \text{where } \delta_{ijl_jl_j} = 1, \delta_{ijl_j0} = 0 \quad (2)$$

Here δ_{ijl_jk} is a sigmoid function relating the odds of observing response $r_{ij} = k$ given true label l_j to a linear combination of the relatedness score s_{l_jk}, the worker expertise e_i and the question difficulty d_j:

$$\delta_{ijl_jk} = \frac{1}{1 + \exp(-(s_{l_jk} - e_i + d_j))} \quad (3)$$

Apart from inferring the relatedness matrix $\boldsymbol{S}$ from responses, our model also allows for the encoding of useful prior knowledge about the entries in each row of the matrix corresponding to a particular category as true labels for items to help calibrate the inference. As shown in Fig. 2b, the Normal prior $\mathcal{N}(\mu_s, \sigma_s^2)$ in Fig. 2a shared by all the entries in the matrix is now replaced by individual priors centered on the product results between a global coefficient β and the observed relatedness matrix $\boldsymbol{X}$ after it is *log-transformed* followed by *standardization*, added with Normally distributed noise following $\mathcal{N}(0, \sigma_s^2)$. Correspondingly, step 2(a) of the above generative process of the DELRA model is now changed to:

2. For each pair of categories (k,k') where $k \neq k'$:
   (a) if $k' > k$ then draw $s_{kk'} \sim \mathcal{N}(\beta x_{kk'}, \sigma_s^2)$;
       else $s_{kk'} \leftarrow s_{k'k}$;

The global term $\beta \sim \mathcal{N}(\mu_\beta, \sigma_\beta^2)$. The term $x_{kk'}$ goes through the transformation:

$$x_{kk'} \leftarrow \frac{\log(x_{kk'}) - \hat{\mu}_{\log(\boldsymbol{X})}}{\hat{\sigma}_{\log(\boldsymbol{X})}} \quad (4)$$

where $\hat{\mu}_{\log(\boldsymbol{X})}$ and $\hat{\sigma}_{\log(\boldsymbol{X})}$ are respectively the sample mean and the sample standard deviation of the logarithm of all the original terms in $\boldsymbol{X}$. The reason behind the logarithm operation is that the outputs from the relatedness function specified by Eq. (1) are very skewed and we do not want such skewness to impact

---

[7] The expression "$a \leftarrow b$" stands for assigning b to a or equivalently replacing a with b.

the estimation of the relatedness matrix. The reason behind the standardization operation is that every log-transformed $x_{kk'}$ is negative, thus having a negative mean. We want to adjust them to be centered on zero with scale one to allow for easier setups of priors for other model parameters. After the transformation by Eq. (4), the prior mean $\beta x_{kk'}$ for the relatedness score $s_{kk'}$ suggests how the relatedness between categories according to the semantic knowledge tree tends to correlate with their latent relatedness in crowdsourcing a priori.

4   Parameter Estimation

In this section, we describe how the model parameters are estimated. More specifically, in each iteration of the estimation, we alternate between the *Collapsed Gibbs sampling* for inferring the true labels of items $\mathcal{L}$ given the current estimates of the other model parameters including the worker expertise e_i, the item difficulty d_j and the relatedness matrix S, and the *LBFGS-B* till its convergence for updating these parameters given the current assignment of $\mathcal{L}$.

Collapsed Gibbs Sampling for $\mathcal{L}$: At this stage, we obtain posterior samples for item true labels $\mathcal{L}$ given the current estimates of all the other parameters. The conditional probabilities of true label l_j of item j is obtained by marginalizing out the multinomial probability vector $\boldsymbol{\theta}$, which ends up being:

$$P(l_j = k | \mathcal{L}_{\neg j}, \mathcal{R}_j, \{e_i\}_{i \in \mathcal{I}_j}, d_j, \boldsymbol{s}_k, \boldsymbol{\gamma}) \propto \frac{N_{\neg jk} + \gamma_k}{\sum\limits_{z \in \mathcal{K}} (N_{\neg jz} + \gamma_z)} \prod_{i \in \mathcal{I}_j} \pi_{ijkr_{ij}} \qquad (5)$$

where $\mathcal{I}_j$ is the set of workers who responded item j with a set of responses $\mathcal{R}_j$, $\mathcal{L}_{\neg j}$ is the set of current true label assignments to all the items except j, and $N_{\neg jk}$ is the number of items except j whose true labels are now inferred as k.

Gradient Descent for Other Parameters: The conditional probability distributions of the other model parameters including $e_i, d_j,$ and S are hard to compute analytically due to the presence of the sigmoid function. Instead, we run the LBFGS-B till its convergence on the following objective function Q:

$$Q = -\log\left(p(\boldsymbol{e}, \boldsymbol{d}, \boldsymbol{S} | \mathcal{R}, \mathcal{L}, \mu_{\{e,d,s\}}, \sigma^2_{\{e,d,s\}})\right) = -\sum_{j \in \mathcal{J}} \sum_{i \in \mathcal{I}_j} \log(\pi_{ijl_j r_{ij}})$$
$$+ \frac{1}{2}\left[\sum_{i \in \mathcal{I}} \frac{(e_i - \mu_e)^2}{\sigma_e^2} + \sum_{j \in \mathcal{J}} \frac{(d_j - \mu_d)^2}{\sigma_d^2} + \sum_{k \in \mathcal{K}} \sum_{k' \in \mathcal{K} \& k' > k} \frac{(s_{kk'} - \mu_s)^2}{\sigma_s^2} \right] \qquad (6)$$

The gradient with respect to the label-relatedness term s_{lk} is computed as:

$$\frac{\partial Q}{\partial s_{lk}} = -\sum_{j \in \mathcal{J}} \sum_{i \in \mathcal{I}_j} \frac{\partial \log(\pi_{ijl_j r_{ij}})}{\partial s_{lk}} + \frac{s_{lk}}{\sigma_s^2} \qquad (7)$$

where for the true label $l = l_j$ and observed response $k = r_{ij}$ we have:

$$\frac{\partial \log(\pi_{ijl_jr_{ij}})}{\partial s_{lk}} = \frac{\delta_{ijl_jk}(1 - \delta_{ijl_jk})}{\pi_{ijl_jr_{ij}}} \tag{8}$$

And for other responses $k \neq r_{ij}$ we have:

$$\frac{\partial \log(\pi_{ijl_jr_{ij}})}{\partial s_{lk}} = \frac{-\delta_{ijl_jk}(1 - \delta_{ijl_jk})}{\pi_{ijl_jr_{ij}}} \text{ if } k = \arg \max_{k':\delta < \delta_{ijl_jr_{ij}}} \delta_{ijl_jk'} \text{ else } 0 \tag{9}$$

Note that we also impose symmetry on the label relatedness terms $s_{kk'} = s_{k'k}$. The gradients with respect to e_i and d_j are similarly easy to derive and thus omitted due to space limitations.

When observed matrix X is introduced into the model, the coefficient β is updated by maximum a posteriori estimation for a linear regression over X.

5   Experiments and Results

Datasets: We have collected four new crowdsourcing datasets from Crowd-Flower for our experiments. Table 1 summarizes these datasets.

- **Dog breed identification (Dog).** The images and the set of categories used in this task originate from the Stanford Dog dataset [10]. There are 120 breeds of dogs involved in the task with 10 images for each dog breed randomly sampled from the Stanford dataset. We collected 5 labels for each image about the breed crowd-workers think appearing in that image. The 120 dog breeds are organized under the subtree "Dog" of the WordNet.
- **Bird species identification (Bird).** The categories involved are species of birds from the Caltech-UCSD Birds 200 dataset [13]. Originally, there are 200 bird species in this dataset, only 72 of which are present in the WordNet. As a result, we have only used these categories for the experiments and randomly sampled 10 images for each of them from the Caltech-UCSD dataset. Since this task is quite difficult, we collect on average 8 labels for each of the images.
- **Classification of Webpages about string instruments (Instrument).** This task asks for judgements about the sub-directories under which Web-pages about string instruments should be put. All the sub-directories share one root directory "Arts/Music/Instruments/String Instruments" from DMOZ. We have collected 5 judgements for each of the 1,323 Webpages across the 193 sub-directories corresponding to different aspects of string instruments.
- **Classification of Webpages about movies (Movie).** The judgements collected are about the sub-directories from DMOZ under which Webpages about movies should be put. All the sub-directories involved share the root directory "Arts/Movies". We have collected 5 judgements for each of the 737 Webpages across the 148 sub-directories about different aspects of movies.

Table 1. Dataset summary. The headers correspond to the notations introduced in Sect. 3.

Dataset	$|\mathcal{I}|$	$|\mathcal{J}|$	$|\mathcal{K}|$	$|\mathcal{R}|$
Dog	136	1,200	120	6,000
Bird	428	707	72	5,660
Instrument	334	1,323	193	7,233
Movie	169	737	148	3,539

Table 2. The accuracy of different models on inferring the true labels of the items across the four datasets.

Methods	Datasets			
	Dog	Bird	Instrument	Movie
DELRA	0.4803	0.4278	0.4489	0.3367
DELRA+X	**0.4833**	**0.4331**	**0.4561**	**0.3433**
SEEK	0.4688	0.4046	0.4406	0.3217
SEEK+X	0.4752	0.4256	0.4453	0.3342
DASM	0.4720	0.4229	0.4448	0.3274
MV	0.4742	0.4170	0.4414	0.3256
GLAD	0.4675	0.4017	0.4450	0.3229
DS	0.4341	0.3219	0.3900	0.2931
MdWC	0.4742	0.4041	0.4409	0.3311
PM	0.4367	0.3621	0.4002	0.2999
Minimax	0.4770	0.4224	0.4456	0.3202

5.1 True Label Prediction

To verify the capability of our model on predicting item true labels, we compare it with the following state-of-the-art crowdsourcing quality control methods.

- **Generative model of Labels, Abilities, & Difficulties (GLAD)** [5]. This model endows every crowd-worker and every item respectively with a latent variable about the worker's expertise and a variable about the item's difficulty. The expertise variable is divided by the difficulty variable to account for the probability of the label given by the worker to the item being correct.
- **Multi-dimensional Wisdom of Crowds (MdWC)** [9]. This model extends the concept of GLAD that worker expertise interacts with item difficulty by making the interaction factorized over latent variable vectors respectively about workers and items. It also adds another variable for each worker to account for their individual biases in choosing label categories.
- **Dawid-Skene (DS)** [1]. Unlike GLAD and MdWc which estimate the marginal correctness probability of a label, this model estimates the conditional probability of every label with which a worker can respond given each true label.
- **Minimax entropy (Minimax)** [8]. The same conditional probabilities are estimated in this model. In this case, the total entropy of the conditional probabilities over all the categories as the responses to the items given their true labels is optimized according to the minimax principle with constraints.
- **Participant-Mine voting (PM)** [4]. The accuracy of each worker and the true label of each item are inferred together using HITS [14] algorithm. An item is treated as a Webpage as in HITS with the total accuracy of the workers

responding it as its authority level and the total difference between the true label estimate of the item and its received worker labels as its hub level.

Apart from these baselines, we also compare our model with the Majority Vote (**MV**), and the original SEEK model discussed in Sect. 2. Moreover, we have also changed the external knowledge matrix input to the SEEK model to be the matrix X input to the DELRA model (called **DELRA+X**) with each entry transformed by Eq. (4) in both cases. We call this model **SEEK+X** and use it as another baseline. Likewise, we adapt the DASM model by calculating the distance between any pair of label categories using our distance definition specified in Eq. (1) rather than theirs as we do not have any observed feature about label categories. To measure the performance of our model and all the baselines, we use the *true label prediction accuracy*, defined as $\frac{1}{|\mathcal{J}|}\sum_{j\in\mathcal{J}}\mathbb{1}\{l_j = \hat{l}_j\}$.

Table 2 shows the results of the true label prediction of both DELRA and all the baselines. We can see that *with* and *without* the knowledge matrix X incorporated, the DELRA model respectively outperforms all the baselines by at least 0.6% and 0.3% over the Dog dataset, 0.8% and 0.22% over the Bird dataset, 1.1% and 0.36% over the Instrument dataset and 0.9% and 0.25% over the Movie dataset. Especially, SEEK+X has the exact same knowledge matrix input as DELRA+X, but has yielded lower performance even compared to the DELRA model without incorporating X. This suggests that not only our model is able to better leverage the external knowledge about semantic relationships between label categories, but also it is a better model in explaining how responses are generated from the interactions among the expertise of workers, the difficulty of items and the relationships between label categories in crowdsourcing.

5.2 True Label Prediction Under Response Sparsity

We now proceed to investigating how DELRA performs under various degrees of sparsity in crowdsourced responses. To do this, we randomly sample different proportions (i.e. between 10% and 50%) of the responses from each of the datasets and average the performance over 10 runs for each model (on each proportion). Figure 3a–d show the results of the true label prediction of all the models under varying degrees of response sparsity across the four datasets. The DELRA model incorporating the knowledge matrix X clearly beats all the baselines with convincing margins across 10% to 50% of the total responses from each dataset. Moreover, even without access to the external knowledge X, DELRA still performs closely to the performance of SEEK+X and outperform the other baselines when the sampling proportion is greater than 10%. When the sampling proportion is only 10%, DELRA without X seems to suffer from the response sparsity as any other baseline that has not leveraged X.

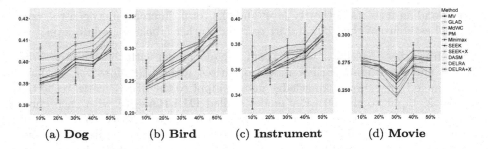

Fig. 3. The accuracy of different models on inferring the true labels of the items from 10% to 50% of the total responses across the four datasets. Note that x-axis and y-axis in each figure are respectively the sampling proportions of responses and the average true label prediction accuracy over 10 runs.

5.3   Consistency of Learned Relatedness Between Categories

In this experiment, we evaluate how consistent the estimates of the relatedness between categories from DELRA *without* X are with the relatedness scores in X pre-computed using Eq. (1) followed by the transformation in Eq. (4). More specifically, for each label category, we calculate the *Pearson correlation coefficients* between the Top-N most related category rank of the other label categories in terms of the estimates of their relatedness to the category, and the Top-N rank of the same set of categories in terms of their pre-computed transformed relatedness scores for that category. We set N to be 2, 3, 5, 10 and 15 to obtain the respective average Pearson correlation coefficients across all the label categories. We also implement two *supervised* baselines (in terms of knowing true labels) for obtaining the Top-N rank of the most related categories:

– **Top-N rank by frequency - asymmetric.** For each label category, the relatedness of the other label categories to it is their frequencies as the responses to the items with the label category as their true labels.
– **Top-N rank by frequency - symmetric.** For each label category, the relatedness of the other label categories to it is their frequencies as either the responses to the items with that label category as their true labels, or the true labels of the items which receive that label category as the responses.

Figure 4a–d show how the Top-N most related category ranks by both the DELRA model without X and the two baselines are correlated with the Top-N ranks by the ground-truth relatedness scores calculated by Eq. (1). Showing overall higher average correlation with the ground-truth relatedness scores across the four datasets, our model clearly yields more consistent category relatedness estimates than the two baselines even though it is unsupervised. We conjecture this is attributed to the ability of our model in distinguishing responses of different quality by accounting for the interaction between worker expertise and item difficulty, while the baselines treat all the responses as the same in their quality.

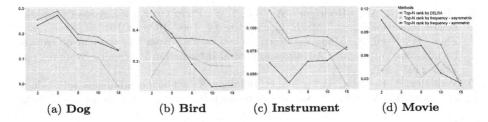

(a) **Dog**   (b) **Bird**   (c) **Instrument**   (d) **Movie**

Fig. 4. Average Pearson correlation coefficients between the Top-N most related category rank yielded by different methods, and the ground-truth Top-N rank yielded by the pre-computed related scores based on Eq. (1). Note that x-axis and y-axis in each figure are N and average correlation, respectively.

6   Conclusion

We propose DELRA, a quality control framework for crowdsourcing that leverages the semantic relationships between label categories. It features a latent real-valued matrix that captures the relatedness between response categories alongside variables for worker expertise, item difficulty and item true labels. DELRA encodes the joint interaction among these variables to refine estimation of conditional probabilities of responses given true labels. This leads DELRA to outperform numerous state-of-the-art quality control methods in terms of true label prediction. Moreover, DELRA allows for elegant encoding of a priori knowledge regarding the relationships between categories for calibrating the estimation of the latent relatedness matrix. This leads to its further improvements in the prediction. Finally, the relatedness matrix learned solely from response data by DELRA shows convincing consistency with the relatedness matrix pre-computed from the external semantic relationships between the categories.

References

1. Dawid, A.P., Skene, A.M.: Maximum likelihood estimation of observer error-rates using the EM algorithm. Appl. Stat. **28**, 20–28 (1979)
2. Liu, Q., Peng, J., Ihler, A.T.: Variational inference for crowdsourcing. In: Advances in Neural Information Processing Systems, pp. 692–700 (2012)
3. Wauthier, F.L., Jordan, M.I.: Bayesian bias mitigation for crowdsourcing. In: Advances in Neural Information Processing Systems, pp. 1800–1808 (2011)
4. Aydin, B.I., Yilmaz, Y.S., Li, Y., Li, Q., Gao, J., Demirbas, M.: Crowdsourcing for multiple-choice question answering (2014)
5. Whitehill, J., Ruvolo, P., Wu, T., Bergsma, J., Movellan, J.R.: Whose vote should count more: optimal integration of labels from labelers of unknown expertise. In: 23rd Annual Conference on Neural Information Processing Systems, NIPS 2009, pp. 2035–2043 (2009)
6. Bachrach, Y., Graepel, T., Minka, T., Guiver, J.: How to grade a test without knowing the answers—a bayesian graphical model for adaptive crowdsourcing and aptitude testing. arXiv preprint arXiv:1206.6386 (2012)

7. Moreno, P.G., Artés-Rodríguez, A., Teh, Y.W., Perez-Cruz, F.: Bayesian nonparametric crowdsourcing. J. Mach. Learn. Res. **16**, 1607–1628 (2015)
8. Zhou, D., Basu, S., Mao, Y., Platt, J.C.: Learning from the wisdom of crowds by minimax entropy. In: Advances in Neural Information Processing Systems, pp. 2195–2203 (2012)
9. Welinder, P., Branson, S., Perona, P., Belongie, S.J.: The multidimensional wisdom of crowds. In: Advances in Neural Information Processing Systems, pp. 2424–2432 (2010)
10. Khosla, A., Jayadevaprakash, N., Yao, B., Li, F.F.: Novel dataset for fine-grained image categorization: stanford dogs. In: Proceedings of CVPR Workshop on Fine-Grained Visual Categorization (FGVC), vol. 2, p. 1 (2011)
11. Fang, Y.L., Sun, H.L., Chen, P.P., Deng, T.: Improving the quality of crowdsourced image labeling via label similarity. J. Comput. Sci. Technol. **32**(5), 877–889 (2017)
12. Han, T., Sun, H., Song, Y., Fang, Y., Liu, X.: Incorporating external knowledge into crowd intelligence for more specific knowledge acquisition. In: IJCAI, pp. 1541–1547 (2016)
13. Welinder, P., Branson, S., Mita, T., Wah, C., Schroff, F., Belongie, S., Perona, P.: Caltech-UCSD birds 200. Technical report CNS-TR-2010-001, California Institute of Technology (2010)
14. Kleinberg, J.M.: Authoritative sources in a hyperlinked environment. J. ACM (JACM) **46**(5), 604–632 (1999)

Embedding Knowledge Graphs Based on Transitivity and Asymmetry of Rules

Mengya Wang, Erhu Rong, Hankui Zhuo$^{(\boxtimes)}$, and Huiling Zhu

Sun Yat-sen University, No. 132, Waihuan East Road, Guangzhou, China
{wangmy38,rongerhu}@mail2.sysu.edu.cn,
{zhuohank,zhuhling6}@mail.sysu.edu.cn

Abstract. Representation learning of knowledge graphs encodes entities and relation types into a continuous low-dimensional vector space, learns embeddings of entities and relation types. Most existing methods only concentrate on knowledge triples, ignoring logic rules which contain rich background knowledge. Although there has been some work aiming at leveraging both knowledge triples and logic rules, they ignore the transitivity and asymmetry of logic rules. In this paper, we propose a novel approach to learn knowledge representations with entities and ordered relations in knowledges and logic rules. The key idea is to integrate knowledge triples and logic rules, and approximately order the relation types in logic rules to utilize the transitivity and asymmetry of logic rules. All entries of the embeddings of relation types are constrained to be non-negative. We translate the general constrained optimization problem into an unconstrained optimization problem to solve the non-negative matrix factorization. Experimental results show that our model significantly outperforms other baselines on knowledge graph completion task. It indicates that our model is capable of capturing the transitivity and asymmetry information, which is significant when learning embeddings of knowledge graphs.

Keywords: Knowledge graph · Logic rules
Non-negative matrix factorization · Transitivity · Asymmetry

1 Introduction

Knowledge graphs (KGs) store rich information of the real world in the form of graphs, which consist of nodes (entities) and labelled edges (relation types between entities) (e.g., *(Trump, PresidentOf, USA)*). This sort of structured data can be interpreted by computers and applied in various fields such as information retrieval [4] and word sense disambiguation [2]. Although powerful in representing structured data, the symbolic nature of relations makes KGs, especially large-scale KGs, difficult to manipulate. Predicting missing entries (known as link prediction) is of great importance in knowledge graph. To do this task, vector space embeddings of knowledge graphs have been widely adopted.

© Springer International Publishing AG, part of Springer Nature 2018
D. Phung et al. (Eds.): PAKDD 2018, LNAI 10938, pp. 141–153, 2018.
https://doi.org/10.1007/978-3-319-93037-4_12

The key idea is to embed entities and relation types of a KG into a continuous vector space.

Many approaches have been proposed to learn embeddings of entities and relations, such as TransE [5], NTN [22], HOLE [6], ComplEx [7], and so on. They, however, only focus on knowledge triples, ignoring rich knowledge from logic rules. Logic rules are taken as complex formulae constructed by combining atoms with logical connectives. To leverage logic rules in knowledge graph embeddings, [18–20] propose to utilize both knowledge triples and logic rules for KB completion. In their works, however, logic rules need to be grounded. Each rule needs to be instantiated with concrete entities. A rule can be grounded into plenty of ground rules, since there are many entities connected by the same relation type in logic rules. As a result, the works cannot scale well to larger KGs since rules will be grounded with more entities. And also, they all neglect the properties of transitivity and asymmetry of rules.

*For example, suppose we have two rules: "CapitalOf ⇒ LocatedIn" and "LocatedIn ⇒ ContainedBy", which indicates the relation that x is a **capital of** y implies another relation that x is **located in** y, and the relation that x is **located in** y implies x is **contained by** y. Provided that we know Paris is a **capital of** France from the knowledge base, we can infer that Paris is **located in** France and **contained by** France as well, according to the **transitivity** of rules. In addition, even though we know Paris is **located in** France, we cannot infer that Paris is a **capital of** France according to the **asymmetry** of rules: "CapitalOf ⇒ LocatedIn" ≠ "LocatedIn ⇒ CapitalOf".*

To leverage the properties of transitivity and asymmetry of rules, we propose a novel knowledge representation learning model to capture the ordering of relations, and infer potential new relations based on the ordering of existing relations and properties of transitivity and asymmetry of rules. We integrate knowledge graphs, existing relations and logic rules together to learn the knowledge graph embeddings. Logic rules are incorporated into relation type representations directly, rather than instantiated with concrete entities. The embeddings learned are therefore compatible not only with triples but also with rules, and the embeddings of relation types are approximately ordered. We call our learning approach TARE, short for **E**mbedding knowledge graphs based on **T**ransitivity and **A**symmetry of **R**ules.

In the remainder of the paper, we first review previous work related to our work. Then we formulate our problem and present our learning algorithm in detail. After that we evaluate our approach by comparing our approach with exiting state-of-the-art algorithms. Finally we conclude the paper with future work.

2   Related Work

Many works have made great efforts on modelling knowledge graphs. Some works explain triples via latent representation of entities and relations such as tensor factorization [8,9,29] and multiway neural networks [22]. The key of relational latent feature models is that the relationship between entities can be derived from interactions of their latent features. Many ways are discovered to model these interactions. RESCAL [29] is a relational latent feature model which explains triples via pairwise interactions of latent features. However, it requires a large number of parameters. TransE [5] translates the latent feature representations via a relation-specific offset. Both entities and relations are projected into the same continuous low-dimensional vector space and relations are interpreted as translating operations between head and tail entities. TransE is efficient when modelling simple relations. To improve the performance of TransE on complicated relations, TransH [10], TransR [11] and TransD [17] are proposed. Unfortunately, these models miss simplicity and efficiency of TransE. To combine the power of tensor product with the efficiency and simplicity of TransE, HOLE [6] uses the circular correlation of vectors to represent pairs of entities. Circular correlation has the advantage, comparing to tensor product, that it does not increase the dimensionality of the composite representation. However, due to the asymmetry of circular correlation, HOLE is unable to deal with symmetric relation. Complex [7] makes use of embeddings with complex value and is able to handle a large number of binary relations, in particular symmetric and antisymmetric relations. Some works such as [12,30,31] learn embeddings by sampling long paths($e0 \rightarrow e_1 \rightarrow e_2... \rightarrow e_n$) in KG. They learn the transitivity of relation triples but not the transitivity of rules.

These models perform the embedding task based solely on triples contained in a KG. Recent work put growing interest in logic rules. [18] tries to utilize rules via integer linear programming or Markov logical networks. However, rules are modeled separately from embedding models and will not help obtaining better embeddings. [19] proposes a joint model which injects first-order logic into embeddings. This work focus on the relation extraction task and created vector embeddings for entity pairs rather than individual entities. As a result, relations between unpaired entities cannot be effectively discovered. KALE-Joint [20] proposes a new approach which learns entity and relation embeddings by jointly modelling knowledge triples and logic rules. However, all logic rules need to be grounded in these works. Since each relation type is linked to plenty of entities, each rule can be grounded into plenty of triples. The more original triples KG have, the more triples grounded from rules need to be used for learning, keeping grow exponentially. And thus they do not scale well to larger KGs. Also, the embeddings of the relation types in rules are not ordered, and thus the transitivity and asymmetry of logic rules are missed.

To address above issues, we propose a novel approach which learns embeddings by combining logic rules with knowledge triples. Logic rules are incorporated into relation type representations directly and the embeddings of relation types in logic rules are approximately ordered to leverage the transitivity and asymmetry of rules.

3   Problem Definition

In this section, we give a formal definition of the problem. A knowledge graph $\mathcal{G}$ is defined as a set of triples of the form (s, r, o). $s, o \in \mathcal{E}$ denote the subject and object entity, respectively. $r \in \mathcal{R}$ denotes the relation type. $\mathcal{E}$ denotes the set of all entities and $\mathcal{R}$ denotes the set of all relation types in $\mathcal{G}$.

Create from $\mathcal{G}$ a set of logic rules: $r_a \Rightarrow r_b$ (in form of (r_a, r_b)) denotes that r_a logically implies r_b, which means that any two entities linked by relation r_a should also be linked by relation r_b; $r_a \wedge r_b \Rightarrow r_c$ (in form of (r_a, r_b, r_c)) denotes that the conjunction of r_a and r_b logically implies r_c: if e_0 and e_1 are linked by r_a, e_1 and e_2 are linked by r_b, then e_0 and e_2 are linked by r_c. $r_a, r_b, r_c \in \mathcal{LR}$, where $\mathcal{LR} \subseteq \mathcal{R}$ is the subset of relation types observed in logic rules.

Our objective is to learn embeddings of entities, relations more precisely by approximately ordering the embeddings of relation types in logic rules, to predict relation types between entities. The embeddings are set in $\mathbb{R}^d$ and denoted with the same letters in boldface.

4 Our Model

4.1 Restricted Triple Model (RTM)

In RTM, we aim to embed entities and relation types to capture the correlations between them. The embeddings of relation types are restricted to be non-negative. Given two entities $s, o \in \mathcal{E}$, the log-odd of the probability of the truth of fact (s, r, o) is:

$$P(Y_{sro} = 1|\Theta) = \sigma(\phi(s, r, o)) \tag{1}$$

where $\sigma(x) = 1/(1 + exp(-x))$ denotes the logistic function that maps x to $(0,1)$, which is just the range of probability; $\phi()$ is the energy function which is based on a factorization of the observed knowledges and indicates the correlation of relation r and the entity pair (s, o). $\Theta = \{e_i\}_{i=1}^{n_e} \cup \{r_k\}_{k=1}^{n_r}$ denotes the the set of all embeddings $\mathbf{v}_e, \mathbf{v}_r \in \mathbb{R}^d$ of the corresponding model, n_e and n_r is the number of entities and relation types in the given KG respectively. $\{Y_{sro}\}_{(s,r,o) \in \Omega} \in \{-1, 1\}^{|\Omega|}$ is a set of labels (true or false) of the triples, where $\Omega \in \mathcal{E} \otimes \mathcal{R} \otimes \mathcal{E}$. $Y_{sro} = 1$ if (s, r, o) is positive. Otherwise, $Y_{sro} = -1$.

The energy function $\phi(s, r, o; \Theta)$ in our model is based on existing model Complex [7], in which complex vectors $\mathbf{v}_e, \mathbf{v}_r \in \mathbb{R}^d$ are learned for each entity $e \in \mathcal{E}$ and each relation type $r \in \mathcal{R}$. It models the score of a triple as:

$$\begin{aligned}
\phi(s, r, o) &= Re(\langle \mathbf{r}\mathbf{s}\overline{\mathbf{o}} \rangle) \\
&= Re(\sum_{i=0}^{d-1} \mathbf{r}_i \mathbf{s}_i \overline{\mathbf{o}}_i) \\
&= \sum_{i=0}^{d-1} Re(\mathbf{r}_i) Re(\mathbf{s}_i) Re(\mathbf{o}_i) + \sum_{i=0}^{d-1} Re(\mathbf{r}_i) Im(\mathbf{s}_i) Im(\mathbf{o}_i) \\
&+ \sum_{i=0}^{d-1} Im(\mathbf{r}_i) Re(\mathbf{s}_i) Im(\mathbf{o}_i) - \sum_{i=0}^{d-1} Im(\mathbf{r}_i) Im(\mathbf{s}_i) Re(\mathbf{o}_i)
\end{aligned} \tag{2}$$

where $\mathbf{r}$, $\mathbf{s}$, $\mathbf{o}$ are complex vector embeddings for relation types, subject and object respectively. $Re(\mathbf{x})$ and $Im(\mathbf{x})$ represent the real part and imaginary part of the complex vector embedding $\mathbf{x}$, respectively. It separates the embedding vector into real part and imaginary part to obtain the symmetric and antisymmetric relations. This function is antisymmetric when $\mathbf{r}$ is purely imaginary (i.e. its real part is zero), and symmetric when $\mathbf{r}$ is real.

To approximately order the relation types in logic rules, we constrain the real part and imaginary part of the vector embeddings of relation types to be non-negative and reduce the problem to Non-negative Matrix Factorization (**NMF**). And also, non-negative constraints make the embeddings interpretable. For most embedding methods, a critical issue is that, we are unaware of what each dimension represent in embeddings. Hence, the dimensions are difficult to interpret. This makes embeddings like a black-box, and prevents them from being human-readable and further manipulation. NMF learn embeddings with good interpretabilities. There are many ways to solve Non-negative Matrix Factorization such as Multiplicative Update [25], Gradient based Update [23] and Alternating Non-negative Least Squares [26, 27].

In our model, we adopt the approach which updates the embeddings based on gradient. Translate the general constrained optimization problem into an unconstrained optimization problem. The embeddings of relation types are translated into unconstrained complex vectors as follows:

$$Re(\mathbf{r}) = Re(\mathbf{q}_a)^{(2)}, \quad Im(\mathbf{r}) = Im(\mathbf{q}_b)^{(2)} \tag{3}$$

where $\mathbf{q}_a$ and $\mathbf{q}_b$ denote the vectors which are initialized randomly(not constrained to be non-negative) and updated during learning, $\mathbf{x}^{(2)}$ denotes the element-wise square of the vector $\mathbf{x}$. In other words, $Re(\mathbf{r}_i) = Re(\mathbf{q}_{ai})^2$ and $Im(\mathbf{r}_i) = Im(\mathbf{q}_{bi})^2$. Plug Eq.(3) into Eq.(2), we get a new energy function:

$$\phi(s, r, o) = \sum_{i=0}^{d-1} Re(\mathbf{q}_{ai})^2 Re(\mathbf{s}_i) Re(\mathbf{o}_i) + \sum_{i=0}^{d-1} Re(\mathbf{q}_{ai})^2 Im(\mathbf{s}_i) Im(\mathbf{o}_i)$$

$$+ \sum_{i=0}^{d-1} Im(\mathbf{q}_{bi})^2 Re(\mathbf{s}_i) Im(\mathbf{o}_i) - \sum_{i=0}^{d-1} Im(\mathbf{q}_{bi})^2 Im(\mathbf{s}_i) Re(\mathbf{o}_i) \tag{4}$$

We train the triples by minimizing the negative log-likelihood of the logistic model with L^2 regularization on the parameters Θ:

$$\mathcal{L}_K = \min_{\Theta} \sum log(1 + exp(-Y_{sro}\phi(s, r, o))) + \lambda_1 \|\Theta\|^2 \tag{5}$$

where λ_1 is the regularization parameter.

The real part and imaginary part of complex vector $\mathbf{q}$ are both unconstrained. Therefore, the novel objective function can be solved by applying Stochastic Gradient Descent (SGD) directly. The negative set of knowledges is generated by local closed world assumption (LCWA) proposed in [28].

4.2 Approximate Order Logic Model (AOLM)

In AOLM, We aim to embed relation types to capture the ordering between relations. We work on two kinds of logic rules: $r_a \Rightarrow r_b$ and $r_a \wedge r_b \Rightarrow r_c$. Since there is no natural linear ordering on the set of complex numbers, we approximately order the complex vector embeddings by ordering the real part and imaginary part of the embeddings respectively. For their vector representations we require that the component-wise inequality holds:

$$r_a \Rightarrow r_b \ if \ and \ only \ if \ \bigwedge_{i=0}^{d-1} Re(\mathbf{r}_{ai}) \leqslant Re(\mathbf{r}_{bi})$$

$$and \ \bigwedge_{i=0}^{d-1} Im(\mathbf{r}_{ai}) \leqslant Im(\mathbf{r}_{bi}) \qquad (6)$$

and

$$r_a \wedge r_b \Rightarrow r_c \ if \ and \ only \ if \ \bigwedge_{i=0}^{d-1} Re(\mathbf{r}_{ai})Re(\mathbf{r}_{bi}) \leqslant Re(\mathbf{r}_{ci})$$

$$and \ \bigwedge_{i=0}^{d-1} Im(\mathbf{r}_{ai})Im(\mathbf{r}_{bi}) \leqslant Im(\mathbf{r}_{ci}) \qquad (7)$$

for all vectors with non-negative coordinates, where $\bigwedge$ denotes the conjunction. Smaller coordinates imply higher position: $r_a \Rightarrow r_b$ if and only if all entries of the real part and imaginary part of the vector embedding of r_a are less than or equal to that of r_b.

The penalty for an ordered pair (r_a, r_b) of a given logic rule $r_a \Rightarrow r_b$ is defined as follows:

$$F(r_a, r_b) = \|max(\mathbf{0}, Re(\mathbf{r}_a) - Re(\mathbf{r}_b)) + max(\mathbf{0}, Im(\mathbf{r}_a) - Im(\mathbf{r}_b))\|^2 \qquad (8)$$

where $max(\mathbf{0}, \mathbf{x})$ returns the greater one by element between $\mathbf{0}$ and $\mathbf{x}$.

Crucially, if $r_a \Rightarrow r_b$, $F(r_a, r_b) = 0$. $F(r_a, r_b)$ is positive if $r_a \Rightarrow r_b$ is not satisfied. $F(r_a, r_b) = 0$ if and only if $max(\mathbf{0}, Re(\mathbf{r}_a) - Re(\mathbf{r}_b)) = \mathbf{0}$ and $max(\mathbf{0}, Im(\mathbf{r}_a) - Im(\mathbf{r}_b)) = \mathbf{0}$. That is, the real part and imaginary part of the embeddings of r_a is less than or equal to that of r_b respectively. This satisfies the condition in Eq. (6), and encourages the learned embeddings of relation types to satisfy the order properties of transitivity and asymmetry. For transitivity, if $r_a \Rightarrow r_b$ and $r_b \Rightarrow r_c$, $\bigwedge_{i=0}^{d-1} Re(\mathbf{r}_{ai}) \leqslant Re(\mathbf{r}_{bi}) \leqslant Re(\mathbf{r}_{ci})$ and $\bigwedge_{i=0}^{d-1} Im(\mathbf{r}_{ai}) \leqslant Im(\mathbf{r}_{bi}) \leqslant Im(\mathbf{r}_{ci})$, then $F(r_a, r_c) = 0$, and thus $r_a \Rightarrow r_c$ is satisfied. For asymmetry, if $r_a \Rightarrow r_b$, $\bigwedge_{i=0}^{d-1} Re(\mathbf{r}_{ai}) \leqslant Re(\mathbf{r}_{bi})$ and $\bigwedge_{i=0}^{d-1} Im(\mathbf{r}_{ai}) \leqslant Im(\mathbf{r}_{bi})$, then $F(r_a, r_b) >= 0$, and thus $r_b \Rightarrow r_a$ is not necessarily satisfied.

For logic rule $r_a \wedge r_b \Rightarrow r_c$ the penalty for (r_a, r_b, r_c) is:

$$F(r_a, r_b, r_c) = \|max(\mathbf{0}, Re(\mathbf{r}_a) * Re(\mathbf{r}_b) - Re(\mathbf{r}_c))$$
$$+ max(\mathbf{0}, Im(\mathbf{r}_a) * Im(\mathbf{r}_b) - Im(\mathbf{r}_c))\|^2 \qquad (9)$$

where $*$ denotes the element-wise multiplication of two vectors: $[a * b]_i = a_i b_i$.

Similarly, if $r_a \wedge r_b \Rightarrow r_c$, $F(r_a, r_b, r_c) = 0$. $F(r_a, r_b, r_c)$ is positive otherwise. $F(r_a, r_b, r_c) = 0$ if and only if $max(\mathbf{0}, Re(\mathbf{r}_a) * Re(\mathbf{r}_b) - Re(\mathbf{r}_c)) = \mathbf{0}$ and $max(\mathbf{0}, Im(\mathbf{r}_a) * Im(\mathbf{r}_b) - Im(\mathbf{r}_c)) = \mathbf{0}$. That is, the multiplication of the real part and imaginary part of the embeddings of r_a and r_b is less than or equal to that of r_c respectively. This satisfies the condition in Eq. (7).

The set of all relation types in logic rules is the subset of all relation types in the given KG. Therefore, the relation types in logic rules are translated similarly to Eq. (3):

$$Re(\mathbf{r}_{a|b|c}) = Re(\mathbf{q}_{a|b|c})^{(2)}, Im(\mathbf{r}_{a|b|c}) = Im(\mathbf{q}_{a|b|c})^{(2)} \tag{10}$$

To learn the approximate order-embedding of relation types in logic rules, we could use a max-margin loss. For rule $r_a \Rightarrow r_b$:

$$\mathcal{L}_R = \min \sum_{(r_a, r_b) \in P} F(r_a, r_b) + \sum_{(r'_a, r'_b) \in N} max(0, \alpha - F(r'_a, r'_b)) \tag{11}$$

If the rule is $r_a \wedge r_b \Rightarrow r_c$:

$$\mathcal{L}_R = \min \sum_{(r_a, r_b, r_c) \in P} F(r_a, r_b, r_c) + \sum_{(r'_a, r'_b, r'_c) \in N} max(0, \alpha - F(r'_a, r'_b, r'_c)) \tag{12}$$

where P and N denote the positive and negative sets of logic rules. If $r_a \Rightarrow r_b$, we construct negatives by replacing r_b in the consequent with a random relation $r \in R$. If $r_a \wedge r_b \Rightarrow r_c$, we construct negatives by replacing r_c in the consequent with a random relation $r \in R$. $\alpha \geq 0$ is a hyper-parameter of margin. $F(r_a, r_b), F(r_a, r_b, r_c)$ is the penalty function score of positive logic rule, and $F(r'_a, r'_b), F(r'_a, r'_b, r'_c)$ is that of negative logic rule calculated by Eqs. (8) or (9). This loss encourages positive examples to have zero penalty, and negative examples to have penalty greater than a margin.

4.3   Global Objective

With both knowledge triples and logic rules modelled, embeddings are learned by minimizing a global loss over this general representation:

$$\mathcal{L} = \mathcal{L}_K + \lambda_2 \mathcal{L}_R \tag{13}$$

where $\mathcal{L}_K$ is calculated by Eq. (5) and $\mathcal{L}_R$ is calculated by Eqs. (11) or (12). The embeddings of relation types are constrained to be non-negative, and are translated into unconstrained complex vector embeddings in loss function. Therefore, stochastic gradient descent (SGD) in mini-batch mode and AdaGrad [13] for tuning the learning rate can be used to carry out the minimization directly. Embeddings learned are able to be compatible with both triples and logic rules. And the embeddings of relation types in logic rules are approximately ordered to capture the transitivity and asymmetry of rules.

Table 1. Complexity of our model and some other models

Model	Space	Time
TransE	$n_e d + n_r d$	$n_t d$
TransR	$n_e d + n_r (d^2 + d)$	$n_t d^2$
HOLE	$n_e d + n_r d$	$n_t \log d$
Complex	$n_e d + n_r d$	$n_t d$
KALE-Joint	$n_e d + n_r d$	$n_t d + n_g d$
TARE	$n_e d + n_r d$	$n_t d + n_l d$

4.4 Discussions

Complexity. We compare our model with several state-of-the-art models in space and time complexity. Table 1 lists the complexity, where d is the dimension of the embedding vectors, n_e, n_r, n_t, n_g, n_l is the number of entities, relations, triples, ground triples, logical rules respectively. It can be seen that our model does not significantly increase the space or time complexity. Note that KALE-Joint [20] needs to ground rules with entities, which further requires $O(n_g d) = O(n_l n_t / n_r d)$ in time complexity, where n_t / n_r is the averaged number of observed triples per relation. Our model only requires $O(n_l d)$ which is n_r / n_t of $O(n_g d)$ KALE-Joint required.

5 Experiments

5.1 Datasets and Experiment Settings

Datasets. We evaluate our model on knowledge graph completion using two commonly used large-scale knowledge graph datasets and a relational learning dataset:

WN36. WordNet is a large lexical database of English. Nouns, verbs, adjectives and adverbs are grouped into sets of cognitive synonyms called synsets. It provides short definitions and usage examples, and records a number of relations among these synonym sets or their members. WordNet can thus be seen as a combination of dictionary and thesaurus. The WN18 dataset is a subset of Word-Net which contains 40,943 entities, 18 relation types and 151,442 binary triples. Since there are no logic rules among all relation types in WN18, we first add the reversed relations into training set. For example, *_hypernym* is the reversed relation type of _hypernym*. We add the triple $(e_1, *\_hypernym, e_0)$ into training set according to the positive triple observed $(e_0, \_hypernym, e_1)$. Then we can find some rules in newly generated training set. We create 14 implication rules.

FB15k. Freebase is a large-scale and growing collaborative KG which provides general facts of the real world. For example, the triple (Barack Obama, Spouse, Michelle Obama) describes there is a relation Spouse between Barack Obama and Michelle Obama. The FB15k dataset is a subset of Freebase which contains 14,951 entities, 1,345 relation types, and 592,213 triples. We use original training, validation and test set splits as provided by [5]. We create 200 implication rules.

Countries. The countries dataset provided by [14] consists of 244 countries, 22 subregions and 5 regions. Each country is located in exactly one region and subregion, each subregion is located in exactly one region, and each country can have a number of neighbour countries. We construct a set of triple relations from the raw data of two relations *LocatedIn* and *NeighborOf*. We create 2 conjunction rules.

The statistics of WN36 and FB15k are listed in Table 2. Examples of rules created are shown in Table 3.

Table 2. Statistics of WN36 and FB15k

Model	Entities	Relations	Train	Valid	Test	Rules
Wn36	40,943	36	282,884	5,000	5,000	14
FB15k	14,951	1,345	483,142	50,000	59,071	200

Table 3. Examples of rules created

WN18	_hyponym ⇒ *_hypernym _member_meronym ⇒ *_member_holonym _part_of ⇒ *_has_part
FB15k	/award/award_honor/award_winner ⇒ /award/award_nomination/award_nominee /location/country/administrative_divisions ⇒ /location/location/contains /ice_hockey/hockey_roster_position/position ⇒ /sports/sports_team_roster/position
Countries	NeighborOf ∧ LocatedIn ⇒ LocatedIn LocatedIn ∧ LocatedIn ⇒ LocatedIn

Experiment Settings. We use a grid search among the following parameters: $d \in \{20, 50, 100, 150, 200\}$, $n \in \{1, 2, 5, 10\}$, $a \in \{1.0, 0.5, 0.2, 0.1, 0.05, 0.02, 0.01\}$, $\lambda_1 \in \{0.1, 0.03, 0.01, 0.003, 0.001, 0.0, 0.0003\}$, $\lambda_2 \in \{0.1, 0.03, 0.01, 0.003, 0.001, 0.0, 0.0003\}$, $m \in \{2.0, 1.0, 0.5, 0.2, 0.05, 0.01\}$ to find the optimal parameters, where d denotes the embedding size of the vectors of the entity and relation type representations; n denotes the number of negatives sampled for per positive triple observed in training set or logic rule; a denotes the initial learning rate which will be tuned during AdaGrad; λ_1 denotes the L^2 regularization parameter, λ_2 is the weight of logic rules, and m is the margin between the positive logic rules and the negative logic rules.

5.2 Knowledge Base Completion

Knowledge base completion aims to complete a triple (s, r, o) when one of s, r, o is missing. In the task of knowledge base completion, we compare our model with several state-of-art models including TransE [5], TransR [11], HOLE [6], ComplEx [7] and KALE-Joint [20]. The former four models only focus on knowledge triples, and KALE-Joint learns embeddings by jointly modelling knowledge triples and logic rules. Rules need to be grounded and are not ordered in KALE-Joint.

We evaluate the performance of our model with Mean Reciprocal Rank (MRR) and top n (Hits@n) which have been widely used for evaluation in previous works. Replace the subject or object entity of each triple (s, r, o) in the testing set with each entity in the whole dataset: (s', r, o) and (s, r, o'), where $\forall s', \forall o' \in \mathcal{E}$. Afterwards, rank all candidate entities in the dataset according to their scores calculated by Eq. (4) in ascending order. Mean Reciprocal Rank (MRR) and the ratio of correct entities ranked in top n (Hits@n) are the standard evaluation measures, which measure the quality of the ranking. They fall into two categories: raw and filtered. The filtered rankings are computed after filtering all other positive triples observed in the whole dataset, whereas the raw rankings do not filter these. We report both filtered and raw MRR, and filtered Hits@10, 3, 1 in Tables 4 and 5 for the models.

It can be seen that TARE is able to outperform TransE, TransR, HOLE, ComplEx on MRR and Hits@ on WN36 and FB15k. This demonstrates the effectiveness of joint logic rules into knowledges. TARE largely outperforms KALE-Joint, with a filtered MRR of 0.955 and 91.4% of Hits@1, compared to 0.662 and 85.5% for KALE-Joint. This demonstrates the effectiveness of considering the transitivity and asymmetry of logic rules.

Table 4. KG completion on WN36

Model	MRR		Hits@		
	Filter	Raw	1	3	10
TransE	0.495	0.351	11.3	88.8	94.3
TransR	0.605	0.427	33.5	87.6	94.0
HOLE	0.938	0.616	93.0	**94.5**	94.9
Complex	0.941	**0.587**	**93.6**	**94.5**	94.7
KALE-Joint	0.662	0.478	85.5	90.1	93.0
TARE	**0.955**	0.545	91.4	94.2	**94.7**

Table 5. KG completion on FB15k

Model	MRR		Hits@		
	Filter	Raw	1	3	10
TransE	0.463	0.222	29.7	57.8	74.9
TransR	0.346	0.198	21.8	40.4	58.2
HOLE	0.524	0.232	40.2	61.3	73.9
Complex	0.692	0.242	59.9	**75.9**	84.0
TARE	**0.781**	**0.292**	**61.7**	72.8	**84.2**

5.3   Relational Learning

We test the relational learning capabilities of our model on the countries dataset. Most of the test triples in the countries dataset can be inferred by directly applying logic rules on the training set. However, to evaluate our model, we do not use the pure logical inference. We split all countries randomly in train (80%), validation (10%), and test (10%) countries, then training, validation, and test set is composed of the relations which start from all countries in the training validation, and test countries respectively.

Remove all triples of the form $(c, LocatedIn, r)$ for each country c in the validation and test set. In the new set S_1, $(c, LocatedIn, r)$ can be predicted by $LocatedIn \wedge LocatedIn \Rightarrow LocatedIn$.

Based on S_1, remove $(c, LocatedIn, s)$ for all countries in the validation and test set. In the new set S_2, $(c, LocatedIn, r)$ can be predicted by $NeighborOf \wedge LocatedIn \Rightarrow LocatedIn$.

Based on S_2, remove $(c_n, LocatedIn, r)$ for all neighbour countries c_n of all countries in the validation and test set. In the new set S_3, $(c, LocatedIn, r)$ can be predicted by $NeighborOf \wedge LocatedIn \Rightarrow LocatedIn$ and $LocatedIn \wedge LocatedIn \Rightarrow LocatedIn$.

The prediction quality is measured by the area under the precision-recall curve (AUC-PR), we compute the mean AUC-PR after 10 fold cross-validation. The results are shown in Table 6. It can be seen that our model performs well in this task. It achieves 13.4% improvement on S2 and 19.3% improvement on S3.

Table 6. Link prediction on Countries dataset

Model	S_1	S_2	S_3
Random	0.323	0.323	0.323
Frequency	0.323	0.323	0.308
ER-MLP	0.960	0.745	0.650
Rescal	**0.997**	0.745	0.650
HOLE	**0.997**	0.772	0.697
TARE	0.994	**0.906**	**0.890**

6 Conclusion and Future Work

In this paper, we propose TARE model for representation learning of knowledge graphs by integrating existing relations and logic rules together. Logic rules are incorporated into relation type representations directly, rather than instantiated with concrete entities. We model logic rules by approximately ordering the relation types in logic rules to leverage the transitivity and asymmetry of rules, and thus obtain better embeddings for entities and relation types. To be ordered, the vector embeddings of relation types are constrained to be non-negative, the general constrained optimization problem is translated into an unconstrained optimization problem in our model. In experiments, we evaluate our models on knowledge base completion and relational learning tasks. Experimental results show that TARE brings significant and consistent improvements over exiting state-of-the-art methods.

For future work, we would like to explore the following research directions:
(1) more complex types of logic rules such as ¬ and ∨ would be modelled to
obtain better performance. (2) logic rules can be extracted from text. There is
richer information in text than triples, more logic rules can be obtained if we
joint the information in text. (3) TARE only consider the order of relation types,
the order over entities would also be helpful to obtain better embeddings.

Acknowledgements.. We thank the National Key Research and Development Program of China (2016YFB0201900), National Natural Science Foundation of China (U1611262), Guangdong Natural Science Funds for Distinguished Young Scholar (2017A030306028), Pearl River Science and Technology New Star of Guangzhou, and Guangdong Province Key Laboratory of Big Data Analysis and Processing for the support of this research.

References

1. Auer, S., Bizer, C., Kobilarov, G., Lehmann, J., Cyganiak, R., Ives, Z.: DBpedia: a nucleus for a web of open data. In: Aberer, K., Choi, K.-S., Noy, N., Allemang, D., Lee, K.-I., Nixon, L., Golbeck, J., Mika, P., Maynard, D., Mizoguchi, R., Schreiber, G., Cudré-Mauroux, P. (eds.) ASWC/ISWC -2007. LNCS, vol. 4825, pp. 722–735. Springer, Heidelberg (2007). https://doi.org/10.1007/978-3-540-76298-0_52
2. Pritsker, E.W., Cohen, W., Minkov, E.: Learning to identify the best contexts for knowledge-based WSD. In: EMNLP, pp. 1662–1667 (2015)
3. Bollacker, K., Evans, C., Paritosh, P., Sturge, T., Taylor, J.: Freebase: a collaboratively created graph database for structuring human knowledge. In: SIGMOD Conference, pp. 1247–1250 (2008)
4. Hoffmann, R., Zhang, C., Ling, X., Zettlemoyer, L., Weld, D.S.: Knowledge-based weak supervision for information extraction of overlapping relations. In: ACL, pp. 541–550 (2011)
5. Bordes, A., Usunier, N., Weston, J., Yakhnenko, O.: Translating embeddings for modeling multi-relational data. In: NIPS, pp. 2787–2795. Curran Associates Inc. (2013)
6. Nickel, M., Rosasco, L., Poggio, T.: Holographic embeddings of knowledge graphs, pp. 1955–1961 (2015)
7. Welbl, J., Riedel, S., Bouchard, G.: Complex embeddings for simple link prediction. In: ICML, pp. 2071–2080 (2016). JMLR.org
8. Riedel, S., Yao, L., Mccallum, A., Marlin, B.M.: Relation extraction with matrix factorization and universal schemas (2013)
9. Chang, K.W., Yih, W.T., Yang, B., Meek, C.: Typed tensor decomposition of knowledge bases for relation extraction. In: EMNLP (2014)
10. Wang, Z., Zhang, J., Feng, J., Chen, Z.: Knowledge graph embedding by translating on hyperplanes. In: AAAI, pp. 1112–1119 (2014)
11. Lin, Y., Liu, Z., Sun, M., Liu, Y., Zhu, X.: Learning entity and relation embeddings for knowledge graph completion. In: AAAI, pp. 2181–2187 (2015)
12. Lin, Y., Liu, Z., Luan, H., Sun, M., Rao, S., Liu, S.: Modeling relation paths for representation learning of knowledge bases. Computer Science (2015)
13. Duchi, J., Hazan, E., Singer, Y.: Adaptive subgradient methods for online learning and stochastic optimization. JMLR 257–269 (2011)

14. Bouchard, G., Singh, S., Trouillon, T.: On approximate reasoning capabilities of low-rank vector spaces (2015)
15. Dong, X., Gabrilovich, E., Heitz, G., Horn, W., Lao, N., Murphy, K., et al.: Knowledge vault: a web-scale approach to probabilistic knowledge fusion. In: ACM SIGKDD, pp. 601–610 (2014)
16. Neelakantan, A., Roth, B., Mccallum, A.: Compositional vector space models for knowledge base completion. Computer Science, 1–16 (2015)
17. Ji, G., He, S., Xu, L., Liu, K., Zhao, J.: Knowledge graph embedding via dynamic mapping matrix. In: ACL, IJCNLP, pp. 687–696 (2015)
18. Wang, Q., Wang, B., Guo, L.: Knowledge base completion using embeddings and rules. In: International Conference on Artificial Intelligence, pp. 1859–1865. AAAI Press (2015)
19. Rocktschel, T., Singh, S., Riedel, S.: Injecting logical background knowledge into embeddings for relation extraction. In: NAACL, pp. 648–664 (2014)
20. Shu, G., Quan, W., Wang, L., Wang, B., Li, G.: Jointly embedding knowledge graphs and logical rules. In: EMNLP, pp. 192–202 (2016)
21. Vendrov, I., Kiros, R., Fidler, S., Urtasun, R.: Order-embeddings of images and language (2015)
22. Socher, R., Chen, D., Manning, C.D., Ng, A.Y.: Reasoning with neural tensor networks for knowledge base completion. In: NIPS, pp. 926–934. Curran Associates Inc. (2013)
23. Lin, C.J.: Projected gradient methods for nonnegative matrix factorization. Neural Comput. 19(10), 2756 (2007)
24. Guan, N., Tao, D., Luo, Z., Yuan, B.: NeNMF: an optimal gradient method for nonnegative matrix factorization. TSP 60(6), 2882–2898 (2012)
25. Lee, D.D., Seung, H.S.: Algorithms for non-negative matrix factorization. In: NIPS, pp. 535–541. MIT Press (2000)
26. Kim, H., Park, H.: Sparse non-negative matrix factorizations via alternating non-negativity-constrained least squares for microarray data analysis. Bioinformatics 23(12), 1495 (2007)
27. Cichocki, A., Zdunek, R.: Regularized alternating least squares algorithms for non-negative matrix/tensor factorization. In: Liu, D., Fei, S., Hou, Z., Zhang, H., Sun, C. (eds.) ISNN 2007. LNCS, vol. 4493, pp. 793–802. Springer, Heidelberg (2007). https://doi.org/10.1007/978-3-540-72395-0_97
28. Krompa, D., Baier, S., Tresp, V.: Type-constrained representation learning in knowledge graphs, vol. 107(8), pp. 640–655 (2015)
29. Nickel, M., Tresp, V., Kriegel, H.P.: A three-way model for collective learning on multi-relational data. In: ICML, pp. 809–816. Omnipress (2011)
30. Guu, K., Miller, J., Liang, P.: Traversing knowledge graphs in vector space. Computer Science (2015)
31. Garca-Durn, A., Bordes, A., Usunier, N.: Composing relationships with translations (2015)

Representation Learning and Embedding

SIGNet: Scalable Embeddings for Signed Networks

Mohammad Raihanul Islam$^{(\boxtimes)}$, B. Aditya Prakash, and Naren Ramakrishnan

Department of Computer Science, Virginia Tech., Blacksburg, USA
{raihan8,badityap,naren}@cs.vt.edu

Abstract. Recent successes in word embedding and document embedding have motivated researchers to explore similar representations for networks and to use such representations for tasks such as edge prediction, node label prediction, and community detection. Such network embedding methods are largely focused on finding distributed representations for unsigned networks and are unable to discover embeddings that respect polarities inherent in edges. We propose SIGNet, a fast scalable embedding method suitable for signed networks. Our proposed objective function aims to carefully model the social structure implicit in signed networks by reinforcing the principles of social balance theory. Our method builds upon the traditional word2vec family of embedding approaches and adds a new targeted node sampling strategy to maintain structural balance in higher-order neighborhoods. We demonstrate the superiority of SIGNet over state-of-the-art methods proposed for both signed and unsigned networks on several real world datasets from different domains. In particular, SIGNet offers an approach to generate a richer vocabulary of features of signed networks to support representation and reasoning.

1 Introduction

Social and information networks are ubiquitous today across a variety of domains; as a result, a large body of research has developed to help construct discriminative and informative features for network analysis tasks such as classification [2], prediction [11], and visualization [12].

Classical approaches to find features and embeddings are motivated by dimensionality reduction research and extensions, e.g., approaches such as Laplacian eigenmaps [1], non-linear dimension reduction [17], and spectral embedding [7]. More recent research has focused on developing network analogs to distributed vector representations such as word2vec [13,14]. In particular, by viewing sequences of nodes encountered on random walks as documents, methods such as DeepWalk [15], node2vec [5], and LINE [16] learn similar representations for nodes (viewing them as words). Although these approaches are scalable to large networks, they are primarily applicable to only unsigned networks. Signed networks are becoming increasingly important in online media,

© Springer International Publishing AG, part of Springer Nature 2018
D. Phung et al. (Eds.): PAKDD 2018, LNAI 10938, pp. 157–169, 2018.
https://doi.org/10.1007/978-3-319-93037-4_13

trust management, and in law/criminal applications. As we will show, applying the above methods to signed networks results in key information loss in the resulting embedding. For instance, if the sign between two nodes is negative, the resulting embeddings could place the nodes in close proximity, which is undesirable.

An attempt to fill this gap is the work of Wang et al. [19] wherein the authors learn node representations by optimizing an objective function through a multilayer neural network based on structural balance theory. This work, however, models only local connectivity information through 2-hop paths and fails to capture global balance structures prevalent in a network. Our contributions are:

1. We propose SIGNet, a scalable node embedding method for feature learning in signed networks that maintains structural balance in higher order neighborhoods. SIGNet is generic by design and can handle both directed and undirected networks, including weighted or unweighted (binary) edges.
2. We propose a novel node sampling method as an improvement over traditional negative sampling. The idea is to maintain a cache of nodes during optimization integral for maintaining the principles of structural balance in the network. This targeted node sampling technique can be treated as an extension of the negative sampling strategy used in word2vec models.
3. Through extensive experimentation, we demonstrate that SIGNet generates better features suitable for a range of prediction tasks such as edge and node label prediction. SIGNet[1] is able to generate embeddings for networks with millions of nodes in a scalable manner.

2   Problem Formulation

Definition 1. *Signed Network: A signed network can be defined as $G = (V, E)$, where V is the set of vertices and E is the set of edges between the vertices. Each element v_i of V represents an entity in the network and each edge $e_{ij} \in E$ is a tuple (v_i, v_j) associated with a weight $w_{ij} \in \mathbb{Z}$. The absolute value of w_{ij} represents the strength of the relationship between v_i and v_j, and the sign represents the nature of relationship (e.g., friendship or antagonism). A signed network can be either directed or undirected. If G is undirected then the order of vertices is not relevant (i.e. $(v_i, v_j) \equiv (v_j, v_i)$). On the other hand, if G is directed then order becomes relevant (i.e. $(v_i, v_j) \not\equiv (v_j, v_i)$ and $w_{ij} \neq w_{ji}$).*

Because the weights in a signed network carry a combined interpretation (sign denotes polarity and magnitude denotes strength), conventional proximity assumptions used in unsigned network representations (e.g., in [5]) cannot be applied for signed networks. Consider a network wherein the nodes v_i and v_j are positively connected and the nodes v_k and v_i are negatively connected (see Fig. 1(a)). Suppose the weights of the edges e_{ij} and e_{ik} are $+w_{ij}$ and $-w_{ik}$ respectively. Now if $|+w_{ij}| < |-w_{ik}|$, conventional embedding methods will

---

[1] The implementation is available at: https://github.com/raihan2108/signet.

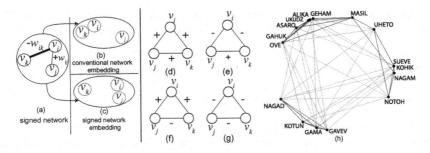

Fig. 1. Given a signed network (a), a conventional network embedding (b) does not take signs into account and can result in faulty representations. (c) SIGNet learns embeddings that respect sign information between edges. Of the possible signed triangles, (d) and (e) are considered balanced but (f) and (g) are not. (h) shows a 2-dimensional embedding of alliances among New Guinea tribes using SIGNet. Alliance (hostility) between the tribes is shown in solid blue (dashed red) edges. We can see that edges representing alliances are comparatively shorter than the edges representing hostility. (Color figure online)

place v_i and v_k closer than v_i and v_j owing to the stronger influence of the weight (Fig. 1(b)). This problem remains unresolved even if we consider the weight of a negative edge as zero, because even though it may place node v_i and v_j closer, node v_k may be relatively closer to v_i because we ignore the adverse relation between node v_i and v_k. This may comprise the quality of embedding space. Ideally, we would like a representation wherein nodes v_i and v_j are closer than nodes v_i and v_k, as shown in Fig. 1(c). This example shows that modeling the polarity is as important as modeling the strength of the relationship.

To accurately model the interplay between the vertices in signed networks we use the theory of *structural balance* proposed in [6]. This theory posits that triangles with an odd number of positive edges are more plausible than an even number of positive edges (see Fig. 1(d–g)). Although different adaptations of and alternatives to balance theory exist in the literature, here we focus on the original notion of structural balance to create the embedding space since it applies naturally to the experimental contexts considered here (e.g., networks constructed from adjectives, described in Sect. 4).

Problem Statement: *Scalable Embedding of Signed Networks (SIGNet):* Given a signed network G, compute a low-dimensional vector $\mathbf{d}_i \in \mathbb{R}^K$, $\forall v_i \in V$, where positively related vertices reside in close proximity to each other and negatively related vertices are distant from each other.

To explain the interpretability of the signed network embedding we utilize a small dataset denoting relations between 16 tribes in New Guinea. This is a signed network depicting alliances and hostility between the tribes. We learned the embeddings using SIGNet in 2 dimensional space as an undirected network as shown in Fig. 1(h). We can see that in general solid blue edges (alliance) are shorter than the dashed red edges (hostility) confirming that allied tribes are closer than the hostile tribes. Therefore the embedding space learned by SIGNet clearly depicts alliances and relationships among the tribes as intended.

3   Scalable Embedding of Signed Networks (SIGNet)

3.1   SIGNet for Undirected Networks

Consider a weighted signed network defined as in Sect. 2. Now suppose each v_i is represented by a vector $\mathbf{x}_i \in \mathbb{R}^K$. Then a natural way to compute the proximity between v_i and v_j is by the following function (ignoring the sign for now):

$$p_u(v_i, v_j) = \sigma(\mathbf{x}_j^T \cdot \mathbf{x}_i) = \frac{1}{1 + \exp(-\mathbf{x}_j^T \cdot \mathbf{x}_i)} \tag{1}$$

where $\sigma(a) = \frac{1}{1+\exp(-a)}$. Now let us breakdown the weight of edge w_{ij} into two components: r_{ij} and s_{ij}. $r_{ij} \in \mathbb{N}$ represents the absolute value of w_{ij} (i.e. $r_{ij} = |w_{ij}|$) and $s_{ij} \in \{-1, 1\}$ represents the sign of w_{ij}. Given this breakdown of w_{ij}, $p_u(v_i, v_j) = \sigma(s_{ij}(\mathbf{x}_j^T \cdot \mathbf{x}_i))$. Now incorporating the weight information, the objective function for undirected signed network can be written as:

$$\mathcal{O}_{un} = \sum_{e_{ij} \in E} r_{ij}\sigma(s_{ij}(\mathbf{x}_j^T \cdot \mathbf{x}_i)) = \sum_{e_{ij} \in E} r_{ij}p_u(v_i, v_j) \tag{2}$$

By maximizing Eq. 2 we obtain a vector $\mathbf{x}_i$ of dimension K for each node $v_i \in V$ (we also use $\mathbf{d}_i$ to refer to this embedding, for reasons that will become clear in the next section).

3.2   SIGNet for Directed Networks

Computing embeddings for directed networks is trickier due to the asymmetric nature of neighborhoods (and thus, contexts). For instance, if the edge e_{ij} is positive, but e_{ji} is negative, it is not clear if the respective representations for nodes v_i and v_j should be proximal or not. We solve this problem by treating each vertex as itself plus a specific context; for instance, a positive edge e_{ij} is interpreted to mean that given the context of node v_j, node v_i should be closer. This enables us to treat all nodes consistently without worrying about reciprocity relationships. To this end, we introduce another vector $\mathbf{y}_i \in \mathbb{R}^K$ besides $\mathbf{x}_i$, $\forall v_i \in V$. For a directed edge e_{ij} the probability of context v_j given v_i is:

$$p_d(v_j | v_i) = \frac{\exp(s_{ij}(\mathbf{y}_j^T \cdot \mathbf{x}_i))}{\sum_{k=1}^{|V|} \exp(s_{ik}(\mathbf{y}_k^T \cdot \mathbf{x}_i))} \tag{3}$$

Treating the same entity as itself and as a specific context is very popular in the text representation literature [13]. The above equation defines a probability distribution over all context space w.r.t. node v_i. Now our goal is to optimize the above objective function for all the edges in the network. However we also need to consider the weight of each edge in the optimization. Incorporating the absolute weight of each edge we obtain the objective function for a directed network as:

$$\mathcal{O}_{dir} = \sum_{e_{ij} \in E} r_{ij} p_d(v_j | v_i) \tag{4}$$

By maximizing Eq. 4 we will obtain two vectors $\mathbf{x}_i$ and $\mathbf{y}_i$ for each $v_i \in V$. The vector $\mathbf{x}_i$ models the outward connection of a node whereas $\mathbf{y}_i$ models the inward connection of the node. Therefore the concatenation of $\mathbf{x}_i$ and $\mathbf{y}_i$ represents the final embedding for each node. We denote the final embedding of node v_i as $\mathbf{d}_i$. It should be noted that for undirected network $\mathbf{d}_i = \mathbf{x}_i$ whereas for a directed network $\mathbf{d}_i$ is the concatenation of $\mathbf{x}_i$ and $\mathbf{y}_i$. This means $|\mathbf{x}_i| = |\mathbf{y}_i| = \frac{K}{2}$ in the case of directed graph (for the same representational length).

3.3 Efficient Optimization by Targeted Node Sampling

The denominator of Eq. 3 is very hard to compute as this requires marginalizing the conditional probability over the entire vertex set V. We adopt the classical negative sampling approach [14] wherein negative examples are selected from some distribution for each edge e_{ij}. However, for signed networks, conventional negative sampling does not work. For example consider the network from Fig. 2(a). Viewing this example as an unsigned network, while optimizing for edge e_{ij}, we will consider v_i and v_y as negative examples and thus they will be placed distantly from each other. However, in a signed network context, v_i and v_y have a friendlier relationship (than with, say, v_x) and thus should be placed closer to each other. We propose a new sampling approach, referred to as simply *targeted node sampling* wherein we first create a cache of nodes for each node with their estimated relationship according to structural balance theory and then sample nodes accordingly.

Constructing the Cache for Each Node: We aim to construct a cache of positive and negative examples for each node v_i where the positive (negative) example cache η_i^+ (η_i^-) contains nodes which should have a positive (negative) relationship with v_i according to structural balance theory. To construct these caches for each node v_i, we apply random walks of length l starting with v_i to obtain a sequence of nodes. Suppose the sequence is $\Omega = < v_i, v_{n_0}, \cdots, v_{n_{l-1}} >$. Now we add each node v_{n_p} to either η_i^+ or η_i^- by observing the estimated sign between v_i and v_{n_p}. The estimated sign is computed using the following recursive formula $\tilde{s}_{in_p} = \tilde{s}_{in_{p-1}} \times s_{n_{p-1}n_p}$. Here $\tilde{s}_{in_{p-1}}$ is the estimated sign between node v_i and node $v_{n_{p-1}}$, which can be computed recursively. The base case for this formula is $\tilde{s}_{in_1} = s_{in_0} \times s_{n_0 n_1}$. If node v_{n_p} is not a neighbor of node v_i and $\tilde{s}_{in_p}$ is positive then we add v_{n_p} to η_i^+. On the other hand if $\tilde{s}_{in_p}$ is negative and v_{n_p} is not a neighbor of v_i then we add it to η_i^-. For example for the graph shown in Fig. 2(a), suppose a random walk starting with node v_i is $< v_i, v_j, v_k, v_z >$. Here node v_k will be added to η_i^+ as $\tilde{s}_{ik} = s_{ij} \times s_{jk} > 0$ (base case) and v_k is not a neighbor of v_i. However, v_z will be added to node η_i^- as $\tilde{s}_{iz} = \tilde{s}_{ik} \times s_{kz} < 0$ and v_z is not a neighbor of v_i.

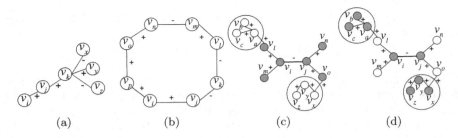

Fig. 2. (a) depicts a small network to illustrate why conventional negative sampling does not work. v_i and v_y might be considered too distant for their representations to be placed close to each other. Targeted node sampling solves this problem by constructing a cache of nodes which can be used as sampling. (b) shows how we resolve *conflict*. Although there are two ways to proceed from node v_i to v_l the shortest path is v_i, v_j, v_k, v_l, which estimates a net positive relation between v_i and v_l. As a result v_l will be added to η_i^+. However for node v_m there are two shortest paths from v_i, with the path v_i, v_p, v_o, v_n, v_m having more positive edges but with a net negative relation, so v_m will be added to η_i^- in case of a *conflict*. (c) and (d) shows a comparative scenario depicting the optimization process inherent in both SiNE and SIGNet. The shaded vertices represent the nodes both methods will consider while optimizing the edge e_{ij}. We can see that SiNE only considers the immediate neighbors because it optimizes edges in 2-hop paths having opposite signs. On the other hand, SIGNet considers higher order neighbors ($v_a, v_b, v_c, v_x, v_y, v_z$) for targeted node sampling.

The one problem with this approach is that a node v_j may be added to both η_i^+ and η_i^-. We denote this phenomena as *conflict* and define the reason for this *conflict* in Theorem 1. We resolve this situation by computing the shortest path between v_i and v_j and compute $\tilde{s}_{ij}$ between them using the shortest path, then add to either η_i^+ or η_i^- based on $\tilde{s}_{ij}$. To compute the shortest path we have to consider the network as unsigned since negative weight has a different interpretation for shortest path algorithms. If there are multiple shortest paths with equal length in case of a *conflict*, then we pick the path with the highest number of positive edges to compute $\tilde{s}_{ij}$. A scenario is shown in Fig. 2(b).

Theorem 1. *(Reason for conflict): Node v_j will be added to both η_i^+ and η_i^- if there are multiple paths from v_i to v_j and the union of these paths has at least one unbalanced cycle.*

Targeted Edge Sampling During Optimization: Now after constructing the cache $\eta_i = \eta_i^+ \bigcup \eta_i^-$ for each node v_i, we can apply the targeted sampling approach for each node. Here our goal is to extend the objective of negative sampling from classical word2vec approaches [14]. In traditional negative sampling, a random word-context pair is negatively sampled for each observed word-context pair. In a signed network both positive and negative edges are present, and thus we aim to conduct both types of sampling while sampling an edge observing its sign. Therefore when sampling a positive (negative) edge e_{ij}, we aim to sample multiple negative (positive) nodes from η_i^- (η_i^+). Therefore the objective function for each edge becomes (taking log):

$$\mathcal{O}_{ij} = \log[\sigma(s_{ij}(\mathbf{y}_j^T \cdot \mathbf{x}_i))] + \sum_{c=1}^{\mathcal{N}} E_{v_n \sim \tau(s_{ij})} \log[\sigma(\tilde{s}_{in}(\mathbf{y}_n^T \cdot \mathbf{x}_i))] \tag{5}$$

Here $\mathcal{N}$ is the number of targeted node examples per edge and τ is a function which selects from η_i^+ or η_i^- based on the sign s_{ij}. τ selects from η_i^+ (η_i^-) if $s_{ij} < 0$ ($s_{ij} > 0$).

The benefit of targeted node sampling in terms of global balance considerations across the entire network is shown in Fig. 2(c) and (d). Here we compare how our proposed approach SIGNet and SiNE [19] maintain structural balance. For simplicity suppose only edge e_{ij} has negative sign. Now SiNE optimizes w.r.t. pairs of edges in 2-hop paths each having different signs. Therefore optimizing the edge e_{ij} involves only the immediate neighbors of node v_i and v_j, i.e. v_l, v_m, v_n, v_o (Fig. 2(c)). However SIGNet skips the immediate neighbors while it uses higher order neighbors (i.e., $v_a, v_b, v_c, v_x, v_y, v_z$). Note that SIGNet actually uses immediate neighbors as separate examples (i.e. edge e_{il}, e_{im} etc.). In this way SIGNet covers more nodes to optimize the embedding space than SiNE.

4 Experiments

Experimental Setup: We compare our algorithm against both the state-of-the-art methods proposed for signed and unsigned network embedding. The description of the methods are below:

- node2vec [5]: This method, not specific to signed networks, computes embeddings by optimizing the neighborhood structure using informed random walks.
- SNE [20]: This method computes the embedding using a log bilinear model; however it does not exploit any specific theory of signed networks.
- SiNE [19]: This method uses a multi-layer neural network to learn the embedding by optimizing an objective function satisfying structural balance theory. SiNE only concentrates on the immediate neighborhood of vertices rather than on the global balance structure.
- SIGNet-NS: This method is similar to our proposed method SIGNet except it uses conventional negative sampling instead of targeted node sampling.
- SIGNet: This is our proposed SIGNet method which uses random walks to construct a cache of positive and negative examples for targeted node sampling.

We skip hand crafted feature generation method for link prediction like [9] because they can not be applied in node label prediction and already shows inferior performance compared to SiNE. For node2vec the weight of negative edges are treated zero since node2vec can not handle negative edges.

In the discussion below, we focus on five real world signed network datasets. Out of these five, two datasets are from social network platforms—Epinions and Slashdot—courtesy the Stanford Network Analysis Project (SNAP). The details on how the signed edges are defined are available at the project website. The third dataset is a voting records of Wikipedia adminship election (Wiki), also from SNAP. The fourth dataset we study is an adjective network (ADJNet) constructed from the synonyms and antonyms collected from Wordnet database. Label information about whether the adjective is positive or negative comes from SentiWordNet. The last dataset is a citation network we constructed from written case opinions of the Supreme Court of the United States (SCOTUS). We expand the notion of SCOTUS citation network into a signed network.

Unless otherwise stated, for directed networks we set $|\mathbf{x}_i| = |\mathbf{y}_i| = \frac{K}{2} = 20$ for both SIGNet-NS and SIGNet; therefore $|\mathbf{d}_i| = 40$. For a fair comparison, the final embedding dimension for others methods is set to 40. For undirected network (ADJNet) $|\mathbf{d}_i| = 40$ for all the methods. We also set the total number of samples (examples) to 100 million, $\mathcal{N} = 5$, $l = 50$ and $r = 1$ for SIGNet-NS and SIGNet. For all the other parameters for node2vec, SNE and SiNE we use the settings recommended in their respective papers.

Does the Embedding Space Learned by SIGNet Support Structural Balance Theory? Here we present our analysis on whether the embedding space learned by SIGNet follows the principles of structural balance theory. We calculate the mean Euclidean distance between representations of nodes connected by positive versus negative edges, as well as their standard deviations (see Table 1). The lower value of positive edges suggests positively connected nodes stay closer together than the negatively connected nodes indicating that SIGNet has successfully learned the embedding using the principles of structural balance theory. Moreover, the ratio of average distance between the positive and negative edges is at most 67% over all the datasets suggesting that SIGNet grasps the principles very effectively.

Table 1. Average Euclidean distance between node representations connected by positive edges versus negative edges with std. deviation. We can see that the avg. distance between positive edge is significantly lower than negative edges indicating that SIGNet preserves the conditions of structural balance theory.

Type of edges	Epinions	Slashdot	Wiki	SCOTUS	ADJNet
Positive	0.86 (0.37)	0.98 (0.31)	1.06 (0.27)	0.84 (0.25)	0.71 (0.16)
Negative	1.64 (0.23)	1.60 (0.19)	1.56 (0.19)	1.64 (0.21)	1.77 (0.08)
Ratio	0.524	0.613	0.679	0.512	0.401

Are Representations Learned by SIGNet Effective at Edge Label Prediction? We now explore the utility of SIGNet for edge label prediction. For all the datasets we sample 50% of the edges as a training set to learn the node embedding. Then we train a logistic regression classifier using the embedding as features and the sign of the edges as label. This classifier is used to predict the sign of the remaining 50% of the edges. Since edges involve two nodes we explore several scores to compute the features for edges from the node embedding. They are **Concatenation:** (concat): $\mathbf{f}_{ij} = \mathbf{d}_i \oplus \mathbf{d}_j$, **Average** (avg): $\mathbf{f}_{ij} = \frac{\mathbf{d}_i + \mathbf{d}_j}{2}$, **Hadamard** (had): $\mathbf{f}_{ij} = \mathbf{d}_i * \mathbf{d}_j$, $\mathcal{L}_1$: $\mathbf{f}_{ij} = |\mathbf{d}_i - \mathbf{d}_j|$ and $\mathcal{L}_2$: $\mathbf{f}_{ij} = |\mathbf{d}_i - \mathbf{d}_j|^2$.

Here $\mathbf{f}_{ij}$ is the feature vector of edge e_{ij} and $\mathbf{d}_i$ is the embedding of node v_i. Except for the method of concatenation (which has a feature vector dimension of 80) other methods use 40-dimensional vectors. We use the micro-F1 scores to evaluate our method. We repeat this process five times and report the average results (see Table 2). Some key observations from this table are as follows:

1. SIGNet, not surprisingly, outperforms node2vec across all datasets. For datasets that contain relatively fewer negative edges (e.g., 14% for Epinions and 22% for Slashdot), the improvements are modest (around 7%). For Wiki the gains are moderate (around 12%) where 25% of edges are negative. For ADJNet and SCOTUS where the sign distribution is less skewed, SIGNet outperforms node2vec by a huge margin (19% for ADJNet and 39% for SCOTUS).
2. SIGNet demonstrates a consistent advantage over SiNE and SNE, with gains ranging from 6–12% (for the social network datasets) to 17–36% (for ADJNet and SCOTUS).
3. SIGNet also outperforms SIGNet-NS in almost all scenarios demonstrating the effectiveness of targeted node sampling over negative sampling.
4. Performance measures (across all scores and across all algorithms) are comparatively better for Epinions over other datasets because almost 83% of the nodes in Epinions satisfy the structural balance condition [3]. As a result in Epinions edge label prediction is comparatively easier than in other datasets.
5. The feature scoring method has a noticeable impact w.r.t. different datasets. The avg. and concat. methods subsidize differences whereas the hadamard, $\mathcal{L}$-1 and $\mathcal{L}$-2 methods promote differences. To understand why this makes a difference, consider networks like ADJNet and SCOTUS where connected components denote strong polarities (e.g., denoting synonyms or justice leanings, respectively). In such networks, the Hadamard, $\mathcal{L}$-1 and $\mathcal{L}$-2 methods provide more discriminatory features. However, Epinions and Slashdot are relatively large datasets with diversified communities and so all these methods perform nearly comparably.

Are Representations Learned by SIGNet Effective at Node Label Prediction? For datasets like SCOTUS and ADJNet (where nodes are annotated with labels), we learn a logistic regression classifier to map from node representations to corresponding labels (with a 50–50 training-test split). We also repeat this five times and report the average. See Table 3 for results. As can be seen, SIGNet consistently outperforms all the other approaches. In particular, in the case of SCOTUS which is a citation network, some cases have a huge number of citations (i.e. landmark cases) in both ideologies. Targeted node sampling,

Table 2. Comparison of edge label prediction in all datasets. We show the micro F1 score for each feature scoring method. The best score across all the scoring method is shown in boldface. SIGNet outperforms node2vec, SNE, and SiNE in every case. The results are statistically significant with $p < 0.01$.

Eval.	Dataset name	Epinions	Slashdot	Wiki	ADJNet	SCOTUS
concat	node2vec	0.831	0.776	0.749	0.594	0.543
	SNE	0.854	0.778	0.751	0.602	0.528
	SiNE	0.856	0.779	0.752	0.598	0.605
	SIGNet-NS	0.911	0.793	0.816	0.599	0.56
	SIGNet	**0.920**	**0.832**	**0.845**	0.573	0.557
avg	node2vec	0.853	0.775	0.747	0.603	0.516
	SNE	0.853	0.776	0.748	0.601	0.532
	SiNE	0.853	0.774	0.749	0.599	0.608
	SIGNet-NS	0.837	0.771	0.769	0.620	0.509
	SIGNet	0.879	0.809	0.801	0.574	0.512
had	node2vec	0.852	0.773	0.748	0.600	0.562
	SNE	0.851	0.775	0.745	0.604	0.541
	SiNE	0.854	0.772	0.748	0.589	0.609
	SIGNet-NS	0.846	0.757	0.741	0.705	**0.793**
	SIGNet	0.883	0.782	0.754	**0.722**	0.792
l1	node2vec	0.852	0.775	0.747	0.601	0.559
	SNE	0.854	0.774	0.749	0.605	0.582
	SiNE	0.853	0.773	0.746	0.609	0.608
	SIGNet-NS	0.851	0.764	0.743	0.639	0.723
	SIGNet	0.901	0.787	0.751	0.703	0.723
l2	node2vec	0.852	0.773	0.747	0.601	0.569
	SNE	0.852	0.774	0.748	0.606	0.547
	SiNE	0.787	0.776	0.745	0.612	0.611
	SIGNet-NS	0.848	0.763	0.743	0.659	0.742
	SIGNet	0.903	0.809	0.753	0.716	0.745
Gain over node2vec		7.85	7.22	12.82	19.73	39.19
Gain over SNE		7.73	6.94	12.52	19.14	36.08
Gain over SiNE		7.48	6.80	12.37	17.97	29.62
Gain over SIGNet-NS		0.99	4.92	3.55	2.41	−0.13

by adding such cases to either η_i^+ or η_i^-, situates the embedding space close to the landmark cases if they are in η_i^+ or away from them if they are in η_i^-, thus supporting accurate node prediction.

Table 3. Comparison of methods for node label prediction on real world datasets. SIGNet outperforms other methods in all datasets.

Metric	micro f1		macro f1	
Datasets	ADjNet	SCOTUS	ADjNet	SCOTUS
node2vec	0.5284	0.5392	0.4605	0.4922
SNE	0.5480	0.5432	0.4840	0.5335
SiNE	0.6257	0.6131	0.6247	0.5796
SIGNet-NS	0.7292	0.8004	0.7261	0.7997
SIGNet	0.8380	0.8419	0.8374	0.8415
Gain over node2vec	58.5920	56.1387	81.8458	70.9671
Gain over SNE	52.9197	54.9890	73.0165	57.7320
Gain over SiNE	33.9300	37.3185	34.0483	45.1863
Gain over SIGNet-NS	14.9205	5.1849	15.3285	5.2270

How Much More Effective Is Our Sampling Strategy in the Presence of Partial Information? To evaluate the effectiveness of our targeted node sampling versus negative sampling, we remove all outgoing edges of a certain percent of randomly selected nodes (test nodes), learn an embedding, and then aim to predict the labels of the test nodes. We show the micro f1 scores for ADJNet (treating it as directed) and SCOTUS in Fig. 3(a) and (b). As seen here, SIGNet consistently outperforms SIGNet-NS. Withholding the outgoing edges of test nodes implies that both methods will miss the same edge information in learning the embedding. However due to targeted node sampling many of these test nodes will be added to η_i^+ or η_i^- in SIGNet (recall only the outgoing edges are removed, but not incoming edges). Because of this property, SIGNet is able to make an informed choice while optimizing the embedding space.

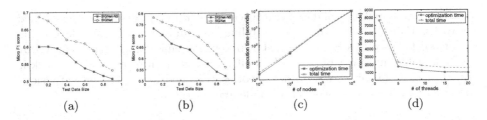

(a)   (b)   (c)   (d)

Fig. 3. Micro F1 of ADJNet (a) and SCOTUS (b) datasets varying the percent of nodes used for training. SIGNet outperforms SIGNet-NS in all cases. (c) and (d) show execution time of SIGNet varying the number of nodes and threads.

How Scalable Is SIGNet for Large Networks? To assess the scalability of SIGNet, we learn embeddings for an Erdos-Renyi random network for upto one million nodes. The average degree for each node is set to 10 and the total number of samples is set to 100 times the number of edges in the network. The size of the dimension is also set to 100 for this experiment. We make the network signed by randomly changing the sign of 20% edges to negative. The optimization time and the total execution time (targeted node sampling + optimization) is compared in Fig. 3(c) for different vertex sizes. On a regular desktop, an unparallelized version of SIGNet requires less than 3 h to learn the embedding space for over 1 million nodes. Moreover, the sampling times is negligible compared to the optimization time (less than 15 min for 1 million nodes). This actually shows SIGNet is very scalable for real world networks. Additionally, SIGNet uses an asynchronous stochastic gradient approach, so it is trivially parallelizable and as Fig. 3(d) shows, we can obtain a 3.5 fold improvement with just 5 threads, with diminishing returns beyond that point.

5   Other Related Work

The concept of unsupervised learning in networks follow the trend opened up originally by Skip-gram models [13,14]. Skip-gram models can be extended to learn feature representations for documents [8], diseases [4] etc. Recently deep learning based models have been proposed for representation learning on graphs to perform prediction in unsigned networks [10,18]. Although these models provide high accuracy by optimizing several layers of non-linear transformations, they are computationally expensive, require a significant amount of training time and are only applicable to unsigned networks as opposed to our proposed method SIGNet.

6   Conclusion

We have presented a scalable feature learning framework suitable for signed networks. Using a targeted node sampling for random walks, and leveraging structural balance theory, we have shown how the embedding space learned by SIGNet yields interpretable as well as effective representations. Future work is aimed at experimenting with other theories of signed networks and extensions to networks with a heterogeneity of node and edge tables.

Acknowledgments. This research has been supported by US NSF grants IIS-1750407 (CAREER), IIS-1633363, and DGE-1545362, by the NEH (HG-229283-15), by ORNL (Task Order 4000143330), by the Maryland Procurement Office (H98230-14-C-0127), and by a Facebook faculty gift.

References

1. Belkin, M., Niyogi, P.: Laplacian eigenmaps and spectral techniques for embedding and clustering. In: NIPS (2001)
2. Bhagat, S., Cormode, G., Muthukrishnan, S.: Node classification in social networks. In: Aggarwal, C. (ed.) Social Network Data Analytics, pp. 115–148. Springer, Boston (2011). https://doi.org/10.1007/978-1-4419-8462-3_5
3. Facchetti, G., Iacono, G., Altafini, C.: Computing global structural balance in large-scale signed social networks. PNAS **108**(52), 20953–20958 (2011)
4. Ghosh, S., et al.: Characterizing diseases from unstructured text: a vocabulary driven word2vec approach. In: CIKM (2016)
5. Grover, A., Leskovec, J.: node2vec: scalable feature learning for networks. In: KDD (2016)
6. Heider, F.: Attitudes and cognitive organization. J. Psychol. **21**, 107–112 (1946)
7. Kunegis, J., et al.: Spectral analysis of signed graphs for clustering, prediction and visualization. In: SDM (2010)
8. Le, Q., Mikolov, T.: Distributed representations of sentences and documents. In: ICML (2014)
9. Leskovec, J., Huttenlocher, D., Kleinberg, J.: Predicting positive and negative links in online social networks. In: WWW (2010)
10. Li, K., et al.: LRBM: a restricted Boltzmann machine based approach for representation learning on linked data. In: ICDM (2014)
11. Liben-Nowell, D., Kleinberg, J.: The link prediction problem for social networks. In: CIKM (2003)
12. van der Maaten, L., Hinton, G.: Visualizing high-dimensional data using t-SNE. JMLR **9**, 2579–2605 (2008)
13. Mikolov, T., Chen, K., Corrado, G., Dean, J.: Efficient estimation of word representations in vector space. CoRR abs/1301.3781 (2013)
14. Mikolov, T., Sutskever, I., Chen, K., Corrado, G.S., Dean, J.: Distributed representations of words and phrases and their compositionality. In: NIPS (2013)
15. Perozzi, B., Al-Rfou, R., Skiena, S.: DeepWalk: online learning of social representations. In: KDD (2014)
16. Tang, J., et al.: LINE: large-scale information network embedding. In: WWW (2015)
17. Tenenbaum, J., de Silva, V., Langford, J.: A global geometric framework for nonlinear dimensionality reduction. Science **290**(5500), 2319–2323 (2000)
18. Wang, D., Cui, P., Zhu, W.: Structural deep network embedding. In: KDD (2016)
19. Wang, S., Tang, J., Aggarwal, C., Chang, Y., Liu, H.: Signed network embedding in social media. In: SDM (2017)
20. Yuan, S., Wu, X., Xiang, Y.: SNE: signed network embedding. In: Kim, J., Shim, K., Cao, L., Lee, J.-G., Lin, X., Moon, Y.-S. (eds.) PAKDD 2017. LNCS (LNAI), vol. 10235, pp. 183–195. Springer, Cham (2017). https://doi.org/10.1007/978-3-319-57529-2_15

Sub2Vec: Feature Learning for Subgraphs

Bijaya Adhikari[✉], Yao Zhang, Naren Ramakrishnan, and B. Aditya Prakash

Department of Computer Science, Virginia Tech, Blacksburg, USA
{bijaya,yaozhang,naren,badityap}@cs.vt.edu

Abstract. Network embeddings have become very popular in learning effective feature representations of networks. Motivated by the recent successes of embeddings in natural language processing, researchers have tried to find network embeddings in order to exploit machine learning algorithms for mining tasks like node classification and edge prediction. However, most of the work focuses on distributed representations of nodes that are inherently ill-suited to tasks such as community detection which are intuitively dependent on subgraphs. Here, we formulate subgraph embedding problem based on two intuitive properties of subgraphs and propose Sub2Vec, an unsupervised algorithm to learn feature representations of arbitrary subgraphs. We also highlight the usability of Sub2Vec by leveraging it for network mining tasks, like community detection and graph classification. We show that Sub2Vec gets significant gains over state-of-the-art methods. In particular, Sub2Vec offers an approach to generate a richer vocabulary of meaningful features of subgraphs for representation and reasoning.

1 Introduction

Graphs are a natural abstraction for representing relational data from multiple domains such as social networks, protein-protein interaction networks, the World Wide Web, and so on. Analysis of such networks include classification [3], detecting communities [4,8], and so on. Many of these tasks can be solved using machine learning algorithms. Unfortunately, since most machine learning algorithms require data to be represented as features, applying them to graphs is challenging due to their high dimensionality and structure. In this context, learning discriminative feature representation of subgraphs can help in leveraging existing machine learning algorithms more widely on graph data.

Apart from classical dimensionality reduction techniques (see related work), recent works [9,14,19,21] have explored various ways of learning feature representation of nodes in networks exploiting relationships to vector representations in NLP (like word2vec [11]). However, application of such methods are limited to binary and multi-class node classification and edge-prediction. It is not clear how one can exploit these methods for tasks like community detection which are inherently based on subgraphs and node embeddings result in loss of information of the subgraph structure. Embedding of subgraphs or neighborhoods themselves seem to be better suited for these tasks. Surprisingly, learning feature

© Springer International Publishing AG, part of Springer Nature 2018
D. Phung et al. (Eds.): PAKDD 2018, LNAI 10938, pp. 170–182, 2018.
https://doi.org/10.1007/978-3-319-93037-4_14

representation of networks themselves (subgraphs and graphs) has not gained much attention. Here, we address this gap by studying the problem of learning distributed representations of subgraphs in a low dimensional continuous vector space. Figure 1(a–b) gives an illustration of our framework. Given a set of subgraphs (Fig. 1 (a)), we learn a low-dimensional feature representation of each subgraph (Fig. 1(b)).

(a) Input subgraphs (b) Embeddings | (c) Ground Truth   (d) Node2Vec   (e) Sub2Vec

Fig. 1. (a) and (b) An overview of our Sub2Vec. Our input is a set of subgraphs S. Sub2Vec learns d dimensional feature embedding of each subgraph. (c)–(e) Leveraging embeddings learned by Sub2Vec for community detection. (c) Communities in School network (different colors represent different communities). (d) Communities discovered via Node2Vec deviates from the ground truth, (e) while those discovered via Sub2Vec closely matches the ground truth. (Color figure online)

As shown later, the embeddings of the subgraphs enable us to apply off-the-shelf machine learning algorithms directly to solve subgraph mining tasks. For example, for community detection, we can first embed the ego-nets of each node using sub2vec, and then apply clustering algorithms like k-means on the embeddings (see Sect. 4.1 later for more details). Figure 1(c–e) shows a visualization of ground-truth communities in a network (c), communities found by using just node embeddings (d), and those found by our method Sub2Vec (e). Clearly our result matches the ground-truth well while the other is far from it. Our contributions are:

– We identify two intuitive properties of subgraphs (Neighborhood and Structural). Then we formulate two novel *Subgraph Embedding* problems, and propose Sub2Vec, a scalable subgraph embedding framework to learn features for arbitrary subgraphs that maintains the properties.
– We conduct multiple experiments over diverse datasets to show correctness, scalability, and utility of Sub2Vec in several tasks. We get upto a gain of 123.5% in community detection and upto 33.3% in graph classification compared to closest competitors.

2   Problem Formulation

We begin with the setting of our problem. Let $G(V, E)$ be a graph where V is the vertex set and E is the associated edge-set (we assume unweighted undirected graphs here, but our framework can be easily extended to weighted and/or directed graphs as well). A graph $g_i(v_i, e_i)$ is said to be a subgraph of a larger graph $G(V, E)$ if $v_i \subseteq V$ and $e_i \subseteq E$. For simplicity, we write $g_i(v_i, e_i)$ as g_i. As input, we require a set of subgraphs $\mathcal{S} = \{g_1, g_2, \ldots, g_n\}$, typically extracted from the same graph $G(V, E)$. Our goal is to embed each subgraph $g_i \in \mathcal{S}$ into d-dimensional feature space $\mathbb{R}^d$, where $d << |V|$.

Main Idea: Intuitively, our goal is to learn a feature representation of each subgraph $g_i \in \mathcal{S}$ such that the likelihood of preserving certain *properties* of each subgraph, defined in the network setting, is maximized in the latent feature space. In this work, we provide a framework to preserve two different properties— namely *Neighborhood* and *Structural*— properties of subgraphs.

Neighborhood Property: Intuitively, the Neighborhood property of a subgraph captures the neighborhood information within a subgraph itself for each node in it. For illustration, consider the following example. In the figure below, let g_1 be the subgraph induced by nodes $\{a, e, c, d\}$. The Neighborhood property of g_1 should be able to capture the information that the nodes a, c are in the neighborhood of node e, that nodes d, e are in the neighborhood of node c. To capture the neighborhood information of all the nodes in a given subgraph, we consider paths annotated by ids of the nodes. We refer to such paths as the *Id-paths* and define the Neighborhood property of a subgraph g_i as the set of all Id-paths in g_i.

The Id-paths capture the neighborhood information in subgraphs and each succession of nodes in Id-paths reveals how the neighborhood in the subgraph is evolving. For example in g_1 described above, the id-path $a \rightarrow c \rightarrow d$ shows that nodes a and c are neighbors of each other. 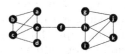 Moreover, this path along with $a \rightarrow e \rightarrow d$ indicate that nodes a and d are in neighborhood of each other (despite not being direct neighbors). Hence, the set of all Id-paths captures important connectivity information of the subgraph.

Structural Property: The Structural property of a subgraph captures the *overall* structure of the subgraph as opposed to just the local connectivity information as captured by the Neighborhood property. Several prior works have leveraged degree of nodes and their neighbors to capture structural information in network representation learning [16,18]. While degree of a node captures its local structural information within a subgraph, it fails in characterizing the similarity between the structures of two nodes in different subgraphs. Note that the nodes of two subgraphs with the same structure but of different sizes will have different degrees. For example, nodes in clique of size 10 have degree of 9, whereas nodes in clique of size 6 have degree 5. Therefore this suggests that instead, the *ratio* of degree to the size of the subgraph, of a node and its neighbors better identifies subgraph structure. Hence we rely on paths in g_i annotated by the ratio of node degrees to the subgraph size. We refer to the set of all such paths as Degree-paths. Degree-paths capture the structure by tracking how the density of edges changes in a neighborhood. While our method is simple, it is effective as shown by the results. One can build upon our framework by leveraging other techniques like rooted subgraphs [13] and predefined motifs [17].

As an example, consider the subgraph g_2 induced by nodes $\{a, b, c, e\}$ and subgraph g_3 induced by nodes $\{f, h, i, g\}$ in the graph shown above. As it is a clique, the ratio of degree to the size of the subgraph for each node in g_2 is 0.75. Hence any Degree-paths of length 3 in g_2 is $0.75 \rightarrow 0.75 \rightarrow 0.75$. Similarly, g_3 is a star and a Degree-path in g_3 from i to g is $0.25 \rightarrow 0.75 \rightarrow 0.25$. The consistent high values in the paths in cliques show that each node in the path is densely connected to the rest of the graph, while the fluctuation in values in stars show that the two spokes in the path are sparsely connected to the rest of the network while the center is densely connected. In practice, since we cannot treat each real value distinctly, we generate labels for each node from a fixed alphabet (see Sect. 3).

Our Problems: Having defined the Neighborhood and Structural properties of subgraphs, we want to learn vector representations in $\mathbb{R}^d$, such that the likelihood of preserving these properties in the feature space is maximized. Formally the two versions of our Subgraph Embedding problem are:

Problem 1. Given a graph $G(V, E)$, d, and set of $\mathcal{S}$ subgraphs (of G) $\mathcal{S} = \{g_1, g_2, \ldots, g_n\}$, learn an embedding function $f : g_i \rightarrow \mathbf{y_i} \in \mathbb{R}^d$ such that the Neighborhood property of each $g_i \in \mathcal{S}$ is preserved.

Problem 2. Given a graph $G(V, E)$, d, and set of $\mathcal{S}$ subgraphs (of G) $\mathcal{S} = \{g_1, g_2, \ldots, g_n\}$, learn an embedding function $f : g_i \rightarrow \mathbf{y_i} \in \mathbb{R}^d$ such that the Structural property of each $g_i \in \mathcal{S}$ is preserved.

3 Our Methods

A common framework leveraged by most prior works in network embedding is to exploit Word2vec [11] to learn feature representation of nodes in the network. Word2vec learns similar feature representations for words which co-appear frequently in the same context. Network embedding methods, such as Deep-Walk [14] and Node2vec [9], generate 'context' around each node based on random walks and embed nodes using Word2vec. These embeddings are known to preserve various node properties. However, such methods lack the global view of subgraphs, hence they are inherently unable to preserve the properties of entire subgraphs and fail in solving our problems.

3.1 Overview

A major challenge in solving our problems is to design an architecture which has global view of subgraphs and is able to capture similarities and differences between the properties of entire subgraphs. Our idea to overcome this challenge is to leverage the Paragraph2vec models for our subgraph embedding problems. Paragraph2vec [10] models learn latent representation of entire paragraphs while maximizing similarity between paragraphs which have similar word co-occurrences. Note that these models have the global view of entire paragraphs. Intuitively, such a model is suitable for solving Problems 1 and 2. Thus, we extend Paragraph2vec to learn subgraph embedding while preserving distance between subgraphs that have similar 'node co-occurrences'. We extend both Paragraph2vec models (PV-DBOW and PV-DM). We call our models *Distributed Bag of Nodes version of Subgraph Vector* (**Sub2Vec**-DBON) and *Distributed Memory version of Subgraph Vector* (**Sub2Vec**-DM) respectively. We discuss **Sub2Vec**-DM and **Sub2Vec**-DBON in detail in Subsects. 3.3 and 3.4.

In addition, another challenge is to generate meaningful context of 'node co-occurrences' which preserve Neighborhood and Structural properties of subgraphs. We tackle this challenge by using our Id-paths and Degree-paths. As discussed earlier, Id-paths and Degree-paths capture the Neighborhood and Structural property respectively. We discuss on efficiently generating samples of Id-paths and Degree-paths next.

3.2 Subgraph Truncated Random Walks

Both Neighborhood and Structural properties of subgraphs require us to enlist paths in the subgraph (Id-paths for Neighborhood and Degree-paths for Structural). Since there are an unbounded number of paths, it is not feasible to enumerate all of them. Hence, we resort to random walks to generate samples of the paths efficiently. Next we describe the random walks for Id-paths and Degree-paths respectively. Note that the random walks are performed *inside* each subgraph g_i separately, and not on the entire graph G.

Id-paths: Our random walk for Id-paths in subgraph $g_i(v_i, e_i)$ starts from a node $n_1 \in v_i$ chosen uniformly at random. We choose a neighbor of n_1 uniformly at random as the next node to visit in the random walk. Specifically, if i^{th} node in the random walk is n_i, then the jth node in the random walk is a node n_j, such that $(n_i, n_j) \in e_i$, chosen uniformly at random among such nodes.

We generate random walk of fixed length l for each subgraph g_i in the input set of subgraphs $\mathcal{S} = \{g_1, g_2, \ldots, g_n\}$. At the end of process, for each subgraph $g_i \in \mathcal{S}$, we obtain a random walk of length l, annotated by the ids of the nodes.

Degree-paths: As mentioned in Sect. 2, we generate labels for each node based on the ratio of degree of a node to the number of nodes in the subgraph for Degree-paths. We formally define the label generation for nodes next.

Let θ_a be the degree of the node n_a. Given a subgraph $g(v_i, e_i)$, we get the ratio $\gamma_a = \frac{\theta_a}{|v_i|}$ for each node $n_a \in v_i$. Note that $\gamma \in [0, 1]$. Roughly, we define various bins of range of values and associate a character with it (e.g. [0,0.2) is 'a', [0.2,0.5) is 'b' and so on). As real networks have very skewed degree distributions, we use logarithmic binning [12] to determine bin-widths. Formally, let Σ be a finite alphabet over distinct characters; and we define a many-to-one relabeling function $\tau : [0, 1] \to \Sigma$.

After generating the labels for each node, we follow the same procedure as in Id-paths, and do a random walk on each subgraph. The only difference is that we generate a sequence of characters $\alpha \in \Sigma$ for Degree-paths as opposed to node-ids for Id-paths.

3.3 Sub2Vec-DM

In the Sub2Vec-DM model, assuming we want to preserve Neighborhood property of subgraphs, we seek to predict a node-id u that appears in an Id-path, given other node-ids that co-occur with u in the path, and the subgraph that u belongs to. By co-occurrence, we mean that two ids co-appear in a sliding window of a fixed length w, i.e., they appear within distance w of each other. Consider a subgraph g_1 (a subgraph induced by nodes $\{a, b, c, e\}$) in the toy graph shown earlier. Suppose the random walk simulation of Id-paths in g_1 returns $a \rightarrow b \rightarrow c$, and we consider $w = 3$, then the model asks to predict node-id c given subgraph g_1, and the node's 2 predecessors (ids a and b), i.e., $\Pr(c|g_1, \{a, b\})$. Note that if our goal is to preserve the Structural property, we use Degree-paths instead.

Here we give a formal formulation of Sub2Vec-DM. Let $V = \bigcup_i v_i$ be the union of node-set of all the subgraphs. Let $\mathbf{W}_1$ be a $|\mathcal{S}| \times d$ matrix, where each row $\mathbf{W}_1(i)$ represents the embedding of a subgraph $g_i \in \mathcal{S}$. Similarly, let $\mathbf{W}_2$ be a $|V| \times d$ matrix, where each column $\mathbf{W}_2(n)$ is the vector representation of node $n \in V'$. Let the set of node-ids that appear within distance w of a node a be θ_a.

In Sub2Vec-DM, we predict a node-id a given θ_a and the subgraph g_i, from which a and θ_a are drawn. Formally, the objective of Sub2Vec-DM is to maximize:

$$\max \sum_{g_i \in \mathcal{S}} \sum_{a \in g_i} \log(\Pr(a|\mathbf{W}_2(\theta_a), \mathbf{W}_1(i))),$$

where $\Pr(a|\mathbf{W}_2(\theta_a), \mathbf{W}_1(i))$ is the probability of predicting node a given the vector representations of θ_a and g_i. It is defined using the softmax function:

$$\Pr(a|\mathbf{W}_2(\theta_a), \mathbf{W}_1(i)) = \frac{e^{\mathbf{W}_3(a) \cdot h(\mathbf{W}_2(\theta_a), \mathbf{W}_1(i))}}{\sum_{v \in V} e^{\mathbf{W}_3(v) \cdot h(\mathbf{W}_2(\theta_a), \mathbf{W}_1(g_i))}} \tag{1}$$

where matrix $\mathbf{W}_3$ is a softmax parameter and $h(\mathbf{x}, \mathbf{y})$ is average or concatenation of vectors $\mathbf{x}$ and $\mathbf{y}$ [10]. In practice, to compute Eq. 1, we use negative sampling or hierarchical softmax [11].

3.4 Sub2Vec-DBON

In the Sub2Vec-DBON model, we want to predict a set θ of co-occurring node-ids in an Id-path sampled from subgraph g_i, given only the sugraph g_i. Note that Sub2Vec-DBON does not explicitly rely on the embeddings of node-ids as in Sub2Vec-DM. As shown in Sect. 3.3, the 'co-occurrence' means that two ids co-appear in a sliding window of a fixed length w. For example, consider the same example as in Sect. 3.3: the subgraph g_1 and the node sequence $a \rightarrow b \rightarrow c$ generated by random walks. Now in the Sub2Vec-DBON model, for $w = 3$, the goal is to predict the set $\{a, b, c\}$ given the subgraph g_1. This model is parallel to the popular skip-gram model. The matrices $\mathbf{W}_1$ and $\mathbf{W}_2$ are the same as in Sect. 3.3.

Formally, given a subgraph g_i, and the θ drawn from g_i, the objective of Sub2Vec-DBON is the following:

$$\max \sum_{g_i \in \mathcal{S}} \sum_{\theta \in g_i} \log(\Pr(\theta | \mathbf{W}_1(i))), \tag{2}$$

where $\Pr(\theta | \mathbf{W}_1(i))$ is a softmax function, i.e.,

$$\Pr(\theta | \mathbf{W}_1(i)) = \frac{e^{\mathbf{W}_2(\theta).\mathbf{W}_1(i)}}{\sum_{\theta' \in G} e^{\mathbf{W}_2(\theta').\mathbf{W}_1(i)}},$$

Since computing Eq. 2 involves summation over all possible sets of co-occuring nodes, we use approximation techniques such as negative sampling [11].

3.5   Algorithm

Our algorithm Sub2Vec works as follows: we first generate the samples of Id-paths/Degree-paths in each subgraph by running random walks. Then we optimize the SV-DBON/SV-DM objectives using the stochastic gradient descent (SGD) method [5] by leveraging the random walks. We used the Gensim package for implementation [15]. The complete pseudocode is presented in Algorithm 1.

Algorithm 1. Sub2Vec

Require: Graph G, subgraph set $\mathcal{S} = \{g_1, g_2, \ldots, g_n\}$, length of the context window w, dimension d
1: walkSet = {}
2: **for each** g_i in s **do**
3:     walk = RandomWalk (g_i)
4:     walkSet$[g_i]$ = walk
5: **end for**
6: f = StochasticGradientDescent(walkSet, d, w)
7: **return** f

---

4   Experiments

We leverage Sub2Vec[1] for two applications, namely Community Detection and Graph Classification, and perform a case-study on a real subgraph data. All experiments are conducted using a 4 Xeon E7-4850 CPU with 512 GB 1066 Mhz RAM. We set the length of the random walk as 1000 and following literature [9], we set dimension of the embedding as 128 unless mentioned otherwise for both parameters.

---

[1] Code in Python available at: https://goo.gl/Ef4q8g.

4.1   Community Detection

Setup. Here we show how to leverage Sub2Vec for the well-known community detection problem [4,8]. A community in a network is a coherent group of nodes which are densely connected among themselves and sparsely connected with the rest of the network. One expects many nodes within the same community to have similar neighborhoods. Hence, we can use Sub2Vec to embed subgraphs while preserving the Neighborhood property and cluster the embeddings to detect communities.

Approach. We propose to use Sub2Vec for community detection by embedding the surrounding neighborhood of each node. First, we extract the neighborhood C_v of each node $v \in V$ from the input graph $G(V, E)$. For each node v, we extract its neighborhood C_v only once. Hence, we get a set $\mathcal{S} = \{C_v | v \in V\}$ of $|V|$ neighborhoods are extracted from G. Since each C_v is a subgraph, we can leverage Sub2Vec to embed each $C_v \in \mathcal{S}$. The idea is that similar C_vs will be embedded together, which can then be clustered to detect communities. We use a clustering algorithm (K-Means) to cluster the feature vectors $f(C_v)$ of each C_v. For datasets with overlapping communities (like Youtube), we use the Neo-Kmeans algorithm [23] to obtain overlapping clusters. Cluster membership of $f(C_v)$ determines the community membership of node v. The complete pseudocode is in Algorithm 2.

In Algorithm 2, we define the neighborhood of each node to be its ego-network for dense networks (School and Work) and its 2-hop ego-network for others. The ego-network of a node is the subgraph induced by the node and its neighbors. The 2-hop ego-network is the subgraph induced by the node, its neighbors, and neighbors' neighbors.

Algorithm 2 . Community Detection using Sub2Vec

Require: A network $G(V, E)$, Sub2Vec parameters, k number of communities
 1: neighborhoodSet = {}
 2: **for each** v in V **do**
 3:     neighborhoodSet = neighborhoodSet $\cup$ neighbordhood of v in G.
 4: **end for**
 5: vecs = Sub2Vec (neighborhoodSet, w, d)
 6: clusters = Clustering(vecs, k)
 7: return clusters

Datasets. We use multiple real world datasets from multiple domains like social-interactions, co-authorship, social networks and so on of varying sizes. See Table 1.

Baselines. We compare Sub2Vec with various traditional community detection algorithms and network embedding based methods. Newman [8] is a well-known community detection algorithm based on betweenness. Louvian [4] is a popular greedy optimization method. DeepWalk and Node2Vec are recent network

Table 1. Information on Datasets for Community Detection (Left) and Graph Classification (Right). # com denotes the number of ground truth communities in each dataset. # nodes denotes the average number of nodes in each graph classification dataset.

Dataset	$|V|$	$|E|$	# com	Domain
Work	92	757	5	contact
Cornell	195	304	5	web
School	182	2221	5	contact
Texas	187	328	5	web
Wash.	230	446	5	web
Wisc.	265	530	5	web
PolBlogs	1490	16783	2	web
Youtube	1.13M	2.97M	5000	social

Dataset	# graphs	# classes	# nodes	# labels
MUT	188	2	17.9	7
PTC	344	2	25.5	19
ENZ	600	6	32.6	3
PRT	1113	2	39.1	3
NC1	4110	2	29.8	37
NC109	4127	2	29.6	38

Table 2. Sub2Vec easily out-performs all baselines in all datasets. Average F-1 score is shown for each method. Winners have been bolded for each dataset. G stands for gain obtained by Sub2Vec in percentage.

Method	Work	School	PolBlogs	Texas	Cornell	Wash.	Wisc.	Youtube
Newman	0.32	0.34	0.58	0.17	0.33	0.21	0.16	0.04
Louvian	0.25	0.31	0.50	0.20	0.20	0.13	0.19	0.01
DeepWalk	0.40	0.48	0.80	0.25	0.32	0.29	0.29	0.15
Node2Vec	0.64	0.79	**0.86**	0.27	0.33	0.28	0.30	0.17
Sub2Vec-DM	**0.77**	**0.93**	0.85	**0.35**	**0.36**	**0.38**	0.32	**0.38**
Sub2Vec-DBON	0.65	0.57	0.82	**0.35**	0.34	0.37	**0.32**	0.36
G	**20.3**	**17.7**	**−1.2**	**29.6**	**9.1**	**31.0**	**6.6**	**123.5**

embedding methods which learn feature representations of nodes in the network which we then cluster (in the same way as us) to obtain communities.

Results. We measure the performance of all the algorithms by computing the Average F1 score [25] against the ground-truth. See Table 2. Both versions of Sub2Vec significantly and consistently outperform all the baselines. We achieve a significant gain of *123.5%* over the closest competitor (Node2Vec) for Youtube. We do better than Node2Vec and DeepWalk because intuitively, we learn the feature vector of the neighborhood of each node for the community detection task; while they just do random probes of the neighborhood. Performance of Newman and Louvian is considerably poor in Youtube as these methods output non-overlapping communities. Performance of Node2Vec is satisfactory in sparse networks like Wash. and Texas. Node2Vec does slightly better (∼1%) than Sub2Vec in PolBlogs—the network consists of homogeneous neighborhoods, which favors it. However, the performance of Node2Vec is significantly worse for dense networks like Work and School. On the other hand, performance of Sub2Vec is even more impressive in these dense networks (where the task is more challenging).

4.2   Graph Classification

Setup. Here, we show an application of our method in the Graph Classification task [18,24]. In the graph classification problem, the data consists of multiple (g_i, Y_i) tuples, where each g_i is a graph and Y_i is its class-label. Moreover, the nodes in each graph g_i are labeled. The goal is to predict the class Y_i for a given graph G_i. Since we can generate embeddings of each graph, we can train any off-the-shelf classifier to classify the graphs. In this experiment, we set the dimension of embedding as 300 and set the length of the random walk as 100000.

Approach. Our approach is to learn the embedding of each graph by treating them as a subgraph of a union of all the graphs. First we learn the feature representation of the graphs such that either the Neighborhood (Sub2Vec-N) or Structural (Sub2Vec-S) property is preserved. We then leverage four off-the-shelf classifiers: Decision Tree, Random Forest, SVM, and Multi-Layered Perceptron, to solve the classification task.

Datasets. We test on classic graph classification benchmark datasets. All the datasets are publicly available[2]. List of datasets is presented in Table 1.

Baselines. We used two state-of-the-art methods as our competitors. WL-Kernel [18]: This is a graph kernel method based on the Weisfeiler-Lehman test of graph-isomorphism. DG-Kernel is a deep-learning version of WL-kernel [24], which relies on latent representation of sub-structures of the graphs. It uses the popular skip-gram model.

Table 3. Testing accuracy of graph classification. G is the % gain obtained by Sub2Vec.

Method	MUT	PTC	ENZ	PRT	NC1	NC109
WL-Kernel	0.80	0.56	0.27	0.72	0.80	0.80
DG-Kernel	0.82	0.60	0.53	0.71	0.80	0.80
Sub2Vec-N	0.74	0.59	**0.89**	0.92	**0.95**	0.90
Sub2Vec-S	**0.85**	**0.62**	0.85	**0.96**	**0.95**	**0.91**
G	**3.7**	**3.3**	**67.9**	**33.3**	**18.8**	**13.8**

Results. We report the testing accuracy of a 5-fold cross validation. For both Sub2Vec-N and Sub2Vec-S, we run both of our models Sub2Vec-DM and Sub2Vec-DBON. We then train all four classifiers and show the best of them. See Table 3. The results show that both of our methods consistently outperform competitors. The gain of Sub2Vec over the state-of-the-art DG-Kernel is upto a significant 67.9%. The better performance of Sub2Vec-S over Sub2Vec-N indicates that structural properties in these datasets are more discriminative. This is intuitive as different bonds between the elements results in different structure

---

[2] http://mlcb.is.tuebingen.mpg.de/Mitarbeiter/Nino/Graphkernels/.

and also determine the chemical properties of a compound [6]. Interestingly, the Neighborhood property outperforms the Structural property in ENZ dataset. This suggests that, in ENZ dataset, which interestingly also has higher number of classes, the neighborhood property is more important than structural property in determining the graph class.

4.3   Case Studies

We perform case-studies on MemeTracker[3] dataset to investigate if the embeddings returned by Sub2Vec are interpretable. Here, we run Sub2Vec to preserve the Neighborhood property. The MemeTracker consists of a series of cascades caused by memes spreading on the network of linked web pages. Each meme-cascade induces a subgraph in the underlying network. We first embed these subgraphs in a continuous vector space by leveraging Sub2Vec. We then cluster these vectors to explore what kind of meme cascade-graphs are grouped together and what characteristics of memes determine their similarity and distance to each other. For this case-study, we pick the top 1000 memes by volume, and cluster them into 10 clusters.

We find coherent clusters which are meaningful groupings of memes based on topics. For example we find cluster of memes related to different topics such as entertainment, politics, religion, technology and so on. Visualization of these clusters is presented above. In the entertainment cluster, we find memes which are names of popular songs and movies such as "sweet home alabama", "Madagascar 2" and so on. Similarly, we also find a cluster of religious memes. These memes are quotes from the Bible. In the politics cluster, we find popular quotes from the 2008 presidential election season e.g. Barack Obama's popular slogan "yes we can" along with his controversial quotes like "you can put lipstick on a pig" in the cluster. Interestingly, we find that all the memes in Spanish language were clustered together. This indicates that memes in different language travel though separate websites, which matches with the reality as most webpages use one primary language.

(a) Politics

yes we can yes we can
the chant is drill baby drill
tax and spend
i barack hussein obama do solemnly swear
you can put lipstick on a pig

(b) Religion

let there be light and there was light
do unto others as you would have them do unto you
god with us
truly you are the son of god
you shall love your neighbor as yourself

(c) Entertainment

single ladies put a ring on it
dr seuss horton hears a who
around the world in 80 days
star trek the next generation technical manual
eternal sunshine of the spotless mind

(d) Spanish

alicia en el pa s de las maravillas
el ni o con el pijama de rayas
viva la rep blica muerte al borb n
en los pr ximos d as
tirar del carro en la misma direcci n

5   Related Work

The network embedding problem, which seeks to generate low dimensional feature representation of nodes, has been well studied. Early work includes [1, 2, 20]. However, these methods are slow and do not scale to large networks. Recently, several deep learning based network embeddings algorithms were proposed.

---

[3] snap.stanford.edu.

DeepWalk [14] and Node2Vec [9] learn feature representation based on contexts generated by random walks. SDNE [21] and LINE [19] learn feature representation of nodes while preserving first and second order proximity. Other recent works include [7,16,22]. However, all of them node embeddings, while our goal is to embed subgraphs.

The most similar network embedding literature includes [13,17,24]. Risen and Bunke [17] propose to learn vector representations of graphs based on edit distance to a set of pre-defined prototype graphs. Yanardag et al. [24] and Narayanan et al. [13] learn vector representation of the subgraphs using the Word2Vec by generating "corpus" of subgraphs where each subgraph is treated as a word. The above works focuses on some specific subgraphs like graphlets and rooted subgraphs. None of them embed subgraphs with arbitrary structure.

6   Conclusions and Discussion

We focus on the embedding problem for a set of subgraphs by formulating two intuitive properties (Neighborhood and Structural). We developed Sub2Vec, a scalable embedding framework which gives interpretable embeddings such that these properties are preserved. We also demonstrate via detailed experiments that Sub2Vec outperforms traditional algorithms as well as node-level embedding algorithms in various applications, more so in challenging settings. Extending our framework to handle more properties of interest would be fruitful. For example, one may think of a 'Positional property' which relates to the *position or role* of nodes in the subgraph w.r.t. the overall graph.

Acknowledgements. This paper is based on work partially supported by the NSF (CAREER-IIS-1750407, DGE-1545362, and IIS-1633363), the NEH (HG-229283-15), ORNL (Task Order 4000143330) and from the Maryland Procurement Office (H98230-14-C-0127), and a Facebook faculty gift.

References

1. Bach, F.R., Jordan, M.I.: Learning spectral clustering. In: NIPS, vol. 16 (2003)
2. Belkin, M., Niyogi, P.: Laplacian eigenmaps and spectral techniques for embedding and clustering. In: NIPS, vol. 14, pp. 585–591 (2001)
3. Bhagat, S., Cormode, G., Muthukrishnan, S.: Node classification in social networks. In: Aggarwal, C. (ed.) Social Network Data Analytics, pp. 115–148. Springer, Boston (2011). https://doi.org/10.1007/978-1-4419-8462-3_5
4. Blondel, V.D., Guillaume, J.L., Lambiotte, R., Lefebvre, E.: Fast unfolding of communities in large networks. J. Stat. Mech. Theor. Exp. (2008)
5. Bousquet, O., Bottou, L.: The tradeoffs of large scale learning. In: NIPS, pp. 161–168 (2008)
6. Carey, F.A., Sundberg, R.J.: Advanced Organic Chemistry. Part A: Structure and Mechanisms. Springer, New York (2007). https://doi.org/10.1007/978-0-387-44899-2
7. Cheng, K., Li, J., Liu, H.: Unsupervised feature selection in signed social networks. In: KDD 2017, pp. 777–786. ACM (2017)

8. Girvan, M., Newman, M.E.: Community structure in social and biological networks. Proc. National Acad. Sci. **99**(12), 7821–7826 (2002)
9. Grover, A., Leskovec, J.: node2vec: scalable feature learning for networks. In: SIGKDD, pp. 855–864. ACM (2016)
10. Le, Q.V., Mikolov, T.: Distributed representations of sentences and documents. In: ICML, vol. 14, pp. 1188–1196 (2014)
11. Mikolov, T., Sutskever, I., Chen, K., Corrado, G.S., Dean, J.: Distributed representations of words and phrases and their compositionality. In: NIPS, pp. 3111–3119 (2013)
12. Milojević, S.: Power law distributions in information science: making the case for logarithmic binning. J. Am. Soc. Inf. Sci. Technol. **61**, 2417–2425 (2010)
13. Narayanan, A., Chandramohan, M., Chen, L., Liu, Y., Saminathan, S.: subgraph2vec: Learning distributed representations of rooted sub-graphs from large graphs. arXiv preprint arXiv:1606.08928 (2016)
14. Perozzi, B., Al-Rfou, R., Skiena, S.: Deepwalk: Online learning of social representations. In: SIGKDD, pp. 701–710. ACM (2014)
15. Rehurek, R., Sojka, P.: Software framework for topic modelling with large corpora. In: LREC 2010 Workshop on New Challenges for NLP Frameworks. Citeseer (2010)
16. Ribeiro, L.F., Saverese, P.H., Figueiredo, D.R.: struc2vec: Learning node representations from structural identity. In: KDD 2017, pp. 385–394. ACM (2017)
17. Riesen, K., Bunke, H.: Graph Classification and Clustering Based on Vector Space Embedding. World Scientific Publishing Co. Inc., River Edge (2010)
18. Shervashidze, N., Schweitzer, P., Leeuwen, E.J.V., Mehlhorn, K., Borgwardt, K.M.: Weisfeiler-lehman graph kernels. J. Mach. Learn. Res. **12**(Sep), 2539–2561 (2011)
19. Tang, J., Qu, M., Wang, M., Zhang, M., Yan, J., Mei, Q.: Line: large-scale information network embedding. In: WWW, pp. 1067–1077. ACM (2015)
20. Tenenbaum, J.B., De Silva, V., Langford, J.C.: A global geometric framework for nonlinear dimensionality reduction. science **290**(5500), 2319–2323 (2000)
21. Wang, D., Cui, P., Zhu, W.: Structural deep network embedding. In: SIGKDD, pp. 1225–1234. ACM (2016)
22. Wang, X., Cui, P., Wang, J., Pei, J., Zhu, W., Yang, S.: Community Preserving Network Embedding, pp. 203–209 (2017)
23. Whang, J.J., Dhillon, I.S., Gleich, D.F.: Non-exhaustive, overlapping k-means. In: SDM, pp. 936–944. SIAM (2015)
24. Yanardag, P., Vishwanathan, S.: Deep graph kernels. In: SIGKDD, pp. 1365–1374. ACM (2015)
25. Yang, J., Leskovec, J.: Overlapping community detection at scale: a nonnegative matrix factorization approach. In: WSDM, pp. 587–596. ACM (2013)

Interaction Content Aware Network Embedding via Co-embedding of Nodes and Edges

Linchuan Xu[1(✉)], Xiaokai Wei[2], Jiannong Cao[1], and Philip S. Yu[3,4]

[1] The Hong Kong Polytechnic University, Hung Hom, Kowloon, Hong Kong
{cslcxu,csjcao}@comp.polyu.edu.hk
[2] Facebook Inc., 1 Hacker Way, Menlo Park, CA, USA
weixiaokai@gmail.com
[3] University of Illinois at Chicago, Chicago, IL, USA
psyu@uic.edu
[4] Institute for Data Science, Tsinghua University, Beijing, China

Abstract. Network embedding has been a hot topic as it can learn node representations that encode the network structure resulting from node interactions. In this paper, besides the network structure, the interaction content within which each interaction arises is also embedded because it reveals interaction preferences of the two nodes involved. Specifically, we propose interaction content aware network embedding (ICANE) via co-embedding of nodes and edges. The embedding of edges is to learn edge representations that preserve the interaction content, which then can be incorporated into node representations through edge representations. Experiments demonstrate ICANE outperforms five recent network embedding models in visualization, link prediction and classification.

1 Introduction

Network embedding has been a hot topic recently. Existing methods [7,13,16,23,24] basically embed the network structure in a Euclidean space of interest. In this way, however, they fair to consider the content within which node interactions arise. In practice, the content can be observed in various networks:

- In academic co-authorship networks as illustrated in Fig. 1, the particular paper is the interaction content associated with co-authorships.
- In gene co-expression networks where genes co-express functional gene products, such as protein, the functional products are the interaction content.
- In social interaction networks where users interact under social media, e.g., discussing under images and documents, the media is the interaction content.

X. Wei—The work was done when the author was a Ph.D. student at University of Illinois at Chicago.

© Springer International Publishing AG, part of Springer Nature 2018
D. Phung et al. (Eds.): PAKDD 2018, LNAI 10938, pp. 183–195, 2018.
https://doi.org/10.1007/978-3-319-93037-4_15

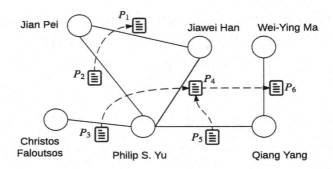

Fig. 1. A co-authorship network sampled from a DBLP dataset [17] where nodes denote researchers and rectangles with lines inside them denote papers that researchers co-authored. Some papers may be missing due to the sampling process.

Interaction content has been shown helpful in network analysis, such as community detection [14]. In the scenario of network embedding, we can see interaction content contains node interaction preferences. Specifically in the co-authorship network, the interaction content indicates research interests. Similarly, the content in the social interaction networks reveals the events or activities that users are interested in. These two cases together indicate interaction preferences are specific for the social environment.

Moreover, some nodes may have multiple distinct interaction preferences, and each interaction may only arise within a single content. For example, a researcher may have interests in three research areas, such as Database, Machine Learning, and Data Mining. Different papers of the researcher and co-authors may belong to different areas. Not distinguishing different co-authorships in terms of the areas while embedding the co-authorship network, hence, is not appropriate.

To achieve this goal, the major challenge is that interaction content cannot be concatenated to node representations because it may not be affiliated to nodes. For example, in social networks where users interact under images or documents, the content may be belong to third parties not involved in the interactions.

In this paper, we propose interaction content aware network embedding (**ICANE**) via co-embedding of nodes and edges. Specifically, ICANE embeds the network structure in node representations, and embeds interaction content in edge representations. Moreover, ICANE incorporates interaction content into node representations via jointly learning representations for nodes and edges.

In some scenarios, interaction content can have relationships, e.g.,

- In co-authorship networks, the interaction content, i.e., papers, usually has citation relationships as illustrated in Fig. 1.
- In gene co-expression networks, the interaction content can be proteins, which can have protein-protein interactions in various biological processes.
- In social interaction networks, the interaction content is the media, such as documents, which can have references to each other.

Because the interaction content is affiliated to edges, we name the network resulting from content relationships as an edge network. Hence, ICANE encodes interaction content into edge representations by embedding the edge network.

In other scenarios, the interaction content has text information, e.g., paper content in co-authorship networks. Collectively, there may exist both an edge network and text information in some scenarios.

It is worthy of noting that node representations may also benefit the learning of edge representations which explicitly preserve node interaction preferences. Since node representations encode the network structure, i.e., interactions between nodes, node representations implicitly preserve interaction preferences of nodes. Hence, node representations and edge representations actually preserve similar characteristics but from different views.

2   Related Work

The development of recent network embedding starts with DeepWalk [13], which employs Skip-gram to present pairs of nodes reached in the same truncated random walks to be close in the embedding space. There are other Skip-gram based models, such as TADW [25] to embed both network structure and node attributes, and node2vec [7] to explore diverse neighborhoods in random walks.

There are also many methods not based on Skip-gram. LINE [16] is proposed to embed large-scale networks by directly presenting pairs of nodes with first-order or second order connections to be close. GraRep [4] models first-order up to a pre-defined k-order proximities into transition matrices. A recent study [26] concludes that modelling high-order proximities can improve the quality of node representations. Besides simply preserving the network structure, some methods also preserve network properties, such as HOPE [12] preserving asymmetric transitivities and M-NMF [21] preserving communities. Some methods [5,6,15,22] even embed heterogeneous information networks. Deep learning has also been applied for network embedding [20]. Most methods above are unsupervised learning methods. Semi-supervised methods [8,19,27] have also been studied.

3   Preliminaries

Definition 1. $G_v(V_v, E_v, C)$ *denotes a* **network with interaction content,** *where V_v is a set of nodes, E_v is a set of weighted or unweighted, directed or undirected edges, and C is a set of interaction content.*

Definition 2. $G_e(V_e, E_e)$ *denotes an* **edge network.** *V_e is a set of nodes which are the concept of edges in $G_v(V_v, E_v, C)$, E_e is a set of weighted or unweighted, directed or undirected edges among the interaction content C.*

Note that $|V_e|$ corresponds to the number of interaction content, and it may not be equal to $|E_v|$ due to two reasons. Firstly, multiple nodes may interact within the same content, e.g., multi-author papers, which results in multiple edges. Secondly, a pair of nodes may interact under multiple content. Multiple interactions are treated as a weighted edge like existing embedding models do.

As an embedding method, ICANE presents nodes connected by edges to be close in an Euclidean space. The closeness of two nodes is quantified as follows:

Definition 3. *The closeness of two nodes is quantified as the probability of an edge between them, where the probability is defined as follows:*

$$p(\boldsymbol{v}_i, \boldsymbol{v}_j) = \frac{1}{1 + exp\{-\boldsymbol{v}_i^\top \boldsymbol{v}_j\}}, \tag{1}$$

where $\boldsymbol{v}_i \in \mathbb{R}^D$ and $\boldsymbol{v}_j \in \mathbb{R}^D$ are column vectors of representations for nodes i and j, respectively, and D is the dimension of the Euclidean space of interest.

The closeness is reasonable as larger probabilities indicate larger inner product of two vectors, which is a measurement of closeness in Euclidean space.

4   Model Development

4.1   Node Representation Learning

To embed the network structure, ICANE not only presents pairs of nodes connected by edges to be close but also presents pairs of nodes not connected to be apart in the embedding space because non-linkage information is also an important part of network structure. Since the closeness is quantified as probability, the network structure preserving can be formulated into an optimization objective according to maximum likelihood estimation as follows:

$$\max_{V \in \mathbb{R}^{|V_v| \times D}} \prod_{(i,j) \in E_v, (h,k) \notin E_v} p(\boldsymbol{v}_i, \boldsymbol{v}_j)(1 - p(\boldsymbol{v}_h, \boldsymbol{v}_k)), \tag{2}$$

which maximizes the probabilities of both linkage and non-linkage relationships.

The multiplication maximization is usually transformed to an equivalent minimization by taking negative natural logarithm, which is denoted as follows:

$$\min_V - \left[\sum_{(i,j) \in E_v} (w_v)_{ij} \log p(\boldsymbol{v}_i, \boldsymbol{v}_j) + \sum_{(h,k) \notin E_v} \log(1 - p(\boldsymbol{v}_h, \boldsymbol{v}_k)) \right], \tag{3}$$

where $(w_v)_{ij} \in \mathbb{R}$ is the weight of edge (i, j) added to reflect to relationship strength. The loss function is referred to as $\mathcal{L}_v$ in the rest of the paper.

4.2   Edge Representation Learning

For interaction content that can produce an edge network, edge representations can be learned by embedding the edge network structure, which can be performed in the same way as embedding the node network structure. Hence, the loss function is referred to as $\mathcal{L}_e$, which is the same as $\mathcal{L}_v$ except that node representations are replaced with edge representations.

For interaction content with text information, the content can be embedded into edge representations via regularizing edge representations to accord with the text information. The regularization is reasonable because the text information is the ground truth about the interaction preferences, e.g., paper content denotes research topics. The regularization can be performed by projecting the representations to corresponding content, which is formulated as follows:

$$\min_{M \in \mathbb{R}^{D \times Q}} ||EM - A||_F^2, \tag{4}$$

where M is a projection matrix to be estimated, Q is the number of terms in text, $E \in \mathbb{R}^{|V_e| \times D}$, $A \in \mathbb{R}^{|V_e| \times Q}$ is a term-frequency matrix extracted from text, and $|| \cdot ||_F^2$ is Frobenius norm. The intuition behind Eq. (4) is that the content is well represented by edge representations through the projection matrix.

4.3   Joint Learning

The key to joint learning is how to relate edge representations to node representations so that interaction content can be incorporated into node representations. As mentioned in the introduction, node representations encoding the network structure implicitly preserve node interaction preferences and edge representations explicitly preserve interaction preferences. Hence, node representations should be similar to representations of their incident edges. To make the problem simple, nodes are presented to be close to their incident edges, which can be achieved in a similar way to encode linkage relationships among nodes.

Hence, the overall loss function for joint learning can be obtained as follows:

$$\mathcal{L}(V, E, M) = \mathcal{L}_v + \mathcal{L}_e - \left[\sum_{v_i \to e_m} \log p(v_i, e_m) + \sum_{v_i \mapsto e_l} \log(1 - p(v_i, e_l)) \right] \tag{5}$$
$$+ ||EM - A||_F^2 + \lambda(||V||_F^2 + ||E||_F^2 + ||M||_F^2),$$

which directly adds loss functions for node representation learning, edge representation learning and joint learning. More sophisticated ways for the combination is left as future work. $v_i \to e_m$ denotes e_m is an incident edge of v_i while $v_i \mapsto e_l$ denotes the opposite. $p(v_i, e_m)$ is the closeness measurement between a node and an edge, which is defined similarly to the closeness among nodes as mentioned above. Specifically, $p(v_i, e_m)$ is quantified as follows:

$$p(v_i, e_m) = \frac{1}{1 + \exp\{-v_i^\top e_m\}}, \tag{6}$$

Equation (5) assumes that there exist both an edge network and text information. In some cases where there may be only one type of content information, we can safely remove the corresponding component from Eq. (5). Hence, for cases where there is only an edge network, we name the model as INCAE(E) while for cases where there is only text information, we name the model as ICANE(A).

Algorithm 1. The optimization algorithm

Input : $G_v(V_v, E_v, C)$, D, λ, and negative ratio
Output: V and E

Pre-training V and E with gradient descent;
while *(not converge)* **do**
  Fix V and E, find the optimal M with the Eq. (10);
  Fix other variable(s), find the optimal E with gradient descent;
  Fix other variable(s), find the optimal V with gradient descent;
return V and E

---

5 The Optimization

$\mathcal{L}(V, E, M)$ is not jointly convex over the three variables. We thus solve it by an alternating algorithm [3] which replaces a complex optimization problem with a sequence of easier sub-problems, and then solves the sub-problems alternatingly. In our case, the sub-problems w.r.t v_i and e_i can be solved by gradient-based algorithms, e.g., steepest descent or L-BFGS. The derivative for minimizing $\mathcal{L}(V, E, M)$ with respect to v_i is computed as follows:

$$\frac{\partial \mathcal{L}(V, E)}{\partial v_i} = - \sum_{(i,j) \in E_v} \left[\frac{(w_v)_{ij} \exp\{-v_i^\top v_j\}}{1 + \exp\{-v_i^\top v_j\}} v_j \right] + \sum_{(i,k) \notin E_v} \left[\frac{v_k}{1 + \exp\{-v_i^\top v_k\}} \right]$$
$$- \sum_{v_i \to e_m} \left[\frac{\exp\{-v_i^\top e_m\}}{1 + \exp\{-v_i^\top e_m\}} e_m \right] + \sum_{v_i \mapsto e_l} \left[\frac{e_l}{1 + \exp\{-v_i^\top e_l\}} \right] + 2\lambda(v_i), \tag{7}$$

The derivative with respect to e_m is computed as follows: $\frac{\partial \mathcal{L}(V, E, M)}{\partial e_m} =$

$$- \sum_{(m,n) \in E_e} \left[\frac{(w_e)_{mn} \exp\{-e_m^\top e_n\}}{1 + \exp\{-e_m^\top e_n\}} e_n \right] + \sum_{(m,l) \notin E_e} \left[\frac{e_l}{1 + \exp\{-e_m^\top e_l\}} \right]$$
$$- \sum_{v_i \to e_m} \frac{\exp\{-e_m^\top v_i\}}{1 + \exp\{-e_m^\top v_i\}} v_i + \sum_{v_k \mapsto e_m} \frac{v_k}{1 + \exp\{-e_m^\top v_k\}} \tag{8}$$
$$+ 2(e_m^T M - a_m^T) M^T + 2\lambda(e_m^d),$$

To minimize $\mathcal{L}(V, E, M)$ with respect to M, the optimization objective actually turns into solving the following problem:

$$\min_M ||EM - A||_2^2 + \lambda ||M||_2^2. \tag{9}$$

It is easy to see that the optimal M can be obtained by setting the derivative of Eq. (9) w.r.t M to zero. Hence, the optimal M is obtained as follows:

$$M = (E^T E + \lambda I)^{-1} E^T A, \tag{10}$$

where $I \in \mathbb{R}^{D \times D}$ is an identity matrix.

The pseudo-codes of the alternating optimization algorithm are presented in Algorithm 1. Negative ratio is the ratio of the number positive edges to that of negative edges as used in LINE [16]. With the negative ration, the scalability to large-scale networks can be guaranteed. ***Pre-training*** is performed to initialize the model to a point in parameter space that renders the learning process more effective [2]. The pre-training on V or E is performed by solely preserving the network structure of $G_v(V_v, E_v, C)$ or $G_e(V_e, E_e)$, i.e., minimizing $\mathcal{L}_v$ or $\mathcal{L}_e$ by gradient descent. The learning rates of gradient descent are obtained by back-tracking line search [1]. If there is no $G_e(V_e, E_e)$, the pre-training of E can be performed by factorizing the term-frequency matrix A using SVD [4].

Algorithm 1 is essentially a block-wise coordinate descent algorithm [18] whose convergence can be guaranteed.

Table 1. Network statistics

Network	Co-authorship	Paper citation	User interaction	Photo(group)
# Nodes	12407	8208	5342	2613
# Edges	27714	10532	230123	38841
# Attributes	6934		4070	

6 Empirical Evaluation

6.1 Datasets

- DBLP [17]: A co-authorship network in Table 1 is sampled with papers as the interaction content. Papers are selected from conferences of four fields, which are SIGMOD, VLDB, ICDE, EDBT, and PODS for Database, KDD, ICDM, SDM, and PAKDD for Data Mining, ICML, NIPS, AAAI, IJCAI and ECML for Machine Learning, SIGIR, WSDM, WWW, CIKM, and ECIR for Information Retrieval. Publication time span is set as 17 years from 1990 to 2006.
- CLEF [11]: From CLEF, we sample a user interaction network where interactions are established between users commenting on the same photo. Hence, the photos are the interaction content. Photos can be categorized into different groups, such as scenery, explore, etc. The groups can be used to construct a photo network where edges are established between two photos belonging to the same group. We refer to this photo network as photo(group) network.

6.2 Experiment Settings

Five recent network embedding models, DeepWalk [13], LINE [16], TADW [25], node2vec [7], and EOE [22] are used as baselines. Both TADW and EOE embed networks with node content. They can be applied to the DBLP co-authorship network because paper content can also be used as node content. However, the tags of images of the Flickr user interaction network cannot be used as node content in that the tags belong to a third party. For the implementation of Algorithm 1, we set the embedding dimension as 128, which is used in all the baselines, negative ratio as 5, which is used in LINE.

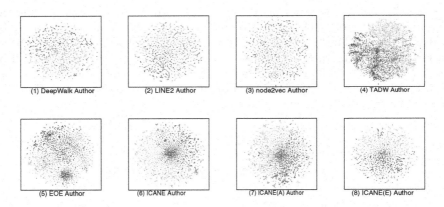

Fig. 2. Visualization of representations for the DBLP dataset, where green points are for authors from DB, light blue for IR, dark blue for DM, and red for ML. The filed of an author is chosen as the one where he/she published the most papers. (Color figure online)

6.3   Representation Visualization

This section visually presents how effectively the representations encode the network structure. The DBLP data is used as the illustration. t-SNE [10] is employed to visualize the author representations in Fig. 2. From Fig. 2(1) through Fig. 2(3) of baselines (LINE(1st) is omitted due to space limitation because it performs worse than LINE(2nd)), we see that a considerably large number of authors from different fields are mixed up. This may be because the selected four fields, i.e., DB, DM, ML, and IR, are closely related, and there are many cross-field co-authorships. Hence, the network structure along is not enough to distinguish authors from one research field to another.

TADW, EOE and the proposed ICANE work better by utilizing the paper content as illustrated in Fig. 2(4) though Fig. 2(6) where data points of the same color are distributed together. This is because research focus of each field is distinct, which is reflected on paper content. To make fair comparison with TADW and EOE, we visualize representations learned by ICANE(A) in Fig. 2(7). We see that ICANE(A) is also comparable with TADW and EOE. Moreover, we visualize representations learned by ICANE(E) in Fig. 2(8). We can see Fig. 2(8) performs better than the baselines only embedding the co-authorship network.

It might not be easy to visually tell which one of TADW, EOE, and ICANE performs better, but ICANE can jointly learn author representations and paper representations while all the baselines can only learn author representations.

Learning paper representations can lend strengths to learn author representations, and vice versa because paper representations capture research interests of authors. As a result, the data mining applications with respect to either nodes or edges may benefit from each other. The advantage of the joint learning is demonstrated in the following link prediction and classification.

6.4 Link Prediction

Link prediction is usually performed by measuring similarities between two nodes [9]. Here, the inner product of two node representations normalized by sigmoid function is employed as the similarity measurement. We first perform Flickr user interaction prediction, and conduct 9 runs of experiments where training interactions range from 10% to 90% of the total interactions and the rest are used as test interactions. Moreover, for each experiment, the same number of negative interactions are randomly sampled for the evaluation purpose. AUC is employed as the evaluation metric, and the results are presented in Table 2.

Table 2. AUC scores(100%) for Flickr interaction prediction when different ratios of interactions are used in the training phase.

Model	10%	20%	30%	40%	50%	60%	70%	80%	90%
DeepWalk	79.30	86.74	89.52	90.59	91.30	91.49	91.73	91.75	91.88
LINE(1st)	74.26	85.45	89.62	91.87	93.17	93.96	94.61	95.00	95.36
LINE(2nd)	78.22	84.03	86.75	88.47	89.34	90.00	90.19	89.81	89.22
node2vec	77.07	80.57	81.67	81.79	81.17	81.64	82.01	81.46	81.73
ICANE(E)	90.25	93.21	94.39	**95.95**	**96.38**	**96.50**	**97.02**	97.25	**97.42**
ICANE(A)	**93.82**	**94.93**	**95.73**	95.22	96.22	96.29	96.96	97.10	97.29
ICANE	92.40	93.58	94.27	94.78	95.64	95.06	96.89	**97.32**	97.39

Table 2 shows ICANE consistently outperform all the baselines no matter what kind of content information is utilized (TADW and EOE are not applicable since photos belong to a third party instead of the nodes of the interaction network). Moreover, ICANE still work well given very limited training interactions, such as 10% and 20%. The superior performance of ICANE benefits from the extra information provided by photo tags and the photo(group) network where photos are the interaction content within which users have interactions.

For DBLP co-authorship prediction, the co-authorships arise from 2007 to 2013 are used as test links, and experiment results are presented in Table 3. ICANE outperforms all the baselines except for LINE(2nd). As we examine the node similarities of links computed on the representations learned by LINE(2nd), all the positive and negative links have node similarities close to 1.0, which is not that interpretable. The reason behind the phenomenon may be that LINE(2nd)

Table 3. AUC scores(100%) for DBLP co-authorship prediction, where (E) and (A) denote ICANE(E) and ICANE(A), respectively.

Model	DeepWalk	LINE(1st)	**LINE(2nd)**	node2vec	TADW	EOE	(E)	(A)	ICANE
AUC	76.62	72.38	**83.05**	76.68	73.82	80.95	78.02	81.30	82.83

presents nodes with second-order link to be close, which is not consistent with the fist-order link prediction. Hence, LINE(2nd) is omit in the following discussion.

ICANE(E) outperforms baselines that only embed the network structure but underperforms EOE which also embeds node content. This may be because the paper citation network brings less useful information than paper content which has explicit information about node interaction preferences. The useful information brought by the paper citation network can be seen in the better performance of ICANE than that of ICANE(A). Moreover, ICANE(A) and ICANE outperform EOE. Recall that ICANE(A) and ICANE use the paper content as interaction content while EOE uses it to construct an author-word coupled network. Hence, the interaction content is explicitly embedded into edge representations, and then is incorporated into node representations in ICANE. In EOE, paper content is fragmentarily embedded in word representations by embedding the word network. We can see the mechanism for incorporating paper content into node representations of ICANE is more effective than that of EOE.

TADW performs even worse than DeepWalk that only embeds the network structure. It is worthy of noting that TADW concatenates node representations learned from node content and node representations learned from the network structure. As a result, the node similarities of links are largely determined by their node content, which indicates research interests of researchers. It is intuitive that it is not necessary for researchers with similar interests to collaborate.

6.5 Multi-label Classification

For DBLP, a research field as a label is assigned to authors if they published papers in this field. For Flickr, there are 99 labels for the photos in the dataset. It is worthy of noting that for Flickr, the photo representations are edge representations in the proposed models while they are node representations learned by baselines from the photo(group) network. The photo tags are the node content of the photo(group) network. We employ Micro-F1 and Macro-F1 as the performance metrics, and results of ten-fold cross validation are presented in Table 4, which is produced by binary-relevance SVM with polynomial kernel. For Flickr, ICANE performs better than all the baselines because ICANE can utilize not only the photo tags and the photo(group) network, but also the user interaction network. The user interaction network results from the photos so that it can provide auxiliary information to the photo representations.

For DBLP, all the models utilizing both the co-authorship network and paper content perform significantly than those only utilizing the co-authorship network.

Table 4. Micro-F1(100%) & Macro-F1(100%) for multi-label classification

Flickr	DeepWalk	LINE(1st)	LINE(2nd)	node2vec	TADW	EOE	(E)	(A)	**ICANE**
Micro-F1	65.1	58.6	57.2	72.5	69.5	70.3	73.2	73.1	**73.7**
Macro-F1	65.3	58.5	57.2	72.3	69.1	70.0	73.0	72.8	**73.2**
DBLP	DeepWalk	LINE(1st)	LINE(2nd)	node2vec	TADW	EOE	(E)	(A)	**ICANE**
Micro-F1	60.2	59.9	59.5	62.9	79.6	79.9	67.3	79.7	**80.8**
Macro-F1	37.6	35.8	36.2	43.7	74.5	74.6	56.4	74.8	**75.3**

ICANE(A) obtains similar performance as TADW and EOE, but ICANE performs better than TADW and EOE because it can even utilize the paper citation network. The benefits brought by the paper citation network can be seen in the superior performance of ICANE(E) to that of DeepWalk, LINE and node2vec.

6.6 Multi-class Classification

The research field as a label is assigned to each paper. We employ SVM with polynomial kernel as the classifier, and present the accuracy obtained by 10-fold cross validation in Table 5. Similarly, ICANE outperforms all the baselines. To this point, we have demonstrated not only the interaction content can help improve node representations, but also the nodes can in turn help improve content representations. Particularly in this case, the authors largely determine the fields of their papers because they have particular expertise.

Table 5. Accuracy on DBLP paper multi-class classification

Model	DeepWalk	LINE(1st)	LINE(2nd)	node2vec	TADW	EOE	(E)	(A)	**ICANE**
Accuracy	68.62	61.40	55.36	70.84	71.06	69.03	72.83	73.56	**73.66**

Acknowledgement. The work described in this paper was partially supported by the funding for Project of Strategic Importance provided by The Hong Kong Polytechnic University (Project Code: 1-ZE26), RGC General Research Fund under Grant PolyU 152199/17E, NSFC Key Grant with Project No. 61332004, NSF through grants IIS-1526499, and CNS-1626432, and NSFC 61672313.

References

1. Armijo, L.: Minimization of functions having Lipschitz continuous first partial derivatives. Pac. J. Math. **16**(1), 1–3 (1966)
2. Bengio, Y., Lamblin, P., Popovici, D., Larochelle, H., et al.: Greedy layer-wise training of deep networks. Adv. Neural Inf. Process. Syst. **19**, 153 (2007)
3. Bezdek, J.C., Hathaway, R.J.: Some notes on alternating optimization. In: Pal, N.R., Sugeno, M. (eds.) AFSS 2002. LNCS (LNAI), vol. 2275, pp. 288–300. Springer, Heidelberg (2002). https://doi.org/10.1007/3-540-45631-7_39

4. Cao, S., Lu, W., Xu, Q.: GraRep: learning graph representations with global structural information. In: Proceedings of the 24th ACM International on Conference on Information and Knowledge Management, pp. 891–900. ACM (2015)
5. Chang, S., Han, W., Tang, J., Qi, G.J., Aggarwal, C.C., Huang, T.S.: Heterogeneous network embedding via deep architectures. In: Proceedings of the 21th ACM SIGKDD International Conference on Knowledge Discovery and Data Mining, pp. 119–128. ACM (2015)
6. Dong, Y., Chawla, N.V., Swami, A.: metapath2vec: scalable representation learning for heterogeneous networks. In: Proceedings of the 23rd ACM SIGKDD International Conference on Knowledge Discovery and Data Mining, pp. 135–144. ACM (2017)
7. Grover, A., Leskovec, J.: node2vec: scalable feature learning for networks. In: Proceedings of the 22nd ACM SIGKDD International Conference on Knowledge Discovery and Data Mining (2016)
8. Li, J., Zhu, J., Zhang, B.: Discriminative deep random walk for network classification. In: ACL, vol. 1 (2016)
9. Liben-Nowell, D., Kleinberg, J.: The link-prediction problem for social networks. J. Am. Soc. Inf. Sci. Technol. 58(7), 1019–1031 (2007)
10. Van der Maaten, L., Hinton, G.: Visualizing data using t-SNE. J. Mach. Learn. Res. 9(2579–2605), 85 (2008)
11. McAuley, J., Leskovec, J.: Image labeling on a network: using social-network metadata for image classification. In: Fitzgibbon, A., Lazebnik, S., Perona, P., Sato, Y., Schmid, C. (eds.) ECCV 2012. LNCS, Part IV, vol. 7575, pp. 828–841. Springer, Heidelberg (2012). https://doi.org/10.1007/978-3-642-33765-9_59
12. Ou, M., Cui, P., Pei, J., Zhang, Z., Zhu, W.: Asymmetric transitivity preserving graph embedding. In: KDD, pp. 1105–1114 (2016)
13. Perozzi, B., Al-Rfou, R., Skiena, S.: DeepWalk: online learning of social representations. In: Proceedings of the 20th ACM SIGKDD International Conference on Knowledge Discovery and Data Mining, pp. 701–710. ACM (2014)
14. Qi, G.J., Aggarwal, C.C., Huang, T.: Community detection with edge content in social media networks. In: 2012 IEEE 28th International Conference on Data Engineering, pp. 534–545. IEEE (2012)
15. Tang, J., Qu, M., Mei, Q.: PTE: predictive text embedding through large-scale heterogeneous text networks. In: Proceedings of the 21th ACM SIGKDD International Conference on Knowledge Discovery and Data Mining, pp. 1165–1174. ACM (2015)
16. Tang, J., Qu, M., Wang, M., Zhang, M., Yan, J., Mei, Q.: Line: large-scale information network embedding. In: Proceedings of the 24th International Conference on World Wide Web, pp. 1067–1077. International World Wide Web Conferences Steering Committee (2015)
17. Tang, J., Zhang, J., Yao, L., Li, J., Zhang, L., Su, Z.: ArnetMiner: extraction and mining of academic social networks. In: Proceedings of the 14th ACM SIGKDD International Conference on Knowledge Discovery and Data Mining, pp. 990–998. ACM (2008)
18. Tseng, P.: Convergence of a block coordinate descent method for nondifferentiable minimization. J. Optim. Theory Appl. 109(3), 475–494 (2001)
19. Tu, C., Zhang, W., Liu, Z., Sun, M.: Max-margin DeepWalk: discriminative learning of network representation. In: IJCAI, pp. 3889–3895 (2016)
20. Wang, D., Cui, P., Zhu, W.: Structural deep network embedding. In: Proceedings of the 22nd ACM SIGKDD International Conference on Knowledge Discovery and Data Mining, pp. 1225–1234. ACM (2016)

21. Wang, X., Cui, P., Wang, J., Pei, J., Zhu, W., Yang, S.: Community preserving network embedding. In: AAAI, pp. 203–209 (2017)
22. Xu, L., Wei, X., Cao, J., Yu, P.S.: Embedding of Embedding (EOE): joint embedding for coupled heterogeneous networks. In: Proceedings of the Tenth ACM International Conference on Web Search and Data Mining, pp. 741–749. ACM (2017)
23. Xu, L., Wei, X., Cao, J., Yu, P.S.: Multi-task network embedding. In: Proceedings of the Fourth IEEE International Conference on Data Science and Advanced Analytics, pp. 571–580. IEEE (2017)
24. Xu, L., Wei, X., Cao, J., Yu, P.S.: Multiple social role embedding. In: Proceedings of the Fourth IEEE International Conference on Data Science and Advanced Analytics, pp. 581–589. IEEE (2017)
25. Yang, C., Liu, Z., Zhao, D., Sun, M., Chang, E.Y.: Network representation learning with rich text information. In: Proceedings of the 24th International Joint Conference on Artificial Intelligence, Buenos Aires, Argentina, pp. 2111–2117 (2015)
26. Yang, C., Sun, M., Liu, Z., Tu, C.: Fast network embedding enhancement via high order proximity approximation. In: Proceedings of the Twenty-Sixth International Joint Conference on Artificial Intelligence, IJCAI, pp. 19–25 (2017)
27. Yang, Z., Cohen, W.W., Salakhutdinov, R.: Revisiting semi-supervised learning with graph embeddings. arXiv preprint arXiv:1603.08861 (2016)

MetaGraph2Vec: Complex Semantic Path Augmented Heterogeneous Network Embedding

Daokun Zhang[1]([✉]), Jie Yin[2], Xingquan Zhu[3], and Chengqi Zhang[1]

[1] Centre for Artificial Intelligence, FEIT,
University of Technology Sydney, Ultimo, Australia
Daokun.Zhang@student.uts.edu.au, Chengqi.Zhang@uts.edu.au
[2] Discipline of Business Analytics, The University of Sydney, Sydney, Australia
jie.yin@sydney.edu.au
[3] Department of CEECS, Florida Atlantic University, Boca Raton, USA
xqzhu@cse.fau.edu

Abstract. Network embedding in heterogeneous information networks (HINs) is a challenging task, due to complications of different node types and rich relationships between nodes. As a result, conventional network embedding techniques cannot work on such HINs. Recently, metapath-based approaches have been proposed to characterize relationships in HINs, but they are ineffective in capturing rich contexts and semantics between nodes for embedding learning, mainly because (1) metapath is a rather strict single path node-node relationship descriptor, which is unable to accommodate variance in relationships, and (2) only a small portion of paths can match the metapath, resulting in sparse context information for embedding learning. In this paper, we advocate a new metagraph concept to capture richer structural contexts and semantics between distant nodes. A metagraph contains multiple paths between nodes, each describing one type of relationships, so the augmentation of multiple metapaths provides an effective way to capture rich contexts and semantic relations between nodes. This greatly boosts the ability of metapath-based embedding techniques in handling very sparse HINs. We propose a new embedding learning algorithm, namely MetaGraph2Vec, which uses metagraph to guide the generation of random walks and to learn latent embeddings of multi-typed HIN nodes. Experimental results show that MetaGraph2Vec is able to outperform the state-of-the-art baselines in various heterogeneous network mining tasks such as node classification, node clustering, and similarity search.

1 Introduction

Recent advances in storage and networking technologies have resulted in many applications with interconnected relationships between objects. This has led to the forming of gigantic inter-related and multi-typed heterogeneous information networks (HINs) across a variety of domains, such as e-government, e-commerce,

D. Phung et al. (Eds.): PAKDD 2018, LNAI 10938, pp. 196–208, 2018.
https://doi.org/10.1007/978-3-319-93037-4_16

biology, social media, etc. HINs provide an effective graph model to characterize the diverse relationships among different types of nodes. Understanding the vast amount of semantic information modeled in HINs has received a lot of attention. In particular, the concept of metapaths [10], which connect two nodes through a sequence of relations between node types, is widely used to exploit rich semantics in HINs. In the last few years, many metapath-based algorithms are proposed to carry out data mining tasks over HINs, including similarity search [10], personalized recommendation [6,9], and object clustering [11].

Despite their great potential, data mining tasks in HINs often suffer from high complexity, because real-world HINs are very large and have very complex network structure. For example, when measuring metapath similarity between two distant nodes, all metapath instances need to be enumerated. This makes it very time-consuming to perform mining tasks, such as link prediction or similarity search, across the entire network. This inspires a lot of research interests in network embedding that aims to embed the network into a low-dimensional vector space, such that the proximity (or similarity) between nodes in the original network can be preserved. Analysis and search over large-scale HINs can then be applied in the embedding space, with the help of efficient indexing or parallelized algorithms designed for vector spaces.

Conventional network embedding techniques [1,4,8,12–16], however, focus on homogeneous networks, where all nodes and relations are considered to have a single type. Thus, they cannot handle the heterogeneity of node and relation types in HINs. Only very recently, metapath-based approaches [2,3], such as MetaPath2Vec [3], are proposed to exploit specific metapaths as guidance to generate random walks and then to learn heterogeneous network embedding. For example, consider a DBLP bibliographic network, Fig. 1(a) shows the HIN schema, which consists of three node types: Author (A), Paper (P) and Venue (V), and three edge types: an author writes a paper, a paper cites another paper, and a paper is published in a venue. The metapath $\mathcal{P}_1$: $A \to P \to V \to P \to A$ describes the relationship where both authors have papers published in the same venue, while $\mathcal{P}_2$: $A \to P \to A \to P \to A$ describes that two authors share the same co-author. If $\mathcal{P}_1$ is used by MetaPath2Vec to generate random walks, a possible random walk could be: $a_1 \to p_1 \to v_1 \to p_2 \to a_2$. Consider a window size of 2, authors a_1 and a_2 would share the same context node v_1, so they should be close to each other in the embedding space. This way, semantic similarity between nodes conveyed by metapaths is preserved.

Due to difficulties in information access, however, real-world HINs often have sparse connections or many missing links. As a result, metapath-based algorithms may fail to capture latent semantics between distant nodes. As an example, consider the bibliographic network, where many papers may not have venue information, as they may be preprints submitted to upcoming venues or their venues are simply missing. The lack of paper-venue connection would result in many short random walks, failing to capture hidden semantic similarity between distant nodes. On the other hand, besides publishing papers on same venues, distant authors can also be connected by other types of relations, like sharing

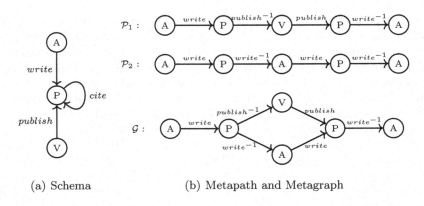

(a) Schema   (b) Metapath and Metagraph

Fig. 1. Schema, metapath and metagraph

common co-authors or publishing papers with similar topics. Such information should be taken into account to augment metapath-based embedding techniques.

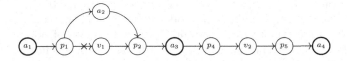

Fig. 2. An example of random walk from a_1 to a_4 based on metagraph $\mathcal{G}$, which cannot be generated using metapaths $\mathcal{P}_1$ and $\mathcal{P}_2$. This justifies the ability of MetaGraph2Vec to provide richer structural contexts to measure semantic similarity between distant nodes.

Inspired by this observation, we propose a new method for heterogeneous network embedding, called MetaGraph2Vec, that learns more informative embeddings by capturing richer semantic relations between distant nodes. The main idea is to use metagraph [5] to guide random walk generation in an HIN, which fully encodes latent semantic relations between distant nodes at the network level. Metagraph has its strength to describe complex relationships between nodes and to provide more flexible matching when generating random walks in an HIN. Figure 1(b) illustrates a metagraph $\mathcal{G}$, which describes that two authors are relevant if they have papers published in the same venue or they share the same co-authors. Metagraph $\mathcal{G}$ can be considered as a union of metapaths $\mathcal{P}_1$ and $\mathcal{P}_2$, but when generating random walks, it can provide a superset of random walks generated by both $\mathcal{P}_1$ and $\mathcal{P}_2$. Figure 2 gives an example to illustrate the intuition behind. When one uses metapath $\mathcal{P}_1$ to guide random walks, if paper p_1 has no venue information, the random walk would stop at p_1 because the link from p_1 to v_1 is missing. This results in generating too many short random walks that cannot reveal semantic relation between authors a_1 and a_3. In contrast, when metagraph $\mathcal{G}$ is used as guidance, the random walk $a_1 \to p_1 \to a_2 \to p_2 \to a_3$, and $a_3 \to p_4 \to v_2 \to p_5 \to a_4$ is generated by taking the path en route A and V in $\mathcal{G}$, respectively. This testifies the ability of MetaGraph2Vec to provide richer structural contexts to measure semantic similarity between distant nodes, thereby enabling more informative network embedding.

Based on this idea, in MetaGraph2Vec, we first propose metagraph guided random walks in HINs to generate heterogeneous neighborhoods that fully encode rich semantic relations between distant nodes. Second, we generalize the Skip-Gram model [7] to learn latent embeddings for multiple types of nodes. Finally, we develop a heterogeneous negative sampling based method that facilitates the efficient and accurate prediction of a node's heterogeneous neighborhood. MetaGraph2Vec has the advantage of offering more flexible ways to generate random walks in HINs so that richer structural contexts and semantics between nodes can be preserved in the embedding space.

The contributions of our paper are summarized as follows:

1. We advocate a new *metagraph* descriptor which augments metapaths for flexible and reliable relationship description in HINs. Our study investigates the ineffectiveness of existing metapath based node proximity in dealing with sparse HINs, and explains the advantage of metagraph based solutions.
2. We propose a new network embedding method, called MetaGraph2Vec, that uses metagraph to capture richer structural contexts and semantics between distant nodes and to learn latent embeddings for multiple types of nodes in HINs.
3. We demonstrate the effectiveness of our proposed method through various heterogeneous network mining tasks such as node classification, node clustering, and similarity search, outperforming the state-of-the-art.

2   Preliminaries and Problem Definition

In this section, we formalize the problem of heterogeneous information network embedding and give some preliminary definitions.

Definition 1. *A **heterogeneous information network (HIN)** is defined as a directed graph $G = (V, E)$ with a node type mapping function $\phi : V \to \mathcal{L}$ and an edge type mapping function $\psi : E \to \mathcal{R}$. $T_G = (\mathcal{L}, \mathcal{R})$ is the network schema that defines the node type set $\mathcal{L}$ with $\phi(v) \in \mathcal{L}$ for each node $v \in V$, and the allowable link types $\mathcal{R}$ with $\psi(e) \in \mathcal{R}$ for each edge $e \in E$.*

Example 1. For a bibliographic HIN composed of authors, papers, and venues, Fig. 1(a) defines its network schema. The network schema contains three node types, author (A), paper (P) and venue (V), and defines three allowable relations, $A \xrightarrow{write} P$, $P \xrightarrow{cite} P$ and $V \xrightarrow{publish} P$. Implicitly, the network schema also defines the reverse relations, i.e., $P \xrightarrow{write^{-1}} A$, $P \xrightarrow{cite^{-1}} P$ and $P \xrightarrow{publish^{-1}} V$.

Definition 2. *Given an HIN G,* **heterogeneous network embedding** *aims to learn a mapping function* $\Phi : V \rightarrow \mathbb{R}^d$ *that embeds the network nodes* $v \in V$ *into a low-dimensional Euclidean space with* $d \ll |V|$ *and guarantees that nodes sharing similar semantics in G have close low-dimensional representations* $\Phi(v)$.

Definition 3. *A* **metagraph** *is a directed acyclic graph (DAG)* $\mathcal{G} = (N, M, n_s, n_t)$ *defined on the given HIN schema* $T_G = (\mathcal{L}, \mathcal{R})$, *which has only a single source node* n_s *(i.e., with 0 in-degree) and a single target node* n_t *(i.e., with 0 out-degree). N is the set of the occurrences of node types with* $n \in \mathcal{L}$ *for each* $n \in N$. *M is the set of the occurrences of edge types with* $m \in \mathcal{R}$ *for each* $m \in M$.

As metagraph $\mathcal{G}$ depicts complex composite relations between nodes of type n_s and n_t, N and M may contain duplicate node and edge types. To clarify, we define the *layer* of each node in N as its topological order in $\mathcal{G}$ and denote the number of layers by $d_{\mathcal{G}}$. According to nodes' layer, we can partition N into disjoint subsets $N[i]$ ($1 \leq i \leq d_{\mathcal{G}}$), which represents the set of nodes in layer i. Each $N[i]$ does not contain duplicate nodes. Now each element in N and M can be uniquely described as follows. For each n in N, there exists a unique i with $1 \leq i \leq d_{\mathcal{G}}$ satisfying $n \in N[i]$ and we define the layer of node n as $l(n) = i$. For each $m \in M$, there exist unique i and j with $1 \leq i < j \leq d_{\mathcal{G}}$ satisfying $m \in N[i] \times N[j]$.

Example 2. Given a bibliographic HIN G and a network schema T_G shown in Fig. 1(a), (b) shows an example of metagraph $\mathcal{G} = (N, M, n_s, n_t)$ with $n_s = n_t = A$. There are 5 layers in $\mathcal{G}$ and node set N can be partitioned into 5 disjoint subsets, one for each layer, where $N[1] = \{A\}, N[2] = \{P\}, N[3] = \{A, V\}, N[4] = \{P\}, N[5] = \{A\}$.

Definition 4. *For a metagraph* $\mathcal{G} = (N, M, n_s, n_t)$ *with* $n_s = n_t$, *its* **recursive metagraph** $\mathcal{G}^\infty = (N^\infty, M^\infty, n_s^\infty, n_t^\infty)$ *is a metagraph formed by tail-head concatenation of an arbitrary number of* $\mathcal{G}$. $\mathcal{G}^\infty$ *satisfies the following conditions:*

1. $N^\infty[i] = N[i]$ *for* $1 \leq i < d_{\mathcal{G}}$, *and* $N^\infty[i] = N[i \bmod d_{\mathcal{G}} + 1]$ *for* $i \geq d_{\mathcal{G}}$.
2. *For each* $m \in N^\infty[i] \times N^\infty[j]$ *with any* i *and* j, $m \in M^\infty$ *if and only if one of the following two conditions is satisfied:*
   (a) $1 \leq i < j \leq d_{\mathcal{G}}$ *and* $m \in M \bigcap (N[i] \times N[j])$;
   (b) $i \geq d_{\mathcal{G}}$, $1 \leq j - i \leq d_{\mathcal{G}}$ *and* $m \in M \bigcap (N[i \bmod d_{\mathcal{G}} + 1] \times N[j \bmod d_{\mathcal{G}} + 1])$.

In the recursive metagraph $\mathcal{G}^\infty$, *for each node* $n \in N^\infty$, *we define its layer as* $l^\infty(n)$.

Definition 5. *Given an HIN G and a metagraph* $\mathcal{G} = (N, M, n_s, n_t)$ *with* $n_s = n_t$ *defined on its network schema* T_G, *together with the corresponding recursive metagraph* $\mathcal{G}^\infty = (N^\infty, M^\infty, n_s^\infty, n_t^\infty)$, *we define the random walk node sequence constrained by metagraph* $\mathcal{G}$ *as* $\mathcal{S}_{\mathcal{G}} = \{v_1, v_2, \cdots, v_L\}$ *with length L satisfying the following conditions:*

1. *For each v_i ($1 \leq i \leq L$) in $\mathcal{S}_{\mathcal{G}}$, $v_i \in V$ and for each v_i ($1 < i \leq L$) in $\mathcal{S}_{\mathcal{G}}$, $(v_{i-1}, v_i) \in E$. Namely, the sequence $\mathcal{S}_{\mathcal{G}}$ respects the network structure in G.*
2. *$\phi(v_1) = n_s$ and $l^{\infty}(\phi(v_1)) = 1$. Namely, the random walk starts from a node with type n_s.*
3. *For each v_i ($1 < i \leq L$) in $\mathcal{S}_{\mathcal{G}}$, there exists a unique j satisfying $(\phi(v_{i-1}), \phi(v_i)) \in M^{\infty} \bigcap (N^{\infty}[l^{\infty}(\phi(v_{i-1}))] \times N^{\infty}[j])$ with $j > l^{\infty}(\phi(v_{i-1}))$, $\phi(v_i) \in N^{\infty}[j]$ and $l^{\infty}(\phi(v_i)) = j$. Namely, the random walk is constrained by the recursive metagraph $\mathcal{G}^{\infty}$.*

Example 3. Given metagraph $\mathcal{G}$ in Fig. 1(b), a possible random walk is $a_1 \rightarrow p_1 \rightarrow v_1 \rightarrow p_2 \rightarrow a_2 \rightarrow p_3 \rightarrow a_3 \rightarrow p_4 \rightarrow a_5$. It describes that author a_1 and a_2 publish papers in the same venue v_1 and author a_2 and a_5 share the common co-author a_3. Compared with metapath $\mathcal{P}_1$ given in Fig. 1(b), metagraph $\mathcal{G}$ captures richer semantic relations between distant nodes.

3   Methodology

In this section, we first present metagraph-guided random walk to generate heterogeneous neighborhood in an HIN, and then present the MetaGraph2Vec learning strategy to learn latent embeddings of multiple types of nodes.

3.1   MetaGraph Guided Random Walk

In an HIN $G = (V, E)$, assuming a metagraph $\mathcal{G} = (N, M, n_s, n_t)$ with $n_s = n_t$ is given according to domain knowledge, we can get the corresponding recursive metagraph $\mathcal{G}^{\infty} = (N^{\infty}, M^{\infty}, n_s^{\infty}, n_t^{\infty})$. After choosing a node of type n_s, we can start the metagraph guided random walk. We denote the transition probability guided by metagraph $\mathcal{G}$ at ith step as $\Pr(v_i | v_{i-1}; \mathcal{G}^{\infty})$. According to Definition 5, if $(v_{i-1}, v_i) \notin E$, or $(v_{i-1}, v_i) \in E$ but there is no link from node type $\phi(v_{i-1})$ at layer $l^{\infty}(\phi(v_{i-1}))$ to node type $\phi(v_i)$ in the recursive metagraph $\mathcal{G}^{\infty}$, the transition probability $\Pr(v_i | v_{i-1}; \mathcal{G}^{\infty})$ is 0. The probability $\Pr(v_i | v_{i-1}; \mathcal{G}^{\infty})$ for v_i that satisfies the conditions of Definition 5 is defined as

$$\Pr(v_i | v_{i-1}; \mathcal{G}^{\infty}) = \frac{1}{T_{\mathcal{G}^{\infty}}(v_{i-1})} \times \frac{1}{|\{u | (v_{i-1}, u) \in E, \phi(v_i) = \phi(u)\}|}. \tag{1}$$

Above, $T_{\mathcal{G}^{\infty}}(v_{i-1})$ is the number of edge types among the edges starting from v_{i-1} that satisfy the constraints of the recursive metagraph $\mathcal{G}^{\infty}$, which is formalized as

$$T_{\mathcal{G}^{\infty}}(v_{i-1}) = |\{j | (\phi(v_{i-1}), \phi(u)) \in M^{\infty} \bigcap (N^{\infty}[l^{\infty}(\phi(v_{i-1}))] \times N^{\infty}[j]), (v_{i-1}, u) \in E\}|, \tag{2}$$

and $|\{u | (v_{i-1}, u) \in E, \phi(v_i) = \phi(u)\}|$ is the number of v_{i-1}'s 1-hop forward neighbors sharing common node type with node v_i.

At step i, the metagraph guided random walk works as follows. Among the edges starting from v_{i-1}, it firstly counts the number of edge types satisfying the constraints and randomly selects one qualified edge type. Then it randomly walks across one edge of the selected edge type to the next node. If there are no qualified edge types, the random walk would terminate.

3.2   MetaGraph2Vec Embedding Learning

Given a metagraph guided random walk $\mathcal{S}_{\mathcal{G}} = \{v_1, v_2, \cdots, v_L\}$ with length L, the node embedding function $\Phi(\cdot)$ is learned by maximizing the probability of the occurrence of v_i's context nodes within w window size conditioned on $\Phi(v_i)$:

$$\min_{\Phi} - \log \Pr(\{v_{i-w}, \cdots, v_{i+w}\} \setminus v_i | \Phi(v_i)), \tag{3}$$

where,

$$\Pr(\{v_{i-w}, \cdots, v_{i+w}\} \setminus v_i | \Phi(v_i)) = \prod_{j=i-w, j \neq i}^{i+w} \Pr(v_j | \Phi(v_i)). \tag{4}$$

Following MetaPath2Vec [3], the probability $\Pr(v_j | \Phi(v_i))$ is modeled in two different ways:

1. **Homogeneous Skip-Gram** that assumes the probability $\Pr(v_j | \Phi(v_i))$ does not depend on the type of v_j, and thus models the probability $\Pr(v_j | \Phi(v_i))$ directly by softmax:

$$\Pr(v_j | \Phi(v_i)) = \frac{\exp(\Psi(v_j) \cdot \Phi(v_i))}{\sum_{u \in V} \exp(\Psi(u) \cdot \Phi(v_i))}. \tag{5}$$

2. **Heterogeneous Skip-Gram** that assumes the probability $\Pr(v_j | \Phi(v_i))$ is related to the type of node v_j:

$$\Pr(v_j | \Phi(v_i)) = \Pr(v_j | \Phi(v_i), \phi(v_j)) \Pr(\phi(v_j) | \Phi(v_i)), \tag{6}$$

where the probability $\Pr(v_j | \Phi(v_i), \phi(v_j))$ is modeled via softmax:

$$\Pr(v_j | \Phi(v_i), \phi(v_j)) = \frac{\exp(\Psi(v_j) \cdot \Phi(v_i))}{\sum_{u \in V, \phi(u) = \phi(v_j)} \exp(\Psi(u) \cdot \Phi(v_i))}. \tag{7}$$

To learn node embeddings, the MetaGraph2Vec algorithm first generates a set of metagraph guided random walks, and then counts the occurrence frequency $\mathbb{F}(v_i, v_j)$ of each node context pair (v_i, v_j) within w window size. After that, stochastic gradient descent is used to learn the parameters. At each iteration, a node context pair (v_i, v_j) is sampled according to the distribution of $\mathbb{F}(v_i, v_j)$, and the gradients are updated to minimize the following objective,

$$\mathcal{O}_{ij} = - \log \Pr(v_j | \Phi(v_i)). \tag{8}$$

To speed up training, negative sampling is used to approximate the objective function:

$$\mathcal{O}_{ij} = \log \sigma(\Psi(v_j) \cdot \Phi(v_i)) + \sum_{k=1}^{K} \log \sigma(-\Psi(v_{N_{j,k}}) \cdot \Phi(v_i)), \tag{9}$$

where $\sigma(\cdot)$ is the sigmoid function, $v_{N_{j,k}}$ is the kth negative node sampled for node v_j and K is the number of negative samples. For Homogeneous Skip-Gram, $v_{N_{j,k}}$ is sampled from all nodes in V; for Heterogeneous Skip-Gram, $v_{N_{j,k}}$ is sampled from nodes with type $\phi(v_j)$. Formally, parameters Φ and Ψ are updated as follows:

$$\Phi = \Phi - \alpha \frac{\partial \mathcal{O}_{ij}}{\partial \Phi}; \quad \Psi = \Phi - \alpha \frac{\partial \mathcal{O}_{ij}}{\partial \Psi}, \tag{10}$$

where α is the learning rate.

The pseudo code of the MetaGraph2Vec algorithm is given in Algorithm 1.

Algorithm 1. The MetaGraph2Vec Algorithm

Input:
  (1) A heterogeneous information network (HIN): $G = (V, E)$;
  (2) A metagraph: $\mathcal{G} = (N, M, n_s, n_t)$ with $n_s = n_t$;
  (3) Maximum number of iterations: $MaxIterations$;
Output:
  Node embedding $\Phi(\cdot)$ for each $v \in V$;
1: $\mathbb{S} \leftarrow$ generate a set of random walks according to $\mathcal{G}$;
2: $\mathbb{F}(v_i, v_j) \leftarrow$ count frequency of node context pairs (v_i, v_j) in $\mathbb{S}$;
3: $Iterations \leftarrow 0$;
4: **repeat**
5:   $(v_i, v_j) \leftarrow$ sample a node context pair according to the distribution of $\mathbb{F}(v_i, v_j)$;
6:   $(\Phi, \Psi) \leftarrow$ update parameters using (v_i, v_j) and Eq. (10);
7:   $Iterations \leftarrow Iterations + 1$;
8: **until** *convergence* or $Iterations \geq MaxIterations$
9: **return** Φ;

---

4 Experiments

In this section, we demonstrate the effectiveness of the proposed algorithms for heterogeneous network embedding via various network mining tasks, including node classification, node clustering, and similarity search.

4.1 Experimental Settings

For evaluation, we carry out experiments on the DBLP[1] bibliographic HIN, which is composed of papers, authors, venues, and their relationships. Based on paper's venues, we extract papers falling into four research areas: *Database, Data Mining, Artificial Intelligence, Computer Vision*, and preserve the associated authors and venues, together with their relations. To simulate the paper-venue sparsity, we randomly select 1/5 papers and remove their paper-venue relations. This results in a dataset that contains 70,910 papers, 67,950 authors, 97 venues, as well as 189,875 paper-author relations, 91,048 paper-paper relations and 56,728 venue-paper relations.

To evaluate the quality of the learned embeddings, we carry out multi-class classification, clustering and similarity search on author embeddings. Metapaths and metagraph shown in Fig. 1(b) are used to measure the proximity between authors. The author's ground true label is determined by research area of his/her major publications.

We evaluate MetaGraph2Vec with Homogeneous Skip-Gram and its variant MetaGraph2Vec++ with Heterogeneous Skip-Gram. We compare their performance with the following state-of-the-art baseline methods:

- DeepWalk [8]: It uses the uniform random walk that treats nodes of different types equally to generate random walks.

---

[1] https://aminer.org/citation (Version 3 is used).

- LINE [12]: We use two versions of LINE, namely LINE_1 and LINE_2, which models the first order and second order proximity, respectively. Both neglect different node types and edge types.
- MetaPath2Vec and MetaPath2Vec++ [3]: They are the state-of-the-art network embedding algorithms for HINs, with MetaPath2Vec++ being a variant of MetaPath2Vec that uses heterogeneous negative sampling. To demonstrate the strength of metagraph over metapath, we compare with different versions of the two algorithms: $\mathcal{P}_1$ MetaPath2Vec, $\mathcal{P}_2$ MetaPath2Vec and Mixed Meta-Path2Vec, which uses $\mathcal{P}_1$ only, $\mathcal{P}_2$ only, or both, to guide random walks, as well as their counterparts, $\mathcal{P}_1$ MetaPath2Vec++, $\mathcal{P}_2$ MetaPath2Vec++, and Mixed MetaPath2Vec++.

For all random walk based algorithms, we start random walks with length $L = 100$ at each author for $\gamma = 80$ times, for efficiency reasons. For the mixed MetaPath2Vec methods, $\gamma/2 = 40$ random walks are generated by following metapaths $\mathcal{P}_1$ and $\mathcal{P}_2$, respectively. To improve the efficiency, we use our optimization strategy for all random walk based methods: After random walks are generated, we first count the co-occurrence frequencies of node context pairs using a window size $w = 5$, and according to the frequency distribution, we then sample one node context pair to do stochastic gradient descent sequentially. For fair comparisons, the total number of samples (iterations) is set to 100 million, for both random walk based methods and LINE. For all methods, the dimension of learned node embeddings d is set to 128.

4.2   Node Classification Results

We first carry out multi-class classification on the learned author embeddings to compare the performance of all algorithms. We vary the ratio of training data from 1% to 9%. For each training ratio, we randomly split training set and test set for 10 times and report the averaged accuracy.

Table 1. Multi-class author classification on DBLP

Method	1%	2%	3%	4%	5%	6%	7%	8%	9%
DeepWalk	82.39	86.04	87.16	88.15	89.10	89.49	90.02	90.25	90.56
LINE_1	71.25	79.25	83.11	85.60	87.17	88.29	89.05	89.45	89.63
LINE_2	75.70	80.80	82.49	83.88	84.83	85.71	86.58	86.90	86.93
$\mathcal{P}_1$ MetaPath2Vec	83.24	87.70	88.42	89.05	89.26	89.46	89.51	89.76	89.69
$\mathcal{P}_1$ MetaPath2Vec++	82.14	86.02	87.04	87.96	88.47	88.66	88.90	88.91	89.02
$\mathcal{P}_2$ MetaPath2Vec	49.59	52.12	53.76	54.67	55.68	55.49	55.83	55.68	56.07
$\mathcal{P}_2$ MetaPath2Vec++	50.31	52.50	53.72	54.47	55.53	55.78	56.30	56.36	57.02
Mixed MetaPath2Vec	83.86	87.34	88.37	89.22	89.70	90.01	90.37	90.42	90.71
Mixed MetaPath2Vec++	83.08	86.91	88.13	89.07	89.69	90.09	90.58	90.68	90.87
MetaGraph2Vec	**85.76**	**89.00**	89.79	90.55	91.02	91.30	91.72	92.13	92.25
MetaGraph2Vec++	85.20	88.97	**89.99**	**90.78**	**91.42**	**91.65**	**92.13**	**92.42**	**92.46**

Table 1 shows the multi-class author classification results in terms of accuracy (%) for all algorithms, with the highest score highlighted by **bold**. Our Meta-Graph2Vec and MetaGraph2vec++ algorithms achieve the best performance in all cases. The performance gain over metapath based algorithms proves the capacity of MetaGraph2Vec in capturing complex semantic relations between distant authors in sparse networks, and the effectiveness of the semantic similarity in learning informative node embeddings. By considering methpaths between different types of nodes, MetaPath2Vec can capture better proximity properties and learn better author embeddings than DeepWalk and LINE, which neglect different node types and edge types.

4.3   Node Clustering Results

We also carry out node clustering experiments to compare different embedding algorithms. We take the learned author embeddings produced by different methods as input and adopt K-means to do clustering. With authors' labels as ground truth, we evaluate the quality of clustering using three metrics, including Accuracy, F score and NMI. From Table 2, we can see that MetaGraph2Vec and MetaGraph2Vec++ yield the best clustering results on all three metrics.

Table 2. Author clustering on DBLP

Method	Accuracy (%)	F (%)	NMI (%)
DeepWalk	73.87	67.39	42.02
LINE_1	50.26	46.33	17.94
LINE_2	52.14	45.89	19.55
$\mathcal{P}_1$ MetaPath2Vec	69.39	63.05	41.72
$\mathcal{P}_1$ MetaPath2Vec++	66.11	58.68	36.45
$\mathcal{P}_2$ MetaPath2Vec	47.51	43.30	6.17
$\mathcal{P}_2$ MetaPath2Vec++	47.65	41.48	6.56
Mixed MetaPath2Vec	77.20	69.50	49.43
Mixed MetaPath2Vec++	72.36	65.09	42.40
MetaGraph2Vec	**78.00**	**70.96**	**51.40**
MetaGraph2Vec++	77.48	70.69	50.60

4.4   Node Similarity Search

Experiments are also performed on similarity search to verify the ability of Meta-Graph2Vec to capture author proximities in the embedding space. We randomly select 1,000 authors and rank their similar authors according to cosine similarity score. Table 3 gives the averaged precision@100 and precision@500 for different embedding algorithms. As can be seen, our MetaGraph2Vec and Meta-Graph2Vec++ achieve the best search precisions.

Table 3. Author similarity search on DBLP

Methods	Precision@100 (%)	Precision@500 (%)
DeepWalk	91.65	91.44
LINE_1	91.18	89.88
LINE_2	91.92	91.38
$\mathcal{P}_1$ MetaPath2Vec	88.21	88.64
$\mathcal{P}_1$ MetaPath2Vec++	88.68	88.58
$\mathcal{P}_2$ MetaPath2Vec	53.98	44.11
$\mathcal{P}_2$ MetaPath2Vec++	53.39	44.11
Mixed MetaPath2Vec	90.94	90.27
Mixed MetaPath2Vec++	91.49	90.69
MetaGraph2Vec	92.50	**92.17**
MetaGraph2Vec++	**92.59**	91.92

4.5 Parameter Sensitivity

We further analyze the sensitivity of MetaGraph2vec and MetaGraph2Vec++ to three parameters: (1) γ: the number of metagraph guided random walks starting from each author; (2) w: the window size used for collecting node context pairs; (3) d: the dimension of learned embeddings. Figure 3 shows node classification performance with 5% training ratio by varying the values of these parameters. We can see that, as the dimension of learned embeddings d increases, Meta-Graph2Vec and MetaGraph2Vec++ gradually perform better and then stay at a stable level. Yet, both algorithms are not very sensitive to the the number of random walks and window size.

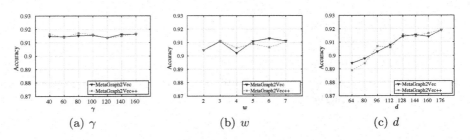

(a) γ (b) w (c) d

Fig. 3. The effect of parameters γ, w, and d on node classification performance

5   Conclusions and Future Work

This paper studied network embedding learning for heterogeneous information networks. We analyzed the ineffectiveness of existing *metapath* based approaches in handling sparse HINs, mainly because metapath is too strict for capturing relationships in HINs. Accordingly, we proposed a new *metagraph* relationship descriptor which augments metapaths for flexible and reliable relationship description in HINs. By using metagraph to guide the generation of random walks, our new proposed algorithm, MetaGraph2Vec, can capture rich context and semantic information between different types of nodes in the network. The main contribution of this work, compared to the existing research in the field, is twofold: (1) a new metagraph guided random walk approach to capturing rich contexts and semantics between nodes in HINs, and (2) a new network embedding algorithm for very sparse HINs, outperforming the state-of-the-art.

In the future, we will study automatic methods for efficiently learning metagraph structures from HINs and assess the contributions of different metagraphs to network embedding. We will also evaluate the performance of MetaGraph2Vec on other types of HINs, such as heterogeneous biological networks and social networks, for producing informative node embeddings.

Acknowledgments. This work is partially supported by the Australian Research Council (ARC) under discovery grant DP140100545, and by the Program for Professor of Special Appointment (Eastern Scholar) at Shanghai Institutions of Higher Learning. Daokun Zhang is supported by China Scholarship Council (CSC) with No. 201506300082 and a supplementary postgraduate scholarship from CSIRO.

References

1. Cao, S., Lu, W., Xu, Q.: Grarep: learning graph representations with global structural information. In: Proceedings of CIKM, pp. 891–900. ACM (2015)
2. Chen, T., Sun, Y.: Task-guided and path-augmented heterogeneous network embedding for author identification. In: Proceedings of WSDM, pp. 295–304. ACM (2017)
3. Dong, Y., Chawla, N.V., Swami, A.: Metapath2vec: scalable representation learning for heterogeneous networks. In: Proceedings of SIGKDD, pp. 135–144. ACM (2017)
4. Grover, A., Leskovec, J.: node2vec: scalable feature learning for networks. In: Proceedings of SIGKDD, pp. 855–864. ACM (2016)
5. Huang, Z., Zheng, Y., Cheng, R., Sun, Y., Mamoulis, N., Li, X.: Meta structure: computing relevance in large heterogeneous information networks. In: Proceedings of SIGKDD, pp. 1595–1604. ACM (2016)
6. Jamali, M., Lakshmanan, L.: HeteroMF: recommendation in heterogeneous information networks using context dependent factor models. In: Proceedings of WWW, pp. 643–654. ACM (2013)
7. Mikolov, T., Sutskever, I., Chen, K., Corrado, G.S., Dean, J.: Distributed representations of words and phrases and their compositionality. In: NIPS, pp. 3111–3119 (2013)
8. Perozzi, B., Al-Rfou, R., Skiena, S.: Deepwalk: online learning of social representations. In: Proceedings of SIGKDD, pp. 701–710. ACM (2014)
9. Shi, C., Zhang, Z., Luo, P., Yu, P.S., Yue, Y., Wu, B.: Semantic path based personalized recommendation on weighted heterogeneous information networks. In: Proceedings of CIKM, pp. 453–462. ACM (2015)

10. Sun, Y., Han, J., Yan, X., Yu, P.S., Wu, T.: Pathsim: meta path-based top-k similarity search in heterogeneous information networks. In: Proceedings of VLDB, pp. 992–1003. ACM (2011)
11. Sun, Y., Norick, B., Han, J., Yan, X., Yu, P.S., Yu, X.: Integrating meta-path selection with user-guided object clustering in heterogeneous information networks. In: Proceedings of SIGKDD, pp. 1348–1356. ACM (2012)
12. Tang, J., Qu, M., Wang, M., Zhang, M., Yan, J., Mei, Q.: LINE: large-scale information network embedding. In: Proceedings of WWW, pp. 1067–1077. ACM (2015)
13. Wang, D., Cui, P., Zhu, W.: Structural deep network embedding. In: Proceedings of SIGKDD, pp. 1225–1234. ACM (2016)
14. Zhang, D., Yin, J., Zhu, X., Zhang, C.: Homophily, structure, and content augmented network representation learning. In: Proceedings of ICDM, pp. 609–618. IEEE (2016)
15. Zhang, D., Yin, J., Zhu, X., Zhang, C.: User profile preserving social network embedding. In: Proceedings of IJCAI, pp. 3378–3384. AAAI Press (2017)
16. Zhang, D., Yin, J., Zhu, X., Zhang, C.: Network representation learning: a survey. arXiv preprint arXiv:1801.05852 (2018)

Multi-network User Identification via Graph-Aware Embedding

Yang Yang, De-Chuan Zhan$^{(\boxtimes)}$, Yi-Feng Wu, and Yuan Jiang

National Key Laboratory for Novel Software Technology,
Nanjing University, Nanjing 210023, China
{yangy,zhandc,wuyf,jiangy}@lamda.nju.edu.cn

Abstract. User identification is widely used in anomaly detection, recommendation system and so on. Previous approaches focus on extraction of features describing users, and the learners try to emphasize the differences between different user identities. However, one applicable user identification scenario occurs in the circumstance of social network, where features of users are not acquirable while only relationships between users are provided. In this paper, we aim at the later situation, i.e., the Network User Identification, where features of users cannot be extracted in social network applications. We consider the information limitation of the single network and focus on utilizing the multiple relationships between identities from multi-networks. Different from the existing common subspace methods in Cross-Network User Identification, we propose a more discriminative Graph-Aware Embedding (GAEM) method for modeling the relationships as well as the transformation between different social networks explicitly in one unified framework. As a consequence, we can get more accurate predictions of the user identities directly based on the learned transferring model with GAEM. The experimental evaluations on real-world data demonstrate the superiorities of our proposed method comparing to the state-of-the-art.

Keywords: Multiple networks · User identification
Graph-Aware Embedding

1 Introduction

With the increasing popularity of social media platforms, more and more people are encouraged to participate in online social networks. Meanwhile, the problem of user identification becomes attractive and has been widely researched recently [9]. However, traditional user identification aims to predict links between user identities and in this case, the essential task turns into recovering the similar closures among identities. These paradigms are almost performed in single social network while neglecting the fact that people take participate in many networks, such as Facebook, Instagram and Twitter, simultaneously. How to identify the

---

This work was supported by NSFC (61773198).

accounts of the same user across different social platforms is a desirable, newly proposed problem. In this paper, we named the later task as "multi-network user identification" (MUI). MUI has obviously practical significance in many web applications [8,26].

Features used in user identification become a crucial fact affecting the performance, especially in single network user identification. Many researchers have been devoted to solving the problem of feature learning/engineering in UI [19]. However, these approaches adopted by existing solutions are always based on the collected profile features or content features, leaving the following essential challenges without considering [22]: Difficulties on obtaining profile features for privacy policies; Incompletion of profile features, owing to many reasons, i.e., law terms, users willings, distributed data storage; Sparsities of content features, due to the divergence of user activity patterns.

The structural information of the social networks, alternatively, can be utilized directly and efficiently for user identification across multiple online social networks and this information can be relatively easy to be obtained (usually permitted by carriers API). There are some structural information based methods proposed, which discover unmatched pair-wise user identities in an iterative way from seed matched pair-wise user identities [30,33]. This type of structural information utilization generally use iterative strategy for spreading over linkages, and named by "propagation" methods. The propagation methods, however, are time-consuming and require more parameters on controlling the information spreading over linkages and is sensitive to linkage noises. To better address the mentioned issues, researchers proposed the embedding methods which first learn the latent features with the information of structure preserved, including TSVM [13], LINE [24], and then identify users based on different distance metric. Nevertheless, most existing graph embedding algorithms are step-by-step. Therefore, [23] built a hypergraph to model high-order relations, and proposed a novel subspace learning algorithm to project seed matching pairs to a node to ensure the aforementioned constraint. However, it needs auxiliary profile information.

In this paper, we focus on the second paradigm of structural information utilization approaches and propose a more discriminative Graph-Aware Embedding (GAEM) method aim at the MUI problem. The proposed GAEM models the relationships as well as the transformation between different social networks in one unified framework. As a consequence, we can identify the user identities across different social platforms using the network structural information only. Especially, rather than utilizing the raw data, we construct more discriminative weighted graphs by exploring the shared neighborhood structures of the vertices globally. Meanwhile, we can predict the transformation among different weighted graphs and ensure the consistency between the transferred weighted graphs, simultaneously. Besides, inspired by the rank constraint, we utilize the rank of the transformation as a rank regularization to improve the construction of the transformation and weighted graph. We empirically validate the effectiveness of our framework, and our model achieves significantly better performance on various tasks.

The rest of this paper starts from the introduction of related work. Then we propose our approach, followed by experiments and conclusion.

2 Related Work

The GAEM approach can identify the user identity across multiple networks in semi-supervised scenarios via the weighted graph embedding. Therefore our work is closely related to: user identity identification and graph embedding.

User identity identification problem was first formalized as connecting corresponding identities across communities in [28]. Considering social network diversity and information asymmetry, previous research can be categorized into three types considering the different feature extraction: profile based, content based and network structure based. User profile based methods aim to collect tagging information provided by users or user profiles from several social networks (e.g., user-name, profile picture, description, location, occupation, etc.), and represent the profiles in vectors [29], then construct models with the new feature representation [16,17]; Content based methods aim to utilize the personally identifiable information from public pages of user-generated content [1,14]. However, previous profile or content based methods always collect specific information of the users, and face serious challenges if required data are not available, i.e., missing features, data sparsity or false data, etc. Alternatively, recent methods have been focused on utilizing the structural network information, [15] unifies learning the latent features of user identities collectively in source and target networks. Nevertheless, the solution is iterative one-to-one mapping method.

Naturally, the local structures are represented by the observed links in the networks, which capture the *first-order* proximity between the vertices [24]. Most existing graph embedding algorithms are designed to preserve this kind of proximity [3,25]. However, the observed *first-order* proximity in real-world data is always not sufficient for preserving the global network structures, while many legitimate linkages in the real-world network are actually not observed, which can be denoted as *second-order* proximity. As a complement, many works explore the *second-order* proximity between the vertices, which is not determined through the observed linkage but through the shared neighborhood structures of the vertices [21,24]. Nevertheless, these graph embedding methods can not directly handle the user identification.

In this paper, we utilize the network information across different social platforms, which can be directed, undirected or weighted, to identify holistic unmatched user identities simultaneously. Specially, we construct the discriminative weighted graph by exploring the shared neighborhood structures of the vertices globally, and learn the transformation to make the transferred weighted graphs consistent. Meanwhile, a rank regularization is also proposed, and the implementation can be optimized effectively.

3 Proposed Method

3.1 Notations

Our method can predict the user identities across multiple networks from different social platform, and we consider the case of two networks for simplicity here. Suppose the two relation networks are represented as $X^1 \in \mathbb{R}^{d_1 \times d_1}$, i.e., the Facebook network, and $X^2 \in \mathbb{R}^{d_2 \times d_2}$, i.e., the Twitter network, where X_{ij}^v denotes the link value between the i-th instance and j-th instance of the v-th network, and $X_{ij}^v \neq 0$ if there is a linkage between i-th instance and j-th instance, $X_{ij}^v = 0$ otherwise, it is notable that X^v can be directed, undirected or weighted. Meanwhile, the problem in our setting can be seen as a transductive problem, we have N_1 matched pair-wise users denoted by $\{(\mathbf{x}_i^1, \mathbf{x}_i^2)\}$ in advance, i.e., the blue dotted lines, and N_2 unmatched pair-wise users, i.e., the red dotted lines, the number of all practical matched pair-wise users is $N = N_1 + N_2$, $N \leq \min(d_1, d_2)$, and GAEM aims to identify the unmatched user identities.

3.2 Graph-Aware Embedding (GAEM)

Naturally, the user linkage network X^v can be seen as a graph $G^v = (V^v, E^v)$, $V^v = \{\mathbf{x}_i^v\}_{i=1}^{d_v}$ corresponds to the set of vertices and $E^v = \{(\mathbf{x}_i^v, \mathbf{x}_j^v)\}$ denotes to the set of edges from $\mathbf{x}_i^v$ to $\mathbf{x}_j^v$ iff $X_{ij}^v \neq 0$. The observed links in the network can be considered as the local structures, which denote the *first-order* proximity between the vertices. However, the *first-order* proximity is insufficient for preserving the global network structures. As a complement, the *second-order* proximity between the vertices are used, which determined by shared neighborhood structures of the vertices, and the nodes with shared neighbors are likely to be similar.

Therefore, given the raw network X^v, inspired from [31], a weighted graph $\hat{G}^v = (V^v, \hat{E}^v, W^v)$ can be reconstructed to characterize the global structure of the raw relation network, $V^v = \{\mathbf{x}_i^v\}_{i=1}^{d_v}$ corresponds to the set of vertices as above, and the $\hat{E}^v = \{(\mathbf{x}_i^v, \mathbf{x}_j^v)\}_{\mathbf{x}_i^v \in KNN(x_j^v)}$ denotes the set of edges from $\mathbf{x}_i^v$ to $\mathbf{x}_j^v$ iff x_i^v is among the K-nearest neighbors of x_j^v. Furthermore, $W^v = [W_{ij}^v] \in \mathbb{R}^{d_v \times d_v}$ represents the nonnegative weight matrix, where $W_{ij}^v = 0$ iff $(\mathbf{x}_i^v, \mathbf{x}_j^v) \notin \hat{E}^v$. Meanwhile, the j-th column $W_{\cdot j}^v = \{W_{1j}^v, W_{2j}^v, \cdots, W_{d_v j}^v\}$ is determined by following problem:

$$\min_{W_{\cdot j}^v} \left\| \mathbf{x}_j^v - \sum_{(\mathbf{x}_i^v, \mathbf{x}_j^v) \in E^v} W_{ij}^v \mathbf{x}_i^v \right\|^2$$

$$s.t. \quad \sum_{(\mathbf{x}_i^v, \mathbf{x}_j^v) \in \hat{E}^v} W_{ij}^v = 1, W_{ij}^v \geq 0 \tag{1}$$

Conceptually, the W_{ij}^v characterizes the relative importance of neighbor example $\mathbf{x}_i^v$ in reconstructing x_j^v. Here the loss term can take any convex forms and we use linear least square loss here for simplicity.

With the embedded weighted matrix W^v in Eq. 1, and the known matched links $\{\mathbf{x}_i^1, \mathbf{x}_j^2\}$, we aim to predict the unmatched user identities. Specially, considering that different networks share the similar global structure, the objective is to learn a transfer matching matrix $M \in \mathbb{R}^{\min(d_v) \times \max(d_v)}$, in which $M_{ij} = 1$ iff the link $\{\mathbf{x}_i^1, \mathbf{x}_j^2\}$ is matched, $M_{ij} = 0$ otherwise. Without any loss of generalizations, we assume $d_1 \leq d_2$ in the remaining of this paper, thus $M \in \mathbb{R}^{d_1 \times d_2}$. The transformation of the larger weighted matrix can be written as $M(MW^2)^T$, note that the M transfers the disordered weighted matrix W^2 in rows and columns. After the transformation, the weighted matrices of different networks should be similar, which can be represented as $\ell(W^v, M)$.

In practical case, the ideal transformation M, gives identical outputs for consistent weighted matrix, and consequently make the rank of M equal to N, which is the number of practical matched pair-wise users, and we define $RC(M) = rank(M)$ here, it is notable that RC(F) reflects the prediction compatibility among the weighted matrices. Thus, rank consistency can be used as a regularization in the learning framework, which is helpful to achieve compatible and consistent transformation upon the achieved weighted matrix. The keys of the proposed method are the reconstructed weighted matrices, the transfer matching matrix and rank regularization, which boost the performance of weighted matrices construction and the learning transfer matching matrix simultaneously. Benefited from these, we can bridge the loss of weighted matrices construction and the gap of transferred weighted matrices in a unified framework:

$$\min_{W^v, M} \sum_v \ell_v(X^v, W^v) + \ell(W^v, M) + \lambda RC(M)$$

$$s.t. \quad M_{i,j} \in \{0, 1\}, \quad \sum_{(\mathbf{x}_i^v, \mathbf{x}_j^v) \in \hat{E}^v} W_{ij}^v = 1, W_{ij}^v \geq 0 \tag{2}$$

The first term $\ell_v(X^v, W^v)$ denotes the loss of the construction of each weighted matrix. Furthermore, W^1 and W^2 are disordered while the construction on each network is self-adaptive. The second term $\ell(W^v, M)$, is the loss of transferred weighted matrices, which leverages the consistency constraint of the weighted matrices on each linkage network. The last term $RC(M)$, is the rank regularization on transfer matrix M, which constrains the degree of freedom. $\lambda > 0$ is the balance parameter.

Specifically, objective function $\ell_v(X^v, W^v)$ in Eq. 2 can be generally represented as the form in Eq. 1, the $\ell(W^v, M)$ can be any convex loss function here, and we use square loss here for simplicity. Thus, the Eq. 2 can be re-formed as:

$$\min_{W^v, M} \sum_v \sum_{j=1}^{d_v} \|\mathbf{x}_j^v - \sum_{\mathbf{x}_i^v \in KNN(x_j^v)} W_{ij}^v \mathbf{x}_i^v\|^2$$

$$+ \|W^1 - M(MW^2)^T\|_F^2 + \lambda RC(M) \tag{3}$$

$$s.t. \quad M_{i,j} \in \{0, 1\}, \quad \sum_{(\mathbf{x}_i^v, \mathbf{x}_j^v) \in \hat{E}^v} W_{ij}^v = 1, W_{ij}^v \geq 0$$

3.3   Optimization

The 2nd term in Eq. 3 involves the product of the weighted matrix W^2 and the transformation M, which makes the formulation not joint convex. Consequently, the formulation cannot be optimized easily. We provide the optimization process below:

Fix M, Optimize W^v: According to the constraint $\sum_{(\mathbf{x}_i^v, \mathbf{x}_j^v) \in \hat{E}^v} W_{ij} = 1$, the Eq. 1 can be re-written as:

$$\min_{W_{\cdot j}^v} W_{\cdot j}^{v^T} G_j^v W_{\cdot j}^v$$

$$s.t. \quad \mathbf{1}^\top W_{\cdot j}^v = 1, W_{i,j} \geq 0$$

Here, $G_j^v \in \mathbb{R}^{d_v \times d_v}$ is the local Gram matrix for $\mathbf{x}_j$ with elements $(G_j^v)_{mn} = (\mathbf{x}_j - \mathbf{x}_m)^\top (\mathbf{x}_j - \mathbf{x}_n)$. Apparently, when M and W^2 are fixed, the 3rd term of Eq. 3 is not related to W^1, besides, it is NP-hard to directly learn the binary $M_{i,j} \in \{0, 1\}$, thus, we relax to $M_{i,j} \in [0, 1]$. Eventually, Eq. 3 can be equivalently written as:

$$\min_{W_{\cdot j}^1} W_{\cdot j}^{1^T} (G_j^1 + I) W_{\cdot j}^1 - 2M (MW^2)_{\cdot j}^{T^T} W_{\cdot j}^1$$

$$s.t. \quad M_{i,j} \in [0, 1], \mathbf{1}^\top W_{\cdot j}^1 = 1, W_{i,j} \geq 0 \tag{4}$$

Here I is the identity matrix with the same size of G_j^1. Equation 4 corresponds to a standard quadratic programming (QP) problem whose optimal solution can be obtained by any off-the-shelf QP solver. The weighted matrix W^1 is constructed by solving column-wisely. Similarly, when M and W^1 are fixed, the Eq. 3 can be equivalently written as:

$$\min_{W_{\cdot j}^2} W_{\cdot j}^{2^T} (G^2 + \hat{M}) W_{\cdot j}^2 - 2 (M^T W^1 M)_{\cdot j}^T W_{\cdot j}^2$$

$$s.t. \quad M_{i,j} \in [0, 1], \mathbf{1}^\top W_{\cdot j}^2 = 1, W_{i,j} \geq 0 \tag{5}$$

Where $\hat{M} = M^T M M^T M$. Equation 5 also corresponds to a standard quadratic programming (QP) problem as Eq. 4.

From the aspect of weighted matrix construction, we treat the extra transferred weighted matrix as supervision to help to construct the discriminative weighted matrix W^v, i.e., we consider the transferred weighted matrices should have consistently global structure in this step.

Fix W^v, Optimize M: Apparently, when W^v are fixed, the 1st term of Eq. 3 is not related to M, thus Eq. 3 can be equivalently written as:

$$\min_M ||W^1 - M(MW^2)^T||_F^2 + \lambda RC(M)$$

$$s.t. \quad M_{i,j} \in [0, 1] \tag{6}$$

Note that the rank norm minimization is NP-hard, and inspired by [6], the nuclear norm usually acts as a convex surrogate. Specifically, given a matrix

$X \in \mathbb{R}^{m \times n}$, its singular value are assumed as $\sigma_i, i = 1, \cdots, \min(m,n)$, which are ordered from large to small. Thus, the nuclear norm can be defined as $\|X\|_* = \sum_{i=1}^{\min(m,n)} \sigma_i$, and the nuclear norm has been widely used in various scenarios in rank norm minimization problem [10].

Nevertheless, blindly minimize the rank will break the natural structure of M. Therefore, a directional optimization approach, which conduct the RC(M) until converging to N during the minimization procedure is desired. According to [2,27], we use truncated nuclear norm as a surrogate function of the $RC(M)$ operator:

Definition 1. *Given a matrix $X \in \mathbb{R}^{m \times n}$, the truncated nuclear norm $\|X\|_r$ is defined as the sum of $\min(m,n) - r$ minimum singular values, i.e., $\|X\|_r = \sum_{i=r+1}^{min(m,n)} \sigma_i(X)$.*

Different from traditional nuclear norm minimization, which preserves all the singular values, truncated nuclear norm minimizes the singular values with first r largest ones unchanged, which is more close to the true rank definition. Specially, if $\|X\|_r = 0$, there are only r non-zero singular values for X, and this explicitly indicates the rank of X is less than or equals to r. Practically, in order to impel the RC(F) directional to the practical matched users, it is clear to set $r = N$.

The truncated nuclear norm can be re-formulated as the equivalent form by the following theorem [11]:

Theorem 1. *Given a matrix $X \in \mathbb{R}^{m \times n}$ and any non-negative integer r ($r \leq \min(m,n)$), for any matrix $A \in \mathbb{R}^{r \times m}$ and $B \in \mathbb{R}^{r \times n}$ such that $AA^\top = I_r, BB^\top = I_r$, where $I_r \in \mathbb{R}^{r \times r}$ is identity matrix. Truncated nuclear norm can be reformulated as:*

$$\|X\|_r = \|X\|_* - \max \mathrm{Tr}(AXB^\top) \tag{7}$$

If the singular value decomposition of matrix X is $X = U\Sigma V^\top$ where Σ is the diagonal matrix of singular values sorted in descending order and $U \in \mathbb{R}^{m \times n}, V \in \mathbb{R}^{n \times n}$. The optimal solution for the trace term in the above equation has a closed form solution: $A = (\mathbf{u}_1, \mathbf{u}_2, \cdots, \mathbf{u}_r)^\top$ and $B = (\mathbf{v}_1, \mathbf{v}_2, \cdots, \mathbf{v}_r)^\top$, corresponds to the first r columns of left and right singular vectors.

With Theorem 1, we can reformulate the Eq. 6 as:

$$\arg \min_M \|W^1 - M(MW^2)^T\|_F^2 + \lambda(\|M\|_* - maxTr(AMB^T))$$
$$s.t. \quad A^T A = I, B^T B = I \tag{8}$$

Because of the non-convexity of truncated nuclear norm, alternative approach can be utilized for the optimization. A simple solution to Eq. 8 is alternating descent method. We can fix M and optimize A, B via SVD on M first, and then fix A and B to optimize M. When A and B are fixed, the subproblem is convex.

A and B can be obtained by SVD on M, which are the left and right singular vectors corresponding to the maximum N singular values. As the number of actually required singular vectors is rather small, partial SVD can be used [4]. The most computational cost step, however, is the subproblem for solving M in Eq. 8. We will give a detailed investigation on employing Accelerate Proximal Gradient Descent Method (APG) [2] for solving this subproblem in the following.

Note that when A and B are fixed, the problem is composed of two convex parts, i.e., a smooth loss term $P_1(M)$ and a non-smooth trace norm $P_2(M)$:

$$P_1(M) = L(M) - \text{Tr}(AMB^\top), P_2(M) = \|M\|_* \tag{9}$$

APG is suitable for solving Eq. 9 [12], which optimizes on a linearized approximation version of the original problem. In the t−th iteration, if we denote the current optimization variable as M^t, then we can linearize the smooth part $P_1(\cdot)$ respect to M^t as:

$$\begin{aligned} Q(M) &= P_1(M^t) + \text{Tr}(<\nabla P_1(M^t), M - M^t>) + \frac{L}{2}\|M - M^t\|_F^2 + P_2(F) \\ &= L(M) + \frac{L}{2}\|M - M^t\|_F^2 + \lambda\|M\|_* - \text{Tr}(AMB^\top) \\ &\quad + \text{Tr}(<\nabla P_1(M^t), M - M^t>) \end{aligned} \tag{10}$$

where $\nabla P_1(M^t) = -((W^1 - M^t(M^tW^2)^\top)M^t(W^2 + W^{2^\top}) + \lambda A^\top B)$. Here L is the Lipschitz coefficient, which can be estimated by line search strategy [2]. Minimizing Q(M) w.r.t. M is equivalent to solving:

$$\hat{M} = \arg\min_M \lambda\|M\|_* + \frac{L}{2}\|M - (M^t - \frac{1}{L}\nabla P_1(M^t))\|_F^2 \tag{11}$$

APG updates the optimal solution in Eq. 11 at each iteration. Given the following theorem [6] about the proximal operator for nuclear norm:

Theorem 2. *For each $\tau \geq 0$ and $Y \in \mathbb{R}^{m \times n}$, we have*

$$D_\tau(Y) = \arg\min_X \frac{1}{2}\|X - Y\|_F^2 + \tau\|X\|_* \tag{12}$$

Here, $D_\tau(Y)$ is a matrix shrinkage operator for matrix Y, which can be calculated by SVD of Y. If SVD of Y is $Y = U\Sigma V^\top$, then

$$D_\tau(Y) = UD_\tau(\Sigma)V^\top \quad D_\tau(\Sigma) = diag(max(\sigma_i - \tau, 0)). \tag{13}$$

we can solve Eq. 11 in a closed form:

$$\hat{M} = \mathfrak{D}_L(M^t) \stackrel{\text{def}}{=} D_{\frac{\lambda}{L}}(M^t - \frac{1}{L}\nabla P_1(M^t)) \tag{14}$$

4 Experiment

4.1 Datasets and Configurations

Data Sets: In this paper, we use the datasets from webpage networks and the social networks in our empirical investigations.

The WebKB dataset [5] contains webpages collected from 4 universities: Wisconsin, Washington, Cornell and Texas (denoted as Wins., Wash., Corn. and Texas in tables) and described with two networks: the content and the citation. The content represents the documents-words matrix, containing 0/1 values, and can be indicated as the similarity relationships between the documents. On the other hand, the citation denotes the number of citation links between documents, which can be acted as another network structure. The ground-truth mapping of WebKB across these two networks is in the documents-mapping. To demonstrate the generalization ability, we also demonstrate our method on different social networks. The social networks collection consists of four popular online social networking sites: LiveJournal (LJ), Flickr (FL), Last.fm (LF), and MySpace (MS) as [32]. We use the linked user accounts dataset from [7,20] as the ground truth. The data was originally collected by [20] through Google Profiles service by allowing users to integrate different social network services. Five subsets are constructed from social networks, i.e., flickr-lastfm; flickr-myspace; livejournal-lastfm, livejournal-myspace and livejournal-flickr.

For all datasets in our experiments, we randomly select $\{20\%, 40\%, 60\%, 80\%\}$ for matched pair-wise examples, and the remains are unmatched for prediction. We repeat this for 30 times, the acc. and std. of predictions are recorded as classification performance. The parameter λ in the training phase is tuned in $\{10^{-2}, \cdots, 10^2\}$. The number of k-nearest neighbors is set 15. Empirically, when the variations between the objective value of Eq. 3 is less than 10^{-5} in iteration, we treat GAEM converged.

Compare Algorithms: Our method solves the problem of user identities identification across networks. Thus, we choose five state-of-the-art user linkage identify classifiers: HYDRA [16], COSNET [32], ULink [17], NS [18], IONE [15]. Note that the HYDRA and COSNET utilize both the profile information and network structure, and we consider the raw network structure information as the profile features in our setting, on the other hand, ULink is difficult to handle the sparse network information, and we use the embedded features by Isomap as the input. Moreover, NS and IONE directly take the network structure as input. Besides, our method is also related to graph embedding. Thus, we construct the weighted matrix with four graph embedding methods: Baseline (BL), Isomap [25], Deepwalk [21], Weight [31], the first two methods, i.e., BL, Isomap, only consider the local structure, while remaining two methods, i.e., Deepwalk, Weight, consider the global structure. It is notable that these methods can not unify the weighted matrix construction and transformation together, thus, we calculates the weighted matrices and then optimize the transformation M separately.

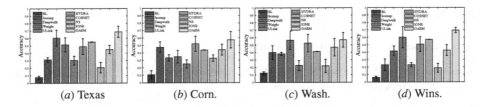

Fig. 1. Webpage linkage comparison for batch data setting (with ratio between between the ground-truth matching pairs to non-matching pairs being 4:1)

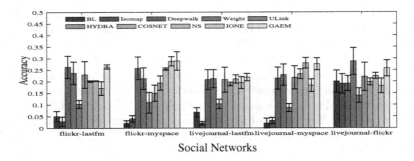

Fig. 2. Social Networks comparison for batch data setting (with ratio between the ground-truth matching pairs to non-matching pairs being 3:1)

4.2   Experiment Results

Multiple-Network Identity Identification: To demonstrate the effectiveness of our proposed method. For both the webpage networks and social networks datasets, We fix the ratio of the number of matched pair-wise users at 80% firstly, and record the avg. $\pm$ std. of the GAEM and compared methods in Figs. 1 and 2.

Figure 1 clearly reveals that on all webpage datasets, the average accuracies of GAEM are the best. Further more, while comparing to the graph embedding methods, the methods considering the global structure proximity, i.e., Deep-Walk, Weight, are superior to the local structure proximity based methods, i.e., BL, Isomap, on the majority of datasets, which indicates that global structure proximity is more efficient and different social networks share the similar global structure, and it confirms to the real significance. On the other hand, the performance of previous user identity linkage methods are not well performed, note that these methods require the specific collected or designed features, specifically, ULink is a supervised method, which can not utilize the network structure information; HYDRA maximizes the structure consistency by modeling the core social network behavior consistency, and the performance hinges heavily upon the availability of the consistent structure, where the consistency is calculated by extra profile features, it performs worse than GAEM in the cases where raw features can not possess consistent structure, COSNET is found to be in a similar situation as HYDRA, as for NS and IONE methods, these methods also require additional information and are sensitive to the parameters. Thus, GAEM is well performed considering the global structure proximity with the network information. To demonstrate the generalization ability, we conduct more experiments, Fig. 2 records the prediction accuracies (avg. ± std.) of the GAEM and compared methods on five social network datasets, and Fig. 2 reveals that on the social network datasets, the average accuracies of GAEM are also competitive with the compared methods, the average accuracies are the best on three datasets, i.e., flickr-lastfm; flickr-myspace; livejournal-lastfm.

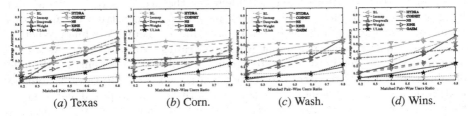

(a) Texas   (b) Corn.   (c) Wash.   (d) Wins.

Fig. 3. Influence of number of matched pair-wise users of Webpage linkage comparison

Influence of Number of Matched Pair-Wise Users: In order to explore the influence of the number of initial matched pair-wise users on performance, more experiments are conducted. In this section, the parameters in each investigation are fixed as the optimal values, the λ in GAEM is set 1, while the ratio of initial matched pair-wise users varies in $\{20\%, 40\%, 60\%, 80\%\}$. Due to the page limits, results on only 4 datasets, i.e., Wins., Wash., Corn., and Texas, and the results are recorded in Fig. 3. From these figures, it clearly shows that GAEM achieves the best performance when the ratio is larger than 40% on most datasets. Besides, we can also find that GAEM achieves an optimal performance fast, and the accuracy of GAEM increases faster than compared methods.

Empirical Investigation on Convergence: To investigate the convergence empirically, the objective function value, i.e., the value of Eq. 3 and the classification performance of GAEM in each iteration are recorded. Due to the page limits, results on only 4 datasets mentioned above, are plotted in Fig. 4. It clearly reveals that the objective function value decreases as the iterations increase, and the classification performance is stable after several iterations. Moreover, these additional experimental results indicate that our GAEM can converge very fast, i.e., GAEM converges after 3 rounds.

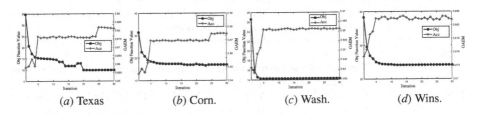

(a) Texas   (b) Corn.   (c) Wash.   (d) Wins.

Fig. 4. Objective function value convergence and corresponding classification accuracy vs. number of iterations of GAEM with matched pair-wise users ratio at 80%

5   Conclusion

The user identification problem in the multi-network environment is a challenging problem. Previous efforts mainly focus on the predefined profile or content features in the learning approaches, while leaving the data incompleteness and sparsities unconsidered. These approaches, meanwhile, are difficult to handle the information of structures provided by multi-networks. In this paper, we propose the Graph-Aware Embedding (GAEM) approach, which utilizes the more general social networks information and identifies the accounts of the same user by exploiting useful information from the networks. We construct the more discriminative weighted graph instead of the raw linkage network, while predicting the transformation among different weighted graphs simultaneously. As a consequence, we can get more accurate predictions of the user identities directly obtained from the learned transformation matrix, experimental evaluations on real-world applications demonstrate the superiority of our proposed method over the compared methods. How to extend multiple platforms and the scalability with improved performance are interesting future works.

References

1. Backstrom, L., Leskovec, J.: Supervised random walks: predicting and recommending links in social networks. In: WSDM, pp. 635–644 (2011)
2. Beck, A., Teboulle, M.: A fast iterative shrinkage-thresholding algorithm for linear inverse problems. SIIMS **2**(1), 183–202 (2009)
3. Belkin, M., Niyogi, P.: Laplacian eigenmaps and spectral techniques for embedding and clustering. In: NIPS, pp. 585–591 (2001)
4. Berkhin, P.: A survey on pagerank computing. Internet Math. **2**(1), 73–120 (2005)
5. Blum, A., Mitchell, T.: Combining labeled and unlabeled data with co-training. In: COLT, pp. 92–100 (1999)
6. Cai, J.F., Candes, E.J., Shen, Z.: A singular value thresholding algorithm for matrix completion. SIOPT **20**(4), 1956–1982 (2008)
7. Chen, W., Liu, Z., Sun, X., Wang, Y.: A game-theoretic framework to identify overlapping communities in social networks. Data Min. Knowledge Disc. **21**(2), 224–240 (2010)
8. Deng, Z., Sang, J., Xu, C.: Personalized video recommendation based on cross-platform user modeling. In: ICME, pp. 1–6 (2013)
9. Han, S., Xu, Y.: Link prediction in microblog network using supervised learning with multiple features. J. Comput. Phys. **11**(1), 72–82 (2016)
10. Harchaoui, Z., Douze, M., Paulin, M., Dudik, M., Malick, J.: Large-scale image classification with trace-norm regularization. In: CVPR, pp. 3386–3393 (2012)
11. Hu, Y., Zhang, D., Ye, J., Li, X., He, X.: Fast and accurate matrix completion via truncated nuclear norm regularization. TPAMI **35**(9), 2117 (2013)
12. Ji, S., Ye, J.: An accelerated gradient method for trace norm minimization. In: ICML, pp. 457–464 (2009)
13. Joachims, T.: Transductive inference for text classification using support vector machines. In: ICML, pp. 200–209 (1999)
14. Liu, J., Zhang, F., Song, X., Song, Y.I., Lin, C.Y., Hon, H.W.: What's in a name?: an unsupervised approach to link users across communities. In: ICWSM, pp. 495–504 (2013)

15. Liu, L., Cheung, W.K., Li, X., Liao, L.: Aligning users across social networks using network embedding. In: IJCAI, pp. 1774–1780 (2016)
16. Liu, S., Wang, S., Zhu, F., Zhang, J., Krishnan, R.: Hydra: large-scale social identity linkage via heterogeneous behavior modeling. In: SIGMOD, pp. 51–62 (2014)
17. Mu, X., Zhu, F., Wang, J., Zhou, Z.H.: User identity linkage by latent user space modelling. In: SIGKDD, pp. 1775–1784 (2016)
18. Narayanan, A., Shmatikov, V.: De-anonymizing social networks. In: SP, Oakland, California, pp. 173–187 (2009)
19. Nie, Y., Jia, Y., Li, S., Zhu, X., Li, A., Zhou, B.: Identifying users across social networks based on dynamic core interests. Neurocomputing 210, 107–115 (2016)
20. Perito, D., Castelluccia, C., AliKaafar, M., Manils, P.: How unique and traceable are usernames? In: PET, pp. 1–17 (2011)
21. Perozzi, B., Al-Rfou, R., Skiena, S.: Deepwalk: Online learning of social representations. In: SIGKDD, pp. 701–710 (2014)
22. Shu, K., Wang, S., Tang, J., Zafarani, R., Liu, H.: User identity linkage across online social networks: a review. SIGKDD Explor. 18(2), 5–17 (2016)
23. Tan, S., Guan, Z., Cai, D., Qin, X., Bu, J., Chen, C.: Mapping users across networks by manifold alignment on hypergraph. In: AAAI, pp. 159–165 (2014)
24. Tang, J., Qu, M., Wang, M., Zhang, M., Yan, J., Mei, Q.: LINE: large-scale information network embedding, pp. 1067–1077 (2015)
25. Tenenbaum, J.B., de Silva, V., Langford, J.C.: A global geometric framework for nonlinear dimensionality reduction. Science 290(5500), 2319–2323 (2000)
26. Wei, Y., Singh, L.: Using network flows to identify users sharing extremist content on social media. In: Kim, J., Shim, K., Cao, L., Lee, J.-G., Lin, X., Moon, Y.-S. (eds.) PAKDD 2017. LNCS (LNAI), vol. 10234, pp. 330–342. Springer, Cham (2017). https://doi.org/10.1007/978-3-319-57454-7_26
27. Ye, H.J., Zhan, D.C., Miao, Y., Jiang, Y., Zhou, Z.H.: Rank consistency based multi-view learning: a privacy-preserving approach. In: CIKM, pp. 991–1000 (2015)
28. Zafarani, R., Liu, H.: Connecting corresponding identities across communities. In: ICWSM, pp. 354–357 (2009)
29. Zhang, J., Kong, X., Yu, P.S.: Transferring heterogeneous links across location-based social networks. In: WSDM. pp. 495–504 (2014)
30. Zhang, J., Yu, P.S.: Integrated anchor and social link predictions across social networks. In: IJCAI, pp. 1620–1626 (2015)
31. Zhang, M.L., Zhou, B.B., Liu, X.Y.: Partial label learning via feature-aware disambiguation. In: SIGKDD, pp. 1335–1344 (2016)
32. Zhang, Y., Tang, J., Yang, Z., Pei, J., Yu, P.S.: Cosnet: Connecting heterogeneous social networks with local and global consistency. In: SIGKDD, pp. 1485–1494 (2015)
33. Zhou, X., Liang, X., Zhang, H., Ma, Y.: Cross-platform identification of anonymous identical users in multiple social media networks. TKDE 28(2), 411–424 (2016)

Knowledge-Based Recommendation with Hierarchical Collaborative Embedding

Zili Zhou[1,2], Shaowu Liu[1], Guandong Xu[1(✉)], Xing Xie[3], Jun Yin[1], Yidong Li[4], and Wu Zhang[2]

[1] Advanced Analytics Institute, University of Technology Sydney, Ultimo, Australia
{zili.zhou,jun.yin-2}@student.uts.edu.au,
{shaowu.liu,guandong.xu}@uts.edu.au
[2] School of Computer Engineering and Science, Shanghai University, Shanghai, China
wzhang@shu.edu.cn
[3] Microsoft Research Asia, Beijing, China
xingx@microsoft.com
[4] School of Computer and Information Technology, Beijing Jiaotong University, Beijing, China
ydli@bjtu.edu.cn

Abstract. Data sparsity is a common issue in recommendation systems, particularly collaborative filtering. In real recommendation scenarios, user preferences are often quantitatively sparse because of the application nature. To address the issue, we proposed a knowledge graph-based semantic information enhancement mechanism to enrich the user preferences. Specifically, the proposed **H**ierarchical **C**ollaborative **E**mbedding (**HCE**) model leverages both network structure and text info embedded in knowledge bases to supplement traditional collaborative filtering. The **HCE** model jointly learns the latent representations from user preferences, linkages between items and knowledge base, as well as the semantic representations from knowledge base. Experiment results on *GitHub* dataset demonstrated that semantic information from knowledge base has been properly captured, resulting improved recommendation performance.

1 Introduction

Recommendation has been widely used in today's business. By observing user past behaviors, recommender systems can identify items with potential to be interested by users. A popular technique in recommendations is collaborative filtering (CF) which is based on the intuition that preference history can be transferred across like-minded users. However, CF suffers from the cold-start problem in which users usually provide limited amount of preferences, i.e., preference data is quantitatively sparse, making recommendation inaccurate. Particularly, in some real recommendation scenarios, user preferences are often quantitatively sparse because of the application nature. For example, unlike watching many movies, users typically can only study a few subjects in *Coursera*[1] or contribute to a few repositories (projects) on *GitHub*[2], both of which contain very high numbers of subjects or projects.

To address the sparsity problem, researchers have proposed to extend the spare data by connecting to external knowledge graphs [1–4]. This approach leverages both network structure and text info embedded in knowledge bases to supplement traditional CF. To be specific, a knowledge base is a data repository containing interlinked entities across different domains. Since knowledge base is often represented in a graph way, it is also called as knowledge graph (KG). The beauty of knowledge graph is not only the textual knowledge representations, but also the linked structure of knowledge entities. Recently, knowledge graph has emerged as a new method in recommender systems research. For example, latent features are often extracted from heterogeneous information network to represent users and items [2–4]. More recently, Zhang et al. [1] proposed the first work to build a hybrid heterogeneous information network containing both recommender system and knowledge graph.

On the other hand, although the above mentioned recommendation tasks exhibit the significant sparsity, we argue that the user choices/behaviors carry on rich semantics info which has not been fully utilized in recommendation. For example, knowing a user's interest in a subject or repository reveals lots of information about this user, such as preferences over *programming language, operating system, field of study, research topic*. From knowledge graph view, such information pieces are not isolated and fragmented, instead, interrelated, forming a comprehensive view of this author. This rich semantic information can play an important role in alleviating such cold-start and sparsity problem, therefore using KG-based approaches becomes an idea solution to this kind of tasks. However, previous studies on KG-based CF suffer from one or more of the following limitations: (1) rely on tedious feature engineering; (2) the high data sparsity; and (3) recommended items need to be one exact entity within knowledge base.

To address the above issues, we propose a novel collaborative recommendation framework to integrate recommender system and knowledge graph with extensible connection between items and knowledge entities. Overall, our method constructs a multi-level network via knowledge graph to enhance sparse semantic information between users and items. Let's take example of *GitHub* recommendation task shown in Fig. 1 to show how our model works. In order to reveal the latent correlations between *GitHub* repositories, the names of which are not existent in knowledge graph, we frame the integrated system into 3-level, where users, repositories, knowledge graph entities are placed in different levels, with edges between users and repositories indicating the user has interest in the repository, edges between repositories and entities indicating that the entity is

---

[1] Online course platform https://www.coursera.org/.

[2] Project hosting platform https://github.com/.

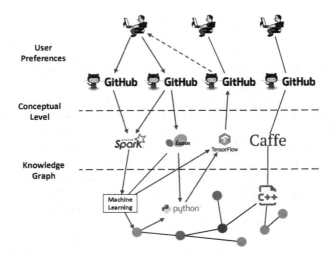

Fig. 1. Conceptual level

possibly related to the repository, and edges between entities meaning there is at least one specific relation between this pair of entities. A hierarchical structure heterogeneous network, which contains multiple types of nodes and multiple types of edges, is built for automatic collaborative learning. Particularly, to link recommender system and knowledge graph properly, knowledge conceptual level is proposed to indirectly map item-entities, different from previous works of direct mapping. Serving as middle level of three-level hierarchical structure model, the knowledge conceptual level can fully interconnect the whole system in a proper way, tackling the restriction that recommendation items need to be within knowledge base.

The main contributions of this paper are as follows:

- A novel KG-based recommender system with knowledge conceptual level is proposed to properly encode the correlation amongst items which are non-existent knowledge graph entities.
- A new collaborative learning algorithm is devised to deal with the proposed three-level network for sparse user preference data.
- We conducted extensive experiments on *GitHub* recommendation task, which is extremely sparse but rich semantics in user preference, to evaluate the effectiveness of our model. To the best of our knowledge, this is the first trial work of using knowledge graph embedding, to deal with semantic enhancement for entities not existent in conventional knowledge graph.

The rest of the paper first introduces the basic concepts of collaborative filtering and knowledge graph, followed by a detailed discussion of the proposed hierarchical collaborative embedding model. The proposed model is compared with several baselines on *GitHub* dataset.

2   Preliminary

This section briefly summarizes necessary background of *Implicit Feedback Recommendation* and *Knowledge Graph* that form the basis of this paper.

2.1   Implicit Feedback Recommendation

This paper considers the implicit feedback recommendation problem [5], i.e., analyzing interactions among users and items instead of explicit ratings. The implicit user feedback is encoded as a matrix $\mathbf{R} \in \mathbb{R}^{m*n}$, where $R_{ij} = 1$ if user i has interacted with item j and $R_{ij} = 0$ otherwise. The user-item interactions are defined per application scenario, e.g., a *GitHub* user "stars" (follows) a repository, or a *Coursera* user "enrolled" in a subject. Generally speaking, an interaction $R_{ij} = 1$ implies the user is interested in the item, however, the meaning of $R_{ij} = 0$ is not necessarily to be not interested. In fact, the matrix $\mathbf{R}$ is often sparse and most entries will be 0, where the 0 value indicates that the user either has no interest in the item or has interest but not interacted with the item yet. The goal of implicit feedback recommendation is to identify which 0 entries in $\mathbf{R}$ have the potential to become 1.

2.2   Knowledge Graph

The implicit feedback matrix $\mathbf{R}$ can be extreme sparse as some users may have only interacted with one or two items. Although modeling moderately sparse data has been considered by traditional CF methods, it remains a challenging problem of utilizing extreme sparse data. Fortunately, if the items contain rich semantic information, then only a few items will be able to connect the user to knowledge graph, such that more complete user profiles can be built. To be specific, knowledge graph is a semantic web consist of *entities* and *relations*, where entities represent anything in the world including people, things, events, etc., and relations connect entities that have interactions with each other. For example, in *GitHub* repository recommendation, entities can be software development concepts such as programming language $C++$, operating system *Linux*, development framework *TensorFlow*, etc. The entities are connected through relations such as "is programming language of", "have dependency on", "is operating system of", etc. Denoting entities as nodes and relations as edges, knowledge graph can be represented by a heterogeneous network with multiple types of nodes and multiple types of edges.

Although using knowledge graph in recommendation is promising, it is assumed that the recommended items are entities in knowledge graph. This assumption may hold for recommending movies or tourism destinations where the items are already entities in knowledge graph, but it becomes invalid for items that are non-existent in knowledge graph, such as repositories in *GitHub*. Therefore, the link between non-existent items and knowledge graph entities must be identified together with reliability and importance measures.

For ease of reference, notations used throughout this paper are summarized as following. U_i represents vector of user i; V_j represents vector of item j; E_i represents vector of entity i; B_j represents bias vector of item j; W_k represents weigh of entity k; I_j represents possibly related entities set of item j; R represents factorized matrix of relation r; r represents vector of relation r; M_r represents subspace mapping matrix of relation r; $p(i, j, j')$ represents preference function of triple (user i, item j, item j'); $\mathcal{X}_{i,j,j'}$ represents user preference term of training function; $\mathcal{Y}_{h,r,t,t'}$ represents knowledge graph embedding term of training function; $\mathcal{Z}$ represents regularization term of training function; f_r represents knowledge graph triple score function used in training function; f_r^{TransR} represents TransR score function; f_r^{RESCAL} represents RESCAL score function.

2.3   Problem Definition

The research problem of this paper can be defined as follows: given quantitatively sparse but semantically dense user feedback data, how to leverage knowledge graph to perform semantic enhancement for items that do not exist in knowledge graph, such that the item recommendation quality can be improved.

3   Hierarchical Collaborative Embedding

In this section, we propose the **H**ierarchical **C**ollaborative **E**mbedding model (**HCE**) to bridge knowledge graph to CF, which jointly learns the embedding of elements, including users, items, entities, and relations.

3.1   Knowledge Graph Structured Embedding

With large amount of knowledge being extracted from open source, knowledge graph was proposed to store the knowledge with graph structure. The knowledge facts are represented by triples, each triple has two entities (head entity and tail entity) and one relation in between. Given all triples, the entities and relations can be considered as nodes and edges, respectively, resulting a large scale of heterogeneous knowledge graph. To capture the latent semantic information of entities and relations, several embedding based methods [6–9] were proposed. These methods embed entities and relations into a continuous vector space, in which the latent semantic information can be reasoned automatically according to vector space position of entities and relations.

Two state-of-the-art knowledge graph structured embedding methods are employed in this paper: RESCAL [8] and TransR [9]. One important advantage of these two methods is the capability of modeling multi-relational data where more than one relation may exist between two entities.

RESCAL uses three-way tensor to represent triples set, each element of a triple (head entity, relation, or tail entity) is represented by one dimension, and tensor factorization is used to obtain the entity and relation representations. To be specific, each entity is represented by a vector and each relation is represented by a matrix. Y is the three-way tensor which represent all the triples, Y_k is a matrix picked up from Y, it only contains triples with relation k. E_i and E_j are the representation vectors of entity h and t, W_k is the representation matrix of relation r.

The representations of entities and relations are constructed by minimizing the following objective function:

$$\min_{E,W_k} \sum_k \|Y_k - EW_k E^T\|_F^2. \tag{1}$$

In a triple (h, r, t), each entity is represented by a vector, E_h for head entity h, E_t for tail entity t, and relation r is represented by a matrix R. The RESCAL score function of a triple (h, r, t) is defined as:

$$f_r^{\text{RESCAL}}(h, t) = \|E_h R E_t\|_2^2, \tag{2}$$

TransR uses a different score function for triples. Given a triple (h, r, t), the head and tail entities are represented by vectors E_h and E_t, respectively. Each relation is represented by a vector $\mathbf{r}$ together with a matrix M_r. TransR firstly maps entity h and t into subspace of relation r by using matrix M_r:

$$E_h^r = M_r E_h, E_t^r = M_r E_t. \tag{3}$$

The score function of TransR is defined as follows:

$$f_r^{\text{TransR}}(h, t) = \|E_h^r + \mathbf{r} - E_t^r\|_2^2. \tag{4}$$

In learning process, we pick up a true triple (h, r, t) and generate a false triple by replacing one entity of the triple by another entity: (h, r, t'). Then we make the score value of true triple larger than that of false triple: $f_r(h, t) > f_r(h, t')$.

3.2   Knowledge Conceptual Level Connection

This work focuses on recommender systems without direct connection to knowledge graph, i.e., most recommendation items do not exist in knowledge graph. For example, a *GitHub* project with a customized name is not an entity in knowledge graph. Consequently, methods such as [1] that rely on direct mapping between items and knowledge graph entities are not applicable. However, by extracting content information from items, such as item description and user reviews, potential links between items and entities can be constructed. To bridge recommender system and knowledge graph with item-entity links, we propose a collaborative learning model with hierarchical structure of three levels: the recommender system level, the knowledge graph level, and the knowledge conceptual level (KCL). The KCL plays a key role in the model to connect the other two levels and enables collaborative learning.

Creating the knowledge conceptual level has two challenges. The first challenge is how to filter irrelevant linkages. The automated extraction of item content introduces lots of irrelevant information for the recommendation task.

For example, in *GitHub* project recommendation, the project description may include vocabulary of specific areas such as biology and chemistry, which are off-topic of general purpose coding recommendation. While this information is irrelevant, it is actually linked to knowledge graph entities, thus introducing noise data.

The second challenge is how to measure the influences of knowledge graph entities on recommendation items. The entities have different influences on items, thus the links between items and entities must be weighed in order to represent an item precisely. To tackle these two challenges, the proposed Knowledge Conceptual Level implements the filtering and weighing functionalities. To be specific, a weighed link function is used to represent each item with their possibly related entities (automatically extracted from side information). The representation of an item is the weighted sum of vectors of possibly related entities plus a bias term B_j. Maximizing the weighed link function of each item is one of the targets in collaborative learning process. The weighed link function of an item is defined as follows:

$$V_j = B_j + \sum_{k \in I_j} W_k E_k \tag{5}$$

where V_j is the representation of item j, E_k is the representation of entity k, W_k is the weigh parameter of entity k, and I_j is the set containing all the entities which are possibly related to item j. If the entity is unrelated to current recommendation task, the weigh parameter should be lowered to near zero during learning process, if the influential degree of the entity is minor, the weigh parameter should be lowered accordingly. The filtering and weighing are both achieved by knowledge conceptual level.

3.3 Collaborative Learning

To integrate recommender system with knowledge graph, the proposed collaborative learning framework learns the embedding representations of both recommender system elements (users and items) and knowledge graph elements (entities and relations).

Because of user feedback is implicit, similar to some previous works [1,10], we use pairwise ranking of items in our learning approach. Given user i, item j and item j', using $F_{i,j}$ to represent the feedback of user i for item j, if $F_{i,j} = 1$ and $F_{i,j'} = 0$, then we consider user i prefer item j over item j', we use preference function $p(i, j, j')$ to represent this pairwise preference relation, and $p(i, j, j') > 0$. More specifically, in our model, we use same-dimension vector representation for user and item, the preference function is defined as following,

$$p(i, j, j') = \ln \sigma(U_i^T V_j - U_i^T V_{j'}) \tag{6}$$

U_i is the vector representing user i, V_j is the vector representing item j, $V_{j'}$ is the vector representing item j', σ is sigmoid function.

Algorithm 1. *HCE* Algorithm

Input: User preferences, Knowledge graph, Item&entity links.
Training:
Step 1: Draw pairwise user-entity triple set $\mathcal{D}$.
Step 2: Repeat
for each $(u_i, v_j, v_{j'}) \in \mathcal{D}$ **do**
    Draw pairwise entity-relation quadruple set $\mathcal{S}_{j,j'}$
    Draw possibly related entities set I_j and $I_{j'}$
    for each $(h, r, t, t') \in \mathcal{S}_{j,j'}$ **do**
        Represent item by embedding of entities:
        $V_j = B_j + \sum_{k \in I_j} W_k * E_k$
        $V_{j'} = B_{j'} + \sum_{k \in I_{j'}} W_k * E_k$
        Compute interaction of user and item:
        $\mathcal{X}_{i,j,j'} = \ln \sigma(U_i^T V_j - U_i^T V_{j'})$
        Compute score of entity-relation quadruple:
        $\mathcal{Y}_{h,r,t,t'} = \ln \sigma(f_r(h, t) - f_r(h, t'))$
        Compute regularization:
        $\mathcal{Z} = \|U_i\|, \|E_{\{h,t,t'\}}\|, \|R_r\|, \|B_{\{j,j'\}}\|, \|W_{\{j,j'\}}\|$
        maximize $\mathcal{X}_{i,j,j'} + \mathcal{Y}_{h,r,t,t'} + \mathcal{Z}$
Predictions:
for each $u_i \in U$ **do**
    Recommend items for user i in order
    $j_1 > j_2 > ... > j_n$ $(U_i^T V_{j1} > U_i^T V_{j2} > ... > U_i^T V_{jn})$.

Integrating knowledge graph embedding and knowledge conceptual level, the collaborative learning leverage the information from both user feedback and knowledge graph. by repeating following procedure. Jointly, we aim to maximize the likelihood function in Eq. 7 and the overall learning algorithm is summarized in Algorithm 1.

$$
\begin{aligned}
\mathcal{L} &= \sum_{(i,j,j') \in D} \mathcal{X}_{i,j,j'} + \sum_{(h,r,t,t') \in S} \mathcal{Y}_{h,r,t,t'} + \mathcal{Z} \\
\mathcal{X}_{i,j,j'} &= \ln \sigma(U_i^T V_j - U_i^T V_{j'}) \\
\mathcal{Y}_{h,r,t,t'} &= \ln \sigma(f_r(h, t) - f_r(h, t')) \\
\mathcal{Z} &= \frac{\lambda_U}{2} \|U\|_2^2 + \frac{\lambda_E}{2} \|E\|_2^2 + \frac{\lambda_R}{2} \|R\|_2^2 + \frac{\lambda_B}{2} \|B\|_2^2 + \frac{\lambda_W}{2} \|W\|_2^2 \\
V_j &= \sum_{k \in I_j} W_k E_k
\end{aligned}
\tag{7}
$$

4   Experiment

In this section, we introduce the dataset, the baselines and the results of comparison experiments.

4.1   Dataset

To demonstrate the effectiveness of proposed method, we collected *GitHub* dataset and conduct experiments on it. The *GitHub* dataset is chosen for several reasons. Firstly, the user feedback is implicit which is more realistic in real-world recommendation. Besides, the *GitHub* dataset is quantitatively sparse but semantically dense, the dataset consists of 3,798 users, 2,477 items and 22,096 interactions. Defining the density ratio as $iteration\_num/(user\_num * item\_num)$, the ratio of *GitHub* dataset is 0.0026. In contrast, the popular MovieLens-1M dataset has a density ratio of 0.0119 even if only 5-star ratings are considered. Though the *GitHub* dataset is quantitatively sparse, it is semantically dense, the repositories are highly related with each other based on their semantic information including some simple entity-based interactions, such as some repositories use programming language "C++" or some repositories use toolkit "TensorFlow". We also leverage some complex interactions from knowledge graph, for example, a repository uses toolkit "TensorFlow" which is implemented by "C++" which is the programming language of another repository. We do recommendation on *GitHub* dataset not only based on historical cooccurrence of items but also based on semantic information enhanced by knowledge graph. The other reason we use *GitHub* dataset is that the items (repositories) can't be directly mapped to entities of knowledge graph because of its highly customized item name. Although directly mapping is used in some previous works, it fails in recommendation tasks where item names are customized.

4.2   Baselines

We choose following methods as baselines of our experiment, **BPRMF** (Bayesian Personalized Ranking based Matrix Factorization), **BPRMF+TransE** and **FM** (Factorization Machines).

BPRMF ignores the knowledge graph information, it only focuses on historical user feedback, the results are learnt by using pairwise item ranking based matrix factorization.

BPRMF+TransE uses almost the same setting as our proposed models (RESCAL-based HCE and TransR-based HCE), while it only considers part of knowledge graph information. By using TransE knowledge graph embedding method, it ignores the multi-relational data.

FM [11,12] is another popular solution for integrating side information into recommendation tasks. While it is limited by only considering the entities as items' features and ignoring the semantic structural relation between entities.

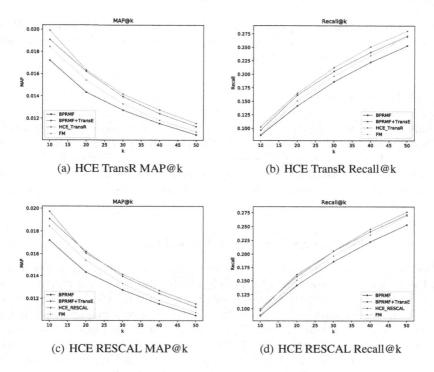

Fig. 2. MAP@k and Recall@k result.

4.3 Comparison

To measure both the precision and recall of recommendation results, we use MAP@k (mean average precision) [13] and Recall@k [14] in our experiments. Due to utilizing two knowledge graph embedding methods, RESCAL and TransR, in proposed Hierarchical Collaborative Embedding (HCE) model, we compare RESCAL-based HCE and TransR-based HCE with baselines (BPRMF, FM, BPRMF+TransE) respectively.

Each experiment is repeated five times with different random seeds and we report the MAP and Recall values by varying the position k in Fig. 2. The results can be summarized as follows: (1) Results of FM model are better than BPRMF, because BPRMF totally ignores knowledge graph information, knowledge graph information is useful to improve the recommendation results. (2) The improvement of FM is limited, less than BPRMF+TransE model, because FM model doesn't consider relation structure of knowledge graph, integrating knowledge graph structured embedding in our proposed model by using knowledge conceptual level effectively elevates MAP@k and Recall@k scores. (3) Although BPRMF+TransE model is effective, it is still outperformed by both RESCAL-based HCE model and TransE-based HCE model, because the latter two models consider the multi-relational data of knowledge graph.

The effectiveness of proposed Hierarchical Collaborative Embedding (HCE) framework is presented. Knowledge Conceptual Level serves as the core component of HCE framework appropriately.

5   Related Work

In this section, we introduce two related works, Knowledge graph structured embedding and Implicit Collaborative Filtering. Knowledge graph structured embedding leverages relational learning methods [15] to extract the latent semantic information of knowledge graph elements including entities and relations. Collaborative filtering learns users' interests from their feedback, either explicit or implicit.

5.1   Knowledge Graph Structured Embedding

Based on different assumptions, each structured embedding method proposes a model to represent knowledge graph triple which consists of head entity, relation and tail entity. There are three categories of models, direct vector space translating, vector space translating with relation subspace or hyperplane mapping, and tensor factorization. Considering of knowledge graph is a multi-relational heterogeneous network, Bordes et al. [7] then proposed another model using direct vector space translating model, which ignore multi-relation problem but make the model much more efficient in training speed. Nickel et al. [8] proposed a new type of relational learning methods based on tensor factorization, which is efficient in both speed and accuracy. Lin et al. [9] use relation subspace mapping instead of hyperplane mapping. Except the models mentioned above, there are some other structured embedding models [6,16]. In this work, we integrate [7] into our framework as one baseline, and we use [8,9] as important components of our proposed model.

5.2   Collaborative Filtering Using Implicit Feedback

Popularized by the *Netflix prize*[3], traditional methods focus on explicit feedback such as *ratings*. However, the last decade has seen a growing trend towards exploiting implicit feedback such as *clicks* and *purchases*. Implicit feedback has a major advantage of eliminating the needs of asking users explicitly. Instead, user feedback is collected silently, resulting more user-friendly recommender systems. Hu et al. [17] and Pan et al. [18] investigated item recommendation from implicit feedback and propose to impute all missing values with zeros. More recently, Shi et al. [19] and Bayer et al. [20] extended Bayesian Personalized Ranking (BPR) [10] for optimizing parameters from implicit feedback. In this paper, we employ an optimization strategy similar to BPR, but with semantic information modeling. Standard BPR methods are also used as baselines for our experiments.

---

[3] http://www.netflixprize.com/.

6   Conclusions

In this paper, we proposed Hierarchical Collaborative Embedding framework, which integrates recommender system with knowledge graph into a three-level model. The information of knowledge graph is leveraged to improve the results of quantitatively sparse but semantically dense recommendation scenarios. Experiment was conducted on real-world *GitHub* dataset showing that semantic information from knowledge graph has been properly captured, resulting improved recommendation performance. To the best of our knowledge, this is the first attempt of using knowledge graph embedding to perform semantic enhancement for items that do not exist in knowledge graph, by using the proposed Knowledge Conceptual Level. LS as well. For future work, we would like to add additional layers to the network to capture higher order interactions of items and entities.

Acknowledgement. The authors thank the reviewers for their helpful comments. This work was partially supported by the Major Research Plan of National Science Foundation of China [No. 91630206].

References

1. Zhang, F., Yuan, N.J., Lian, D., Xie, X., Ma, W.Y.: Collaborative knowledge base embedding for recommender systems. In: Proceedings of the 22nd ACM SIGKDD International Conference on Knowledge Discovery and Data Mining, pp. 353–362. ACM (2016)
2. Yu, X., Ren, X., Sun, Y., Gu, Q., Sturt, B., Khandelwal, U., Norick, B., Han, J.: Personalized entity recommendation: a heterogeneous information network approach. In: Proceedings of the 7th ACM International Conference on Web Search and Data Mining, pp. 283–292. ACM (2014)
3. Shi, C., Zhang, Z., Luo, P., Yu, P.S., Yue, Y., Wu, B.: Semantic path based personalized recommendation on weighted heterogeneous information networks. In: Proceedings of the 24th ACM International Conference on Information and Knowledge Management, pp. 453–462. ACM (2015)
4. Burke, R., Vahedian, F., Mobasher, B.: Hybrid recommendation in heterogeneous networks. In: Dimitrova, V., Kuflik, T., Chin, D., Ricci, F., Dolog, P., Houben, G.-J. (eds.) UMAP 2014. LNCS, vol. 8538, pp. 49–60. Springer, Cham (2014). https://doi.org/10.1007/978-3-319-08786-3_5
5. Yu, X., Ren, X., Sun, Y., Sturt, B., Khandelwal, U., Gu, Q., Norick, B., Han, J.: Recommendation in heterogeneous information networks with implicit user feedback. In: ACM Conference on Recommender Systems, pp. 347–350 (2013)
6. Bordes, A., Weston, J., Collobert, R., Bengio, Y., et al.: Learning structured embeddings of knowledge bases. In: AAAI, vol. 6, p. 6 (2011)
7. Bordes, A., Usunier, N., Garcia-Duran, A., Weston, J., Yakhnenko, O.: Translating embeddings for modeling multi-relational data. In: Advances in Neural Information Processing Systems, pp. 2787–2795 (2013)
8. Nickel, M.: Tensor factorization for relational learning. Ph.D. thesis, LMU (2013)
9. Lin, Y., Liu, Z., Sun, M., Liu, Y., Zhu, X.: Learning entity and relation embeddings for knowledge graph completion. In: AAAI, pp. 2181–2187 (2015)

10. Rendle, S., Freudenthaler, C., Gantner, Z., Schmidt-Thieme, L.: BPR: Bayesian personalized ranking from implicit feedback. In: Proceedings of the Twenty-Fifth Conference on Uncertainty in Artificial Intelligence, pp. 452–461. AUAI Press (2009)
11. Rendle, S.: Factorization machines. In: 2010 IEEE 10th International Conference on Data Mining (ICDM), pp. 995–1000. IEEE (2010)
12. Rendle, S.: Factorization machines with libFM. ACM Trans. Intell. Syst. Technol. (TIST) **3**(3), 57 (2012)
13. Dieleman, S., Schrauwen, B.: Deep content-based music recommendation. In: International Conference on Neural Information Processing Systems, pp. 2643–2651 (2013)
14. Davis, J., Goadrich, M.: The relationship between precision-recall and ROC curves. In: International Conference on Machine Learning, pp. 233–240 (2006)
15. Nickel, M., Murphy, K., Tresp, V., Gabrilovich, E.: A review of relational machine learning for knowledge graphs. Proc. IEEE **104**(1), 11–33 (2015)
16. Zhou, Z., Xu, G., Zhu, W., Li, J., Zhang, W.: Structure embedding for knowledge base completion and analytics. In: International Joint Conference on Neural Networks, pp. 737–743 (2017)
17. Hu, Y., Koren, Y., Volinsky, C.: Collaborative filtering for implicit feedback datasets. In: Eighth IEEE International Conference on Data Mining, ICDM 2008, pp. 263–272. IEEE (2008)
18. Pan, R., Zhou, Y., Cao, B., Liu, N.N., Lukose, R., Scholz, M., Yang, Q.: One-class collaborative filtering. In: Eighth IEEE International Conference on Data Mining, ICDM 2008, pp. 502–511. IEEE (2008)
19. Shi, Y., Karatzoglou, A., Baltrunas, L., Larson, M., Oliver, N., Hanjalic, A.: Climf: learning to maximize reciprocal rank with collaborative less-is-more filtering. In: Proceedings of the Sixth ACM Conference on Recommender Systems, pp. 139–146. ACM (2012)
20. Bayer, I., He, X., Kanagal, B., Rendle, S.: A generic coordinate descent framework for learning from implicit feedback. In: Proceedings of the 26th International Conference on World Wide Web, pp. 1341–1350. International World Wide Web Conferences Steering Committee (2017)

DPNE: Differentially Private Network Embedding

Depeng Xu[1], Shuhan Yuan[1], Xintao Wu[1(✉)], and HaiNhat Phan[2]

[1] University of Arkansas, Fayetteville, AR, USA
{depengxu,sy005,xintaowu}@uark.edu
[2] New Jersey Institute of Technology, Newark, NJ, USA
phan@njit.edu

Abstract. Learning the low-dimensional representations of the vertices in a network can help users understand the network structure and perform other data mining tasks efficiently. Various network embedding approaches such as DeepWalk and LINE have been developed recently. However, how to protect the individual privacy in network embedding has not been exploited. It is challenging to achieve high utility as the sensitivity of stochastic gradients in random walks and that of edge sampling are very high, thus incurring high utility loss when applying Laplace mechanism and exponential mechanism to achieve differential privacy. In this paper, we develop a differentially private network embedding method (DPNE). In this method, we leverage the recent theoretical findings that network embedding methods such as DeepWalk and LINE are equivalent to factorization of some matrices derived from the adjacency matrix of the original network and apply objective perturbation on the objective function of matrix factorization. We evaluate the learned representations by our DPNE from three different real world datasets on two data mining tasks: vertex classification and link prediction. Experiment results show the effectiveness of DPNE. To our best knowledge, this is the first work on how to preserve differential privacy in network embedding.

1 Introduction

Network embedding learns the lower dimensional representations of the vertices in a high-dimensional social network [7]. The latent representations encode the social relations in a continuous vector space, which can be used to conduct a variety of applications such as vertex classification and link prediction. The first network embedding model, DeepWalk [14], uses the sequences of vertices generated by random walks to learn the vertex representations. It adopts SkipGram [12], which was previously used to learn word representations in natural language processing. Several models based on the neural language model have been proposed, such as node2vec [8], Discriminative Deep Random Walk (DDRW) [10], Large-scale Information Network Embedding (LINE) [18] and Signed Network Embedding (SNE) [23].

© Springer International Publishing AG, part of Springer Nature 2018
D. Phung et al. (Eds.): PAKDD 2018, LNAI 10938, pp. 235–246, 2018.
https://doi.org/10.1007/978-3-319-93037-4_19

However, releasing the representations of vertices in a social network gives malicious attackers a potential way to infer the sensitive information of individuals. For example, the widely-used network embedding methods like DeepWalk [14] and LINE [18] train the vertex representations based on the linkage information between vertices. Hence, the released vertex representations may potentially breach link privacy of social network users. Currently, it is under-exploited how to preserve differential privacy in network embedding.

Differential privacy is a formal standard for protecting individual privacy in data analysis [5]. Differential privacy ensures that the inclusion or exclusion of a single record from a dataset makes no statistical difference when we perform a data analysis task on the dataset. The mechanisms to achieve differential privacy mainly include the classic approach of adding Laplacian noise [5], the exponential mechanism [11], the objective perturbation approach [4], the functional perturbation approach [25] and the sample and aggregate framework [13]. There have been many studies on the application of differential privacy in some particular analysis tasks, e.g., data collection [6,19], stochastic gradient descents [17], regression [3], spectral graph analysis [20], causal graph discovery [21] and deep learning models [1,15]. In this work, we aim to ensure no "privacy loss" in case of the inclusion or exclusion of an edge between two vertices from a network, a.k.a. link privacy.

In DeepWalk, the inputs of the two embedding models are generated by random walks and edge sampling. The objective function derived from random walks has an uncertain and complex mapping from edges. Edge sampling directly discloses the presence or absence of an edge between a vertex pair, which leads to a high sensitivity for privacy protection. Thus, it is difficult to directly incorporate well-studied differential privacy mechanisms [17,24] onto DeepWalk. Meanwhile, the matrix factorization based method, which is proven to be equivalent to DeepWalk, learns representations through factorizing a matrix with the pointwise mutual information of the vertices pairs in a network. It is convenient to achieve differential privacy in network embedding via small perturbation on matrix factorization.

In this work, we focus on developing a differential privacy preserving method of network embedding based on the equivalent matrix factorization method. We propose a differentially private network embedding method (DPNE). In this method, we leverage the findings that network embedding methods such as DeepWalk and LINE are equivalent to factorization of some matrices derived from the adjacency matrix of the original network and apply objective perturbation on the objective function of matrix factorization. We show that with only adding a small amount of noise onto the objective function, the learned low-dimensional representations satisfy differential privacy. Experimental results show that the embedded representations learned by DPNE achieve good utility with a small privacy budget on both vertex classification and link prediction tasks. To our best knowledge, this is the first work on how to preserve differential privacy in network embedding.

2 Preliminaries

2.1 Network Embedding

A social network is defined as $\mathcal{G} = (\mathcal{V}, \mathcal{E})$, where $\mathcal{V}$ is a set of vertices and $\mathcal{E}$ is the set of edges. We use v_i to denote a vertex, and we use e_{ij} to denote an edge between a pair of vertices v_i and v_j. The goal of the network embedding is to learn the low-dimensional representations $\mathbf{X} \in \mathbb{R}^{|\mathcal{V}| \times k}$ for all vertices in $\mathcal{V}$, where $k \ll |\mathcal{V}|$. The i-th row of $\mathbf{X}$ (denoted as $\mathbf{x}_i$) is the k-dimensional latent representation of vertex v_i.

DeepWalk. DeepWalk adopted SkipGram [12], which was previously used to learn word representations, to learn vertex representations according to the network structure. DeepWalk first generates short random walks for each vertex. Then the model uses the sequences of vertices S generated by random walks as the input of SkipGram function to learn the vertex representations. In particular, for each target vertex $v_i \in \mathcal{V}$ and a context vertex v_j within t window size of v_i in a walk sequence ($v_j \in C_i = \{v_{i-t}, \ldots, v_{i+t}\} \setminus \{v_i\}$), DeepWalk optimizes the co-occurrence probability between v_i and its context vertices within S:

$$\mathcal{L}(\mathcal{S}) = \frac{1}{|S|} \sum_{v_i \in S} \sum_{v_j \in C_i} \log \Pr(v_j | v_i), \tag{1}$$

where the probability function $\Pr(v_j | v_i)$ is defined by the softmax function:

$$\Pr(v_j | v_i) = \frac{\exp(\mathbf{x}_i \mathbf{y}_j^T)}{\sum_{v_{j'} \in \mathcal{V}_c} \exp(\mathbf{x}_i \mathbf{y}_{j'}^T)}, \tag{2}$$

where $\mathbf{x}_i$ and $\mathbf{y}_j$ are the k-dimensional representations of the target vertex v_i and the context vertex v_j, respectively; and $\mathcal{V}_c$ denotes the set of context vertices.

DeepWalk as Matrix Factorization. DeepWalk using SkipGram with Negative Sampling (SGNS) has been proven that it is equivalent to factorize a matrix $\mathbf{M}$ derived from $\mathcal{G}$ $\underset{|\mathcal{V}| \times |\mathcal{V}_c|}{\mathbf{M}} = \underset{|\mathcal{V}| \times k}{\mathbf{W}} \underset{k \times |\mathcal{V}_c|}{\mathbf{H}^T}$ [22]. The factorized matrices $\mathbf{W}, \mathbf{H}$ are equivalent to the vertex/context representations, as $\mathbf{w}_i = \mathbf{x}_i$ and $\mathbf{h}_j = \mathbf{y}_j$. Each value m_{ij} in $\mathbf{M}$ represents logarithm of the average probability that vertex v_i randomly walks to vertex v_j within fixed t steps. Formally, m_{ij} is defined as:

$$m_{ij} = \log \frac{[I_i(\mathbf{P} + \mathbf{P}^2 + \cdots + \mathbf{P}^t)]_j}{t}, \tag{3}$$

where $\mathbf{P}$ is the transition matrix of $\mathcal{G}$ with $p_{ij} = \frac{1}{d_i}$ if $e_{ij} \in \mathcal{E}$; d_i is the degree of v_i in $\mathcal{G}$; and I_i denotes an indicator vector, in which the i-th entry is 1 and the others are all 0.

Hence, we formalize the DeepWalk as matrix factorization $\mathbf{M} = \mathbf{W}\mathbf{H}^T$. Let Ω be the set of vertex pairs referenced by each entry m_{ij} of $\mathbf{M}$. The model aims to find matrices $\mathbf{W}$ and $\mathbf{H}$ to minimize the objective function as follows:

$$\mathcal{L}(\mathbf{w}_i, \mathbf{h}_j, \mathcal{G}) = \sum_{(i,j)\in\Omega} ||m_{ij} - \mathbf{w}_i\mathbf{h}_j^T||_2^2 + \lambda\sum_{i\in\mathcal{V}} ||\mathbf{w}_i||_2^2 + \mu\sum_{j\in\mathcal{V}_c} ||\mathbf{h}_j||_2^2, \quad (4)$$

where λ and μ are the weights of regularization terms.

We adopt the stochastic gradient descent (SGD) approach to minimize Eq. 4. SGD iteratively learns $\mathbf{W}$ and $\mathbf{H}$. The partial derivatives of $\mathcal{L}(\mathbf{w}_i, \mathbf{h}_j, \mathcal{G})$ with respect to $\mathbf{w}_i$ and $\mathbf{h}_j$ are as follows:

$$\nabla_{\mathbf{w}_i} \mathcal{L}(\mathbf{w}_i, \mathbf{h}_j, \mathcal{G}) = -2\sum_{j\in\mathcal{V}_c} \mathbf{h}_j(m_{ij} - \mathbf{w}_i\mathbf{h}_j^T), \quad (5)$$

$$\nabla_{\mathbf{h}_j} \mathcal{L}(\mathbf{w}_i, \mathbf{h}_j, \mathcal{G}) = -2\sum_{i\in\mathcal{V}} \mathbf{w}_i(m_{ij} - \mathbf{w}_i\mathbf{h}_j^T). \quad (6)$$

2.2 Differential Privacy

Definition 1 *Differential privacy* [5]. *A graph analysis mechanism $\mathcal{M}$ satisfies ϵ-differential privacy, if for all neighboring graphs $\mathcal{G}$ and $\mathcal{G}'$ and all subsets Z of $\mathcal{M}$'s range:*

$$\Pr(\mathcal{M}(\mathcal{G}) \in Z) \leq \exp(\epsilon) \cdot \Pr(\mathcal{M}(\mathcal{G}') \in Z), \quad (7)$$

where $\mathcal{G} = (\mathcal{V}, \mathcal{E})$, $\mathcal{G}' = (\mathcal{V}, \mathcal{E}')$, $\mathcal{E}' = \mathcal{E}\bigcup\{e_{rs}\}$, e_{rs} is the differed edge between $\mathcal{G}$ and $\mathcal{G}'$.

The parameter ϵ denotes the privacy budget (smaller values indicates a stronger privacy guarantee).

Definition 2 *Global sensitivity* [5]. *Given a function $f : \mathcal{G} \to \mathbb{R}^k$. The sensitivity $S_f(\mathcal{G})$ is defined as*

$$S_f(\mathcal{G}) = \max_{\mathcal{G},\mathcal{G}'} ||f(\mathcal{G}) - f(\mathcal{G}')||. \quad (8)$$

Definition 3 *Laplace mechanism* [5]. *Given a graph $\mathcal{G}$ and a query f, a mechanism $\mathcal{M}(\mathcal{G}) = f(\mathcal{G}) + (Y_1, \cdots, Y_k)$ satisfies ϵ-differential privacy, where Y_i is drawn i.i.d. from $Lap(S_f(\mathcal{G})/\epsilon)$.*

Laplace mechanism ensures differential privacy for any function f by adding random noises generated from a Laplace distribution onto the true answer of $f(\mathcal{G})$. The global sensitivity of f controls the magnitude of the noise distribution.

For many data mining and machine learning algorithms, we usually optimize some objective functions (e.g., cross entropy) to derive coefficients of released models. Rather than adding noise to coefficients of the released model, Chaudhuri et al. [4] proposed an **objective perturbation** approach by perturbing the objective function $\mathcal{L}$ and then optimizing the perturbed objective function,

$$\mathcal{L}_{priv}(\boldsymbol{\omega}, \mathcal{G}) = \mathcal{L}(\boldsymbol{\omega}, \mathcal{G}) + \boldsymbol{\omega}\boldsymbol{\eta}^T, \tag{9}$$

where $\boldsymbol{\eta}$ is a random noise vector and its probability density is given by

$$\Pr(\boldsymbol{\eta}) \propto e^{-\beta||\boldsymbol{\eta}||}, \tag{10}$$

and the parameter β is a function of privacy budget ϵ and the scale of $||\nabla_\omega \mathcal{L}(\boldsymbol{\omega}, e_{ij})||$. To implement this, we pick the norm of $\boldsymbol{\eta}$ from the $\Gamma(k, \beta)$ distribution and the direction of $\boldsymbol{\eta}$ uniformly at random. Then we compute the private output $\widehat{\boldsymbol{\omega}}$, where $\widehat{\boldsymbol{\omega}} = \arg\min_\omega \mathcal{L}_{priv}(\boldsymbol{\omega}, \mathcal{G})$ satisfies ϵ-differential privacy.

3 Differentially Private Network Embedding

3.1 Differentially Private Network Embedding (DPNE)

DPNE adopts the objective perturbation mechanism on matrix factorization to protect the individual's link privacy in the social network. Note that $\mathbf{M}$ represents logarithm of the average probability that one vertex randomly walks to another vertex within fixed steps. When an edge e_{ij} in $\mathcal{G}$ is added or removed, the entire entries in $\mathbf{M}$ are changed accordingly. Hence, although there are some works [9] to protect the privacy in terms of a value in a matrix, it is not straightforward to adapt the existing models to our scenario. We need to derive the scale of the objective function in terms of changing one edge in $\mathcal{G}$.

In DPNE, we define the perturbed objective function of matrix factorization in Eq. 4 as follows:

$$\mathcal{L}_{priv}(\mathbf{w}_i, \mathcal{G}) = \sum_{(i,j)\in\Omega} ||m_{ij} - \mathbf{w}_i\mathbf{h}_j^T||_2^2 + \lambda\sum_{i\in\mathcal{V}} ||\mathbf{w}_i||_2^2 + \mu\sum_{j\in\mathcal{V}_c} ||\mathbf{h}_j||_2^2 + \sum_{i\in\mathcal{V}} \mathbf{w}_i\boldsymbol{\eta}_i^T, \tag{11}$$

where $\mathbf{N} = [\boldsymbol{\eta}_i]_{|\mathcal{V}|\times k}$ is a noise matrix with each row $\boldsymbol{\eta}_i$ of $\mathbf{N}$ as a k-dimensional noise vector. In practice, the context representation matrix $\mathbf{H}$ is kept confidential. We first minimize Eq. 4 to update $\mathbf{H}$, then fix $\mathbf{H}$ and learn $\mathbf{W}$ by minimizing Eq. 11. Hence, $\mathbf{W}$ is the only matrix variable in Eq. 11.

Theorem 1. *Let $\mathbf{M}$ be a matrix where each of its entry m_{ij} is defined by Eq. 3. Let $\boldsymbol{\eta}_i$ in Eq. 11 be a k-dimensional noise vector that is independently and randomly picked for each vertex v_i from the density function $\Pr(\boldsymbol{\eta}_i) \propto \exp(-\dfrac{\epsilon||\boldsymbol{\eta}_i||}{2\Delta})$, where $\Delta = \max||\mathbf{M}' - \mathbf{M}||$. The derived $\widehat{\mathbf{W}} = \arg\min_\mathbf{W} \mathcal{L}_{priv}(\mathbf{w}_i, \mathcal{G})$ by minimizing Eq. 11 satisfies ϵ-differential privacy.*

Proof. Let $\mathcal{G}$ and $\mathcal{G}'$ be two neighboring graphs differing by one edge, where $\mathcal{G} = (\mathcal{V}, \mathcal{E})$, $\mathcal{G}' = (\mathcal{V}, \mathcal{E}')$, $\mathcal{E}' = \mathcal{E}\bigcup\{e_{rs}\}$. Let $\mathbf{M}$ and $\mathbf{M}'$ be two matrices derived from $\mathcal{G}$ and $\mathcal{G}'$ following Eq. 3. Let $\mathbf{N}$ and $\mathbf{N}'$ be the noise matrices in Eq. 11 when training with $\mathcal{G}$ and $\mathcal{G}'$. Meanwhile, $\mathcal{L}_{priv}(\mathbf{w}_i, \mathcal{G})$ is differentiable anywhere.

Let $\bar{\mathbf{W}} = \arg\min_{\mathbf{W}} \mathcal{L}_{priv}(\mathbf{w}_i, \mathcal{G}) = \arg\min_{\mathbf{W}} \mathcal{L}_{priv}(\mathbf{w}_i, \mathcal{G}')$, we have $\forall v_i \in \mathcal{V}$, $\bigtriangledown_{\mathbf{w}_i} \mathcal{L}_{priv}(\bar{\mathbf{w}}_i, \mathcal{G}) = \bigtriangledown_{\mathbf{w}_i} \mathcal{L}_{priv}(\bar{\mathbf{w}}_i, \mathcal{G}') = 0$. Thereby,

$$\boldsymbol{\eta}_i - 2\sum_{j \in \mathcal{V}_c} \mathbf{h}_j (m_{ij} - \bar{\mathbf{w}}_i \mathbf{h}_j^T) = \boldsymbol{\eta}_i' - 2\sum_{j \in \mathcal{V}_c} \mathbf{h}_j (m_{ij}' - \bar{\mathbf{w}}_i \mathbf{h}_j^T); \qquad (12)$$

We can derive from Eq. 12 that:

$$\boldsymbol{\eta}_i - \boldsymbol{\eta}_i' = 2\sum_{j \in \mathcal{V}_c} \mathbf{h}_j (m_{ij}' - \bar{\mathbf{w}}_i \mathbf{h}_j^T) - 2\sum_{j \in \mathcal{V}_c} \mathbf{h}_j (m_{ij} - \bar{\mathbf{w}}_i \mathbf{h}_j^T) = 2\sum_{j \in \mathcal{V}_c} \mathbf{h}_j (m_{ij} - m_{ij}');$$

$$\sum_{i \in \mathcal{V}} (\boldsymbol{\eta}_i - \boldsymbol{\eta}_i') = 2\sum_{i \in \mathcal{V}} \sum_{j \in \mathcal{V}_c} \mathbf{h}_j (m_{ij} - m_{ij}');$$

We normalize $\|\mathbf{h}_j\| \leq 1$. Since $\|\mathbf{M}' - \mathbf{M}\| \leq \Delta$, we have $\|\mathbf{N} - \mathbf{N}'\| \leq 2\Delta$ regardless of $\mathbf{H}$. Then for $\mathcal{G}$ and $\mathcal{G}'$,

$$\frac{\Pr[W = \bar{W}|\mathcal{G}]}{\Pr[W = \bar{W}|\mathcal{G}']} = \frac{\prod_{i \in \mathcal{V}} \Pr(\boldsymbol{\eta}_i)}{\prod_{i \in \mathcal{V}} \Pr(\boldsymbol{\eta}_i')} = \exp(-\frac{\epsilon(\sum_{i \in \mathcal{V}} \|\boldsymbol{\eta}_i\| - \sum_{i \in \mathcal{V}} \|\boldsymbol{\eta}_i'\|)}{2\Delta})$$

$$\leq \exp(-\frac{\epsilon(\|\mathbf{N} - \mathbf{N}'\|)}{2\Delta}) \leq \exp(\epsilon). \qquad (13)$$

The above proof also holds when we use $\left|\sum_{v_i \in \mathcal{V}} \mathbf{w}_i \boldsymbol{\eta}_i^T\right|$ as the noise term in the perturbed objective function. In our implementation, we use the absolute noise term to get a better performance on the optimization. Next, we show the upper bound of $\max \|\mathbf{M}' - \mathbf{M}\|$, which will be used for adding noise to the objective function.

Lemma 1. *The L_2-sensitivity of* $\mathbf{M}$ *is* $\max \|\mathbf{M}' - \mathbf{M}\| \leq \sqrt{2}$.

Proof. When window size $t = 1$, $\|\mathbf{M}' - \mathbf{M}\|$ decreases as degrees of v_r and v_s (the vertices linked by the edge e_{rs} in $\mathcal{G}'$) increase. $\|\mathbf{M}' - \mathbf{M}\|$ takes the maximum value $\sqrt{2}$ when degrees of v_r and v_s are both 0 in $\mathcal{G}$ and 1 in $\mathcal{G}'$. When window size $t \geq 2$, $\|\mathbf{P}'^t - \mathbf{P}^t\| \leq \|\mathbf{P}' - \mathbf{P}\|$.

$$\|\mathbf{M}' - \mathbf{M}\|_{(t \geq 2)} = \left\|\frac{\mathbf{P}' + \ldots + \mathbf{P}'^t}{t} - \frac{\mathbf{P} + \ldots + \mathbf{P}^t}{t}\right\| \leq \left\|\frac{t(\mathbf{P}' - \mathbf{P})}{t}\right\| = \|\mathbf{M}' - \mathbf{M}\|_{(t=1)}.$$

3.2   DPNE vs. Other DP-Preserving Embedding Approaches

A naive way to achieve differential privacy in network embedding is to get a differentially private matrix $\mathbf{M}$ (dpM) and then to apply matrix factorization, where $\mathbf{M}$ is calculated with the transition matrix $\mathbf{P}$ and its powers $\mathbf{P}^t$. To get a differentially private matrix $\mathbf{M}$, we can use the Laplace mechanism to add a noise matrix $\mathbf{N}$ on $\mathbf{M}$, where each entry n_{ij} of $\mathbf{N}$ is drawn i.i.d. from $Lap(S_f(\mathcal{G})/\epsilon)$, let $S_f(\mathcal{G})$ be the sensitivity of $\mathbf{M}$, $S_f(\mathcal{G}) = \max ||\mathbf{M'} - \mathbf{M}|| \leq \sqrt{2}$. The worst case for $||\mathbf{M'} - \mathbf{M}||$ is when adding an edge to two isolated vertices in $\mathcal{G}$. We will use dpM as a baseline in our empirical evaluation.

Another way to achieve differential privacy in network embedding is to enforce differential privacy in the process of network embedding models. Let D be a vertex-context set generated from the random walk sequences, where each member of D is a vertex-context pair (v_i, v_j). For a "walk step" from v_i to v_j in random walks, it is uncertain how many times in total the same "walk step" appears in D. For example, the number of times that an edge e_{ij} gets walked through by all the random walks sequences has the upper bound $\sum_{i=0}^{\min\{b, \lceil \log_a |\mathcal{V}| \rceil\}} a^i \times c \times (b - i)$, where a is the maximal degree of $\mathcal{G}$, b is the walk length, c is the number of walks starting at each vertex. Stochastic Gradient Descent (SGD) is used in SkipGram to learn the embedding vectors based on the objective function. The input for SGD is the vertex-context set D. Although there are existing works on how to apply objective perturbation [17] or exponential mechanism [24] on SGD to make private updates, the privacy of D and the privacy of $\mathcal{G}$ are not the same. For example, if we want to apply the functional mechanism [25] on DeepWalk with hierarchical softmax [14], in terms of the privacy of D, the hierarchical softmax function iterates one time over each vertex-context pair in D; the sensitivity of the hierarchical softmax function on D is about $\lceil \log_2 |\mathcal{V}| \rceil (k/2 + k^2/8)$. But in terms of the privacy of $\mathcal{G}$, for each edge in $\mathcal{G}$, the hierarchical softmax function iterates an unset number of times; thus, the sensitivity of the hierarchical softmax function on $\mathcal{G}$ is very large after multiplying the iteration times.

Also, negative sampling is used for SkipGram function in DeepWalk or LINE. The processes of positive sampling and negative sampling already indicate link privacy. It is viable to make the sampling process private by applying the exponential mechanism [11] or the Laplace mechanism [5]. However, the sensitivity of edge sampling domain is also large. Hence, it is challenging to intuitively apply the existing differentially private approaches in DeepWalk to achieve privacy protection on $\mathcal{G}$.

The equivalent matrix factorization approach avoids random walks and negative sampling. The effect of window size in DeepWalk is expressed as the powers of the transition matrix. It considers the expected times that the vertex pairs appear rather than the randomly generated times. It is more convenient to apply differential privacy mechanisms on the matrix factorization based network embedding method. The only remaining challenge is to bridge the "gap" between the privacy of D and the privacy of $\mathcal{G}$. Theorem 1 addresses this problem.

4   Evaluation

In this section, we evaluate the performance of DPNE on two tasks: vertex classification and link prediction. For vertex classification, we predict the label of each vertex in the network by using vertex embeddings as inputs to build

Table 1. Comparing the accuracy for vertex classification with different privacy budget ϵ [non-priv: (a) Wiki 0.555 (b) Cora 0.700 (c) Citeseer 0.505]

Dataset	(a) Wiki		(b) Cora		(l) Citeseer	
ϵ	dpM	DPNE	dpM	DPNE	dpM	DPNE
0.01	0.094	0.093	0.187	0.181	0.179	0.178
0.1	0.091	0.096	0.189	0.213	0.182	0.186
1	0.088	0.521	0.190	0.662	0.177	0.452
10	0.090	0.527	0.196	0.669	0.180	0.459
100	0.355	0.537	0.500	0.677	0.311	0.466
1000	0.545	0.552	0.679	0.700	0.497	0.490

a classifier. For link prediction, we aim to use vertex embeddings to predict whether there is an edge between two vertices.

Baselines. We compare our differentially private network embedding method (**DPNE**) with the naive method (**dpM**) and the non-private network embedding as matrix factorization method (**non-priv**).

Datasets. We adopt the following three datasets to evaluate the proposed model. (1) **Wiki** contains 2,405 documents from 19 classes and 17,981 links between them. (2) **Cora** is a research paper set which contains 2,708 machine learning papers from 7 classes and 5,429 links between them. (3) **Citeseer** is another research paper set which contains 3,312 publications from 6 classes and 4,732 links between them.

Parameter Settings. In our experiments, we set the window size $t = 2$ and $\mathbf{M} = (\mathbf{P} + \mathbf{P}^2)/2$. For all three datasets, we choose a series of values for the embedding size $k = \{10, 20, 50, 100, 200, 500, 1000\}$ for vertex representations, and a series of values for the privacy budget $\epsilon = \{0.01, 0.1, 1, 10, 100, 1000\}$. We set the regularization coefficients $\lambda = \mu = 0.001$ and the learning rate $\gamma = 0.015$. The train ratios of SVM and logistic regression in the two tasks are both 10%. For each parameter setting, we report the average result over 10 different runs.

4.1   Vertex Classification Task

For the vertex classification task, we first get the vertex representation $\mathbf{W}$ based on the matrix factorization method. Then, we train a multi-class support vector machine (SVM) on a training dataset L based on a subset of vertex representations $\mathbf{W}_L$ and further predict the labels of a testing dataset U by the SVM classifier based on $\mathbf{W} \backslash \mathbf{W}_L$.

Different Privacy Budgets. We evaluate the performance of two private algorithms on all three datasets with embedding size $k = 100$ and different privacy

Table 2. Comparing the accuracy for vertex classification with different embedding size k and privacy budget ϵ (dataset=Wiki)

	ϵ						
	non-priv	DPNE					
k		0.01	0.1	1	10	100	1000
10	0.324	0.080	0.108	0.255	0.265	0.272	0.350
20	0.466	0.072	0.095	0.420	0.428	0.429	0.457
50	0.537	0.080	0.087	**0.526**	0.520	0.515	0.539
100	**0.555**	0.093	0.096	0.521	**0.527**	**0.537**	**0.552**
200	0.530	0.096	0.101	0.470	0.511	0.515	0.539
500	0.474	0.105	0.110	0.285	0.437	0.424	0.482
1000	0.429	**0.109**	**0.112**	0.169	0.270	0.295	0.405

budgets ϵ. Table 1 shows the comparison results of each method for vertex classification on three datasets. We can observe that both DPNE and dpM have the similar trend on three datasets in terms of accuracy while we increase the privacy budget. The classification results of DPNE are close to the non-private method when $\epsilon \geq 1$. However, the performance of dpM method has a big improvement when $\epsilon \geq 100$. It indicates that, compared with dpM, DPNE can achieve the same performance with a much smaller privacy budget. Meanwhile, when the $\epsilon \leq 0.1$, both private methods have poor performance. It indicates that when the privacy budget is low, the matrix factorization method cannot be converged due to the large noisy injected to the objective function or matrix itself.

Different Embedding Sizes. Based on Eq. 10, η_i increases with embedding size k. There is potentially a compromised performance of DPNE at a larger k. We evaluate the performance of the DPNE algorithm on the Wiki dataset with different embedding size k and privacy budget ϵ. For non-priv, as shown in the second column of Table 2, the highest accuracy 55.5% is achieved when $k = 100$. When increasing or decreasing embedding size k, the accuracy decreases. For DPNE, with relatively large privacy budget, $\epsilon = 10, 100, 1000$, high accuracy is also achieved when $k = 100$. When $\epsilon = 1$, high accuracy is achieved when $k = 50$. However, with very small privacy budget, $\epsilon = 0.01, 0.1$, DPNE has significantly lower accuracy comparing to non-priv no matter how we choose k.

4.2 Link Prediction Task

For the link prediction task, we first use vertex representations to compose edge representations. Given a pair of vertices v_i, v_j connected by an edge, we use an Hadamard operator to combine the vertex vectors $\mathbf{w}_i$ and $\mathbf{w}_j$ to compose the edge vector $\widehat{\mathbf{e}}_{ij} = \mathbf{w}_i * \mathbf{w}_j$ [8]. Then, we use the constructed edge vectors as inputs to train a logistic regression classifier and adopt the classifier to predict the presence or absence of an edge.

Different Privacy Budgets. We evaluate the performance of two private algorithms on three datasets for link prediction with embedding size $k = 100$ and different privacy budgets ϵ. Table 3 shows the link prediction accuracy of DPNE and dpM. We can observe that when $\epsilon \leq 0.1$, both private algorithms have only about 50% accuracy on three datasets. The accuracy of DPNE has a big improvement while ϵ increases to 1. However, dpM can achieve the comparable results when the $\epsilon \geq 100$. It indicates DPNE can achieve better performance with a small privacy budget.

Different Embedding Sizes. We also evaluate the performance of the DPNE algorithm on the Wiki dataset with different embedding size k and privacy budget ϵ. Table 4 shows our result. For non-priv, it achieves the highest accuracy 74.7% when $k = 200$. For DPNE, it often achieves the highest accuracy with $k = 200$ for large privacy budget values and with $k = 1000$ for small privacy budget values. More interestingly, DPNE even outperforms non-priv at large k for $\epsilon = 10$ or 100.

Table 3. Comparing the accuracy for link prediction under different privacy budget ϵ [non-priv: (a) Wiki 0.734 (b) Cora 0.697 (c) Citeseer 0.699]

Dataset	(a) Wiki		(b) Cora		(c) Citeseer	
ϵ	dpM	DPNE	dpM	DPNE	dpM	DPNE
0.01	0.502	0.520	0.501	0.532	0.498	0.535
0.1	0.502	0.524	0.501	0.541	0.501	0.523
1	0.503	0.743	0.500	0.698	0.503	0.690
10	0.527	0.713	0.552	0.673	0.566	0.666
100	0.708	0.720	0.729	0.666	0.751	0.662
1000	0.725	0.725	0.706	0.703	0.695	0.696

Table 4. Comparing the accuracy for link prediction with different embedding size k and privacy budget ϵ (dataset = Wiki)

	ϵ						
	non-priv	DPNE					
k		0.01	0.1	1	10	100	1000
10	0.607	0.499	0.552	0.614	0.612	0.609	0.605
20	0.648	0.500	0.538	0.661	0.650	0.654	0.649
50	0.717	0.510	0.525	0.718	0.694	0.695	0.718
100	0.734	0.520	0.524	0.743	0.713	0.720	0.725
200	**0.747**	0.532	0.541	**0.763**	0.734	**0.741**	**0.746**
500	0.670	0.550	0.566	0.733	0.778	0.721	0.695
1000	0.650	**0.568**	**0.595**	0.698	**0.784**	0.726	0.679

5   Conclusions and Future Work

In this work, we proposed a differentially private network embedding method (DPNE) based on DeepWalk as matrix factorization. We applied the objective perturbation approach on the objective function of matrix factorization. Our evaluation shows that on both vertex classification and link prediction tasks our DPNE achieves satisfactory performance. DPNE can be easily employed in other network embedding methods if there exists an equivalent matrix factorization of a certain matrix. For example, LINE is proven to be factorizing a similar matrix to $\mathbf{M}$ [16]. We would derive differential privacy preserving LINE similarly. One potential limitation of the matrix factorization based methods is that they are not scalable to large networks. Graph factorization [2] uses a streaming algorithm for graph partitioning to improve factorization based embedding methods. In future work, we will extend our DPNE to deal with large networks.

Acknowledgments. This work is supported in part by U.S. National Institute of Health (1R01GM103309) and National Science Foundation (1564250, 1523115 and 1502273).

References

1. Abadi, M., Chu, A., Goodfellow, I., McMahan, H.B., Mironov, I., Talwar, K., Zhang, L.: Deep learning with differential privacy. In: Proceedings of the 2016 ACM SIGSAC Conference on Computer and Communications Security, pp. 308–318 (2016)
2. Ahmed, A., Shervashidze, N., Narayanamurthy, S.M., Josifovski, V., Smola, A.J.: Distributed large-scale natural graph factorization. In: Proceedings of the 22nd International Conference on World Wide Web, pp. 37–48 (2013)
3. Chaudhuri, K., Monteleoni, C.: Privacy-preserving logistic regression. In: Advances in Neural Information Processing Systems, pp. 289–296 (2008)
4. Chaudhuri, K., Monteleoni, C., Sarwate, A.D.: Differentially private empirical risk minimization. J. Mach. Learn. Res. **12**, 1069–1109 (2011)
5. Dwork, C., McSherry, F., Nissim, K., Smith, A.: Calibrating noise to sensitivity in private data analysis. In: Halevi, S., Rabin, T. (eds.) TCC 2006. LNCS, vol. 3876, pp. 265–284. Springer, Heidelberg (2006). https://doi.org/10.1007/11681878_14
6. Erlingsson, U., Pihur, V., Korolova, A.: RAPPOR: Randomized aggregatable privacy-preserving ordinal response. In: Proceedings of the 2014 ACM SIGSAC Conference on Computer and Communications Security, pp. 1054–1067 (2014)
7. Goyal, P., Ferrara, E.: Graph embedding techniques, applications, and performance: A survey (2017). CoRR abs/1705.02801
8. Grover, A., Leskovec, J.: node2vec: Scalable feature learning for networks. In: Proceedings of the 22nd ACM SIGKDD International Conference on Knowledge Discovery and Data Mining, pp. 855–864 (2016)
9. Hua, J., Xia, C., Zhong, S.: Differentially private matrix factorization. In: IJCAI, pp. 1763–1770 (2015)
10. Li, J., Zhu, J., Zhang, B.: Discriminative deep random walk for network classification. In: Proceedings of the 54th Annual Meeting of the Association for Computational Linguistics (2016)

11. McSherry, F., Talwar, K.: Mechanism design via differential privacy. In: 48th Annual IEEE Symposium on Foundations of Computer Science, pp. 94–103 (2007)
12. Mikolov, T., Sutskever, I., Chen, K., Corrado, G.S., Dean, J.: Distributed representations of words and phrases and their compositionality. In: Advances in Neural Information Processing Systems, pp. 3111–3119 (2013)
13. Nissim, K., Raskhodnikova, S., Smith, A.: Smooth sensitivity and sampling in private data analysis. In: Proceedings of the Thirty-ninth Annual ACM Symposium on Theory of Computing, pp. 75–84. ACM (2007)
14. Perozzi, B., Al-Rfou, R., Skiena, S.: Deepwalk: online learning of social representations. In: Proceedings of the 20th ACM SIGKDD International Conference on Knowledge Discovery and Data Mining, pp. 701–710 (2014)
15. Phan, N., Wang, Y., Wu, X., Dou, D.: Differential privacy preservation for deep auto-encoders: an application of human behavior prediction. In: AAAI, pp. 1309–1316 (2016)
16. Qiu, J., Dong, Y., Ma, H., Li, J., Wang, K., Tang, J.: Network embedding as matrix factorization: unifying deepwalk, LINE, PTE, and node2vec. In: Proceedings of the Eleventh ACM International Conference on Web Search and Data Mining, pp. 459–467 (2018)
17. Song, S., Chaudhuri, K., Sarwate, A.D.: Stochastic gradient descent with differentially private updates. In: 2013 IEEE Global Conference on Signal and Information Processing, pp. 245–248 (2013)
18. Tang, J., Qu, M., Wang, M., Zhang, M., Yan, J., Mei, Q.: LINE: large-scale information network embedding. In: Proceedings of the 24th International Conference on World Wide Web, pp. 1067–1077 (2015)
19. Wang, Y., Wu, X., Hu, D.: Using randomized response for differential privacy preserving data collection. In: Proceedings of the 9th International Workshop on Privacy and Anonymity in the Information Society (2016)
20. Wang, Y., Wu, X., Wu, L.: Differential privacy preserving spectral graph analysis. In: Pei, J., Tseng, V.S., Cao, L., Motoda, H., Xu, G. (eds.) PAKDD 2013. LNCS (LNAI), vol. 7819, pp. 329–340. Springer, Heidelberg (2013). https://doi.org/10.1007/978-3-642-37456-2_28
21. Xu, D., Yuan, S., Wu, X.: Differential privacy preserving causal graph discovery. In: 2017 IEEE Symposium on Privacy-Aware Computing (PAC), pp. 60–71 (2017)
22. Yang, C., Liu, Z., Zhao, D., Sun, M., Chang, E.Y.: Network representation learning with rich text information. In: IJCAI, pp. 2111–2117 (2015)
23. Yuan, S., Wu, X., Xiang, Y.: SNE: signed network embedding. In: Kim, J., Shim, K., Cao, L., Lee, J.-G., Lin, X., Moon, Y.-S. (eds.) PAKDD 2017. LNCS (LNAI), vol. 10235, pp. 183–195. Springer, Cham (2017). https://doi.org/10.1007/978-3-319-57529-2_15
24. Zhang, J., Xiao, X., Yang, Y., Zhang, Z., Winslett, M.: PrivGene: Differentially private model fitting using genetic algorithms. In: Proceedings of the 2013 ACM SIGMOD International Conference on Management of Data, pp. 665–676 (2013)
25. Zhang, J., Zhang, Z., Xiao, X., Yang, Y., Winslett, M.: Functional mechanism: regression analysis under differential privacy. Proc. VLDB Endowment 5(11), 1364–1375 (2012)

A Generalization of Recurrent Neural Networks for Graph Embedding

Xiao Han[(✉)], Chunhong Zhang, Chenchen Guo, and Yang Ji

Key Laboratory of Universal Wireless Communications, Ministry of Education,
Beijing University of Posts and Telecommunications, Beijing, China
{hanxiao1007,zhangch,orangegcc,jiyang}@bupt.edu.cn

Abstract. Due to the ubiquity of graphs, machine learning on graphs facilitates many AI systems. In order to incorporate the rich information of graphs into machine learning models, graph embedding has been developed, which seeks to preserve the graphs into low dimensional embeddings. Recently, researchers try to conduct graph embedding via generalizing neural networks on graphs. However, most existing approaches focus on node embedding, ignoring the heterogeneity of edges. Besides, the similarity relationship among random walk sequences has been rarely discussed. In this paper, we propose a generalization of Recurrent Neural Networks on Graphs (G-RNN) for graph embedding. More specifically, first we propose to utilize edge embedding and node embedding jointly to preserve graphs, which is of great significance in multi-relational graphs with heterogeneous edges. Then we propose the definition of subgraph level high-order proximity to preserve the inter-sequence proximity into the embeddings. To verify the generalization of G-RNN, we apply it to the embedding of knowledge graph, a typical multi-relational graph. Empirically we evaluate the resulting embeddings on the tasks of link prediction and node classification. The results show that the embeddings learned by G-RNN are powerful on both tasks, producing better performance than the baselines.

1 Introduction

Graph is a primary abstraction for various physical worlds such as social networks, protein-protein interaction networks, and knowledge graph [11], where nodes model individual units and edges capture the relations between them. As a result, machine learning on graphs is a ubiquitous task with applications ranging from drug design to friendship recommendation in social networks. In order to incorporate the information about graph structure into machine learning models, representation learning has been developed to encode graph structure into low-dimensional embeddings. A "good" graph embedding representation should preserve both topological and semantic structure of the graph.

Existing neural network based graph embedding approaches focus on node embedding, which preserves graph topological structure and node semantics [12,21,25,26]. All of these approaches take edges as topological connections between nodes, ignoring their semantic information. It might work well for

© Springer International Publishing AG, part of Springer Nature 2018
D. Phung et al. (Eds.): PAKDD 2018, LNAI 10938, pp. 247–259, 2018.
https://doi.org/10.1007/978-3-319-93037-4_20

single-relational graph with homogeneous edges, where all the edges are treated identically. In contrast in multi-relational graphs, the rich semantic information of heterogeneous edges is also essential for graph representation. For example, given two nodes *Steve Jobs* and *USA* of a graph, there are multiple directed edges connecting them with different edge labels such as *BornIn* and *LiveIn*, providing different semantic dependency for the two nodes. Unfortunately, previous node embedding approaches tend to ignore the two edge labels and regard them identically. Therefore, only node embedding can not encode and distinguish the different edges between two nodes. To address this issue, we propose to perform edge embedding and node embedding jointly, considering that the semantic information of edges are indispensable for the structure of multi-relational graphs. As a result, we can predict the specific edge type between two nodes in the link prediction task, while previous node embedding approaches can only predict whether the two nodes are directly linked. To sum up, edge embedding is an important complement to node embedding approaches.

Generally, graph embedding approaches model the structure of graph with the proximity relationships. To begin with, the first-order proximity in [1,23] describes the pairwise similarity only between the directly linked nodes, which characterizes the local relationship of nodes. Since many legitimate links are missing, the first-order proximity is not sufficient to model the graph structure. Therefore, [25,26] propose the second-order proximity, describing the similarity of nodes according to their shared neighborhood structures. Furthermore, [12,21] preserve the high-order proximity within a random-walk multi-hop sequence. However, even the high-order proximity used in existing approaches only reflects the **intra-sequence proximity**, incapable of capturing the similarity among different sequences. As in Fig. 1, given a subgraph derived from two nodes, the sequences within the subgraph generally imply similar pattern. To preserve the **inter-sequence proximity** into embedding space, we define a **subgraph level high-order proximity**. A related work is PTransE [17], where the similarity between the multi-hop edge sequences and the direct edge within a subgraph is maximized. However, PTransE only considers the edges of the sequences and omits all the intermediate nodes, which will result in ambiguity in some cases. To avoid this problem, we compose both nodes and edges of the sequences so as to represent the exact sequence semantic meanings.

To model the subgraph level high-order proximity, we propose a generalization framework of Recurrent Neural Network on Graphs (G-RNN). As Fig. 1 shows, G-RNN is composed of two stages: subgraph extraction and G-RNN training. First, we extract the subgraphs from a graph via finding the one-hop and multi-hop sequences between arbitrary two nodes. Then, the sequences of each subgraph are sent into G-RNN. After G-RNN alternately preforms the same calculation on the nodes and edges along each sequence, we obtain the sequence embeddings by averaging all of the hidden states. Finally, we optimize model parameters and the embeddings of nodes and edges by maximizing the subgraph level high-order proximity. Empirically, we use G-RNN to learn the embeddings of knowledge graph, a typical multi-relational graph. We evaluate the learned embeddings on the tasks of link prediction and node classification. The results show that our approach provides better performance than the baselines.

The main contributions of this paper are summarized as follows:

1. In addition to graph topology and node semantics, we preserve the semantic information of edges into the embedding space, which is critical for multi-relational graphs with heterogeneous edges.

2. We model the inter-sequence proximity within a subgraph by preserving the subgraph level high order proximity in the embeddings.

3. We verify the generalization of G-RNN by applying it to knowledge graph embedding and evaluating the resulting embeddings on the tasks of link prediction and node classification.

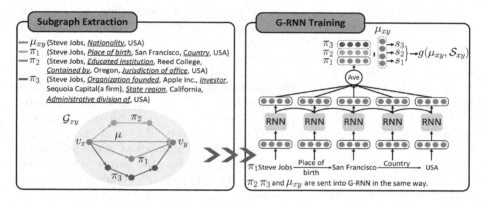

Fig. 1. Framework of G-RNN. $\mathcal{G}_{xy}$ is the subgraph derived from v_x *Steve Jobs* and v_y *USA*. The one-hop sequence μ_{xy} and the multi-hop sequences $\pi_1, \pi_2, \pi_3 \in \mathcal{S}_{xy}$ are sent into G-RNN. G-RNN recurrently processes the components of the sequences. After obtaining the embeddings of each sequence by averaging the hidden states, we calculate the similarity score between μ_{xy} and $\mathcal{S}_{xy}$ at the output.

2   Framework of Graph Recurrent Neural Network

In this section, we describe the generalization framework of G-RNN for the embedding of multi-relational graphs, whose definition is as follows:

Definition 1. *Multi-relational graph $\mathcal{G}$ is defined as $\mathcal{G} = (\mathcal{V}, \mathcal{E}, \mathcal{R})$, where $\mathcal{V}$ represents a set of $|\mathcal{V}| = n$ nodes, $\mathcal{E}$ is a set of $|\mathcal{E}| = m$ edges, and $\mathcal{R}$ is a finite set of $|\mathcal{R}| = k$ edge types corresponding to the semantic meanings of edges with $k \ll m$. Each edge $e \in \mathcal{E}$ belongs to a particular edge type $r \in \mathcal{R}$: $\psi(e) \in \mathcal{R}$, where $\psi(\cdot)$ is a edge type mapping function $\psi : \mathcal{E} \to \mathcal{R}$.*

Single-relational graphs can be viewed as a special case of multi-relational graphs, where the edge type set $\mathcal{R}$ has unique element r_0, and all the edges in $\mathcal{E}$ are mapped to r_0, i.e. $\psi(e) = r_0 \ \forall e \in \mathcal{E}$.

We propose G-RNN to learn a mapping: $v_i \rightarrow \mathbf{v}_i \in \mathbb{R}^d \ \forall i \in [n]$ and $r_j \rightarrow \mathbf{r}_j \in \mathbb{R}^d \ \forall j \in [k]$, so as to preserve the graph in a low-dimensional embedding space. The framework of G-RNN is composed of two stages: subgraph extraction and G-RNN training. Next, we detail the two stages as follows.

2.1 Subgraph Extraction

Given two nodes $v_x, v_y \in \mathcal{V}$, a subgraph $\mathcal{G}_{xy} \in \mathcal{G}$ is defined as $\mu_{xy} \cup \mathcal{S}_{xy}$, where $\mu_{xy} = (v_x, r, v_y)$ is the **one-hop sequence** with direct edge type r and $\mathcal{S}_{xy} = \{\pi_1, \cdots, \pi_q\}$ is the **multi-hop sequence** set. The multi-hop sequence $\pi_i = (v_x, c_1, \cdots, c_j, c_{j+1}, \cdots, v_y)$, where c_j and c_{j+1} are successive components the sequence passes through. If there is no direct edge between v_x and v_y, $\mathcal{G}_{xy} = \emptyset$.

Different instantiation of the component $c_j \in \pi_i$ suggests different consideration of the edge semantic significance to graph embedding. When c_j corresponds only to nodes in $\mathcal{V}$ as in [12,21], π_i is merely a **Node Sequence** with no semantic information of edges. If c_j is selected only from the edge type set $\mathcal{R}$ as in [16,19], this pure **Edge Sequence** $\pi_i = \{r_1, \cdots, r_l\}$ may lead to ambiguity since all the intermediate nodes joining the sequence are omitted. To address the semantic loss of the above two sequence types, we resort a **Node & Edge Sequence** $\pi_i = \{v_x, r_1, v_1, \cdots, r_p, v_p, \cdots, r_l, v_y\}$ where c_j and c_{j+1} alternatively correspond to an edge type and its contiguous node respectively (if $c_j = r_p$, $c_{j+1} = v_p$; if $c_j = v_p$, $c_{j+1} = r_{p+1}$). The hop number l is the number of edges joining the sequence. The composition of nodes and edges enables the sequence to preserve the semantic information as much as possible, and thus can be taken as the input of G-RNN to facilitate both the node and edge embedding learning.

Since obtaining the one-hop sequence μ_{xy} is trivial, the key problem of the subgraph extraction is finding the multi-hop sequences. We use bidirectional random walk [10] to find the multi-hop sequences, considering the random walk in Path Ranking Algorithm (PRA) [15] will become extremely time-consuming when the graph has a large scale or the hop number increases. Although we can get all the multi-hop sequences between v_x and v_y theoretically, it is inefficient and unnecessary for the following two reasons: First, the number of sequences will exponentially increase with the number of hops. Second, when the multi-hop sequences become too long, their correlation with the corresponding one-hop sequence will recede. Thus we set a maximum hop number l_{max} in practice.

2.2 G-RNN Training

In this work, the training objective of G-RNN is to preserve the subgraph level high-order proximity in embedding space, which is defined as follows:

Definition 2. *The subgraph level high-order proximity describes the proximity between the one-hop sequence and multi-hop sequences within a subgraph. Given the subgraph $\mathcal{G}_{xy} = \mu_{xy} \cup \mathcal{S}_{xy}$ derived from $v_x, v_y \in \mathcal{V}$, the subgraph level high-order proximity is determined by the similarity between μ_{xy} and $\mathcal{S}_{xy}$.*

In order to compare the similarity between μ_{xy} and $\mathcal{S}_{xy}$ in embedding space, we send μ_{xy} and each $\pi_i \in \mathcal{S}_{xy}$ into G-RNN. At step t, G-RNN calculates the hidden state according to the function $\mathbf{h}_t = \text{sigmoid}(\mathbf{U}\mathbf{x}_t + \mathbf{W}\mathbf{h}_{t-1})$, where $\mathbf{x}_t$ is the embedding vector of node $v_p \in \pi_i$ or edge type $r_p \in \pi_i$, $\mathbf{U}$ and $\mathbf{W}$ are weight matrixes. We obtain the embedding of π_i by averaging the hidden states of each step: $\boldsymbol{\pi}_i = (\boldsymbol{h}_1 + \cdots + \boldsymbol{h}_T)/T$, where T is the number of components in π_i. Then, the similarity between μ_{xy} and $\pi_i \in \mathcal{S}_{xy}$ is measured by:

$$s\left(\mu_{xy}, \pi_i\right) = \boldsymbol{\mu}_{xy}^{\mathsf{T}} \boldsymbol{\pi}_i. \tag{1}$$

Let $\{s_1, s_2, ..., s_q\}$ be the similarity scores for q multi-hop sequences in $\mathcal{S}_{xy}$. We calculate the similarity between μ_{xy} and $\mathcal{S}_{xy}$ according the following Log-SumExp (LSM) function:

$$g\left(\mu_{xy}, \mathcal{S}_{xy}\right) = \log\left(\sum_{i=1}^{q} \exp(s_i)\right). \tag{2}$$

We formulate the subgraph level high-order proximity using the conditional probability of one-hop sequence μ_{xy} given the multi-hop sequence set $\mathcal{S}_{xy}$:

$$p_\theta(\mu_{xy}|\mathcal{S}_{xy}) = \frac{\exp\left(g\left(\mu_{xy}, \mathcal{S}_{xy}\right)\right)}{\sum_{\mathcal{G}_{ij} \in \mathcal{G}} \exp(g(\mu_{ij}, \mathcal{S}_{xy}))} \tag{3}$$

where μ_{ij} represents the one-hop sequence of each subgraph $\mathcal{G}_{ij} \in \mathcal{G}$, θ represents the parameters to be optimized, including the embeddings and G-RNN weight matrixes. To estimate the parameters of $p_\theta(\mu_{xy}|\mathcal{S}_{xy})$, we use maximum likelihood estimation (MLE) to maximize the empirical (log-)likelihood of one hop sequence given the multi-hop sequences within the same subgraph:

$$\theta^{\text{MLE}} = \arg\max_\theta \sum_{\mathcal{G}_{xy} \in \mathcal{G}} \log p_\theta\left(\mu_{xy}|\mathcal{S}_{xy}\right). \tag{4}$$

The problem defined by Eq. (4) is a standard classification problem, whose training objective function is cross entropy ideally. However, due to the numerous one-hop sequences (or edges) in the graph, Eq. (4) is computationally expensive. Faced with this problem, hinge loss with negative samples [7] is a classic solution. Specifically, the training objective is to minimize the following loss function:

$$\mathcal{L} = \sum_{\mathcal{G}_{xy} \in \mathcal{G}} \sum_{\mu'_{xy} \in \Delta'_{\mu_{xy}}} \max\left[0, \gamma - g\left(\mu_{xy}, \mathcal{S}_{xy}\right) + g\left(\mu'_{xy}, \mathcal{S}_{xy}\right)\right] \tag{5}$$

where γ is the predefined margin, $\max(0, \cdot)$ is the hinge loss, and $\Delta'_{\mu_{xy}}$ is the set of negative one-hop sequences by replacing one component of μ_{xy}.

3   Application of G-RNN in Knowledge Graph

To verify the generalization of G-RNN, we apply it to knowledge graph (KG) embedding, encoding the rich symbolic information of KG into a low-dimensional space. KG is a typical multi-relational graph, where edges have different types and nodes have different types and attributes. The customary notations in KG are slightly different from those in general graphs. Given a KG $\mathcal{G}$, the entities in $\mathcal{G}$ denote the nodes in general graphs, and the relation types denote edge types. KG is often stored as triples in the form of (s, r, o), where r is the relation type between subject entity s and object entity o. The triple (s, r, o) is also a one-hop sequence μ_{so} as defined in Sect. 2. In addition, the multi-hop sequences are called multi-hop paths in KG so as to be consistent with the previous work [16].

The reasons why G-RNN can be used for KG embedding are as follows: First, G-RNN learn both entity embeddings and relation type embeddings, thus can preserve the topology as well as semantic structure of KG. Second, the subgraph level high-order proximity defined in G-RNN is also valid for KG. Consequently, we can directly use the loss function of Eq. (5) to optimize the embeddings of KG. For triple $\mu_{so} = (s, r, o)$, the training objective is

$$\mathcal{L}(\mu_{so}, \mathcal{S}_{so}) = \sum_{\mu'_{so} \in \Delta'_{\mu_{so}}} \max \left[0, \gamma_1 - g\left(\mu_{so}, \mathcal{S}_{so}\right) + g\left(\mu'_{so}, \mathcal{S}_{so}\right)\right] \qquad (6)$$

where $\mathcal{S}_{so}$ is the corresponding multi-hop sequence set of μ_{so}. The set of negative triples $\Delta'_{\mu_{so}}$ are the original triple μ_{so} with one of three components replaced:

$$\Delta'_{\mu_{so}} = \{(s', r, o)\} \cup \{(s, r', o)\} \cup \{(s, r, o')\}. \qquad (7)$$

The framework of G-RNN is flexible and can be combined with the existing approaches. For example, we can incorporate the widely used constraint based on translation, which is first introduced in [5]:

$$f(\mu_{so}) = -\left\|\mathbf{s} + \mathbf{r} - \mathbf{o}\right\|_{L_n} \qquad (8)$$

where $\|\cdot\|_{L_n}$ measures the L_n-distance between a translated subject entity $\mathbf{s} + \mathbf{r}$ and the object entity $\mathbf{o}$. To be consistent with the variation trend of score function $g(\mu_{so}, \mathcal{S}_{so})$, we add a minus before $\|\cdot\|_{L_n}$. So, $f(\mu_{so})$ gets a high score when (s, r, o) holds, and low otherwise. Based on Eq. (8), we obtain the following objective function:

$$\mathcal{L}(\mu_{so}) = \sum_{\mu'_{so} \in \Delta'_{\mu_{so}}} \max \left[0, \gamma_2 - f(\mu_{so}) + f(\mu'_{so})\right]. \qquad (9)$$

Let Δ be the triple set of the KG $\mathcal{G}$, the final loss function for KG embedding is the combination of (6) and (9):

$$\mathcal{L} = \sum_{\mu_{so} \in \Delta} \left(\mathcal{L}(\mu_{so}, \mathcal{S}_{so}) + \mathcal{L}(\mu_{so})\right). \qquad (10)$$

The training procedure is summarized in Algorithm 1. In line 1, all the entities and relation types are initialized as suggested in [5]. Lines 2–11 are the major part of training. We randomly sample a batch of triples and corresponding multi-hop sequences. According to the loss function in line 10, the embeddings of involved

---

Algorithm 1. Training G-RNN for KG embedding

Input: Training triple set $\Delta = \{\mu\}$, multi-hop path set $\{\mathcal{S}\}$, entities and rel. sets $\mathcal{V}$
   and $\mathcal{R}$, margins γ_1 and γ_1, embeddings dim. d.
Output: Embeddings of $v \in \mathcal{V}$ and $r \in \mathcal{R}$, weight matrix parameters of G-RNN.
   1: **initialize:** $\mathbf{v}, \mathbf{r} \leftarrow$ uniform$(-\sqrt{6/d}, \sqrt{6/d})$ for $v \in \mathcal{V}$, $r \in \mathcal{R}$
   2: **loop**
   3:    $\mathbf{v} \leftarrow \frac{v}{||v||}$, $r \leftarrow \frac{r}{||r||}$ for each entity $v \in \mathcal{V}$ and relation type $r \in \mathcal{R}$
   4:    $\Delta_{batch}, \mathcal{S}_{batch} \leftarrow$ sample$(\Delta, \{\mathcal{S}\}, b)$ // b is the size of mini batch
   5:    $T_{batch} \leftarrow \emptyset$ // initialize the set of pairs of valid and negative triples
   6:    **for** $\mu \in \Delta_{batch}$ **do**
   7:       $\mu' \leftarrow$ sample(Δ'_μ)
   8:       $T_{batch} \leftarrow T_{batch} \cup \{(\mu, \mu')\}$
   9:    **end for**
  10:    Update embeddings and parameters w.r.t. $\displaystyle\sum_{\substack{(\mu,\mu') \in T_{batch} \\ \mathcal{S} \in \mathcal{S}_{batch}}} \nabla [\mathcal{L}(\mu, \mathcal{S}) + \mathcal{L}(\mu)]$
  11: **end loop** until convergence

---

entities and relation types as well as G-RNN weight matrixes are optimized by
the Adam optimizer [14].

In this paper, we think of Eq. (8) as a modified first-order proximity. According to the original first-order proximity in [11], two nodes are similar if they are
linked by a direct edge. But in multi-relational graphs, the similarity of two nodes
should be measured under the condition of the edge semantics, especially for two
nodes of different types. For example, in triple (*Steve Jobs, nationality, USA*), it
is illogical to judge that the linked nodes *Steve Jobs* and *USA* are similar. Their
similarity should be characterized under the edge type *nationality*.

4   Experiments

In order to evaluate the embeddings learned by our proposed approach, we conduct the following two tasks: Link Prediction for predicting missing links and
Node Classification for predicting entity types. Both of the two tasks are conducted on FB15k, a public benchmark dataset with 14,951 entities and 1,345
relation types extracted from a typical large-scale knowledge graph Freebase [3].

4.1   Link Prediction

Owing to edge embedding, the proposed approach can predict the specific type
of the missing edge in addition to predicting whether a direct edge exists between
two nodes as in the previous node embedding approaches. To be specific, our
task of link prediction aims to complete the triple (s, r, o) when one of the components is missing, including two sub-tasks: **entity prediction** and **relation
type prediction**. Given an incomplete test triple, we fill up the missing position with each candidate entity or relation type in the dataset and calculate the

scores according to the following formula:

$$G(s, r, o) = \log \left(\sum_{\pi_i \in \mathcal{S}_{so}} \exp(\boldsymbol{\mu}_{so}^{\mathrm{T}} \boldsymbol{\pi}_i) \right) - \|\mathbf{s} + \mathbf{r} - \mathbf{o}\|_{L_n} . \tag{11}$$

Then, we rank the scores in descending order and record the ranking value of the correct candidate. Following [5], we use two measures as our evaluation metrics, including the mean rank of correct entities and the proportion of the correct candidates ranked in top 10 (for entity) and top 1 (for relation type). We also follow the two evaluation settings named as "Raw" and "Filter".

Entity Prediction. Entity prediction completes each test triple with missing head or tail entity. According to Eq. (11), all multi-hop paths between the known entity of the test triple and each candidate entity are necessary for the score calculation, which makes the sub-task time consuming. For simplification, we adopt the re-ranking method in [16]. First, we rank all candidate entities according to the scores from Eq. (8). Then, the top-500 candidate entities are re-ranked according to the scores from Eq. (11).

We decide the best hyper-parameters according to the mean rank of validation dataset. The exact values of the hyper-parameters are: learning rate $\lambda = 0.001$, margins $\gamma_1 = 0.3$, $\gamma_2 = 0.3$, dimension $d = 50$, and the dissimilarity measure is L_2.

The experiment results are shown in Table 1. For PTransE, we list the ADD composition with at most 2-hop and 3-hop, which provide better performance than other composition operations. From Table 1 we observe that: (1) G-RNN

Table 1. Entity prediction results.

Metric	Mean Rank		Hits@10(%)	
	Raw	Filter	Raw	Filter
RESCAL [20]	828	683	28.4	44.1
SE [6]	273	162	28.8	39.8
SME (linear) [4]	274	154	30.7	40.8
SME (bilinear) [4]	284	158	31.3	41.3
LFM [13]	283	164	26.0	33.1
TransE [5]	243	125	34.9	47.1
TransH [27]	212	87	45.7	64.4
TransR [17]	198	77	48.2	68.7
PTransE (2-hop) [16]	200	54	51.8	83.4
PTransE (3-hop) [16]	207	58	51.4	84.6
G-RNN (3-hop)	161	55	51.7	84.7
G-RNN (4-hop)	**156**	**50**	**52.1**	**85.2**

outperforms other baselines on all metrics, especially with a wide margin on "Raw" Mean Rank. It indicates that the graph embeddings learnt by G-RNN preserve the graph structure better. (2) For the paths with at most 3-hop, G-RNN performs better than PTransE. The reason is that G-RNN encodes the exact meanings of the paths by incorporating both entities and relation types of the paths, while PTransE only considers the relation types. Besides, the embedding dimension of G-RNN is 50 while it is 100 in PTransE, which indicates that G-RNN can preserve the structure of KG in lower-dimensional embeddings. (3) G-RNN with at most 4-hop performs better than 3-hop, which is a straightforward outcome as more paths can be used with the maximum hop number increasing. But as we have analyzed in Sect. 2, the hop number should not be increased continuously. Generally, we restrict the maximum hop number to 4.

According to the classification criteria in [5], relation types in KG can be categorized into four classes depending on their mapping properties: *1-to-1, 1-to-M, M-to-1, M-to-M*. The detailed results with respect to different classes of relation types are given in Table 2. We observe that on most classes of relation types, G-RNN performs better than the baselines.

Table 2. Detailed results on FB15k by mapping properties of relation types. (%)

Tasks	Predicting head entities (Hits@10)				Predicting tail entities (Hits@10)			
	1-to-1	1-to-M	M-to-1	M-to-M	1-to-1	1-to-M	M-to-1	M-to-M
SE	35.6	62.6	17.2	37.5	34.9	14.6	68.3	41.3
SME (linear)	35.1	53.7	19.0	40.3	32.7	14.9	61.6	43.3
SME (bilinear)	30.9	69.6	19.9	38.6	28.2	13.1	76.0	41.8
TransE	43.7	65.7	18.2	47.2	43.7	19.7	66.7	50.0
TransH	66.8	87.6	28.7	64.5	65.5	39.8	83.3	67.2
TransR	78.8	89.2	34.1	69.2	79.2	37.4	**90.4**	72.1
PTransE (2-hop)	91.0	**92.8**	60.9	83.8	91.2	74.0	88.9	86.4
PTransE (3-hop)	90.1	92.0	58.7	86.1	90.7	70.7	87.5	88.7
G-RNN (3-hop)	**92.3**	91.5	61.1	86.5	**92.2**	73.9	88.1	89.0
G-RNN (4-hop)	91.7	92.7	**62.5**	**87.7**	91.8	**74.2**	89.3	**89.7**

Relation Type Prediction. Relation type prediction aims to predict the relation type between two entities. In this sub-task, we directly use the score function in Eq. (11) to rank the candidate relation types. The KG embeddings used in this sub-task is the same as in entity prediction.

We evaluate G-RNN on relation type prediction by comparing it with PTransE [16] and TransE [5]. Following PTransE, we report Hits@1 rather than Hits@10, considering that all approaches provide pretty good results for Hits@10. We use the reported results in PTransE directly since the evaluation dataset is identical. To evaluate the effect of multi-hop paths, we conduct the experiment

Table 3. Relation type prediction results.

Table 4. Node classification results.

Metric	Mean Rank		Hits@1(%)	
	Raw	Filter	Raw	Filter
TransE	2.8	2.5	65.1	84.3
PTransE (2-hop)	1.7	**1.2**	69.5	93.6
-TransE	135.8	135.3	51.4	78.0
-Path	2.0	1.6	69.7	89.0
PTransE (3-hop)	1.8	1.4	68.5	94.0
G-RNN (3-hop)	1.7	1.3	69.8	93.8
-TransE	67.6	60.3	56.3	80.7
-Path	2.1	1.8	69.7	91.8
G-RNN (4-hop)	**1.5**	**1.2**	**70.2**	**94.1**

Metric	FB15k
TransE	87.9
DKRL	**90.1**
PTransE (ADD, 2-hop)	87.5
PTransE (ADD, 3-hop)	86.7
node2vec	63.2
G-RNN (3-hop)	88.5
G-RNN (4-hop)	89.7

with only paths (-TransE), with only score function in Eq. (8) (-Path) and with both.

From Table 3 we observe that: (1) G-RNN outperforms the baselines on both Mean Rank and Hits@1, which indicates that G-RNN preserves more information of relation types by incorporating both entities and relation types of the paths. (2) G-RNN with only considering path outperforms PTransE with the same setting by reducing the predict error more than half. It indicates that the entity embeddings also contain helpful information for relation type prediction. (3) When only use the score function of Eq. (8), the performance of G-RNN is better than PTransE on Hits@1.

4.2 Node Classification

Node classification is a multi-label classification problem aiming to predict the node types (entity types in KG), which has been used as an evaluation by many graph embedding approaches [11].

The dataset is extracted from FB15k according to the settings of DKRL [29]. We select the top 50 entity types according to their frequency, covering 13,445 entities. These entities are randomly split into training set (12,113 entities) and test set (1,332 entities).

We decompose the multi-label classification problem to a binary classification problem based on the one-versus-rest setting [18]. For a fair comparison, we use Logistic Regression as classifier as in DKRL [29]. We use MAP for evaluation, which is defined as the mean of average precision over all entity types [18].

The results of node classification are shown in Table 4. The baselines include TransE [5], DKRL [29], PTransE [16] and node2vec [12]. The observation and analysis are as follows: (1) G-RNN performs largely better than the node embedding approach node2vec. It indicates that the edge embeddings are also informative for entity type predicting. (2) G-RNN outperforms TransE and PTransE since G-RNN utilize the complete information of the paths. (3) The performance of DKRL is slightly better than G-RNN, considering that the entity descriptions may contain more information about entity types.

5   Related Work

Recently, researchers have developed a number of graph embedding approaches to preserve the graph structure into low dimensional embeddings. Many works utilize node embedding to achieve this goal. Among them, factorization based approaches [1,2] represent the first-order proximity between nodes in a matrix and obtain the node embeddings by matrix factorization; random walk based approaches [12,21,22] extend skip-gram architecture to graphs, preserving the high-order proximity of nodes along single sequence; besides, first and second-order proximities are jointly considered in [25,26]. Although the above node embedding approaches can preserve the topological linking between nodes, they are incapable of identifying the specific edge types. To address this issue, [24] distinguishes edge types by weight matrixes while embeds nodes in vector space, using graph convolutional neural networks for graph embedding. However, the weight matrixes of edge types are predefined and can not be optimized during training. Unlike the previous approaches, we leverage node and edge embedding jointly to preserve the graph structure. Furthermore, we generalize RNNs to graphs to model their subgraph level high-order proximity.

Another line of related works are the approaches for KG completion, which compose most of the baselines. The basic translation based approach is introduced in [5]. In order to deal with the polysemous entities and relations, [17,27] project the entities or relation types or both into hyperplanes and then perform translation at the new space. Some approaches improve the basic translation based approach with the help of supplementary information, such as entity descriptions [29], entity types [8] and related images [28]. Some other approaches utilize topological structure as an improvement, such as [9,16]. Besides, collective matrix factorization approach [20] and neural network based approaches [4,6,13] are also our baselines.

6   Conclusion

In this paper, we propose a generalization of RNNs on graphs to learn graph embedding. Via representing edge types in the same embedding space as nodes, we preserve the semantic information of heterogeneous edges. Furthermore, we define a subgraph level high-order proximity to preserve the similarity among different sequences within a subgraph. Besides, we incorporate both nodes and edge types of the sequence to obtain its semantic information as much as possible. We apply G-RNN to KG embedding and evaluate resulting embeddings on the tasks of link prediction and node classification. Experimental results show that G-RNN performs better than the baselines.

Acknowledgments. This work is supported by National Natural Science Foundation of China, 61602048, 61520106007, BUPT-SICE Excellent Graduate Students Innovation Fund, 2016.

References

1. Ahmed, A., Shervashidze, N., Narayanamurthy, S., Josifovski, V., Smola, A.J.: Distributed large-scale natural graph factorization. In: Proceedings of the 22nd International Conference on World Wide Web, pp. 37–48. ACM (2013)
2. Belkin, M., Niyogi, P.: Laplacian eigenmaps and spectral techniques for embedding and clustering. In: Advances in Neural Information Processing Systems, pp. 585–591 (2002)
3. Bollacker, K., Evans, C., Paritosh, P., Sturge, T., Taylor, J.: Freebase: a collaboratively created graph database for structuring human knowledge. In: Proceedings of the 2008 ACM SIGMOD International Conference on Management of Data, pp. 1247–1250. ACM (2008)
4. Bordes, A., Glorot, X., Weston, J., Bengio, Y.: A semantic matching energy function for learning with multi-relational data. Mach. Learn. **94**(2), 233–259 (2014)
5. Bordes, A., Usunier, N., Garcia-Duran, A., Weston, J., Yakhnenko, O.: Translating embeddings for modeling multi-relational data. In: Advances in Neural Information Processing Systems, pp. 2787–2795 (2013)
6. Bordes, A., Weston, J., Collobert, R., Bengio, Y., et al.: Learning structured embeddings of knowledge bases. In: AAAI, vol. 6, p. 6 (2011)
7. Collobert, R., Weston, J.: A unified architecture for natural language processing: deep neural networks with multitask learning. In: Proceedings of the 25th International Conference on Machine Learning, pp. 160–167. ACM (2008)
8. Das, R., Neelakantan, A., Belanger, D., McCallum, A.: Chains of reasoning over entities, relations, and text using recurrent neural networks. arXiv preprint arXiv:1607.01426 (2016)
9. Garcia-Duran, A., Bordes, A., Usunier, N.: Composing relationships with translations. Ph.D. thesis, CNRS, Heudiasyc (2015)
10. Gardner, M., Mitchell, T.M.: Efficient and expressive knowledge base completion using subgraph feature extraction. In: EMNLP, pp. 1488–1498 (2015)
11. Goyal, P., Ferrara, E.: Graph embedding techniques, applications, and performance: a survey. arXiv preprint arXiv:1705.02801 (2017)
12. Grover, A., Leskovec, J.: node2vec: scalable feature learning for networks. In: Proceedings of the 22nd ACM SIGKDD International Conference on Knowledge Discovery and Data Mining, pp. 855–864. ACM (2016)
13. Jenatton, R., Roux, N.L., Bordes, A., Obozinski, G.R.: A latent factor model for highly multi-relational data. In: Advances in Neural Information Processing Systems, pp. 3167–3175 (2012)
14. Kingma, D., Ba, J.: Adam: a method for stochastic optimization. arXiv preprint arXiv:1412.6980 (2014)
15. Lao, N., Mitchell, T., Cohen, W.W.: Random walk inference and learning in a large scale knowledge base. In: Proceedings of the Conference on Empirical Methods in Natural Language Processing, pp. 529–539. Association for Computational Linguistics (2011)
16. Lin, Y., Liu, Z., Luan, H., Sun, M., Rao, S., Liu, S.: Modeling relation paths for representation learning of knowledge bases. arXiv preprint arXiv:1506.00379 (2015)

17. Lin, Y., Liu, Z., Sun, M., Liu, Y., Zhu, X.: Learning entity and relation embeddings for knowledge graph completion. In: AAAI, pp. 2181–2187 (2015)
18. Neelakantan, A., Chang, M.W.: Inferring missing entity type instances for knowledge base completion: new dataset and methods. arXiv preprint arXiv:1504.06658 (2015)
19. Neelakantan, A., Roth, B., McCallum, A.: Compositional vector space models for knowledge base inference. In: 2015 AAAI Spring Symposium Series (2015)
20. Nickel, M., Tresp, V., Kriegel, H.P.: A three-way model for collective learning on multi-relational data. In: Proceedings of the 28th International Conference on Machine Learning (ICML 2011), pp. 809–816 (2011)
21. Perozzi, B., Al-Rfou, R., Skiena, S.: Deepwalk: online learning of social representations. In: Proceedings of the 20th ACM SIGKDD International Conference on Knowledge Discovery and Data Mining, pp. 701–710. ACM (2014)
22. Ristoski, P., Paulheim, H.: RDF2Vec: RDF graph embeddings for data mining. In: Groth, P., Simperl, E., Gray, A., Sabou, M., Krötzsch, M., Lecue, F., Flöck, F., Gil, Y. (eds.) ISWC 2016. LNCS, vol. 9981, pp. 498–514. Springer, Cham (2016). https://doi.org/10.1007/978-3-319-46523-4_30
23. Roweis, S.T., Saul, L.K.: Nonlinear dimensionality reduction by locally linear embedding. Science 290(5500), 2323–2326 (2000)
24. Schlichtkrull, M., Kipf, T.N., Bloem, P., Berg, R.v.d., Titov, I., Welling, M.: Modeling relational data with graph convolutional networks. arXiv preprint arXiv:1703.06103 (2017)
25. Tang, J., Qu, M., Wang, M., Zhang, M., Yan, J., Mei, Q.: Line: large-scale information network embedding. In: Proceedings of the 24th International Conference on World Wide Web, pp. 1067–1077. International World Wide Web Conferences Steering Committee (2015)
26. Wang, D., Cui, P., Zhu, W.: Structural deep network embedding. In: Proceedings of the 22nd ACM SIGKDD International Conference on Knowledge Discovery and Data Mining, pp. 1225–1234. ACM (2016)
27. Wang, Z., Zhang, J., Feng, J., Chen, Z.: Knowledge graph embedding by translating on hyperplanes. In: AAAI, pp. 1112–1119 (2014)
28. Xie, R., Liu, Z., Chua, T.s., Luan, H., Sun, M.: Image-embodied knowledge representation learning. arXiv preprint arXiv:1609.07028 (2016)
29. Xie, R., Liu, Z., Jia, J., Luan, H., Sun, M.: Representation learning of knowledge graphs with entity descriptions. In: AAAI, pp. 2659–2665 (2016)

NE-FLGC: Network Embedding Based on Fusing Local (First-Order) and Global (Second-Order) Network Structure with Node Content

Hongyan Xu, Hongtao Liu, Wenjun Wang, Yueheng Sun[(✉)], and Pengfei Jiao

Tianjin Key Laboratory of Advanced Networking (TANK),
School of Computer Science and Technology, Tianjin University, Tianjin, China
{hongyanxu,htliu,wjwang,yhs,pjiao}@tju.edu.cn

Abstract. This paper studies the problem of Representation Learning for network with textual information, which aims to learn low dimensional vectors for nodes by leveraging network structure and textual information. Most existing works only focus on one aspect of network structure and cannot fuse network first-order proximity, second-order proximity and textual information. In this paper, we propose a novel network embedding method NE-FLGC: Network Embedding based on Fusing Local (first-order) and Global (second-order) network structure with node Content. Especially, we adopt context-enhance method that obtains node embedding by concatenating the vector of itself and the context vectors. In experiments, we compare our model with existing network embedding models on four real-world datasets. The experimental results demonstrate that NE-FLGC is stable and significantly outperforms state-of-the-art methods.

Keywords: Network embedding · Attributed network · Node content

1 Introduction

Nowadays, networks are more and more ubiquitous in the real world, e.g. social networks, paper citation networks and so on. At the same time, network analysis becomes even more difficult along with the size of the network larger and larger. How to represent network effectively is the fundamental and critical problem in network analysis, and thus has attracted much attention recently. Network Representation Learning(NRL) or Network Embedding(NE) generally aims to represent each node of networks into a continuous low-dimensional space. In recent years, network embedding has shown highly effective in node classification, link prediction, network visualization etc. [3,14,16,19].

Most previous NRL works mainly study the network structure. Some works represent a network as a set of random walks. For example, Deep Walk [14] adopts truncated random walks to preserve network structure information, and utilize Skip-Gram [12] model to learn node embedding. Instead of sampling paths from a network, LINE [16] designs objective function that preserves both the local and global network structures. In real world networks, nodes usually contain rich information. To incorporate textual information associated with nodes, TADW [20] incorporates the feature matrix into low-rank matrix factorization. However, it can't scale to large networks due to high computation cost.

Those methods above achieves significant improvement compared to traditional models [1,15], but they either only encode the network structure into node embeddings, or preserve first-order proximity and second-order proximity separately when they learn network representation. To address these issues, we propose a novel NRL method NE-FLGC: Network Embedding based on Fusing Local (first-order) and Global (second-order) network structure with node Content. Especially, a context-enhanced method is proposed, which obtains the embedding for each node by concatenating the vector of itself and the context vectors.

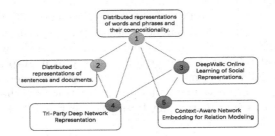

Fig. 1. Example of a citation network with title or abstract as textual information.(Green, blue and red font represent key words in Natural Language Processing, Network, and both respectively.) (Color figure online)

In order to make full use of the information hidden in the network structure, we introduce first-order proximity and second-order proximity. First-order proximity in network is the similarity between nodes linked by observed links. For example, in Fig. 1, there is a link between vertex 3 and vertex 5, meaning that the first-order proximity between them are large. However, real-world networks are usually sparse, only a small ratio links observed [9]. Only first-order proximity is not enough to preserve network structure. Thus, we introduce second-order proximity which is the similarity between node's neighbor structure. If two nodes have common nodes, then they tend to be similar with each other. For example, in Fig. 1, vertex 4 and vertex 5 have no link between them, meaning that the first-order proximity between them are small. In contrast, the second-order proximity is large due to that they both link with vertex 1, vertex 3. They are also similar in terms of node content. As we all know, they should be close in embedding space because these papers are from the same area. This shows that node content, first-order and second order proximity are all essential parts to learn network embedding. Specifically, we adopt context-enhanced method to recover global structure in network, which gets node embedding by concatenating the vector of it self and the average of context vectors.

We evaluate our model against several baselines on four networks. The effectiveness of the learned embeddings is evaluated within multiple data mining tasks, including multi-class classification, link prediction and visualization. The results suggest that NE-FLGC outperforms baselines in terms of both effectiveness and efficient. There are two contributions of this paper: (1) we propose a novel network embedding method NE-FLGC, which fuses textual information associated with nodes, first-order proximity and second-order proximity of network structure to learn network embedding; (2) We conduct experiments on four real-world networks. Experimental results prove the effectiveness of our model.

2   Related Work

Recently, there has been a lot of Network representation learning models proposed to learn effective embeddings [3,14,19] etc. Node2vec [3] performs a biased random walk over network and gets node embedding by using skip-gram model, proposed in word2vec [12]. GraRep [2] learns representations by integrating different k-step local relational information into a global graph representation. SDNE [19] takes adjacency matrix as input to auto encoder and uses Laplacian Eigenmap [1] to preserve first-order proximity. However, these methods only consider the structure information, neglecting the textual information associated with nodes.

To address this problem, researchers make great efforts to incorporate textual information into structure-based representation models. DeepWalk has been proved to be equivalent to PMI matrix factorization [8]. TADW [20] incorporates feature matrix into NRL under the framework of matrix factorization and gets considerable improvement compared to baselines. Tri-DNR [13] is a tri-party NRL model with three DeepWalk: network structure(node-to-node), content of nodes(node-to-word), label of nodes(label-to-word). To the best of our knowledge, we are the first to preserve node text and the first-order, second-order proximity of network structure simultaneously in network embedding. We propose NE-FLGC, which learn network embedding by fusing textual information associated with nodes, first-order and second-order proximity in network.

3   Problem Formulation

We first give basic notations and definitions in our work.

Definition 1 (Attributed network). Let G =(V, E, T) denote an attributed network. $V = \{u_i\}_{i=0,1,2,\ldots,N-1}$ is the set of nodes. $E \in (V \times V)$ is the set of edges and each $e_{ij} \in E$ represents the relation between node u_i and u_j. And, $t_i \in T$ denotes the textual information of node u_i.

Definition 2 (Network Embedding). Given an attribute network, network embedding aims to learns a distributed low-dimension vector $\vec{u}_i \in \mathbb{R}^d$ for each node u_i according to its network structure and associated textual information, where $d \ll |V|$.

Definition 3 (Embedding proximity). We introduce first-order proximity and second-proximity in network. (1)The **first-order proximity** is local pairwise proximity between two nodes, based on node embedding $\vec{u}_i$ and $\vec{u}_j$. The weight w_{ij} in edge e_{ij} indicates the first-order proximity. (2)The **second-order proximity** of node pair (u_i, u_j) is the similarity between their neighbor structure, based on node embedding $\vec{u}_i$ and context embedding $\vec{u}'_j$. If two nodes have common nodes, then they are more likely to be similar to each other.

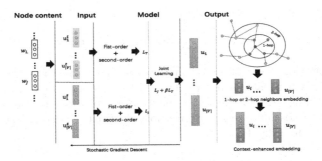

Fig. 2. Framework overview.

4   Our Model

To make full use of the network information, our model contains two part: structure-based module and text-based module. Afterwards, in order to integrate the neighbor information of nodes we propose a context-enhance method in the inference stage. The framework of our model is illustrated in Fig. 2.

4.1   Structure-Based Module

In social networks, people who share similar interests are more likely to be friends with each other. On the other hand, people who have common friends tend to have similar interests and thus become friends. To exploit the observation above, we introduce first-order proximity and second-order proximity. According to the definition of embedding proximity, each node has two roles: as node itself $\vec{u}^s$ and as context node of other node $\vec{u}^{s'}$. We assume edges in network is directed and undirected edge can be considered as two directed edges. Given a pair of nodes (i, j), we define the probability of node j generated by node i as follows:

$$p(u_j^s | u_i^s) = \frac{exp(\vec{u}_i^{sT} \cdot \vec{u}_j^s + \alpha \vec{u}_i^{sT} \cdot \vec{u}_j^{s'})}{Z}, \tag{1}$$

where $Z = \sum_{\{p,q\} \in V \times V, p \neq q} exp(\vec{u}_p^{sT} \cdot \vec{u}_q^s + \alpha \vec{u}_p^{sT} \cdot \vec{u}_q^{s'})$ is a normalization term and α is the balance weight which indicates the importance of second-order proximity. If node i and j have similar embedding vectors, meaning they have larger first-order similarity, $\vec{u}_i^{sT} \cdot \vec{u}_j^s$ will be large, leading to a large conditional probability. On the other hand, if the embedding vector of node i is similar to the context vector of node j, meaning they have larger second-order similarity, $\vec{u}_i^{sT} \cdot \vec{u}_j^{s'}$ will be large, leading to a large conditional probability.

To learn node embedding, we expect estimated distribution $p(u_j|u_i)$ to be close to the empirical distribution $\hat{p}(u_j|u_i)$, which is defined as $\hat{p}(u_j|u_i) = \frac{w_{ij}}{d_i}$, where $d_i = \sum_{k=1}^{|V|} w_{ik}$ is the out-degree of node u_i and w_{ij} is the weight of edge e_{ij}. We minimize the KL-divergence [5] of two probability distributions: $L_J = KL(\hat{p}(.|.), p(.|.))$. After removing some constants, the above equation can be written as:

$$L_J = - \sum_{(i,j) \in E} w_{ij} log(p(u_j^s|u_i^s)). \tag{2}$$

Optimizing Eq. (2) is computationally expensive, which requires summation over all nodes when calculating Eq. (1). Therefore, we adopt the approach of negative sampling [12] to speed up the training process. It samples negative edges according to some noise distribution. So Eq. (2) is transformed into the following form (for edge (u_i, u_j)):

$$log\sigma(\vec{u}_i^{sT} \cdot \vec{u}_j^s + \alpha \vec{u}_i^{sT} \cdot \vec{u}_j^{s'}) + \sum_{k=1}^{K} E_{u_n \sim P_n(u)}[log\sigma(-\vec{u}_i^{sT} \cdot \vec{u}_n^s - \alpha \vec{u}_i^{sT} \cdot \vec{u}_n^{s'})], \tag{3}$$

where $\sigma(x) = \frac{1}{(1+exp(-x))}$ is the sigmoid function and K is number of negative edges. The first term models the positive edge that can be observed in network and the second term models the negative and invisible edges, which are randomly sampled from the noise distribution. According to Word2vec [12], we set $P_n(u) \propto d_u^{3/4}$, d_u is the degree of node u.

4.2   Text-Based Module

To incorporate textual information into network embedding model, here we apply a simple but effective approach to encode the textual information. Similar as [4], we use the average of words embedding to represent sentences/document associated with nodes. For each node i, we define text embedding vector $\vec{u}_i^t$, initialized by the average of word embedding (if available) or zeros. $\vec{u}_i^t = \frac{1}{|t_i|} \sum_{k=0}^{|t_i|-1} \vec{e}_{W_k}$ where t_i is the node content of node i and $\vec{e}_{W_k}$ is the vector of the k-th word in it.

The embedding proximity defines not only about the network structure but also the textual information. Therefore, following the structure proximity above, we denote the text proximity with a conditional probability:

$$p(u_j^t|u_i^t) = \frac{exp(\vec{u}_i^{tT} \cdot \vec{u}_j^t)}{\sum_{\{p,q\} \in V \times V, p \neq q} exp(\vec{u}_p^{tT} \cdot \vec{u}_q^t)}. \tag{4}$$

We formulate the text-based objective function L_T according to Eq. (3) and adopt the approach of negative sampling. After omitting some constrains, we have:

$$L_T = \sum_{(i,j)\in E} log\sigma(\vec{u}_i^{tT} \cdot \vec{u}_j^t) + \sum_{i=1}^{K} E_{v_n \sim P_n(v)}[log\sigma(-\vec{u}_i^{tT} \cdot \vec{u}_n^t))]. \qquad (5)$$

4.3   Training

Finally, we optimize the following joint objective function:

$$L = L_J + \beta L_T. \qquad (6)$$

where L_J, L_T is the structure-based and text-based objective function respectively and hyper-parameter β is the balance weight that indicates the importance of node content. The gradients are computed with back-propagation and optimized with stochastic gradient descent (SGD). In each step, SGD algorithm samples a mini-batch edges and then updates the model parameters. Given an edge (u_i, u_j), the structure embedding vector $\vec{u}_i^s$ of u_i will be calculated as:

$$\frac{\partial L}{\partial \vec{u}_i^s} = w_{ij} \frac{\partial log(p(u_j^s|u_i^s))}{\partial \vec{u}_i^s}. \qquad (7)$$

Algorithm 1. NE-FLGC

Input: G=(V, E, T), Parameter α, β, DIM, number of epoch T
Output: node embedding $\vec{u}^s$; node context embedding $\vec{u}^{s'}$; node text embedding $\vec{u}^t$
1: Pre-train node content and get $\vec{u}_p^t$;
2: Randomly Initialize $\vec{u}^s$ and $\vec{u}^{s'}$, initialize $\vec{u}^t$ with node content vector$\vec{u}_p^t$
3: For epc *gets* 0 to T do
       Lookup embedding of node structure/text from $\vec{u}^s$, $\vec{u}^{s'}$ and $\vec{u}^t$ respectively
       Loss $= -\sum_{(i,j)\in E} w_{ij}(log(p(u_j^s|u_i^s) + \beta log(p(u_j^t|u_i^t))$
       Update embeddings with SGD
4: End for
5: **return** node embedding $\vec{u}^s$, node attribute embedding $\vec{u}^t$

4.4   Context-Enhance

We optimize our model with independence assumption. It means that the conditional probability $p(u_j|u_i)$ is independent to $p(u_k|u_i)$ for any nodes u_i, u_j, u_k. However, coherent structure in the world leads to strong correlations between inputs (such as between neighboring pixels in images) [15], meaning that independence assumption will weaken the representation power of network embedding models. To address this problem, we propose context-enhanced method to recover global structure in network, which gets node embedding by concatenating the vector of it self and the average of context vectors.

The k-hop neighborhood of node u_i is defined as $N_i = \{j \in V | i \neq j, sp(i,j) = k\}$, where V is the set of all nodes, $sp(i,j)$ returns the length of the shortest path from node i to node j ($sp(i,j) = \infty$ if node j is not reachable). We define the embedding of nodes as follows.

$$\vec{u}_i = (\vec{u}_i^s \oplus \vec{u}_i^t) \oplus \frac{1}{|N_i|} \sum_{u_k \in N_i} \vec{u}_k^s \oplus \vec{u}_k^t, \tag{8}$$

where u^s and u^t is the node embedding and text embedding respectively.

5 Experiments

5.1 Dataset

We evaluate our model on four public benchmark datasets. To construct network of textual information, we select four citation network dataset. **Cora** is a paper citation network [11] with 2,277 machine learning papers which are divided into 7 categories, and 5,214 links between nodes. **HepTh** is a citation network [7] from arXiv, in which there are 1,038 papers with textual information, and 1,990 links between nodes. **DBLP** is a citation network [17] with 60,744 papers which are divided into 4 categories and 52,890 links between papers. **CiteSeer-M10** is a subxf [10] of CiteSeerX data. There are 38997 nodes in network and only 10310 nodes with textual information which are divided into 10 distinct research areas, and 77,218 links between papers. We take the abstract of academic papers as node content.

5.2 Experimental Settings

We perform node classification, link prediction and network visualization to evaluate the quality of different models and implement NE-FLGC based on Pytorch.
Baselines

Structure-Only: **DeepWalk** [14] adopts random walk on network structure and uses the Skip-Gram [12] model to learn node embedding. **LINE** [16] preserves both the local and global network structures and can scale to large network. **node2vec** [3] is based on DeepWalk, preserving neighborhoods with BFS, DFS random walk strategies.

With node content and/or node type: **Doc2Vec** [6] represents each document by a dense vector and trains model by predicting words in the documents. **TADW** [20] is based on matrix factorization and incorporates text features of vertices into network embedding learning. **CANE** [18] learns context-aware embeddings for vertices with mutual attention mechanism. **Tri-DNR** [13] uses coupled DeepWalk model to jointly learn network embedding with multi-type information: node structure, node content, node labels.

Parameter Setting

In multi-label classification task, we use SVM classifier. We train our model and baseline methods with structure/text embedding dimension d = 100 and set other parameters in baselines like [13,18] do. Besides, we pre-train text with Word2vec [12] model and set the number of negative samples k = 1 and the learning rate $\rho = 0.02$. We also apply grid search to set the hyper-parameters α and β, and we set $\alpha = 6$, $\beta = 10$ in NE-FLGC. We use 2-hop neighborhood for CiteSeer-M10 dataset and use 1-hop neighborhood for other datasets.

5.3 Multi-class Classification

Similar as [13], we conduct multi-class classification on DBLP and Citeseer M10 datasets. We vary the percentages of training samples t% from 10% to 70% and report the result in Tables 1 and 2.

Table 1. Average Macro f1 score values on dblp

t%	DeepWalk	LINE	node2vec	Doc2Vec	DW+D2V	TADW	Tri-DNR	NE-FLGC
10%	0.398	0.427	0.424	0.605	0.653	0.676	0.687	**0.730**
30%	0.423	0.438	0.425	0.617	0.681	0.689	0.727	**0.738**
50%	0.426	0.438	0.430	0.620	0.686	0.692	0.738	**0.742**
70%	0.428	0.439	0.432	0.623	0.690	0.695	0.744	**0.748**

Table 2. Average Macro f1 score values on M10

t%	DeepWalk	LINE	node2vec	Doc2Vec	DW+D2V	TADW	Tri-DNR	NE-FLGC
10%	0.354	0.531	0.528	0.432	0.495	0.600	0.626	**0.729**
30%	0.411	0.569	0.548	0.477	0.586	0.652	0.715	**0.775**
50%	0.425	0.581	0.560	0.494	0.614	0.671	0.753	**0.792**
70%	0.434	0.589	0.561	0.503	0.628	0.681	0.777	**0.802**

From Tables 1 and 2, we have the following observations: (1) DeepWalk, LINE and node2vec performs poorly on these two datasets. It is mainly because that structure-based algorithms learn network representation without node content. Doc2Vec performs poorly in these two datasets too, because it only learns document or sentence embedding to represent nodes without network structure. The naive combination of the DeepWalk embedding and Doc2Vec embedding performs better than any one of them. However, NE-FLGC performs better than all these baselines, mainly because we jointly train network structure and node content. (2) NE-FLGC performs better than TADW and is applying to large scale network. For a tradeoff between speed and accuracy, TADW factorizes the approximate Matrix $M = (A + A^2)/2$, which will weaken its representation power. (3) As Tables 1 and 2 show, our model performs better than Tri-DNR, even though we do not incorporate label information. It is mainly because that our model learn NRL by fusing first-order, second-order proximity and node content together. Even though Tri-DNR contains network structure, textual information, and node labels, it only uses simply coupled neural network model with three DeepWalk [14] models.

To summarize, all the above observations demonstrate that NE-FLGC achieves better performance compared to baselines. Especially, NE-FLGC beats LINE and Tri-DNR by 37.2% and 16.4% under 10% training ratios on M10 dataset. It demonstrates that our model perform well even with small training ratio.

Table 3. AUC values on Cora

Training%	DeepWalk	LINE	node2vec	TADW	CANE	NE-FLGC
15%	56.0	55.0	55.9	86.6	**86.8**	85.9
25%	63.0	58.6	62.4	88.2	**91.5**	90.4
35%	70.2	66.4	66.1	90.2	92.2	**92.3**
45%	75.5	73.0	75.0	90.8	93.9	**93.9**
55%	80.1	77.6	78.7	90.0	94.6	**94.6**
65%	85.2	82.8	81.6	93.0	94.9	**95.9**
75%	85.3	85.6	85.9	91.0	95.6	**96.9**
85%	87.8	88.4	87.3	93.4	96.6	**97.3**

5.4 Link Prediction

Link prediction is an important application of network embedding. Similar as [18], we evaluate the AUC values with ratios of training data ranking from 15% to 85% on Cora and HepTh dataset.

As shown in Tables 3 and 4, our model achieves comparable performance than baselines on link prediction. DeepWalk, LINE, node2vec always perform poorly in small ratio training data. It is mainly because they don't incorporate textual information. By adopting context-enhanced method, our model obtains considerable improvements than baselines. Our model performs better than TADW and CANE, because we preserve first-order and second-order proximity simultaneously. CANE learns context-aware embeddings for vertices with mutual attention mechanism, whose objective function is based on the first-order proximity in network.

5.5 Parameter Sensitivity

In this subsection, we investigate how the parameter dimension d and weights β, indicating the importance of node content, affects NE-FLGC. We test the

Table 4. AUC values on HepTh

Training%	DeepWalk	LINE	node2vec	TADW	CANE	NE-FLGC
15%	55.2	53.7	57.1	87.0	90.0	**92.7**
25%	66.0	60.4	63.6	89.5	91.2	**93.6**
35%	70.0	66.5	69.9	91.8	92.0	**94.4**
45%	75.7	73.9	76.2	90.8	93.0	**95.3**
55%	81.3	78.5	84.3	91.1	94.2	**95.8**
65%	83.3	83.8	87.3	92.6	94.6	**96.2**
75%	87.6	87.5	88.4	93.5	95.4	**96.6**
85%	88.9	87.7	89.2	91.9	95.7	**97.0**

classification average F1-score with dimension d ranging from 10 to 100 on DBLP
and Citeseer M10 datasets. When d is under test, other parameters are set to
their default values. Figure 3(a) and (b) shows the classification performance of
our model on DBLP and M10 respectively, compared with, Tri-DNR, DeepWalk
etc. We can see the performance of our model is very stable. Especially, our
model performs well even under small training ratios. Besides, we test the link
prediction AUC with β ranging from 0 to 1 on cora dataset. In Fig. 3(c), We can
see that the performance of our method is stable when β is bigger than 0.4.

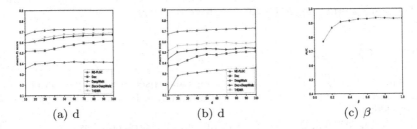

Fig. 3. Parameter sensitivity analysis on dimension d and β

5.6   Network Visualization

Network visualization is a basic application of network representation. Similar as
[16], we visualize a citation network extracted from M10 dataset by mapping the
embedding vectors of nodes into 2D space. We represent papers in different areas
with different colors. As shown in Fig. 4(a), in the visualization of Doc2vec, nodes
with the same color are not clustered together. It is mainly because that Doc2vec
model neglects network structure. In Fig. 4(b), Tri-DNR model performs better
than Doc2vec, but the border of different color is not clear. NE-FLGC performs
well and produces meaningful layout for each community(nodes with different
colors are distributed far in Fig. 4(c)).

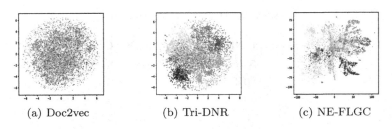

| (a) Doc2vec | (b) Tri-DNR | (c) NE-FLGC |

Fig. 4. Visualization on M10 dataset (Color figure online)

6   Conclusion and Future Work

In this paper, we propose a novel network embedding method NE-FLGC, which learns network embedding by fusing textual information, first-order and second-order proximity of network structure simultaneously. Especially, a context-enhanced method is proposed to capture the correlations between node and it's context. Experiments on a variety of different networks illustrate the effectiveness of our approach on multi-class classification, link prediction and network visualization tasks. In the future, we plan to investigate the embedding of heterogeneous information networks with node content/types.

Acknowledgments. This work was supported by the Major Project of National Social Science Fund(14ZDB153),the major research plan of the National Natural Science Foundation (91746205,91746107,91224009,51438009).

References

1. Belkin, M., Niyogi, P.: Laplacian eigenmaps and spectral techniques for embedding and clustering. In: Advances in Neural Information Processing Systems, pp. 585–591 (2002)
2. Cao, S., Lu, W., Xu, Q.: GraRep: learning graph representations with global structural information. In: Proceedings of the 24th ACM International on Conference on Information and Knowledge Management, pp. 891–900. ACM (2015)
3. Grover, A., Leskovec, J.: node2vec: scalable feature learning for networks. In: Proceedings of the 22nd ACM SIGKDD International Conference on Knowledge Discovery and Data Mining, pp. 855–864. ACM (2016)
4. Joulin, A., Grave, E., Bojanowski, P., Mikolov, T.: Bag of tricks for efficient text classification. arXiv preprint arXiv:1607.01759 (2016)
5. Kullback, S., Leibler, R.A.: On information and sufficiency. Ann. Math. Stat. **22**(1), 79–86 (1951)
6. Le, Q., Mikolov, T.: Distributed representations of sentences and documents. In: International Conference on Machine Learning, pp. 1188–1196 (2014)
7. Leskovec, J., Kleinberg, J., Faloutsos, C.: Graphs over time: densification laws, shrinking diameters and possible explanations. In: Proceedings of the Eleventh ACM SIGKDD International Conference on Knowledge Discovery in Data Mining, pp. 177–187. ACM (2005)

8. Levy, O., Goldberg, Y.: Neural word embedding as implicit matrix factorization. In: Advances in Neural Information Processing Systems, pp. 2177–2185 (2014)
9. Liben-Nowell, D., Kleinberg, J.: The link-prediction problem for social networks. J. Assoc. Inf. Sci. Technol. **58**(7), 1019–1031 (2007)
10. Lim, K.W., Buntine, W.: Bibliographic analysis with the citation network topic model. arXiv preprint arXiv:1609.06826 (2016)
11. McCallum, A.K., Nigam, K., Rennie, J., Seymore, K.: Automating the construction of internet portals with machine learning. Inf. Retrieval **3**(2), 127–163 (2000)
12. Mikolov, T., Sutskever, I., Chen, K., Corrado, G.S., Dean, J.: Distributed representations of words and phrases and their compositionality. In: Advances in Neural Information Processing Systems, pp. 3111–3119 (2013)
13. Pan, S., Wu, J., Zhu, X., Zhang, C., Wang, Y.: Tri-party deep network representation. Network **11**(9), 12 (2016)
14. Perozzi, B., Al-Rfou, R., Skiena, S.: DeepWalk: online learning of social representations. In: Proceedings of the 20th ACM SIGKDD International Conference on Knowledge Discovery and Data Mining, pp. 701–710. ACM (2014)
15. Roweis, S.T., Saul, L.K.: Nonlinear dimensionality reduction by locally linear embedding. Science **290**(5500), 2323–2326 (2000)
16. Tang, J., Qu, M., Wang, M., Zhang, M., Yan, J., Mei, Q.: LINE: large-scale information network embedding. In: Proceedings of the 24th International Conference on World Wide Web, pp. 1067–1077. International World Wide Web Conferences Steering Committee (2015)
17. Tang, J., Zhang, J., Yao, L., Li, J., Zhang, L., Su, Z.: Arnetminer: extraction and mining of academic social networks. In: Proceedings of the 14th ACM SIGKDD International Conference on Knowledge Discovery and Data Mining, pp. 990–998. ACM (2008)
18. Tu, C., Liu, H., Liu, Z., Sun, M.: Cane: Context-aware network embedding for relation modeling. In: Proceedings of the 55th Annual Meeting of the Association for Computational Linguistics (Volume 1: Long Papers), vol. 1, pp. 1722–1731 (2017)
19. Wang, D., Cui, P., Zhu, W.: Structural deep network embedding. In: Proceedings of the 22nd ACM SIGKDD International Conference on Knowledge Discovery and Data Mining, pp. 1225–1234. ACM (2016)
20. Yang, C., Liu, Z., Zhao, D., Sun, M., Chang, E.Y.: Network representation learning with rich text information. In: IJCAI, pp. 2111–2117 (2015)

Semi-structured Data and NLP

Category Multi-representation: A Unified Solution for Named Entity Recognition in Clinical Texts

Jiangtao Zhang[1]($\boxtimes$), Juanzi Li[1], Shuai Wang[1], Yan Zhang[1], Yixin Cao[1], Lei Hou[1], and Xiao-Li Li[2]

[1] Department of Computer Science and Technology, Tsinghua University, Beijing 100084, China
`zhang-jt13@mails.tsinghua.edu.cn, lijuanzi@tsinghua.edu.cn, 18813129752@163.com, zhangyan9988@qq.com, caoyixin2011@gmail.com, greener2009@gmail.com`
[2] Institute for Infocomm Research, A*STAR, Singapore 138632, Singapore
`xlli@i2r.a-star.edu.sg`

Abstract. Clinical Named Entity Recognition (CNER), the task of identifying the entity boundaries in clinical texts, is essential for many applications. Previous methods usually follow the traditional NER methods that heavily rely on language specific features (i.e. linguistics and lexicons) and high quality annotated data. However, due to the problem of *Limited Availability of Annotated Data* and *Informal Clinical Texts*, CNER becomes more challenging. In this paper, we propose a novel method that learn multiple representations for each category, namely *category-multi-representation* (CMR) that captures the semantic relatedness between words and clinical categories from different perspectives. CMR is learned based on a large scale unannotated corpus and a small set of annotated data, which greatly alleviates the burden of human effort. Instead of the language specific features, our proposed method uses more evidential features without any additional NLP tools, and enjoys a lightweight adaption among languages. We conduct a series of experiments to verify our new CMR features can further improve the performance of NER significantly without leveraging any external lexicons.

1 Introduction

Electronic Medical Records (EMR) contains valuable and detailed medical information of patients accessed and modified in a digital format [15]. Identifying the boundaries of clinically relevant entities in *clinical texts* from EMR and classifying them into predefined categories such as *disease*, *treatment* and *symptom*, namely *Clinical Named Entity Recognition* (CNER) is a fundamental task both in medical data mining and information extraction. CNER could benefit many applications in medical domain such as comorbidity analyses, syndromic

© Springer International Publishing AG, part of Springer Nature 2018
D. Phung et al. (Eds.): PAKDD 2018, LNAI 10938, pp. 275–287, 2018.
https://doi.org/10.1007/978-3-319-93037-4_22

surveillance, adverse drug event detection and the analysis of drug-drug interaction [12], as well as the NLP related tasks like information retrieval, relation
extraction and question answering [18].

Most existing work of NER in medical domain [1,4,6,10,21] simply follows
the conventional NER methods in general domain which focus on identifying
general named entities such as *person, location* and *organization.* They usually
utilize linguistic features based on syntactics and lexicons[1] to feed a supervised
model, e.g. SVM [24], CRF [21] or a hybrid of several classification models [6,10].
However, these methods may achieve poor performance in realistic applications
because (i) they heavily depend on linguistic features and lexicons, which varies
greatly among different datasets or across various languages, and (ii) the annotated data for the supervised model is not always available.

Despite the success of traditional NER, CNER receives relatively few studies
which has the following challenges:

Limited Availability of Annotated Data. As mentioned above, previous
works following traditional NER rely on a supervised model over a high quality training data. However, in the clinical domain, annotated data are not only
expensive (usually requires domain expertise) but also often unavailable due to
patient privacy and confidentiality requirements. Even though there are a few
public available annotated datasets for CNER task, such as i2b2 2010 [25] and
ShARe/CLEF eHealth 2013 [23], they are usually insufficient for training an
applicable system. For example, ShARe dataset contains only 300 documents
including 9,768 entity mentions annotated. On the other hand, the gap among
different languages always requires new language-specific annotated data. Therefore, we need to use the unlabeled data, usually available in clinical domain, such
as MIMC III [11], to alleviate the burden of human effort involved in creating
annotated resources and improve the performance of CNER.

Informal Clinical Texts. A clinical text is dictated by a doctor (and tran-
scribed later by a third-party) to capture the proceedings of a doctor-patient
interaction, or to document the results of a medical procedure or test [12]. It is
usually far different from general texts and even scholarly medical literatures.
Clinical texts have the following unique characteristics: (1) incomplete sentences,
(2) informal grammar, and (3) littered with misspellings and non-standard short-
hand, abbreviations and acronyms. All these characteristics result in the unre-
liability of the linguistic based features used in the traditional NER and the
effectiveness of NLP tools (e.g. POS tagging). Therefore, we need to explore
more evidential features with good generalization and independent of language
to cope with characteristics of clinical texts.

To address these challenges, our solution is to learn semantic features by
taking advantage of large scale unannotated corpora, instead of the language
specific features, such as syntactic and lexicon. The semantic features will be

---

[1] These methods extract lexicons from UMLS [3] or MeSH: https://www.nlm.nih.gov/
mesh/meshhome.html.

trained in an unsupervised way, and measure the similarity between the words in clinical texts and CNER categories. Our solution doesn't rely on any additional NLP tools which can avoid the unreliable linguistic features, and alleviate the burden of language specific annotated data.

In this paper, we propose *a unified solution* for CNER without leveraging any language specific features. It induces multiple representations for each category, namely *category-multi-representation* (CMR) that is used to measure the semantic similarity between words and categories. Specifically, we first construct a semantic space of clinical texts by employing a model of distributed representation (word embedding) over a large unannotated clinical corpus (e.g. MIMC III). As each entity mention has been classified into a certain predefined category in the annotated dataset, each category could be regarded as a *vector cluster* in the semantic space. Then we learn multiple representations for each category from 4 different aspects by leveraging the statistics and context information derived from the large unlabeled data to holistically capture the meaning of each category. That is, CMR shares a common semantic space with words in clinical texts which could easily be used to measure the semantic similarity between words and categories. Based on these representations, our proposed model only requires a small annotated dataset for training a sequence labeling model due to the good generalization ability of CMR. For inference, we adopt a heuristic method to assign a threshold for each CMR, which aims to filter out irrelevant noise (words) belonging to the corresponding category.

Contributions. Our main contributions are summarized as follows.

- To the best of our knowledge, this is the first work for CNER that represents category from multiple perspectives, which is based on unlabeled clinical corpus without any additional NLP tools.
- Our CNER model is a *united* method, which is independent of language-specific features (i.e. lexicons and linguistic features), and lightweight for adaption to identify clinical entities in another language and another dataset.
- Extensive experiments are conducted on two public datasets and the results demonstrate that our new CMR features can further improve the performance of CNER by 2.05% in terms of F1 score.

2   Problem Definition

Given a clinical text $\mathbf{s} = \langle w_1, w_2, ...w_{|\mathbf{s}|} \rangle$ and a set of predefined categories $C = \{c_1, ...c_{|C|}\}$, the output of our task is to generate a list of tags t_i for each word $w_i \in \mathbf{s}$. $t_i \in \mathcal{T} = \{cp | c \in C, p \in \mathcal{P} - \{O\}\} \cup \{O\}$ is a *category-position* combinatorial tag for w_i, where $\mathcal{P} = \{B, I, O\}$ is a set of *position* tags indicating the position information of a word located in an entity mention. B and I stand for beginning, intermediate positions of a *multi-word* entity respectively and O denotes outside of any entity mention. In short, our task is to identify every entity mention $m = \langle w_i, \cdots, w_j \rangle, i \leq j$ (perhaps including multiple words) occurring in the clinical text $\mathbf{s}$ and then classify it into a predefined category

$c_i \in C$. Figure 1 gives an example of sequence labeling for CNER, in which $C = \{Pr, Tr, Te\}$, represents *Problem*, *Treatment* and *Test* respectively.

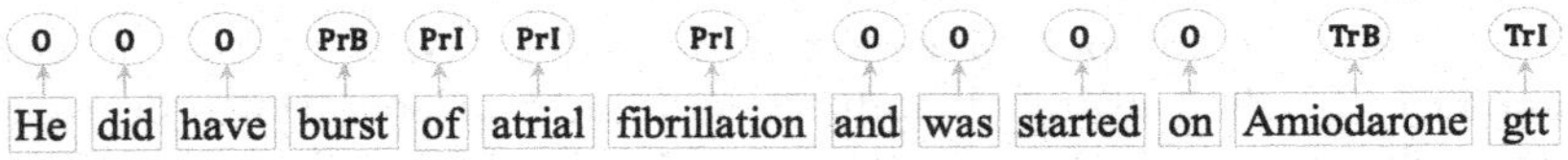

O	O	O	PrB	PrI	PrI	PrI	O	O	O	O	TrB	TrI
He	did	have	burst	of	atrial	fibrillation	and	was	started	on	Amiodarone	gtt

Fig. 1. An example of labeling process for CNER.

3   Our Proposed Approach

Our proposed method presupposes the existence of two resources: (1) an annotated corpus $\mathcal{L}$ in which each word has been annotated as a predefined category $c \in C$; (2) a much larger unannotated clinical corpus $\mathcal{U}$. The main steps of our method are as follows. Firstly, we construct a semantic clinical space by training a word embedding model over $\mathcal{U}$. Each predefined category can be seen as a word cluster in this space. Secondly, we learn abstract representations for each category from many different perspectives (CMR) derived from $\mathcal{U}$. Thirdly, we generate a bundle of novel features for the target word based on its distance to each of CMR. Lastly, an appropriate learning algorithm is applied to $\mathcal{L}$ with the generated new features to evaluate our method. The focus of this paper is primarily on the first three steps.

3.1   Generating Semantic Space

We first construct a semantic space by learning word embeddings (e.g. GloVe [17] and Word2vec [16]) on $\mathcal{U}$ to obtain low-dimensional, real-valued vector representation for each word in clinical texts. Each word $w \in \mathcal{U} \cup \mathcal{L}$ is represented as a point (vector) $\mathbf{v}_w$ in this semantic space. If an entity mention $m = \langle w_i, \cdots, w_j \rangle$ contains more than one word $(i < j)$, we simply represent it as the mean vector of its component words, i.e. $\mathbf{v}_m = (\frac{1}{j-i+1}) \sum_{k=i}^{j} \mathbf{v}_{w_k}$.

3.2   Category Multi-representation

We first build a *category-words* set for each predefined category based on $\mathcal{L}$ and $\mathcal{U}$. That is, $\forall c_i \in C$, we get $\mathbf{c}_i = \{w_{i1}, \cdots, w_{ij}, \cdots, w_{i|\mathbf{c}_i|}\}$ where each w_{ij} has been annotated as c_i in training dataset of $\mathcal{L}$ and occurs at least 100 times in $\mathcal{U}$. Then each predefined category can be regarded as a word cluster in the semantic space. The key point is how to represent the cluster of each category in order to more holistically capture the meaning of it.

One-Center Representation. Since the distance between vectors indicates the strength of the semantic relatedness of their corresponding words in the semantic space, we regard each category as a *hypersphere* constructed by the vectors in its category-words set. Each word located in the hypersphere of a category is more likely classified into it without considering any orthographic and syntactic features. In another word, the closer a word w is to the centre of the hypersphere of a category c_i, the more likely the word w belongs to the category c_i. Then we represent the category c_i as the *centroid vector* of the semantic vectors of its category-words set $\mathbf{c}_i$ as follows:

$$\mathbf{R_o}(c_i) = centroid(\mathbf{c}_i) = \langle median_{i1}, \cdots, median_{in} \rangle, i = 1, \cdots, |C| \quad (1)$$

where the centroid vector is defined as the median value of each dimension of the semantic vectors of words in $\mathbf{c}_i$ and n is the dimension size of the embedding vectors.

Multi-sub-center Representation. Each predefined category usually can be subdivided into several sub-categories in clinical texts. For example, category *Disease* can be classified as *Mental or Behavioral Dysfunction* and *Neoplastic Process*. Words in the same sub-category are more similar (closer in semantic space) to each other than to those in other sub-categories. In other words, the category may not be a normal hypersphere, it could be represented as several smaller sub-hyperspheres. Therefore, we use a clustering algorithm (Affinity propagation used in this paper which does not predefine the number of clusters) to group all words in each category $\mathbf{c}_i$ into K_i clusters $\{\mathbf{s}_{i1}, \cdots, \mathbf{s}_{ij}, \cdots, \mathbf{s}_{iK_i}\}$ where $\mathbf{s}_{ij}$ is a subset of words in $\mathbf{c}_i$. Then, we represent each category c_i as the *set of centroids* of its sub-hyperspheres. The premise of this representation is that some words which are a bit far from the centroid of the category are probably close to the centroids of some sub-hyperspheres.

$$\mathbf{R_m}(c_i) = \{centroid(\mathbf{s}_{i1}), \cdots, centroid(\mathbf{s}_{iK_i})\}, i = 1, \cdots, |C| \quad (2)$$

Influence Representation. The first two representations do not consider the importance of component words of categories. However, different words belonging to a certain category may have different influence on the category. Those mentions occurring more frequently in $\mathcal{U}$ generally are more prominent and representative for their categories. For example, since mentions *cancer* and *tumor* representing certain diseases occur in $\mathcal{U}$ frequently, we consider they are more representative for category *Disease* and those mentions related to them closely such as *Carcinoma* are more likely be recognized as *Disease*. We define the influence factor $if(w_{ij})$ of each word $w_{ij} \in \mathbf{c}_i$ as the normalized frequency of the mention that it belongs to[2] occurring in $\mathcal{U}$. Then we represent each category as the *weighted mean vector* of word embeddings of its category-words set:

$$\mathbf{R_i}(c_i) = \frac{1}{|\mathbf{c}_i|} \sum_{j=1}^{|\mathbf{c}_i|} if(w_{ij}) \cdot \mathbf{v}_{w_{ij}}, i = 1, \cdots, |C| \quad (3)$$

---

[2] If one word belongs to multiple mentions, we simply choose the one with highest frequency.

Context Representation. Our last category representation bases on following assumption: contexts of each mention occurring in $\mathcal{U}$ embrace rich information and patterns which are helpful to recognize the entity mention. For example, *"the effect of $\cdots$"* is always followed by a *drug name*. Therefore, adding context information into category representation will be useful. We consider a fixed length of window for each mention: two previous words and two following words in $\mathcal{U}$. Then we construct a set of context words for each category $\mathbf{cw}_i = \{cw_{i1}, \cdots, cw_{i|\mathbf{cw}_i|}\}$ where cw_{ij} denotes a bigram or unigram context word occurring over a certain number of times in $\mathcal{U}$ (e.g. 50). Then we represent each category c_i as the *mean vector* of the set of its context words:

$$\mathbf{R_c}(c_i) = \frac{1}{|\mathbf{cw}_i|} \sum_{j=1}^{|\mathbf{cw}_i|} \mathbf{v}_{cw_{ij}}, i = 1, \cdots, |C| \tag{4}$$

where $\mathbf{v}_{cw_{ij}}$ denotes the embedding vector of a context word cw_{ij}.

In summary, we learn 4 representations $\mathbf{R}_*(c_i), * \in \{\mathbf{o}, \mathbf{m}, \mathbf{i}, \mathbf{c}\}$ for each predefined category c_i which capture the four different semantic information of it.

3.3 Generating CMR Features

We first calculate 4 kinds of semantic relatedness between target word w_j and a category c_i based on CMR by leveraging a *distance function* such as *Euclidean distance* as follows.

$$d_\mathbf{o}(w_j, \mathbf{R_o}(c_i)) = dist(\mathbf{v}_{w_j}, centroid(\mathbf{s}_j))$$
$$d_\mathbf{m}(w_j, \mathbf{R_m}(c_i)) = \min_{k \in [1, \cdots, K_i]} dist(\mathbf{v}_{w_j}, centroid(\mathbf{s}_{ik}))$$
$$d_\mathbf{i}(w_j, \mathbf{R_i}(c_i)) = \frac{1}{|\mathbf{c}_i|} \sum_{k=1}^{|\mathbf{c}_i|} if(w_{ik}) \cdot dist(\mathbf{v}_{w_j}, \mathbf{v}_{w_{ik}}) \tag{5}$$
$$d_\mathbf{c}(w_j, \mathbf{R_c}(c_i)) = \frac{1}{|\mathbf{cw}_i|} \sum_{k=1}^{|\mathbf{cw}_i|} dist(\mathbf{v}_{w_j}, \mathbf{v}_{cw_{ik}})$$

Then we define a *threshold* of each category for each CMR based on the distances between the annotated word and each representation of its corresponding category, which is selected with the optimization objective to maximize F_β-score.

$$\tau_*(c_i) = \arg\max_{t^* \in V}((1 + \beta^2)\frac{P^{(t^*)} \cdot R^{(t^*)}}{(\beta^2 \cdot P^{(t^*)}) + R^{(t^*)}}), * \in \{\mathbf{o}, \mathbf{m}, \mathbf{i}, \mathbf{c}\} \tag{6}$$

where P is precision and R is recall; $V = (0, 0.01, 0.02, \cdots, 1)$; β determines the weight that should be given to recall relative to precision. The lowest threshold $\tau_*(c_i)$ is chosen that optimizes the F_β-score.

Finally, we generate one feature per representation of each predefined category. The value of the feature is either *True* or *False* depending on whether the calculated *distance* is above the threshold $\tau_*(c_i)$ or not.

$$f_*^{c_i}(w_j) = \begin{cases} 0 \text{ if } f_*(w_j, \mathbf{R}_*(c_i)) > \tau_*(c_i) \\ 1 \text{ if } f_*(w_j, \mathbf{R}_*(c_i)) \leq \tau_*(c_i) \end{cases}, * \in \{\mathbf{o}, \mathbf{m}, \mathbf{i}, \mathbf{c}\} \tag{7}$$

4 Experiments

4.1 Data Sets

To the best of our knowledge, the annotated corpora of the i2b2/VA 2010 shared task (i2b2) and ShARe/CLEF eHealth 2013 Shared Task (ShARe) are the only two public available datasets for CNER. Tables 1 and 2 show the statistics of these two datasets respectively. In i2b2, 3 different categories have been annotated: *Problem* (Pr), *Treatment* (Tr), *Test* (Te) from discharge summaries and progress notes. ShARe involves annotation of disorder mentions in a set of narrative clinical reports. Since ShARe does not provide the exact category of each disorder mention, we map each disorder mention into a category by ourselves according to its linking UMLS CUI (Concept Unique Identifier)[3]. Then we get 11 different semantic types for this dataset and merge them into 5 categories: *Anatomical Abnormality* (AA), *Pathologic Function* (PF), *Injury or Poisoning* (IP), *Signs and Symptoms* (SS) and *Others*[4](O) according to hierarchies of semantic types in UMLS. Notice these two datasets have totally different categories and our proposed method could work well on both of them which will be demonstrated in following subsections. Two public available corpora are used as unannotated clinical data: the 378,000 *Medline abstracts* that are indexed as pertaining to clinical trials and *MIMIC III* that comprises de-identified health data associated with 40,000 critical care patients. Then we build a semantic space by training a word embedding model — GloVe [17] used in this paper — on these two corpora (merged).

Table 1. The statistics of i2b2

Dataset	Pr	Tr	Te	All
Training	11968	8500	7369	27837
Test	18550	13560	12899	45009

Table 2. The statistics of ShARe

Dataset	AA	PF	IP	SS	O	All
Training	250	2304	221	838	1525	5138
Test	157	2107	96	735	1535	4630

4.2 Our Models and Parameter Settings

In our experiments, We apply two state-of-the-art sequence labeling models: CRF and BLSTM+CRF (BLSTM for short) with the generated new CMR features to evaluate our method. We implement CRF employing *CRFsuite*[5] and BLSTM using *theano* library.[6] The parameter settings of these two models are showed in Tables 3 and 4 respectively.

---

[3] In UMLS, each concept (entity) is represented by its CUI and is semantically classified into one of semantic types.

[4] For those mentions mapping to unknown CUI, i.e. CUI-less.

[5] http://www.chokkan.org/software/crfsuite/.

[6] http://deeplearning.net/software/theano/.

Table 3. CRF settings

C-value	Context window	Regularization
5	2+2	L1&L2

Table 4. BLSTM settings

Layers	Layer size	Batch size	Activation function	Learning rate	Drop out	Epochs	Optimizer
2	100	64	RELU	1E-04	0.5	100	adam

The considered performance metrics are precision, recall and F1-score and we adopt the *strict metrics* for evaluation used in both tasks. Performance scores are macro-averaged over classes, giving equal weight to all classes.

4.3   Threshold Settings for Determining CMR Features

We first investigate the impact of providing threshold of CMR that determine the feature values on NER performance. Figure 2 shows the threshold setting procedure for different CMR in which threshold is set by finding the distance that maximizes F1-score on i2b2. It can be seen that the thresholds are generally lower and the F1-scores higher for Multi-sub-center Representation and One-center Representation (also observed on ShARe). It indicates these two representations are better to separate the categories and important to capture the meaning of a category. This is confirmed in the subsequent experiments, the results of which show that the highest performance is obtained with these two representations.

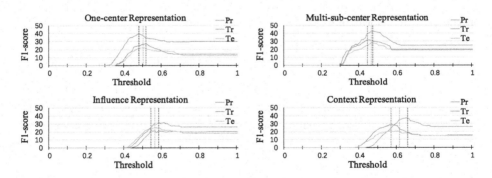

Fig. 2. Threshold setting procedure for CMR on i2b2.

To study the impact of changing the optimization objective to various F_β-scores on NER performance, experiments are conducted with the following β values: 0.5, 1.0, 2.0 and 5.0. The highest F1-score are observed when β is set to 1.0 in i2b2 and 2.0 in ShARe. Then in our following experiments, we set $\beta = 1$ for i2b2 and $\beta = 2$ for ShARe.

4.4   Comparison with Different CMR Features

In order to study and verify the effectiveness of the proposed new CMR features to the learning algorithms, our four groups of CMR features are evaluated and compared one by one. For CRF model, we combine our CMR features with a set of traditional features — orthographic and syntactic features[7] — as our baseline which is the most traditional method in CNER. For BLSTM, we take the word embedding concatenating character embedding as the baseline — which is state-of-the-art in general domain of NER task. Tables 5 and 6 show the comparison of different feature combinations on two datasets respectively.

Table 5. Comparison with different CMR features on i2b2

CMR	CRF				BLSTM			
	P	R	F1	ΔF1	P	R	F1	ΔF1
Baseline	82.05	78.86	80.42		82.20	81.57	81.88	
$+f_o$	84.69	78.35	81.40	+0.97	84.29	81.70	82.97	+1.09
$+f_m$	83.45	80.01	81.69	+1.27	83.92	83.07	83.50	+1.61
$+f_i$	82.77	79.11	80.90	+0.48	83.40	81.67	82.53	+0.65
$+f_c$	82.35	79.54	80.92	+0.50	83.26	82.48	82.87	+0.98
$+f_{all}$	83.92	80.12	81.98	+1.55	**84.48**	**83.39**	**83.93**	+2.05

Table 6. Comparison with different CMR features on ShARe

CMR	CRF				BLSTM			
	P	R	F1	ΔF1	P	R	F1	ΔF1
Baseline	74.22	61.16	67.06		73.93	66.11	69.80	
$+f_o$	75.95	61.06	67.70	+0.64	75.32	66.31	70.53	+0.73
$+f_m$	75.34	62.12	68.09	+1.04	75.31	67.35	71.11	+1.31
$+f_i$	74.99	61.34	67.48	+0.42	74.83	66.36	70.34	+0.54
$+f_c$	74.79	61.58	67.55	+0.49	74.25	66.95	70.41	+0.61
$+f_{all}$	75.71	62.32	68.37	+1.31	**75.74**	**67.93**	**71.62**	+1.82

Two traditional state-of-the-art models without leveraging lexicons (baselines) perform not well on both datasets. When we add our CMR features to these models one by one, the experiment results show each group of CMR features achieves improvement on both datasets. We can see the Multi-sub-center Representation features achieve the best improvement among all CMR features while the improvement obtained from Context Representation and Influence Representation features are relatively small. This indicates that Multi-sub-center Representation is more representative than other CMR and could more holistically capture the meaning of the category. When we combine all CMR features,

---

[7] The same as the ones used in [10] except lexical features extracted from existing annotated tools.

we achieve further significant improvement on both datasets (1.55% improvement of CRF and 2.05% improvement of BLSTM on i2b2 as well as 1.31% and 1.82% on ShARe) that indicates the four groups of CMR features could compensate each other and combination of them could further improve the performance. We also find our CMR features get more improvement for BLSTM (2.05% on i2b2 and 1.82% on SHARe) than CRF and achieve the best performance on both datasets. The possible reason is that our CMR features are derived from word embedding and could work better when combining with it. Furthermore, in addition to powerful capability of BLSTM model, features used in BLSTM including word embedding and character embedding and CMR features are semantic, without considering orthographic and syntactic features, which could potentially more effectively address the challenge of informal clinical texts.

4.5   Comparison with Previous Systems

Our evaluation show that the performance of NER significantly improves after adding our new CMR features. However, how much it contributes toward improving the state-of-the-art determines the practical significance of the improvement. Thus, we compare the performance of our method to the top systems in the i2b2/VA 2010 concept extraction task and ShARe/CLEF eHealth 2013 Shared Task.

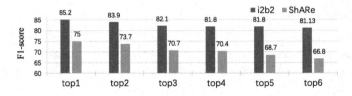

Fig. 3. Comparison with top 6 systems in two shared tasks.

Figure 3 shows the results of the top 6 systems in these two tasks. Almost all systems use hybrid models integrating several models such as CRF and SSVM with a set of rich features. Furthermore, all systems leverage the output of existing annotation tools such as cTAKEs, MetaMap and rely lexicons derived from UMLS to improve the NER performance. Our best method (BLSTM + 4 CMR) is better than system 3 and equal to system 2 in i2b2 and ranks the third in ShARe. The results suggest that by integrating CMR features derived from a large scaled unlabeled corpus into one single model our work can achieve state-of-the-art without leveraging any outside lexicons and any existing annotated tools. In addition, our CMR features can easily integrate with other models such as discriminative semi-Markov HMM Models used by the best system in i2b2. It may further improve the performance of these systems.

5   Related Works

Most early existing NER techniques in medical domain typically focus on traditional machine learning methods such as Support Vector Machine (SVM) [24], Hidden Markov Model (HMM) [22] and Conditional Random Fields (CRF) [21] integrating a set of complicated hand-crafted features. Some other methods leverage hybrid models [6,10] to improve the performance of NER. However, their performance may be affected by some common drawbacks: (1) with the change of corpora and languages, the process to reconstruct feature set is difficult; (2) some complex features with syntactic information rely on the performance of other NLP modules; (3) these features with expert knowledge are expensive to acquire.

There also exist some well-known annotation tools in clinical texts such as cTAKEs [20], MetaMap [2] and ConText [7]. Most of them can extract various types of named entities from clinical texts and link them to concepts in UMLS. However, these tools heavily rely on external dictionaries such as SNOMED-CT [9] and are only suitable for English. A large amount of works [5,6,10] usually leverage the annotating results of these tools as a part of features to feed into their models and achieve further improvement of performance.

Another thread of NER in medical domain focuses on recognizing one single named entity, such as [27] finding anatomies from discharge summaries, [14, 19,28] recognizing drug names and [26] extracting disease names from clinical texts. Different with these clinical NER works addressing single entity type, we are addressing a comprehensive set of challenges in identifying multiple named entities to analysis the clinical texts.

Recently, some attempts [8,13,29] focus on applying deep neural network to NER in clinical texts. Most of these concatenate word-level embedding, character-level embedding and lexicon embedding as input. Then multiple convolutional layers are stacked over the input to extract useful features automatically and then fed into RNN models. Although these methods claim no feature engineering, their performance are heavily rely on the training dataset (also rely on lexicons) and usually not satisfied when the training set is small. Since our proposed CMR features are derived from large scale unannotated corpus, our method reduce the limitation of small training set and is easy to be adapted to new domains while large scale unannotated corpora are often readily available.

6   Conclusion and Future Work

The existing CNER systems simply follow the traditional NER methods used in general domain which usually leverage the linguistic features including syntactic and lexicon features. Compared with successful performance of NER in general domain, CNER achieves relatively pool performance due to the issues of *Limited Availability of Annotated Data* and *Informal clinical texts*. In this paper, we propose a novel unified method for CNER without considering any linguistic features. It learned multiple representations for each category to capture the

semantic similarity between words and categories from 4 different perspectives
In the future, we will evaluate our method in other domains, such as biomedical
domain. In addition, we will explore new unsupervised methods that is useful
when training dataset is not available.

Acknowledgements. The work is supported by major national research and development projects (2017YFB1002101), NSFC key project (U1736204, 61661146007), Fund of Online Education Research Center, Ministry of Education (No. 2016ZD102), and THU-NUS NExT Co-Lab.

References

1. Abacha, A.B., Zweigenbaum, P.: Medical entity recognition: a comparison of semantic and statistical methods. In: BioNLP, pp. 56–64 (2011)
2. Aronson, A.R., Lang, F.M.: An overview of metamap: historical perspective and recent advances. JAMIA **17**, 229–236 (2010)
3. Bodenreider, O.: The unified medical language system (UMLS): integrating biomedical terminology (2004)
4. Bodnari, A., Deléger, L., Lavergne, T., Névéol, A., Zweigenbaum, P.: A supervised named-entity extraction system for medical text. In: Working Notes for CLEF Conference (2013)
5. Bodnari, A., Deléger, L., Lavergne, T., Névéol, A., Zweigenbaum, P.: A supervised named-entity extraction system for medical text. In: CLEF (2013)
6. de Bruijn, B., Cherry, C., Kiritchenko, S., Martin, J., Zhu, X.: Machine-learned solutions for three stages of clinical information extraction: the state of the art at i2b2 2010. J. Am. Med. Inform. Assoc. **18**(5), 557 (2011)
7. Chapman, W.W., Chu, D., Dowling, J.N.: Context: an algorithm for identifying contextual features from clinical text. In: BioNLP 2007, pp. 81–88 (2007)
8. Dernoncourt, F., Lee, J.Y., Uzuner, Ö., Szolovits, P.: De-identification of patient notes with recurrent neural networks. JAMIA **24**(3), 596–606 (2017)
9. Donnelly, K.: SNOMED-CT: the advanced terminology and coding system for eHealth. Stud. Health Technol. Inform. **121**, 279–90 (2006)
10. Jiang, M., Chen, Y., Liu, M., Rosenbloom, S.T., Mani, S., Denny, J.C., Xu, H.: A study of machine-learning-based approaches to extract clinical entities and their assertions from discharge summaries. JAMIA **18**, 601–606 (2011)
11. Johnson, A.E.W., Pollard, T.J., Shen, L., Lehman, L.H., Feng, M., Ghassemi, M., Moody, B., Szolovits, P., Celi, L.A., Mark, R.G.: MIMIC-III, a freely accessible critical care database. Sci. Data **3**, 160035 (2016)
12. Kundeti, S.R., Vijayananda, J., Mujjiga, S., Kalyan, M.: Clinical named entity recognition: challenges and opportunities. In: 2016 IEEE International Conference on Big Data, pp. 1937–1945 (2016)
13. Li, L., Jin, L., Jiang, Z., Song, D., Huang, D.: Biomedical named entity recognition based on extended recurrent neural networks. In: BIBM, pp. 649–652 (2015)
14. Liu, S., Tang, B., Chen, Q., Wang, X.: Effects of semantic features on machine learning-based drug name recognition systems: word embeddings vs. manually constructed dictionaries. Information **6**(4), 848–865 (2015)
15. Meystre, S.M., Savova, G.K., Kipper-Schuler, K.C., Hurdle, J.F.: Extracting information from textual documents in the electronic health record: a review of recent research. Yearb. Med. Inform. **35**, 128–144 (2008)

16. Mikolov, T., Sutskever, I., Chen, K., Corrado, G.S., Dean, J.: Distributed representations of words and phrases and their compositionality. Adv. Neural Inf. Proc. Syst. **26**, 3111–3119 (2013)
17. Pennington, J., Socher, R., Manning, C.D.: Glove: global vectors for word representation. In: EMNLP, vol. 14, pp. 1532–1543 (2014)
18. Ratinov, L., Roth, D.: Design challenges and misconceptions in named entity recognition. In: CoNLL (2009)
19. Sadikin, M., Fanany, M.I., Basaruddin, T.: A new data representation based on training data characteristics to extract drug name entity in medical text. Comput. Intell. Neurosci. **2016**, 16 (2016)
20. Savova, G.K., Masanz, J.J., Ogren, P.V., Zheng, J., Sohn, S., Kipper-Schuler, K.C., Chute, C.G.: Mayo clinical text analysis and knowledge extraction system (cTAKES): architecture, component evaluation and applications. JAMIA **17**(5), 507–513 (2010)
21. Settles, B.: Biomedical named entity recognition using conditional random fields and rich feature sets. In: JNLPBA, pp. 104–107 (2004)
22. Shen, D., Zhang, J., Zhou, G., Su, J., Tan, C.L.: Effective adaptation of a hidden Markov model-based named entity recognizer for biomedical domain. In: BioMed, pp. 49–56 (2003)
23. Suominen, H., Salanterä, S., Velupillai, S., Chapman, W.W., Savova, G., Elhadad, N., Pradhan, S., South, B.R., Mowery, D.L., Jones, G.J.F., Leveling, J., Kelly, L., Goeuriot, L., Martinez, D., Zuccon, G.: Overview of the ShARe/CLEF eHealth evaluation lab 2013. In: Forner, P., Müller, H., Paredes, R., Rosso, P., Stein, B. (eds.) CLEF 2013. LNCS, vol. 8138, pp. 212–231. Springer, Heidelberg (2013). https://doi.org/10.1007/978-3-642-40802-1_24
24. Takeuchi, K., Collier, N.: Bio-medical entity extraction using support vector machines. In: BioMed, pp. 57–64 (2003)
25. Uzuner, Ö., South, B.R., Shen, S., DuVall, S.L.: 2010 i2b2/VA challenge on concepts, assertions, and relations in clinical text. J. Am. Med. Inform. Assoc. **18**(5), 552 (2011)
26. Wei, Q., Chen, T., Xu, R., He, Y., Gui, L.: Disease named entity recognition by combining conditional random fields and bidirectional recurrent neural networks. In: Database 2016 (2016)
27. Xu, Y., Hua, J., Ni, Z., Chen, Q., Fan, Y., Ananiadou, S., Chang, E.I.C., Tsujii, J.: Anatomical entity recognition with a hierarchical framework augmented by external resources. Plos One **9**, 1–13 (2014)
28. Zeng, D., Sun, C., Lin, L., Liu, B.: Enlarging drug dictionary with semi-supervised learning for drug entity recognition. In: BIBM, pp. 1929–1931 (2016)
29. Zhao, Z., Yang, Z., Luo, L., Zhang, Y., Wang, L., Lin, H., Wang, J.: ML-CNN: a novel deep learning based disease named entity recognition architecture. In: BIBM, p. 794 (2016)

A Heterogeneous Information Network Method for Entity Set Expansion in Knowledge Graph

Xiaohuan Cao[1], Chuan Shi[1(✉)], Yuyan Zheng[1], Jiayu Ding[1], Xiaoli Li[2], and Bin Wu[1]

[1] Beijing Key Lab of Intelligent Telecommunications Software and Multimedia, Beijing University of Posts and Telecommunications, Beijing 100876, China
devil_baba@126.com, {shichuan,wubin}@bupt.edu.cn,
zyy0716_source@163.com, username.djy@gmail.com
[2] Institute for Infocomm Research, A*STAR, Singapore, Singapore
xlli@i2r.a-star.edu.sg

Abstract. Entity Set Expansion (ESE) is an important data mining task, e.g. query suggestion. It aims to expand an entity seed set to obtain more entities which have traits in common. Traditionally, text and Web information are widely used for ESE. Recently, some ESE methods employ Knowledge Graph (KG) to extend entities. However, these methods usually fail to sufficiently and efficiently utilize the rich semantics contained in KG. In this paper, we use the Heterogeneous Information Network (HIN) to represent KG, which would effectively capture hidden semantic relations between seed entities. However, the complex KG introduces new challenges for HIN analysis, such as generation of meta paths between entities and addressing ambiguity caused by multiple types of objects. To solve these problems, we propose a novel Concatenated Meta Path based Entity Set Expansion method (CoMeSE). With the delicate design of the concatenated meta path generation and multi-type-constrained meta path, CoMeSE can quickly and accurately detect important path features in KG. In addition, heuristic learning and PU learning are employed to learn the weights of extracted meta paths. Extensive experiments on real dataset show that the CoMeSE accurately and quickly expands the given small entity set.

Keywords: Heterogeneous Information Network · Knowledge Graph
Entity Set Expansion · Meta path

1 Introduction

Entity Set Expansion (ESE) is mainly about, given a small set of seed entities, finding out other entities belonging to it and expanding the set to a more complete one. For example, given a few seeds like "New York", "Los Angeles" and "Chicago", ESE will discover the relation among them and obtain other city

© Springer International Publishing AG, part of Springer Nature 2018
D. Phung et al. (Eds.): PAKDD 2018, LNAI 10938, pp. 288–299, 2018.
https://doi.org/10.1007/978-3-319-93037-4_23

instances in America, such as "Houston". ESE has been widely used in many applications, e.g., dictionary construction [2] and query suggestion [1].

Plenty of ESE works have been done, and most of them discover distribution information or context pattern of seeds in text or Web resources to infer the intrinsic relation for ESE [8]. Recently, Knowledge Graph (KG), a kind of structured data source, has been more and more important for knowledge mining. So some ESE works use KG as a supplement for text to improve performance [7].

Fig. 1. A snapshot in knowledge graph.

However, few researches utilize KG as individual data source for ESE. Owing to rich semantics and structural representation of KG, it is feasible to employ KG to extend entities. KG, constructed by triples $<Subject, Property, Object>$, can be considered as a Heterogeneous Information Network (HIN) that contains different types of objects and relations [9]. In HIN, meta path [9], a sequence of relations connecting two objects, is widely used for semantic capture. For example, in Fig. 1, the fact triples can form a network and the path in it can show semantics. Therefore, we can consider KG as an HIN and employ meta path features to solve ESE problem. However, this idea faces challenges as follows.

- It is impossible to enumerate meta paths in KG. In traditional HIN with only a few types of objects and relations, it is easy to enumerate useful meta paths. However, it is not the case for KG because of its complexity. For example, DBpedia has 3 billion facts. It is impossible to find meta paths by manual.
- In KG, objects connected by a relation may affiliate to multiple types, which will cause ambiguity. In traditional HIN, objects have a unique type, which makes meta path have definite semantics. But in KG, objects may affiliate to multiple types which will lead to uncertain semantics. For example, in Fig. 1, the objects connecting to the $\xrightarrow{created}$ relation affiliate to the types of executive, painter and etc. The $\xrightarrow{created}$ between different pairs has different meaning. The relation between (Bill Gates, Microsoft) means establishing, while for (Vincent van Gogh, Starry Night), it means painting.
- It is not easy to combine path features for ESE, though we extract path features among entities. ESE problem usually has few seeds, so it is difficult to use traditional supervised method to build a ranking or classification model.

It is not a trivial task to solve these challenges. A very recent attempt by Zheng et al. [14] illustrates limited performance improvement on the problems but cost huge time and space. In this paper, we propose a novel *Concatenated Meta Path based Entity Set Expansion* method (CoMeSE) for ESE problem in KG. The CoMeSE includes three steps. Firstly, in order to reduce time and reuse visited paths, CoMeSE designs a novel random walk based Concatenated Meta Path Generation (RWCP) method to quickly and accurately discover useful meta paths by concatenating meta paths that have been visited in KG. Secondly, in order to solve the ambiguity problem caused by multiple types of objects, we propose a multi-type-constrained meta path concept to subtly capture semantics in KG, and further design a novel similarity measure based on it. Thirdly, for solving the problem of very limited positive samples, besides a heuristic weight strategy, we employ a PU learning method (Learning from Positive and Unlabeled Example) to effectively learn the weights of meta paths. Plenty of experiments on real dataset have been done to validate the effectiveness and efficiency of CoMeSE. The experiments show that, compared to the state of the arts, CoMeSE can quickly and accurately extend entities because of its delicate designs.

2   Preliminary

In this section, we describe some main concepts and preliminary knowledge.

Heterogeneous Information Network [4] is a kind of information network defined as a directed graph $G = (\mathcal{O}, \mathcal{R})$, which consists of either different types of objects $\mathcal{O}$ or different types of relations $\mathcal{R}$. In an HIN, there can be different paths connecting two objects and these paths are called **meta path** [12]. A meta path $\mathcal{P}$ is defined as $\mathcal{P}^{R_1 \circ \cdots \circ R_n} = \mathcal{T}(o_1) \xrightarrow{R_1} \cdots \xrightarrow{R_n} \mathcal{T}(o_{n+1})$, where o_i presents the object at position i in $\mathcal{P}$, $\mathcal{T}(o_i)$ is the type of o_i, and R_i is a type of relation. Note that $\mathcal{T}(o_i)$ corresponds to a unique entity type.

Knowledge Graph [10] is a knowledge base system with semantic properties and derived from text data of knowledge sources. KG is conducted by triples $<Subject, Property, Object>$. In this paper, we model KG as an HIN, "Subject" and "Object" as nodes, "Property" as links. This HIN is not like general HIN with simple schema but with thousands of node and link types. Besides, in KG, one entity is subordinate to multiple types and the meanings of links may introduce ambiguity. So traditional meta path would capture exact semantics badly. In order to solve the ambiguity problem, we propose a novel concept of Multi-Type-Constrained Meta Path to more subtly capture semantic relations.

Definition 1. *Multi-Type-Constrained Meta Path (MuTyPath) is a special meta path where each object position is constrained by a set of entity types. A MuTyPath $\tilde{\mathcal{P}}$ is represented as $\tilde{\mathcal{P}}^{R_1 \circ \cdots \circ R_n} = \mathcal{TS}(o_1) \xrightarrow{R_1} \cdots \xrightarrow{R_n} \mathcal{TS}(o_{n+1})$, where $\mathcal{TS}(o_i)$ represents the type set of object o_i at position i in $\tilde{\mathcal{P}}$. Different from $\mathcal{T}(o_i)$ in meta path, $\mathcal{TS}(o_i)$ can correspond to multiple entity types. When the cardinality of every $\mathcal{TS}(o_i)$ is 1, the MuTyPath is equal to meta path.*

Let's give an example in Fig. 1 to show the difference between MuTyPath and meta path. As the fact that Steve Jobs established Apple, meta path can not normally show their relation because entities have multiple types, like Jobs belonging to Executive, Person, and Inventor. Or meta path can only describe this fact as "$Obj \xrightarrow{Created} Obj$" through ignoring the node types, which may cause semantic ambiguity. However, we can use MuTyPath "$\{Person, Inventor, Excusive\} \xrightarrow{Created} \{Company, Manufacturer\}$" to describe the fact more exactly.

3   The Method Description

In order to expand entity set in KG efficiently and accurately, we propose an algorithm named **C**oncatenated **M**eta **P**ath based **E**ntity **S**et **E**xpansion (CoMeSE) to capture semantic relations between seeds. Firstly, for reducing space and time, we design an efficient concatenated meta path generation method. Secondly, to handle ambiguity of meta path, we extract multi-type-constrained meta paths and design a novel similarity measure MuTySim. Thirdly, due to the lack of negative cases, we design a heuristic weight strategy and PU learning method to assign the importance of extracted paths for ESE model.

3.1   Random Walk Based Concatenated Meta Path Generation Method

Meta path is a kind of effective feature to capture semantic relationship among nodes. In traditional HIN with simple schema, meta path is usually predefined, while it is hard in KG owing to its massive types of objects and relations. Thus, we propose an algorithm named **R**andom **W**alk based **C**oncatenated Meta **P**ath generation method (RWCP) to quickly and automatically generate meta paths.

A naive method to generate meta path in KG is, using a walker to wander with one-directional random walk to find meta paths between seed pairs. MP_ESE [14] adopts this idea. However, it has two disadvantages. Firstly, it has a huge space and time cost. If m is the average number of neighbors of a node, discovering n-length paths should visit m^n nodes. Secondly, many paths are duplicately visited and the visiting information is not reused, which makes it inefficient. For saving time and space, we can use a bi-directional random walk where two walkers wander from two sides respectively and meet at an intersection node while wandering. The searching space would reduce to $m^{n/2}$. For reusing visited path, we record the wandering information of walkers, and the walkers can continue to wander or decide to concatenate existing paths.

Specifically, given the seed set, RWCP randomly walks with a half or less of the maximum path length to obtain a set of paths for seeds, and then concatenates different paths with visiting information to get useful meta paths. In fact, RWCP is a repeating process of meta path concatenation and extension. For explaining RWCP clearly, we give some basic structural definitions as follows.

Definition 2. *Recorder is a structure to record visiting information. It includes: the meta path passed by, a series of entity pairs generated along the path and corresponding similarity values, entity lists between the entity pairs along the path, an arriving entity set and the score Sco of meta path defined later.*

Definition 3. *The score of meta path,* **Sco**, *is designed to indicate the importance of the path to seed set. The value of Sco means the priority to handle.*

$$\mathcal{S}co(\mathcal{P}) = \sum_s \frac{1}{K} \sum_t \sigma(s,t|\mathcal{P}), \tag{1}$$

where s and t are source and arriving node respectively on meta path $\mathcal{P}$, K as the number of arriving nodes from s, and $\sigma(s,t|\mathcal{P})$ is the similarity value based on $\mathcal{P}$, which is calculated by MuTySim *for meta path introduced in Sect. 3.2. Moreover,* **AvgS** *is an average value of Sco of two meta paths which are pending to be concatenated. The AvgS shows priority of Recorder pair for path concatenation.*

Besides, some assistant sets are needed. We use Recorder Set ($\mathcal{RS}$) to store Recorders, Extension Backlog ($\mathcal{EB}$) to record serial numbers and $\mathcal{S}co$ of Recorders which would be extended, and Concatenation Backlog ($\mathcal{CB}$) to record serial number pairs and $\mathcal{A}vgS$ of the Recorder pairs which would be concatenated.

Definition 4. *η is a threshold determining whether adding the new generated Recorder into $\mathcal{RS}$, for excluding unimportant paths.*

$$\eta = \epsilon \cdot (|\mathcal{SS}| + |\mathcal{PS}|), \tag{2}$$

where ϵ is a limited coefficient, and $|\mathcal{SS}|$ is the cardinality of seed set $\mathcal{SS}$, $|\mathcal{PS}|$ is the number of meta paths chosen. So η is dynamically increasing according to the number of chosen meta paths to converge faster to terminate the algorithm.

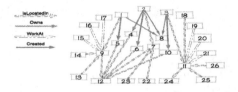

Fig. 2. Subgraph example for RWCP.

Thus, as introduced above, we mainly use $\mathcal{S}co$ and $\mathcal{A}vgS$ together to determine action in RWCP. In detail, comparing $\mathcal{A}vgS$ and $\mathcal{S}co$ values, if there are $\mathcal{A}vgS$ greater than all $\mathcal{S}co$ values in $\mathcal{EB}$, we will do path concatenation for these Recorder pairs. When all $\mathcal{A}vgS$ in $\mathcal{CB}$ are smaller than the largest $\mathcal{S}co$, we extend the Recorder with largest $\mathcal{S}co$ to generate new Recorder pairs. Path concatenation and Recorder extension take place alternately. Figure 2 is a simple network and seed set is {1,2,3} which means the seed pairs are {(1,2),(1,3),(2,1),(2,3),(3,1),(3,2)}. Figure 3 describes how RWCP algorithm works in the sample shown in Fig. 2.

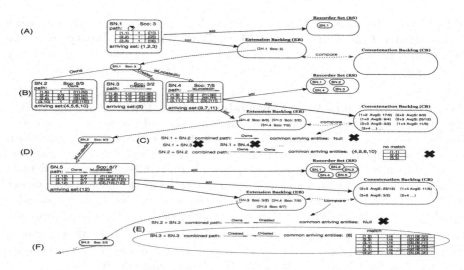

Fig. 3. A sample of RWCP process.

First of all, RWCP builds the initial Recorder Rec_1 and puts it into RS, (serial number SN_1, score Sco_1) into EB in Step A. Then, RWCP is in a loop process of path concatenation and extension. When CB is empty and EB contains Recorder, like in Step B, or all $AvgS$ in CB are smaller than the largest Sco in EB, like in Step D, RWCP pops the Recorder with the max Sco from EB to extend outwards. When extending, like in Step B, the arriving nodes in Rec_1 move one step forward and RWCP generates new Recorders with the serial numbers $SN.2-4$ based on different paths passing by, and $(SN.1)$ is removed from EB. After that, RWCP judges if Sco of each new generated Recorder is larger than the minimum threshold, η. If yes, put it into RS and EB. For each Recorder in RS, pair it with the new Recorder and put this pair with its $AvgS$ into CB. If the largest $AvgS$ of Recorder pairs is larger than all Sco in EB, like in the Step C and E, RWCP will get the Recorder pair with largest $AvgS$ out of CB and concatenate paths. For example, in Step E, handling the Recorder pair $(SN.3, SN.3)$, RWCP concatenates two meta paths of them as "$\xrightarrow{Created}\xleftarrow{Created}$", and judges whether two arriving entity sets have nodes in common or not. If yes, generate new entity pairs based on the common node set $\{8\}$. If the new entity pairs match to seed pairs, the concatenated path would be chosen as a path feature $\mathcal{P}$. Finally, if CB and EB are all empty, RWCP would be terminated.

3.2 Multi-Type-Constrained Meta Path Extraction and Similarity Calculation

Traditional meta path may not well capture subtle semantics in KG, because the type of entities at ends of a relation may be non-unique and entity may be subordinate to multiple types. To avoid ambiguity problem, we design MuTyPath in Definition 1 to make the explored relationships more accurate and a measure named **Multi-Type-Constrained Meta Path-based *Sim*ilarity measure** (MuTySim) to compute similarity of entity pairs in MuTyPath. The similarity value vectors based on MuTyPaths can be used as features for ESE model.

Extraction of Multi-Type-Constrained Meta Path. Applying RWCP algorithm, we get the meta paths with relations only and a series of visited entity lists for seed pairs along the paths. These entity lists are the path instances. With the lists, we can change meta paths to MuTyPaths for more precise semantics.

We design an extraction strategy for MuTyPath. Given an n-length meta path $\mathcal{P}$ and a list of instances $\{p_1, \cdots, p_m\}$, every position o_i of $\mathcal{P}$ has an object set $\{a_{1i}, \cdots, a_{mi}\}$. We check the types of each entity a_{ji} one by one and judge whether current entity type set $\mathcal{TS}(a_{ji})$ has intersection with existing common type sets $\mathcal{TS}'_k(o_i)$(initial common type set $\mathcal{TS}'_1(o_i) = \mathcal{TS}(a_{1i})$). If yes, update with $\mathcal{TS}'_k(o_i) = \mathcal{TS}'_k(o_i) \bigcap \mathcal{TS}(a_{ji})$. Otherwise, create another common type set to store its types. Then we will have one or more common type sets $\mathcal{CTS}_i = \{\mathcal{TS}'_1, \cdots\}$ at every position o_i. After that, we use Cartesian Product, $\mathcal{CTS}_1 \times \cdots \times \mathcal{CTS}_n$, to get multiple common type set combinations, and combine them with the original meta path to finally form one or more MuTyPaths.

Multi-Type-Constrained Meta Path-Based Similarity Measure. Based on the common type set of seeds obtained from selected MuTyPaths, we can get candidates sharing the same common types. To find out the relationship between seeds and candidates, we should re-calculate the similarity of each seed-candidate pair and seed-seed pair along MuTyPaths. Here we propose a novel **Multi-Type-Constrained Meta Path-based *Sim*ilarity measure** (MuTySim).

MuTySim has the following advantages. Firstly, MuTySim supports meeting at any node along the path for RWCP. Current similarity measures have fixed random walk direction and measurement. Secondly, MuTySim considers both conditions of two ends between a link, while existing measures do not. Moreover, MuTySim considers multi-type constraint of MuTyPath.

Given a MuTyPath $\tilde{\mathcal{P}}^{o_1} = \mathcal{TS}(o_1)$, where $\mathcal{TS}(o_1)$ is the common type set at position o_1 constrained by MuTyPath, the similarity of object s and itself is:

$$\sigma(s, s | \tilde{\mathcal{P}}^{o_1}) = 1 - \alpha \cdot \frac{|\mathcal{TS}(o_1) - \mathcal{TS}(s)|)}{|\mathcal{TS}(o_1)|}, \tag{3}$$

where $\mathcal{TS}(s)$ represents the type set of s and α is the impact factor of type set importance degree over similarity with range of [0,1]. The larger α is, the more constraints of types will be, and the clearer the semantic will be.

Given two objects s and t, and a MuTyPath $\tilde{\mathcal{P}}^{o_1\cdots o_n} = \mathcal{TS}(o_1) \xrightarrow{R_1} \cdots \xrightarrow{R_{n-1}} \mathcal{TS}(o_n)$, considering both conditions of two ends of a relation and type set constraint, the MuTySim similarity of two objects along the path $\tilde{\mathcal{P}}$ is defined as:

$$\sigma(s,t|\tilde{\mathcal{P}}^{o_1\cdots o_n}) = \sigma(t,t|\tilde{\mathcal{P}}^{o_n}) \sum_{x\in I(t|R_{n-1})} \frac{2|O(x|R_{n-1}) \bigcap I(t|R_{n-1})|}{|O(x|R_{n-1})| + |I(t|R_{n-1})|} \cdot \sigma(s,x|\tilde{\mathcal{P}}^{o_1\cdots o_{n-1}}),$$

(4)

where $O(x|R_{n-1})$ is the out-neighbors of object x based on relation R_{n-1}, and $I(t|R_{n-1})$ is the in-neighbors of t based on R_{n-1}.

When two objects s and t are connected by the concatenated MuTyPath $\tilde{\mathcal{P}}^{o_1\cdots o_n}$ and meet at any position along the MuTyPath, the MuTySim similarity values of two objects are equal:

$$\sigma(s,t|\tilde{\mathcal{P}}^{o_1\cdots o_n}) = \sum_{x\in C_j} \sigma(s,x|\tilde{\mathcal{P}}^{o_1\cdots o_j}) \cdot \sigma(x,t|\tilde{\mathcal{P}}^{o_j\cdots o_n}),$$

(5)

where o_j is the meeting position and C_j is the object set at position o_j.

In particular, when α equals to 0, the entity type will not be considered and MuTySim can measure similarity based on traditional meta path, so that it can be seen as *MuTySim for meta path* which is useful for random walking in opposite way and meeting at any position. Therefore, applying *MuTySim for meta path* in RWCP can discover meta path feature more accurately.

3.3   Weight Learning of Meta Paths

MuTyPaths should be combined effectively based on their importances to constitute ESE model. ESE is in fact to build a ranking model that calculates the probability of a candidate in the expansion set, and take the top $\mathcal{K}$ entities as the expansion result. The formula of the ranking model can be defined as follows:

$$CSSim(c,\mathcal{SS}) = \sum_{s\in\mathcal{SS}} CSim(c,s),$$

(6)

where c is the candidate node, and $\mathcal{SS}$ is the seed set, and $CSim(c,s)$ represents the matching probability of candidate node c and seed s.

Whether the candidate matches the seed can be seen as a classification problem. We can regard the MuTySim similarity value vector of entity pair on selected MuTyPaths as a feature for classification. Besides, positive data are the seed pairs, while the pairs of candidates and seeds are all unlabeled data. However, there are no effective methods for the automatic selection of the negative data. Without the negative data, we can not use traditional supervised learning method to do classification. To solve the problem, we come up with two weight learning solutions: the heuristic method and PU learning method.

Weight Learning with Heuristic Method. It is easy to understand that the meta path connecting more seed pairs will be more important, and the path with larger similarity value indicates closer relationship. Depending on the importance degree of each MuTyPath connecting the seed set, we calculate the corresponding weight for each path based on the similarity information of seed pairs and linearly combine the weight with similarity value together to form the matching probability equation for the candidates.

$$CSim(c, s) = \sum_{\tilde{\mathcal{P}}_i \in \tilde{\mathcal{P}S}} \varpi_i \cdot \sigma(c, s | \tilde{\mathcal{P}}_i), \tag{7}$$

$$\varpi_i = \frac{f(\tilde{\mathcal{P}}_i) \cdot \sum_{s_m, s_n \in \mathcal{SS}, m \neq n} \sigma(s_m, s_n | \tilde{\mathcal{P}}_i)}{\sum_{\tilde{\mathcal{P}}_t \in \tilde{\mathcal{P}S}} f(\tilde{\mathcal{P}}_t) \cdot \sum_{s_m, s_n \in \mathcal{SS}, m \neq n} \sigma(s_m, s_n | \tilde{\mathcal{P}}_t)}, \tag{8}$$

where $\tilde{\mathcal{P}S}$ is the MuTyPath set generated in Sect. 3.2 and ϖ_i is the weight for path $\tilde{\mathcal{P}}_t$. $f(\tilde{\mathcal{P}}_i) = \xi^{|zeros(\tilde{\mathcal{P}}_i)|}$ is a penalty function for having seed pairs not connected by $\tilde{\mathcal{P}}_i$. So $|zeros(\tilde{\mathcal{P}}_i)|$ is the number of similarity values of seed pairs as 0 based on path $\tilde{\mathcal{P}}_i$ and ξ is the penalty constant for it as $1/2$. $\sigma(s_m, s_n | \tilde{\mathcal{P}}_i)$ is the MuTySim value in seed pair (s_m, s_n) based on $\tilde{\mathcal{P}}_i$.

Weight Learning with PU Learning. PU learning is used to train classifier with positive and unlabeled training data, which is suitable for ESE. In our method, the seed pairs can be seen as positive data while the candidate-seed pairs as unlabeled data. We adopt a novel PU learning method proposed by Elkan et al. [3], which could train a traditional classifier to distinguish the positive and unlabeled examples and get a better result than existing PU learning methods. The main idea is to detect the reliable negative samples and then use the positive and negative cases to do classification training. This method is very flexible to choose any traditional classifier for PU learning, so that we can use suitable classifier to form the matching model for candidate nodes based on the exact situation.

4   Experiment

In order to verify the superiority of CoMeSE for entity set expansion in KG, we validate the effectiveness of CoMeSE with a series of experiments.

4.1   Experiment Settings

Dataset. We use KG Yago [11] to conduct relevant experiments. In experiments, we adopt "*COREFact*" and "*yagoSimpleTypes*" parts of this dataset, which contain 4.4 million facts, 35 relationships and 1.3 million entities of 3455 types.

Four ESE tasks are chosen to evaluate the performance of CoMeSE. These tasks are as follows: (1) in the *Actor* task, the seeds are actors who won Emmy Award, and their spouses are also actors; (2) in the *Company* task, the seeds are companies which own a channel in America; (3) in the *Writer* task, the seeds are writers which are graduated from the universities in New York; (4) in the *Movie* task, the seeds are movies, and their director won National Film Award. The real numbers of instances in these tasks are 193, 76, 60, and 653, respectively.

Criteria. We use precision-at-k ($p@k$) and Mean Average Precision (MAP) to evaluate performance. Here, they are $p@10$, $p@30$, and $p@60$. And MAP is the mean of the Average Precision (AP) of the $p@10$, $p@30$, and $p@60$.

Compared Methods. We denote the CoMeSE with heuristic and PU learning method, as "*CoMeSE_He*" and "*CoMeSE_PU*", respectively. And we use four baselines as follows: (1) Link. According to the pattern-based methods [8], it only considers 1-hop link of an entity, denoted as Link. (2) Neighbor. Inspired by QBEES [6], it considers 1-hop link and 1-hop entity as features, called Neighbor. (3) PCRW. With path constrained random walk based similarity measure PCRW [5], it employs different max path lengths to connect objects. The PCRW within length-2, 3, 4 paths are denoted as PCRW-2, PCRW-3, PCRW-4, respectively. (4) MP_ESE [14]. The KG-based ESE method finds meta paths by an one-directed generated method and uses a heuristic weight learning method.

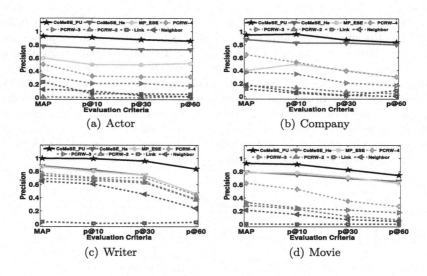

(a) Actor   (b) Company

(c) Writer   (d) Movie

Fig. 4. The results of entity set expansion.

In CoMeSE, we set ϵ as 10^{-6} in Eq. 2, α as 1 in Eq. 3 for calculating similarity based on MuTyPath, based on parameter study. The max length of path is set to be 4 since meta paths with length more than 4 are almost irrelevant [13].

4.2 Effectiveness Experiments

In this section, we validate the effectiveness of CoMeSE in 4 tasks introduced above. For each task, we randomly take three seeds from the instance set to conduct an experiment. We run 20 times and report the average results.

We illustrate the experiment results in Fig. 4. Firstly, the meta path based methods, CoMeSE and MP_ESE, almost have higher accuracy than other methods in all tasks. That is because, in KG, path feature can effectively embody intrinsic relations among seeds. Secondly, the results of CoMeSE_He and CoMeSE_PU are better than MP_ESE in almost all tasks. Traditional meta path fails to capture exact meanings owing to the uncertain types connected by relations, while MuTyPath used in CoMeSE can subtly distinguishes these paths. Thirdly, CoMeSE_PU performs better than CoMeSE_He in all tasks. The reason is that PU learning judges path features more precisely than heuristic method. In all, CoMeSE performs the best.

4.3 Efficiency Study

Here we validate the efficiency of finding meta paths under different seed size. We conduct experiments by varying the seed size from 2 to 6 on the *Actor* and *Movie* tasks. For each seed size, we randomly select the same-scale seeds to run 10 times. We show the average running time in Fig. 5. It is obvious that CoMeSE almost has the smallest running time in both tasks. PCRW-2 only explores 1-hop and 2-hops paths, so it has small running time. But short path exploration also gets bad performance, shown in Fig. 4. We think the bi-directional random walk strategy and the reuse of visited paths make the CoMeSE significant efficiency improvement. In addition, the running time of CoMeSE and PCRW methods near linearly increase with the increment of the seed size. It is reasonable, since these methods need to discover more paths to connect more seed pairs. Some strategies in MP_ESE make it less affected by seed size. In all, CoMeSE has obviously high efficiency of meta path discovery.

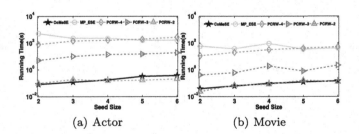

| (a) Actor | (b) Movie |

Fig. 5. Running times of different methods.

5   Conclusion

In this paper, we study the problem of entity set expansion in KG. We model KG as an HIN and propose a novel *Co*ncatenated *Me*ta Path based *E*ntity *Se*t *E*xpansion Method called CoMeSE, which proposes a random walk based concatenated meta path generation method to detect meta paths, a multi-type-constrained meta path algorithm to subtle capture path semantics, and uses two path weight learning methods to determine the importance of paths. Extensive experiments on Yago validate the performance of CoMeSE.

Acknowledgement. This work is supported in part by the National Natural Science Foundation of China (No. 61772082, 61375058), The National Key Research and Development Program of China (2017YFB0803304).

References

1. Cao, H., Jiang, D., Pei, J., He, Q., Liao, Z., Chen, E., Li, H.: Context-aware query suggestion by mining click-through and session data. In: KDD, pp. 875–883 (2008)
2. Cohen, W.W., Sarawagi, S.: Exploiting dictionaries in named entity extraction: combining semi-Markov extraction processes and data integration methods. In: KDD, pp. 89–98. ACM (2004)
3. Elkan, C., Noto, K.: Learning classifiers from only positive and unlabeled data. In: SIGKDD, pp. 213–220 (2008)
4. Jaiwei, H.: Mining heterogeneous information networks: the next frontier. In: SIGKDD, pp. 2–3 (2012)
5. Lao, N., Cohen, W.W.: Relational retrieval using a combination of path-constrained random walks. Mach. Learn. **81**(1), 53–67 (2010)
6. Metzger, S., Schenkel, R., Sydow, M.: QBEES: query by entity examples. In: CIKM, pp. 1829–1832. ACM (2013)
7. Qi, Z., Liu, K., Zhao, J.: Choosing better seeds for entity set expansion by leveraging wikipedia semantic knowledge. In: CCPR, pp. 655–662 (2012)
8. Shi, B., Zhang, Z., Sun, L., Han, X.: A probabilistic co-bootstrapping method for entity set expansion. In: 25th International Conference on Computational Linguistics, COLING 2014, Proceedings of the Conference: Technical Papers, Dublin, Ireland, 23–29 August 2014, pp. 2280–2290 (2014)
9. Shi, C., Li, Y., Zhang, J., Sun, Y., Yu, P.S.: A survey of heterogeneous information network analysis. IEEE Trans. Knowl. Data Eng. **29**(1), 17–37 (2017)
10. Singhal, A.: Introducing the knowledge graph: things, not strings. Official google blog (2012)
11. Suchanek, F.M., Kasneci, G., Weikum, G.: Yago: a core of semantic knowledge. In: WWW, pp. 697–706 (2007)
12. Sun, Y., Norick, B., Han, J., Yan, X., Yu, P.S., Yu, X.: Integrating meta-path selection with user-guided object clustering in heterogeneous information networks. In: KDD, pp. 1348–1356 (2012)
13. Wang, C., Song, Y., Li, H., Zhang, M., Han, J.: Knowsim: a document similarity measure on structured heterogeneous information networks. In: ICDM, pp. 1015–1020 (2015)
14. Zheng, Y., Shi, C., Cao, X., Li, X., Wu, B.: Entity set expansion with meta path in knowledge graph. In: Kim, J., Shim, K., Cao, L., Lee, J.-G., Lin, X., Moon, Y.-S. (eds.) PAKDD 2017. LNCS (LNAI), vol. 10234, pp. 317–329. Springer, Cham (2017). https://doi.org/10.1007/978-3-319-57454-7_25

Identifying In-App User Actions from Mobile Web Logs

Bilih Priyogi[1(✉)], Mark Sanderson[1], Flora Salim[1], Jeffrey Chan[1],
Martin Tomko[2], and Yongli Ren[1]

[1] School of Science, Computer Science and Information Technology,
RMIT University, Melbourne, Australia
{bilih.priyogi,mark.sanderson,flora.salim,
jeffrey.chan,yongli.ren}@rmit.edu.au
[2] Department of Infrastructure Engineering, University of Melbourne,
Melbourne, Australia
tomkom@unimelb.edu.au

Abstract. We address the problem of identifying in-app user actions from Web access logs when the content of those logs is both encrypted (through HTTPS) and also contains automated Web accesses. We find that the distribution of time gaps between HTTPS accesses can distinguish user actions from automated Web accesses generated by the apps, and we determine that it is reasonable to identify meaningful user *actions* within mobile Web logs by modelling this temporal feature. A real-world experiment is conducted with multiple mobile devices running some popular apps, and the results show that the proposed clustering-based method achieves good accuracy in identifying user actions, and outperforms the state-of-the-art baseline by 17.84%.

Keywords: Transaction identification · Mobile Web logs

1 Introduction

Mobile devices have become an important component of people's daily life, allowing near ubiquitous access to services on the Internet. Such accesses can potentially be logged. It is vital to understand users' needs by mining their Web usage logs, e.g. grouping Web accesses into meaningful units to find patterns [6,19]. Although Web log mining has been extensively studied for many years, the focus of such work is on the logs of Web sites [6,20]. Providers of free Internet access, such as shopping malls, airports and train stations, can collect a different sort of log: a mobile Web log capturing accesses from users accessing different services [1]. This log is quite different from those captured from a single Web site. Most of the Web requests will be encrypted (through HTTPS protocol) and there are a mix of user-driven accesses to Web sites/services, as well as automated accesses sent by the apps installed on user phones as observed by [17]. Accurately identifying and understanding in-app user actions from these mobile

© Springer International Publishing AG, part of Springer Nature 2018
D. Phung et al. (Eds.): PAKDD 2018, LNAI 10938, pp. 300–311, 2018.
https://doi.org/10.1007/978-3-319-93037-4_24

Web logs is critical in many fields, ranging from promoting personalised Web services to enforcing user activity monitoring and information security [8,10,11,16]. For example, mobile users will be provided the right information to satisfy their needs at the moment; the providers of free Internet can automatically monitor users' Web behaviours at high level and immediately be alerted when a potential risk user (or group of users) does something dangerous; and researchers in this field can learn lessons and build various models for modelling mobile users, etc.

Users' true actions in apps can be used to understand or infer users' behaviour. Users' behaviour can be extracted from mobile Web logs using various data mining techniques, such as association rule mining [3,12]. However, the mining result might be too coarse if we use a single log entry as a viewpoint, while a single user session might be too broad to give a fine-grained knowledge. Thus, it is necessary to group Web logs into meaningful units, in order to provide proper meaningful granularity for mobile user behaviour research [6,19,21].

We attempt to identify users' true actions from such logs. The challenges include: (1) as the concern of privacy and security become more prominent, more mobile apps are encrypting their Web accesses. This means that only hostname would be visible and it is impossible to know the content requested, including textual content and filename suffixes, which makes existing approaches infeasible [6], e.g. identifying access by using file suffixes. (2) the multitasking nature of a mobile device allows numerous applications to access the Web almost simultaneously, which should be taken into account before identifying in-app user actions within the corresponding sequential logs. (3) Automated URL requests issued by mobile applications also introduce bias in determining user actions, because it is not directly triggered by the user. Then, a new research question appears:

How to identify in-app user actions of Web accesses from encrypted and noisy mobile Web logs?

To tackle this problem, we introduce the concept of a *transaction* in the context of mobile Web logs, and propose a method to identify them. Specifically, a *transaction* is a group of *sequential Web access* (URLs) to one or more relevant Web domains, which correspond to a singular user action in a single mobile app. Moreover, we found that the distribution of time gaps between Web accesses, when there are user actions, is significantly different from that when there are no user actions. This indicates that it is reasonable to identify *transactions* within the Web logs with clustering techniques by modelling this time gap feature. Another reason why we model transaction identification by using clustering technique is because there are generally no labels in mobile Web logs and unsupervised learning is more fitting. Finally, to evaluate the performance of the proposed method, we conducted a controlled real-world experiment with multiple devices and mobile applications. Note the labels gathered in controlled experiment are only used for testing purposes. The experimental results show the proposed method significantly outperforms the state-of-the-art in terms of identifying *transactions* from mobile Web logs.

2   Related Work

Here, we briefly review relevant works about transaction identification in traditional Web site logs and traffic analysis in mobile logs.

Shu-yue et al. [19] and Cooley et al. [6] discuss several preprocessing techniques needed before executing mining algorithms on Web site logs. The techniques include data cleaning, log consolidation, log formatting, user identification, session identification, and transaction identification. These preprocessing techniques aim to provide high-quality data to ensure logs become more credible, accurate, and representative so that log mining and analysis can be conducted effectively. From Web server perspective, [2] investigates automated network request in the query stream of a large search engine provider's logs. [2] also proposes some features to distinguish between queries generated by people actually searching for information and those generated by the autonomous process.

Both [6,19] discuss techniques to identify transactions within a log in order to cluster log entries into meaningful units. Cooley et al. [5] discussed three transaction identification methods: *Reference Length* (RL), *Maximal Forward Reference* (MFR), and *Time Window*. Both RL and MFR categorized each page accessed as either a navigational or content page. A navigational page was considered to be only used by users to locate pages of interest, while a content page is a page containing desired information. The RL technique is based on an assumption that the amount of time a user spends on a page indicates whether the page should be classified as a navigational or content page. Then, in RL, a transaction can be defined as a sequence of navigational pages that lead to a content page. In MFR a transaction is defined as a set of page accesses before a backward reference is made. Such a reference is defined as a page not already in the set of pages from a current transaction. Different from aforementioned approaches, *Time Window* does not try to identify a page as navigational or content, but assumes that meaningful transaction should have an average overall length associated with them. User sessions were partitioned into time intervals no larger than a specified threshold.

Chen et al. [4] and Li et al. [14] proposing their own approach to identifying transactions. However, the technique is not feasible because of encryption in modern mobile Web traffic. In work by [13,15] a device-level point of view is taken to investigate network activity characteristics of mobile devices. Both authors gather and analyse data from users who voluntarily installed an app on their device, which captured their applications' network activity over a period of time. Similar to the work in this paper, [13,15] are also explanatory in nature and serve as an initial step toward a global network utilization model of mobile devices. However, little research was conducted to understand the meaningful *transaction* unit in the context of mobile Web traffic. In this paper, in the context of mobile traffic logs, a transaction is defined as a group of URL requests without considering its content, and a data mining based transaction identification method has been investigated.

3   Transaction Identification

Here, we first define the concept of a *transaction* in mobile Web logs to describe in-app user actions, then propose a method to identify them by modelling temporal features.

3.1   Definition

Based on a transaction definition in traditional Web log mining [6], a *transaction* t is defined as a meaningful group of sequential network activities (URL requests) in one application (*app*) on a user's mobile *device*. Transaction t is a tuple:

$$t = <device_t, app_t, \{(url_1^t, time_1^t), ..., (url_n^t, time_n^t)\}> \tag{1}$$

where $(url_i^t, time_i^t)$ denotes the i-th URL request from app_t on $device_t$ in transaction t. Namely, this indicates that *transactions* are defined per user (*device*) and *app*. Note, *url* could be the full URL request, or only the hostname if it is encrypted (e.g. through HTTPS protocol).

3.2   Identifying Transactions

We first conducted a comprehensive analysis of the time gap feature in mobile Web logs, which is defined as the gap in seconds between consecutive URL requests from the same *device* and *app*. We examine the logs of six representative *apps* (Facebook, Twitter, Instagram, Path, MSN, Sina) on three different devices (Android Tablet, iOS Phone, iOS Tablet), which will be detailed in Sec. 4.1. There are other apps and devices of interest, but we believe these apps and devices provide a good representative sample to analyse on. The gaps are separated into two groups: *idle* that means there are no user actions, and *active* that means there are user actions with the device. A Kolmogorov-Smirnov (KS) test is deployed to examine whether the *idle* vs *active* time gap distributions are significantly different to each other. The KS test results are shown in Table 1, indicating that there is a statistically significant difference between *idle* and *active* time gaps across all tested apps and devices, e.g. for *facebook* on Android Table, the KS statistic $D = 0.340$ and p-value ≤ 0.0001. This indicates that it is possible to identify transactions by modelling this temporal gaps.

Furthermore, Fig. 1 illustrates the example of difference between *idle* and *active* time gaps, in terms of *Expected Cumulative Distribution Function* (ECDF). We can observe that time gaps on *active* device seems to be more rapid than its counterpart. In other words, user actions on *active* device evidently caused a burst of consecutive URL requests. In contrast, consecutive URL requests occurred intermittently when there is no user actions.

Based on this analysis, we treat the problem of *transaction identification* as a clustering problem on the temporal gap feature. Specifically, we deploy the classic DBSCAN clustering method [7] to perform the transaction identification. The main reason is that, when modelling the temporal gaps, DBSCAN takes into account the sequential nature of the data (network activities) via its unique density connectivity concept. Namely, DBSCAN clusters periods of high activity, separated by gaps of idles, which is the characteristics of the transaction identification problem.

Table 1. *KS*-test results of comparing time gap distribution between *idle* and *active* mobile usage. Note D is the two-sample KS statistic, indicating it is greater than the corresponding critical value.

Device	app	D	p-value
Android tablet	Facebook	0.340	<0.0001
	Twitter	0.333	<0.0001
	Instagram	0.364	<0.0001
	Path	0.341	<0.0001
	MSN	0.460	<0.0001
	Sina	0.494	<0.0001
iOS phone	Facebook	0.297	<0.0001
	Twitter	0.311	<0.0001
	Instagram	0.294	0.016
	Path	0.306	<0.0001
	MSN	0.453	<0.0001
	Sina	0.490	<0.0001
iOS tablet	Facebook	0.299	<0.0001
	Twitter	0.299	<0.0001
	Instagram	0.297	<0.0001
	Path	0.299	<0.0001
	MSN	0.404	<0.0001
	Sina	0.389	<0.0001

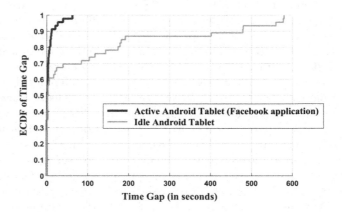

Fig. 1. ECDF of time gap feature in two cases: *Idle* VS. *Active*

4   Experiment

4.1   Experiment Environment

We configured a controlled Wi-Fi network to collect Web logs from connected mobile devices, as shown in Fig. 2. Each of mobile devices connected to the wireless access point via wi-fi, as the only available communication line to the Internet. A computer acts as a proxy server that records all URL requests coming through a wireless access point. Untangle[1] used as network gateway management software, which provides the capability of capturing network traffic. There are four different devices in the experiment: an Android tablet (Nexus 7 3G), iOS phone (iPhone 4S), iOS tablet (iPad 2), and iOS music player (iPod Touch 4th Gen). A factory reset was performed on all devices prior to the experiment to minimize noise introduced by previously installed apps. All devices were installed with the latest version of each application that was available for the device's latest OS version.

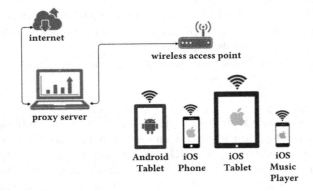

Fig. 2. Experiment configuration

To better simulate the real-world usage, six representative popular applications were installed on each device: Facebook, Twitter, Instagram, Path, MSN, and Sina. However, on iOS music player, MSN application could not be installed due to incompatible older OS version. These applications were selected because presumably, their rich and constantly updated content nature will give an adequate volume of network traffic to observe for. This set-up expected to give controlled environment in terms of unrelated adware's network access that might be embedded in some applications, while still represents real world mobile device usage. All applications were logged into with the same user account across all devices, except for MSN and Sina, to ensure they receive a similar volume of content. Note that, in this experiment, we only consider URL requests issued by

---

[1] www.untangle.com.

applications, and not from the browser. Then, a user usage scenario was simulated to capture network activities on the active (used) device. Specifically, five user actions are defined, conducted, and considered within 1-min time boundary to simulate real world application usage by users, as listed in Table 2. Note that in this scenario, automated and user triggered URL requests might get mixed in the Web logs captured. Thus, only URL requests to Web domains that related with the tested application at the time are considered as URL requests triggered by user, otherwise it will be labelled as automated URL requests. Furthermore, URL requests that occurred outside the time boundary of scheduled usage scenario will also be considered as automated URL requests.

Table 2. User actions examined

Minute	User action	Description
1	Open app	Starting an application session
2	Browsing	Reading content and scrolling through at normal reading speed
3	Dwelling	Reading one post and stop scrolling through
4	Skimming	Reading content and scrolling through at skim reading speed
5	Close app	Closing an application by pressing device's home button

The collected mobile Web logs are divided per device and application. Transactions within the logs are identified and labelled manually according to time boundary of use. We only consider URL requests directly triggered by user action to form a transaction, while the automated URL requests are discarded. The resulting labels will form a ground truth to evaluate the experiment result. We use the *Time Window* [6] approach as a baseline method. For the threshold parameter in *Time Window*, it is calculated from the average duration of transactions occurred on a device caused by an application, where duration means the time interval between the first and last Web accesses of a particular transaction in seconds.

4.2 Measurement Metrics

Accuracy and *purity* are selected to measure the performance of the proposed method. Given the clustering result C, the majority of the log entries in a cluster $c \in C$ determines which transaction it represents. If two clusters represent the same transaction, the large one is selected to calculate *accuracy*. Specifically, for each selected cluster c, we define N_c as the number of the log entries correctly identified to the corresponding represented transaction. Then, *accuracy* is defined as follows:

$$accuracy = \frac{\sum_{i=1}^{j} N_{c_i}}{\# \ of \ all \ log \ entries}, \tag{2}$$

where j denotes the number of selected clusters, representing each transaction. *Purity* is also reported here because *accuracy* is based on an assumption that one transaction will be clustered into exactly one cluster. However, a single transaction might be divided into several clusters in the clustering process, and it is acceptable as long as a cluster is not a mix of log entries from different transactions. Thus, *purity* is defined as follows:

$$purity = \frac{\sum_{i=1}^{|C|} N_{c_i}}{\sum_{i=1}^{|C|} |c_i|},\tag{3}$$

where $|C|$ denotes the number of generated clusters, and $|c_i|$ denotes the number of log entries in cluster c_i.

4.3 Parameters

There are two parameters in the deployed DBSCAN algorithm: *MinPts* and *Epsilon*. *MinPts* is the minimum number of data points (URL log entries) within a transaction, and *Epsilon* is the maximum distance (time gap) from each data point to any other data point within the same cluster. We set $MinPts = 3$ as a fair number of URL requests in a single transaction. This is based on observation that a user action on an application is often triggered by more than one or two URL requests. But we are also aware that a higher number of *MinPts* might have risk of mixing few transactions together because it needs to reach *MinPts* to be considered as a cluster.

Meanwhile, *Epsilon* can be determined for each device-application by using the heuristic suggested in [7]. Specifically, the distances to the k-th nearest neighbour ($k = MinPts$) of all URL requests are sorted and plotted, as shown in Fig. 3. The corresponding distance value on the first valley or knee point in the graph indicates the threshold value for *Epsilon*. This heuristic gives an intuitive way to find a suitable *Epsilon* value, as a smaller *Epsilon* might break a transaction into several clusters. Vice versa, a bigger *Epsilon* value might group several transactions into one cluster. In this work, we use four knee point detection formulas similar to [9,18], in order to automatically find a suitable *Epsilon* value. The formulas are modified to suit our case in which the distance distribution is monotonic and non-increasing, as illustrated in Fig. 3. Then, we compare the DBSCAN clustering performance with four knee point detection approaches, including:

$$j_{i-1} - j_i : difference\ between\ magnitudes\tag{4}$$

$$\frac{j_{i-1} - j_i}{j_i} : relative\ difference\ between\ magnitudes\tag{5}$$

$$\frac{j_{i-1}}{j_i} - \frac{j_i}{j_{i+1}} : difference\ between\ ratios\tag{6}$$

$$\frac{j_{i-1} - j_i}{j_i - j_{i+1}} : difference\ between\ magnitude\ ratios\tag{7}$$

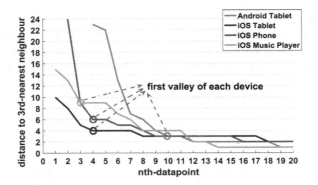

Fig. 3. Example distribution of the sorted 3rd-nearest neighbour distance

Table 3 displays the clustering performance with the four knee point detection approaches, in terms of average clustering accuracy and purity. The result indicates that the difference between magnitudes approach works best with 80.19% of average accuracy. The difference between magnitude's ratios approach gives second best performance, while the other two approaches seems to fail with average accuracy below 50%. Moreover, the four approaches gives similar performance in terms of average purity, which all of them exceed 99%. Based on this result, we use the difference between magnitudes approach to estimate *Epsilon* parameter in our experiment.

Table 3. Comparison of different knee point detection approaches to find k.

Method	$\jmath_{i-1} - \jmath_i$	$\frac{\jmath_{i-1} - \jmath_i}{\jmath_i}$	$\frac{\jmath_{i-1}}{\jmath_i} - \frac{\jmath_i}{\jmath_{i+1}}$	$\frac{\jmath_{i-1} - \jmath_i}{\jmath_i - \jmath_{i+1}}$
Description	Magnitude difference	Relative magnitude difference	Ratio difference	Magnitude's ratio difference
Accuracy	**80.19%**	42.32%	39.36%	72.22%
Purity	**99.27%**	99.72%	99.93%	99.52%

4.4 Experiment Results

Table 4 shows the comparison of experiment results between *clustering with DBSCAN* and *Time Window* (the baseline), in terms of accuracy. It is observed that on average, the proposed clustering-based method achieves better accuracy than the baseline. Specifically, the average accuracy of *clustering with DBSCAN* over all device-applications is 80.19%, with the highest and lowest accuracy are 100% and 51.35%, respectively. In contrast, the average accuracy of *Time Window* over all device-applications is 62.35%, with the highest and lowest accuracy are 85.00% and 50.48%, respectively. The *clustering with DBSCAN* method significantly outperforms *Time Window* in 20 out of 23 configurations, with equivalent or slight lower performance in the rest 3 configurations. The reason for the higher accuracy of the clustering-based method might be due to its flexibility in determining a transaction boundary based on the density of URL requests over time, rather than just a fixed time interval partitioning.

Table 4. Comparison on accuracy

Device	Application	TimeWindow	DBSCAN
Android tablet	Facebook	71.74%	**86.96%**
	Twitter	53.57%	**89.29%**
	Instagram	56.18%	**98.88%**
	Path	62.07%	**86.21%**
	MSN	**76.83%**	67.89%
	Sina	60.31%	**68.70%**
iOS phone	Facebook	76.27%	**88.14%**
	Twitter	51.43%	**87.14%**
	Instagram	54.55%	**90.91%**
	Path	71.76%	**81.18%**
	MSN	65.50%	**89.96%**
	Sina	57.45%	**100%**
iOS tablet	Facebook	**85.00%**	**85.00%**
	Twitter	60.68%	**99.15%**
	Instagram	56.06%	**56.92%**
	Path	50.48%	**51.43%**
	MSN	76.83%	**98.35%**
	Sina	55.48%	**81.51%**
iOS music player	Facebook	56.25%	**71.88%**
	Twitter	61.54%	**65.38%**
	Instagram	**51.35%**	**51.35%**
	Path	57.32%	**73.17%**
	MSN	N/A	N/A
	Sina	74.67%	**74.93%**
Average accuracy		62.35%	**80.19%**

Moreover, we also compare the two methods in terms of *purity*. It is observed that the *purity* is similar between the two methods, with *clustering with DBSCAN* performing slightly better than *Time Window*. Specifically, the average purity of *clustering with DBSCAN* over all device-applications is 99.27%; and the average purity of *Time Window* over all device-applications is 97.84%. In addition, the surprisingly good performance of *Time Window* approach in purity might be caused by our method in choosing suitable time interval for each device-application pair based on adequate prior knowledge or good estimation of average transaction duration, which would not be available in a real life situation.

It is observed that generally clustering *purity* is higher than *accuracy*. In this case, this phenomenon mainly because a single transaction often split into several clusters, which will penalised accuracy as much as mixing different transactions into one cluster. Thus, although each cluster tends to be quite pure as it consists of URLs within a single transaction, the accuracy will seem to be relative low. Nonetheless, high purity shows that inter-transaction time gap is quite significant to distinguish one transaction from another, while future work should also be addressed to recognize the substantial variation in intra-transaction time gap.

Overall, the experiment results indicate that it is possible to solve the challenges in the mobile Web logs, and so as to identify transactions from mobile Web logs by modelling the temporal gaps (via DBSCAN). Note, this is different to previous approaches, as they relied on the full URL information or Web content without considering time gap pattern.

5   Conclusion

The paper introduces the concept of a *transaction* to determine in-app user actions from encrypted and noisy mobile Web logs. A clustering algorithm is deployed to examine the possibility of identifying these transactions by only using time stamp information and without the assumption of knowing Web content, which is infeasible due to Web traffic encryption. The experiment results indicate that the proposed method can identify transactions successfully. In the future, we plan to improve the model to distinguish different user actions in a particular app (e.g. browsing on Facebook).

Acknowledgments. This research is supported by LPDP (Indonesia Endowment Fund for Education) and a Linkage Project grant of the Australian Research Council (LP120200413).

References

1. Arora, D., Neville, S.W., Li, K.F.: Mining wifi data for business intelligence. In: 2013 Eighth International Conference on P2P, Parallel, Grid, Cloud and Internet Computing, pp. 394–398, October 2013. https://doi.org/10.1109/3PGCIC.2013.67
2. Buehrer, G., Stokes, J.W., Chellapilla, K.: A large-scale study of automated web search traffic. In: Proceedings of the 4th International Workshop on Adversarial Information Retrieval on the Web, AIRWeb 2008, pp. 1–8 (2008)
3. Bulut, E., Szymanski, B.K.: Understanding user behavior via mobile data analysis. In: 2015 IEEE International Conference on Communication Workshop (ICCW), pp. 1563–1568, June 2015. https://doi.org/10.1109/ICCW.2015.7247402
4. Chen, L., Li, Z., Ju, S.: Based on forward reference object transaction identification algorithm on web mining. In: ALPIT 2007, pp. 469–473, August 2007
5. Cooley, R., Mobasher, B., Srivastava, J.: Grouping web page references into transactions for mining world wide web browsing patterns. In: Proceedings 1997 IEEE Knowledge and Data Engineering Exchange Workshop, pp. 2–9, November 1997. https://doi.org/10.1109/KDEX.1997.629824

6. Cooley, R., Mobasher, B., Srivastava, J.: Data preparation for mining world wide web browsing patterns. KIS **1**(1), 5–32 (1999)
7. Ester, M., Kriegel, H.P., Sander, J., Xu, X.: A density-based algorithm for discovering clusters in large spatial databases with noise, pp. 226–231. AAAI Press (1996)
8. Fan, Y.C., Chen, Y.C., Tung, K.C., Wu, K.C., Chen, A.L.P.: A framework for enabling user preference profiling through wi-fi logs. IEEE Trans. Knowl. Data Eng. **28**(3), 592–603 (2016). https://doi.org/10.1109/TKDE.2015.2489657
9. Foss, A., Wang, W., Zaïane, O.R.: A non-parametric approach to web log analysis (2001)
10. Gu, Y., Quan, L., Ren, F.: Wifi-assisted human activity recognition. In: 2014 IEEE Asia Pacific Conference on Wireless and Mobile, pp. 60–65, August 2014. https://doi.org/10.1109/APWiMob.2014.6920266
11. Guerbas, A., Addam, O., Zaarour, O., Nagi, M., Elhajj, A., Ridley, M., Alhajj, R.: Effective web log mining and online navigational pattern prediction. Knowl. Based Syst. **49**(Supplement C), 50–62 (2013). https://doi.org/10.1016/j.knosys.2013.04.014. http://www.sciencedirect.com/science/article/pii/S0950705113001263
12. Huang, J., Xu, F., Lin, Y., Li, Y.: On the understanding of interdependency of mobile app usage. In: 2017 IEEE 14th International Conference on Mobile Ad Hoc and Sensor Systems (MASS), pp. 471–475, October 2017. https://doi.org/10.1109/MASS.2017.89
13. Lee, J., Seeling, P.: An overview of mobile device network traffic and network interface usage patterns. In: IEEE EIT 2013, pp. 1–5, May 2013
14. Li, Y., Feng, B.: The construction of transactions for web usage mining. In: CINC 2009, vol. 1, pp. 121–124, June 2009
15. Mead, S., Veeramachaneni, N., Seeling, P.: An overview of mobile device network activities: characteristics of heterogeneous network interfaces. In: CCNC 2016, pp. 305–306 (2016)
16. Morichetta, A., Bocchi, E., Metwalley, H., Mellia, M.: Clue: clustering for mining web urls. In: 2016 28th International Teletraffic Congress (ITC 28), vol. 01, pp. 286–294, September 2016. https://doi.org/10.1109/ITC-28.2016.146
17. Qian, F., Wang, Z., Gao, Y., Huang, J., Gerber, A., Mao, Z., Sen, S., Spatscheck, O.: Periodic transfers in mobile applications: network-wide origin, impact, and optimization. In: Proceedings of the 21st International Conference on World Wide Web, WWW 2012, pp. 51–60. ACM, New York (2012). https://doi.org/10.1145/2187836.2187844, http://doi.acm.org/10.1145/2187836.2187844
18. Sadri, A., Ren, Y., Salim, F.D.: Information gain-based metric for recognizing transitions in human activities. Pervasive Mob. Comput. **38**(Part 1), 92–109 (2017). https://doi.org/10.1016/j.pmcj.2017.01.003.   http://www.sciencedirect.com/science/article/pii/S1574119217300081
19. Shu-yue, M., Wen-cai, L., Shuo, W.: The study on the preprocessing in web log mining. In: KAM 2011, pp. 315–317 (2011)
20. Suadaa, L.H.: A survey on web usage mining techniques and applications. In: 2014 International Conference on Information Technology Systems and Innovation (ICITSI), pp. 39–43, November 2014. https://doi.org/10.1109/ICITSI.2014.7048235
21. Woon, Y.K., Ng, W.K., Lim, E.P.: Online and incremental mining of separately-grouped web access logs. In: Proceedings of the Third International Conference on Web Information Systems Engineering, WISE 2002, pp. 53–62, December 2002. https://doi.org/10.1109/WISE.2002.1181643

Harvesting Knowledge from Cultural Heritage Artifacts in Museums of India

Abhilasha Sancheti[1], Paridhi Maheshwari[2], Rajat Chaturvedi[3],
Anish V. Monsy[4], Tanya Goyal[5], and Balaji Vasan Srinivasan[1(✉)]

[1] Big data Experience Lab, Adobe Research, Bangalore, India
{sancheti,balsrini}@adobe.com
[2] Department of Electrical Engineering, Indian Institute of Technology,
Kanpur, India
1997.paridhi@gmail.com
[3] Department of Computer Science, Indian Institute of Technology,
Bombay, India
chaturvedirajat96@gmail.com
[4] Department of Computer Science, Indian Institute of Technology,
Guwahati, India
anishvmonsy2@gmail.com
[5] Department of Computer Science, University of Texas at Austin,
Austin, TX, USA
tanyagoyal.93@gmail.com

Abstract. Recent efforts towards digitization of cultural heritage artifacts have resulted in a surge of information around these artifacts. However, the organization of these artifacts falls short with respect to accessing the facts across these entities. In this paper, we present a method to harvest the knowledge and form a knowledge graph from the digitized artifacts in the Museums of India repository via distant supervision to enable better accessibility of the facts and ability to extract new insights around the artifacts. Triples extracted from an open information extractor are first canonicalized to a standard taxonomy based on a metric-based scoring. Since a standard taxonomy is insufficient to capture all the relationships, we propose a sequential clustering based approach to add artifact specific relationships to the taxonomy (and to the knowledge graph). The graph is enriched by inferring missing facts based on a probabilistic soft logic approach seeded from a frequent item set framework. Human evaluation of the final knowledge graph showed an accuracy of 75% on par with knowledge bases like DBpedia.

1 Introduction

Cultural heritage represents a legacy of traditions and customs inherited from the past, maintained in the present and preserved for the benefit of future generations. As an attempt to reach wider audiences, various museums across the

Electronic supplementary material The online version of this chapter (https://doi.org/10.1007/978-3-319-93037-4_25) contains supplementary material, which is available to authorized users.

globe have digitized their artifact collections [6, 9, 14] and have made them available on web portals to facilitate better availability of the artifacts data to the public. However, in the absence of a proper organization, the large amount of digital content in these portals can be overwhelming and infeasible to interpret the information associated with the artifacts.

For a cultural enthusiast, a simple keyword search might not always fetch what (s)he is looking for since some of the information can span details from multiple artifacts. Standard information retrieval system cannot satisfy such needs since they serve information from a single source only. For example, Fig. 1(a) shows a sample query "tempera images by Jamini" to Museums of India (MOI) [14], an online portal about cultural artifacts in India. This illustrates that the current organization of the artifacts does not capture the specific style of paintings by an author. There could be several such aspects that could be useful for gaining insights and discovering relationships between the artifacts. This calls for a systematic approach to harvest the knowledge from cultural artifacts in order to enhance the understanding and facilitate the organization around them to enable better accessibility of the facts around these artifacts.

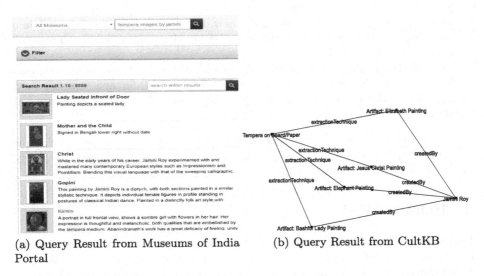

(a) Query Result from Museums of India Portal

(b) Query Result from CultKB

Fig. 1. Search result for query paintings by 'Jamini Roy' using 'Tempera' from the Museums of India web portal and the constructed CultKB

There is a growing body of work that focuses on harvesting knowledge from structured and unstructured data sources [23] towards building a knowledge base. Such knowledge bases/graphs serve as an excellent organization for insightful explorations as well as cross-artifact fact extraction/retrieval. Popular knowledge bases like DBpedia [2], NELL [7], YAGO [24] contain "facts" of the form "subject-predicate-object" and are generally extracted from a generic corpus like Wikipedia and canonicalized based on a standard taxonomy.

While knowledge graphs offer a good solution towards exploration, standard taxonomies are insufficient to capture the facts in cultural artifacts and render the taxonomies from standard knowledge bases inapplicable to the needs of cultural artifacts. However, building such a taxonomy from scratch specific to the cultural artifacts is also infeasible since this requires significant input from domain-experts. To address these challenges, we start with a standard taxonomy and harvest the facts from artifacts canonicalized to this taxonomy. Simultaneously, we also enrich the taxonomy to cover the needs of the cultural heritage artifacts by adding new artifact-specific relationships to the taxonomy (and the corresponding facts to the knowledge graph). The proposed approach addresses 3 major challenges:

1. The meta-data in the digitized cultural artifacts do not always have well-formed text and hence can result in noisy facts. Therefore, processing these noisy facts to extract meaningful facts via appropriate canonicalization is the first major challenge that we address in our approach.
2. Since standard taxonomies are insufficient to canonicalize all the facts that exist in the data from a specific domain, building a systematic approach to enrich the taxonomy with domain-specific relationships is the second challenge that we address. Identifying the new predicates for the taxonomy would require de-duplicating their multiple representations and reducing the overall noise in the extracted facts.
3. Finally, the uniqueness of the data about cultural artifacts provides an opportunity to look for patterns in the extracted facts to infer new/missing facts and enrich the knowledge graph with additional relationships that are not already present in the data.

Figure 1(b) shows a list of paintings made by the artist using tempera technique in response to the same query in Fig. 1(a) extracted from our proposed knowledge graph built on the MOI [14] data.

2   Related Work

Knowledge harvesting deals with extracting meaningful relationships and constructing knowledge graphs from text and other unstructured as well as structured sources [15]. Knowledge graph extraction involves the problem of inferring entities (nodes) and their relations/predicates (edges) from uncertain data while simultaneously incorporating constraints imposed by ontological inferences [23]. Entities in uncertain data might appear in different forms due to mis-spelled usages, use of synonyms or any other factors. Therefore, entities and relations extracted from the uncertain data are canonicalized, which is the process of standardizing the extracted facts to a taxonomy to achieve a consistent knowledge graph. Ontologies in the taxonomy aid in adding constraints to the facts for maintaining consistency and meaningfulness of the extracted facts [7].

There exists a number of large-scale **publicly available knowledge bases** like YAGO [24], DBpedia [2], Freebase [5], etc. While DBpedia [2] builds upon the structured info-boxes of Wikipedia, YAGO [24] automatically derives its facts from Wikipedia and Wordnet using a combination of rule-based and heuristic approaches. But these works deal with knowledge covering a broad range of real-world concepts and are not restricted to any particular domain. There exists very limited work on building knowledge bases for a specific domain. Kobren et al. [16] build a knowledge base of scientists and their affiliation via crowdsourcing. Similarly, Zhao et al. [26] use crowdsourcing to build a software-engineering knowledge base from StackOverflow. However, given the limited expertise available for cultural artifacts, such crowd-sourced approaches are not feasible in our case.

Developments in **digitizing cultural artifacts** have led to a few efforts to understand and organize such cultural artifacts. Agirre et al. [1] developed a system, PATH to aid people in navigating through the Europeana [9] artifacts. PATH measures artifact similarity to Wikipedia articles/entities by comparing the topics generated from each artifacts metadata using Latent Dirichlet Allocation (LDA) with the Wikipedia topics. The matched Wikipedia articles/entities are used to generate hierarchies which help in browsing and exploring the artifacts. Fernando et al. [11] explored techniques to automatically add Wikipedia links to resources in order to provide relevant background information. However, these approaches are not suitable to organize our data owing to the limited information available about Indian cultural heritage on open knowledge sources like Wikipedia. This restrains the use of external data sources.

With these limitations in mind, we propose a novel algorithm to harvest knowledge from a cultural artifact corpus [14], canonicalize it to a standard taxonomy and simultaneously enrich the taxonomy, and finally refine and infer any missing information in the extracted facts. The proposed approach is designed for distant supervision and hence can scale without annotations from a human expert.

3   Harvesting Data from MOI [14] into CultKB

Table 1(a) shows the list of museums and their artifacts from the portal. Table 1(b) shows the distribution of different categories of artifacts in the data. The portal currently hosts information of over $90k$ artifacts including paintings, manuscripts, coins, and sculptures.

Figure 2 shows a sample artifact with the associated meta-data. The artifact meta-data is in the form of field-value pairs which includes the structured data such as the title, creator and year of work along with the unstructured data such as brief and detailed description about the artifact. We build **CultKB** our knowledge base of cultural artifacts from MOI by harvesting this meta-data.

We begin with the canonicalization of the structured data to the YAGO taxonomy [24]. We canonicalize the unstructured data to the YAGO taxonomy using a voting based mechanism across different scoring functions. To account for predicates not in the taxonomy, we use a density-based spatial clustering approach to identify valid predicates and de-noise their multiple manifestations.

Table 1. Statistics around various artifacts in the Museums of India web portal

(a) Artifacts in different museusm

Museums	Artifacts
Salar Jung Museum, Hyderabad	23, 981
National Museum, New Delhi	21, 384
The Allahabad Museum	13, 277
Indian Museum, Kolkata	12, 228
Nagarjunakonda Museum	8, 400
Victoria Memorial Hall, Kolkata	2, 900
National Gallery of Modern Art, New Delhi	5, 423
National Gallery of Modern Art, Mumbai	1, 400
National Gallery of Modern Art, Bengaluru	500
Goa Museum	700

(b) Artifacts in top 10 categories

Category	Artifacts
Painting	10207
Decorative Art	7227
Manuscript	7152
Coin	6714
Terracotta	4010
Miniature Painting	3889
Soldier	3843
Porcelain	3799
Anthropology	3710
Central Asian Antiquities	2930

Fig. 2. A sample artifact along with the associated structured and unstructured information from Museums of India

We finally use a probabilistic soft logic based approach to infer missing facts in the constructed knowledge graph.

Canonicalization of Structured Data: The structured data, in the form of field-value pairs, naturally occurs in the <subject, predicate, object> format with each distinct field representing an edge between the artifact and the corresponding field value. Since the number of predicates in the structured data was small, we identified the predicates in the structured data as a part of preprocessing and manually mapped them to the appropriate predicates in the YAGO taxonomy [24] after an initial set of candidate predicates being extracted via string matching. The triples thus extracted are directly added to our knowledge graph which has the subject/object as its nodes and the predicates as edges.

Canonicalization of Unstructured Data: For canonicalizing the unstructured text, the artifact description is preprocessed to resolve all co-referencing pronouns using the Stanford Co-reference Parser [17]. All possible triples are extracted from the processed text based on an open information extraction (OpenIE) architecture [4,10]. OpenIE architecture identifies relation phrases in sentences based on syntactic and lexical constraints and assigns a pair of noun arguments for each extracted relation. For each triple, the entity type of subject and object are recognized using the Stanford Named Entity Recognizer [12].

The OpenIE triple extraction is based on the sentence structure analysis and therefore tends to be noisy.

To reduce the noisy triples and resolve redundant and ambiguous facts, the entities (subject and object) and the predicates are mapped to the YAGO taxonomy [24]. For the entities, an edit distance is computed from the matching entities in YAGO and the map beyond a threshold (σ_{entity}) is used as the canonicalized entity.

The canonicalization of predicates is constrained on the nature of entities associated in the artifact triples and that of YAGO triples by incorporating the ontological knowledge of the relationships between entity types to remove noisy triples. For example, the domain and range constraints DOMAIN(isWrittenBy, book) and RANGE(isWrittenBy, person) specify that the relation 'isWrittenBy' is a mapping from entities with type book to entities with type person. The appropriate YAGO predicate for a given triple is then identified based on an ensemble of three approaches:

1. *Semantic Mapping:* The first approach captures the semantic similarity of words in the phrase and the YAGO relations using a vector space model. It involves computing the cosine similarity between the Word2Vec embeddings [18,19] of the relationships from artifact triples and those from YAGO. Word2Vec captures the semantic space of the words and therefore such a measure maps the relationships based on their semantic closeness to the relationships in the YAGO taxonomy.
2. *Syntactic Mapping:* In this approach, the resemblance of two predicates is determined by the resemblance of the main verbs. A dependency parser is used to extract the dependency tree from the unstructured source text and a network of "cognitive synonyms" [20] of the root verb of the dependency tree is identified. This network of synonyms is compared with the root verbs of the YAGO relations to establish a correspondence between relations in the syntactic sense.
3. *Pattern based Mapping:* Two verbal phrases are likely to be similar if they share some common pattern of words, with a possible difference of some words like helper verbs and adjectives. With this intuition, the last approach is extended from [21] which obtains textual patterns in binary relations, transforms them into syntactic-ontologic-lexical patterns using frequent item set mining [13] and constructs a taxonomy for these patterns. We match the closest YAGO relationship corresponding to a current pattern taxonomy triple (including the respective POS tags) and assign it as the predicate.

An empirical threshold is used for every approach to find suitable predicate in the taxonomy. A ranked order of target predicates (beyond the threshold) from all the 3 methods is combined based on a voting mechanism to determine the best canonicalized relationship for the current triple.

Enriching the Taxonomy with Cultural Heritage Specific Predicates: Canonicalization based on a standard taxonomy does not standardize all the extracted triples due to the uniqueness of the relationships in the cultural artifacts. Since the OpenIE predicates are extracted from the sentences, there are multiple manifestations of the same relationships in the database. This calls for a novel approach to enrich the initial taxonomy with predicates specific to the cultural heritage.

Starting with the mapped and unmapped relations, cosine similarity between the Word2vec [18,19] embeddings of the relationships is used to perform a density-based spatial clustering (DBSCAN) [8]. DBScan is capable of identifying the number of clusters simultaneously. This resulted in 20,000 different relation types being grouped into 7000 clusters. Incorporating a constraint of maintaining the same NER tags of the subject and object throughout the cluster resulted in partitioning into 9,000 clusters.

For the rest of the clusters, if a predicate from YAGO taxonomy is a part of the cluster, the cluster is tagged with the corresponding YAGO predicate and all the facts are updated with this predicate. In the absence of such a predicate, a representative predicate was chosen based on its frequency of occurrence in the corpus. The NER tags of the subject and object of the associated predicate are used to define the domain and range of the new relationship. Clusters with a significant relation (based on the threshold) are added to the taxonomy and the rest are ignored.

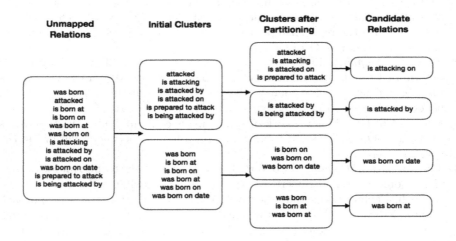

Fig. 3. Illustration of clustering algorithm

Figure 3 shows an illustration of various steps in the clustering process. The unmapped relations are first clustered depending upon the verb 'attack' or 'born'. 'was born at' and 'was born on' are originally in the same cluster but the corresponding NER tags are 'location' and 'date' respectively. Hence, they are partitioned into different clusters.

Inferring Missing Information: Since the facts are extracted from manually curated content, the knowledge graph is subject to the Open World Assumption, which states that *any missing triple is not necessarily false, just unknown.*

The knowledge graph is enriched with new triples based on a probabilistic approach that simultaneously identifies the missing information, strengthens the confidence value of correct facts and resolves conflicts in the data.

Logical rules are extracted via association rule mining [13] taking into account the Partial Completeness Assumption (PCA) which states that if there is at least one object associated with a subject through a relation then the relationship is considered complete. This implies that the PCA makes predictions only for those entities that have an object for the relationship, and remains silent otherwise. Logical rules, of the form,

$$<E_1, R_1, E_2> \wedge <E_2, R_2, E_3> \wedge \ldots \wedge <E_n, R_n, E_{n+1}> \Rightarrow <E_1, R_{n+1}, R_n>,$$

encode frequent correlations in the data. The left-hand side of the implication is called the body and the right-hand side is the head. The rules are assigned a normalized confidence score based on their support in the extracted knowledge base. A support for a rule is defined as the number of distinct subject and object pairs in the head of all the instantiations that appear in the knowledge base. The confidence score is calculated as the ratio of the support of the rule to the number of all the known true facts together with the assumed false facts in the extracted knowledge graph.

A Probabilistic Soft Logic (PSL) model [22] is defined based on the rules from the frequent items that includes the input set of rules along with the predicted triples. PSL minimizes a Markov Hinge-Loss function [3] that uses the input triples and their confidence scores to infer new facts along with their probabilistic confidence. PSL forms a probability distribution over all the interpretations/facts possible out of the derived and extracted facts and then infers the "most likely" facts. The task of "most likely explanation" inference corresponds to finding the confidence of each fact in the knowledge graph that maximizes the probability distribution over the derived facts. Confidence scores of facts endorsed by multiple rules are amplified, thus reinforcing the correct triples in the knowledge graph.

4   Evaluation of CultKB

We extracted 847,547 facts from the structured data input of 90,193 artifacts. The canonicalization of triples from the unstructured data to the YAGO taxonomy yielded 3615 more facts. The unmapped relations from the above step went through the clustering phase and gave us further 147,176 facts and added 5,502 new relationships to the taxonomy. Finally, the enrichment phase added 408,752 more facts to the knowledge base summing up to an overall of 1,407,090 facts.

In the absence of a gold standard dataset, we test the correctness of the facts in the knowledge base via human annotations from Amazon Mechanical Turk (AMT). Each AMT worker was presented with a text snippet from the Museums of India dataset to evaluate the correctness of the facts extracted from them. Each worker annotates 3 facts extracted from the presented passage, with one of the subject, predicate or object missing. The worker is tasked with selecting the appropriate option for the missing part from a list of options while ensuring the correctness of the completed fact. Occasionally, none of the options may correspond to the correct fact. We, therefore, allowed the AMT worker to opt for "None of the above" for such cases. Note that we are not evaluating the correctness of facts itself, but only the "correctness of the facts" as present in the text. Such an evaluation technique allows us to evaluate the correctness of the facts extracted by the algorithm, as well as establish ground truth for future experiments.

Each fact or triple is evaluated by 3 annotators. We used the Cohen's Kappa score to check for inter-annotator agreement which measures the agreement between categorical options, while simultaneously accounting for agreement by chance. Hence it is more robust than simple percentage calculation. We simulated two annotators by randomly selecting 2 turkers for each fact and calculating the agreement between them. This is repeated for 1,000 iterations and we report on the median of these iterations. We obtain a Cohen's Kappa score of 0.763 with 95% confidence interval of 0.0455 indicating high inter-annotator agreement. More details about the evaluation is provided in the supplementary material.

Table 2. Accuracy of facts in CultKB. We also report on the Wilson interval for $\alpha = 5\%$ to ensure that the accuracy values are significant.

Stage	Accuracy-interval
YAGO canonicalized	63.03% ± 18.15%
Sequential clustering	82.16% ± 6.18%
Overall after enrichment	75.50% ± 6.67%

Table 2 shows the accuracy of CultKB facts extracted. The accuracy of the facts canonicalized to YAGO (where both the predicates and the entities are canonicalized) are lower than the rest but is reasonable at 63.03% indicating that the canonicalization to a taxonomy is fruitful when the entire triple can be canonicalized.

The accuracy increases when the predicates are enriched using the clustering technique and this further establishes the need for building a base taxonomy to the needs of the cultural artifacts. The higher accuracy also justifies the ability of the proposed approach to introduce the culture specific predicates thus addressing the inadequacies of the standard taxonomy.

An overall accuracy of 75.50% ± 6.67% of the facts is comparable to that of DBpedia [2] (81% [25]) built from a cleaner and more structured source establishing the integrity of the constructed knowledge base.

Exploring CultKB: Table 3 shows the frequency of facts with a given relationship for the top 20 relationships in CultKB; note that there exists a long tail of relations with lower frequency counts. This count varies from as high as 1,578 for relation "painted" to as low as 69 for the relation "created". The relations 'painted', 'is Fragment of', 'painted from', 'is decorated with', 'has depicted

Table 3. Distribution of different relationships in CultKB

Predicate	Count	Predicate	Count
painted	1578	belongs to	600
is Fragment of	544	is written in	497
placedIn	483	depicts	466
studied art at	376	visited	201
is Head of	194	is A handle of	158
consists of	131	was born in	131
has studied	129	is written by	126
painted from	123	is decorated with	106
is Drawing of	100	is seated in	88
is seated on	86	has depicted a portrait of	74

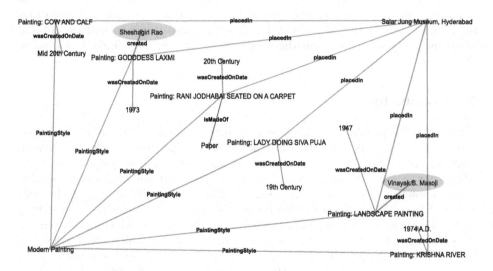

Fig. 4. Random subgraph from CultKB

a portrait of' reflect the facts around the intrinsic details of an artifact itself, while relations like 'studied art at', 'visited', 'created', 'belongs to' reflect the information about the artist involved.

The constructed knowledge graph facilitates a **navigation** through the various artifacts of the Museums of India and allows to hop between different artifacts sharing the same facets. Figure 4 shows such a sub graph of CultKB.

The artifacts are labelled with their corresponding title. The labels on edges are the relationships between nodes. We can visualize information such as 'placedIn', 'created', 'wasCreatedOnDate' for an artifact as well as can also see how different artifacts are related to each other. It is easy to see that the artifacts are placed in the 'Salar Jung Museum, Hyderabad'. Note that such a navigation is much richer than the one proposed in PATHS [1] since PATHS connects related artifacts without providing any reason for connections. But our knowledge graph representation naturally allows for a deeper artifact navigation experience.

A combination of the navigation experience and the retrievability of the organized data in CultKB allows for interesting knowledge discovery from the data. For example, the path between two artists 'Sheshagiri Rao' and 'Vinayak S. Masoji' in Fig. 4, whose paintings are housed in Salar Jung Museum, reveals that the two painters are part of the school of "Modern Paintings". Such connections are impossible without an organized representation like CultKB.

The knowledge graph also aids in easy accessibility of facts in the original data. Recall the example in Fig. 1, where a query on "tempera images by jamini" on the Museums of India portal yields irrelevant results (Fig. 1(a)). The structured knowledge representation in CultKB facilitated the results via a "path query" that connects the entities 'Jamini Roy' and 'Tempera Images' in the graph yielding the result in Fig. 1(b) which shows that there are four paintings by 'Jamini Roy' in the tempera style. Note that algorithms to understand and serve such queries are beyond the scope of this paper, but CultKB can aid in serving such queries.

5   Conclusion

We studied the problems with the accessibility of cultural heritage artifacts and proposed a novel approach to construct a knowledge base for the artifacts in Museums of India. The need for such a domain-specific knowledge base is justified due to the lack of facts supporting Indian cultural artifacts present in global knowledge bases like YAGO [24], DBpedia [2]. Evaluation of the constructed knowledge base with human annotators showed acceptable accuracy along the scales of existing knowledge bases. The structured knowledge graph thus obtained facilitates both knowledge discovery and enhanced retrieval of the cultural artifacts. Although, we had applied the proposed approach to the domain of cultural artifacts, the approach is generic and can be easily extended to other domains as well.

References

1. Agirre, E., Aletras, N., Clough, P.D., Fernando, S., Goodale, P., Hall, M.M., Soroa, A., Stevenson, M.: Paths: a system for accessing cultural heritage collections. In: Conference System Demonstrations, pp. 151–156. ACL (2013)
2. Auer, S., Bizer, C., Kobilarov, G., Lehmann, J., Cyganiak, R., Ives, Z.: DBpedia: a nucleus for a web of open data. In: Aberer, K., et al. (eds.) ASWC/ISWC 2007. LNCS, vol. 4825, pp. 722–735. Springer, Heidelberg (2007). https://doi.org/10.1007/978-3-540-76298-0_52
3. Bach, S.H., Broecheler, M., Huang, B., Getoor, L.: Hinge-loss Markov random fields and probabilistic soft logic. arXiv preprint arXiv:1505.04406 (2015)

4. Banko, M., Cafarella, M.J., Soderland, S., Broadhead, M., Etzioni, O.: Open information extraction from the web. In: IJCAI, vol. 7, pp. 2670–2676 (2007)
5. Bollacker, K., Evans, C., Paritosh, P., Sturge, T., Taylor, J.: Freebase: a collaboratively created graph database for structuring human knowledge. In: Proceedings of the 2008 ACM SIGMOD International Conference on Management of Data, pp. 1247–1250. ACM (2008)
6. British Museum: http://www.britishmuseum.org/
7. Carlson, A., Betteridge, J., Kisiel, B., Settles, B., Hruschka Jr., E.R., Mitchell, T.M.: Toward an architecture for never-ending language learning. In: AAAI, vol. 5, p. 3 (2010)
8. Ester, M., Kriegel, H.P., Sander, J., Xu, X., et al.: A density-based algorithm for discovering clusters in large spatial databases with noise. In: KDD (1996)
9. Europeana Museums: https://www.europeana.eu/portal/en
10. Fader, A., Soderland, S., Etzioni, O.: Identifying relations for open information extraction. In: Proceedings of the Conference on Empirical Methods in Natural Language Processing, pp. 1535–1545. Association for Computational Linguistics (2011)
11. Fernando, S., Stevenson, M.: Adapting wikification to cultural heritage. In: Proceedings of the 6th Workshop on Language Technology for Cultural Heritage, Social Sciences, and Humanities, pp. 101–106. Association for Computational Linguistics (2012)
12. Finkel, J.R., Grenager, T., Manning, C.: Incorporating non-local information into information extraction systems by Gibbs sampling. In: Proceedings of the 43rd Annual Meeting on Association for Computational Linguistics, pp. 363–370. Association for Computational Linguistics (2005)
13. Galárraga, L.A., Teflioudi, C., Hose, K., Suchanek, F.: AMIE: association rule mining under incomplete evidence in ontological knowledge bases. In: Proceedings of the 22nd International Conference on World Wide Web, pp. 413–422. ACM (2013)
14. Museums of India. http://museumsofindia.gov.in
15. Pujara, J., Sameer Singh, B.D.: Knowledge graph construction from text. In: AAAI Tutorial (2017)
16. Kobren, A., Logan, T., Sampangi, S., McCallum, A.: Domain specific knowledge base construction via crowdsourcing. In: Neural Information Processing Systems Workshop on Automated Knowledge Base Construction, AKBC, Montreal, Canada (2014)
17. Lee, H., Peirsman, Y., Chang, A., Chambers, N., Surdeanu, M., Jurafsky, D.: Stanford's multi-pass sieve coreference resolution system at the CONLL 2011 shared task. In: Proceedings of the Fifteenth Conference on Computational Natural Language Learning: Shared Task, pp. 28–34. Association for Computational Linguistics (2011)
18. Mikolov, T., Chen, K., Corrado, G., Dean, J.: Efficient estimation of word representations in vector space. arXiv preprint arXiv:1301.3781 (2013)
19. Mikolov, T., Sutskever, I., Chen, K., Corrado, G.S., Dean, J.: Distributed representations of words and phrases and their compositionality. In: Advances in Neural Information Processing Systems, pp. 3111–3119 (2013)
20. Miller, G.A.: Wordnet: a lexical database for english. Commun. ACM **38**(11), 39–41 (1995)

21. Nakashole, N., Weikum, G., Suchanek, F.: PATTY: a taxonomy of relational patterns with semantic types. In: Proceedings of the 2012 Joint Conference on Empirical Methods in Natural Language Processing and Computational Natural Language Learning, pp. 1135–1145. Association for Computational Linguistics (2012)
22. Pujara, J., Miao, H., Getoor, L., Cohen, W.: Knowledge graph identification. In: Alani, H., et al. (eds.) ISWC 2013. LNCS, vol. 8218, pp. 542–557. Springer, Heidelberg (2013). https://doi.org/10.1007/978-3-642-41335-3_34
23. Pujara, J., Miao, H., Getoor, L., Cohen, W.W.: Using semantics and statistics to turn data into knowledge. AI Mag. **36**(1), 65–74 (2015)
24. Suchanek, F.M., Kasneci, G., Weikum, G.: YAGO: a core of semantic knowledge. In: Proceedings of the 16th International Conference on World Wide Web, pp. 697–706. ACM (2007)
25. Zaveri, A., Kontokostas, D., Sherif, M.A., Bühmann, L., Morsey, M., Auer, S., Lehmann, J.: User-driven quality evaluation of DBpedia. In: Proceedings of the 9th International Conference on Semantic Systems, pp. 97–104. ACM (2013)
26. Zhao, X., Xing, Z., Kabir, M.A., Sawada, N., Li, J., Lin, S.W.: HDSKG: harvesting domain specific knowledge graph from content of webpages. In: 2017 IEEE 24th International Conference on Software Analysis, Evolution and Reengineering (SANER), pp. 56–67. IEEE (2017)

Query-Based Automatic Training Set Selection for Microblog Retrieval

Khaled Albishre[1,2]([✉]), Yuefeng Li[1], and Yue Xu[1]

[1] School of EECS, Queensland University of Technology (QUT),
Brisbane, QLD, Australia
khaled.albishre@hdr.qut.edu.au, {y2.li,yue.xu}@qut.edu.au
[2] Umm Al-Qura University, Makkah, Saudi Arabia

Abstract. Typical pseudo-relevance feedback models assume that the first-pass documents are the most relevant and use those documents to select feedback terms for query expansion. In real applications, however, short documents, such as microblogs, may not have enough information about the searched topic, thus increasing the chance that irrelevant documents will be included in the initial set of retrieved documents. This situation reduces a feedback model's ability to capture information that is relevant to users' needs, which makes determining the best documents for relevant feedback without requiring extra effort from the user a critical challenge. In this paper, we propose an innovative mechanism to automatically select useful feedback documents using a topic modeling technique to improve the effectiveness of pseudo-relevance feedback models. The main idea behind the proposed model is to discover the latent topics in the top-ranked documents that allow for the exploitation of the correlation between terms in relevant topics. To capture discriminative terms for query expansion, we incorporated topical features into a relevance model that focuses on the temporal information in the selected set of documents. Experimental results on TREC 2011–2013 microblog datasets illustrate that the proposed model significantly outperforms all state-of-the-art baseline models.

Keywords: Microblog retrieval · Topic model · Query expansion
Pseudo relevance feedback

1 Introduction

Pseudo-relevance feedback (PRF) is a technique for using top-ranked documents to boost document retrieval performance. However, the top-ranked documents include both relevant and irrelevant documents. In real applications, such as microblog retrieval, terms selected from top-ranked documents do not always represent the users' information needs because that they encounter language challenges, such as mismatched vocabularies and queries that are often too short or ambiguous. Microblog documents also have unique aspects that set them apart

© Springer International Publishing AG, part of Springer Nature 2018
D. Phung et al. (Eds.): PAKDD 2018, LNAI 10938, pp. 325–336, 2018.
https://doi.org/10.1007/978-3-319-93037-4_26

from other document types. These aspects include insufficient information, short length, time sensitivity and unstructured phrases [4]. Therefore, query expansion features (e.g. terms) from top-ranked documents that include microblogs can introduce a lot of irrelevant feature terms.

The challenging issue remains of how to infer the relevant features (e.g., terms, topics or themes) from top-ranked documents. Relevant documents focus more on the topic being searched and are separated from irrelevant documents [27]; however, both relevant and irrelevant documents share many common terms [5,21,35]. Therefore, term-based methods are limited when working to identify relevant features. Recently, we have observed that there are several focused topics in relevant documents, but the topics in irrelevant documents are diverse. This makes it more likely to find relevant topics in the top-ranked documents because the number of top-ranked relevant documents is normally higher than the number of top-ranked irrelevant documents, and the frequency of a focused topic is likely greater than the frequency of all the diverse irrelevant topics. Most existing PRF-based [10,26,29,30,36] methods use the top-ranked documents without fully assessing their contents because it is impossible to determine true relevance feedback. The basic hypothesis is that the top-ranked documents are relevant to a short query, include much relevant information, and may be helpful for query expansion. However, this hypothesis does not always hold in the microblogosphere [30,35] due to the overwhelming amount of extraneous and redundant information it includes.

In this paper, we address this challenging issue by automatically predicting the best top-ranked documents by using a topic modeling technique. The proposed method views parameter k (top-k documents) as a random variable and tries to work out the best k value for each query-based pseudo-relevant feedback document. The hypothesis states that the proposed method provides better query expansion than the approach of using a fixed-k for all feedbacks. The proposed method includes two stages. In the first stage, we automatically discern the top-k documents among the top-ranked documents for a given query. More specifically, the random variable k arranges the top-ranked documents into different small subsets. Unlike term-based approaches, such as TF-IDF [34], BM25 [33], RFD [20] and two-stage model [22], which have difficulty identifying relevant terms in PRF due to the terms' weak discriminative power, topic modeling techniques, such as Latent Dirichlet Allocation (LDA) [7], can find some latent topics for describing the focusing topics in PRF by using a more discriminative power than terms. We then use the latent topics to estimate how precise the top-k documents will be in helping to select the best k value. In the second stage, we use the top-k documents selected from the first stage to do the query expansion which uses both lexical features and latent topics to select possible relevant terms for the original query. It also considers the temporal distribution of the recent documents and provides a model to combine both the lexical features and latent topics effectively.

The main contributions of this paper include the following. (1) We propose a topic-based model to automatically determine the best (most likely) top-k

documents from PRF for a given query (described in Sect. 3.1); and (2) from the selected relevant feedback, we combine the relevance features with topical features to select discriminative features for query expansion taking in account their temporal distribution (described in Sect. 3.2). We conducted multiple experiments in real-time microblog datasets published by TREC 2011–2013. The evaluation was accomplished by comparing the proposed method with the state-of-the-art baseline models. The evaluation results showed that the proposed method performs significantly better than all baseline models for the microblog retrieval task. To the best of our knowledge, the proposed model is the first to integrate automatic query-based selection model into pseudo-relevance feedback for microblog retrieval application without any human efforts or external evidence, such as knowledge base.

2   Related Works

Query expansion approaches based on PRF are widely used in microblog retrieval research [4, 30]. Miyanishi et al. [30] proposed a relevance feedback model based on a two-stage relevance feedback that models search interests through manual tweet selection and the integration of lexical evidence into its relevance model. Lin et al. [24] utilised PRF to get relevant information and the graph-based model-generated storyline for a given query. To emphasise the short-term interests of user information needs, Albakour et al. [2] applied PRF to expand the user profile and tackle the sparsity problem. Using global evidence, such as Freebase or Probase, [15, 26, 36] proposed a two-stage feedback entity model based on a mixture strategy developed by observing the underlying entities in the original users' intents. In fact, utilising external evidence, such as Wikipedia under PRF framework requires a double run for a query, and it can be increased the complexity efficiency [8]. The majority of the above contributions assume that the top-ranked documents are relevant to the user's information needs.

Temporal signal has been extensively utilized in earlier microblog retrieval studies, proving that time is vital to reflect the relevance of the feedback [12, 26]. To discover the relationship between time and relevance, Li and Croft [19] proposed a time-based language model by integrating the time factor and relevance models. Efron and Golovchinksy [13] combined temporal information from first-pass documents to estimate the rate parameter for the query's likelihood model and used PRF to estimate the expansion query and show the effectiveness of recent user information needs. To discover users' behaviour using recency features, Choi and Croft [9] combined temporal evidence from PRF into the relevance model to improve the query expansion based on user attitude (e.g., retweet). Relevant information in a real application, such as microblog retrieval, tends to cluster together in time (i.e., event). Based on this idea, Efron et al. [14] proposed a retrieval model for microblog searchs that used temporal feedback to estimate the density of relevant information.

Recently, probabilistic topic mining models, such as LDA [7], have received much attention in the information retrieval research community [6, 23, 28, 29].

LDA is one of the state-of-the-art topic models and holds the assumption that each document in a given data collection could have multiple topics and the data collection itself could also contain more than one topic. Andrzejewski and Buttler [6] proposed an LDA-based model that distinguishes the potentially relevant latent topics that are manually chosen by the user. It then expands the original query from the selected topic by using the most associated terms. However, the user's manual efforts are expensive in real applications, such microblog retrieval systems. To extend LDA applications in microblogs, Kotov et al. [17] utilised the relevance model expansion for microblog collection that are publishing on geographical area. To infer topic distributions of short-documents over time, a recent study by Liang et al. [23] proposed a dynamic clustering topic model to cluster short documents using a dynamic topic model. The proposed framework utilised LDA to observe the quality of relevant information in the first-pass documents and determine the best documents for each query based under PFR framework that does not rely on user efforts or external knowledge.

3   The Proposed Model

3.1   Determining Top-k Documents

The main purpose of this stage is to discover the topical features of a set of ranked documents using LDA [7]. This stage emphasises the relevant information (i.e. theme) of the first-pass documents and enables the discovery and preservation of essential statistical relationships. LDA is typically a generative probabilistic model to learn the semantic mixture of topics from a set of documents [16]. For the i_{th} word in document d, denoted as w_i, the probability of w_i in a document is defined as:

$$P(w_i|d) = \sum_{j=1}^{V} P(w_i|z_j) \times P(z_j|d) \times P(z_j) \tag{1}$$

where $P(w_i|z_j)$ is the multinomial probability distribution over all words w_i for z_j, $P(z_j|d)$ is the topic assignment in document d, $P(z_j)$ is the topic assignment of topic z_j and V denotes the number of topics.

The LDA result consists of a mixture of latent topics. Each topic z_j is defined by a multinomial distribution over words, denoted as $\phi_j = (\varphi_{j,1}, \varphi_{j,2}, \dots, \varphi_{j,n})$ where $\varphi_{j,h}$ is the probability of a word w_i in the topic z_j, and the sum of all elements in the topic space denotes as $\sum_{h=1}^{n} \varphi_{j,h} = 1$. For all topics Z over all words in a document, $\Phi = (\phi_1, \phi_2, \dots, \phi_V)$ is the composition for each topic z_j. Each document is defined by a multinomial distribution over topics Z as $\Theta_d = (\vartheta_{d,1}, \vartheta_{d,2}, \dots, \vartheta_{d,V})$ where $\vartheta_{d,j}$ means the distribution of topic z_j in document d, and the sum of all elements in the Θ_d denotes as $\sum_{j=1}^{V} \vartheta_{d,j} = 1$. The posterior distribution is estimated for exact inference using the Gibbs sampler [32].

Here, the probability distribution over words in the topics space emphasises the degree to which the ranked documents relate to the users' information needs.

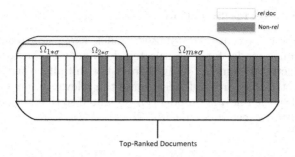

Fig. 1. Selecting pseudo-relevance feedback

Let Ω_k represent a set of top-k documents in the top-ranked documents. Based on the LDA, we can compute the probability distribution of a word w_i in documents by using Eq. 1. In this paper, we assume feature words come from latent topics because we need to work out the probability of feature words in Ω_k rather than in a specific document. We use the following equation to calculate $P(w_i|\Omega_k)$.

$$P(w_i|\Omega_k) = \frac{\sum_{w_i \in d, d \in \Omega_k} P(w_i|d)}{|\{d|d \in \Omega_k, w_i \in d\}|} \tag{2}$$

In next step of the first stage, we assume that the top-k pseudo documents in Ω_k are relevant. Let Ω be the top-ranked documents for a given query, we consider $\Omega = \{\Omega_{\sigma*1}, \Omega_{\sigma*2}, ..., \Omega_{\sigma*m}\}$ where σ is an incremental parameter and the number of subsets are $m = \frac{|\Omega|}{\sigma}$. For each candidate training subset $\Omega_{\sigma*x} \subseteq \Omega$, $\sigma * x$ indicates the size of the subset, (i.e., $|.| = \sigma * x$). For example, if Ω is equal 50 and an incremental parameter σ is equal 10, thus, the number of subsets that we considered is $5 = \frac{50}{10}$. Figure 1 shows the process of selecting the top-k pseudo-relevant documents.

We examine the features in each candidate subset $\Omega_{\sigma*x}$. The discriminative power of word w_i in a subset that has more relevant documents will be significant. To do this task, we observe the probability distribution $P(w_i|\Omega_{\sigma*x})$ from the topic level of features in the candidate subset $\Omega_{\sigma*x}$ as in Eq. 2. When the number of features in the next subset dramatically increase, it indicates more extraneous features may be introduced and thus more uncertain information could be included in the relevance feedback model. Finally, the largest integral of each candidate subset in all the top-k ranked documents is more likely to be relevant feedback when processed as follows:

$$argmax_{1 \le x \le m} \sum_{i=1}^{n} \left(\frac{P(w_i|\Omega_{\sigma*x})}{\sigma * x} \right) \tag{3}$$

where the probability distribution $P(w_i|\Omega_{\sigma*x}) = 0$ if $w_i \notin \Omega_{\sigma*x}$.

3.2 Expansion Terms Selection

During the second stage in our proposed model, expansion terms are selected using the relevance model for each selected document in the top-ranked documents. The selected documents derived from the initially retrieved documents described in Sect. 3.1 are utilised to estimate the relevance model. By following the work in [1,18], the relevance model is re-weighted so that the candidate terms w are on top of the selected documents Ω_k as follows:

$$P(w|Q) \propto P(w|\Omega_k) + \sum_{d\in\Omega_k} P(w|d)P(d) \prod_{i=1}^{n} P(q_i|d) \tag{4}$$

where $P(w|\Omega_k)$ is the topical distribution weight of word w on the relevant topics of selected documents Ω_k (as estimated in Eq. 2), $\prod_{i=1}^{n} P(q_i|d)$ is the query likelihood with Dirichlet smoothing for document d and the document prior $P(d)$ is usually assumed to be uniform. However, the quality of the documents (e.g., microblogs) is not uniform due to the documents' timestamp variation. The user usually favors recent relevant information in a microblog search system. In this paper, we follow the work of [19] to integrate the temporal information into the recency-based document $P(d|T_d)$ in Eq. 4, as follows:

$$P(d|T_d) = r * e^{-r*(T_Q-T_d)} \tag{5}$$

where r is the parameter that controls the temporal information, T_Q is the time the query was issued and T_d is the time the document was published.

The final phase of the proposed model is a linear combination of the relevance model $P(w|Q)$ and the original query model θ_Q we computed as follows:

$$P(w|\theta_{Q'}) = \lambda P(w|\theta_Q) + (1 - \lambda) P(w|Q) \tag{6}$$

where $\lambda \in [0, 1]$ is a free parameter. Then, we estimate the simple form for the original query model as follows:

$$P(w|\theta_Q) = \frac{c(w,Q)}{\sum_{w'\in Q} c(w',Q)} = \frac{c(w,Q)}{|Q|} \tag{7}$$

where $c(w,Q)$ is the count of word w in Q, and $|Q|$ is the length of the query.

4   Experiments and Results

4.1 Datasets

The experiments were conducted using the TREC 2011 microblog collection, called **Tweets2011** [31], and the TREC 2013 microblog collection, called **Tweets2013** [25]. The size of the **Tweets2011** dataset is about 16 million tweets over a period of two weeks (January 23, 2011 to February 8, 2011), which included important events such as, the US Super Bowl and the Egyptian revolution. However, the size of the **Tweets2013** dataset is much larger than **Tweets2011**, which is about 240 million tweets, and covers a period of two months (February 1, 2013 to March 31, 2013). The **Tweets2011** dataset has two topic sets that include 49 (**MB2011**) and 59 (**MB2012**) topics while Tweets2013 has a topic set that contains 60 (**MB2013**) topics. Each official query consist of the query number, title and query time stamp. In this paper, we use all topics for both datasets. The National Institute of Standards and Technology (NIST) assessors used a standard pooling strategy for evaluation, assigning multi-scale judgments to each tweet denoted as highly relevant, relevant and not relevant.

Pre-processing tweets was a critical stage to improve the retrieving model effectiveness [3]. In these experiments, we treated the tweets and queries text based on the following steps. The first step was tweet filtering, Non-English tweets make data noisy, so we discarded these tweets using language detector called *ldig*[1]. We also filtered out tweets that contain non-ASCII characters, such as emojis and symbols. Then, following the microblog tracking guidelines[2], we normalised retweets that started with "RT @". Finally, we used the Porter algorithm to remove stop words and stem word. All tweets in these collections were indexed using Apache Lucene library[3].

4.2 Experiment Metrics

The metrics applied were broadly acknowledged and settled assessment measurements in TREC 2011–2014 Microblog tracks [25, 31]. The evaluation metrics, which are the precision at N (P@N) and the Mean Average Precision (MAP), are extensively utilized in information retrieval and the official microblog metric at the TREC Microblog tracks. Precision p took all retrieved tweets into account. We fix N to 30 to adjust with the cut-off utilized in the previous TREC Microblog tracks. MAP was figured by measuring the precision of each relevant tweet and then averaging the precision over all the quires of the top 1000 tweets. All experiments took into account the standard topic set utilised as a part of the TREC 2011–2013 datasets, called *allrel*. Moreover, a statistical difference in our evaluation was utilised with a two-tailed paired t-test with level of p value. The statistical improvement happened when the value is small enough ($p < 0.05\%$).

4.3 Parameter Tuning

We set $\lambda = 0.6$ in our experiment settings for the relevance model in Eq. 6. For the relevance model settings, we set $|\Omega| = 50$ top-ranked documents and the number of expansion terms was $n = 30$. The value of the temporal rate parameter r in Eq. 5 was 0.01 based on reference [36]. In this experiment, we tuned all the parameters of the proposed model **TBS** with TREC 2011 topics from the **Tweets11** collection.

---

[1] http://github.com/shuyo/ldig/.
[2] https://github.com/lintool/twitter-tools/wiki/.
[3] http://lucene.apache.org/.

Different experimental parameters settings were used with our framework. In our experimental framework, we utilised the Java Machine Learning for Language Toolkit (MALLET)[4] in our experimental framework. The hyperparameters settings for the LDA model was $\alpha = 50/V$ and $\beta = 0.01$ as was recommended in [11].

4.4   Baseline Algorithms

The first baseline was a probabilistic, state-of-the-art retrieval model **BM25** [33] that can estimate the similarity between document d and query Q containing words w. The $k1$ and b parameters were the experimental parameters (in this paper we set $k1 = 0.9$ and $b = 0.4$, respectively). In the second baseline, the query likelihood model with Dirichlet smoothing that was referred as **QL** utilised the Dirichlet smoothing parameter $\mu = 100$ based on the settings in paper [26]. **Recency**, which is one of the simplest techniques that allows time to influence the ranking model, was given by Li and Croft [19], who proposed a document prior that favoured recently published documents. Kernel Density Estimation **KDE** [14] is a state-of-the-art model that estimates the temporal density of relevance feedback for microblog documents. The final baseline is the **RM3** relevance model [1] used to compare with the proposed model. In **RM3**, for a given query the relevance model is estimated and then interpolated with the original query with a control parameter.

4.5   Experimental Results

Table 1 shows the P@30 and MAP performances of the proposed model **TBS** compared with the baseline approaches that had statistically significant test results. The change in percentage is denoted as $ch\%$. The superscript † denotes statistically significant improvements over the state-of-the-art baseline model **RM3** where all p values were less than 0.05. The best result per evaluation metric is marked in bold font. All the parameters were tuned on the **Tweets2011** collection with **MB2011** topics. Then, we tested the proposed model **TBS** on the **Tweets2011** and **Tweets2013** collections with **MB2012** and **MB2013** topics.

It can be clearly seen in our experiment that our model **TBS** outperformed and showed significant improvement over the baseline models in all metrics across all microblog TREC collections 2011-2013. Table 1 shows that for microblog TREC 2011 topics, the **TBS** improved over the P@30 by a maximum improvement of 11.45% compared to **KDE** and improved by a 4.60% minimum compared to **RM3**. The **TBS** improved over the MAP by a maximum of 24.55% compared to **KDE** and a minimum of 14.02% compared to **RM3**. For microblog TREC 2012 topics, the **TBS** improved the P@30 by a maximum of 12.45% compared to **KDE** and a 2.81% minimum compared to **RM3**. **TBS** improved the MAP by a

---

[4] http://mallet.cs.umass.edu/.

Table 1. Results comparison of the proposed method **TBS** and baselines of MB2011, MB2012 and MB2013 where symbol † represents a ($p < 0.05$) statistically significant improvement over **RM3**.

Model	MB2011		MB2012		MB2013	
	P@30	MAP	P@30	MAP	P@30	MAP
TBS	**0.4169†**	**0.4232†**	**0.3729†**	**0.2680†**	**0.5226†**	**0.3506†**
BM25	0.3599	0.3310	0.3270	0.2118	0.4383	0.2603
QL	0.3714	0.3561	0.3327	0.2248	0.4544	0.2825
Recency	0.3776	0.3581	0.3349	0.2255	0.4694	0.2875
ch%	+10.42%	+18.19%	+11.35%	+18.86%	+11.33%	+21.94%
KDE	0.3741	0.3398	0.3316	0.2249	0.4644	0.2791
ch%	+11.45%	+24.55%	+12.45%	+19.18%	+12.53%	+25.61%
RM3	0.3986	0.3712	0.3627	0.2534	0.4467	0.3035
ch%	+4.60%	+14.02%	+2.81%	+5.77%	+16.99%	+15.51%

maximum of 19.18% against compared to **KDE** and a 5.77% minimum against compared to **RM3**.

To confirm the superiority of the proposed model, we tested **TBS** on the microblog TREC 2103 collections, which were much larger than the TREC 2011-2012 collections, and showed the variations in performance. For microblog TREC 2103 topics, Table 1 shows that our model **TBS** improved the MAP by a maximum and minimum of 25.61% and 15.51% over **KDE** and **RM3**, respectively, while the corresponding increments of P@30 were a maximum of 16.99% and minimum of 11.33% over **RM3** and **Recency**, respectively.

4.6   Discussions

The proposed framework's performance can be affected by using several parameters settings. This section shows and analyses the robustness of the proposed model against a coefficients setting that could affect the overall performance. All these experiments in this section were done on the microblog TREC 2011 collection that was utilized for parameters tuning.

Figure 2a shows the proposed model **TBS**'s performance regarding the MAP across a different number of top-ranked documents Ω (from 10 to 200). The proposed model **TBS** achieved the optimal performance when $\Omega = 50$. Figure 2b illustrates the sensitivity for changing the number of expansion features of the **TBS** model. We can observe that, when the number of expansion features was equal to 30, **TBS** achieved the optimal performance regarding the MAP value. As we mentioned in Sect. 3.2, we integrated the topic model with the relevance feedback in Eq. 4 to get the most likely ranked documents and then interpolated the features with the original query in Eq. 6. At this stage, parameter λ controlled the expansion features' weight with the original query features. Figure 2c shows

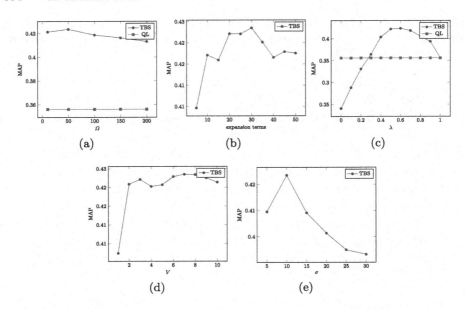

Fig. 2. Sensitivity of (a) the number of feedback documents Ω, (b) expansion terms, (c) interpolation parameters λ, (d) the number of topics V and (e) the incremental factor parameter σ for the **TBS** at microblog TREC 2011 collection.

the performance of **TBS** across different λ. When $\lambda = 1$, we completely ignored using the relevance model features and the **TBS** performance degenerates into the baseline method **QL**. While $\lambda = 0$, we completely ignored the original query and only used the relevance model features. From Fig. 2c, we could observe that, when setting the interpolation coefficient $\lambda = 0.6$, **TBS** obtained its optimal performance regarding the MAP value.

We will demonstrate how the topic numbers can affect the proposed model. More specifically, based on the analysis of the above figures, we can fix a number of feedback documents to 50, the number of expansion terms is to 30 and the interpolation parameter λ is to 0.6. Figure 2d shows the performance across a different number of topics. When the number of topics was greater than one, the performance of the proposed model **TBS** become more stable. The **TBS** performance was stable against different numbers of topics, and when the number of topics was seven, it obtained the optimal performance regarding the MAP value. In Sect. 3.1, we investigated the proposed model **TBS** to see its performance according to the incremental coefficient. Based on the previous parameters analysis, we fixed the feedback number to 50, the expansion terms to 30, the interpolation parameter λ to 0.6, and the number of topics is to 7. Figure 2e shows the **TBS**'s performance across different values of an incremental parameter σ. It is clearly observed that when an incremental parameter $\sigma = 10$, **TBS** obtained the best performance across other values.

5 Conclusions

In real applications, such as microblog retrieval, the challenging task is how to infer user information needs and to retrieve high-quality results without requiring extra effort from user. PRF is an automatic technique that uses top-ranked documents to enhance retrieval performance. However, top-ranked documents in the microblog retrieval application are not optimal for representing the initial user information needs. To tackle this challenge, in this paper, we proposed a method that can automatically determine the best the top-rank documents using a topic modeling technique. The proposed method views the top-k documents as a random variable and then find the best k value for each query-based PRF. Furthermore, lexical features and latent topics are combined to select possible relevant terms for the original query based on temporal evidence. We exhibited that the proposed model's performance was evaluated on real-time datasets from TREC collections 2011–2013, and the experimental results showed that our model achieved significant improvement when compared to all state-of-the-art baseline models for the microblog retrieval task.

References

1. Abdul-Jaleel, N., Allan, J., Croft, W.B., Diaz, F., Larkey, L., Li, X., Smucker, M.D., Wade, C.: UMass at TREC 2004: Novelty and hard. In: TREC (2004)
2. Albakour, M., Macdonald, C., Ounis, I., et al.: On sparsity and drift for effective real-time filtering in microblogs. In: Proceedings of CIKM, pp. 419–428 (2013)
3. Albishre, K., Albathan, M., Li, Y.: Effective 20 newsgroups dataset cleaning. In: Proceedings of the WI-IAT, vol. 3, pp. 98–101 (2015)
4. Albishre, K., Li, Y., Xu, Y.: Effective pseudo-relevance for microblog retrieval. In: Proceedings of ACSW, p. 51 (2017)
5. Algarni, A., Li, Y., Xu, Y.: Selected new training documents to update user profile. In: Proceedings of CIKM, pp. 799–808. ACM (2010)
6. Andrzejewski, D., Buttler, D.: Latent topic feedback for information retrieval. In: Proceedings of KDD, pp. 600–608 (2011)
7. Blei, D.M., Ng, A.Y., Jordan, M.I.: Latent dirichlet allocation. J. Mach. Learn. Res. **3**, 993–1022 (2003)
8. Carpineto, C., Romano, G.: A survey of automatic query expansion in information retrieval. CSUR **44**(1), 1 (2012)
9. Choi, J., Croft, W.B.: Temporal models for microblogs. In: Proceedings of CIKM, pp. 2491–2494 (2012)
10. Choi, J., Croft, W.B., Kim, J.Y.: Quality models for microblog retrieval. In: Proceedings of CIKM, pp. 1834–1838 (2012)
11. Chuang, J., Gupta, S., Manning, C., Heer, J.: Topic model diagnostics: assessing domain relevance via topical alignment. In: Proceedings of ICML, pp. 612–620 (2013)
12. Dong, A., Zhang, R., Kolari, P., Bai, J., Diaz, F., Chang, Y., Zheng, Z., Zha, H.: Time is of the essence: improving recency ranking using twitter data. In: Proceedings of WWW, pp. 331–340 (2010)
13. Efron, M., Golovchinsky, G.: Estimation methods for ranking recent information. In: Proceedings of SIGIR, pp. 495–504 (2011)

14. Efron, M., Lin, J., He, J., De Vries, A.: Temporal feedback for tweet search with non-parametric density estimation. In: Proceedings of SIGIR, pp. 33–42 (2014)
15. Fan, F., Qiang, R., Lv, C., Yang, J.: Improving microblog retrieval with feedback entity model. In: Proceedings of CIKM, pp. 573–582 (2015)
16. Gao, Y., Xu, Y., Li, Y.: Pattern-based topics for document modelling in information filtering. IEEE Trans. Knowl. Data Eng. **27**(6), 1629–1642 (2015)
17. Kotov, A., Wang, Y., Agichtein, E.: Leveraging geographical metadata to improve search over social media. In: Proceedings of WWW, pp. 151–152 (2013)
18. Lavrenko, V., Croft, W.B.: Relevance based language models. In: Proceedings of SIGIR, pp. 120–127 (2001)
19. Li, X., Croft, W.B.: Time-based language models. In: Proceedings of CIKM, pp. 469–475 (2003)
20. Li, Y., Algarni, A., Albathan, M., Shen, Y., Bijaksana, M.A.: Relevance feature discovery for text mining. IEEE Trans. Knowl. Data Eng. **27**(6), 1656–1669 (2015)
21. Li, Y., Algarni, A., Zhong, N.: Mining positive and negative patterns for relevance feature discovery. In: Proceedings of KDD, pp. 753–762 (2010)
22. Li, Y., Zhou, X., Bruza, P., Xu, Y., Lau, R.Y.: A two-stage decision model for information filtering. Decis. Support Syst. **52**(3), 706–716 (2012)
23. Liang, S., Yilmaz, E., Kanoulas, E.: Dynamic clustering of streaming short documents. In: Proceedings of KDD, pp. 995–1004 (2016)
24. Lin, C., Lin, C., Li, J., Wang, D., Chen, Y., Li, T.: Generating event storylines from microblogs. In: Proceedings of CIKM, pp. 175–184 (2012)
25. Lin, J., Efron, M.: Overview of the TREC-2013 microblog track. In: TREC (2013)
26. Lv, C., Qiang, R., Fan, F., Yang, J.: Knowledge-based query expansion in real-time microblog search. In: Zuccon, G., Geva, S., Joho, H., Scholer, F., Sun, A., Zhang, P. (eds.) AIRS 2015. LNCS, vol. 9460, pp. 43–55. Springer, Cham (2015). https://doi.org/10.1007/978-3-319-28940-3_4
27. Lv, Y., Zhai, C.: Adaptive relevance feedback in information retrieval. In: Proceedings of CIKM, pp. 255–264 (2009)
28. Metzler, D., Croft, W.B.: Latent concept expansion using markov random fields. In: Proceedings of SIGIR, pp. 311–318 (2007)
29. Miao, J., Huang, J.X., Zhao, J.: TopPRF: a probabilistic framework for integrating topic space into pseudo relevance feedback. TOIS **34**(4), 22 (2016)
30. Miyanishi, T., Seki, K., Uehara, K.: Improving pseudo-relevance feedback via tweet selection. In: Proceedings of CIKM, pp. 439–448 (2013)
31. Ounis, I., Macdonald, C., Lin, J., Soboroff, I.: Overview of the TREC-2011 microblog track. In: TREC (2011)
32. Porteous, I., Newman, D., Ihler, A., Asuncion, A., Smyth, P., Welling, M.: Fast collapsed gibbs sampling for latent dirichlet allocation. In: Proceedings of KDD, pp. 569–577 (2008)
33. Robertson, S.E., Walker, S., Jones, S., Hancock-Beaulieu, M.M., Gatford, M., et al.: Okapi at trec-3. NIST Special Publication SP **109**, 109 (1995)
34. Salton, G., Buckley, C.: Term-weighting approaches in automatic text retrieval. Inf. Process. Manage. **24**(5), 513–523 (1988)
35. Song, Y., Wang, H., Chen, W., Wang, S.: Transfer understanding from head queries to tail queries. In: Proceedings of CIKM, pp. 1299–1308 (2014)
36. Wang, Y., Huang, H., Feng, C.: Query expansion based on a feedback concept model for microblog retrieval. In: roceedings of WWW, pp. 559–568 (2017)

Distributed Representation of Multi-sense Words: A Loss Driven Approach

Saurav Manchanda[✉] and George Karypis

University of Minnesota, Twin Cities, MN 55455, USA
{manch043,karypis}@umn.edu

Abstract. Word2Vec's Skip Gram model is the current state-of-the-art approach for estimating the distributed representation of words. However, it assumes a single vector per word, which is not well-suited for representing words that have multiple senses. This work presents LDMI, a new model for estimating distributional representations of words. LDMI relies on the idea that, if a word carries multiple senses, then having a different representation for each of its senses should lead to a lower loss associated with predicting its co-occurring words, as opposed to the case when a single vector representation is used for all the senses. After identifying the multi-sense words, LDMI clusters the occurrences of these words to assign a sense to each occurrence. Experiments on the contextual word similarity task show that LDMI leads to better performance than competing approaches.

1 Introduction

Many NLP tasks benefit by embedding the words of a collection into a low dimensional space in a way that captures their syntactic and semantic information. Such NLP tasks include analogy/similarity questions [11], part-of-speech tagging [2], named entity recognition [1], machine translation [12,16] etc. Distributed representations of words are real-valued, low dimensional embeddings based on the distributional properties of words in large samples of the language data. However, representing each word by a single vector does not properly model the words that have multiple senses (i.e., polysemous and homonymous words). For multi-sense words, a single representation leads to a vector that is the amalgamation of all its different senses, which can lead to ambiguity.

To address this problem, models have been developed to estimate a different representation for each of the senses of multi-sense words. The common idea utilized by these models is that if the words have different senses, then they tend to co-occur with different sets of words. The models proposed by Reisinger and Mooney [14], Huang et al. [10] and the Multiple-Sense Skip-Gram (MSSG) model of Neelakantan et al. [13] estimates a fixed number of representations per word, without discriminating between the single-sense and multi-sense words. As a result, these approaches fail to identify the right number of senses per word and estimate multiple representations for the words that have a single sense.

© Springer International Publishing AG, part of Springer Nature 2018
D. Phung et al. (Eds.): PAKDD 2018, LNAI 10938, pp. 337–349, 2018.
https://doi.org/10.1007/978-3-319-93037-4_27

In addition, these approaches cluster the occurrences without taking into consideration the diversity of words that occur within the contexts of these occurrences (explained in Sect. 3). The Non-Parametric Multiple-Sense Skip-Gram (NP-MSSG) model [13] estimates a varying number of representations for each word but uses the same clustering approach and hence, is not effective in taking into consideration the diversity of words that occur within the same context.

We present an extension to the Skip-Gram model of Word2Vec to accurately and efficiently estimate a vector representation for each sense of multi-sense words. Our model relies on the fact that, given a word, the Skip-Gram model's loss associated with predicting the words that co-occur with that word, should be greater when that word has multiple senses as compared to the case when it has a single sense. This information is used to identify the words that have multiple senses and estimate a different representation for each of the senses. These representations are estimated using the Skip-Gram model by first clustering the occurrences of the multi-sense words by accounting for the diversity of the words in these contexts. We evaluated the performance of our model for the contextual similarity task on the Stanford's Contextual Word Similarities (SCWS) dataset. When comparing the most likely contextual sense of words, our model was able to achieve approximately 13% and 10% improvement over the NP-MSSG and MSSG approaches, respectively. In addition, our qualitative evaluation shows that our model does a better job of identifying the words that have multiple senses over the competing approaches.

2   Definitions, Notations and Background

Distributed representation of words quantify the syntactic and semantic relations among the words based on their distributional properties in large samples of the language data. The underlying assumption is that the co-occurring words should be similar to each other. We say that the word w_j co-occurs with the word w_i if w_j occurs within a window around w_i. The context of w_i corresponds to the set of words which co-occur with w_i within a window and is represented by $C(w_i)$.

The state-of-the-art technique to learn the distributed representation of words is Word2Vec. The word vector representations produced by Word2Vec are able to capture fine-grained semantic and syntactic regularities in the language data. Word2Vec provides two models to learn word vector representations. The first is the Continuous Bag-of-words Model that involves predicting a word using its context. The second is called the Continuous Skip-gram Model that involves predicting the context using the current word. To estimate the word vectors, Word2Vec trains a simple neural network with a single hidden layer to perform the following task: Given an input word (w_i), the network computes the probability for every word in the vocabulary of being in the context of w_i. The network is trained such that, if it is given w_i as an input, it will give a higher probability to w_j in the output layer than w_k if w_j occurs in the context of w_i but w_k does not occur in the context of w_i. The set of all words in the vocabulary is represented by V. The vector associated with the word w_i is denoted by $\boldsymbol{w}_i$.

The vector corresponding to word w_i when w_i is used in the context is denoted by $\tilde{\boldsymbol{w}}_i$. The size of the word vector $\boldsymbol{w}_i$ or the context vector $\tilde{\boldsymbol{w}}_i$ is denoted by d.

The objective function for the Skip-Gram model with negative sampling is given by [8]

$$\text{minimize} \ -\sum_{i=1}^{|V|}\left(\sum_{w_j \in C(w_i)} \log \sigma(\langle \boldsymbol{w}_i, \tilde{\boldsymbol{w}}_j\rangle) + \sum_{\substack{k \in R(m,|V|) \\ w_k \notin C(w_i)}} \log \sigma(-\langle \boldsymbol{w}_i, \tilde{\boldsymbol{w}}_k\rangle)\right),$$

where $R(m, n)$ denotes a set of m random numbers from the range $[1, n]$ (negative samples), $\langle \boldsymbol{w}_i, \boldsymbol{w}_j\rangle$ is the dot product of $\boldsymbol{w}_i$ and $\boldsymbol{w}_j$ and $\sigma(\langle \boldsymbol{w}_i, \tilde{\boldsymbol{w}}_j\rangle)$ is the sigmoid function.

The parameters of the model are estimated using Stochastic Gradient Descent (SGD) in which, for each iteration, the model makes a single pass through every word in the training corpus (say w_i) and gathers the context words within a window. The negative samples are sampled from a probability distribution which favors the frequent words. The model also down-samples the frequent words using a hyper-parameter called the sub-sampling parameter.

3   Prior Approaches for Dealing with Multi-sense Words

Various models have been developed to deal with the distributed representations of the multi-sense words. These models presented in this section work by estimating multiple vector-space representations per word, one for each sense. Most of these models estimate a fixed number of vector representations for each word, irrespective of the number of senses associated with a word. In the rest of this section, we review these models and discuss their limitations.

Reisinger and Mooney [14] clusters the occurrences of a word using the mixture of von Mises-Fisher distributions [3] clustering method to assign a different sense to each occurrence of the word. The clustering is performed on all the words even if the word has a single sense. This approach estimates a fixed number of vector representations for each word in the vocabulary. As per the authors, the model captures meaningful variation in the word usage and does not assume that each vector representation corresponds to a different sense. Huang et al. [10] also uses the same idea and estimates a fixed number of senses for each word. It uses spherical k-means [5] to cluster the occurrences.

Neelakantan et al. [13] proposed two models built on the top of the Skip-Gram model: Multiple-Sense Skip-Gram (MSSG), and its Non-Parametric counterpart NP-MSSG. MSSG estimates a fixed number of senses per word whereas NP-MSSG discovers varying number of senses. MSSG maintains clusters of the occurrences for each word, each cluster corresponding to a sense. Each occurrence of a word is assigned a sense based on the similarity of its context with the already maintained clusters, and the corresponding vector representation, as well as the sense cluster of the word is updated. During training, NP-MSSG creates a new sense for a word with the probability proportional to the distance of the context to the nearest sense cluster. Both MSSG and NP-MSSG create an auxiliary vector to represent an occurrence, by taking the average of vectors associated with all the words belonging to its context. The similarity between the two occurrences is computed as the cosine similarity between these auxiliary vectors. This approach does not take into consideration the variation among the words that occur within the same context. Another disadvantage is that the auxiliary vector is biased towards the words having higher L_2 norm. This leads to noisy clusters, and hence, the senses discovered by these models are not robust.

4   Loss Driven Multisense Identification (LDMI)

In order to address the limitations of the existing models, we developed an extension to the Skip-Gram model that combines two ideas. The first is to identify the multi-sense words and the second is to cluster the occurrences of the identified words such that the clustering correctly accounts for the variation among the words that occur within the same context. We explain these parts as follows:

4.1   Identifying the Words with Multiple Senses

For the Skip-Gram model, the loss associated with an occurrence of w_i is

$$L(w_i) = -\left(\sum_{w_j \in C(w_i)} \log \sigma(\langle \boldsymbol{w_i}, \boldsymbol{\tilde{w}_j} \rangle) + \sum_{\substack{k \in R(m, |V|) \\ w_k \notin C(w_i)}} \log \sigma(-\langle \boldsymbol{w_i}, \boldsymbol{\tilde{w}_k} \rangle) \right).$$

The model minimizes $L(w_i)$ by increasing the probability of the co-occurrence of w_j and w_i if w_j is present in the context of w_i and decreasing the probability of the co-occurrence of w_k and w_i if w_k is not present in the context of w_i. This happens by aligning the directions of $\boldsymbol{w_i}$ and $\boldsymbol{\tilde{w}_j}$ closer to each other and aligning the directions of $\boldsymbol{w_i}$ and $\boldsymbol{\tilde{w}_k}$ farther from each other. At the end of the optimization process, we expect that the co-occurring words have their vectors aligned closer in the vector space. However, consider the polysemous word *bat*. We expect that the vector representation of *bat* is aligned in a direction closer to the directions of the vectors representing the terms like *ball, baseball, sports* etc. (the sense corresponding to *sports*). We also expect that the vector representation of *bat* is aligned in a direction closer to the directions of the vectors representing the terms like *animal, batman, nocturnal* etc. (the sense corresponding to *animals*). But at the same time, we do not expect that the directions of the vectors representing the words corresponding to the *sports* sense are closer to the directions of the vectors representing the words corresponding to the *animal* sense. This leads to the direction of the vector representing *bat* lying in between the directions of the vectors representing the words corresponding to the *sports* sense and the directions of the vectors representing the words corresponding to the *animal* sense. Consequently, the multi-sense words will tend to contribute more to the overall loss than the words with a single sense.

Having a vector representation for each sense of the word *bat* will avoid this scenario, as each sense can be considered as a new single-sense word in the vocabulary. Hence, the loss associated with a word provides us information regarding whether a word has multiple senses or not. LDMI leverages this insight to identify a word w_i as multi-sense if the average $L(w_i)$ across all its occurrences is more than a threshold. However, $L(w_i)$ has a random component associated with it, in the form of negative samples. We found that, in general, infrequent words have higher loss as compared to the frequent words. This can be attributed to the fact that given a random negative sample while calculating the loss, there is a greater chance that the frequent words have already seen this negative sample before during the optimization process as compared to the infrequent words. This way, infrequent words end up having higher loss than frequent words. Therefore, for the selection purposes, we ignore the loss associated with negative samples. We denote the average loss associated with the prediction of the context words in an occurrence of w_i as $L^+(w_i)$ and define it as

$$L^+(w_i) = -\frac{1}{|C(w_i)|} \sum_{w_j \in C(w_i)} \log \sigma(\langle \boldsymbol{w}_i, \tilde{\boldsymbol{w}}_j \rangle).$$

We describe $L^+(w_i)$ as the *contextual loss* associated with an occurrence of w_i.

To identify the multi-sense words, LDMI performs a few iterations to optimize the loss function on the text dataset, and shortlist the words with average contextual loss (average $L^+(w_i)$ across all the occurrences of the w_i) that is higher than a threshold. These shortlisted words represent the identified multi-sense words, which form the input of the second step described in the next section.

4.2 Clustering the Occurrences

To assign senses to the occurrences of each of the identified multi-sense words, LDMI clusters its occurrences so that each of the clusters corresponds to a particular sense. The clustering solution employs the $\mathcal{I}_1$ criterion function [15] which maximizes the objective function of the form

$$\text{maximize} \sum_{i=1}^{k} n_i Q(S_i), \tag{1}$$

where $Q(S_i)$ is the quality of cluster S_i whose size is n_i. We define $Q(S_i)$ as

$$Q(S_i) = \frac{1}{n_i^2} \sum_{u,v \in S_i} \text{sim}(u, v),$$

where $\text{sim}(u, v)$ denotes the similarity between the occurrences u and v, and is given by

$$\text{sim}(u, v) = \frac{1}{|C(u)||C(v)|} \sum_{x \in C(u)} \sum_{y \in C(v)} \cos(x, y), \tag{2}$$

According to Eq. (2), LDMI measures the similarity between the two occurrences as the average of the pairwise cosine similarities between the words belonging to the contexts of these occurrences. This approach considers the variation among the words that occur within the same context. We can simplify Eq. (1) to the following equation

$$\text{maximize} \sum_{i=1}^{k} \frac{1}{n_i} \left\| \sum_{u \in S_i} \left(\sum_{x \in C(u)} \frac{x}{\|x\|_2} \right) \right\|_2^2.$$

LDMI maximizes this objective function using a greedy incremental strategy [15].

4.3   Putting Everything Together

LDMI is an iterative algorithm with two steps in each iteration. The first step is to perform a few SGD iterations to optimize the loss function. In the second step, it calculates the contextual loss associated with each occurrence of each word and identifies the words having the average contextual loss that is more than a threshold. It then clusters the occurrences of the identified multi-sense words into two clusters ($k = 2$) as per the clustering approach discussed earlier. The algorithm terminates after a fixed number of iterations. x number of iterations of LDMI can estimate a maximum of 2^x senses for each word.

5   Experimental Methodology

5.1   Datasets

We train LDMI on two corpora of varying sizes: The Wall Street Journal (WSJ) dataset [9] and the Google's One Billion Word (GOBW) dataset [4]. In preprocessing, we removed all the words which contained a number or did not contain any alphabet and converted the remaining words to lower case.

For WSJ, we removed all the words with frequency less than 10 and for GOBW, we removed all the words with frequency less than 100. The statistics of these datasets after preprocessing are presented in Table 1.

Table 1. Dataset statistics.

Dataset	Vocabulary size	Total words
WSJ	88,118	62,653,821
GOBW	73,443	710,848,599

We use Stanford's Contextual Word Similarities (SCWS) dataset [10] for evaluation on the contextual word similarity task. SCWS contains human judgments on pairs of words (2,003 in total) presented in sentential context. The word pairs are chosen so as to reflect interesting variations in meanings.

When the contextual information is not present, different people can consider different senses when giving a similarity judgment. Therefore, having representations for all the senses of a word can help us to find similarities which align better with the human judgments, as compared to having a single representation of a word. To investigate this, we evaluated our model on the WordSim-353 dataset [7], which consists of 353 pairs of nouns, without any contextual information. Each pair has an associated averaged human judgments on similarity.

5.2   Evaluation Methodology and Metrics

Baselines. We compare the LDMI model with the MSSG and NP-MSSG approaches as they are also built on top of the Skip-Gram model. As mentioned earlier, MSSG estimates the vectors for a fixed number of senses per word whereas NP-MSSG discovers varying number of senses per word. To illustrate the advantage of using the clustering with $\mathcal{I}_1$ criterion over the clustering approach used by the competing models, we also compare LDMI with LDMI-SK. LDMI-SK uses the same approach to select the multi-sense words as used by the LDMI, but instead of clustering with the $\mathcal{I}_1$ criterion, it uses spherical K-means [5].

Parameter Selection. For all our experiments, we consider 10 negative samples and a symmetric window of 10 words. The sub-sampling parameter is 10^{-4} for both the datasets. To avoid clustering the infrequent and stop-words, we only consider the words within a frequency range to select them as the multi-sense words. For the WSJ dataset, we consider the words with frequency between 50 and 30,000 while for the GOBW dataset, we consider the words with frequency between 500 and 300,000. For the WSJ dataset, we consider only 50-dimensional embeddings while for the GOBW dataset, we consider 50, 100 and 200 dimensional embeddings. The model checks for multi-sense words after every 5 iterations. We selected our hyperparameter values by a small amount of manual exploration to get the best performing model.

To decide the threshold for the average contextual loss to select the multi-sense words, we consider the distribution of the average contextual loss after running an iteration of Skip-Gram. For example, Fig. 1 shows the average contextual loss of every word in the vocabulary for the GOBW dataset for the 50-dimensional embeddings. We can see that there is an increase in the average contextual loss around 2.0−2.4. We experiment around this range to select a loss threshold for which our model performs best. For the experiments pre-

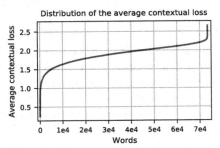

Fig. 1. Distribution of the average contextual loss for all words (Words on the x-axis are sorted in order of their loss)

sented in this paper, this threshold is set to 2.15 for the WSJ (50-dimensional embeddings), and 2.15, 2.10 and 2.05 for the GOBW corresponding to the 50,

100 and 200-dimensional embeddings, respectively. With increasing dimensionality of the vectors, we are able to model the information from the dataset in a better way, which leads to a relatively lower loss.

For the MSSG and NP-MSSG models, we use the same hyperparameter values as used by Neelakantan et al. [13]. For MSSG, the number of senses is set to 3. Increasing the number of senses involves a compromise between getting the correct number of senses for some words while noisy senses for the others. For NP-MSSG, the maximum number of senses is set to 10 and the parameter λ is set to -0.5 (A new sense cluster is created if the similarity of an occurrence to the existing sense clusters is less than λ). The models are trained using SGD with AdaGrad [6] with 0.025 as the initial learning rate and we run 15 iterations.

Metrics. For evaluation, we use the similarities calculated by our model and sort them to create an ordering among all the word-pairs. We compare this ordering against the one obtained by the human judgments. To do this comparison, we use the Spearman rank correlation (ρ). Higher score for the Spearman rank correlation corresponds to the better correlation between the respective orderings. For the SCWS dataset, to measure the similarity between two words given their sentential contexts, we use two different metrics [14]. The first is the maxSimC, which for each word in the pair, identifies the sense of the word that is the most similar to its context and then compares those two senses. It is computed as

$$\text{maxSimC}(w_1, w_2, C(w_1), C(w_2)) = \cos(\hat{\pi}(w_1), \hat{\pi}(w_2)),$$

where, $\hat{\pi}(w_i)$ is the vector representation of the sense that is most similar to $C(w_i)$. As in Eq. (2), we measure the similarity between x and $C(w_i)$ as

$$\text{sim}(x, C(w_i)) = \frac{1}{Z} \left(\sum_{y \in C(w_i)} \sum_{j=1}^{m(y)} \cos(x, V(y, j)) \right),$$

where, $Z = \sum_{y \in C(w_i)} m(y)$, $m(y)$ is the number of senses discovered for the word y and $V(y, i)$ is the vector representation associated with the ith sense of the word y. For simplicity, we consider all the senses of the words in the sentential context for the similarity calculation. The second metric is the avgSimC which calculates the similarity between the two words as the weighed average of the similarities between each of their senses. It is computed as

$$\text{avgSimC}(w_1, w_2, C(w_1), C(w_2)) =$$
$$\sum_{i=1}^{m(w_1)} \sum_{j=1}^{m(w_2)} \left(Pr(w_1, i, C(w_1)) Pr(w_2, j, C(w_2)) \times \cos(V(w_1, i), V(w_2, j)) \right),$$

where $Pr(x, i, C(x))$ is the probability that x takes the ith sense given the context $C(x)$. We calculate $Pr(x, i, C(x))$ as

$$Pr(x, i, C(x)) = \frac{1}{N} \left(\frac{1}{1 - \text{sim}(x, C(x))} \right),$$

where N is the normalization constant so that the probabilities add to 1. Note that, the maxSimC metric models the similarity between two words with respect to the most probable identified sense for each of them. If there are noisy senses as a result of overclustering, maxSimC will penalize them. Hence, maxSimC is a stricter metric as compared to the avgSimC.

For the WordSim-353 dataset, we used the avgSim metric, which is qualitatively similar to the avgSimC, but does not take contextual information into consideration. The avgSim metric is calculated as

$$\text{avgSim}(w_1, w_2) = \frac{1}{m(w_1)m(w_2)} \times \sum_{i=1}^{m(w_1)} \sum_{j=1}^{m(w_2)} \cos(V(w_1, i), V(w_2, j)).$$

For qualitative analysis, we look into the similar words corresponding to different senses for some of the words identified as multi-sense by the LDMI and compare them to the ones discovered by the competing approaches.

6   Results and Discussion

6.1   Quantitative Analysis

Table 2 shows the Spearman rank correlation (ρ) on the SCWS and WordSim-353 dataset for various models and different vector dimensions. For all the vector dimensions, LDMI performs better than the competing approaches on the maxSimC metric. For the GOBW dataset, LDMI shows an average improvement of about 13% over the NP-MSSG and 10% over the MSSG on the maxSimC metric. The average is taken over all vector dimensions. This shows the advantage of LDMI over the competing approaches. For the avgSimC metric, LDMI performs at par with the competing approaches. The other approaches are not as effective in identifying the correct number of senses, leading to noisy clusters and hence, poor performance on the maxSimC metric. LDMI also performs better than LDMI-SK on both maxSimC and avgSimC, demonstrating the effectiveness of the clustering approach employed by LDMI over spherical k-means. Similarly, LDMI performs better than other approaches on the avgSim metric for the WordSim-353 dataset in all the cases, further demonstrating the advantage of LDMI.

In addition, we used the Kolmogorov-Smirnov two-sample test to assess if LDMI's performance advantage over the Skip-Gram is statistically significant. We performed the test on maxSimC and avgSimC metrics corresponding to the 1,000 runs each of LDMI and Skip-Gram on the WSJ dataset. For the null hypothesis that the two samples are derived from the same distribution, the resulting p-value ($\approx 10^{-8}$) shows that the difference is statistically significant for both maxSimC and avgSimC metrics. Similarly, the difference in the LDMI's and LDMI-SK's performance is also found to be statistically significant.

Table 2. Results for the Spearman rank correlation ($\rho \times 100$).

Dataset	Model	d	maxSimC (SCWS)	avgSimC (SCWS)	avgSim (WordSim-353)
WSJ	Skip-Gram	50	57.0	57.0	54.9
WSJ	MSSG	50	41.4	56.3	50.5
WSJ	NP-MSSG	50	33.0	52.2	47.4
WSJ	LDMI-SK	50	57.1	57.9	55.2
WSJ	LDMI	50	**57.9**	**58.9**	**56.8**
GOBW	Skip-Gram	50	60.1	60.1	62.0
GOBW	MSSG	50	50.0	59.6	57.1
GOBW	NP-MSSG	50	48.2	60.0	58.9
GOBW	LDMI-SK	50	60.1	60.6	62.8
GOBW	LDMI	50	**60.6**	**61.2**	**63.8**
GOBW	Skip-Gram	100	61.7	61.7	64.3
GOBW	MSSG	100	53.4	62.6	60.4
GOBW	NP-MSSG	100	47.9	**63.3**	61.7
GOBW	LDMI-SK	100	61.9	62.4	64.9
GOBW	LDMI	100	**62.2**	63.1	**65.3**
GOBW	Skip-Gram	200	63.1	63.1	65.4
GOBW	MSSG	200	54.7	64.0	64.2
GOBW	NP-MSSG	200	51.5	64.1	62.8
GOBW	LDMI-SK	200	63.3	63.9	66.4
GOBW	LDMI	200	**63.9**	**64.4**	**66.8**

6.2 Qualitative Analysis

In order to evaluate the actual senses that the different models identify, we look into the similar words corresponding to different senses for some of the words identified as multi-sense by LDMI. We compare these discovered senses with other competing approaches. Table 3 shows the similar words (corresponding to the cosine similarity) with respect to some of the words that LDMI identified as multi-sense words and estimated a different vector representation for each sense. The results correspond to the 50-dimensional embeddings for the GOBW dataset. The table illustrates that LDMI is able to identify meaningful senses. For example, it is able to identify two senses of the word *digest*, one corresponding to the *food* sense and the other to the *magazine* sense. For the word *block*, it is able to identify two senses, corresponding to the *hindrance* and *address* sense.

Table 4 shows the similar words with respect to the identified senses for the words *digest* and *block* by the competing approaches. We can see that LDMI is able to identify more comprehensible senses for *digest* and *block*, compared to MSSG and NP-MSSG. Compared to the LDMI, LDMI-SK finds redundant

Table 3. Top similar words for different senses of the multi-sense words (different lines in a row correspond to different senses).

Word	Similar words	Sense
figure	status; considered; iconoclast; charismatic; stature; known; calculate; understand; know; find; quantify; explain; how; tell; doubling; tenth; average; percentage; total; cent; gdp; estimate	leader deduce numbers
cool	breezy; gentle; chill; hot; warm; chilled; cooler; sunny; frosty; pretty; liking; classy; quite; nice; wise; fast; nicer; okay; mad;	weather expression
block	amend; revoke; disallow; overturn; thwart; nullify; reject; alley; avenue; waterside; duplex; opposite; lane; boulevard;	hindrance address
digest	eat; metabolize; starches; reproduce; chew; gut; consume; editor; guide; penguin; publisher; compilers; editions; paper;	food magazine
head	arm; shoulder; ankles; neck; throat; torso; nose; limp; toe; assistant; associate; deputy; chief; vice; executive; adviser;	body organization

Table 4. Senses discovered by the competing approaches (different lines in a row correspond to different senses).

digest (Skip-Gram)	block (Skip-Gram)
nutritional; publishes; bittman; reader	annex; barricade; snaked; curving; narrow
digest (MSSG)	**block (MSSG)**
comenu; ponder; catch; turn; ignore	street; corner; brick; lofts; lombard; wall
areat; grow; tease; releasing; warts	yancey; linden; calif; stapleton; spruce; ellis
nast; conde; blender; magazine; edition	bypass; allow; clears; compel; stop
digest (NP-MSSG)	**block (NP-MSSG)**
guide; bible; ebook; danielle; bookseller	acquire; pipeline; blocks; stumbling; owner
snippets; find; squeeze; analyze; tease	override; approve; thwart; strip; overturn
eat; ingest; starches; microbes; produce	townhouse; alley; blocks; street; entrance
oprah; cosmopolitan; editor; conde; nast	mill; dix; pickens; dewitt; woodland; lane
disappointing; ahead; unease; nervousness	slices; rebounded; wrestled; effort; limit
observer; writing; irina; reveals; bewildered	target; remove; hamper; remove; binding
	hinder; reclaim; thwart; hamper; stop
	side; blocks; stand; walls; concrete; front
	approve; enforce; overturn; halted; delay
	inside; simply; retrieve; track; stopping
digest (LDMI-SK)	**block (LDMI-SK)**
almanac; deloitte; nast; wired; guide	cinder; fronted; avenue; flagstone; bricks
sugars; bacteria; ingest; enzymes; nutrients	amend; blocking; withhold; bypass; stall
liking; sort; swallow; find; bite; whole	
find; fresh; percolate; tease; answers	

senses for the word *digest*, but overall, the senses found by the LDMI-SK are comparable to the ones found by the LDMI.

7   Conclusion

We presented LDMI, a model to estimate distributed representations of the multi-sense words. LDMI is able to efficiently identify the meaningful senses of words and estimate the vector embeddings for each sense of these identified words. The vector embeddings produced by LDMI achieves state-of-the-art results on the contextual similarity task by outperforming the other related work.

Acknowledgments. This work was supported in part by NSF (IIS-1247632, IIP-1414153, IIS-1447788, IIS-1704074, CNS-1757916), Army Research Office (W911NF-14-1-0316), Intel Software and Services Group, and the Digital Technology Center at the University of Minnesota. Access to research and computing facilities was provided by the Digital Technology Center and the Minnesota Supercomputing Institute.

References

1. Al-Rfou, R., Kulkarni, V., Perozzi, B., Skiena, S.: POLYGLOT-NER: massive multilingual named entity recognition. In: SDM (2015)
2. Al-Rfou, R., Perozzi, B., Skiena, S.: Polyglot: distributed word representations for multilingual NLP. In: CoNLL (2013)
3. Banerjee, A., Dhillon, I.S., Ghosh, J., Sra, S.: Clustering on the unit hypersphere using von Mises-Fisher distributions. J. Mach. Learn. Res. **6**, 1345–1382 (2005)
4. Chelba, C., Mikolov, T., Schuster, M., Ge, Q., Brants, T., Koehn, P., Robinson, T.: One billion word benchmark for measuring progress in statistical language modeling. Technical report (2013). http://arxiv.org/abs/1312.3005
5. Dhillon, I.S., Modha, D.S.: Concept decompositions for large sparse text data using clustering. Mach. Learn. **42**(1–2), 143–175 (2001)
6. Duchi, J., Hazan, E., Singer, Y.: Adaptive subgradient methods for online learning and stochastic optimization. J. Mach. Learn. Res. **12**, 2121–2159 (2011)
7. Finkelstein, L., Gabrilovich, E., Matias, Y., Rivlin, E., Solan, Z., Wolfman, G., Ruppin, E.: Placing search in context: the concept revisited. In: Proceedings of the 10th International Conference on World Wide Web (2001)
8. Goldberg, Y., Levy, O.: word2vec explained: deriving Mikolov et al.'s negative-sampling word-embedding method. arXiv preprint arXiv:1402.3722 (2014)
9. Harman, D., Liberman, M.: Tipster complete. Corpus number LDC93T3A, Linguistic Data Consortium, Philadelphia (1993)
10. Huang, E.H., Socher, R., Manning, C.D., Ng, A.Y.: Improving word representations via global context and multiple word prototypes. In: Proceedings of the 50th Annual Meeting of the Association for Computational Linguistics (2012)
11. Mikolov, T., Chen, K., Corrado, G., Dean, J.: Efficient estimation of word representations in vector space. arXiv preprint arXiv:1301.3781 (2013)
12. Mikolov, T., Le, Q.V., Sutskever, I.: Exploiting similarities among languages for machine translation. arXiv preprint arXiv:1309.4168 (2013)
13. Neelakantan, A., Shankar, J., Passos, A., McCallum, A.: Efficient non-parametric estimation of multiple embeddings per word in vector space. In: EMNLP (2014)

14. Reisinger, J., Mooney, R.J.: Multi-prototype vector-space models of word meaning. In: NAACL:HLT (2010)
15. Zhao, Y., Karypis, G.: Criterion functions for document clustering. Technical report, Department of Computer Science, University of Minnesota (2005)
16. Zou, W.Y., Socher, R., Cer, D.M., Manning, C.D.: Bilingual word embeddings for phrase-based machine translation. In: EMNLP (2013)

Active Blocking Scheme Learning
for Entity Resolution

Jingyu Shao and Qing Wang[✉]

Research School of Computer Science, Australian National University,
Canberra, Australia
{jingyu.shao,qing.wang}@anu.edu.au

Abstract. Blocking is an important part of entity resolution. It aims to
improve time efficiency by grouping potentially matched records into the
same block. In the past, both supervised and unsupervised approaches
have been proposed. Nonetheless, existing approaches have some limitations: either a large amount of labels are required or blocking quality is
hard to be guaranteed. To address these issues, we propose a blocking
scheme learning approach based on active learning techniques. With a
limited label budget, our approach can learn a blocking scheme to generate high quality blocks. Two strategies called active sampling and active
branching are proposed to select samples and generate blocking schemes
efficiently. We experimentally verify that our approach outperforms several baseline approaches over four real-world datasets.

Keywords: Entity resolution · Blocking scheme · Active learning

1 Introduction

Entity Resolution (ER), which is also called Record Linkage [11,12], Deduplication [6] or Data Matching [5], refers to the process of identifying records which
represent the same real-world entity from one or more datasets [17]. Blocking
techniques are commonly applied to improve time efficiency in the ER process
by grouping potentially matched records into the same block [16]. It can thus
reduce the number of record pairs to be compared. For example, given a dataset
D, the total number of record pairs to be compared is $\frac{|D|*(|D|-1)}{2}$ (i.e. each
record should be compared with all the others in D). With blocking, the number
of record pairs to be compared can be reduced to no more than $\frac{m*(m-1)}{2}*|B|$,
where m is the number of records in the largest block and $|B|$ is the number of
blocks, since the comparison only occurs between records in the same block.

In recent years, a number of blocking approaches have been proposed to
learn blocking schemes [3,13,15]. They generally fall into two categories: (1)
supervised blocking scheme learning approaches. For example, Michelson and

---

Q. Wang–This work was partially funded by the Australian Research Council (ARC)
under Discovery Project DP160101934.

Knoblock presented an algorithm called *BSL* to automatically learn effective blocking schemes [15]; (2) Unsupervised blocking scheme learning approaches [13]. For example, Kejriwal and Miranker proposed an algorithm called *Fisher* which uses record similarity to generate labels for training based on the TF-IDF measure, and a blocking scheme can then be learned from a training set [13].

However, these existing approaches on blocking scheme learning still have some limitations: (1) It is expensive to obtain ground-truth labels in real-life applications. Particularly, match and non-match labels in entity resolution are often highly imbalanced [16], which is called the *class imbalance problem*. Existing supervised learning approaches use random sampling to generate blocking schemes, which can only guarantee the blocking quality when sufficient training samples are available [15]. (2) Blocking quality is hard to be guaranteed in unsupervised approaches. These approaches obtain the labels of record pairs based on the assumption that the more similar two records are, the more likely they can be a match. However, this assumption does not always hold [17]. As a result, the labels may not be reliable and no blocking quality can be promised. A question arising is: Can we learn a blocking scheme with blocking quality guaranteed and the cost of labels reduced?

To answer this question, we propose an active blocking scheme learning approach which incorporates active learning techniques [7,10] into the blocking scheme learning process. In our approach, we actively learn the blocking scheme based on a set of blocking predicates using a balanced active sampling strategy, which aims to solve the class imbalance problem of entity resolution. The experimental results show that our proposed approach yields high quality blocks within a specified error bound and a limited budget of labels.

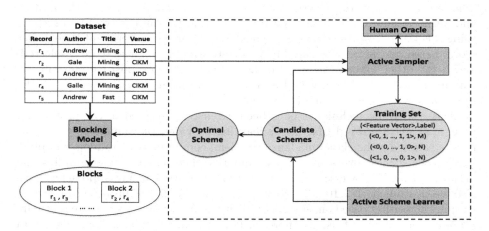

Fig. 1. Overview of the active blocking scheme learning approach

Figure 1 illustrates our proposed approach, which works as follows: Given a dataset D, an active sampler selects samples from D based on a set of candidate schemes, and asks a human oracle for labels. Then an active scheme learner generates a set of refined candidate schemes, enabling the active sampler to adaptively select more samples. Within a limited label budget and an error bound, the optimal scheme will be selected among the candidate schemes.

The contributions of this paper are as follows. (1) We propose a blocking scheme learning approach based on active learning, which can efficiently learn a blocking scheme with less samples while still achieving high quality. (2) We develop two complementary and integrated active learning strategies for the proposed approach: (a) *Active sampling strategy* which overcomes the class imbalance problem by selecting informative training samples; (b) *Active branching strategy* which determines whether a further conjunction/disjunction form of candidate schemes should be generated. (3) We have evaluated our approach over four real-world datasets. Our experimental results show that our approach outperforms state-of-the-art approaches.

2   Related Work

Blocking for entity resolution was first mentioned by Fellegi and Sunter in 1969 [9]. Later, Michelson and Knoblock proposed a blocking scheme learning algorithm called *Blocking Scheme Learner (BSL)* [15], which is the first algorithm to learn blocking schemes, instead of manually selecting them by domain experts. In the same year, Bilenko et al. [3] proposed two blocking scheme learning algorithms called *ApproxRBSetCover* and *ApproxDNF* to learn disjunctive blocking schemes and DNF (i.e. Disjunctive Normal Form) blocking schemes, respectively. Kejriwal et al. [13] proposed an unsupervised algorithm for learning blocking schemes. In their work, a weak training set was applied, where both positive and negative labels were generated by calculating the similarity of two records using TF-IDF. A set of blocking predicates was ranked in terms of their fisher scores based on the training set. The predicate with the highest score is selected, and if the other lower ranking predicates can cover more positive pairs in the training set, they will be selected in a disjunctive form. After traversing all the predicates, a blocking scheme is learned.

Active learning techniques have been extensively studied in the past years. Ertekin et al. [8] proved that active learning provided the same or even better results in solving the class imbalance problem, compared with random sampling approaches such as oversampling the minority class and/or undersampling the majority class [4]. For entity resolution, several active learning approaches have also been studied [1,2,10]. For example, Arasu et al. [1] proposed an active learning algorithm based on the monotonicity assumption, i.e. the more textually similar a pair of records is, the more likely it is a matched pair. Their algorithm aimed to maximize recall under a specific precision constraint.

Different from the previous approaches, our approach uses active learning techniques to select balanced samples adaptively for tackling the class imbalance problem. This enables us to learn blocking schemes within a limited label budget. We also develop a general strategy to generate blocking schemes that can be conjunctions or disjunctions of an arbitrary number of blocking predicates, instead of limiting at most k predicates to be used in conjunctions [3,13].

3   Problem Definition

Let D be a dataset consisting of a set of records, and each record $r_i \in D$ be associated with a set of attributes $A = \{a_1, a_2, ..., a_{|A|}\}$. We use $r_i.a_k$ to refer to the value of attribute a_k in a record r_i. Each attribute $a_k \in A$ is associated with a domain $Dom(a_k)$. A *blocking function* $h_{a_k} : Dom(a_k) \rightarrow U$ takes an attribute value $r_i.a_k$ from $Dom(a_k)$ as input and returns a value in U as output. A *blocking predicate* $\langle a_k, h_{a_k} \rangle$ is a pair of attribute a_k and blocking function h_{a_k}. Given a record pair r_i and r_j, a blocking predicate $\langle a_k, h_{a_k} \rangle$ returns *true* if $h_{a_k}(r_i.a_k) = h_{a_k}(r_j.a_k)$ holds, otherwise *false*. For example, we may have *soundex* as a blocking function for attribute *author*, and accordingly, a blocking predicate $\langle author, soundex \rangle$. For two records with values "Gale" and "Gaile", $\langle author, soundex \rangle$ returns *true* because of $soundex(Gale) = soundex(Gaile) = G4$. A *blocking scheme* is a disjunction of conjunctions of blocking predicates (i.e. in the Disjunctive Normal Form).

A *training set* $T = (X, Y)$ consists of a set of feature vectors $X = \{x_1, x_2, ..., x_{|T|}\}$ and a set of labels $Y = \{y_1, y_2, ..., y_{|T|}\}$, where each $y_i \in \{M, N\}$ is the label of x_i $(i = 1, ..., |T|)$. Given a record pair r_i, r_j, and a set of blocking predicates P, a *feature vector* of r_i and r_j is a tuple $\langle v_1, v_2, ..., v_{|P|} \rangle$, where each v_k $(k = 1, ..., |P|)$ is an equality value of either 1 or 0, describing whether the corresponding blocking predicate $\langle a_k, h_{a_k} \rangle$ returns true or false. Given a pair of records, a *human oracle* ζ is used to provide the label $y_i \in Y$. If $y_i = M$, it indicates that the given record pair refers to the same entity (i.e. *a match*), and analogously, $y_i = N$ indicates that the given record pair refers to two different entities (i.e. *a non-match*). The human oracle ζ is associated with a budget limit $budget(\zeta) \geq 0$, which indicates the total number of labels ζ can provide.

Given a blocking scheme s, a *blocking model* can generate a set of pairwise disjoint blocks $B_s = \{b_1, ..., b_{|B_s|}\}$, where $b_k \subseteq D$ $(k = 1, ..., |B_s|)$, $\bigcup_{1 \leq k \leq |B_s|} b_k = D$ and $\bigwedge_{1 \leq i \neq j \leq |B_s|} b_i \cap b_j = \emptyset$. Moreover, for any two records r_i and r_j in a block $b_k \in B_s$, s must contain a conjunction of block predicates such that $h(r_i.a_k) = h(r_j.a_k)$ holds for each block predicate $\langle a_k, h \rangle$ in this conjunction. For convenience, we use $tp(B_s)$, $fp(B_s)$ and $fn(B_s)$ to refer to *true positives*, *false positives* and *false negatives* in terms of B_s, respectively. Ideally, a good blocking scheme should yield blocks that minimize the number of record pairs to be compared, while preserving true matches at a required level. We thus define the active blocking problem as follows.

Definition 1. *Given a human oracle ζ, and an error rate $\epsilon \in [0, 1]$, the **active blocking problem** is to learn a blocking scheme s in terms of the following objective function, through actively selecting a training set T:*

$$\textit{minimize} \quad |fp(B_s)|$$

$$\textit{subject to} \quad \frac{|fn(B_s)|}{|tp(B_s)|} \leq \epsilon, \textit{ and } |T| \leq budget(\zeta) \tag{1}$$

4   Active Scheme Learning Framework

In our active scheme learning framework, we develop two complementary and integrated strategies (i.e. *active sampling* and *active branching*) to adaptively generate a set of blocking schemes and learn the optimal one based on actively selected samples. The algorithm we propose is called *Active Scheme Learning (ASL)* and described in Sect. 4.3.

4.1   Active Sampling

To deal with the active blocking problem, we need both match and non-match samples for training. However, one of the well-known challenges in entity resolution is the class imbalance problem [18]. That is, if samples are selected randomly, there are usually much more non-matches than matches. To tackle this problem, we have observed, as well as shown in the previous work [1,2], that the more similar two records are, the higher probability they can be a match. This observation suggests that, a balanced representation of similar records and dissimilar records is likely to represent a training set with balanced matches and non-matches. Hence, we define the notion of balance rate.

Definition 2. *Let s be a blocking scheme and X a non-empty feature vector set, the **balance rate** of X in terms of s, denoted as $\gamma(s, X)$, is defined as:*

$$\gamma(s, X) = \frac{|\{x_i \in X | s(x_i) = true\}| - |\{x_i \in X | s(x_i) = false\}|}{|X|} \tag{2}$$

Conceptually, the balance rate describes how balance or imbalance of the samples in X by comparing the number of similar samples to that of dissimilar samples in terms of a given blocking scheme s. The range of balance rate is $[-1, 1]$. If $\gamma(s, X) = 1$, there are all similar samples in T with regard to s, whereas $\gamma(s, X) = -1$ means all the samples are dissimilar samples. In these two cases, X is highly imbalanced. If $\gamma(s, X) = 0$, there is an equal number of similar and dissimilar samples, indicating that X is balanced.

Based on the notion of balance rate, we convert the class imbalance problem into the balanced sampling problem as follows:

Definition 3. *Given a set of blocking scheme S and a label budget $n \leq budget(\zeta)$, the **balanced sampling problem** is to select a training set $T = (X, Y)$, where $|X| = n$, in order to:*

$$minimize \sum_{s_i \in S} \gamma(s_i, X)^2 \tag{3}$$

For two different blocking schemes s_1 and s_2, they may have different balance rates over the same feature vector set X, i.e. $\gamma(s_1, X) \neq \gamma(s_2, X)$ is possible. The objective here is to find a training set that minimizes the balance rates in terms of the given set of blocking schemes. The optimal case is $\gamma(s_i, X) = 0$, $\forall s_i \in S$. However, this is not always possible to achieve in real world applications.

4.2 Active Branching

Given n blocking predicates, we have 2^n possible blocking schemes which can be constructed upon blocking predicates in the form of only conjunctions or disjunctions. Thus, the number of all possible blocking schemes which can be constructued through aribitary combinations of conjunction and disjunction of blocking predicates is more than 2^n. To efficiently learn blocking schemes, we therefore propose a hierarchical blocking scheme learning strategy called *active branching* to avoid enumerating all possible blocking schemes and reduce the number of candidate blocking schemes to $\frac{n(n+1)}{2}$.

Given a blocking scheme s, there are two types of branches through which we can extend s with another blocking predicate: conjunction and disjunction. Let s_1 and s_2 be two blocking schemes, we have the following lemmas.

Lemma 1. *For the conjunction of s_1 and s_2, the following holds:*

$$|fp(B_{s_i})| \geq |fp(B_{s_1 \wedge s_2})|, \ where \ i = 1, 2 \tag{4}$$

Proof. For any true negative record pair $t \notin B_{s_1}$, we have $t \notin B_{s_1 \wedge s_2}$, which means $|tn(B_{s_1})| \leq |tn(B_{s_1 \wedge s_2})|$. Since the sum of true negatives and false positives is constant for a given dataset, we have $|fp(B_{s_1})| \geq |fp(B_{s_1 \wedge s_2})|$. □

Lemma 2. *For the disjunction of s_1 and s_2, the following holds:*

$$\frac{|fn(B_{s_i})|}{|tp(B_{s_i})|} \geq \frac{|fn(B_{s_1 \vee s_2})|}{|tp(B_{s_1 \vee s_2})|}, \ where \ i = 1, 2 \tag{5}$$

Proof. For any true positive record pair $t \in B_{s_1}$, we have $t \in B_{s_1} \cup B_{s_2} = B_{s_1 \vee s_2}$. This is, the number of true positives generated by s_1 cannot be larger than that generated by $s_1 \vee s_2$, i.e. $|tp(B_{s_1})| \leq |tp(B_{s_1 \vee s_2})|$. Since the sum of true positives and false negatives is constant, we have $|fn(B_{s_1})| \geq |fn(B_{s_1 \vee s_2})|$. □

Based on Lemmas 1 and 2, we develop an active branching strategy as follows. First, a locally optimal blocking scheme is learned from a set of candidate schemes. Then, by Lemma 1, the locally optimal blocking scheme is extended. If no locally optimal blocking scheme is learned, the strategy selects the one with minimal error rate and extends it in disjunction with other blocking predicates to reduce the error rate, according to Lemma 2. The extended blocking schemes are then used as a candidate scheme for active sampling to select more samples. Based on more samples, active branching strategy adaptively refines the locally optimal scheme. This process iterates until the label budget is used out.

4.3 Algorithm Description

We present the algorithm called Active Scheme Learning (ASL) used in our framework. A high-level description is shown in Algorithm 1. Let S be a set of blocking schemes, where each blocking scheme $s_i \in S$ is a blocking predicate at the beginning. The budget usage is initially set to zero, i.e. $n = 0$. A set of feature vectors is selected from the dataset as seed samples (lines 1 and 2).

After initialization, the algorithm iterates until the number of samples in the training set reaches the budget limit (line 3). At the beginning of each iteration, the active sampling strategy is applied to generate a training set (lines 4 to 10). For each blocking scheme $s_i \in S$, the samples are selected in two steps: (1) firstly, the balance rate of this blocking scheme s_i is calculated (lines 5 and 7), (2) secondly, a feature vector to reduce this balance rate is selected from the dataset (lines 6 and 8). Then the samples are labeled by the human oracle and stored in the training set T. The usage of label budget is increased, accordingly (lines 9 and 10).

A locally optimal blocking scheme s is searched among a set of blocking schemes S over the training set, according to a specified error rate ϵ (line 11). If it is found, new blocking schemes are generated by extending s to a conjunction with each of the blocking schemes in S_{prev} (lines 12 and 13). Otherwise a blocking scheme with the minimal error rate is selected and new schemes are generated using disjunctions (lines 14 to 16).

Algorithm 1. Active Scheme Learning (ASL)

Input: Dataset: D
Error rate $\epsilon \in [0, 1]$
Human oracle ζ
Set of blocking predicates P
Sample size k
Output: A blocking scheme s

1   $S = S_{prev} = P$, $n = 0$, $T = \emptyset$, $X = \emptyset$
2   $X = X \cup \text{RANDOM\_SAMPLE}(D)$
3   **while** $n < budget(\zeta)$ **do**
4   **for** *each* $s_i \in S$ **do**   // Begin active sampling
5   **if** $\gamma(s_i, X) \leq 0$ **then**
6         $X = X \cup \text{SIMILAR\_SAMPLE}(D, s_i, k)$
7   **else**
8         $X = X \cup \text{DISSIMILAR\_SAMPLE}(D, s_i, k)$
9       $n = |X|$ // End active sampling
10    $T = T \cup \{(x_i, \zeta(x_i)) | x_i \in X\}$ // Add labeled samples into T
11    $s = \text{FIND\_OPTIMAL\_SCHEME}(S, T, \epsilon)$; $S_{prev} = S$ // Begin active branching
12   **if** $\text{FOUND}(s)$ **then**
13      $S = \{s \wedge s_i | s_i \in S_{prev}\}$
14   **else**
15      $s = \text{FIND\_APPROXIMATE\_SCHEME}(S, T, \epsilon)$
16      $S = \{s \vee s_i | s_i \in S_{prev}\}$ // End active branching

17 **Return** s

---

5   Experiments

We have evaluated our approach to answer the following two questions: (1) How do the error rate ϵ and the label budget affect the learning results in our approach? (2) What are the accuracy and efficiency of our active scheme learning approach compared with the state-of-the-art approaches?

5.1   Experimental Setup

Our approach is implemented in Python 2.7.3, and is run on a server with 6-core 64-bit Intel Xeon 2.4 GHz CPUs, 128GBytes of memory.

Datasets: We have used four datasets in the experiments: (1) *Cora*[1] dataset contains bibliographic records of machine learning publications. (2) *DBLP-Scholar* (see footnote 1) dataset contains bibliographic records from the DBLP and Google Scholar websites. (3) *DBLP-ACM* [14] dataset contains bibliographic records from the DBLP and ACM websites. (4) *North Carolina Voter Registration (NCVR)*[2] dataset contains real-world voter registration information of people from North Carolina in the USA. Two sets of records collected in October 2011 and December 2011 respectively are used in our experiments. The characteristics of these data sets are summarized in Table 1, including the number of attributes, the number of records (for each dataset) and the class imbalance ratio.

Table 1. Characteristics of datasets

Dataset	# Attributes	# Records	Class imbalance ratio
Cora	4	1,295	1:49
DBLP-Scholar	4	2,616/64,263	1:31,440
DBLP-ACM	4	2,616/2,294	1:1,117
NCVR	18	267,716/278,262	1:2,692

Baseline Approaches: Since no active learning approaches were proposed on blocking scheme learning, we have compared our approach (ASL) with the following three baseline approaches: (1) *Fisher* [13]: this is the state-of-the-art unsupervised scheme learning approach proposed by Kejriwal and Miranker. Details of this approach have been outlined in Sect. 2. (2) *TBlo* [9]: this is a traditional blocking approach based on expert-selected attributes. In the survey [6], this approach has a better performance than the other approaches in terms of the F-measure results. (3) *RSL (Random Scheme Learning)*: it uses random sampling, instead of active sampling, to build the training set and learn blocking

---

[1] Available from: http://secondstring.sourceforge.net.
[2] Available from: http://alt.ncsbe.gov/data/.

schemes. We run the RSL ten times, and present the average results of blocking schemes it learned.

Measurements: We use the following common measures [6] to evaluate blocking quality: *Reduction Ratio (RR)* is one minus the total number of record pairs in blocks divided by the total number of record pairs without blocks, which measures the reduction of compared record pairs. *Pairs Completeness (PC)* is the number of true positives divided by the total number of true matches in the dataset. *Pairs Quality (PQ)* is the number of true positives divided by the total number of record pairs in blocks. *F-measure (FM)* $FM = \frac{2*PC*PQ}{PC+PQ}$ is the harmonic mean of PC and PQ. In addition to these, we define the notion of *constraint satisfaction* as $CS = \frac{N_s}{N}$, where N_s is the times of learning an optimal blocking scheme by an algorithm and N is the times the algorithm runs.

5.2 Experimental Results

Now we present our experimental results in terms of the constraint satisfaction, blocking quality, blocking efficiency and label cost.

Constraint Satisfaction. We have conducted experiments to evaluate the constraint satisfaction. In Fig. 2, the results are presented under different error rates $\epsilon \in \{0.1, 0.2, 0.4, 0.6, 0.8\}$ and different label budgets ranging from 20 to 500 over four real datasets. We use the total label budget as the training label size for RSL to make a fair comparison on active sampling and random sampling. Our experimental results show that random sampling with a limited label sizes often fails to produce an optimal blocking scheme. Additionally, both error rate and label budget can affect the constraint satisfaction. As shown in Fig. 2(a)–(d), when the label budget increases, the CS value goes up. In general, when ϵ becomes lower, the CS value decreases. This is because a lower error rate is usually harder to achieve, and thus no scheme that satisfies the error rate can be learned in some cases. However, if the error rate is set too high (e.g. the red line), it could generate a large number of blocking schemes satisfying the error rate, and the learned blocking scheme may vary depending on the training set.

Blocking Quality. We present the experimental results of four measures (i.e. RR, PC, PQ, FM) for our approach and the baseline approaches. In Fig. 3(a), all the approaches yield high RR values over four datasets. In Fig. 3(b), the PC values of our approach are not the highest over the four datasets, but they are not much lower than the highest one (i.e. within 10% lower except in DBLP-Scholar). However, out approach can generate higher PQ values than all the other approaches, from 15% higher in NCVR (0.9956 vs 0.8655) to 20 times higher in DBLP-ACM (0.6714 vs 0.0320), as shown in Fig. 3(c). The FM results are shown in Fig. 3(d), in which our approach outperforms all the baselines over all the datasets.

Blocking Efficiency. Since blocking aims to reduce the number of pairs to be compared in entity resolution, we evaluate the efficiency of blocking schemes by the number of record pairs each approach generates. As shown in Table 2, *TBlo*

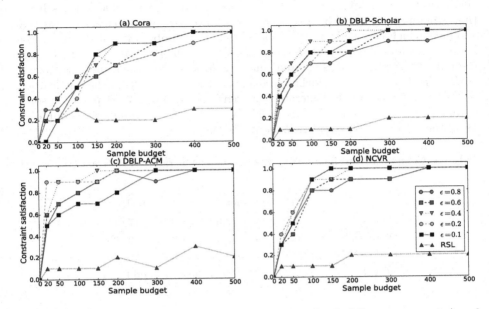

Fig. 2. Comparison on constraint satisfaction by ASL (with different error rates) and RSL under different label budgets over four real datasets (Color figure online)

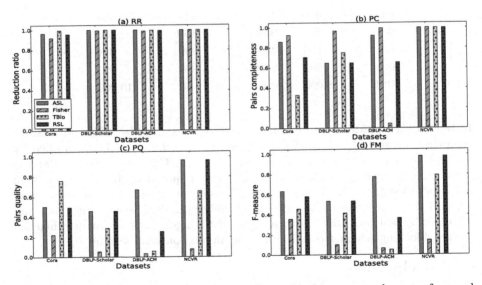

Fig. 3. Comparison on blocking quality by different blocking approaches over four real datasets using the measures: (a) RR, (b) PC, (c) PQ, and (d) FM

generates the minimal number of record pairs in Cora. This is due to the scheme that is manually selected by domain experts. *Fisher* targeted to learn disjunctive schemes, which can lead to large blocks, thus the number of record pairs is the largest over four datasets. *ASL* considers a trade-off between PC and PQ, and the number of record pairs is often small. In *RSL*, we use the same label size as *ASL*, thus it may learn a blocking scheme that is different from the one learned by RSL, and accordingly generates different numbers of record pairs for some datasets such as Cora and DBLP-ACM. When a sufficient number of samples is used, the results of *ASL* and *RSL* would be the same.

Table 2. Comparison on the number of record pairs generated by different approaches

	TBlo	Fisher	ASL	RSL
Cora	**2,945**	67,290	29,306	17,974
DBLP-Scholar	6,163	1,039,242	**3,328**	**3,328**
DBLP-ACM	25,279	69,037	**3,043**	17,446
NCVR	932,239	7,902,910	**634,121**	**634,121**

Table 3. Comparison on label cost by ASL and RSL over four real datasets

Error rate	Cora	DBLP-Scholar	DBLP-ACM	NCVR
0.8	600	500	300	300
0.6	**400**	350	200	350
0.4	450	**250**	**150**	250
0.2	550	300	200	**200**
0.1	500	**250**	300	250
RSL	8,000	10,000+	2,500	10,000+

Label Cost. In order to compare the label cost required by ASL and RSL for achieving the same block quality, we present the numbers of labels needed by our approach to generate a blocking scheme with CS = 100% under different error rates, and compare them with the labels required by RSL in Table 3. In our experiments, the label budget for ASL under a given error rate starts with 50, and then increases by 50. The label budget for RSL starts with 500, and increases by 500 each time. Both ASL and RSL algorithms terminate when the learned blocking schemes remain the same in ten consecutive runs.

6   Conclusions

In this paper, we have used active learning techniques to develop a blocking scheme learning approach. Our approach overcomes the weaknesses of existing works in two aspects: (1) Previously, supervised blocking scheme learning approaches require a large number of labels for learning a blocking scheme, which is an expensive task for entity resolution; (2) Existing unsupervised blocking scheme learning approaches generate training sets based on the similarity of record pairs, instead of their true labels, thus the training quality can not be guaranteed. Our experimental results show that the proposed approach outperforms the baseline approaches under a specific error rate with a sample budget.

References

1. Arasu, A., Götz, M., Kaushik, R.: On active learning of record matching packages. In: Proceedings of the 2010 ACM SIGMOD International Conference on Management of data, pp. 783–794 (2010)
2. Bellare, K., Iyengar, S., Parameswaran, A.G., Rastogi, V.: Active sampling for entity matching. In: Proceedings of the 18th ACM SIGKDD International Conference on Knowledge Discovery and Data mining, pp. 1131–1139 (2012)
3. Bilenko, M., Kamath, B., Mooney, R.J.: Adaptive blocking: learning to scale up record linkage. In: Proceedings of the 6th International Conference on Data Mining, pp. 87–96 (2006)
4. Chawla, N.V., Bowyer, K.W., Hall, L.O., Kegelmeyer, W.P.: SMOTE: synthetic minority over-sampling technique. J. Artif. Intell. Res. **16**, 321–357 (2002)
5. Christen, P.: Data Matching. Concepts and Techniques For Record Linkage, Entity Resolution, and Duplicate Detection. Springer, Heidelberg (2012). https://doi.org/10.1007/978-3-642-31164-2
6. Christen, P.: A survey of indexing techniques for scalable record linkage and deduplication. IEEE Trans. Knowl. Data Eng. **24**(9), 1537–1555 (2012)
7. Dasgupta, S., Hsu, D.: Hierarchical sampling for active learning. In: Proceedings of the 25th International Conference on Machine Learning, pp. 208–215 (2008)
8. Ertekin, S., Huang, J., Bottou, L., Giles, L.: Learning on the border: active learning in imbalanced data classification. In: Proceedings of the 16th ACM Conference on Information and Knowledge Management, pp. 127–136 (2007)
9. Fellegi, I.P., Sunter, A.B.: A theory for record linkage. J. Am. Stat. Assoc. **64**(328), 1183–1210 (1969)
10. Fisher, J., Christen, P., Wang, Q.: Active learning based entity resolution using Markov logic. In: Bailey, J., Khan, L., Washio, T., Dobbie, G., Huang, J.Z., Wang, R. (eds.) PAKDD 2016. LNCS (LNAI), vol. 9652, pp. 338–349. Springer, Cham (2016). https://doi.org/10.1007/978-3-319-31750-2_27
11. Gruenheid, A., Dong, X.L., Srivastava, D.: Incremental record linkage. Proc. VLDB Endowment **7**(9), 697–708 (2014)
12. Hu, Y., Wang, Q., Vatsalan, D., Christen, P.: Improving temporal record linkage using regression classification. In: Kim, J., Shim, K., Cao, L., Lee, J.-G., Lin, X., Moon, Y.-S. (eds.) PAKDD 2017. LNCS (LNAI), vol. 10234, pp. 561–573. Springer, Cham (2017). https://doi.org/10.1007/978-3-319-57454-7_44
13. Kejriwal, M., Miranker, D.P.: An unsupervised algorithm for learning blocking schemes. In: Proceedings of the 13th International Conference on Data Mining, pp. 340–349 (2013)
14. Köpcke, H., Thor, A., Rahm, E.: Evaluation of entity resolution approaches on real-world match problems. Proc. VLDB Endowment **3**(1–2), 484–493 (2010)
15. Michelson, M., Knoblock, C.A.: Learning blocking schemes for record linkage. In: Proceedings of the 21st Association for the Advancement of Artificial Intelligence, pp. 440–445 (2006)

16. Wang, Q., Cui, M., Liang, H.: Semantic-aware blocking for entity resolution. IEEE Trans. Knowl. Data Eng. **28**(1), 166–180 (2016)
17. Wang, Q., Gao, J., Christen, P.: A clustering-based framework for incrementally repairing entity resolution. In: Bailey, J., Khan, L., Washio, T., Dobbie, G., Huang, J.Z., Wang, R. (eds.) PAKDD 2016. LNCS (LNAI), vol. 9652, pp. 283–295. Springer, Cham (2016). https://doi.org/10.1007/978-3-319-31750-2_23
18. Wang, Q., Vatsalan, D., Christen, P.: Efficient interactive training selection for large-scale entity resolution. In: Cao, T., Lim, E.-P., Zhou, Z.-H., Ho, T.-B., Cheung, D., Motoda, H. (eds.) PAKDD 2015. LNCS (LNAI), vol. 9078, pp. 562–573. Springer, Cham (2015). https://doi.org/10.1007/978-3-319-18032-8_44

Mining Relations from Unstructured Content

Ismini Lourentzou[1], Alfredo Alba[3], Anni Coden[2], Anna Lisa Gentile[3(✉)],
Daniel Gruhl[3], and Steve Welch[3]

[1] University of Illinois at Urbana - Champaign, Champaign, USA
lourent2@illinois.edu
[2] IBM Watson Research Laboratory, New York, NY, USA
{anni,aalba,dgruhl,welchs}@us.ibm.com
[3] IBM Research Almaden, San Jose, CA, USA
annalisa.gentile@ibm.com

Abstract. Extracting relations from unstructured Web content is a
challenging task and for any new relation a significant effort is required
to design, train and tune the extraction models. In this work, we investigate how to obtain suitable results for relation extraction with modest
human efforts, relying on a dynamic active learning approach. We propose a method to reliably generate high quality training/test data for
relation extraction - for any generic user-demonstrated relation, starting
from a few user provided examples and extracting valuable samples from
unstructured and unlabeled Web content. To this extent we propose a
strategy which learns how to identify the best order to human-annotate
data, maximizing learning performance early in the process. We demonstrate the viability of the approach (i) against state of the art datasets
for relation extraction as well as (ii) a real case study identifying text
expressing a causal relation between a drug and an adverse reaction from
user generated Web content.

1 Introduction

Recent years have seen the rise of neural networks for addressing many Information Extraction tasks. Particular interest is focused on Relation Extraction
from unstructured text content. While crafting the right model architecture has
gained significant attention, a major and often overlooked challenge is the acquisition of solid training data and reliable gold standard datasets for validation.

Kick-starting a relation extraction process - i.e. acquiring reliable training
and testing data - for an arbitrary user-defined relation presents many hurdles.
This is especially true when the pool of unannotated data to choose from is
virtually infinite, as in the case of Web data and social streams - where one first
needs to identify a relevant corpus from which relations should be extracted.
Systems extracting relations from open data have been described in the literature. Although they may perform well, they are in general quite expensive:
on one hand, supervised methods require an upfront annotation effort for each

© Springer International Publishing AG, part of Springer Nature 2018
D. Phung et al. (Eds.): PAKDD 2018, LNAI 10938, pp. 363–375, 2018.
https://doi.org/10.1007/978-3-319-93037-4_29

relation, while on the other hand unsupervised methods present many drawbacks - mainly the reliance on Natural Language Processing (NLP) tools, which might not be available for all languages and which do not perform that well on the ungrammatical text often found in social posts. Moreover for any approach it is desirable to evaluate the performance: this implies the availability of test data, which in general involves an expensive manual annotation process.

In this paper, we address these challenges by presenting a system for extracting relations from unlabeled data which minimizes the required annotations, which needs no NLP tools and performs well with respect to the available state of the art methods. These relations can be fairly well defined (e.g. given a color and an object, does the author imply the object is that color) to somewhat more subjective ones (e.g. detecting asserted causal relations between drugs and adverse events). Here relation detection is defined in a standard way, i.e. determining if a relation is present in a text sample or not, or in Relation Extraction terms, the goal is to recognize whether a predefined set of relations holds between two or more entities in a sentence. We propose an end-to-end system for extracting relations from unstructured Web content. First, the type of entities involved in the relation, e.g. drugs and adverse events, must be specified - this step can be seen as a blackbox component here. Then we obtain a relevant pool of potential examples of the relations from the Web and social media by selecting sentences where the entities co-occur. We discard parts of the corpus which seems to contain highly ambiguous data, while retaining useful data for the task (Sect. 3).

After collecting relevant unlabeled data, we ask a Subject Matter Expert (SME) to annotate the data in small batches (e.g. 100 examples at a time). The selection of examples that are presented to the SME is dependent on the learning model and the active learning strategy which in itself is dependent on the type of relation and data at hand. In this paper, we show how for a given model the system "learns" a quite successful strategy (in general it is not feasible to determine an "optimal" strategy with reasonable effort): we measure the performance of several neural models and several active learning strategies at the end of each batch, devising a method to only promote the most successful strategies for subsequent steps (Sect. 3.2). By optimizing the order of examples to annotate, the work required by the SME is much less than manually creating labeled training data [29] or building/tuning NLP tools for different languages and styles. We show this with several experiments on standard benchmarking datasets.

The contribution of this work is threefold. First, we propose an end-to-end method for relation extraction with a human-in-the-loop. We design a systematic procedure for generating datasets for relation extraction on any domain and any concept that the user is interested in. This is valuable to kick-start arbitrary extraction tasks for which annotated resources are not yet available. Our method does not have to rely on any NLP tools and hence is independent of document style and language. Second, we experiment using a combination of active learning strategies on neural models and devise a method to prune the ones that are not effective for the task at hand. Besides testing the approach on a real use case, we prove its efficacy on publicly available standard datasets for relation extraction and show that by using our pruning technique - and observing the results a posteriori - we achieve similar performance to the "optimal active learning strategy" for the task and the specific dataset. In addition, as one does not know a priori what the optimal strategy is, our system learns which strategy among the available ones to use. The technique works comparably well regardless of the chosen neural architecture. Finally, we present a real use case scenario where we address the challenging task of extracting the causal relation between drugs and their adverse events from user generated content.

The advantage of the proposed approach is the possibility to rapidly deploy a system able to quickly generate high quality train/test data on any relation of interest, regardless of language and text style of the corpus. Given the fact that the method gives feedback on performance after every small annotation step, the user can decide when to stop annotating when she is satisfied with the level of accuracy (e.g. accuracy above 75%) or decide to stop if she understands that the underlying data might not be useful for the task at hand. Substantially, we are able to early identify high quality train/test data for challenging relation extraction tasks while minimizing the user annotation effort.

2   Related Work

One of the key to success for machine learning tasks is the availability of high quality annotated data, which is often costly to acquire. For the relation extraction task, the definition of a relation is highly dependent on the task at hand and on the view of the user, therefore having pre-annotated data available for any specific case is unfeasible. Various approaches have been proposed to minimize the cost of obtaining labelled data, one of the most prominent being distant supervision, which exploits large knowledge bases to automatically label entities in text [5,12,16,26,27]. Despite being a powerful technique, distant supervision has many drawbacks including poor coverage for tail entities [16], as well as the broad assumption that when two entities co-occur, a certain relation is expressed in the sentence [5]. The latter can be especially misleading for unusual relations, where the entities might co-occur but not fit the desired semantic (e.g. a user wants to classify "positive" or desirable side effects of drugs). One way to tackle the problem is to use targeted human annotations to expand the large pool of examples labelled with distant supervision [3]. This combination approach produced good results in the 2013 KBP English Slot Filling task[1]. Another way is to address it as a noise reduction problem: e.g. Sterckx et al. [30] exploit hierarchical clustering of the distantly annotated samples to select the most reliable ones, while Fu and Grishman [10] propose to interleave self-training with co-testing. Nonetheless, it is nearly impossible to refrain from manual annotation entirely: at the very least test data (that serves as gold standard) needs to be annotated manually. The question then is how to minimize the human annotation effort.

Active Learning (AL) aims at incorporating *targeted* human annotations in the process: the learning strategy interactively asks the user to annotate certain

---

[1] http://surdeanu.info/kbp2013/.

specific data points, using several criteria to identify the best data to annotate next. Some of the most commonly used criteria are: (i) *uncertainty sampling*, which ranks the samples according to the model's belief it will mislabel them [18]; (ii) *density weighted uncertainty sampling*, which clusters the unlabeled instances to pick examples that the model is uncertain for, but also are "representative" of the underlying distribution [9,23]; (iii) *QUIRE*, which measures each instance's informativeness and representativeness by its prediction uncertainty [15]; (iv) Bayesian methods such as *bald* (Bayesian Active Learning by Disagreement) which select examples that maximize the models's information gain [11]. The effectiveness of these criteria is highly dependent on the underlying data and the relation to extract and it is very difficult to identify strong connections between any of the criteria and the task [14]. The open question is then how to decide which technique to use on a new extraction task. Following [14] we argue that it is best to dynamically decide on the criteria on a task-driven basis. The "active learning by learning" method (*albl*) [14] has an initial phase where all criteria are tested extensively and one is chosen. Our intuition is that the technique that seems to perform the best at the beginning might not be best one in the long run. Therefore we propose a method that initially distributes the budget of annotation among all considered criteria and discards the worst performing one at each iteration. We argue that keeping a pool of options for a longer number of iterations will maximize performance on average for a larger number of tasks, especially given the very small sample set, and we support the claim with comparative experiments.

For the sake of completeness, it is worth mentioning that in relation extraction, as in many other machine learning tasks, there is no one-fits-all model and many have been proposed ranging from early solutions based on SVMs and tree kernels [7,8,20,32,34] to most recent ones exploiting neural architectures [24,31,33]. Neither the model nor the active learning strategy or any particular combination is universally (on all relations/all data) "the best" performer - hence our proposal of a data driven approach. The aim of this work is to investigate the influence of different active learning strategies on different extraction tasks (regardless of the underlying neural model) and to devise strategies to effectively annotate data, rather than proposing a new neural architecture per-se. Therefore for our experiments we considered several state of the art deep learning models for relation classification, including Convolutional Neural Networks (CNNs) [24,33], Recurrent Neural Networks (RNNs) [19], such as bi-directional GRUs [35], as well as ensembles [31]. For all models, we do not require any NLP preprocessing (besides tokenization) of the text.

3   Relation Classification

We consider relation extraction as a binary classification task. Given a text snippet s containing one or more target entities e_i[2] our goal is to identify if

---

[2] In our experiments we use pairs of entities, however we should note that our models can handle n-ary relations as well. We leave this to future work.

s expresses a certain relation r among the entities e_i. Our goal is two-fold: (i) create a relation classification system that gradually increases accuracy from each recognized relation, as well as (ii) identifying the sentence snippets for which the system is most/least confident about expressing the desired relation. We first obtain a large pool of relevant unlabeled text from a given social media stream (e.g. the Twitter stream, a social forum etc.), applying the following method. We consider the (two) types of entities involved in the relation, for which we construct dictionaries using any off-the-shelf tool (e.g. [2]) and select sentences where the (two) entities co-occur. Note that this will produce lot of noisy data, therefore noise reduction needs to be in place. For this work we treat entity identification in sentences as a blackbox component with various valid available solutions [16, 26, 30].

We then segment the learning process into small steps of b examples at a time ($b = 100$ in this work[3]) and interactively annotate the data as we train the models. Example refers here to a text snippet expressing the relation between the entities and annotation refers to manually assigning a "true/false" label to each example. We select the first batch of b examples with a *curriculum learning* strategy (details in Sect. 3.2) and manually annotate them. With those we train (i) several neural models, using (ii) two different data representation paradigms and (iii) several active learning strategies to determine the next batch of examples. Our goal is not to specifically improve a particular learning model per-se, but rather (i) identify at an early stage, i.e. with minimal annotation effort, if a specific relation *can* be learned from the available data and (ii) minimize the labelling effort by using first examples that are more likely to boost the learning performance. As no active learning strategy is universally preferable (we show tests on ready-available gold standard datasets for relation extraction in Sect. 4.2) we propose a pruning method (Sect. 3.2) that dynamically selects the best strategy for a given task.

3.1 Models, Data Representations and Parameter Choices

We employ commonly used neural models for relation extraction, specifically CNNs [24, 33] and bi-directional GRUs [35].

As for data representation, we do not rely on lexical features or any other language-dependent information, but after using a simple tokenizer (white spaces, punctuation) we merely exploit distributional semantics - statistical properties of the text data - to ensure portability to different languages, domains and relation types. We explore two different representations for the text: (i) word sequences concatenated with positional features (as in [33]), i.e. we generate three embedding matrices, one initialized with pre-trained word embeddings and two randomly initialized for the positional features; (ii) a context-wise split of the sentence (as in [1]), i.e. using pre-trained word embeddings and using the two

---

[3] The size of the batch is adjustable, the human-in-the-loop can specify it. In our experiments, the involved medical doctor indicated 100 as a good size in terms of keeping focus.

entities in the text as split points to generate three matrices - left, middle and right context.

As for the neural network architectures specifications, following literature, all our models use: 100-dimensional pre-trained GloVe word embeddings [25]; 100-dimensional positional embeddings optimized with Adam [17]; initial learning rate = 0.001; batch size $b = 100$; validation split = 0.2; early stopping [21] to avoid overfitting (if no improvement happens for 5 consecutive iterations). For the CNNs we use: 100 filters; kernels width = 3; ReLU nonlinearities [22] - for CNNs with multiple filter sizes we set the kernels width from 2 to 5. For the GRU we use: ReLU activations and layer size = 100.

3.2 Active Learning by Pruning

At the bootstrapping phase, we have no information on the performance of each model as all data is unlabeled. We used *curriculum learning (CL)* strategies [6], where the order of the data is decided in advance - before starting the learning process using several text based criteria. While we tested several criteria, including random as baseline, the best performance was obtained by maximizing dissimilary. Starting from a random example (sentence) we sort the data as to maximize dissimilarity between the sentences. We calculate sentence similarity exploiting GloVe embeddings as proposed by [4].

For all subsequent steps, we can use previously annotated examples to test the performance of the different active learning strategies. We consider a pool-based active learning scenario [28] in which there exists a small set of labeled data $L = (x_1, y_1), \ldots, (x_{n_l}, y_{n_l})$ (in this case we consider the batch of 100 examples selected by CL) and a large pool of unlabeled data $U = x_1, \ldots, x_{n_u}$. The task for the active learner is to draw examples to be labeled from U, so as to maximize the performance of the classifier (the neural net) while limiting the number of required annotations to achieve a certain accuracy. We train the model on the first batch of annotated examples, using 5-fold validation on the batch itself. At each subsequent iteration we select $\frac{b}{n}$ examples according to each of the n target active learning strategies; after labelling those b examples we calculate the performance for each of them and identify the worst performing AL strategy, which gets discarded in subsequent iterations. After n iterations we remain with one strategy for the particular task. In this particular work, we select $n = 5$ active learning strategies: uncertainty sampling (*us*), density weighted uncertainty sampling (*dwus*), bayesian active learning by disagreement (*bald*), QUIRE and we include as baseline the random selection (*rs*) of examples. The proposed approach is not limited to those - any other strategy can be added without changing the overall framework.

We perform extensive experiments testing all possible combinations of models, data representations and active learning strategies. Results are summarized in Table 2. We show that our proposed "active learning by pruning" strategy is robust across relation extraction tasks and datasets (Sect. 4).

4   Experiments

The relation extraction task is a challenging one. Especially in the case of developing early prototype systems, little can be done with a traditional neural network in the absence of a significant quantity of hand labeled data. While a task specific labeling system can help [29], it makes sense to consider the "best order" to ask the user for input in the hopes of achieving a sufficiently performant system with minimal human effort.

Assuming the existence of a relevant corpus of unlabelled examples for the relation at hand our aim in this work is to identify the best active learning strategy for each extraction task to prioritize the annotation of examples that have a better impact on the models. We exploit existing benchmark datasets on relation extraction and simulate the human-in-the-loop: we treat all examples as unlabelled and "request" the annotations in small batches from the existing labels, as if they were annotated in real-time by a user. This gives us useful insights, as we can compare partial performance (after any given annotation batch) against the best achievable performance (using the whole dataset), as well as run in parallel all active learning strategies to figure out if any of them is "universally" better for all tasks. A post-hoc analysis reveals that in terms of active learning strategy there is no one-fits-all solution (Sect. 4.2) but that our proposed solution is able to promote good performing ones for the task. We test our pruning technique on all the benchmark relations, as well as on our real case scenario on extracting adverse drug events, for a total of 10 different relations (details on the data in Sect. 4.1).

4.1   Datasets

We test our method in a real case experiment, extracting Adverse Drug Events (ADE) relations from a Web forum (http://www.askapatient.com/). Our human-in-the-loop is a medical doctor using our system to annotate the data. We produced annotations in the same style as CADEC (CSIRO Adverse Drug Event Corpus)[4], totaling of 646 positive and 774 negative examples of causal relationships between drugs and ADEs. We name this dataset *causalADEs*[5]. Posts are tagged based on mentions of certain drugs, ADEs, symptoms, findings etc. However, the mere co-occurrence of a drug and an ADE in a sentence does not necessarily imply a causal relation among the two. Figure 1 shows three sentences, one where the drug caused an ADE and others where it did not.

We also test our method on the *Semeval2010-Task8* dataset [13], which consists of 8,000 training and 2,717 test examples on nine relation types: Cause-Effect, Component-Whole, Content-Container, Entity-Destination, Entity-Origin, Instrument-Agency, Member-Collection, Message-Topic, and Product-Producer.

---

[4] http://doi.org/10.4225/08/570FB102BDAD2.
[5] https://github.com/Isminoula/CausalADEs.

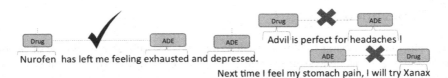

Fig. 1. Examples of causal and non-causal relations between drugs and ADE mentions in sentences.

4.2 Fixed Active Learning Strategy VS Dynamic Selection

The aim of the experiments is to compare all the considered active learning strategies (as used individually) against dynamic selection, either our proposed pruning strategy or the *albl* method [14]. We ran experiments using various different configurations of neural networks, data representations and active learning strategies. For the sake of reporting clarity we use CNN with positional features to plot results on the different active learning strategies - but we summarize results for different configurations in Table 2. Figure 2 shows the accuracy on the *Semeval* extraction tasks for all the strategies as a function of the number of labelled examples (Fig. 2): no single AL strategy is always the best, but we can observe that our pruning strategy has a consistent behavior across all tasks, approximating top performance.

In a real case scenario, where all data is unlabeled and we do not have a designated test set, the feedback that we can provide at each step is the performance calculated with cross-validation on the currently annotated data.

Taking a closer look at individual results (Table 2) one can observe that our proposed pruning strategy (i) obtains top performance - with respect to other strategies - with exhaustive annotation, i.e. when all examples are labelled on most tasks (9 out of 10) and (ii) can consistently "near" top performance (with a loss ≤2% in most cases (7 out of 10) with less than half of the annotated data, for some relations as early as after 400 annotations. For completeness we also compare our pruning strategy to *albl* [14], which is to the best of our knowledge

Table 1. Examples of correct and incorrect predictions on causalADEs

Sentence	y	$\hat{y}$	$P(\hat{y} = y|x)$
I was on **Crestor** for only two months when my knee just flared up in pain followed by **muscle pain**	1	1	0.99
However, I am afraid to discontinue the **Paxil** due to fear of withdrawal symptoms and/or return of **panic attacks**	0	0	0.99
I felt like **Zoloft** turned me into a little bit of a **zombie**	1	0	0.722
I was **crying** at the drop of a hat until I started taking the **Celexa**, so has been a life saver in my opinion	0	1	0.497
Put me on **prozac** and it made me more **jittery**	1	0	0.803

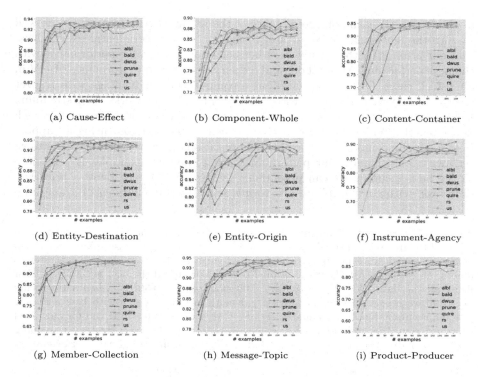

(a) Cause-Effect   (b) Component-Whole   (c) Content-Container

(d) Entity-Destination   (e) Entity-Origin   (f) Instrument-Agency

(g) Member-Collection   (h) Message-Topic   (i) Product-Producer

Fig. 2. Active learning strategies comparison across all *Semeval* relation extraction tasks, fixing CNN as neural model and context-wise splitting as data representation. Accuracy is calculated on the reference test set.

the best performing method for dynamically selecting active learning strategies. Our pruning method outperforms *albl* (or has same performance) in all the runs, with a maximal increase of 6% (the *Entity-Origin* task).

A further observation is about the *causalADEs* dataset, which is our true real-case experiments (while the others are simulated with existing benchmark datasets). The presence of a causal relation between a Drug and an Adverse Drug Event can be very tricky to identify. Table 1 shows examples of correct and incorrect predictions of our models.

Our overarching goal is to be able to identify a pool of examples for which we are highly confident to have been annotated correctly leading to a reliable training/test dataset to train the extraction of a new relation. This is particularly valuable in situations where the unlabeled sample data is particularly large and we can afford to discard examples, as long as the selected ones are of high quality.

In Table 2 we summarize our results and report top performance for each extraction task. Regarding neural architecture we observe that the simple CNN model performed better in most cases, with a preference for the *context-wise split* data representation. Regarding active learning strategies we compare our method with (i) each considered state-of-the-art fixed AL strategy (ii) as well as with the *albl* dynamic AL strategy selection. The experimental results show that our pruning method achieves either better or comparable performance with the best performing AL method, and we surpass the best performance of *albl* in almost all cases. Our method also has a computational advantage with respect to *albl*. While we train and test in small batches, *albl* works in a streaming fashion where a micro-training and performance estimation is done after each new example. While this is affordable in their tested settings (using a Support Vector Machine model) it becomes computationally heavy in neural network settings[6]. Additionally, our performance using only 500 examples is very close to the best accuracy we can potentially achieve (but do not know a priori) in each task, with comparable results to *albl* with the same number of examples.

Table 2. For each extraction **task** we report: the best neural network configuration, in terms of the **model** and the **data** representation - either *context*-wise split or *positional* features - which have produced best results; which of the fixed single AL strategies among *rs (r), quire (q), dwus (u), bald (b), us (u)* produced the best accuracy - either using all the data (**A@all**) or the first 500 examples (**A@500**) - when a tie occurs at A@all we report all tying strategies and mark in bold the one with best accuracy A@500; accuracy for the dynamic selection of AL strategies for **albl** and for our novel proposed **pruning** technique - for which we report the last AL strategy remaining (**selection**) after the pruning is completed. We highlight in bold highest performances.

Task	Best performing NN		Fixed AL strategy			ALBL		Pruning		
	Model	Data	Best AL	A@all	A@500	A@all	A@500	Selection	A@all	A@500
Content-Container	GRU_{att}	positional	*r, q, d, b*	0.95	0.95	0.94	0.93	*dwus*	**0.96**	0.95
Member-Collection	CNN	positional	*us*	**0.96**	0.96	**0.96**	0.94	*us*	**0.96**	0.95
Message-Topic	CNN	positional	*r, q, d, b, u*	**0.94**	0.93	**0.94**	0.90	*rs*	**0.94**	0.93
Cause-Effect	CNN_{msf}	context	*bald*	**0.94**	0.93	0.92	0.90	*QUIRE*	0.93	0.93
Entity-Destination	CNN	context	*r, q, d, b*	**0.94**	0.95	**0.94**	0.93	*dwus*	**0.94**	0.93
Entity-Origin	GRU	context	*rs*	0.92	0.90	0.87	0.86	*QUIRE*	**0.93**	0.89
Component-Whole	CNN	context	*q, r, u*	0.88	0.86	0.86	0.87	*bald*	**0.89**	0.85
Product-Producer	CNN	context	*rs*	**0.88**	0.83	0.84	0.83	*rs*	0.87	0.83
Instrument-Agency	CNN	positional	*dwus*	**0.91**	0.89	0.88	0.86	*bald*	0.88	0.86
causalADEs	GRU_{mp}	positional	*q, r*	**0.80**	0.77	0.78	0.75	*rs*	0.79	0.76

Another observation is that when using the whole available training data, the different active learning strategies tend to converge. On the other hand at the first stages of training some strategies might be "slower" in terms of performance gain. We can observe this in Table 2: after using all available data several AL strategies achieve top performance (as reported in column **Best AL**), while when using only 500 examples (strategies marked in bold in column **Best AL**)

---

[6] On a Linux server with 48 Intel Xeon CPUs @2.20GHz, 231GBs RAM, NVIDIA GeForce GTX 1080 GPU, on *causalADE* task *albl* (the *libact* implementation https://github.com/ntucllab/libact) took 3hrs-10mins, our pruning method took 7 min.

we have less ties. Regarding our **pruning** method we report which AL strategy is selected (column **selection**) after the pruning is completed. It is important to note that this is not equivalent to running the selected strategy alone, because the first stages of training include data selected with various techniques, and this contributes to learning a slightly different model than with a single technique.

5   Conclusions and Future Work

Previous literature on relation extraction has been focusing on improving model performance by either developing new architectures, incorporating additional linguistic features or acquiring additional data. We conjecture that in order to be able to capture any domain specific relation, we need to design models that take into account the effect of the data size and type in addition to the computational cost occurring from training under streamed annotations. To this end, we train neural models with minimal data pre-processing, without using any linguistic knowledge and we propose a novel active learning strategy selection technique. We achieve promising performance on various relation extraction tasks. Moreover, we demonstrate that our method is effective for the rapid generation of train/test data for ambiguous relations and we release a novel dataset for the detection of adverse drug reactions in user generated data. In future work, we will investigate pruning strategies, specifically a hierarchical approach which, given the small amount of data, may result in faster convergence, especially when exploring many AL options.

References

1. Adel, H., Roth, B., Schütze, H.: Comparing convolutional neural networks to traditional models for slot filling. In: NAACL-HLT (2016)
2. Alba, A., Coden, A., Gentile, A.L., Gruhl, D., Ristoski, P., Welch, S.: Language agnostic dictionary extraction. In: ISWC (ISWC-PD-Industry). CEUR Workshop Proceedings, vol. 1963 (2017)
3. Angeli, G., Tibshirani, J., Wu, J., Manning, C.D.: Combining distant and partial supervision for relation extraction. In: EMNLP, pp. 1556–1567 (2014)
4. Arora, S., Liang, Y., Ma, T.: A simple but tough-to-beat baseline for sentence embeddings. In: ICLR (2017)
5. Augenstein, I., Maynard, D., Ciravegna, F.: Distantly supervised web relation extraction for knowledge base population. Semant. Web **7**(4), 335–349 (2016)
6. Bengio, Y.: Curriculum learning. In: ICML (2009)
7. Bunescu, R.C., Mooney, R.J.: A shortest path dependency kernel for relation extraction. In: HLT/EMNLP, pp. 724–731. ACL (2005)
8. Culotta, A., Sorensen, J.: Dependency tree kernels for relation extraction. In: ACL, p. 423. ACL (2004)
9. Donmez, P., Carbonell, J.G., Bennett, P.N.: Dual strategy active learning. In: Kok, J.N., Koronacki, J., Mantaras, R.L., Matwin, S., Mladenič, D., Skowron, A. (eds.) ECML 2007. LNCS (LNAI), vol. 4701, pp. 116–127. Springer, Heidelberg (2007). https://doi.org/10.1007/978-3-540-74958-5_14

10. Fu, L., Grishman, R.: An efficient active learning framework for new relation types. In: IJCNLP, pp. 692–698 (2013)
11. Gal, Y., Islam, R., Ghahramani, Z.: Deep bayesian active learning with image data. In: ICML (2017)
12. Gentile, A.L., Zhang, Z., Augenstein, I., Ciravegna, F.: Unsupervised wrapper induction using linked data. In: K-CAP, pp. 41–48. ACM (2013)
13. Hendrickx, I., Kim, S.N., Kozareva, Z., Nakov, P., Ó Séaghdha, D., Padó, S., Pennacchiotti, M., Romano, L., Szpakowicz, S.: Semeval-2010 task 8: multi-way classification of semantic relations between pairs of nominals. In: DEW Workshop, pp. 94–99. ACL (2009)
14. Hsu, W., Lin, H.: Active learning by learning. In: Bonet, B., Koenig, S. (eds.) AAAI, pp. 2659–2665. AAAI Press (2015)
15. Huang, S.J., Jin, R., Zhou, Z.H.: Active learning by querying informative and representative examples. In: NIPS, pp. 892–900 (2010)
16. Ji, G., Liu, K., He, S., Zhao, J.: Distant supervision for relation extraction with sentence-level attention and entity descriptions. In: AAAI, pp. 3060–3066 (2017)
17. Kingma, D.P., Ba, J.: Adam: a method for stochastic optimization. In: ICLR (2015)
18. Lewis, D.D., Catlett, J.: Heterogeneous uncertainty sampling for supervised learning. In: ICML, pp. 148–156 (1994)
19. Liu, M.X.C.: Semantic relation classification via hierarchical recurrent neural network with attention. In: COLING (2016)
20. Mooney, R.J., Bunescu, R.C.: Subsequence kernels for relation extraction. In: NIPS, pp. 171–178 (2006)
21. Morgan, N., Bourlard, H.: Generalization and parameter estimation in feedforward nets: some experiments. In: NIPS, pp. 630–637 (1990)
22. Nair, V., Hinton, G.E.: Rectified linear units improve restricted boltzmann machines. In: ICML, pp. 807–814 (2010)
23. Nguyen, H.T., Smeulders, A.: Active learning using pre-clustering. In: ICML. ACM (2004)
24. Nguyen, T.H., Grishman, R.: Relation extraction: perspective from convolutional neural networks. In: VS@ HLT-NAACL, pp. 39–48 (2015)
25. Pennington, J., Socher, R., Manning, C.D.: Glove: global vectors for word representation. In: EMNLP, vol. 14, pp. 1532–1543 (2014)
26. Ratner, A.J., Sa, C.D., Wu, S., Selsam, D., Ré, C.: Data programming: creating large training sets, quickly. In: NIPS, pp. 3567–3575 (2016)
27. Roth, B., Barth, T., Wiegand, M., Klakow, D.: A survey of noise reduction methods for distant supervision. In: AKBC, pp. 73–78. ACM (2013)
28. Settles, B.: Active learning literature survey. Univ. Wis. Madison **52**(55–66), 11 (2010)
29. Stanovsky, G., Gruhl, D., Mendes, P.: Recognizing mentions of adverse drug reaction in social media using knowledge-infused recurrent models. In: EACL, pp. 142–151. ACL (2017)
30. Sterckx, L., Demeester, T., Deleu, J., Develder, C.: Using active learning and semantic clustering for noise reduction in distant supervision. In: AKBC at NIPS, pp. 1–6 (2014)
31. Vu, N.T., Adel, H., Gupta, P., et al.: Combining recurrent and convolutional neural networks for relation classification. In: NAACL-HLT, pp. 534–539 (2016)
32. Zelenko, D., Aone, C., Richardella, A.: Kernel methods for relation extraction. J. Mach. Learn. Res. **3**, 1083–1106 (2003)
33. Zeng, D., Liu, K., Lai, S., Zhou, G., Zhao, J., et al.: Relation classification via convolutional deep neural network. In: COLING, pp. 2335–2344 (2014)

34. Zhao, S., Grishman, R.: Extracting relations with integrated information using kernel methods. In: ACL, pp. 419–426. ACL (2005)
35. Zhou, P., Shi, W., Tian, J., Qi, Z., Li, B., Hao, H., Xu, B.: Attention-based bidirectional long short-term memory networks for relation classification. In: ACL - Short Papers, vol. 2, pp. 207–212 (2016)

Incorporating Word Embeddings into Open Directory Project Based Large-Scale Classification

Kang-Min Kim, Aliyeva Dinara, Byung-Ju Choi, and SangKeun Lee[✉]

Korea University, Seoul, Republic of Korea
{kangmin89,dinara_aliyeva,bj1123,yalphy}@korea.ac.kr

Abstract. Recently, implicit representation models, such as embedding or deep learning, have been successfully adopted to text classification task due to their outstanding performance. However, these approaches are limited to small- or moderate-scale text classification. Explicit representation models are often used in a large-scale text classification, like the Open Directory Project (ODP)-based text classification. However, the performance of these models is limited to the associated knowledge bases. In this paper, we incorporate word embeddings into the ODP-based large-scale classification. To this end, we first generate category vectors, which represent the semantics of ODP categories by jointly modeling word embeddings and the ODP-based text classification. We then propose a novel semantic similarity measure, which utilizes the category and word vectors obtained from the joint model. The evaluation results clearly show the efficacy of our methodology in large-scale text classification. The proposed scheme exhibits significant improvements of 10% and 28% in terms of macro-averaging F1-score and precision at k, respectively, over state-of-the-art techniques.

Keywords: Text classification · Word embeddings

1 Introduction

Text classification is the process of determining and assigning topical categories to text. It plays an important role in many web applications, such as contextual advertising [7], topical web search [1], and web search personalization [2]. Usually, text classification requires a sufficiently large taxonomy of topical categories to capture various topics in arbitrary texts. In addition, it is necessary to collect a large amount of training data for each category in the taxonomy.

Many studies have utilized an implicit representation model [14], such as embedding [6,9,10] or a deep neural network [4], which adopts dense semantic encodings and measures semantic similarity accordingly. Implicit representation models have been successfully adopted for text classification task. Such implicit representation models, however, may perform poorly in a large-scale text classification (as we shall show in Sect. 5.4). This is largely attributed to the fact

© Springer International Publishing AG, part of Springer Nature 2018
D. Phung et al. (Eds.): PAKDD 2018, LNAI 10938, pp. 376–388, 2018.
https://doi.org/10.1007/978-3-319-93037-4_30

that the training data for each category is relatively insufficient and distributed unevenly among classification categories. In addition, such approaches are not intuitively interpretable to humans.

In another line of work, many studies have been done with an explicit representation model [14], which uses popular knowledge bases, such as ProBase, Wikipedia, or the Open Directory Project (ODP)[1]. Because the explicit model represents knowledge in terms of vectors that are interpretable to both humans and machines, it is relatively easy for humans to tune and understand it. Another advantage of the explicit representation model is that it enables a large-scale text classification with the direct representation of a large-scale knowledge taxonomy already built-in.

To handle the large-scale text classification, several works [3,7,12] have utilized the ODP, which is a large-scale and taxonomy-structured web directory. These studies have used their explicit representation of text to represent ODP knowledge, based on bag-of-words [3,7] or bag-of-phrases [12] to develop ODP-based text classification techniques. They showed that ODP-based text classification techniques are effective at the large-scale text classification. The performance of previous ODP-based text classification, however, is limited to ODP and/or Wikipedia knowledge bases.

To alleviate the limitation of ODP-based text classification, we incorporate word embeddings into the ODP-based text classification. To this end, we propose two novel joint models of ODP-based classification and word2vec, a representative word embeddings technique. The joint models seek to project both words and ODP categories into the same vector space. Therefore, category vectors of ODP categories successfully identify words learned from external knowledge. In addition, we effectively measure the semantic relatedness between an ODP category and a document by utilizing both category and word vectors. In summary, our contributions are three-fold:

- We propose a novel methodology to handle the large-scale text classification, which utilizes both the explicit and implicit representation.
- We develop two novel joint models of ODP-based classification and word2vec to generate category vectors that represent the semantics of ODP categories. In addition, we develop a new semantic similarity measure that utilizes both the category and word vectors.
- We demonstrate the efficacy of the proposed methodology through extensive experiments on real-world datasets. The performance evaluation clearly shows that our approach significantly outperforms the state-of-the-art techniques in terms of macro-averaging F1-score and precision at k.

The remainder of this paper is organized as follows. We briefly describe the ODP-based knowledge representation and word2vec in Sect. 2. Section 3 describes the joint models of ODP-based classification and word2vec to generate category vectors. Section 4 details the similarity measure between a category and document. We present the performance evaluation results in Sect. 5. We discuss related research and conclude this work in Sects. 6 and 7, respectively.

---

[1] http://www.curlie.org.

2 Preliminary

2.1 ODP-Based Knowledge Representation

We employ the ODP-based text classification scheme [7] as our explicit representation model. To compute the centroid $\overrightarrow{\mu_i}$ of category c_i, we calculate the averaged term vector of all ODP documents as:

$$\overrightarrow{\mu_i} = \frac{1}{\|D_{c_i}\|} \sum_{d \in D_{c_i}} \overrightarrow{d} \tag{1}$$

where D_{c_i} is a set of ODP documents in c_i, and $\overrightarrow{d}$ is a weighted vector represented as a *tf-idf* value. Due to the large-scale taxonomy structure of the ODP, however, each ODP category contains a different number of documents, sometimes resulting in sparsity or unavailability of training documents in a category. This issue is addressed in the works [3,7], in which they merge the centroid $\overrightarrow{\mu_i}$ of the descendant categories to build a classifier. As a result, this approach outperforms all other ODP-based text classifiers, and exhibits a stable performance in large-scale text classification [3,7]. Therefore, we utilize this approach to compute the centroid $\overrightarrow{\mu_i}$ of category c_i.

Table 1. Example of ODP-based representation. A document d, *"Trump became prez"*, to be classified, and a category c_1, *Society/Government/President* are considered.

Vector	Term weights				
	Trump	President	Prez	Government	...
Term vector of d	0.67	0	0.51	0	...
Centroid vector of c_1	0.10	0.44	0.05	0.31	...

For example, as shown in Table 1, the category c_1, *Society/Government/President* is explicitly represented by the centroid vector. Given a document d, however, *"Trump became prez"*, the ODP-based classification may not be able to classify the document d as the category c_1. This is because, this approach cannot capture the semantic relations between words (e.g., *prez* and *president*).

2.2 Word2Vec

To complement the ODP-based classification, we adopt the word2vec [9,10], a popular word embeddings technique. In word2vec, each word vector is trained using a shallow neural networks language model, such as Skip-gram [9]. Skip-gram aims to predict context words given a target word in a sliding window. Mathematically, given a sequence of training words $w_1, w_2, w_3..., w_T$, the objective of Skip-gram is to maximize the following average log probability:

$$\frac{1}{T} \sum_{i=1}^{T} \sum_{c=i-k, c \neq t}^{i+k} log\, p(w_c|w_t) \tag{2}$$

where k is the size of the context window centered at the target word, and w_t and w_c are the target and context words, respectively.

Trained word vectors with similar semantic meanings would be located at high proximity within the vector space. In addition, word vectors can be composed by an element-wise addition of their vector representations, e.g., *Russian + river = Volga River*. This property of the vectors is called "additive compositionality" [10].

3 Joint Models of Explicit and Implicit Representation

In this section, we describe two joint models of ODP-based text classification and word2vec. These joint models generate category vectors, which represent the semantics of ODP categories. Each category vector not only semantically encodes the explicitly expressed ODP category, but also understands semantically related words that do not appear in the ODP knowledge base. This is because they are projected into the same semantic space as word vectors learned in an additional volume of knowledge outside the ODP.

3.1 Generating Category Vector with Algebraic Operation

Given the centroid vector of an ODP category and word vectors of the pre-trained word2vec model, our first approach generates the category vector by using the vector scalar multiplication and vector addition methods, as follows.

First, we multiply the term weights of each word in the ODP category by each word vector of the words. Second, the weighted word vectors are composed as a category vector using element-wise addition. This type of vector algebra is quite simple, but it can also clearly represent the semantics of an ODP category. This is because word vectors are not only multiplied by a precisely trained term weight from the centroid vector, but also have additive compositionality. The logic for generating the category vector of the ODP category is as follows:

$$\overrightarrow{C_i} = \sum_{w \in W_i} \overrightarrow{\mu_i}(w) \cdot \overrightarrow{w} \tag{3}$$

where $\overrightarrow{C_i}$ is the category vector of c_i, W_i is the set of words of c_i, $\overrightarrow{w}$ is the word vector (obtained from the pre-trained word2vec model) of word w, and $\overrightarrow{\mu_i}(w)$ is the term weight of w in c_i. For example, in Fig. 1(a), the word vectors of *president*, *government*, and *trump* are multiplied by 0.44, 0.31, and 0.10, respectively, then the weighted word vectors are added. Finally, we obtain the category vector of the category *Society/Government/President*. Vector representations of documents to be classified are generated in the same manner.

3.2 Generating Category Vector with Embedding

Our second approach extends word2vec to represent category vectors, instead of using the pre-trained word2vec model to compose word vectors in ODP categories. We first assign appropriate ODP categories for each word in a text corpus.

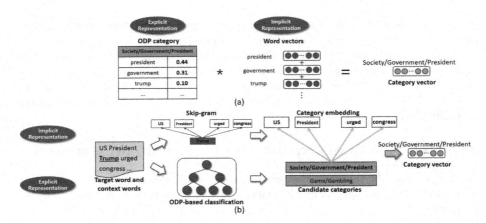

Fig. 1. Illustration of category vector generation with algebraic operation (a) and embedding (b)

Then, we train the category vectors of the assigned ODP categories by applying a modified Skip-gram model. The category vector of an ODP category is expected to represent the collective semantics of words under this category.

The process of generating category vectors with embedding is as follows. First, we identify candidate ODP categories for the target word. If an ODP category is largely associated with the target word, the ODP-based text classification selects this category as a candidate. The ODP-based text classification determines the degree of association by using the term weight of the target word in each ODP category. For example, when *Trump* is the target word, the ODP-based classification identifies categories such as *Game/Gambling* and *Society/Government/President*, as shown in Fig. 1(b). We then select the most appropriate ODP category in the current context by using the ODP-based text classification. For example, when the context is *"US President Trump urged congress"*, the most appropriate category is *Society/Government/President*. Finally, we apply the modified Skip-gram algorithm, which trains the category vector corresponding to the most appropriate category. The objective of category embedding is to maximize the following average log probability:

$$\frac{1}{T}\sum_{i=1}^{T}\sum_{c=i-k,c\neq t}^{i+k} log\, p(w_c|w_t)p(w_c|c_t) \tag{4}$$

Unlike the Skip-gram model, where the target word w_t is used only to predict context words, the category embedding model also uses the ODP category c_t of the target word to predict context words.

4 Semantic Similarity Measure

We develop a novel semantic similarity measure, on the basis of category and word vectors, which captures both the semantic relations between words and the semantics of ODP categories.

4.1 Using Word-Level Semantics

First, we propose a semantic similarity measure that considers word-level semantics by using only the word vectors. The word vectors can be used to calculate the semantic relatedness between two words. The key idea of this measure is to align words with similar meanings in a category and document, although the words represented in this category and document are different.

Before describing the proposed measure, we explain how to compute the similarity between category c_i and document d by means of the existing ODP-based text classification as follows:

$$\cos(c_i,\, d) = \frac{\sum_{j=1}^{n_{c_i}} \sum_{k=1}^{n_d} \delta(w_j - w_k) \cdot \overrightarrow{\mu_i}(w_j) \cdot \overrightarrow{d}(w_k)}{\|\overrightarrow{\mu_i}\| \cdot \|\overrightarrow{d}\|} \tag{5}$$

where w_j and w_k are non-zero terms in centroid vector $\overrightarrow{\mu_i}$ of c_i and term vector $\overrightarrow{d}$, respectively, while n_{c_i} and n_d are the number of non-zero terms in $\overrightarrow{\mu_i}$ and $\overrightarrow{d}$, respectively. $\delta(\cdot)$ is the Dirac function defined by $\delta(0) = 1$ and $\delta(other) = 0$ [13].

The cosine similarity between the centroid vector of category and the term vector of document could increase when w_j and w_k are equal. However, in Table 1, we observe that *prez* has a very similar meaning to *president*, which is a very important word in the category *Society/Government/President*. Therefore, we propose a new measure that increases the similarity between proper $\overrightarrow{\mu_i}$ and $\overrightarrow{d}$ by utilizing word2vec. By substituting the Dirac function $\delta(\cdot)$ with the word similarity $\phi(\cdot)$, it is possible to consider semantic relatedness between two words and calculate the weight more densely:

$$\text{sim}(c_i,\, d) = \frac{\sum_{j=1}^{n_{c_i}} \sum_{k=1}^{n_d} \phi(w_j, w_k) \cdot \overrightarrow{\mu_i}(w_j) \cdot \overrightarrow{d}(w_k)}{\|\overrightarrow{\mu_i}\| \cdot \|\overrightarrow{d}\|} \tag{6}$$

where $\phi(\cdot)$ is the word similarity function. Given two words w_j and w_k, we define the word similarity function $\phi(w_j, w_k)$ in Eq. (6) as follows:

$$\phi(w_j, w_k) = \begin{cases} cos(\overrightarrow{w_j}, \overrightarrow{w_k}) & \text{if } cos(\overrightarrow{w_j}, \overrightarrow{w_k}) > \theta, \\ 0 & \text{otherwise} \end{cases} \tag{7}$$

where $\overrightarrow{w_j}$ and $\overrightarrow{w_k}$ are the word vectors of w_j and w_k, $cos(\overrightarrow{w_j}, \overrightarrow{w_k})$ is the cosine similarity between $\overrightarrow{w_j}$ and $\overrightarrow{w_k}$, and θ is a threshold, which is empirically set to 0.6 in our analysis. The similarity between $\overrightarrow{\mu_i}$ and $\overrightarrow{d}$ increases not only when

w_j and w_k are equal, but also they have similar semantics. For example, *prez* and *president* have highly similar semantics in Table 1. The semantic similarity using word-level semantics, thus, is additionally computed by $0.51 \times 0.44 \times \phi(prez, president)$, unlike the original cosine similarity.

4.2   Using Category- and Word-Level Semantics

In this paper, we develop a robust similarity measure by utilizing both the category and word vectors. A category vector is utilized as a pseudo word in the process of computing semantic similarity. We define the measure as follows:

$$\text{sim}'(c_i, d) = \frac{\sum_{j=1}^{n_{c_i}+1} \sum_{k=1}^{n_d+1} \phi(w_j, w_k) \cdot \overrightarrow{\mu_i}(w_j) \cdot \overrightarrow{d}(w_k)}{\|\overrightarrow{\mu_i}\| \cdot \|\overrightarrow{d}\|} \tag{8}$$

In Eq. (8), the category vector is inserted into the corresponding category as the $(n_{c_i} + 1)^{th}$ word. This is motivated by the fact that category vectors share the same semantic space with word vectors. Similarly, the document vector is inserted into the corresponding document as the $(n_d+1)^{th}$ word. We will examine how to determine the weight (i.e., pseudo term weight) α of the category vector through many parameter experiments in Sect. 5.4.

5   Experiments

5.1   Datasets

Training Datasets. We use the RDF dump from the original ODP dataset released on January 8, 2017, which contains 802,379 categories and 3,624,444 webpages. To obtain a well-organized ODP taxonomy, we apply heuristic rules suggested in [7] and build our own taxonomy with 2,735 categories. Thus, the final training dataset used in our experiments consists of 52,046 webpages. To construct the moderate-scale classification dataset, we use only 13 top-level categories from the ODP taxonomy by excluding two categories, *Top/News* and *Top/Adult*, which contain fewer than 100 webpages. Thus, the training dataset used in the moderate-scale classification consists of 51,856 webpages.

In addition to the ODP dataset, we train our category embedding model and word2vec model on the "One Billion Word Language Modeling Benchmark" dataset released by Google[2]. The word and category vectors are 300-dimensional, while the window size is set to 5 with 15 negative samples.

Test Datasets. We build two test datasets, ODP and NYT, to evaluate our methodology. The ODP test dataset consists of webpages collected from the original ODP. The webpages in each category are randomly divided into a training set and a test set at a ratio of seven to three. In particular, we build two kinds of ODP test datasets. In the large-scale classification task, we collect 24,121

---

[2] https://code.google.com/archive/p/word2vec/.

webpages from 2,735 ODP categories in our taxonomy, while collecting 24,046 webpages from 13 ODP categories in the moderate-scale classification task. In addition to the ODP test datasets, we select six categories related to the New York Times: *art*, *business*, *food*, *health*, *politics*, and *sports*, as the source for our second test dataset. We randomly collect 20 news articles from each of these categories. Table 2 shows the statistics of datasets.

Table 2. Statistics of datasets

		Training dataset	Test dataset
ODP (large-scale/moderate-scale)	No. Categories	2,735/13	2,735/13
	No. Webpages	52,046/51,856	24,121/24,046
NYT	No. Articles	-	120

5.2 Evaluation Metrics

For the ODP test dataset, we use the macro-averaging precision, recall, and F_1-score [15] as the classification performance metric. We adopt the macro-averaging, which assigns equal weights to each category instead of each test document, because the distribution of the ODP training dataset is highly skewed [3,7]. For the NYT test dataset, we use precision at k. Three participants manually assess the top-k ODP categories obtained by text classifiers in three scales: relevant, somewhat relevant, and not relevant.

5.3 Experimental Setup

We evaluate the performance of six methods. We adopt the ODP-based text classifier for our experiments. Other baselines include the paragraph vector [6] and convolutional neural networks-based text classifier [4], which are state-of-the-art methods on multi-class text classification. In our experiments, we compare the following methods:

- *ODP* (baseline): This is the ODP-based text classification only [7].
- *PV* (baseline): This is the text classification method using paragraph vectors [6]. The learned vector representations have 1000 dimensions. We represent ODP categories by averaging the document embeddings for each document in a category.
- *CNN* (baseline): This is the convolutional neural networks-based text classifier [4]. The dimension of word embedding is 300, and the number of filters for the CNN is 900. We use the ReLU for nonlinearity. Optimization is performed using SGD with a mini-batch size of 64.
- ODP_{CV}: This is our proposed text classification method using category vectors, which are generated by the joint model of ODP-based text classification and word2vec.

- ODP_{WV}: This is our proposed ODP-based text classification combined with the similarity measure of word-level semantics.
- ODP_{CV+WV}: This is our proposed ODP-based text classification combined with the similarity measure of both category- and word-level semantics.

5.4   Experimental Results

We first compare the two methods to generate category vectors with the ODP dataset (2,735 categories). In Table 3, $ODP_{CV(Algebra)}$ denotes the text classification utilizing the category vector generated by algebraic operations, while $ODP_{CV(Embedding)}$ denotes the text classification utilizing the category vector generated by embedding. Unexpectedly, we observe that a simple $ODP_{CV(Algebra)}$ clearly outperforms a relatively elaborate $ODP_{CV(Embedding)}$. Thus, we adopt $ODP_{CV(Algebra)}$ in the remaining experiments, which is simply denoted by ODP_{CV}.

Table 3. Comparison of category vector generations on the ODP dataset (2,735 categories)

	Precision	Recall	F1-score
$ODP_{CV(Algebra)}$	**0.449**	**0.458**	**0.453**
$ODP_{CV(Embedding)}$	0.278	0.195	0.230

Next, we perform a parameter setting to determine the term weight α of a category vector as a pseudo word. Figure 2 shows the classification performance obtained by ODP_{CV+WV} based on different α values. We find that the curve reaches a peak at $\alpha = 0.9$. This result shows that the category vector plays a major role in the performance of ODP_{CV+WV}. However, we observe that when the weight of category vector is 1.0, performance drops sharply. This means that the word overlap feature is still helpful. In the remaining experiments, α is set to 0.9 for ODP_{CV+WV}.

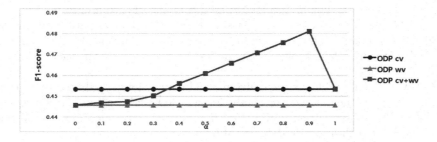

Fig. 2. Classification performance based on different α values

Table 4(a) summarizes the experimental results for text classification on the ODP test dataset with 2,735 target classes. We observe that ODP_{CV+WV} outperforms all the other proposed methods, as well as the baselines. ODP_{CV+WV} performs better than ODP over 9%, 12%, and 10% on average in terms of precision, recall, and F1-score, respectively. Our experimental results show that PV [6] performs worse than ODP. In addition, it turns out that CNN [4] performs the worst among the six methods. This can be explained by the fact the distribution of webpages is skewed toward a few categories in the original ODP [7]. Actually, we observe that 73% of ODP categories contain fewer than five webpages. We also compare the performance of CNN with the ODP baseline on the ODP test dataset with 13 target categories. From Table 4(b), we observe that CNN exhibits a better performance than ODP in the moderate-scale text classification. From Table 4, we confirm that CNN is indeed limited to the moderate-scale text classification.

Table 4. Classification performance on the ODP dataset.

(a) large-scale (2,735 categories)

	Precision	Recall	F1-score
ODP [7]	0.431	0.440	0.436
PV [6]	0.331	0.398	0.361
CNN [4]	0.402	0.232	0.294
ODP_{CV}	0.449	0.458	0.453
ODP_{WV}	0.451	0.440	0.446
ODP_{CV+WV}	**0.468**	**0.494**	**0.481**

(b) moderate-scale (13 categories)

	Precision	Recall	F1-score
ODP [7]	0.667	**0.707**	0.687
CNN [4]	**0.736**	0.661	**0.696**

Table 5 shows the evaluation results on the NYT test dataset. Again, the performance of ODP_{CV+WV} outperforms ODP, PV, CNN, ODP_{CV} and ODP_{WV} over 28%, 119%, 216%, 12%, and 10% in terms of precision at k on average, respectively. We also observe that both ODP_{CV} and ODP_{WV} outperform ODP.

Table 5. Classification performance on the NYT dataset (2,735 categories).

Precision at k					
k	1	2	3	4	5
ODP [7]	0.575	0.496	0.450	0.421	0.403
PV [6]	0.317	0.292	0.261	0.250	0.245
CNN [4]	0.242	0.200	0.186	0.165	0.155
ODP_{CV}	0.583	0.550	0.536	0.510	0.493
ODP_{WV}	0.600	0.583	0.547	0.508	0.482
ODP_{CV+WV}	**0.692**	**0.617**	**0.583**	**0.556**	**0.545**

These results clearly demonstrate that both category and word vectors are effective at text classification. Specifically, ODP_{CV+WV}, which utilizes both category and word vectors, achieves the best performance in all experiments. We also perform the t-test for the classification results, and find that ODP_{CV+WV} results are statistically significant with $p < 0.01$.

5.5   Analysis

We also qualitatively examine the meaning of category vectors to analyze why adding category vectors improves the performance of ODP-based text classification. From Table 6, we observe that the category vector expresses the meaning of category quite well. First, from the parent category *Home/Cooking/Baking_and_Confections* and child category *Home/Cooking/Baking_and_Confections/Breads*, we observe that their category vectors share the core semantically rich words (e.g., Recipe, Baking, Cookies), while they have their own unique semantically rich words (e.g., Dessert, Bread). These observations imply that the category vector actually understands the semantics better than the centroid vector.

Interestingly, we also observe that the category vector identifies semantically related words that do *not* appear in the ODP knowledge base (e.g., Henin, a Belgian former professional tennis player, in the category *Sports/Tennis/Players*). Thus, category vectors combined with the ODP-based classification successfully enable us to improve the performance of text classification.

Table 6. Nearest words of category vector (Explicit + Implicit) and highly weighted words in centroid vector (Explicit) of ODP categories

Category	Nearest words of category vector (Explicit + Implicit)	Highly weighted words in centroid vector (Explicit)
Home/Cooking/ Baking_and_Confections	Recipe, Baking, Cookies, Cake Dessert, Cupcake, Bake, ...	Recipe, Baking, Cookies, Cake Bake, Pastries, Bread, Mix, ...
Home/Cooking/ Baking_and_Confections/Breads	Bread, Recipe, Baking, Flour, Biscuit, Cookies, Pancake, ...	Bread, Recipe, Sourdough, Baking, Yeast, Quick, ...
Sports/Tennis/Players	Tennis, Wimbledon, Nadal, Henin, Federer, Sharapova, ...	Tennis, Wimbledon, Winners, Players, Detailed, Seed, ...

6   Related Work

For the large-scale text classification, many approaches have been developed to handle data sparsity on a knowledge base. Data sparsity on a hierarchical taxonomy was firstly addressed in [8]. This work applied a statistical technique to estimate the parameters of data-sparse child categories with their data-rich ancestor categories. In [3,7], they proposed the merge-centroid (MC) classification that utilizes enriched training data for each category based on webpages classified into their ancestor and/or descendants in the ODP. In another line of work [12], they enriched semantic information in the ODP by incorporating another knowledge base, Wikipedia.

A simple convolutional neural network approach [4] has been proven to be an effective text classifier. Still, it exhibits limitations in the large-scale text classification, which is verified in our analysis. A few work [5,11] has recently studied large-scale multi-label text classification using deep neural networks. However, they do not utilize the explicit representation model built from knowledge base. To the best of our knowledge, our current work is one of only a few works that utilizes both the explicit and implicit knowledge representation, which enables us to perform the large-scale text classification quite well.

7   Conclusion

In this paper, we have proposed novel joint models of the explicit and implicit representation techniques to handle the large-scale text classification. Specifically, we have incorporated the word2vec model into the ODP-based classification framework. Our approach involves two tasks. First, we generate category vectors, which represent the semantics of ODP categories. Second, we develop a new semantic similarity measure that utilizes both category and word vectors. We have verified the large-scale classification performance of our methodology using real-world datasets. The experimental results confirm that our scheme significantly outperforms baseline methods. We plan to apply our methodology to different applications, including contextual and mobile advertising.

Acknowledgment. This research was supported by Basic Science Research Program through the National Research Foundation of Korea (NRF) funded by the Ministry of Science, ICT (number 2015R1A2A1A10052665).

References

1. Broder, A., Fontoura, M., Gabrilovich, E., Joshi, A., Josifovski, V., Zhang, T.: Robust classification of rare queries using web knowledge. In: SIGIR, pp. 231–238 (2007)
2. Chirita, P.A., Nejdl, W., Paiu, R., Kohlschütter, C.: Using ODP metadata to personalize search. In: SIGIR, pp. 178–185 (2005)
3. Ha, J., Lee, J.H., Jang, W.J., Lee, Y.K., Lee, S.: Toward robust classification using the open directory project. In: DSAA, pp. 607–612 (2014)
4. Kim, Y.: Convolutional neural networks for sentence classification. In: EMNLP, pp. 1746–1751 (2014)
5. Kurata, G., Xiang, B., Zhou, B.: Improved neural network-based multi-label classification with better initialization leveraging label co-occurrence. In: NAACL, pp. 521–526 (2016)
6. Le, Q.V., Mikolov, T.: Distributed representations of sentences and documents. In: ICML, pp. 1188–1196 (2014)
7. Lee, J.H., Ha, J., Jung, J.Y., Lee, S.: Semantic contextual advertising based on the open directory project. ACM Trans. Web **7**(4), 24:1–24:22 (2013)

8. McCallum, A., Rosenfeld, R., Mitchell, T.M., Ng, A.Y.: Improving text classification by shrinkage in a hierarchy of classes. In: ICML, pp. 359–367 (1998)
9. Mikolov, T., Chen, K., Corrado, G., Dean, J.: Efficient estimation of word representations in vector space. In: ICLR (Workshop) (2013)
10. Mikolov, T., Sutskever, I., Chen, K., Corrado, G.S., Dean, J.: Distributed representations of words and phrases and their compositionality. In: NIPS, pp. 3111–3119 (2013)
11. Nam, J., Kim, J., Loza Mencía, E., Gurevych, I., Fürnkranz, J.: Large-scale multi-label text classification — revisiting neural networks. In: Calders, T., Esposito, F., Hüllermeier, E., Meo, R. (eds.) ECML PKDD 2014. LNCS (LNAI), vol. 8725, pp. 437–452. Springer, Heidelberg (2014). https://doi.org/10.1007/978-3-662-44851-9_28
12. Shin, H., Lee, G., Ryu, W.J., Lee, S.: Utilizing wikipedia knowledge in open directory project-based text classification. In: SAC, pp. 309–314 (2017)
13. Song, Y., Roth, D.: Unsupervised sparse vector densification for short text similarity. In: NAACL, pp. 1275–1280 (2015)
14. Wang, Z., Wang, H.: Understanding short texts. In: ACL (Tutorial) (2016)
15. Yang, Y.: An evaluation of statistical approaches to text categorization. Inf. Retr. 1(1), 69–90 (1999)

Inference of a Concise Regular Expression Considering Interleaving from XML Documents

Xiaolan Zhang[1,2], Yeting Li[1,2], Fanlin Cui[1,2], Chunmei Dong[1,2],
and Haiming Chen[1(✉)]

[1] State Key Laboratory of Computer Science, Institute of Software,
Chinese Academy of Sciences, Beijing 100190, China
{zhangxl,liyt,cuifl,dongcm,chm}@ios.ac.cn
[2] University of Chinese Academy of Sciences, Beijing, China

Abstract. XML schemas are useful in various applications. However, many XML documents in practice are not accompanied by a schema or by a valid schema. Therefore, it is essential to design efficient algorithms for schema learning. Each element in XML schema has its content model defined by a regular expression. Schema learning can be reduced to the inference of restricted regular expressions. In this paper, we focus on learning restricted regular expressions with interleaving from a set of XML documents. The new subclass is named as *CHAin Regular Expression with Interleaving (ICHARE)*. Then based on single occurrence automaton (SOA) and maximum independent set (MIS), we introduce an inference algorithm *GenICHARE*. The algorithm is proved to infer a descriptive *ICHARE* from a set of given sample. At last, based on the data set crawled from the Web, we compare the coverage proportion of *ICHARE* compared with other existing subclasses. Besides, we analyze the conciseness of regular expressions inferred by *GenICHARE* based on *DBLP*. Experimental results show that *ICHARE* is more concise and useful in practice, and the inference algorithm is promising and effective.

Keywords: XML · Regular expression · Interleaving
Language learning · Algorithms

1 Introduction

As a main data exchange format, eXtensible Markup Language (XML) has been widely used in various applications on the Web [1]. XML schemas are convenient for data processing, automatic data integration, static analysis of transformations and so on [16,18,19]. REgular LAnguage for XML Next Generation (Relax

H. Chen—Work supported by the National Natural Science Foundation of China under Grant Nos. 61472405.

© Springer International Publishing AG, part of Springer Nature 2018
D. Phung et al. (Eds.): PAKDD 2018, LNAI 10938, pp. 389–401, 2018.
https://doi.org/10.1007/978-3-319-93037-4_31

NG) is a simple schema language recommended by OASIS[1]. However, many XML documents are not accompanied by a (valid) schema in practice. A survey [15] showed that XML documents on the Web with corresponding schema definitions only accounted for 24.8% in 2013, with the proportion of 8.9% for valid ones. Therefore, it is essential to devise algorithms for schema inference. And schema inference can be reduced to learning restricted regular expressions from a set of given sample [4,6,12].

Gold [14] proposed a classical language learning model (*learning in the limit or explanatory learning*) and proved that the class of regular expressions can not be identifiable from finite positive data only, even for the class of deterministic regular expressions (proved by Bex et al. [2]). So researchers have turned to the study of restricted subclasses of regular expressions. Different from the emphasis of an exact representation of the target language in Gold-style learning, Freydenberger proposed another learning model called *descriptive generalization* in [10]. For a class $\mathcal{D}$ of language representation mechanisms (e.g., a class of automata, regular expressions, or grammars), a representation $\alpha \in \mathcal{D}$ is called $\mathcal{D}$-descriptive for the sample S if $\mathcal{L}(\alpha)$ is an inclusion-minimal generalization of S. It means that there is no $\beta \in \mathcal{D}$ such that $S \subseteq \mathcal{L}(\beta) \subset \mathcal{L}(\alpha)$.

Popular subclasses such as SORE [4], Simplified CHARE [4] (originally was called CHARE), eSimplified CHARE [9], CHARE [3] (originally was called simple regular expressions), eCHARE [20] were discussed in detail in [17]. They are all based on standard regular expressions. In data-centric applications, there may be no order constraint among siblings [1]. But the relative order within siblings may be still important. In such cases the interleaving is necessary. However, there is only a few work which support interleaving. In [7], Ciucanu and Staworko studied the unordered XML documents and proposed two unordered schema formalisms: *disjunctive multiplicity expressions* (DME) and *disjunction-free multiplicity expressions* (ME). However, they both do not support the concatenation operator. Peng [21] proposed a subclass *SIRE*, which supports the concatenation but not the union operator. Ghelli et al. [13] introduced a grammar considering interleaving: $T ::= \varepsilon |a^{[m,n]}| T + T | T \cdot T | T \& T$ where $m \in N \backslash \{0\}$ and $n \in N \backslash \{0\} \cup \{*\}$. This grammar requires each terminal symbol appear at most once and counter can only occur as a constraint for terminal symbols.

As for inference algorithms for subclasses, Bex et al. [4,5] gave two algorithms *RWR* and *CRX* for SORE and Simplified CHARE. However, regular expressions inferred by these two algorithms are not descriptive generalized. Freydenberger et al. [10] proposed algorithms *Soa2Chare* and *Soa2Sore* for Simplified CHARE and SORE respectively which satisfy *descriptive generalization*. Ciucanu and Staworko introduced an algorithm for *DME* based on *max clique* [7]. Peng et al. [21] developed an approximate and heuristic solution to infer an approximate descriptive generalized *SIRE*.

In present paper, we propose a new subclass of restricted regular expressions called *CHAin regular expression with interleaving (ICHARE)* and give an inference algorithm *GenICHARE* to infer a descriptive generalized *ICHARE* based

---

[1] https://www.oasis-open.org/standards.

on *single occurrence automaton* (SOA) and *maximum independent set* (MIS). First, an SOA is constructed for given sample S. Then each non-trival strongly connected component is renamed by the return value of *Repair()*. Next, assign level numbers for the new graph. Finally, all nodes of each level will be converted to one or more chain factors.

The main contributions of this paper are as follows.

1. We propose a subclass of restricted regular expressions: CHAin regular expression with interleaving (*ICHARE*).
2. We design an inference algorithm *GenICHARE* to infer descriptive generalized *ICHAREs*.
3. A series of experiments were conducted. We compare the coverage proportion of *ICHARE* with existing subclasses. Based on the data set *DBLP*, we analyze the conciseness of regular expressions inferred by *GenICHARE* compared with other methods. The experimental results show that *ICHARE* is more useful and concise in real-world applications.

The paper is organized as follows. Section 2 is the definitions. The inference algorithm *GenICHARE* is introduced in Sect. 3. Experiments are given in Sect. 4. Section 5 is the conclusions.

2 Preliminaries

Definition 1 *Regular Expression with interleaving.* *Let Σ be a finite alphabet. Σ^* is the set of all strings over Σ. A regular expression with interleaving is inductively defined as follows: ε or $a \in \Sigma$ is a regular expression. For any regular expressions E_1 and E_2, the disjunction $E_1|E_2$, the concatenation $E_1 \cdot E_2$, the interleaving $E_1 \& E_2$, or the Kleene-Star E_1^* is also a regular expression. $E^?$ and E^+ are used as abbreviations of $E|\varepsilon$ and EE^*, respectively. $L(E)$ is the language generated by the regular expression E. We use $sym(E)$ to denote the set of terminal symbols occurred in E and $term(v)$ to denote all symbols of the form a or a^+ where $a \in \Sigma$.*

Let $u = au'$ and $v = bv'$ where $a, b \in \Sigma$ and $u', v' \in \Sigma^*$, then $u \& v = a \cdot (u' \& v) \cup b \cdot (u \& v')$. For example, strings accepted by $(abc)\&(d)$ is the set $\{abcd, dabc, adbc, abdc\}$. Regular expressions with interleaving, in which each terminal symbol occurs at most once is called *ISORE* extended from *SORE* in [4]

Definition 2 *Single Occurrence Automaton (SOA)* [5]. *Let Σ be a finite alphabet. src, snk do not occur in Σ. A single occurrence automaton over Σ is a finite directed graph $G = (V, E)$ which satisfies the following two conditions.*

1. *$src, snk \in V$, and $V \subseteq \Sigma \cup \{src, snk\}$. All nodes in G are distinct.*
2. *src is the starting node which only has the outgoing edges while the accepting node snk only has the incoming edges. Each node in the set of $\{V \setminus \{src, snk\}\}$ lies on a path from src to snk.*

A *generalized* SOA over Σ is a finite directed graph in which each node $v \in V \setminus \{src, snk\}$ is an ISORE and all nodes are pairwise alphabet-disjoint ISOREs.

level number (ln) and skip level (sl) [10]. For a given directed acyclic graph $G = (V, E)$ of a SOA, the level number of node v is the longest path from src denoted by $ln(v)$. $ln(src)$ is 0. If there exists a path $v_1 \to_G v_2$ such that $ln(v_1) < ln(v)$ and $ln(v) < ln(v_2)$, then $ln(v)$ is a skip level.

Independent Set (IS). An independent set is a set S of nodes in a graph G, in which there is no edge between every two nodes in S. A *maximum independent set (MIS)* is an independent set with largest possible size for a given graph G where the size means the number of nodes in an *IS*.

Let $POR(S)$ denote the set of all partial order relations of each string in S. Then the Constraint Set (CS) and Non-Constraint Set (NCS) for a set of given sample S are computed as follows.

1. $CS(S) = \{<a_i, a_j> \mid <a_i, a_j> \in POR(S), \text{ and } <a_j, a_i> \in POR(S)\}$;
2. $NCS(S) = \{<a_i, a_j> \mid <a_i, a_j> \in POR(S), \text{ but } <a_j, a_i> \notin POR(S)\}$.

Clearly, for S, its Constraint Set and Non-constraint Set are both unique. If $CS(S_1) \neq CS(S_2)$ (or $NCS(S_1) \neq NCS(S_2)$), then $S_1 \neq S_2$.

Definition 3 PSubstring (P, s). *PSubstring (P, s) is a function in which P is a finite set of symbols and s is a string. It returns a new string each symbol of which is computed as follows: $\pi_s(P, s_i) = s_i$ if $s_i \in P$; $\pi_s(P, s_i) = \varepsilon$ otherwise.*

For example, let $P = \{b, c, r\}$ and $s = ebbdfc$, $PSubstring(P, s) = bbc$.

Definition 4 CHAin Regular Expression with Interleaving (ICHARE). *Let Σ be a finite alphabet. An ICHARE over Σ is a ISORE. It is of the form $(f_1 f_2 \cdots f_n)$ where $n \geq 1$. Each factor f_i has two possible forms. One is $(b_1|b_2| \cdots |b_m)$, $(b_1|b_2| \cdots |b_m)^?$, $(b_1|b_2| \cdots |b_m)^*$ or $(b_1|b_2| \cdots |b_m)^+$ where $m \geq 1$, b_i can be a or a^+ $(a \in \Sigma)$. The other is $s_1^{c_1} \& s_2^{c_2} \& \cdots \& s_t^{c_t}$ where $t \geq 2$, $s_i \in \Sigma^*$ and $c_i \in \{1, ?, *, +\}$.*

For example, $E = (a_1|a_2^+)((a_3^? a_4) \& (a_5 a_6))$ is an *ICHARE*.

3 Inference Algorithm GenICHARE

For a set of given sample S, *GenICHARE* can infer a descriptive generalized *ICHARE* α such that $S \subseteq L(\alpha)$. $sym(A)$ is the set of terminal symbols occur in A. all_mis means the set of all maximum independent sets iteratively obtained from a graph G. The main procedures are illustrated as follows.

1. Construct the graph G $(V, E) = SOA(S)$ for S.
2. For each node v with a self-loop, remove the self-loop and rename it with v^+. For each non-trival strongly connected component (NTSCC), replace it with the return value of *Repair()*. Relations with any node in NTSCC rebuild the relation with the new node.

3. Compute the level numbers for the new graph and find all skip levels.
4. Nodes of each level number ln are turned into one or more chain factors. If there are more than one non-letter node (node with more than one terminal symbols), or if ln is a skip level, then ? is appended to every chain factor on ln.

Now is the pseudo-code inference algorithm $GenICHARE$. $Repair()$ is the subfunction which is to transform each NTSCC into a regular expression with interleaving. $setln(G)$ is to assign level number to each node. $issl(ln)$ returns $true$ if ln is a skip level and returns $false$ otherwise. For a given SOA $G = (V, E)$, the number of NTSCCs, the set of all_mis and level numbers are all finite. Then $Repair()$ will stop after finite steps. Therefore, Algorithm 1 will also stop in finite steps.

Algorithm 1. GenICHARE(S)

Input: A set of given sample S
Output: An $ICHARE$ R
1: Construct $G(V, E) = SOA(S)$ using method $2T\text{-}INF$ [11];
2: For each node v with self-loop, remove the self-loop and rename it with v^+; for each NTSCC, replace it with $Repair(NTSCC, S)$. Relations with any node in NTSCC, rebuild the relation with the new node; Update the graph G;
3: $G.setln(), R \leftarrow \varepsilon, level = 1$;
4: **while** $level \leq (ln(G.sink)) - 1$ **do**
5:    $V_T \leftarrow$ all vertices with $level$ and $length(sym(v)) \geq 2$;
6:    $V_S \leftarrow$ all vertices with $level$ and $length(sym(v)) = 1$;
7:    **for** each $v_T \in V_T$ **do**
8:       **if** $G.issl(level)$ or $(|V_T| + |V_S|) > 1$ **then**
9:          $R \leftarrow R \cdot v_T^?$
10:      **else**
11:         $R \leftarrow R \cdot v_T$
12:      **end if**
13:   **end for**
14:   **if** $|V_S| > 0$ **then**
15:      **if** $G.issl(level)$ or $|V_T| > 0$ **then**
16:         $R \leftarrow R \cdot AIT(V_S)^?$
17:      **else**
18:         $R \leftarrow R \cdot AIT(V_S)$
19:      **end if**
20:   **end if**
21:   $level = level + 1$
22: **end while**
23: **return** R

Algorithm 2. Repair(V,S)

Input: A node set V and a set of given sample S
Output: A repaired regular expression $newRE$
1: Construct the new sample $S' \leftarrow \bigcup_{s \in Strings} PSubstring(V, s)$;
2: **if** $CS(S') == \emptyset$ **then**
3:      **return** RE
4: **else**
5:      $G = Graph(CS(S'))$;
6:      **while** $G.nodes()! = \emptyset$ **do**
7:          $v = clique\_removal(G)$; $all\_mis.append(v)$;
8:      **end while**
9:      **for** each $M_i \in all\_mis$ **do**
10:         $T.append(topological_sort(M_i))$;
11:     **end for**
12:     for nodes in $NCS(S')$ but not in $CS(S')$, add them in M_i with largest size. Determine the orders of symbols in $M_i \in T$ under condition of $NCS(S')$;
13:     add repetition operator $\{1, *, ?, +\}$ for each symbol in T using algorithm CRX[4]
14:     $newRE \leftarrow$ Combine all elements in T using &;
15:     **return** $newRE$
16: **end if**

Time Complexity. Let $G(V, E) = SOA(S)$, $n = |V|$ and $m = |E|$. It costs $O(n)$ to find all nodes with loops and $O(m+n)$ to find all NTSCCs using depth-first search algorithm [8]. The time complexity of $clique\_removal()$ is $O(m+n^2)$. $setln()$ can be finished in time of $O(m + n)$ using breadth-first search [8]. For each NTSCC, computation of all_mis costs time $O(m + n^3)$ and the topological sort for each subgraph M_i costs time $O(m+n)$. The number of NTSCCs is finite. Then computing all_mis for all NTSCCs also cost time $O(m + n^3)$. All nodes will be converted into specific chain factors of $ICHARE$ in $O(n)$. Therefore, the time complexity of $GenICHARE$ is $O(m + n^3)$.

Theorem 1. *For a set of given string sample S, $\alpha = GenICHARE(SOA(S))$. If $S \subseteq L(\beta) \subset L(\alpha)$, then $L(\beta) = L(\alpha)$ holds for every ICHARE β.*

Proof. We construct SOA for S, α, β as G_s, G_α, G_β respectively. Clearly, we have $sym(G_s) = sym(G_\alpha) = sym(G_\beta)$. Let $\alpha = \alpha_1 \alpha_2 \cdots \alpha_n$, $\beta = \beta_1 \beta_2 \cdots \beta_m$. Now we first consider α_1. α_1 contains all nodes with $level = 1$. We use V_S and V_T to denote the sets with only one terminal symbol and multiple terminal symbols, respectively. 6 cases have to be considered. 1. $V_S \neq \emptyset$, $V_T = \emptyset$, 1 is not a skip number. 2. $V_S \neq \emptyset$, $V_T = \emptyset$, 1 is a skip level. 3. $V_S = \emptyset$, $|V_T| = 1$, 1 is not a skip level. 4. $V_S = \emptyset$, $|V_T| = 1$, 1 is a skip number. 5. $V_S = \emptyset$, $|V_T| > 1$. 6. $V_S \neq \emptyset$ and $V_T \neq \emptyset$.
1. $V_S \neq \emptyset$, $V_T = \emptyset$, 1 is not a skip number.

Let $V_S = \{v_1, v_2, \cdots, v_k\}$. According to Algorithm 1, we know that $\alpha_1 = (v_1|v_2| \cdots |v_k)$ and there is no edge between any two nodes. $sym(\alpha_1) = sym(\beta_1)$, otherwise edges from src of G_α and G_β are different which will cause a contradiction with $S \subseteq L(\beta) \subseteq L(\alpha)$. & and $\cdot$ operators must not appear in β_1.

Because these two operators would lead to an edge $e_{<v_i,v_j>}$ that is not in α_1. Therefore, β_1 only have union operator. Because $ln = 1$ is not a skip level, there does not exist an edge from src to a node in $\Sigma \setminus \{src\} \setminus symbol(\alpha_1)$. Now we have $\beta_1 = \alpha_1$ and therefore $L(\beta_1) = L(\alpha_1)$.

2. $V_S = \emptyset$, $V_T \neq \emptyset$, 1 is a skip level.

According to the proof of case 1, we can conclude that $term(\alpha_1) = term(\beta_1)$. β_1 only have union operator. At the same time, there exists an edge from src to some node $v \in \Sigma \setminus \{src\} \setminus symbol(\alpha_1)$. $\alpha = GenICHARE(G_s)$, $src \rightarrow_s v$ holds. Because G_s is a subgraph of G_{β_1}, $src \rightarrow_{\beta_1} v$ holds also. Therefore, $\beta_1 = \alpha_1$. $L(\beta_1) = L(\alpha_1)$.

3. $V_S = \emptyset$, $|V_T| = 1$, 1 is not a skip level.

Let $V_T = \{V_T^1\}$, $V_T^1 = (v_1 \& v_2 \& \cdots \& v_k)^+$. V_T is a NTSCC of G_α. According to the theorem, it is also a NTSCC of G_β and G_s. In this situation, we use Algorithm 2 to decide whether it needs to be repaired.

- α_1 can not be repaired. Then $\alpha_1 = (v_1|v_2|\cdots|v_k)^+$. The CS of $S' = \bigcup_{s \in S} PSubstring(sym(\alpha_1), s)$ is an empty set. If β_1 has $\&$ operator, it must generate two edges $e_{<v_i,v_j>}$ and $e_{<v_j,v_i>}$ at the same time which will lead to a contradiction of $L(G_\beta) \subseteq L(G_\alpha)$. If β_1 has concatenation operator, then the order of two nodes v_i, v_j is fixed. That means in G_β, $v_i \rightarrow^+ v_j$ and $v_j \rightarrow^+ v_i$ can not hold at the same time. Therefore it leads to a contradiction that $G_s \subseteq G_\beta$ because V_T is a NTSCC of G_s.

- α_1 can be repaired. In this situation, β_1 does not has union operator. Otherwise, there is a string $s \in L(G_\beta)$ with a substring aaa but $s \notin L(G_\alpha)$.
  - α_1 only has $\&$ operator. According to $GenICHARE$, the NCS of α_1 is empty. The CS of α_1 contains all nodes of $sym(\alpha_1)$. If β_1 has concatenation operator, the CS of β_1 is different from that of α_1. This will lead to the conclusion that they are generated from different string samples, which is not right. Therefore, β_1 only has $\&$ operator. According to the theorem, the languages of $L(\alpha_1)$ is the minimal-inclusion. Because $L(G_s) \subseteq L(G_\beta) \subset L(G_\alpha)$ and each terminal symbol can only occur once at most, $L(G_{\beta_1}) = L(G_{\alpha_1})$ must hold.
  - α_1 has concatenation and interleaving operators. If β_1 does not has concatenation operator, the CS of β_1 is larger than that of α_1 which leads to the conclusion that they are generated by different string samples, which is wrong. Therefore, β_1 has two operators $\cdot$ and $\&$. And their CS and NCS must be the same because they are both generated from S. Let $\alpha_1 = ABC$ where A and C only have concatenation and all symbols in $sym(A)$ must occur before that of $sym(B)$ and $sym(C)$. Similarly, all symbols in $sym(C)$ must occur after that of $sym(A)$ and $sym(B)$. Due to $L(G_s) \subseteq L(G_\beta) \subset L(G_\alpha)$, we can conclude that there must exist A' and B' with $sym(A) = sym(A')$ and $sym(B) = sym(B')$ in which A' and B' only has concatenation operator. For B in α_1, according to the proof above, $B = B'$. Otherwise, their CSs must be different. Therefore, we can conclude that $\beta_1 = \alpha_1$ and $L(G_{\beta_1}) = L(G_{\alpha_1})$.

4. $V_S = \emptyset$, $|V_T| = 1$, 1 is a skip number.

Similar as the proof in case 3 and case 2, we can easily come to the conclusion that $L(G_{\beta_1}) = L(G_{\alpha_1})$.

5. $V_S = \emptyset$, $|V_T| > 1$.

In this case, there are more than one NTSCC in G_{α_1}. For each NTSCC $C^i_{\alpha_1}$, we can prove that there exists a $\beta_m = \alpha^i_1$ where α^i_1 is the expression of $C^i_{\alpha_1}$ using the same proof in case 3. Each α^i_1 must be added with ?. Otherwise, there exists one node $v \in V_T$ with no outgoing edge. Therefore each β_m must have choice operator ? in order to accept all strings in sample S. This leads to the conclusion that $L(G_{\beta_1}) = L(G_{\alpha_1})$.

6. $V_S \neq \emptyset$ and $V_T \neq \emptyset$.

Obviously $L(G_{\beta_1}) = L(G_{\alpha_1})$ similar with the proof of case 1 and 5.

From the analysis above, we can conclude that there exists a i where $1 \leq i \leq m$ that $L(\alpha_1) = L(\beta_1 \cdots \beta_i)$ holds. Let $\alpha' = \alpha_2 \cdots \alpha_n$ and $\beta' = \beta_{i+1} \cdots \beta_m$. For each string $s \in S$, if $symb(s) \cap symb(\alpha_1) = \emptyset$, then $s \in S'$. If $symb(s) \cap symb(\alpha_1) \neq \emptyset$, then replace all alphabet in $symb(\alpha_1)$ as ε and put the new string in set S'. We construct $G_{S'} = SOA(S')$. α_1 is absorbed by src and the new starting node becomes $src \cdot \alpha_1$. Using the same proof procedure above, we can conclude that there exist a k such that $\alpha_2 = \beta_{j+1} \cdots \beta_{j+k}$. Go on the same operation until the graph G has only two nodes $src \cdot \alpha_1 \cdots \alpha_n$ and snk, and $\alpha_n = \beta_t \cdots \beta_m$, then we have $L(\alpha) = L(\beta)$.

4   Experiments

4.1   Usage of *ICHARE* in Practice

We first investigated the usage and proportion of *ICHARE*(denoted as G in the Fig. 1) compared with SORE(A), Simplified CHARE(B), eSimplified CHARE(C), CHARE(D), eCHARE(E), SIRE(F) based on real-world data set, and then validated our inference algorithm *GenICHARE* on DBLP[2]. All our experiments were conducted on a machine with Intel Core i5-5200U@2.20GHz, 4G memory, OS: Ubuntu 16.04. All codes were written in python 2.7.

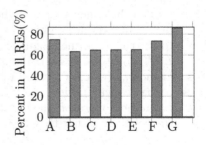

Fig. 1. Coverage proportions of subclasses

Our data set contains 13946 Relax NG files crawled from 268 Web sites, Maven[3], and 171 GitHub repositories with 509267 regular expressions extracted from these files. From Fig. 1, we can find that the coverage proportion of *ICHARE* is the highest (85.78%) among these subclasses.

---

[2] http://dblp.uni-trier.de/xml/.

[3] http://repo1.maven.org/maven2/.

4.2 Analysis of Inference Results Among Different Methods

DBLP is Data-centric database of information on major computer science journals and proceedings. We downloaded *dblp-2015-03-02.xml.gz*.

The results are shown in Tables 1 and 2 inferred by methods: Original Schema, Exact Minimal Schema, Trang[4], InstanceToSchema[5] and *GenICHARE*. The first column is the elements names with corresponding size of strings (considering both large and small data sets). The second column is regular expressions inferred by different methods. In both tables, we use "," to denote the concatenation operator. Method of Exact Minimal Schema uses the exact MIS to infer a regular expression. Trang can infer a schema from one or more XML documents. InstanceToSchema is a Relax NG schema generator from XML documents.

We analyze the inferred regular expressions from two aspects: conciseness and descriptive generalization. For conciseness, regular expressions inferred by the Original Schema, Trang and *GenICHARE* are in accordance with chain regular expressions. This form is concise and has a better readability while the rest two have no notable structure features. For descriptive generalization, we discuss the experimental inference results from the following four cases.

(1) Language generated by the inferred regular expression must contain all strings of given sample. Original Schema, Exact Minimal Schema, Trang and *GenICHARE* all satisfy this requirement while InstanceToSchema does not. Take *inproceedings* for example, the regular expression inferred by Instance-ToSchema is $ee^*\&note^?\&editor^*\&year\&author^*$ $\&title\&cdrom\&url^?$ $\&number^?\&pages^?\&month^?\&cite^*\&booktitle\&crossref^*$. *cdrom* in this regular expression can occur only once. But we find out that the word *cdrom* appeared more than once in 1610138 strings i.e. $w = author \cdot title \cdot pages \cdot year \cdot booktitle \cdot url \cdot ee \cdot cdrom \cdot cdrom \cdot cite \cdot cite \cdot cite \cdot cite \cdot cite \cdot cite \cdot cite$.

(2) It is better for inferred regular expressions to support all binary operators ($\cdot$, $|$, $\&$) which will have a higher practicality. However, regular expressions inferred by Exact Minimal Schema do not support union operator. Trang supports the generation of Relax NG schema using XML documents. But in fact it is almost equivalent to DTD. In other words, it does not support the regular expressions with interleaving. InstanceToSchema supports the interleaving operator but not concatenation and union. *GenICHARE* supports all operators which could generate real Relax NG schemas.

(3) Improper use of union will lead to the over-generalization for the given sample, which will generate many invalid XML documents. This shortage could be found in Original Schema and Trang. Take element of *article* for example. Regular expression inferred by Trang is $editor^*\cdot(author|booktitle|cdrom|cite|cros\ sref|ee|journal|month|note| number|pages|publisher|title|url|volume|year)^+$. But the title and published year for an actual article are both unique. This shortage is also

---

[4] http://www.thaiopensource.com/relaxng/trang.html.
[5] http://www.xmloperator.net/i2s/.

Table 1. Inference results using different methods on DBLP

name	1.Original Schema; 2.Exact Minimal Schema
size	3.Result of Trang; 4.Result of InstanceToSchema; 5.Result of GenICHARE

phdt-	(author|editor|title|booktile|pages|year|address|journal|volume|number|month
hesis	|url|ee|cdrom|cite|publisher|note|crossref|isbn|series|school|chapter)*
6953	school?&author+,title,pages?,publisher?,series?,number?,month?,volume?,
	ee*,url?&year,isbn?,note*
6953	author+,title,(isbn|month|number|pages|publisher|school|series|volume|year)
	+,(ee|note|url)*
6953	ee*&note*&year&author+&isbn?&title&url?&number?&volume?&pages?
	&month?&school?&series?&publisher?
6953	author+,title,(year,month?,volume?,isbn?&school?&pages?,publisher?,
	series?,number?),(ee*,url&note?)
master-	(author|editor|title|booktile|pages|year|address|journal|volume|number|month
sthesis	|url|ee|cdrom|cite|publisher|note|crossref|isbn|series|school|chapter)*
9	ee?&author,title,year,school,url?
9	author,title,year,school,(ee|url)*
9	ee?&year&author&title&url?&school
9	author,title,year,school,(ee?&url?)
www	(author|editor|title|booktile|pages|year|address|journal|volume|number|month
	|url|ee|cdrom|cite|publisher|note|crossref|isbn|series|school|chapter)*
1557527	title?&url*,cite*&crossref?,author*,note*,editor*,booktitle?,ee?,year?
1557527	(crossref|editor*|author*),(booktiltle|cite|ee|note|title|url|year)*
1557527	ee?&editor*&note*&year?&author*&cite*&title?&booktitle?&crossref?&url*
1557527	(editor+|author+|crossref)?,(note*,year?&ee?,booktitle?,url*,cite*&title?)

reflected in the result of Original Schema but rest methods do not have this problem.

(4) Improper use of & without considering the actual orders in real-world data will also lead to over-generalization. Take the element of *masterthesis* for example. Regular expressions inferred by Exact Minimal Schema, Trang, InstanceToSchema and *GenICHARE* are $ee^?\&author \cdot title \cdot year \cdot school \cdot url^?$; $author \cdot title \cdot year \cdot school \cdot (ee|url)^*$; $ee^?\&year\&author\&title\&url^?\&school$ and $author \cdot title \cdot year \cdot school \cdot (ee^?\&url^?)$. Compared with the first result, it is clear that the form of third result will lead to more redundant XML documents. The results of Trang and *GenICHARE* reveal that orders of first four elements are fixed. Therefore, the result inferred by InstanceToSchema is not good enough. From the investigation of *mastersthesis*, the combinations of ee and url ($ee \cdot url$, $url \cdot ee$, ee, url or ε) can only occur in the last position. Therefore, regular expression inferred by *GenICHARE* is reasonable and descriptive generalized.

The proper use of interleaving & can ensure the actual orders of elements at the beginning. Therefore, the optimization is the significant feature of our algorithm *GenICHARE*.

Table 2. Inference results using different methods on DBLP

name size	1.Original Schema; 2.Exact Minimal Schema 3.Result of Trang; 4.Result of InstanceToSchema; 5.Result of *GenICHARE*
article	(author|editor|title|booktile|pages|year|address|journal|volume|number|month |url|ee|cdrom|cite|publisher|note|crossref|isbn|series|school|chapter)*
1270262	booktitle?,cdrom?,cite*,note?&number?&publisher?,month?,ee*&title ,journal?&editor*,volume?&pages?&author*,url?&year?,crossref?
1270262	editor*,(author|booktitle|cdrom|cite|crossref|ee|journal|month|note|number |pages|publisher|title|url|volume|year)+
1270262	ee*&note?&editor*&year?&author*&title&cdrom?&url?&volume?&number? &journal?&pages?&month?&cite*&publisher?&booktitle?&crossref?
1270262	editor*,(title,booktitle?,volume?,cite*&note?,cdrom?&year?&author*, publisher?,month?,url?&journal?&pages?&number?,crossref?&ee*)
inproce- edings	(author|editor|title|booktile|pages|year|address|journal|volume|number|month |url|ee|cdrom|cite|publisher|note|crossref|isbn|series|school|chapter)*
1610138	title,year&ee*&editor*,month?,booktitle,number?&url?&crossref*,cite* &pages?&author*,note?,cdrom*
1610138	editor*,(author|booktitle|cdrom|cite|crossref|ee|month|note|number|pages|title |url|year)+
1610138	ee*&note?&editor*&year&author*&title&cdrom&url?&number?&pages? &month?&cite*&booktitle&crossref*
1610138	editor*,(pages?&booktitle&ee*&crossref*,month?,number?&author*,note?, cdrom*&title,url?,cite*&year)
proce- edings	(author|editor|title|booktile|pages|year|address|journal|volume|number|month |url|ee|cdrom|cite|publisher|note|crossref|isbn|series|school|chapter)*
26502	isbn*&series*&title,note*&publisher?&editor*,pages?,crossref?,ee*,number?& booktitle?&year?&url*&author*,journal?,address?,volume?,cite*
26502	author*,(booktitle|crossref|editor|ee|isbn|journal|note|number|pages|publisher |series|title|url|volume|year|address)+,cite*
26502	ee*&note*&editor*&year?&author*&isbn*&title&url*&volume?&number?& journal?&pages?&cite*&series*&publisher?&booktitle?&crossref?&address?
26502	author*,(journal?&volume?&number?&pages?&note*&address?&title,crossref ?,ee*&publisher?&isbn*&series*&editor*&booktitle?&year?&url*),cite*
book	(author|editor|title|booktile|pages|year|address|journal|volume|number|month |url|ee|cdrom|cite|publisher|note|crossref|isbn|series|school|chapter)*
11600	series*&publisher?&booktitle?,url*,school?&ee*&pages?&isbn*,cdrom?,cite* ,note*&author*,editor*,title,volume?,month?&year
11600	(editor*|author*),title,(booktitle|cdrom|cite|ee|isbn|month|note|pages |publisher|school|series|url|volume|year)+
11600	ee*&editor*&note*&year&author*&isbn*&title&cdrom?&url*&volume? &pages?&month?&school?&series*&publisher?&cite*&booktitle?
11600	(editor+|author+)?,title,(series*&booktitle?,url*,school?&year&publisher? &isbn*&ee*&cdrom?,cite*,volume?,note*&pages?,month?)
incoll- ection	(author|editor|title|booktile|pages|year|address|journal|volume|number|month |url|ee|cdrom|cite|publisher|note|crossref|isbn|series|school|chapter)*
31260	ee?&author*,title,year,chapter?,number?,isbn?,cite*,note?,cdrom?&booktile ,publisher?,url?&pages?&crossref?
31260	author*,title,(booktitle|crossref|ee|isbn|number|pages|publisher|url|year| chapter)+,(cdrom|note|cite*)
31260	ee?&note?&year&author*&isbn?&title&cdrom?&url?&number?&pages?& cite*&publisher?&booktile&crossref?&chapter?
31260	author*,title,(pages?&ee?&booktitle,publisher?,number?,isbn?,chapter?, url?&crossref?&year),(cite+|note|cdrom)?

From the analysis above, the subclass *ICHARE* has a higher coverage proportion based on real-world data. This means that it will have a better practicality in actual applications. Regular expression inferred by *GenICHARE* are more concise and satify descriptive generalization.

5   Conclusion and Future Work

Based on real-world data, we proposed a new subclass *ICHARE* of restricted regular expressions with interleaving. Then we introduced an efficient algorithm *GenICHARE* to infer a descriptive generalized *ICHARE*. A series of experiments were conducted to compare the coverage of *ICHARE* with other existing subclasses based on real-world data. Besides, based on the data of *DBLP*, we analyse the conciseness of regular expressions inferred by *GenICHARE* and compare with other methods. Experimental results show that *ICHARE* is more useful in practice with higher coverage proportion than other subclasses. Regular expressions inferred by *GenICHARE* perform better from the aspects of conciseness and descriptive generalization than methods of Original Schema, Exact Minimal Schema, Trang and InstanceToSchema.

In the future, we will consider constructing the automata for regular expressions with interleaving and study the inference algorithms for subclass *k-ORE* (each terminal symbol can occur k times at most).

References

1. Abiteboul, S., Buneman, P., Suciu, D.: Data on the Web: from Relations to Semistructured Data and XML. Morgan Kaufmann, San Francisco (2000)
2. Bex, G.J., Gelade, W., Neven, F., Vansummeren, S.: Learning deterministic regular expressions for the inference of schemas from XML data. ACM Trans. Web **4**(4), 1–32 (2010)
3. Bex, G.J., Neven, F., Bussche, J.V.D.: DTDs versus XML schema: a practical study. In: International Workshop on the Web and Databases, pp. 79–84 (2004)
4. Bex, G.J., Neven, F., Schwentick, T., Tuyls, K.: Inference of concise DTDs from XML data. In: International Conference on Very Large Data Bases, Seoul, Korea, September, pp. 115–126 (2006)
5. Bex, G.J., Neven, F., Schwentick, T., Vansummeren, S.: Inference of concise regular expressions and DTDs. ACM Trans. Database Syst. **35**(2), 1–47 (2010)
6. Bex, G.J., Neven, F., Vansummeren, S.: Inferring XML schema definitions from XML data. In: International Conference on Very Large Data Bases, University of Vienna, Austria, September, pp. 998–1009 (2007)
7. Boneva, I., Ciucanu, R., Staworko, S.: Simple schemas for unordered XML. In: International Workshop on the Web and Databases (2015)
8. Cormen, T.H., Leiserson, C.E., Rivest, R.L., Stein, C.: Introduction to Algorithms, 2nd edn., pp. 1297–1305 (2001)
9. Feng, X.Q., Zheng, L.X., Chen, H.M.: Inference Algorithm for a Restricted Class of Regular Expressions, vol. 41. Computer Science (2014)
10. Freydenberger, D.D., Kötzing, T.: Fast learning of restricted regular expressions and DTDs. Theory Comput. Syst. **57**(4), 1114–1158 (2015)

11. Garcia, P., Vidal, E.: Inference of K-testable languages in the strict sense and application to syntactic pattern recognition. IEEE Trans. Pattern Anal. Mach. Intell. **12**(9), 920–925 (2002)
12. Garofalakis, M., Gionis, A., Shim, K.: XTRACT: learning document type descriptors from XML document collections. Data Min. Knowl. Discov. **7**(1), 23–56 (2003)
13. Ghelli, G., Colazzo, D., Sartiani, C.: Efficient inclusion for a class of XML types with interleaving and counting. Inf. Syst. **34**(7), 643–656 (2009)
14. Gold, E.M.: Language Identification in the limit. Inf. Control **10**(5), 447–474 (1967)
15. Grijzenhout, S., Marx, M.: The quality of the XML web. Web Semant. Sci. Serv. Agents World Wide Web **19**, 59–68 (2013)
16. Koch, C., Scherzinger, S., Schweikardt, N., Stegmaier, B.: Schema-based scheduling of event processors and buffer minimization for queries on structured data streams. In: Proceedings of the Thirtieth International Conference on Very Large Data Bases, vol. 30, pp. 228–239. VLDB Endowment (2004)
17. Li, Y., Zhang, X., Peng, F., Chen, H.: Practical study of subclasses of regular expressions in DTD and XML schema. In: Li, F., Shim, K., Zheng, K., Liu, G. (eds.) APWeb 2016. LNCS, vol. 9932, pp. 368–382. Springer, Cham (2016). https://doi.org/10.1007/978-3-319-45817-5_29
18. Manolescu, I., Florescu, D., Kossmann, D.: Answering XML queries on heterogeneous data sources. VLDB **1**, 241–250 (2001)
19. Martens, W., Neven, F.: Typechecking top-down uniform unranked tree transducers. In: Calvanese, D., Lenzerini, M., Motwani, R. (eds.) ICDT 2003. LNCS, vol. 2572, pp. 64–78. Springer, Heidelberg (2003). https://doi.org/10.1007/3-540-36285-1_5
20. Martens, W., Neven, F., Schwentick, T.: Complexity of decision problems for XML schemas and chain regular expressions. Siam J. Comput. **39**(4), 1486–1530 (2009)
21. Peng, F., Chen, H.: Discovering restricted regular expressions with interleaving. In: Cheng, R., Cui, B., Zhang, Z., Cai, R., Xu, J. (eds.) APWeb 2015. LNCS, vol. 9313, pp. 104–115. Springer, Cham (2015). https://doi.org/10.1007/978-3-319-25255-1_9

Spatial-Temporal, Time-Series and Stream Mining

Accelerating Adaptive Online Learning
by Matrix Approximation

Yuanyu Wan and Lijun Zhang[(⊠)]

National Key Laboratory for Novel Software Technology, Nanjing University,
Nanjing 210023, China
{wanyy,zhanglj}@lamda.nju.edu.cn

Abstract. Adaptive subgradient methods are able to leverage the second-order information of functions to improve the regret, and have become popular for online learning and optimization. According to the amount of information used, adaptive subgradient methods can be divided into diagonal-matrix version (ADA-DIAG) and full-matrix version (ADA-FULL). In practice, ADA-DIAG is the most commonly adopted rather than ADA-FULL, because ADA-FULL is computationally intractable in high dimensions though it has smaller regret when gradients are correlated. In this paper, we propose to employ techniques of matrix approximation to accelerate ADA-FULL, and develop two methods based on random projections. Compared with ADA-FULL, at each iteration, our methods reduce the space complexity from $O(d^2)$ to $O(\tau d)$ and the time complexity from $O(d^3)$ to $O(\tau^2 d)$ where d is the dimensionality of the data and $\tau \ll d$ is the number of random projections. Experimental results about online convex optimization show that both methods are comparable to ADA-FULL and outperform other state-of-the-art algorithms including ADA-DIAG. Furthermore, the experiments of training convolutional neural networks show again that our method outperforms other state-of-the-art algorithms including ADA-DIAG.

1 Introduction

Adaptive subgradient methods (ADAGRAD) dynamically integrate knowledge of the geometry of data observed in earlier iterations to guide the direction of updating [1]. Different from the conventional online methods [2], ADAGRAD employ adaptive proximal functions to control the learning rate for each dimension, and the proximal functions are iteratively modified by the algorithm instead of tuning manually. There are two versions of adaptive subgradient methods: ADA-DIAG which uses a diagonal matrix to define the proximal function, and ADA-FULL which uses a full matrix to define the proximal function. Because ADA-FULL is computationally intractable in high dimensions, ADA-DIAG is the most commonly studied and adopted version in practice.

Electronic supplementary material The online version of this chapter (https://doi.org/10.1007/978-3-319-93037-4_32) contains supplementary material, which is available to authorized users.

However, compared with ADA-FULL, ADA-DIAG cannot capture the correlation in gradients. As a result, the regret bound of ADA-DIAG may be worse than that of ADA-FULL when the high-dimensional data is dense and has a low-rank structure. This dilemma prompts a question as to whether we can design algorithms that possess the merits of two versions: i.e., the light computation of ADA-DIAG and the small regret of ADA-FULL. In a recent work [3], Krummenacher et al. presented two approximation algorithms to accelerate ADA-FULL, namely ADA-LR and RADAGRAD. Although ADA-LR is equipped with a regret bound, its space and time complexities are quadratic in the dimensionality d, which is unacceptable when d is large. In contrast, the space and time complexities of RADAGRAD are linear in d, but it lacks theoretical guarantees.

Along this line of research, this paper aims to attain theoretical guarantees, and at the same time keeping the computations light. Note that ADA-FULL is computationally impractical mainly due to the fact it needs to maintain a matrix of gradient outer products, and compute its square root and inverse in each round. Actually, similar problems have been encountered in online Newton step (ONS) for exponentially concave functions [4]. Recently, Luo et al. proposed to accelerate ONS using matrix sketching methods including random projections [5]. Motivated by previous work, we first propose to employ random projections to construct a low-rank approximation of gradient outer products, and manipulate this low-rank matrix in subsequent calculations. In this way, the new algorithm, named ADA-GP, reduces the space complexity from $O(d^2)$ to $O(\tau d)$ and the time complexity from $O(d^3)$ to $O(\tau^2 d)$, implying both the space and time complexities have a linear dependence on the dimensionality d.

ADA-GP achieves excellent empirical performance in our experiments. However, due to subtle independence issues, it is difficult to analyze ADA-GP theoretically. To circumvent this problem, we propose to replace the outer product matrix of gradients in ADA-FULL with the outer product matrix of data, and then develop a similar method, named ADA-DP, that applies random projections to the outer product matrix of data. The space and time complexities of ADA-DP are similar to those of ADA-GP. Moreover, we present theoretical analysis for ADA-DP when the outer product matrix of data is low-rank, and further extend to the full-rank case. In the experiments, we first examine the performance of our methods on online convex optimization, and the results demonstrate that they are highly comparable to ADA-FULL and are much more efficient. Furthermore, we conduct experiments on training convolutional neural networks, and show that ADA-GP outperforms ADA-DIAG and RADAGRAD.

Finally, we would like to emphasize the difference between this work and the recent work [3]. First, although both studies make use of random projections, our ADA-GP and ADA-DP are much more simple than ADA-LR and RADAGRAD. Second, our ADA-GP and ADA-DP are very efficient in the sense that their computational complexities are linear in the dimensionality d, and ADA-DP is equipped with theoretical guarantees. In contrast, although RADAGRAD has a similar computational cost, it does not have theoretical justifications.

2 Related Work

ADAGRAD. Adaptive subgradient methods use the second-order information to tune the step size of gradient descent adaptively [1]. For sparse data, the regret guarantee of ADAGRAD could be exponentially smaller in the dimension d than the non-adaptive regret bound. In the following, we provide a brief introduction of ADAGRAD. Note that the idea of ADAGRAD can be incorporated into either primal-dual subgradient method [6] or composite mirror descent [7]. For brevity, we take the composite mirror descent as an example.

In the t-th round, the learner needs to determine an action $\boldsymbol{\beta}_t \in \mathbb{R}^d$ and then observes a composite function $F_t(\boldsymbol{\beta}) = f_t(\boldsymbol{\beta}) + \varphi(\boldsymbol{\beta})$ where f_t and φ are convex. The learner suffers loss $F_t(\boldsymbol{\beta}_t)$, and the goal is to minimize the accumulated loss over T iterations. Let $\partial f_t(\boldsymbol{\beta})$ denote the subdifferential set of function f_t evaluated at $\boldsymbol{\beta}$ and $\mathbf{g}_t \in \partial f_t(\boldsymbol{\beta}_t)$ be a particular vector in the subdifferential set. Define the outer product matrix of gradients $G_t = \sum_{i=1}^t \mathbf{g}_i \mathbf{g}_i^\top$. Then, we use the square root of G_t to construct a positive definite matrices H_t, and have the following two choices:

$$H_t = \begin{cases} \sigma I_d + \operatorname{diag}(G_t)^{1/2} & \text{ADA-DIAG} \\ \sigma I_d + G_t^{1/2} & \text{ADA-FULL} \end{cases}$$

where $\sigma > 0$ is a parameter. The proximal term is given by $\Psi_t(\boldsymbol{\beta}) = \frac{1}{2}\langle \boldsymbol{\beta}, H_t \boldsymbol{\beta}\rangle$ and the Bregman divergence associated with Ψ_t is

$$B_{\Psi_t}(\mathbf{x}, \mathbf{y}) = \Psi_t(\mathbf{x}) - \Psi_t(\mathbf{y}) - \frac{1}{2}\langle \nabla \Psi_t(\mathbf{y}), \mathbf{x} - \mathbf{y}\rangle.$$

In each iteration, the composite mirror descent method updates by

$$\boldsymbol{\beta}_{t+1} = \underset{\boldsymbol{\beta}}{\operatorname{argmin}} \left\{ \eta\langle \mathbf{g}_t, \boldsymbol{\beta}\rangle + \eta\varphi(\boldsymbol{\beta}) + B_{\Psi_t}(\boldsymbol{\beta}, \boldsymbol{\beta}_t) \right\}$$

$$= \boldsymbol{\beta}_t - \eta H_t^{-1}\mathbf{g}_t, \quad \text{if } \varphi = 0$$

where $\eta > 0$ is a fixed step size. When the dimensionality d is large, ADA-FULL is impractical because the storage cost of G_t and the running time of finding its square root and inverse of H_t are unacceptable.

To make ADA-FULL scalable, Krummenacher et al. proposed two methods that approximate the proximal term $\Psi_t(\boldsymbol{\beta})$ [3]. Based on the fast randomized singular value decomposition (SVD) [8], they presented an algorithm ADA-LR that performs the following updates:

$$\begin{aligned} G_t &= G_{t-1} + \mathbf{g}_t \mathbf{g}_t^\top \\ \widetilde{G}_t &= G_t \Pi \quad\quad \text{Random Projection} \\ QR &= \widetilde{G}_t \quad\quad \text{QR-decomposition} \\ B &= Q^\top G_t \\ U\Sigma V^\top &= B \quad\quad \text{SVD} \\ \boldsymbol{\beta}_{t+1} &= \boldsymbol{\beta}_t - \eta V(\Sigma^{1/2} + \sigma I_\tau)^{-1} V^\top \mathbf{g}_t \end{aligned} \quad (1)$$

where $\Pi \in \mathbb{R}^{d \times \tau}$ is the random matrix of the subsampled randomized Fourier transform. We note that random projections are utilized in the 2nd step to generate a smaller matrix $\widetilde{G}_t \in \mathbb{R}^{d \times \tau}$. It is easy to verify that the space and time complexities of ADA-LR are respectively $O(d^2)$ and $O(\tau d^2)$, which are still unacceptable when the d is large. To further improve the efficiency, they presented algorithm RADAGRAD by introducing more randomized approximations, the space and time complexities of which are respectively $O(\tau d)$ and $O(\tau^2 d)$. Unfortunately, RADAGRAD is a heuristic method and lacks theoretical guarantees.

As previous mentioned, in [5], Luo et al. adopted matrix sketching methods to accelerate ONS that also encounters the similar problems as ADA-FULL. Specifically, their ONS updates by $\beta_{t+1} = \beta_t - A_t^{-1} g_t$ where $A_t = \alpha I_d + \sum_{i=1}^{t} \eta_i g_i g_i^{\top}$, $\alpha > 0$ and $\eta_i = O(1/\sqrt{i})$ for general convex functions. We can reformulate this update rule as

$$\beta_{t+1} = \beta_t - \eta H_t^{-1} g_t$$

where $H_t = \delta I_d + \sum_{i=1}^{t} \frac{1}{\sqrt{i}} g_i g_i^{\top}$. To accelerate ONS, they use matrix sketching methods to calculate a low-rank approximation of $\sum_{i=1}^{t} \frac{1}{\sqrt{i}} g_i g_i^{\top}$. Motivated by [5], our work employs random projections to calculate a low-rank approximation of full matrix. But there are obviously differences between our work and this related work. First, although both our methods and RP-SON in [5] use random projections to approximate the full matrix, we further propose to use the outer product matrix of data to replace the outer product matrix of gradients which leads to ADA-DP. Note that this simple change can avoid the dependence issue that the gradient g_t depends on the random vectors. Second, the theoretical analysis in our work is obviously different from [5]. The only common part is the property of the random projections for low-rank data. But we further exploit the property of the random projection for full-rank data. Third, our methods and this related work are designed for different tasks. Our paper aims to accelerate ADA-FULL, and this related work aims to accelerate ONS. Note that ADA-FULL is a data dependent algorithm for general convex function and ONS is proposed to derive a logarithmic regret for exponentially concave functions.

Random Projection. Random projection [9–11] is a simple yet powerful dimensionality reduction method. For a data point $x \in \mathbb{R}^n$, random projection reduces its dimensionality to τ by $R^{\top} x$, where $R \in \mathbb{R}^{n \times \tau}$ is a random matrix. It has been successfully applied to many machine learning tasks including classification [12,13], regression [14], clustering [15,16], manifold learning [17,18] and optimization [19,20]. Random projection can be implemented in various different ways [21,22], and the most classical one is the Gaussian random projection, where each entry of R is sampled from a Gaussian distribution. In this paper, we focus on Gaussian random projection due to its nice theoretical properties and easy implementations.

3   Main Results

In this section, we introduce our proposed methods and theoretical results. Due to the limitation of space, we defer the proof of theoretical results to the supplementary material.

3.1   Problem Setting

To facilitate presentations, we consider the case $\varphi = 0$, and our methods can be directly extended to the general case $\varphi \neq 0$. The goal of the learner is to minimize the regret, defined as $R(T) = \sum_{t=1}^{T} f_t(\beta_t) - \sum_{t=1}^{T} f_t(\beta^*)$ where β^* is a fixed optimal predictor.

3.2   The Proposed ADA-GP Method

From previous discussions, we know that if one can find a low-rank matrix to approximate G_t, then both space and time complexities of ADA-FULL can be reduced dramatically. Random projections provide an elegant way for low-rank matrix approximations, as explained below.

Define

$$A_t^\top = [\mathbf{g}_1, ..., \mathbf{g}_t] \in \mathbb{R}^{d \times t}, \ R_t = [\mathbf{r}_1, ..., \mathbf{r}_t] \in \mathbb{R}^{\tau \times t}$$

where the i-th column of $A_t^\top$ is gradient $\mathbf{g}_i$, and each entry of R_t is a Gaussian random variable drawn from $\mathcal{N}(0, 1/\tau)$ independently. Then, we have

$$G_t = A_t^\top A_t, \ \mathbb{E}[R_t^\top R_t] = I_d.$$

To accelerate the computation, we define

$$S_t = R_t A_t = \sum_{i=1}^{t} \mathbf{r}_i \mathbf{g}_i^\top \in \mathbb{R}^{\tau \times d}.$$

Note that S_t can be calculated on the fly as $S_t = S_{t-1} + \mathbf{r}_t \mathbf{g}_t^\top$. When τ is large enough, we expect $R_t^\top R_t \approx I_d$, implying

$$S_t^\top S_t = A_t^\top R_t^\top R_t A_t \approx A_t^\top A_t = G_t.$$

Thus, $S_t^\top S_t$ could be used as a low-rank approximation of G_t. The matrix H_t in the proximal term can be redefined as

$$H_t = \sigma I_d + (S_t^\top S_t)^{1/2}.$$

Let SVD of S_t be $S_t = U\Sigma V^\top$, then we have $S_t^\top S_t = V\Sigma^2 V^\top$ and $H_t = \sigma I_d + V\Sigma V^\top$. According to Woodbury Formula [23], we have

$$H_t^{-1} = (\sigma I_d + V\Sigma V^\top)^{-1}$$
$$= \frac{1}{\sigma}\big(I_d - V(\sigma I_\tau + \Sigma)^{-1}\Sigma V^T\big).$$

Algorithm 1. ADA-GP

1: **Input:** $\eta > 0, \sigma > 0, \tau,\ S = \mathbf{0}_{\tau \times d}, \boldsymbol{\beta}_1 = \mathbf{0}$;
2: **for** $t = 1, ..., T$ **do**
3:     Receive $\mathbf{g}_t = \nabla f_t(\boldsymbol{\beta}_t)$
4:     $S_t = S_{t-1} + \mathbf{r}_t \mathbf{g}_t^\top$ {Random Projections}
5:     $U\Sigma V^\top = S_t$ {SVD}
6:     $\boldsymbol{\beta}_{t+1} = \boldsymbol{\beta}_t - \frac{\eta}{\sigma}\big(\mathbf{g}_t - V(\sigma I_\tau + \Sigma)^{-1}\Sigma V^\top \mathbf{g}_t\big)$
7: **end for**

As a result, in the t-th round, our algorithm performs the following updates

$$S_t = S_{t-1} + \mathbf{r}_t \mathbf{g}_t^\top \quad \text{Random Projection}$$
$$U\Sigma V^\top = S_t \qquad \text{SVD} \tag{2}$$
$$\boldsymbol{\beta}_{t+1} = \boldsymbol{\beta}_t - \frac{\eta}{\sigma}\big(\mathbf{g}_t - V(\sigma I_\tau + \Sigma)^{-1}\Sigma V^\top \mathbf{g}_t\big).$$

The detailed procedure is summarized in Algorithm 1, and named as adaptive online learning with gradient projection (ADA-GP).

Remark. First, it is easy to verify the time and space complexities of our ADA-GP is $O(\tau d)$ and $O(\tau^2 d)$, respectively. Thus, both of them are linear in the dimensionality d. Second, comparing (2) with (1), we observe that our updating rules are much more simple than those of ADA-LR. Note that the RADAGRAD algorithm of [3] is even more complicated than ADA-LR.

3.3 The Proposed ADA-DP Method

Although ADA-GP performs very well in our experiments, it is difficult to establish a regret bound due to dependence issues. To be specific, the gradient $\mathbf{g}_t$ depends on the random vectors $[\mathbf{r}_1, \cdots, \mathbf{r}_{t-1}]$, and as a result, standard concentration inequalities cannot be directly applied [24].

To avoid the aforementioned problem, we propose a strategy to get ride of the dependence issues and the new algorithm is equipped with theoretical guarantees. We consider the case $f_t(\boldsymbol{\beta}_t) = l(\boldsymbol{\beta}_t^\top \mathbf{x}_t)$ where $\mathbf{x}_t$ is a data vector. Then, we assume the data points $\mathbf{x}_1, ..., \mathbf{x}_t$ are independent from our algorithm. The key idea is to replace the outer product matrix of gradients G_t with the outer product matrix of data $X_t = \sum_{i=1}^t \mathbf{x}_i \mathbf{x}_i^\top$. Accordingly, H_t will be defined as $\sigma I_d + X_t^{1/2}$. To accelerate computations, our problem becomes finding a low-rank approximation of X_t.

Let $C_t^\top = [\mathbf{x}_1, ..., \mathbf{x}_t] \in \mathbb{R}^{d \times t}$, where each column is a data vector. Similar to ADA-GP, we define

$$S_t = R_t C_t = \sum_{i=1}^t \mathbf{r}_i \mathbf{x}_i^\top \in \mathbb{R}^{\tau \times d}$$

Algorithm 2. ADA-DP

1: **Input:** $\eta > 0, \sigma > 0, \tau, \mathrm{S} = \mathbf{0}_{\tau \times d}, \beta_1 = \mathbf{0}$;
2: **for** $t = 1, ..., T$ **do**
3:     Receive $\mathbf{x}_t$ and $\mathbf{g}_t = \nabla f_t(\beta_t) = l'(\beta_t^\top \mathbf{x}_t)\mathbf{x}_t$
4:     $\mathrm{S}_t = \mathrm{S}_{t-1} + \mathbf{r}_t \mathbf{x}_t^\top$ {Random Projections}
5:     $\mathrm{U}\Sigma\mathrm{V}^\top = \mathrm{S}_t$ {SVD}
6:     $\beta_{t+1} = \beta_t - \frac{\eta}{\sigma}\big(\mathbf{g}_t - \mathrm{V}(\sigma\mathrm{I}_\tau + \Sigma)^{-1}\Sigma\mathrm{V}^\top\mathbf{g}_t\big)$
7: **end for**

---

where $\mathrm{R}_t \in \mathbb{R}^{\tau \times t}$ is the Gaussian random matrix. In this case, since R_t is independent of C_t, we have

$$\mathbb{E}[\mathrm{S}_t^\top \mathrm{S}_t] = \mathrm{C}_t^\top \mathbb{E}[\mathrm{R}_t^\top \mathrm{R}_t]\mathrm{C}_t = \mathrm{C}_t^\top \mathrm{C}_t = \mathrm{X}_t$$

which means $\mathrm{S}_t^\top \mathrm{S}_t$ is an unbiased estimation of X_t.

The rest steps are similar to that of ADA-GP. The detailed procedure is summarized in Algorithm 2, named as adaptive online learning with data projection (ADA-DP). It is obvious that the computation cost of ADA-DP is almost the same as that of ADA-GP. Thus, both the space and time complexities of ADA-DP are linear in d.

The main advantage of ADA-DP is that it has formal theoretical guarantees. We first consider the case that the data matrix C_T is low-rank, and have the following theorem regarding the regret of Algorithm 2.

Theorem 1. *Let $r \ll d$ be the rank of C_T, and $0 < \delta < 1$ be the confidence parameter. Assume each entry of $\mathbf{r}_t \in \mathbb{R}^\tau$ is a Gaussian random variable drawn from $\mathcal{N}(0, 1/\tau)$ independently, $\tau = \Omega(\frac{r + \log(T/\delta)}{\epsilon^2})$ and $\sigma \geq 0$, then ADA-DP ensures*

$$R(T) \leq \frac{\sigma}{2\eta}\|\beta_*\|_2^2 + \frac{1}{2\eta}\max_{t \leq T}\|\beta^* - \beta_t\|_2^2 tr(\mathrm{X}_T^{1/2})$$

$$+ \frac{2\eta}{\sqrt{1-\epsilon}}\max_{t \leq T}l'(\beta_t^\top \mathbf{x}_t)^2 tr(\mathrm{X}_T^{1/2}) + \frac{\epsilon}{2\eta}\max_{t \leq T}\|\beta^* - \beta_t\|_2^2 \sum_{t=1}^T \|\mathrm{X}_t^{1/2}\|$$

with probability at least $1 - \delta$.

Remark. Theorem 1 means that we can set the number of random projections as $\tau = \widehat{\Omega}(r)$ when the data matrix is low-rank.

When the data matrix is full-rank, Theorem 1 is inappropriate because it implies the number of random projections is on the order of the dimensionality. Let $\lambda_i(\cdot)$ be the i-th largest eigenvalue of a matrix. For the full-rank case, we further establish the following theorem to bound the regret of Algorithm 2.

Theorem 2. *Let $c \geq 1/32$, $\sigma \geq \sqrt{\alpha} > 0$, $\sigma_{ti}^2 = \lambda_i(\mathrm{C}_t^\top \mathrm{C}_t)$, $\tilde{r}_t = \sum_i \frac{\sigma_{ti}^2}{\alpha + \sigma_{ti}^2}$, $\tilde{r}_* = \max\limits_{1 \leq t \leq T}\tilde{r}_t$, $\sigma_{*1}^2 = \max\limits_{1 \leq t \leq T}\sigma_{t1}^2$, and $0 < \delta < 1$. Assume each entry of $\mathbf{r}_t \in \mathbb{R}^\tau$*

is an independent random Gaussian variable drawn from $\mathcal{N}(0, 1/\tau)$, $\tau \geq$ $\frac{\tilde{r}_* \sigma_{*1}^2}{c\epsilon^2(\alpha+\sigma_{*1}^2)} \log \frac{2dT}{\delta}$ and then ADA-RP ensures

$$R(T) \leq \frac{\sigma}{2\eta}\|\boldsymbol{\beta}_*\|_2^2 + \frac{1}{2\eta}\max_{t \leq T}\|\boldsymbol{\beta}^* - \boldsymbol{\beta}_t\|_2^2 tr(\mathbf{X}_T^{1/2})$$

$$+ \frac{2\eta}{\sqrt{1-\epsilon}}\max_{t \leq T} l'(\boldsymbol{\beta}_t^\top \mathbf{x}_t)^2 tr(\mathbf{X}_T^{1/2}) + \frac{\epsilon}{2\eta}\max_{t \leq T}\|\boldsymbol{\beta}^* - \boldsymbol{\beta}_t\|_2^2 \sum_{t=1}^{T}\|\mathbf{X}_t^{1/2}\|$$

$$+ \frac{\sqrt{\epsilon\alpha}T}{\eta}\max_{t \leq T}\|\boldsymbol{\beta}^* - \boldsymbol{\beta}_t\|_2^2.$$

with probability at least $1 - \delta$.

Remark. Following [20], we introduce the quantity $\tilde{r}_t$ to measure the effective rank of the data matrix $\mathbf{C}_t$. When the eigenvalues of $\mathbf{C}_t^\top \mathbf{C}_t$ decrease rapidly, $\tilde{r}_t$ could be significantly smaller than d, even when $\mathbf{C}_t$ is full-rank. Compared with Theorem 1, the upper bound in this theorem contains an additional term caused by the approximation error of full-rank matrices. Note that Theorem 2 means that we can set the number of random projections as $\tau = \widehat{\Omega}(\max_t \tilde{r}_t)$ when the data matrix has low effective rank.

Note that our methods and theories can be extended to the general case $\varphi \neq 0$. We just need to replace the updating rule as

$$\boldsymbol{\beta}_{t+1} = \operatorname*{argmin}_{\boldsymbol{\beta}}\left\{\eta\langle\mathbf{g}_t, \boldsymbol{\beta}\rangle + \eta\varphi(\boldsymbol{\beta}) + B_{\Psi_t}(\boldsymbol{\beta}, \boldsymbol{\beta}_t)\right\}.$$

Although the updating of $\boldsymbol{\beta}_{t+1}$ may not have closed-form solution, the computational cost of $\mathbf{H}_t^{-1}$ can still be reduced dramatically. The regret bound remains on the same order.

4   Experiments

In this section, we conduct numerical experiments to demonstrate the efficiency and effectiveness of our methods.

4.1   Online Convex Optimization

First, we compare our two methods against ADA-FULL, ADA-DIAG, RADA-GRAD [3] and RP-SON [5] on a synthetic data, which is approximately low-rank. Let $\boldsymbol{\beta}_* = \hat{\boldsymbol{\beta}}_*/\|\hat{\boldsymbol{\beta}}_*\|_2$ where each entry of $\hat{\boldsymbol{\beta}}_*$ is drawn independently from $\mathcal{N}(0, 1)$. We consider the problem of online regression where $f_t(\boldsymbol{\beta}) = |\boldsymbol{\beta}^\top \mathbf{x}_t - y_t|$ and $y_t = \boldsymbol{\beta}_*^\top \mathbf{x}_t$. We generate a regression dataset with $T = 10000$ and $d = 500$. In order to meet the requirement of low-rankness, each data point $\mathbf{x}_t$ is sampled independently from a Gaussian distribution $\mathcal{N}(\boldsymbol{\mu}, \Sigma)$ where $\boldsymbol{\mu} = \mathbf{1}$ and Σ has rapidly decaying eigenvalues $\lambda_j(\Sigma) = \lambda_0 j^{-\alpha}$ with $\alpha = 2$ and $\lambda_0 = 100$.

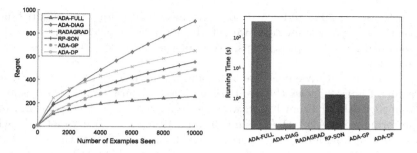

Fig. 1. The left is the comparison of regret among different algorithms on the synthetic data, and the right is the comparison of running time

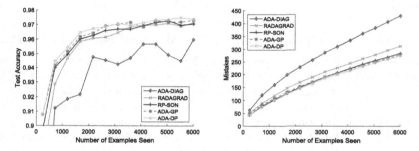

Fig. 2. The left is the comparison of test accuracy among different algorithms on Gisette dataset, and the right is the comparison of mistakes during training

The parameters η and σ are searched in $\{1e-4, 1e-3, \cdots, 100\}$, and we choose the best values for each algorithm. For fairness, all the algorithms are running with the same permutations of the function sequence. For ADA-GP, ADA-DP, RADAGRAD and RP-SON, their results are averaged over 5 runs. Figure 1 shows the regret and running time of different algorithms where we set $\tau = 10$ for methods using random projections. The regret of our two methods are very close and better than ADA-DIAG, RADAGRAD and RP-SON, which indicates our methods approximate ADA-FULL very well. Moreover, our two methods are obviously faster than ADA-FULL according to the comparison of running time.

Second, following [1], we perform online classification with the squared hinge loss (i.e., $f_t(\boldsymbol{\beta}) = \frac{1}{2}\left(\max\left(0, 1 - y_t\boldsymbol{\beta}^\top \mathbf{x}_t\right)\right)^2$) to evaluate the performance of our methods. In each round, the learning algorithm receives a single example and ends with a single pass through the training data. There are two metrics to measure the performance: the online mistakes and the offline accuracy on the testing data.

We conduct numerical experiments on a real world dataset from LIBSVM repository [25]: Gisette which is high-dimensional (i.e. $d = 5000$) and dense. Similar as before, parameters η and σ are searched in $\{1e-4, 1e-3, \cdots, 10\}$ and

$\{2e-4, 2e-3, \cdots, 20\}$, and we choose the best values for each algorithm. To reduce the computational cost, we set the number of projections $\tau = 10$ for methods using random projections. We omit the result of ADA-FULL, because it is too slow.

For training data, we generate 5 random permutations, and report the average result. Figure 2 shows the comparison of test accuracy and mistakes among different algorithms. From Fig. 2, we have some conclusions as following. First, the performance of ADA-DIAG is much worse than all the other methods, which means only keeping a diagonal matrix is insufficient to capture the second-order information. Second, our two methods, ADA-GP and ADA-DP, are better than RADAGRAD and RP-SON. Third, ADA-GP and ADA-DP are close to ADA-FULL, which indicates that random projections cause little adverse affect on the performance.

4.2 Non-convex Optimization in Convolutional Neural Networks

Recently, ADA-DIAG becomes popular for non-convex optimization such as training neural networks, and Krummenacher et al. also show that RADAGRAD performs well for training neural networks [3]. Therefore, we also examine the performance of our method on training convolutional neural networks (CNN). Because the convolutional layer does not meet the case $f_t(\boldsymbol{\beta}_t) = l(\boldsymbol{\beta}_t^\top \mathbf{x}_t)$, we only perform ADA-GP on training CNN. We use the simple and standard architecture shown in Fig. 3 to perform classification on the MNIST [26], CIFAR10 [27] and SVHN [28] datasets.

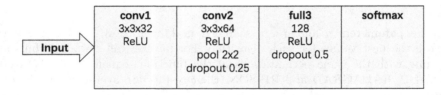

Fig. 3. The 4-layer CNN architecture used in our experiment

Parameters η of all algorithms and δ of ADA-GP and RADAGRAD are searched in $\{1e-4, 1e-3, \cdots, 1\}$. For ADA-DIAG, δ is set to $1e-8$ as it is typically recommended. We choose the best values for each algorithm. Following as [3], we only consider applying ADA-GP and RADAGRAD to the convolutional layer, and other layers are still trained with ADA-DIAG for all datasets. For all algorithms, we run 5 times with batch size 128 and report the average results. Figure 4 shows the comparison of training loss and test accuracy during training among different algorithms where we set $\tau = 20$. We find that ADA-GP obviously improves the performance of ADA-DIAG on all datasets, and note that RADAGRAD is outperformed by ADA-DIAG in term of training loss on CIFAR10. This results shows that ADA-GP is a better approximation of ADA-FULL than RADAGRAD.

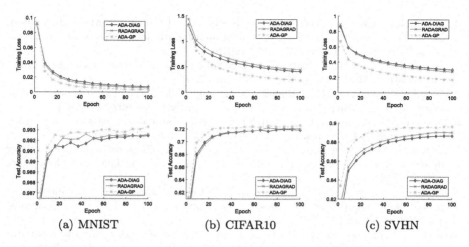

Fig. 4. The comparison of training loss (top row) and test accuracy (bottom row) among different algorithms

5 Conclusions and Future Work

In this paper, we present ADA-GP and ADA-DP to approximate ADA-FULL using random projections. The time and space complexities of both algorithms are linear in the dimensionality d, and thus they are able to accelerate the computation significantly. Furthermore, according to our theoretical analysis, the number of random projections in ADA-DP is on the order of the low rank or low effective rank. Numerical experiments on online convex optimization show that our methods outperform ADA-DIAG, RADAGRAD and RP-SON and are close to ADA-FULL. And experiments on training convolutional neural networks show that ADA-GP outperforms ADA-DIAG and RADAGRAD.

Besides random projection, there exist other ways for low-rank matrix approximations, such as matrix sketching [11]. In the future, we will investigate different techniques to approximate ADA-FULL.

Acknowledgements. This work was partially supported by the NSFC (61603177), JiangsuSF (BK20160658), YESS (2017QNRC001), and the Collaborative Innovation Center of Novel Software Technology and Industrialization.

References

1. Duchi, J., Hazan, E., Singer, Y.: Adaptive subgradient methods for online learning and stochastic optimization. J. Mach. Learn. Res. **12**, 2121–2159 (2011)
2. Zinkevich, M.: Online convex programming and generalized infinitesimal gradient ascent. In: Proceedings of the 20th International Conference on Machine Learning, pp. 928–936 (2003)

3. Krummenacher, G., McWilliams, B., Kilcher, Y., Buhmann, J.M., Meinshausen, N.: Scalable adaptive stochastic optimization using random projections. Adv. Neural Inf. Process. Syst. **29**, 1750–1758 (2016)
4. Hazan, E., Agarwal, A., Kale, S.: Logarithmic regret algorithms for online convex optimization. Mach. Learn. **69**(2), 169–192 (2007)
5. Luo, H., Agarwal, A., Cesa-Bianchi, N., Langford, J.: Efficient second order online learning by sketching. Adv. Neural Inf. Process. Syst. **29**, 902–910 (2016)
6. Xiao, L.: Dual averaging method for regularized stochastic learning and online optimization. Adv. Neural Inf. Process. Syst. **22**, 2116–2124 (2009)
7. Duchi, J., Shalev-Shwartz, S., Singer, Y., Tewari, A.: Composite objective mirror descent. In: Proceedings of the 23rd Annual Conference on Learning Theory, pp. 14–26 (2010)
8. Nalko, N., Martinsson, P.G., Tropp, J.A.: Finding structure with randomness: probabilistic algorithms for constructing approximate matrix decompositions. SIAM Rev. **53**(2), 217–288 (2011)
9. Kaski, S.: Dimensionality reduction by random mapping: Fast similarity computation for clustering. In: Proceedings of the 1998 IEEE International Joint Conference on Neural Networks, vol. 1, pp. 413–418 (1998)
10. Magen, A., Zouzias, A.: Low rank matrix-valued Chernoff bounds and approximate matrix multiplication. In: Proceedings of the 22nd Annual ACM-SIAM Symposium on Discrete Algorithms, pp. 1422–1436 (2011)
11. Woodruff, D.P.: Sketching as a tool for numerical linear algebra. Found. Trends Mach. Learn. **10**(1–2), 1–157 (2014)
12. Fradkin, D., Madigan, D.: Experiments with random projections for machine learning. In: Proceedings of the 9th ACM SIGKDD International Conference on Knowledge Discovery and Data Mining, pp. 517–522 (2003)
13. Rahimi, A., Recht, B.: Random features for large-scale kernel machines. Adv. Neural Inf. Process. Syst. **21**, 1177–1184 (2008)
14. Maillard, O.A., Munos, R.: Linear regression with random projections. J. Mach. Learn. Res. **13**, 2735–2772 (2012)
15. Fern, X.Z., Brodley, C.E.: Random projection for high dimensional data clustering: a cluster ensemble approach. In: Proceedings of the 20th International Conference on Machine Learning, pp. 186–193 (2003)
16. Boutsidis, C., Zouzias, A., Drineas, P.: Random projections for k-means clustering. Adv. Neural Inf. Process. Syst. **23**, 298–306 (2010)
17. Dasgupta, S., Freund, Y.: Random projection trees and low dimensional manifolds. In: Proceedings of the 40th Annual ACM Symposium on Theory of Computing, pp. 537–546 (2008)
18. Freund, Y., Dasgupta, S., Kabra, M., Verma, N.: Learning the structure of manifolds using random projections. Adv. Neural Inf. Process. Syst. **21**, 473–480 (2008)
19. Gao, W., Jin, R., Zhu, S., Zhou, Z.H.: One-pass AUC optimization. In: Proceedings of the 30th International Conference on Machine Learning, pp. 906–914 (2013)
20. Zhang, L., Mahdavi, M., Jin, R., Yang, T., Zhu, S.: Recovering the optimal solution by dual random projection. In: Proceedings of the 26th Annual Conference on Learning Theory, pp. 135–157 (2013)
21. Achlioptas, D.: Database-friendly random projections: Johnson-Lindenstrauss with binary coins. J. Comput. Syst. Sci. **66**(4), 671–687 (2003)
22. Liberty, E., Ailon, N., Singer, A.: Dense fast random projections and lean walsh transforms. Discrete Computat. Geom. **45**(1), 34–44 (2011)
23. Hager, W.W.: Updating the inverse of a matrix. SIAM Rev. **31**(2), 221–239 (1989)

24. Tropp, J.A.: An introduction to matrix concentration inequalities. Found. Trends Mach. Learn. **8**(1–2), 1–230 (2015)
25. Chang, C.C., Lin, C.J.: LIBSVM: a library for support vector machines. ACM Trans. Intell. Syst. Technol. **2**(3), 1–27 (2011)
26. LeCun, Y., Bottou, L., Bengio, Y., Haffner, P.: Gradient-based learning applied to document recognition. Proc. IEEE **86**, 2278–2324 (1998)
27. Krizhevsky, A.: Learning multiple layers of features from tiny images. Technical report, University of Toronto (2009)
28. Netzer, Y., Wang, T., Coates, A., Bissacco, A., Wu, B., Ng, A.Y.: Reading digits in natural images with unsupervised feature learning. In: NIPS Workshop on Deep Learning and Unsupervised Feature Learning 2011 (2011)

Cruising or Waiting: A Shared Recommender System for Taxi Drivers

Xiaoting Jiang, Yanyan Shen$^{(\boxtimes)}$, and Yanmin Zhu$^{(\boxtimes)}$

Shanghai Jiao Tong University, Shanghai, China
{anhuijxt,shenyy,yzhu}@sjtu.edu.cn

Abstract. Recent efforts have been made on mining mobility of taxi trajectories and developing recommender systems for taxi drivers. Existing systems focused on recommending seeking routes to the place with the highest passenger pick-up possibility. They mostly ignore that waiting at nearby taxi stands may also help increase the profit. Furthermore, the recommended results seldom consider potential competitions among drivers and real-time traffic. In this paper, we propose a shared recommender system for taxi drivers by including waiting as one kind of seeking policy. We model a seeking process as a Markov Decision Process, and propose a novel Q-learning algorithm to train the model based on massive trajectory data efficiently. During online recommendation, we update the model using feedbacks from drivers and recommend the optimal seeking policy by taking competitions among drivers and real-time traffic into account. Experimental results show that our system achieves better performance than the state-of-the-art approaches.

Keywords: Recommender system · Reinforcement learning · MDP

1 Introduction

Taxi plays an important role in public transportation service. By analyzing about 1500 million records of taxi trips collected from Shanghai over one year, we observe that different seeking policies may cause huge daily income differences. As shown in Fig. 1a, we can see that (1) the difference between minimum and maximum profits per hour can be up to 60 yuan; (2) on average, only 10% drivers earn more than 40 yuan per hour and up to 10% drivers earn lower than 20 yuan per hour. We further plot the average profit earned by top 10%, middle 30% and bottom 10% drivers over different hours in Fig. 1b, and find that the average profit earned by top 10% drivers in any of the 24 h is roughly twice than that earned by bottom 10% drivers. We raise a question: can we identify effective passenger seeking policies with potentially maximum profit in real time?

Recent efforts have been devoted to recommending effective passenger seeking policies for taxi drivers [1–3]. Most existing systems formulate the problem of discovering the best seeking policy as an optimization problem, with the objective of maximizing potential profit for the next trip. While the proposed

© Springer International Publishing AG, part of Springer Nature 2018
D. Phung et al. (Eds.): PAKDD 2018, LNAI 10938, pp. 418–430, 2018.
https://doi.org/10.1007/978-3-319-93037-4_33

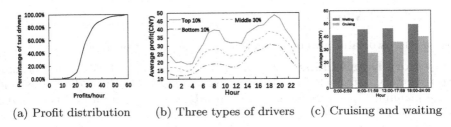

| (a) Profit distribution | (b) Three types of drivers | (c) Cruising and waiting |

Fig. 1. Profit analysis results from trajectory data

approaches are effective in achieving the objective, there still exist several limitations. First, finding optimal seeking policy is known to be NP-hard [5]. Qu et al. [2] proposed a recursion tree method, but it has to make a trade-off between optimality and efficiency. Second, the recommended results often have nothing to deal with real-time environments such as traffic jams and the potential competitions between drivers. Last but not least, taxi drivers may choose to queue at taxi stands waiting for passengers rather than cruise along the roads, but existing systems typically ignore the advantages of waiting. To illustrate the benefits of waiting, we conduct a case study on Shanghai taxi dataset. We identify drivers that wait at airport terminals and those cruise to seek passengers near airports. Figure 1c shows the average profit earned by two groups of drivers in different hours. Apparently, waiting for passengers sometimes achieves higher average profits and hence should also be considered as an effective passenger seeking policy.

To address the above limitations, we develop a recommender system to provide effective seeking policy, with the consideration of real-time environments. Waiting is also included as one of the seeking policies. We model a passenger seeking process as a Markov Decision Process (MDP). We propose a novel Q-learning algorithm to compute the seeking policy with maximum expected profit. The main contributions of this paper are summarized as follows:

- We develop a shared recommender system for taxi drivers that suggests a series of passenger seeking actions in real time with the objective of maximizing the expected profit. We propose a Q-learning approach to the problem efficiently. To the best of our knowledge, we are the first to include waiting as one kind of seeking policy.
- We consider potential competitions among drivers and real-time traffic during online recommendation. We produce seeking actions based on a weighted round robin algorithm to avoid recommending routes run into traffic jams.
- We conduct extensive experiments on real-world trajectories. Results show that our approach makes higher profit than the state-of-the-art solutions.

The remainder of this paper is organized as follows. Section 2 discusses the related works. Section 3 formulates the seeking policy recommendation problem and introduces preliminaries. Section 4 provides details of our system. Section 5 illustrates the experimental results. We conclude this paper in Sect. 6.

2   Related Work

We discuss two categories of works that are related to this paper.

Trajectory Data Mining. Many researches have proposed techniques for mining trajectory data including classification [6], clustering [7] prediction [8] and route planning [12,13]. Some works learn urban mobility from taxi traces [10], while others leverage trajectories for urban planning [9]. Zhang et al. [11] built a model to represent driving patterns and predict drivers' revenue. Chen et al. [13] addressed the bus route planning issue by analyzing taxi traces. Our work also tries to find driving patterns from trajectory data, but we focus on using them to recommend routes for taxi drivers.

Seeking Policy Recommendation for Taxi Drivers. Several works have focused on recommending seeking policy for taxi drivers. The first type is pick-up point recommendation. Existing systems for pick-up point recommendation are developed based on probabilistic models [1], which return the places with high probabilities to pick up passengers. The second type is route recommendation [2–4]. Instead of suggesting several pick-up points, many systems recommend a connected trajectory to increase the profits for taxi drivers. In [2], Qu et al. modeled the road network with a graph and developed effective pruning rules to produce the best seeking routes efficiently. Other works consider the influence of other drivers. Wang et al. [4] proposed a novel approach to calculating the expected revenue of possible routes for individual taxicabs while considering the influence of others. SCRAM [3] provided recommendation fairness for a group of competing taxi drivers, without sacrificing driving efficiency. But none of them considered waiting as a seeking policy.

3   Definition and Problem

Consider a city that is divided into grids of the same size. A taxi trip T is a sequence of spatial points logged by a working taxi from one pick-up point to the corresponding drop-off point, where each point p has the following fields: timestamp $p.t$, latitude $p.lat$, longitude $p.lon$.

Definition 1 (Cruising cost). *We define the cruising cost $Cost_c(T)$ of a trip T as: $Cost_c(T) = l(T) \times GasFee + t_c(T) \times \beta$, where $l(T)$ is the cruising (i.e., a taxi is seeking passengers) distance of T, $t_c(T)$ is the cruising time of T and β is a constant cost per time unit, e.g. vehicle depreciation or company fee.*

Definition 2 (Waiting cost). *We define the waiting cost $Cost_w(T)$ of trip T as: $Cost_w(T) = t_w(T) \times \beta$, where $t_w(T)$ is the waiting time of T and β is a constant cost per time unit.*

Definition 3 (Profit). *We define the average profit of trips in which pick-up points are in grid g as $Profit(g,t)$, as follows.*

$$Profit(g,t) = \frac{\sum_{i=1}^{N_s(g,t)} Earn(T_i;g)}{N_s(g,t)}$$

where $N_s(g, t)$ is the number of taxi trips that have pick-up events in grid g during time period t. $Earn(T; g) = \frac{Fee(T) - Cost(T)}{t(T)}$. $t(T)$ is the total time of T. $Fee(T)$ is the taxi fee getting from trip T. $Cost(T)$ contains gas fee caused by the trip and the cruising/waiting costs before the trip starts.

Definition 4 (Action). *An action of a taxi is its moving direction from one grid to another. We consider 9 actions: {up, down, left, right, upper left, upper right, bottom left, bottom right, waiting nearby}, where waiting nearby is a novel action in this paper.*

Definition 5 (Road network). *The road network can be represented by a graph $G = \langle V, E \rangle$, where $V = \{g_i\}$ consists of all grids and E is the edge set which satisfies $\exists e_{ij} \in E$ iff g_i, g_j are neighboring grids and there exists a path connecting them.*

Given a road network, the location of a taxi and time t, our goal is to provide a seeking policy in the current grid at time t to maximize the expected profit. To formulate the problem, we first model the passenger seeking process as a Markov Decision Process (MDP) [14], which is defined as a four-field tuple $(S, A, P(s'|s, a), R)$. In our context, (1) S is a set of states and each state corresponds to a pair of gird g and time t; (2) $A(s)$ is the set of valid actions that can be taken in grid s; (3) $P(s'|s, a)$ is the transition probability from state s to state s' when action a is taken; (4) $R_a(s, s')$ represents the immediate reward received after transition from state s to state s', due to action a.

We define the seeking policy as π, which is a mapping from S to A. $\pi : S \rightarrow A$. $\pi(s)$ is action to do at state s. We use a value function $V : S \rightarrow R$ to measure the reward accumulated by π, i.e., the accumulated profit obtained when a taxi moves among states according to the policy. Let $V^\pi(s)$ denotes value of policy at state s. $V^\pi(s)$ not only depends on immediate reward, but also the reward subsequently by following π which is referred to delayed reward.

$$V^\pi(s) = \sum_{i=1}^{n} \gamma^i R_{a_i}(s_i, s_{i+1}) \tag{1}$$

where $0 \leq \gamma \leq 1$ is a discount factor represents the importance of delayed reward and immediate reward. n is the number of actions taken following policy π.

We now formalize the optimal seeking policy discovery problem as follows.

Definition 6 (Optimal Seeking Policy Discovery). *The Optimal Seeking Policy Discovery problem is to find an optimal policy π that maximizes the cumulative expected profits $V^\pi(s)$.*

$$\pi^* \equiv \underset{\pi}{\mathrm{argmax}}\, V^\pi(s) \tag{2}$$

Since our problem is a finite-horizon MDP problem, the optimal seeking policy on our problem exists [15]. However, the MDP problem is known to be NP-hard, and there exist many heuristic strategies to solve the problem including dynamic programming methods [16], Monte Carlo methods [17] and temporal difference learning [18]. As the computation complexity of dynamic programming is very high and the convergence rate of Monte Carlo methods is known to be low, we propose to use Q-learning which is a kind of temporal difference learning methods. Q-learning combines the advantages of Monte Carlo methods and dynamic programming and can handle optimization problems without the knowledge of the environment.

4   Online Shared Recommender System

4.1   Overview

Figure 2 shows the framework of our shared recommender system. It contains two major components:

- Offline optimal policy learning: given historical trajectory data, we develop an MDP model to simulate passenger seeking process and leverage Q-learning [22] method to learn model parameters efficiently. The resultant model is able to deliver optimal seeking policy that maximizes the expected profit for drivers.
- Competition-aware online recommendation: given a location and current time from a taxi driver, our shared recommendation system provides the best seeking policy that considers both traffic condition and potential competitions from other drivers.

4.2   Offline Optimal Policy Learning

Taxi Stands Detection. In our MDP model, $A(g)$ is the set of available movements from grid g to another grid. For each grid g, we use road network to decide valid actions that can be taken in the grid. However, whether waiting at nearby

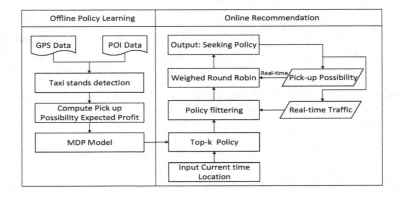

Fig. 2. System framework

action is allowed depends on the existence of taxi stands in the current grid. A straightforward way is to use POI information. If POIs such as airports, shopping malls and railway stations exist in a grid, it is very likely to have a taxi stand in that grid. However, this approach may not be able to identify all the taxi stands accurately. In this section, we propose a data-driven algorithm to detect taxi stands that leverages both POI information and taxi trajectories. Our algorithm consists of three steps: waiting trajectory detection, filtering and clustering.

Waiting Trajectory Detection. Given a trajectory T, we try to detect whether it contains a subsequence that is a *waiting trajectory*. If so, we add the subsequence into a set C. The algorithm of detecting waiting trajectories as follows:

- Stage 1: Finding start waiting point. Given a trajectory T, if the distance between two consecutive points p_j and p_{j+1} is smaller than a distance threshold τ, p_j is considered as a start waiting point.
- Stage 2: Finding end waiting point. We compute the distances between every two adjacent points after p_s, and find the first two points $(p_j, p_{j+1})(j > s)$ whose distance is large than τ. p_j is identified as the end waiting point p_e.
- Stage 3: Getting waiting trajectory. We calculate the time difference t between start waiting point p_s and end waiting point p_e. If t is within the range of $[\delta_{min}, \delta_{max}]$, the sub-trajectory from p_s to p_e is regarded as a waiting trajectory and put into set C.
- Stage 4: Termination. When all trajectories $T \in \mathbb{T}$ are checked, the algorithm terminates and outputs candidate set C.

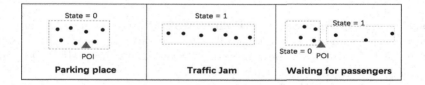

Fig. 3. Three types of trajectories.

Waiting Trajectory Filtering. Given a set C of waiting trajectories, we have to distinguish whether the taxi is waiting at a taxi stand or experiencing traffic jams. Figure 3 shows the differences between parking places, traffic jams and waiting for passengers. If a waiting trajectory satisfies the following two conditions, we consider it to be waiting at taxi stands and add it to C'. (1) The distance from p_c (the geometrical center of the trajectory) to its nearest POI with a taxi stand is within a given threshold. (2) Taxi state in the end of waiting trajectory changes from unoccupied to occupied.

Waiting Trajectory Clustering. Different waiting trajectories may actually point to the same waiting place. Hence, we perform clustering over waiting trajectories

---

Algorithm 1. Discovering optimal seeking policy

    Input: Trajectory T; Learning rate α; Discount rate γ
    Output: Policy Q
1 Initialize Q;
2 repeat
3   Random choose a start state s;
4   **repeat**
5   greedy = random(0,1);
6   **if** *greedy* $< \varepsilon$ **then**
7 | Select an action a with the highest $Q(s, a)$
8   **else**
9 | Select possible action a at random
10   **end**
11   Take action a, observe next state s';
12   Compute expected revenue as reward $R_a(s, s')$);
13 $Q(s_i, a_i) \leftarrow$
                $Q(s_i, a_i) + \alpha \cdot (R_{a_i}(s_i, s_{i+1}) + \gamma \cdot \max\limits_{a} Q(s_{i+1}, a) - Q(s_i, a_i))$;
14 $s \leftarrow s'$;
15   **until** *Find passengers*;
16 until *Converage*;
17 Return Q

---

in C' to eliminate duplicated taxi stands. To do this, we first calculate the geometrical center p_c^i of a waiting trajectory T_i. We then use DBSCAN [19] to cluster all p_c^i and each cluster center is a taxi stand.

Discovering Optimal Seeking Policy. We now present our Q-learning algorithm that finds an optimal seeking policy for any given (finite) Markov decision process (MDP). And it has been proved [22] that for any finite MDP, Q-learning eventually converges to an optimal policy.

Our method works by learning an action-value function $Q(s, a)$ that ultimately gives the expected utility of taking an action in a given state s following the optimal policy. After the function is learned, optimal policy can be constructed by selecting the action with the highest value in each state.

Before learning starts, Q is first initialized by a fixed value. Each time the agent selects an action, observes a reward and a new state based on the previous state and the selected action. Then Q is updated. The update of action-value function follows the Bellman equation as follows [20].

$$Q(s_i, a_i) \leftarrow Q(s_i, a_i) + \alpha \cdot (R_{a_i}(s_i, s_{i+1}) + \gamma \cdot \max_{a} Q(s_{i+1}, a) - Q(s_i, a_i)) \quad (3)$$

where α is the learning rate ($0 < \alpha < 1$) and $R_a(s, s')$ represents the immediate reward in our MDP model. Here, γ is the discount rate that indicates the trade-off between the sooner rewards and later rewards.

The taxi seeking policy recommendation algorithm is outlined in Algorithm 1. We use $\varepsilon - greedy$ policy algorithm to make a balance between exploration and exploitation. ε is the possibility the agent selects the action with the highest value and $1 - \varepsilon$ is the possibility the agent randomly selects an action.

After taking the selected action(moving to a certain grid), the taxi gets a reward. Meanwhile, the action-value function is updated. An episode of the algorithm finishes when the driver finds passengers.

Simulating Seeking Process. Different from traditional reinforcement learning for routing problem in which taxi can learn from real driving experience, we cannot let taxi drivers to drive for Q-learning. Furthermore, it is difficult to use historical trajectory data directly. The reasons are two-fold. First, at the beginning of learning process, the agent sometimes cannot make a reasonable recommendation. In fact, it is almost impossible to find historical trajectories that match our recommended routes. Second, our algorithm may need a large number of iterations to converge and it may not be likely to find enough historical trajectories for each grid.

To address the problems, we design a method to simulate seeking process based on historical trajectories. In the Q-learning algorithm, we can assume that there is a taxi following our policy to drive using the simulating seeking process. Given the time period t_c, current location g_i and action a_i, we first calculate the possibility $P(g_i, t_c)$ for a taxi to find passengers in grid g_i. $P(g_i, t_c)$ equals to the total numbers of taxi pass through grid g_i divided by the pick-up events in g_i in time period t_c. If the taxi finds passengers, the process will terminate. Else the taxi will continue moving to the next grid according to the action a_i in our policy until it finds passengers.

4.3 Online Recommendation

We use action-value function Q to find the best action for each state. When a taxi asks for a recommendation, its current grid and time will be forwarded to the system. Our system responses to the request by continuously providing a seeking policy to the driver until she finds passengers successfully. When providing seeking policies, we also consider road condition and the potential competitions between drivers. Our algorithm performs the following three steps.

Top k Policy Candidate Generation. Let the current state be s. For all actions in $A(g)$, we first rank them by the values of $Q(s, a)$. We then obtain the first k actions with the highest Q-values and add them into the candidate set K.

Candidate Policy Filtering. This step aims to filter actions which will lead taxis into traffic jams based on the candidate policies. We decide whether a grid at time t involves traffic jams by computing the traffic performance index $idx(g, t)$ for the grid. $idx(g, t) = \frac{V(g)}{V_c(g,t)}$, where $V_c(g, t)$ is the average speed of all taxis in grid g during the time period t, and $V(g)$ is the average speed of all taxis in grid g during all day time. After analyzing historical data in peak hours and according to our experience, we use traffic performance index to remove the actions from the candidate set which lead taxis to grids with traffic index above 2.5.

Situation-Aware Policy Recommendation. When more than one taxi drivers are asking for seeking policy at the same place during the same time period t, we have to consider the competitions between them. We adopt a weighted round robin algorithm [21] to solve this problem.

Consider a taxi asks for recommendation in grid g at time period t, we first calculate the weight w of each action in candidate set C. $w_a = P(g_a, t)$, where g_a is the grid the taxi will move to after action a. Then the taxi will select the action a with the highest weight in C. Finally, the traffic index $idx(g_a, t)$ and the pick up possibility of $P(g_a, t)$ will be updated based on the feedback information.

At last, we want to discuss the computational complexity of online recommendation to make sure it is fast enough to handle all the online traffic. Given the number of all actions in $A(g)$ as N, the computational complexity of online recommendation is $O(NlogN)$, where $N < 10$, which means that our system is fast enough to handle all the online traffic.

5   Experiments

5.1   Experimental Setup

Datasets. *Taxi GPS traces.* We use the real-world taxi GPS traces in Shanghai within one year. This dataset contains about 1.5 billion records where each record consists of the longitude, latitude, speed, status (occupied or not), time stamp and direction angle for a taxi. After preprocessing, we obtain about 11 million pick-up activities. We use the first 270-day data to build our system and evaluate the method using the remaining 95-day data.

POI data. We use the POI data obtained from Baidu Map which has 17 categories including transport facilities, living quarters and so on.

Compared Methods. We compare our proposed waiting strategy considered MDP(WMDP) method with the following two methods.

MNP approach [2]. Given a current location of a taxi, the MNP approach recommends the next five road segments with the maximum total expected profit. If the taxi cannot find passengers, the algorithm will recommend the next five road segments based on the new location.

Pick-up possibility based method. Given a current location of a taxi, this method recommends the next grid with the largest passenger pick-up possibility.

Parameter Settings. In our experiment, we set $Gasfee = 0.4\,yuan/km$ and $\beta = 43\,yuan/day$ by default. For a trip T with the total journey of x kilometers,

$$fee = \begin{cases} 14yuan & x \leq 3 \\ (14 + 2.5(x-3))yuan & 3 < x \leq 15 \\ (44 + 2.5(x-15))yuan & x \geq 15 \end{cases}$$

We follow the discussions in Sect. 5.2 for setting the parameters in the Q-learning algorithm.

5.2   Parameter Tuning

We evaluate the effects of learning rate α, discount rate γ and the probability of exploration ε. We measure the average passing grids before finding passengers in Table 1.

Table 1. Impact of parameters

Fixed parameters	Unfixed parameter	Average grids
$\gamma = 0.3, \varepsilon = 0.2$	$\alpha = 0.75$	10.40
	$\alpha = 0.5$	10.11
	$\alpha = 0.25$	9.53
$\alpha = 0.5, \varepsilon = 0.2$	$\gamma = 0.3$	10.11
	$\gamma = 0.5$	10.26
	$\gamma = 0.7$	10.19
$\alpha = 0.5, \gamma = 0.3$	$\varepsilon = 0.1$	10.63
	$\varepsilon = 0.2$	10.11
	$\varepsilon = 0.3$	10.34

We first discuss the impact of α by fixing $\gamma = 0.3$ and $\varepsilon = 0.1$. The average number of passing grids decreases when α becomes smaller which means that lower learning-rate value performs better. We then fixed $\alpha = 0.5$ and $\varepsilon = 0.1$, and vary the value of γ. When $\gamma = 0.3$, the algorithm performs slightly better than other values of gamma. This is because in our problem, immediate rewards have more impact than future rewards. Finally, we compare different ε by fixing $\alpha = 0.5$ and $\gamma = 0.3$. Smaller value of ε gets better results, but at first, ε should be larger, because the agent need more exploration to find better policy at the beginning of the algorithm. Therefore, in the following experiment, we choose $\alpha = 0.25, \gamma = 0.3$. ε is changed every 10,000 episodes. For the N-th episode, if $N < 5,000,000$, $\varepsilon = 0.3 - 0.0004 \frac{N}{10000}$. Otherwise, we set ε to 0.1.

5.3   Comparison Results

To evaluate the influence of load balance method, we consider our algorithm with load-balance method (LWMDP) and without load-balance method (WMDP). We simulate queries in Shanghai with all the drop-off points (in total 166,241 points) during five days. Assuming that half of the taxi drivers use our system, and they will query seeking policy when they finish a trip, and assume our system to be queried every 5.22 s. The results of the two algorithm are shown in Fig. 4. Comparing to WMDP, LWMDP can improve 21.28% profits.

We compare the driving routes based on the recommendation results with those without recommendation. Figure 5 presents the results for average profits of different driving routes. Obviously, our method outperforms the competing methods for all time slots. Moreover, as shown in Fig. 6, the advantage of our recommender system is more significant in off-peak hours than peak hours. The reason is that the passenger pickup demands during peak hours are typically higher than off-peak hours and it is easier for ordinary taxi drivers to find passengers even without recommendation. During off-peak hours, it is difficult for inexperienced drivers (whose profit is lower than 50% drivers) to find passengers because of the limited pickup demands. And our method can help inexperienced drivers to find passengers and make higher profits.

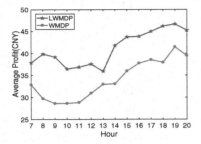

Fig. 4. Load balance results

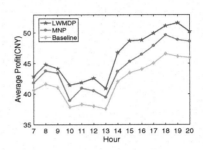

Fig. 5. Average profit comparison

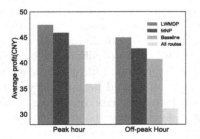

Fig. 6. Average profit results

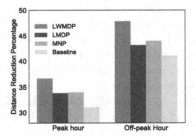

Fig. 7. Improvement achieved by waiting strategy

We further discuss the effects of waiting strategy on recommendation. We compare distance and profit with and without using waiting strategy in our algorithm. As shown in Fig. 7, during off-peak hours, considering waiting as a seeking policy can reduce the total seeking distance significantly. During peak hours, waiting strategy can also reduce the total seeking distance, although the improvement is not as much as that in off-peak hours.

6   Conclusion

In this paper, we develop a shared recommender system for taxi drivers. We model the seeking policy recommendation problem as an MDP problem. We propose a Q-learning algorithm to find an optimal solution efficiently. During the real-time recommendation, our system produces seeking policies by taking both the competitions between drivers and traffic jams into account. We conduct extensive experiments using real-world taxi trajectories. The results show that our method can increase the profits by 36.59%. Compared to the existing solutions, we are the first to include waiting as one kind of seeking policy. As our system is considering competitions among drivers, we can use game theory to find an equilibrium the drivers can cooperate to reach in order to increase driving efficiency in our future work.

Acknowledgments. This research is supported in part by 973 Program (No. 2014CB340303), NSFC (No. 61772341, 61472254, 61170238, 61602297 and 61472241), and Singapore NRF (CREATE E2S2). This work is also supported by the Program for Changjiang Young Scholars in University of China, the Program for China Top Young Talents, and the Program for Shanghai Top Young Talents.

References

1. Yuan, N., Zheng, Y., Zhang, L., Xie, X.: T-finder: a recommender system for finding passengers and vacant taxis. In: TKDE, vol. 25(10), pp. 2390–2403 (2013)
2. Qu, M., Zhu, H., Liu, J., Liu, G., Xiong, H.: A cost-effective recommender system for taxi drivers. In: KDD, pp. 45–54 (2014)
3. Qian, S., Cao, J., et al.: SCRAM: a sharing considered route assignment mechanism for fair taxi route recommendations. In: KDD, pp. 955–964 (2015)
4. Wang, Y., Liang, B., et al.: The development of a smart taxicab scheduling system: a multi-source data fusion perspective. In: ICDM, pp. 1275–1280 (2016)
5. Qian, S., Zhu, Y., Li, M.: Smart recommendation by mining large-scale GPS traces. In: WCNC, pp. 3267–3272 (2012)
6. Lee, J., Han, J., et al.: TraClass: trajectory classification using hierarchical region-based and trajectory-based clustering. In: PVLDB, vol. 1(1), pp. 1081–1094 (2008)
7. Liu, S., Liu, Y., Ni, L., Fan, J., Li, M.: Towards mobility-based clustering. In: KDD, pp. 919–928 (2010)
8. Chen, C., Zhang, D., Castro, P., Li, N., Sun, L., Li, S.: Real-time detection of anomalous taxi trajectories from GPS traces. In: MobiQuitous, pp. 63–74 (2011)
9. Zheng, Y., Liu, Y., Yuan, J., Xie, X.: Urban computing with taxicabs. In: MobiQuitousg, pp. 89–98 (2011)
10. Veloso, M., Phithakkitnukoon, S., Bento, C.: Urban mobility study using taxi traces. In: TDMA, pp. 23–30 (2011)
11. Zhang, D., Sun, L., Li, B., Chen, C.: Understanding taxi service strategies from taxi GPS traces. In: ITITS, vol. 16(1), pp. 123–135 (2015)
12. Yuan, J., Zheng, Y., Xie, X., Sun, G.: Driving with knowledge from the physical world. In: KDD, pp. 316–324 (2011)
13. Chen, C., Zhang, D., Zhou, Z., Li, N., Atmaca, T., Li, S.: B-Planner: night bus route planning using large-scale taxi GPS traces. In: PerCom, pp. 225–233 (2013)

14. Bellman, R.: Science, vol. 153, no. 3731, pp. 34–37 (1966)
15. Puterman, M.L.: Taxi-RS: Markov Decision Processes: Discrete Stochastic Dynamic Programming, p. 353 (1994)
16. Bellman, R: The Theory of Dynamic Programming, pp. 716–719 (1952)
17. Hastings, W.K: Monte Carlo sampling methods using Markov chains and their applications, pp. 665–685. Springer, Heidelberg (1998)
18. Sutton, R.S.: Reinforcement learning. Biometrika **57**(1), 97–109 (1970)
19. Ester, M., Kriegel, H.P., Xu, X., et al.: A density-based algorithm for discovering clusters in large spatial databases with noise. In: KDD, pp. 226–231 (1996)
20. Sutton, R.S.: Learning to predict by the methods of temporal differences. Mach. Learn. **3**(1), 9–44 (1988)
21. Shimonishi, H., et al.: An improvement of weighted round robin cell scheduling in ATM networks. In: GLOBECOM, vol. 2, pp. 1119–1123 (1997)
22. Watkins, C.J.C.H., Dayan, P.: Technical note Q-learning. In: Machine Learning, vol. 8, pp. 279–292 (1992)

A Local Online Learning Approach for Non-linear Data

Xinxing Yang[✉], Jun Zhou, Peilin Zhao, Cen Chen,
Chaochao Chen, and Xiaolong Li

Ant Financial Services Group, Hangzhou, China
{xinxing.yangxx,jun.zhoujun,chencen.cc,chaochao.ccc,xl.li}@antfin.com,
peilinzhao@hotmail.com

Abstract. The efficiency and scalability of online learning methods make them a popular choice for solving the learning problems with big data and limited memory. Most of the existing online learning approaches are based on global models, which consider the incoming example as linear separable. However, this assumption is not always valid in practice. Therefore, local online learning framework was proposed to solve non-linear separable task without kernel modeling. Weights in local online learning framework are based on the first-order information, thus will significantly limit the performance of online learning. Intuitively, the second-order online learning algorithms, e.g., Soft Confidence-Weighted (SCW), can significantly alleviate this issue. Inspired by the second-order algorithms and local online learning framework, we propose a **S**oft **C**onfidence-**W**eighted **L**ocal **O**nline **L**earning (SCW-LOL) algorithm, which extends the single hyperplane SCW to the case with multiple local hyperplanes. Those local hyperplanes are connected by a common component and will be optimized simultaneously. We also examine the theoretical relationship between the single and multiple hyperplanes. The extensive experimental results show that the proposed SCW-LOL learns an online convergence boundary, overall achieving the best performance over almost all datasets, without any kernel modeling and parameter tuning.

Keywords: Online learning · Optimization · Non-linear data

1 Introduction

In recent years, online learning algorithms play more important role in data mining and machine learning. Opposed to traditional batch learning techniques which generate the best prediction by learning on the entire training data, online learning just processes one instance at a time and updates the current model with streaming data repeatedly. The key idea of online learning is to update current

© Springer International Publishing AG, part of Springer Nature 2018
D. Phung et al. (Eds.): PAKDD 2018, LNAI 10938, pp. 431–443, 2018.
https://doi.org/10.1007/978-3-319-93037-4_34

model by retraining the new model close to the current one, while imposing a margin separation on the most recent examples [21]. As online learning does not need to track any history examples, it avoids expensive re-training cost and reduces large memory consumption. These advantages of online learning make it increasingly popular in applications: such as malicious URL detection [25], anomaly detection [20], image retrieval [24], learning to rank [19], collaborative filtering [15], and etc. Hence, the efficiency and scalability of online learning methods make them a popular choice for solving the learning problems with big data and limited memory.

A variety of global online learning algorithms have been developed by machine learning community to process sequential data, which handle the examples with a global hyperplane. Some first-order online learning algorithms were proposed, such as Perceptron Algorithm [7,18] and Passive Aggressive (PA) algorithm [3]. Because first-order online learning algorithm only uses the first-order information, a large number of second-order algorithms have emerged to improve the performance on sequential data. The most representative second-order online learning algorithms are Confidence-Weighted (CW) [6], Adaptive Regularization of Weights (AROW) [4] and Soft Confidence-Weighted (SCW-I, SCW-II) algorithms [21]. These second-order algorithms maintain a Gaussian distribution over some linear classifier hypotheses and apply it to control the direction and scale of parameter updates [6]. Both first-order and second-order online learning algorithms assume that incoming examples are almost linearly separable. However, the examples are linearly non-separable or even non-linear in practice. To address the non-linear separation problem in online learning, kernel algorithms were studied. These algorithms include direct application of kernel tricks on online linear classifiers [7,13], budget-based online models [1,5,8,22], and kernel approximation mapping based models [16]. Because of keeping wrong classified examples as support vectors, kernel online learning methods require more computation and memory.

On the other hand, in order to address linearly non-separable tasks, some local classifiers have been proposed in offline learning. Locally linear support vector machine [14] and local deep kernel learning [12] assume that in a sufficiently small region the decision boundary is approximately linear and the data is locally linearly separable. Those local classifiers avoid kernel modeling and are significantly faster than kernel methods, however they are not specifically designed for online learning tasks. As an improvement of local prototype methods [9,10,14,23], a novel local online learning framework [26] was proposed, which jointly learns multiple local hyperplanes to process non-linear sequential data in a one-pass manner. The weights in local online learning framework are based on first-order information, thus limits its performance. Meanwhile, the second-order based online learning algorithms, e.g., Soft Confidence-Weighted (SCW), can overcome this limitation.

Inspired by both second-order algorithms and local online learning framework, we propose a **S**oft **C**onfidence-**W**eighted **L**ocal **O**nline **L**earning (SCW-LOL) algorithm, which extends the single hyperplane SCW method to the case with multiple local hyperplanes. The SCW-LOL algorithm consists of two components: a global hyperplane and multiple local hyperplanes. All the local hyperplanes are connected by one global hyperplane and will be optimized simultaneously. We also show the theoretical relationship between the single and multiple hyperplanes and proof the online convergence boundary is not higher than SCW method. Experiment results show that our proposed method consistently achieves the best performance on various tasks compared to other online learning methods, especially on multi-class classification tasks.

The paper is outlined as follows. Section 2 reviews the related literature. Section 3 details the SCW-LOL algorithm and the theoretical analysis. Experiment results are shown in Sect. 4, and finally Sect. 5 concludes our paper.

2   Related Work

2.1   Global Methods

One of the well-known online learning algorithms for learning a linear threshold function is Perceptron Algorithm [7,18]. Another popular thread of first-order online learning approach is Passive-Aggressive algorithm [3], which only updates the classifier when a new example's prediction loss score is less than a predefined margin. Rosenblatt [18] empirically shows that the maximum margin based online learning algorithms are more effective than the Perceptron Algorithm.

Because first-order based online learning algorithms only use the first-order information, a large number of second-order algorithms have emerged to improve online learning performance on sequential data. The most representative second-order online learning algorithms are Confidence-Weighted (CW) [6], Adaptive Regularization of Weights (AROW) [4] and Soft Confidence-Weighted (SCW-I, SCW-II) algorithms [21]. These second-order algorithms maintain a Gaussian distribution over some linear classifier hypotheses and apply it to control the direction and scale of parameter updates [6]. Experimental results show that SCW algorithms significantly outperform the original CW algorithm and perform better than other second-order online learning methods with greater efficiency [21].

Both first-order and second-order online learning algorithms assume that incoming examples are almost linearly separable. However the examples are linearly non-separable or even non-linear in practice. To address the non-linear separation problem in online learning, some non-linear algorithms were studied. These algorithms include direct application of kernel tricks on online linear classifiers [7,13], budget-based online models [1,5,8,22], and kernel approximation mapping based models [16] by using random Fourier features or method. However, some limitations exist in these methods, including memory overflow after processing a large amount of data while keeping the historical examples [7,13] and huge computational burden caused by processing support vectors in the budget [1,5,8,22].

2.2   Local Methods

Local classifiers, such as locally linear support vector machine [14] and local deep kernel learning [12], have been proposed recently to solve the linearly non-separable tasks in offline learning, which avoid kernel modeling and are significantly faster than kernel methods. In those local classifiers, data examples are assigned to a set of prototypes and used for updating the weights for model combination of local classifiers. Those classifiers assume that in a sufficiently small region the decision boundary is approximately linear and data is locally linearly separable. However, they are not specifically designed for online learning tasks.

As an improvement of local prototype methods [9,10,14,23], a novel local online learning framework [26] was proposed, which jointly learns multiple local hyperplanes to process non-linear sequential data in a one-pass manner. This approach considers the non-linear separable task as multiple linear separable tasks and aims to learn multiple local hyperplanes on one-pass examples. It achieves notably better performance without using any kernel modeling in one-pass manner. The weights in local online learning framework are based on first-order information, thus could significantly limit the performance of online learning.

3   SCW Local Online Learning

First, we present the preliminary in Subsect. 3.1. Then we describe the objective function and the inference for weight updating in SCW-LOL algorithm in Subsect. 3.2. As online clustering is the key difference between local and traditional global online learning methods, we further describe our online clustering strategy in Subsect. 3.3. Finally, we present the pseudo-code of SCW-LOL algorithm and proof the online convergence boundary in Subsect. 3.4.

3.1   Preliminary

Online learning focuses on processing streaming examples with time steps. At time step t, the algorithm meets an incoming example $x_t \in R^d$ and predicts its label $\overline{y}_t \in \{-1, +1\}$. After the prediction, the true label $y_t \in \{-1, +1\}$ is received and the suffered loss $\ell(y_t, \overline{y}_t)$ is calculated, which is the difference between its prediction and true label. The loss is updated to weights w_t of the model at each time step t and the goal of online learning is to minimize the cumulative error over the whole streaming examples.

Soft Confidence-Weighted (SCW) learning is one of widely used online learning methods. It extends the confidence-weighted learning for soft margin learning and assumes that weights w follow the Gaussian distribution with mean vector μ and covariance matrix Σ, which makes it more robust than other online learning methods. The optimization of SCW learning is formulated as:

$$(\mu_{t+1}, \Sigma_{t+1}) = \arg\min_{\mu, \Sigma} D_{KL}(N(\mu, \Sigma), N(\mu_t, \Sigma_t)),$$

$$s.t. \qquad y_t \mu^T x_t \geq \phi \sqrt{x_t^T \Sigma x_t}, \tag{1}$$

where D_{KL} stands for the Kullback-Leibler divergence, $\phi = \Phi^{-1}(\eta)$ (Φ is the cumulative function of the normal distribution and η is the hyper-parameter) and the loss function is stated below [21]:

$$\ell^{SCW}(N(\boldsymbol{\mu}, \boldsymbol{\Sigma}); (\boldsymbol{x}_t, y_t)) = \max\left(0, \phi\sqrt{\boldsymbol{x}_t^T \boldsymbol{\Sigma} \boldsymbol{x}_t} - y_t \boldsymbol{\mu}^T \boldsymbol{x}_t\right). \tag{2}$$

Similar to Passive-Aggressive algorithms, SCW-I includes slack variable ξ and reformulates optimization (1) for learning soft-margin classifiers as follows:

$$(\boldsymbol{\mu}_{t+1}, \boldsymbol{\Sigma}_{t+1}) = \arg\min_{\boldsymbol{\mu}, \boldsymbol{\Sigma}} D_{KL}(N(\boldsymbol{\mu}, \boldsymbol{\Sigma}), N(\boldsymbol{\mu}_t, \boldsymbol{\Sigma}_t))$$
$$+ C\ell^{SCW}(N(\boldsymbol{\mu}, \boldsymbol{\Sigma}); (\boldsymbol{x}_t, y_t)). \tag{3}$$

The above optimization (3) has the closed-form solution:

$$\boldsymbol{\mu}_{t+1} = \boldsymbol{\mu}_t + \alpha_t y_t \boldsymbol{\Sigma}_t \boldsymbol{x}_t, \quad \boldsymbol{\Sigma}_{t+1} = \boldsymbol{\Sigma}_t - \beta_t \boldsymbol{\Sigma}_t \boldsymbol{x}_t \boldsymbol{x}_t^T \boldsymbol{\Sigma}_t. \tag{4}$$

The updating coefficients are as follows:

$$\alpha_t = \max\{C, \max\{0, \frac{1}{v_t \zeta}(-m_t \psi + \sqrt{m_t^2 \frac{\phi^4}{4} + v_t \phi^2 \zeta})\}\}, \beta_t = \frac{\alpha_t \phi}{\sqrt{u_t} + v_t \alpha_t \phi},$$

where $u_t = \frac{1}{4}\left(-\alpha_t v_t \phi + \sqrt{\alpha_t^2 v_t^2 \phi_t^2 + 4v_t}\right)^2$, $v_t = \boldsymbol{x}_t^T \boldsymbol{\Sigma}_t \boldsymbol{x}_t$, $m_t = y_t \boldsymbol{\mu}^T \boldsymbol{x}_t$, $\phi = \Phi^{-1}(\eta)$, $\psi = 1 + \frac{\phi^2}{2}$ and $\zeta = 1 + \phi^2$.

3.2   Model

Although SCW algorithm can handle some noisy and non-separable cases, it has limitations on linearly non-separable or even non-linear classification tasks. We propose SCW-LOL algorithm to address classification tasks for non-linear data.

In SCW-LOL model, there are two components: a global component $(\boldsymbol{\mu}, \boldsymbol{\Sigma})$ shared by all local components and k specifical local components $(\boldsymbol{\theta}_i, \boldsymbol{\Gamma}_i)$, $i = 1, ..., k$. The optimization of SCW-LOL is formulated as follows:

$$((\boldsymbol{\mu}_{t+1}, \boldsymbol{\Sigma}_{t+1}), (\boldsymbol{\theta}_{1,t+1}, \boldsymbol{\Gamma}_{1,t+1}), ..., (\boldsymbol{\theta}_{k,t+1}, \boldsymbol{\Gamma}_{k,t+1})) =$$
$$\arg\min_{(\boldsymbol{\mu}, \boldsymbol{\Sigma}), (\boldsymbol{\theta}_1, \boldsymbol{\Gamma}_1), ..., (\boldsymbol{\theta}_k, \boldsymbol{\Gamma}_k)} \lambda D_{KL}(N(\boldsymbol{\mu}, \boldsymbol{\Sigma}) \| N(\boldsymbol{\mu}_t, \boldsymbol{\Sigma}_t))$$
$$+ \sum_{i=1}^{k} D_{KL}(N(\boldsymbol{\theta}, \boldsymbol{\Gamma}) \| N(\boldsymbol{\theta}_{i,t}, \boldsymbol{\Gamma}_{i,t})),$$
$$s.t.\ y_t\left(\boldsymbol{\mu}^T \boldsymbol{x}_t + \sum_{i=1}^{k} \theta_i^T \boldsymbol{x}_t\right) \geq \phi\sqrt{\boldsymbol{x}_t^T \boldsymbol{\Sigma} \boldsymbol{x}_t + \sum_{i=1}^{k} \boldsymbol{x}_t^T \boldsymbol{\Gamma}_i \boldsymbol{x}_t}. \tag{5}$$

Let the incoming example $\tilde{\boldsymbol{x}}_t$ be redefined as:

$$\tilde{\boldsymbol{x}}_t = \left[\frac{\boldsymbol{x}_t^T}{\sqrt{\lambda}}, \boldsymbol{0}_1^T, ..., \boldsymbol{0}_{j-1}^T, \boldsymbol{x}_t^T, \boldsymbol{0}_{j+1}^T, ..., \boldsymbol{0}_k^T\right]^T, \tag{6}$$

where x_t is clustered to the jth local component. Similarly, the weights of global and local components could be represented as follows:

$$\tilde{\mu} = \left[\sqrt{\lambda}\mu^T, \theta_1^T, ..., \theta_{j-1}^T, \theta^T, \theta_{j+1}^T, ..., \theta_k^T \right]^T,$$ (7)

$$\tilde{\Sigma} = \begin{bmatrix} \sqrt{\lambda}\Sigma, & 0, & ..., & 0 \\ 0, & \Gamma_1, & ..., & 0 \\ \vdots & 0, & \ddots & 0 \\ 0, & ..., & ..., & \Gamma_k \end{bmatrix}.$$ (8)

We use SCW-I optimization function as an example and other methods can be extended easily in our framework.

Plugging Eqs. (6), (7) and (8) into the optimization model (5), the objective function could be rewritten as follow:

$$\left(\tilde{\mu}_{t+1}, \tilde{\Sigma}_{t+1} \right) = \arg\min_{\tilde{\mu}, \tilde{\Sigma}} D_{KL} \left(N \left(\tilde{\mu}, \tilde{\Sigma} \right) || N \left(\tilde{\mu}_t, \tilde{\Sigma}_t \right) \right),$$

$$s.t. \qquad y_t \tilde{\mu}^T \tilde{x}_t \geq \phi \sqrt{\tilde{x}_t^T \tilde{\Sigma} \tilde{x}_t}.$$ (9)

In this way, SCW-LOL algorithm learns locally sensitive online classifier. The above formulation (9) also has the closed-form solution:

$$\tilde{\mu}_{t+1} = \tilde{\mu}_t + \tilde{\alpha}_t y_t \tilde{\Sigma}_t \tilde{x}_t, \quad \tilde{\Sigma}_{t+1} = \tilde{\Sigma}_t - \tilde{\beta}_t \tilde{\Sigma}_t \tilde{x}_t \tilde{x}_t^T \tilde{\Sigma}_t.$$ (10)

The updating coefficients are calculated as Eq. (10):

$$\tilde{\alpha}_t = \max\{C, \max\{0, \frac{1}{\tilde{v}_t \tilde{\zeta}}(-\tilde{m}_t \tilde{\psi} + \sqrt{\tilde{m}_t^2 \frac{\tilde{\phi}^4}{4} + \tilde{v}_t \tilde{\phi}^2 \tilde{\zeta})}\}\},$$

$$\tilde{\beta}_t = \frac{\tilde{\alpha}_t \tilde{\phi}}{\sqrt{\tilde{u}_t} + \tilde{v}_t \tilde{\alpha}_t \tilde{\phi}},$$ (11)

where $\tilde{u}_t = \frac{1}{4}\left(-\tilde{\alpha}_t \tilde{v}_t \tilde{\phi} + \sqrt{\tilde{\alpha}_t^2 \tilde{v}_t^2 \tilde{\phi}_t^2 + 4\tilde{v}_t} \right)^2$, $\tilde{v}_t = \tilde{x}_t^T \tilde{\Sigma}_t \tilde{x}_t$, $\tilde{m}_t = y_t \tilde{\mu}^T \tilde{x}_t$, $\tilde{\phi} = \Phi^{-1}(\eta)$, $\tilde{\psi} = 1 + \frac{\tilde{\phi}^2}{2}$ and $\tilde{\zeta} = 1 + \tilde{\phi}^2$.

3.3 Online Clustering

Different from traditional online learning algorithm, local online learning (LOL) needs to assign the incoming example to specific local prototype component and updates the prototype weights online. In order to process example x_t for prototyping in a sequential order, we include a sequential version of K-means [17] which uses the first k examples as initial prototypes at time step $t <= k$. When at time step $t > k$, it assigns the new incoming example x_t to the closest local prototype which is learned from time step $t - 1$, and update the assigned local prototype weights [26]. So the rule of online clustering could be defined as:

$$P_{i,t+1} = P_{i,t} + \frac{x_t - P_{i,t}}{n_{i,t}}, \tag{12}$$

where $P_{i,t}$ is the ith local prototype and $n_{i,t}$ is the total number of previous examples assigned to the ith local prototype at time step t. It is possible that some noisy examples could affect the online clustering performance, but experiments show that the gaps between sequential K-means and the offline K-means are small enough to be tolerated after sufficient training [26].

3.4   Algorithm

Algorithm 1 shows the details of SCW-LOL algorithm.

Algorithm 1. SCW Local Online Learning (SCW-LOL)

Input: parameters: $C > 0, \eta > 0, \lambda > 0$
   the number of local prototypes: $k > 0$
1 Initialization: $\tilde{\mu}_0 = (0, ..., 0)^T, \tilde{\Sigma}_0 = I$.
2 **for** $t = 1$ to T **do**
3   receive example: $x_t \in R^d$
4   **if** $t \le k$ **then**
5 | local prototype initialization: $P_t \leftarrow x_t, n_t = 1$
6   **end**
7   **else**
8   find the closest local prototype: $i = \arg\min_j Distance(P_j, x_t)$
9   update assigned local prototype: $P_i \leftarrow P_i + \frac{x_t - P_i}{n_i}, n_i \leftarrow n_i + 1$
10   **end**
11   build example: $\tilde{x}_t = \left[\frac{x_t^T}{\sqrt{\lambda}}, 0_1^T, ..., 0_{i-1}^T, x_t^T, 0_{i+1}^T, ..., 0_k^T \right]^T, \tilde{x}_t \in R^{kd}$
12   make prediction: $\overline{y}_t = sgn\left(\tilde{\mu}^T \tilde{x}_t\right)$
13   receive true label: y_t
14   calculate loss: $\ell = \max\left(0, \tilde{\phi}\sqrt{\tilde{x}_t^T \tilde{\Sigma}\tilde{x}_t} - y_t\tilde{\mu}^T \tilde{x}_t\right), \tilde{\phi} = \Phi^{-1}(\eta)$
15   **if** $\ell > 0$ **then**
16       $\tilde{\mu}_{t+1} = \tilde{\mu}_t + \tilde{\alpha}_t y_t \tilde{\Sigma}_t \tilde{x}_t, \ \tilde{\Sigma}_{t+1} = \tilde{\Sigma}_t - \tilde{\beta}_t \tilde{\Sigma}_t \tilde{x}_t \tilde{x}_t^T \tilde{\Sigma}_t$
17   where $\tilde{\alpha}_t$ and $\tilde{\beta}_t$ are calculated by Eq. (11)
18   **end**
19 **end**

In SCW-LOL algorithm, we project the weights $\tilde{\mu}$ by Eq. (7) and $\tilde{\Sigma}$ by Eq. (8). Based on the rebuilt example $\tilde{x}_t$ in Eq. (6), we have $\ell^{SCW}(N(\mu, \Sigma); (x_t, y_t)) = \ell^{LOL}\left(N(\tilde{\mu}, \tilde{\Sigma}); (\tilde{x}_t, y_t)\right)$, where ℓ^{SCW} and ℓ^{LOL}

represent the same loss function. Thus, local online learning is actually equivalent to linear SCW method, which indicates that SCW-LOL method will not obtain a higher loss bound than SCW algorithm.

4   Evaluation

4.1   Environment Setup

Datasets: We evaluate our approach on 10 benchmark datasets [2] from different domains: (1) MIT cbcl face data[1]; (2) machine learning datasets from LIBSVM, such as "ijcnn1", "splice", "cod-rna", "svmguide1", "usps", "letter", "mnist", "pendigits" and "shuttle"[2]. Table 1 shows the details of the datasets.

Table 1. Datasets details

Dataset	#Training	#Testing	#Features	#Classes
ijcnn1	49,990	91,701	22	2
splice	1,000	2,175	60	2
cod-rna	59,535	271,617	8	2
cbcl_face	6,977	24,045	361	2
svmguide1	3,089	4,000	4	2
ups	7,291	2,007	256	10
letter	15,000	5,000	16	26
mnist	60,000	10,000	780	10
pendigits	7,494	3,498	16	10
shuttle	43,500	14,500	9	7

Baselines: We compare our approach with various baselines: (1) first-order Passive Aggressive (PA) [3], (2) second-order confidence weighted family methods, such as Confidence-Weighted (CW) [6], Adaptive Regularization of Weight (AROW) [4], and Soft Confidence-Weighted (SCW-I, SCW-II) algorithms [11,21], (3) kernel approximation online learning methods [16] such as FOGD and NOGD and (4) Passive Aggressive Local Online Learning (PA-LOL) algorithm [26] (Table 2).

Parameter Setting: We follow the similar parameter settings in [21]. The parameter C in PA, SCW-I, and SCW-II as well as the parameter r in AROW are all determined by cross validation to select the best one from $\{2^{-4}, 2^{-4}, ..., 2^3, 2^4\}$; the parameter η in CW, SCW-I, and SCW-II are determined by cross validation to select the best one from $\{0.5, 0.55, ..., 0.9, 0.95\}$. For

---

[1] http://cbcl.mit.edu/software-datasets/FaceData2.html.

[2] https://www.csie.ntu.edu.tw/~cjlin/libsvmtools/datasets/.

Table 2. Properties of compared methods

Methods	Second order	Using kernel	Non-linear	Local
PA	No	No	No	No
CW	Yes	No	No	No
AROW	Yes	No	No	No
SCW-I	Yes	No	No	No
SWC-II	Yes	No	No	No
FOGD	No	Yes	Yes	No
NOGD	No	Yes	Yes	No
PA-LOL	No	No	Yes	Yes
SCW-LOL	Yes	No	Yes	Yes

kernel approximation online learning methods[3], the number of fourier components D in FOGD is set as 2000 and k in NOGD is set as 200 and learning
rate is 0.001. After parameters are determined, all experiments are conducted
over 10 random shuffles for each dataset and all results are averaged over these
10 runs. For local online learning method, similar to parameter setting in [26],
we fix the number of local prototypes k to 60, balancing parameter λ to 1 and
the aggressive parameter C to 1. In multi-class classification task, we adopt
one-vs-all strategy to predict result.

Metrics: We evaluate the performance on the following metrics: (1) cumulative
error rate, which is defined as $\frac{\sum_{i=1}^{t} I(y_t \neq \overline{y}_t)}{t}$ at time step t. It shows the accuracy
of prediction for sequential examples and (2) test error rate: the model is trained
by the train set and tested using the test set. It indicates the generalization
ability of the model on unseen examples.

Table 3. Test error of the methods in binary-class classification

Methods	ijcnn1	splice	cod-rna	cbcl_face	svmguide1
PA	0.0763 ± 0.0023	0.1921 ± 0.0496	0.1132 ± 0.0029	0.0364 ± 0.0145	0.2173 ± 0.0001
CW	0.1043 ± 0.0209	0.2342 ± 0.0101	0.1080 ± 0.0346	0.0311 ± 0.0022	0.2162 ± 0.0000
AROW	0.0803 ± 0.0002	0.1473 ± 0.0014	0.1111 ± 0.0001	0.0341 ± 0.0005	0.2041 ± 0.0003
SCW-I	0.0587 ± 0.0011	0.1530 ± 0.0022	**0.0486 ± 0.0000**	0.0286 ± 0.0061	0.2099 ± 0.0005
SCW-II	0.0589 ± 0.0007	0.1520 ± 0.0018	0.0487 ± 0.0002	0.0260 ± 0.0015	0.1990 ± 0.0001
FOGD	0.0437 ± 0.0061	0.2095 ± 0.0769	0.1259 ± 0.0001	0.0401 ± 0.0069	**0.0357 ± 0.0009**
BOGD	0.0950 ± 0.0001	0.3178 ± 0.0999	0.1217 ± 0.0030	0.0654 ± 0.0231	0.0359 ± 0.0021
PA-LOL	0.0489 ± 0.0048	0.1811 ± 0.0806	0.1672 ± 0.0032	0.0338 ± 0.0048	0.1537 ± 0.0151
SCW-LOL	**0.0235 ± 0.0018**	**0.1282 ± 0.0101**	0.1098 ± 0.0021	**0.0219 ± 0.0019**	0.1351 ± 0.0106

---

[3] https://github.com/LIBOL/KOL.

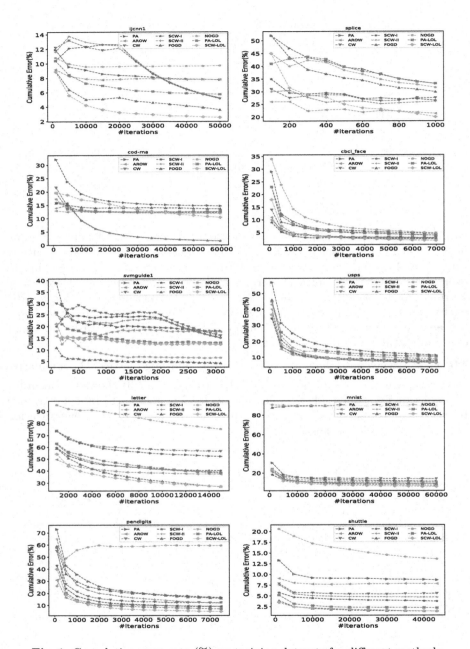

Fig. 1. Cumulative error rate (%) on training datasets for different methods.

Table 4. Test error of the methods in multi-class classification

Methods	usps	letter	mnist	pendigits	shuttle
PA	0.1141 ± 0.0054	0.5109 ± 0.0240	0.0943 ± 0.0040	0.1685 ± 0.0153	0.0979 ± 0.0248
CW	0.1107 ± 0.0029	0.5646 ± 0.0239	0.1275 ± 0.0037	0.1459 ± 0.0076	0.0733 ± 0.0372
AROW	0.0872 ± 0.0016	0.3724 ± 0.0015	0.0867 ± 0.0008	0.1411 ± 0.0028	0.0738 ± 0.0005
SCW-I	0.0868 ± 0.0015	0.3915 ± 0.0052	0.0772 ± 0.0006	0.1277 ± 0.0028	0.0255 ± 0.0027
SCW-II	0.0844 ± 0.0022	0.3758 ± 0.0044	0.0789 ± 0.0006	0.1351 ± 0.0062	0.0285 ± 0.0161
FOGD	0.0758 ± 0.0096	0.2416 ± 0.0027	0.9083 ± 0.0124	0.1654 ± 0.01066	0.1588 ± 0.0026
BOGD	0.0968 ± 0.0041	0.5961 ± 0.0227	0.9012 ± 0.0317	0.5965 ± 0.01348	0.1047 ± 0.0278
PA-LOL	0.0948 ± 0.0048	0.3118 ± 0.0132	0.0591 ± 0.0041	0.0764 ± 0.0094	0.0228 ± 0.0086
SCW-LOL	**0.0725 ± 0.0029**	**0.2097 ± 0.0022**	**0.04829 ± 0.0013**	**0.0568 ± 0.0089**	**0.0138 ± 0.0028**

4.2 Results

Table 3 shows the test error for different methods on binary-class classification task over 5 different datasets. SCW-LOL performs the best for 3 datasets and achieves lower error rate than PA, CW, AROW, and PA-LOL on all datasets. The SCW family algorithms perform well on "cod-rna" dataset and is far ahead of other methods. Online kernel methods, i.e., FOGD and NOGD, outperform the rest of the methods on "svmguide1" dataset by one order of magnitude.

Table 4 shows the test error for multi-class classification on 5 different datasets. SCW-LOL performs the best among all datasets. FOGD and NOGD are inferior on "mnist" dataset as opposed to others. Both PA-LOL and SCW-LOL perform better than other baselines in multi-class classification.

Figure 1 shows the cumulative error rate (%) on all datasets. Overall, the error rates decrease over time, except for FOGD and NOGD on "mnist" dataset. SCW-LOL method achieves the best cumulative error rate in most datasets.

Overall, we can see that SCW-LOL significantly outperforms other methods in most datasets. Compared with the second-order algorithms, such as CW and AROW methods, SCW consistently outperforms the traditional second-order online learning algorithms, while SCW-LOL algorithm still performs better than SCW method in all datasets except for "cod-rna" dataset. This demonstrates that local hyperplane technique is a good way to address non-linear separable task. For online kernel methods, both FOGD and NOGD, have good performance in binary classification and close to other online learning algorithms. But in multi-class task, the kernel methods performance are worse. Meanwhile, local online learning methods PA-LOL and SCW-LOL perform well in multi-class classification. As SCW-LOL method can learn the second-order information, it outperforms PA-LOL on all datasets and has stable improvement.

5 Conclusion

In order to solve non-linear separable task on streaming data, we proposed a soft Confidence-Weighted local online learning (SCW-LOL) algorithm, which learns from streaming data and updates local model. The incoming samples will be assigned to the right local model, thus solving the non-linear separable problem. Moreover, our method keeps the second-order information and outperforms the first-order online learning methods. From the loss bound analysis, SCW-LOL will not obtain a higher loss bound than SCW. Our experimental results show that SCW-LOL has outstanding performance in streaming data, especially in multi-class classification tasks. In the near future, we will improve our proposed sequential version of the K-means clustering with an online Gaussian Mixture Model (GMM), where the parameter k will be automatically learned.

References

1. Cavallanti, G., Cesa-Bianchi, N., Gentile, C.: Tracking the best hyperplane with a simple budget perceptron. Mach. Learn. **69**(2–3), 143–167 (2007)
2. Chang, C.C., Lin, C.J.: LIBSVM: a library for support vector machines. TIST **2**(3), 27 (2011)
3. Crammer, K., Dekel, O., Keshet, J., Shalev-Shwartz, S., Singer, Y.: Online passive-aggressive algorithms. JMLR **7**, 551–585 (2006)
4. Crammer, K., Kulesza, A., Dredze, M.: Adaptive regularization of weight vectors. In: NIPS, pp. 414–422 (2009)
5. Dekel, O., Shalev-Shwartz, S., Singer, Y.: The forgetron: a kernel-based perceptron on a budget. SIAM J. Comput. **37**(5), 1342–1372 (2008)
6. Dredze, M., Crammer, K., Pereira, F.: Confidence-weighted linear classification. In: ICML, pp. 264–271. ACM (2008)
7. Freund, Y., Schapire, R.E.: Large margin classification using the perceptron algorithm. Mach. Learn. **37**(3), 277–296 (1999)
8. Friedman, J.H., Tukey, J.W.: A projection pursuit algorithm for exploratory data analysis. IEEE Trans. Comput. **100**(9), 881–890 (1974)
9. Gou, J., Zhan, Y., Rao, Y., Shen, X., Wang, X., He, W.: Improved pseudo nearest neighbor classification. KBS **70**, 361–375 (2014)
10. Gu, Q., Han, J.: Clustered support vector machines. In: AISTATS, pp. 307–315 (2013)
11. Hoi, S.C., Wang, J., Zhao, P.: LIBOL: a library for online learning algorithms. JMLR **15**(1), 495–499 (2014)
12. Jose, C., Goyal, P., Aggrwal, P., Varma, M.: Local deep kernel learning for efficient non-linear SVM prediction. In: ICML, pp. 486–494 (2013)
13. Kivinen, J., Smola, A.J., Williamson, R.C.: Online learning with kernels. IEEE Trans. Sig. Process. **52**(8), 2165–2176 (2004)
14. Ladicky, L., Torr, P.: Locally linear support vector machines. In: ICML, pp. 985–992 (2011)
15. Lu, J., Hoi, S., Wang, J.: Second order online collaborative filtering. In: ACML, pp. 325–340 (2013)
16. Lu, J., Hoi, S.C., Wang, J., Zhao, P., Liu, Z.Y.: Large scale online kernel learning. JMLR **17**(47), 1 (2016)
17. MacQueen, J., et al.: Some methods for classification and analysis of multivariate observations. In: Proceedings of the Fifth Berkeley Symposium on Mathematical Statistics and Probability, Oakland, CA, USA, vol. 1, pp. 281–297 (1967)
18. Rosenblatt, F.: The perceptron: a probabilistic model for information storage and organization in the brain. Psychol. Rev. **65**(6), 386 (1958)

19. Wang, J., Wan, J., Zhang, Y., Hoi, S.C.: Solar: scalable online learning algorithms for ranking. In: ACL (2015)
20. Wang, J., Zhao, P., Hoi, S.C.: Cost-sensitive online classification. TKDE **26**(10), 2425–2438 (2014)
21. Wang, J., Zhao, P., Hoi, S.C.: Soft confidence-weighted learning. ACM Trans. Intell. Syst. Technol. (TIST) **8**(1), 15 (2016)
22. Wang, Z., Crammer, K., Vucetic, S.: Breaking the curse of kernelization: budgeted stochastic gradient descent for large-scale SVM training. JMLR **13**, 3103–3131 (2012)
23. Weinberger, K., Dasgupta, A., Langford, J., Smola, A., Attenberg, J.: Feature hashing for large scale multitask learning. In: ICML, pp. 1113–1120. ACM (2009)
24. Wu, P., Hoi, S.C., Xia, H., Zhao, P., Wang, D., Miao, C.: Online multimodal deep similarity learning with application to image retrieval. In: MM, pp. 153–162 (2013)
25. Zhao, P., Hoi, S.C.: Cost-sensitive online active learning with application to malicious URL detection. In: SIGKDD, pp. 919–927. ACM (2013)
26. Zhou, Z., Zheng, W.S., Hu, J.F., Xu, Y., You, J.: One-pass online learning: a local approach. Pattern Recogn. **51**, 346–357 (2016)

Contextual Location Imputation for Confined WiFi Trajectories

Elham Naghizade[1]([✉]), Jeffrey Chan[2], Yongli Ren[2], and Martin Tomko[1]

[1] Department of Infrastructure Engineering,
University of Melbourne, Melbourne, Australia
{enaghi,mtomko}@unimelb.edu.au
[2] School of Science, RMIT University, Melbourne, Australia
{jeffrey.chan,yongli.ren}@rmit.edu.au

Abstract. The analysis of mobility patterns from large-scale spatio-temporal datasets is key to personalised location-based applications. Datasets capturing user location are, however, often incomplete due to temporary failures of sensors, deliberate interruptions or because of data privacy restrictions. Effective location imputation is thus a critical processing step enabling mobility pattern mining from sparse data. To date, most studies in this area have focused on coarse location prediction at city scale. In this paper we aim to infer the missing location information of individuals tracked within structured, mostly *confined* spaces such as a university campus or a mall. Many indoor tracking datasets may be collected by sensing user presence via WiFi sensing and consist of timestamped associations with the network's access points (APs). Such coarse location information imposes unique challenges to the location imputation problem. We present a contextual model that combines the regularity of individuals' visits to enable accurate imputation of missing locations in sparse indoor trajectories. This model also considers implicit social ties to capture similarities between individuals, applying Graph-regularized Nonnegative Matrix Factorization (GNMF) techniques. Our findings suggest that people' movement in confined spaces is largely habitual and their social ties plays a role in their less frequently visited locations.

Keywords: Spatio-temporal trajectories · Data imputation
Matrix factorization

1 Introduction

The uptake in sensor-enabled smartphones and devices has enabled capturing human mobility data at a fine-grained scale. Such rich spatio-temporal datasets facilitate the (i) understanding of individuals' movement patterns and potentially their intention, (ii) delivery of tailored recommendations and services to individuals or groups as well as (iii) understanding the space use patterns and provision of public and social applications that benefit from people-centric sensing.

© Springer International Publishing AG, part of Springer Nature 2018
D. Phung et al. (Eds.): PAKDD 2018, LNAI 10938, pp. 444–457, 2018.
https://doi.org/10.1007/978-3-319-93037-4_35

In data mining, *data quality* is a key predictor of the success of analyses. Large spatio-temporal datasets are, however, often incomplete due to sensor failure or, low sampling rates to maximise a sensor's life. Location data imputation has been suggested as a solution to tackle this issue [8].

Research has focused primarily on large-scale outdoor mobility datasets. Such trajectories cover a larger space and are the location is captured at fine precision using GPS measurement at fine intervals. These approaches are not suitable to coarse trajectories based on proximity sensing (symbolic trajectories) [5]. We explore location imputation for trajectories recorded using WiFi sensing across multiple indoor or well defined confined spaces. These trajectories capture a user's proximity to a single access point. This imposes unique challenges to trajectory sensing: the captured locations are coarse and may be sparse. A range of efforts has focused on using additional information such as Received Signal Strength Indicators to improve location points, or used crowd-sourced data to calibrate the location information and interpolate missing points [7,16].

Social ties between visitors to these environments represent a significant source of information that remains largely unexplored in location prediction. Human physical behaviour is strongly mediated by social interactions. A key question is *how can social ties between individuals improve location imputation models in coarse trajectories?* Several studies explored the ability to predict social links between users given their trajectories [12,15], primarily over large areas and using explicit social links (e.g., using Facebook data). Here, we explore physical co-presence to explore implicit social ties.

Factorization techniques has been successfully used in recommender systems – systems that estimate users' preferences (e.g., rankings) for unrated items. As such, they address a similar problem – the inference of missing values associated with a user's behaviour. Here, we utilise the graph-regularized non-negative matrix factorization technique (GNMF) to impute missing locations., based on the intuition that individuals with stronger social ties express higher trajectory similarity and are also more likely to visit specific locations together. Motivated by the approach proposed in [9], we build user profiles that capture users spatio-temporal behaviour (regularity, temporal entropy and gaps of visits). We further implicitly capture social ties between users. These are inferred based on physical co-presence, as actual information about users' relationships may not be known. A co-presence graph captures the strength of users' associations through edge weights (frequency, duration and location of co-presence between pairs of users). Finally, we build an affinity graph between locations, with edges capturing the number of times each pair of locations has been visited together in a trajectory.

Using two real-world datasets (a university campus and a shopping mall) we explore two location imputation scenarios: (i) the case of partially missing location information in individual trajectories (i.e., due to sensor failure), and (ii) the case of location imputation for new users, never observed in the environment in the past (cold start). Our experiments show that the consideration of social co-presence improves the performance of location imputation for, in particular, less frequently visited locations of a user, e.g., top 10 to 20 locations.

This is significant, as these are individually important locations that are not shared with others. We also observe that the history of users' visits plays an important role in predicting missing locations.

2   Related Work

2.1   Collaborative filtering in location recommendation

Collaborative filtering techniques have been successfully used for recommending movies, books, points of interest to users whose rating for a certain item is not known. This is why we focus on it for our location imputation problem.

Zheng et al. [19] provide an example of successfully using tensor factorization to decompose a user-location-activity tensor and provide suggestions on points of interest or activities using GPS trajectories of the individuals and their social network profiles. Yin et al. [17] use a user's location history as well as local preference to provide location recommendations. This approach particularly focus on recommending location/event to users with known profiles, but no location history. The model in [11], takes the the distance between regions, their popularity and user mobility patterns to build a geographical probabilistic latent model for location recommendation.

Similar to [6], this work assume an implicit feedback based on the frequency of users' visit to each places. The collaborative filtering model proposed in [9], leverages users' mobility to retrieve user similarities, however, it assumes an implicit positive feedback for locations that the user has visited rather than considering a negative feedback to the unvisited locations. It also uses the retrieved user profile or location features as the additional contextual information to improve the model. Further, the authors in [10,18] have focused on time-aware location recommendation problem and utilise the temporal affects of visiting places as well as spatial influences when providing location recommendations. Unlike the above-mentioned studies, we focus on a confined area and proximity-based trajectories rather than explicit location check-ins or GPS trajectories within large areas. Also, most of these studies assume the availability of other sources of data to extract additional information and boost the performance of the algorithm, while we do not use any information other than the mobility data.

2.2   Social Ties and Mobility Patterns

The study in [4] is one of the early attempts to find the relationship between individuals' movement patterns and their online social network. Similarly, the authors in [2] explore the spatial and social influences on individuals' mobility patterns. Mobility patterns has been used to predict users' social ties, including using an entropy-based measure of the diversity and frequency of their co-locations [13] and similarities between users' trajectories [14]. Similar to [14], we use how specific the location of co-presence is.

On the other hand, social ties have been consequently used to predict users' location. In [12], McGee et al. build a network based on social ties to estimate the location of users. In particular, they used the links between users in Twitter as well as users' profile information to estimate the distance between any pair of users. The authors in [15], proposed a hybrid model to capture both the regularity of individuals' repetitive patterns as well as the influence of their social network, to predict their location. This approach also uses additional sources of information and ensures improved location information through heterogeneous data sources. These studies mainly focus on low resolution location predictions, e.g., city level resolutions.

3   Problem Description

Consider a mobility dataset D consisting of observations o of tracked individual users. Each observation o, $o = \{id, t, l\}$, is a tuple of a user, identified by a unique id, at (symbolic) location $l \in L$ at timestamp t. Assuming time to be discrete offsets from an arbitrary origin, the time of the observations belongs to a range between $[0, T_{max}]$, where T_{max} is the timestamp of the last observation in the dataset. Dividing the timestamps into intervals of window size w, results in a set of distinct trajectories of an individual:

For all users u_i from the set of users U ($\forall u_i \in U$) we have a set of sequences, s_j, in the form of $s_j = \{(l_1, t_1^d), (l_2, t_2^d), ..., (l_m, t_m^d)\}$ where $1 \leq i \leq M$ and $t^d \in \frac{T_{max}}{w}$. Assuming N_i is the number of u_i's trajectories in the dataset, we denote the set of distinct trajectories of u_i as $S_i = \{s_j | j = 1, 2, ... N_i\}$. In this paper we assume a day as the window size w, i.e., a user has at most one trajectory per day (Table 1).

Table 1. Frequently used notations.

Symbol	Description	Symbol	Description
A	User/Location matrix	α	Confidence weighting function
C	User/Location confidence matrix	G_s, G_l	Implicit social and spatial(location) graphs
P	User profile matrix	λ_s, λ_l	Social and location graph weighting parameters
M	Number of users	ω	Social strength parameter
N	Number of locations	ρ	Regularization parameter
k	Number of latent features	K	top location query parameter

We assume some sessions are either partially or completely missing in D. We focus on the location information in the data and aim to impute users' location. One possible formalisation of this problem is to represent individuals' visited locations in a matrix:

Definition 1 - User-location Matrix (A): From the set of users' sequences, S_i, one can derive a matrix of size $M \times N$, where N is the cardinality of the location set and $A_{i,n}$ corresponds to the number of times u_i has been observed at l_n over all his/her trajectories. For any location n that is not visited by the i_{th} user, $A_{i,n} = 0$. It is noteworthy that elements in A do not consider the sequence of a user's movement, but rather the frequency of being seen by a certain access point during all visits.

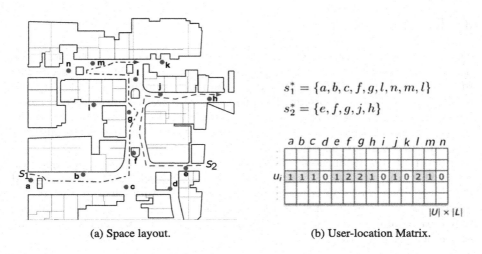

$$s_1^* = \{a, b, c, f, g, l, n, m, l\}$$
$$s_2^* = \{e, f, g, j, h\}$$

(a) Space layout.　　　　(b) User-location Matrix.

Fig. 1. Two sample trajectories of a user in an indoor space. The red circles indicate the location of access points. Note that $s_{1,2}^*$ represent the footprint of a session and the time of the observations is not captured in the user-location matrix (Color figure online).

Intuition: Individuals movement behaviour in an indoor space largely follows a frequent pattern. Intuitively, we would like to use the similarities between such movement pattern (reflected in the frequency of visits to certain locations) as well as user similarities that is potentially reflected in their temporal profile in order to impute the missing location frequencies. Furthermore, being socially connected to others can be reflected in an individual's movement pattern [13–15]. We aim to leverage the knowledge on potential social ties to improve our location imputation model.

Figure 1a depicts two sample trajectories of a user (dashed lines). The red circles represent the access points and Fig. 1b shows the corresponding row in A for this scenario. Thus, assuming some location information of u_i in Fig. 1b is missing and the respective row in A is then $[1, 1, ?, 0, 1, ?, 2, ?, 0, 1, 0, ?, 1, 0]$, the aim of our location imputation technique is to predict the missing values, shown as question marks.

4 Matrix Factorization for Location Imputation

4.1 Location Imputation with Implicit Feedback

One possible approach to infer unknown values in A is to apply matrix factorization (MF) techniques and decompose A into two matrices. The underlying idea is that MF uses the data of other users to discover the latent features that govern the interaction between uses and locations.

Given that matrix A has missing values, i.e., missing location information for some users, and assuming that we have k latent features, the aim of MF is to find two matrices $U_{M \times k}$ and $V_{k \times N}$ that approximate A in the following way:

$$U \times V = A' \approx A \quad where \quad a'_{ij} = \sum_{k=1}^{k} u_{ik} v_{kj}$$

U intuitively captures user similarities in the k-dimensional latent feature space and V reflects location similarities. The product of the obtained U and V provides the estimate values for missing location information.

As discussed in [9], in a mobility dataset, a visit to a certain location can reflect an individual's positive feedback to that location. However, not visiting a location does not necessarily imply negative feedback. In our setting, an individual can actually be present at a location, but not accessing the WiFi network would exacerbate this issue, e.g., having the mobile phone off. Hence, similar to [6], we define a confidence matrix, C, whose elements are computed as follows:

$$c_{i,j} = \begin{cases} \alpha(a_{i,j}) & \text{if } a_{i,j} > 0 \\ 1 & \text{otherwise,} \end{cases} \tag{1}$$

where α is a monotonically increasing function. Having a binary matrix R, where $r_{u,l} = 1$ if u_i has visited l_j at least once during his/her visits, the objective function can be expressed as minimizing the following error (ρ is the regularization parameter):

$$\mathcal{O} = \sum_{i,j} c_{i,j} (r_{ij} - u_{ik} v_{kj})^2 + \frac{\rho}{2} (\|U\|_F^2 + \|V\|_F^2)$$

4.2 Contextual Imputation

The direct factorization does not use the contextual information to improve the location predictions. In this paper we assume no additional source of information about the users is available and propose to create a user profile matrix, P, which mainly incorporates temporal features of users' visits. We focus on the following factors:

– *The average duration* of each visit, which is determined as $1/N_i \sum_{j=1}^{N_i} (t_{end_j} - t_{start_j})$, where t_{start_j} and t_{end_j} are the first and last point of trajectory s_j of user i.

- *Regularity* of the visits captures the predictability of an individual's visits with respect to time. We compute the entropy of a user's visits given specific days of the week, hour of the day as well as the combination of both, i.e., $\{D\}, \{H\}, \{D, H\}$ to estimate whether or not they follow some repetitive temporal pattern.
- *Temporal Gap* between the visits reflects how often a user is observed in the dataset. The regularity measure cannot differentiate regular visits that happen daily from those that happen monthly, hence we compute the temporal gap, in terms of the intervening number of days between a user's consecutive visit.

The above-mentioned factors are estimated for each individual and form a $M \times k'$ matrix, where k' is the number of derived factors in P. We augment U with this information and create $\hat{U}_{(M+k') \times k}$ matrix. Following the proposed algorithm in [9], we can rewrite the objective function as:

$$\mathcal{O} = \sum_{i,j} c_{i,j} (r_{ij} - \hat{u}_{ik} v_{kj})^2 + \frac{\rho}{2} (\|\hat{U}\|_F^2 + \|V\|_F^2)$$

4.3   Implicit Social Ties and GNMF

The Contextual Imputation approach aims to capture users' similarity to assign locations to users, however, the features discussed in the previous section does not consider potential social ties. As discussed in Sect. 2, mobility patterns of individuals can be strongly correlated to their social dependencies. As a result, we propose to leverage such connections to improve our location imputation task. Similar to the previous section, we assume that no additional source of data, e.g., social networks, is available and aim to build an implicit social graph using the available movement data.

To create our implicit social graph, G_s, we build a co-presence graph where there is an edge between any pair of users who have been present at a given location at least for one timestamp. Given that such graph may be noisy (since it adds an edge between random users who are at a popular location at the same time), we use the following parameters in addition to the co-presence frequency to estimate the social strength between any pair of individual:

- *Relative Average Duration:* The duration of co-occurence at each session can be an indicator of the strength of the relationship, however, considering the absolute duration of the co-presence has two weaknesses as i) it does not consider the overall length of users' trips, i.e., a co-presence of for instance 10 min for a 10-min trip is much more intense compared to a 60-min trip and ii) it has not an absolute upper-bound. Hence, the relative average duration computes the duration of co-presence divided by the duration of users' trips:

$$d_{i,j} = 1/N \sum_{n=1}^{f_{i,j}} \frac{d_{ij,n}}{min(d_{i,n}, d_{j,n})}$$

where $1 \leq n \leq f_{i,j}$ is the frequency of co-presence between u_i and u_j and $d_{i,n}, d_{j,n}, d_{ij,n}$ denote the duration of u_ith, duration of u_jth and the duration of their co-presence, at the nth session respectively.

- *Specificity of the co-presence:* Specificity aims to capture the popularity of the location where the co-presence has occurred. This aims to differentiate between popular locations such as main entrances or food courts, where random users may be seen at the same time, compared to less popular locations that can better reflect potential social ties. For any pair of users, (u_i, u_j) we compute the location specificity, i.e., $z_{i,j}$, at location l as the number of times u_i and u_j co-presence has been observed at l divided by the total number of co-presences occurred at l. We further average the location specificity for all the co-presences of u_i and u_j.

We define a decay function, $score(u_i, u_j) = \omega e^{f_{i,j} * d_{i,j} * log(\hat{z}_{i,j})}$, to compute the social strength between any pair of users. This function assigns lower weights to the edges that are likely to be noisy.

We use the GNMF technique to incorporate potential social ties in our model. GNMF builds a nearest neighbour grap, W, using $score(.)$ to determine the neighbours, and aims to decompose the original matrix in a way that connected points in the graph are closer in the latent space. Hence, similar to [1], we define our objective function as:

$$\mathscr{O} = \|A - UV\|_F^2 + \lambda Tr(U^T L U) + \frac{\rho}{2}(\|U\|_F^2 + \|V\|_F^2)$$

where L is the graph Laplacian [3] of the nearest neighbour graph and $Tr(.)$ is the trace of the matrix. L can be estimated as $D - W$, where D is a diagonal matrix whose entries are column sums of W.

Furthermore, we can build a graph, G_l to capture the similarity between the location points, where the edges indicates how frequently any pair of locations are visited together. The objective function can then be expressed as:

$$\mathscr{O} = \|A - UV\|_F^2 + \lambda_s Tr(U^T L_s U) + \lambda_l Tr(V^T L_l V) + \frac{\rho}{2}(\|U\|_F^2 + \|V\|_F^2)$$

and the update rules to minimize $\mathscr{O}$ are as follows:

$$u_{ij} \leftarrow u_{ij} \frac{(AV + \lambda_s W_s U)_{ij}}{(UV^T V + \lambda_s D_s U)_{ij}} \quad v_{ij} \leftarrow v_{ij} \frac{(A^T U + \lambda_l W_l V)_{ij}}{(VU^T U + \lambda_l D_l V)_{ij}}$$

5 Experiments

5.1 Dataset and Experimental Setup

Dataset: In this work we evaluate the performance of our location imputation algorithm using the following two mobility datasets:

D_1: This dataset is the collection of users connecting to the WiFi network of a University over a period of 22 days (10 days in May and 22 days in September) in 2016. We focused on a coarser location granularity (building level) to address the sparsity issue. We focused on individuals that have at least 2 sessions in the dataset and further filtered users whose overall footprint perfectly matches their daily visited locations (As shown in Fig. 2c, we compute the mutual information between daily sessions of each user and the set of all their visited locations. For more than 18% of users in the dataset, the average NMI is equal to 1, i.e., they visited the same set of locations in every session). Hence, we obtained a dataset of ≈120,000 users visiting 118 buildings.

D_2: This dataset has been collected from over 120,000 anonymized users connecting to the operating Wi-Fi network of a shopping mall in the city of X between September 2012 to October 2013. There are 67 Wi-Fi access points in an area of around 90,000 square meters. We focus on users who have visited the mall at least 5 times (5% of the users).

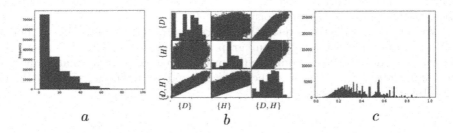

a b c

Fig. 2. Properties of the sessions in D_1. Figure 2a shows the distribution of number of unique locations visited by users. Figure 2b shows the correlation between Day, Hour and Day/Hour entropy. Figure 2c shows the distribution of normalised mutual information between users' daily trips and the overall footprint.

Methods: To evaluate our approach consider the following scenarios:

f-fold missing: We adopt an f fold cross validation framework where for each user we randomly split their mobility data into f folds and for each fold we train our model based on $f - 1$ folds and test its performance to predict the visited locations on the remaining fold. Our default is a 2 fold cross validation which assumes half of the location information of all users is missing. For D_1, we train our model based on the sessions in one period and test it on the other.

cold start: In this scenario we evaluate the perfomance of our model in finding the top locations of new users. We use a 10 fold cross validation framework where the model is trained using the entire location history of 90% fold of the users and is evaluated based on its success in inferring the location of the remaining users.

We consider four varieties of the GNMF-based models: a model that only uses the location graph ($\lambda_s = 0$), L, a model that only considers the users' social graph, called U, a model that uses both graphs to make the prediction, called LU, and a model that uses users' profile when building the users' graph, called $LU + P$.

We compare the performance of the GNMF-based model with approaches proposed in [6,9], denoted as WRMF and ICCF respectively. For the f-fold missing scenario, we assume a model, called $HIST$, that returns the observed set of locations in the train set as the estimates of the missing locations, i.e., it relies on the history of users' location. We also implemented a Frequency-based Imputation technique (FI) for the cold start scenario that selects the top K locations visited by all users who have complete location information as the result of the top K query for any new user.

Metric: Similar to [6,9] we used the precision at K and recall at K to evaluate the performance of our algorithm. For each user, we sort the score for each location and select the top K predicted locations. The precision at K and recall at K, denoted as $p@K$ and $r@K$ respectively are therefore computed as:

$$p@K = \frac{1}{M} \sum_{i=1}^{M} \frac{|T_i(K) \cap V_i|}{K}, \quad r@K = \frac{1}{M} \sum_{i=1}^{M} \frac{|T_i(K) \cap V_i|}{|V_i|}$$

where $T_i(K)$ is the set of top K predicted locations and V_i is the set of actual visited locations for u_i. Note that $r@K$ does not necessarily reach 1.

Setting: Our default number of latent features, k is set to 40 for D_1 and 20 for D_2 since the location set in D_2 is much smaller. Also, the number of factors that are considered when building P is 6 in our work since we add the average and maximum gap between users' visits to our feature set. Prior to feeding P to the model, we perform PCA to reduce the dimensionality of P to 3 since we observed that the derived temporal features in our datasets are moderately correlated (Fig. 3b). When building the affinity graph, we set $n = 20\% * M$ and $\omega = 10$. We also used a similar function suggested in [9] to build our confidence matrix: $\alpha(c_{i,j}) = 1 + log(1 + c_{i,j} * 10^{\epsilon})$, where $\epsilon = 300$. We also set λ_s and λ_l to 0.5, 100 for D_1 and 0.05, 0.1 for D_2. Also, $\rho = 0.05$.

5.2   Results

Figure 3 shows the performance of our proposed model with regard to the first scenario. Figure 3a and b consider the case where the imputation models are trained based on sessions in May and are then tested on sessions in September (D_1). As can be seen, the naive baseline outperforms all varieties of our proposed model as well as ICCF and WRMF for smaller Ks. However, with an increase in K, i.e., more specific or less frequent locations, the model that uses both location and user information outperforms the other methods. This suggests that in confined spaces, individuals' habits plays the most important role in predicting their top visited locations. Another interesting observation is that when user profile is used as the extra side information is a similarity graph, it can actually help to improve the performance of the GNMF-based model. However,

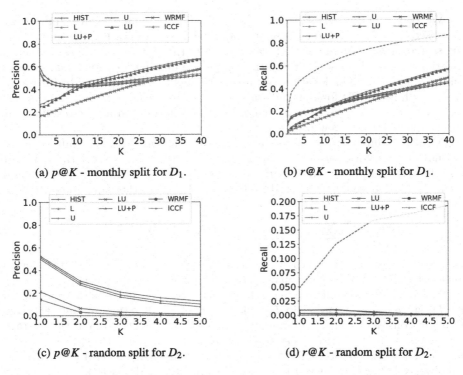

(a) $p@K$ - monthly split for D_1.

(b) $r@K$ - monthly split for D_1.

(c) $p@K$ - random split for D_2.

(d) $r@K$ - random split for D_2.

Fig. 3. The performance of our location imputation models in the case that 50% of the location data for all users is missing. The dashed grey line shows the highest achievable recall@K. Also, note that we changed the y-axis scale in Fig. 3d for visibility. (Color figure online)

the same trend is not observed when it is added as extra feature columns in the matrix, and is mixed with the location data (ICCF versus WRMF). We do not include the performance of WRMF in Fig. 4 since the same observation holds in that scenario. As can be seen in Fig. 3c and d, the same trend is not observed for D_2, which may be due to the limited number of options in a shopping mall.

Figure 4 depicts the performance of our proposed model in the cold case scenario. It can be observed that considering spatial and social influences as well as users' profile, i.e., $LU + P$, improves the performance of the model, however, FI performs almost as good as $LU + P$. This may be due to the fact that the new users are predicted to visit the most popular places, such as libraries in a campus and food court in a shopping mall. This assumption seems to be largely true in confined spaces.

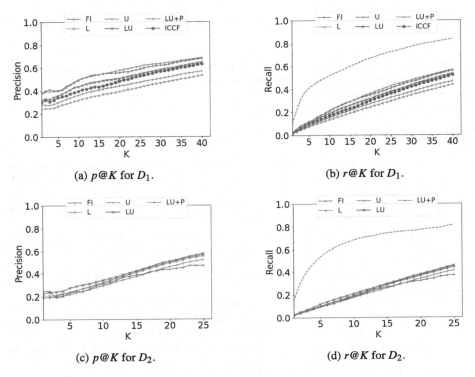

(a) $p@K$ for D_1.

(b) $r@K$ for D_1.

(c) $p@K$ for D_2.

(d) $r@K$ for D_2.

Fig. 4. The performance of our location imputation models in the cold start scenario.

6   Conclusion

In this paper, we proposed a data imputation model for sparse spatio temporal trajectories. In contrast to most previous research, we have focused on constrained and indoor environments, such as a large, multi-building campus environment or a complex shopping mall. The highly regular user behaviour and sparsity of the data capturing their behaviour poses a challenge to sophisticated imputation algorithms. First, the highly regular nature of the movements renders the prediction of the top K locations (here, up to $K = 10$) trivial. We see that the naïve baseline matching the users to the most frequently visited locations in the dataset performs surprisingly well, and significantly outperforms previous state-of-the-art-models, WRMF and ICCF. Yet, the inclusion of physical associations between visitors, captured by their physical co-occurrences outperforms all models for the less frequently visited locations. Our findings is consistent with the observations of the authors in [2]: 'social relationships explain about 10% to 30% of all human movement, while periodic behaviour explains substantially more, 50% to 70% of the behaviour.' As the value in sophisticated location prediction lies in the ability to reflect on the less frequently visited locations, our

$LU + P$ model improves on the state-of-the-art and can support personalised services.

Acknowledgement. This research is supported by a Linkage Project grant of the Australian Research Council (LP120200413) and a Discovery Project grant of the Australian Research Council (DP170100153).

References

1. Cai, D., He, X., Han, J., Huang, T.S.: Graph regularized nonnegative matrix factorization for data representation. IEEE Trans. Pattern Anal. Mach. Intell. **33**, 1548–1560 (2011)
2. Cho, E., Myers, S.A., Leskovec, J.: Friendship and mobility: user movement in location-based social networks. In: KDD (2011)
3. Chung, F.R.: Spectral Graph Theory, No. 92. American Mathematical Soc. (1997)
4. Cranshaw, J., Toch, E., Hong, J., Kittur, A., Sadeh, N.: Bridging the gap between physical location and online social networks. In: UBICOMP (2010)
5. Guting, R.H., Valdés, F., Damiani, M.L.: Symbolic trajectories. ACM Trans. Spat. Algorithms Syst. (2015)
6. Hu, Y., Koren, Y., Volinsky, C.: Collaborative filtering for implicit feedback datasets. In: Proceedings of ICDM, pp. 263–272. IEEE (2008)
7. Kim, Y., Shin, H., Cha, H.: Smartphone-based Wi-Fi pedestrian-tracking system tolerating the RSS variance problem. In: 2012 IEEE International Conference on Pervasive Computing and Communications (PerCom) (2012)
8. Li, W., Hu, Y., Fu, X., Lu, S., Chen, D.: Cooperative positioning and tracking in disruption tolerant networks. IEEE Trans. Parallel Distrib. Syst. **26**, 382–391 (2015)
9. Lian, D., Ge, Y., Zhang, F., Yuan, N.J., Xie, X., Zhou, T., Rui, Y.: Content-aware collaborative filtering for location recommendation based on human mobility data. In: Proceedings of ICDM, pp. 261–270. IEEE (2015)
10. Lian, D., Zhang, Z., Ge, Y., Zhang, F., Yuan, N.J., Xie, X.: Regularized content-aware tensor factorization meets temporal-aware location recommendation. In: ICDM (2016)
11. Liu, B., Fu, Y., Yao, Z., Xiong, H.: Learning geographical preferences for point-of-interest recommendation. In: Proceedings of KDD, New York, NY, USA, pp. 1043–1051 (2013)
12. McGee, J., Caverlee, J., Cheng, Z.: Location prediction in social media based on tie strength. In: Proceedings of CIKM (2013)
13. Pham, H., Shahabi, C., Liu, Y.: Inferring social strength from spatiotemporal data. ACM Trans. Database Syst. (2016)
14. Wang, D., Pedreschi, D., Song, C., Giannotti, F., Barabasi, A.L.: Human mobility, social ties, and link prediction. In: ACM SIGKDD (2011)
15. Wang, Y., Yuan, N.J., Lian, D., Xu, L., Xie, X., Chen, E., Rui, Y.: Regularity and conformity: location prediction using heterogeneous mobility data. In: Proceedings of KDD (2015)
16. Ye, A., Sao, J., Jian, Q.: A robust location fingerprint based on differential signal strength and dynamic linear interpolation. Secur. Commun. Netw. **9**(16), 3618–3626 (2016). sCN-15-0656.R2

17. Yin, H., Sun, Y., Cui, B., Hu, Z., Chen, L.: LCARS: a location-content-aware recommender system. In: Proceedings of KDD (2013)
18. Yuan, Q., Cong, G., Sun, A.: Graph-based point-of-interest recommendation with geographical and temporal influences. In: Proceedings of CIKM (2014)
19. Zheng, V.W., Cao, B., Zheng, Y., Xie, X., Yang, Q.: Collaborative filtering meets mobile recommendation: a user-centered approach. In: Proceedings of AAAI (2010)

Low Redundancy Estimation
of Correlation Matrices for Time Series
Using Triangular Bounds

Erik Scharwächter[1,2]([✉]), Fabian Geier[2], Lukas Faber[2],
and Emmanuel Müller[1,2]

[1] GFZ German Research Centre for Geosciences, Potsdam, Germany
[2] Hasso Plattner Institute, Potsdam, Germany
{erik.scharwaechter,fabian.geier,emmanuel.mueller}@hpi.de,
lukas.faber@student.hpi.de

Abstract. The dramatic increase in the availability of large collections of time series requires new approaches for scalable time series analysis. Correlation analysis for all pairs of time series is a fundamental first step of analysis of such data but is particularly hard for large collections of time series due to its quadratic complexity. State-of-the-art approaches focus on efficiently approximating correlations larger than a hard threshold or compressing fully computed correlation matrices in hindsight. In contrast, we aim at estimates for the *full pairwise correlation structure without computing and storing all pairwise correlations*. We introduce the novel problem of low redundancy estimation for correlation matrices to capture the complete correlation structure with as few parameters and correlation computations as possible. We propose a novel estimation algorithm that is very efficient and comes with formal approximation guarantees. Our algorithm avoids the computation of redundant blocks in the correlation matrix to drastically reduce time and space complexity of estimation. We perform an extensive empirical evaluation of our approach and show that we obtain high-quality estimates with drastically reduced space requirements on a large variety of datasets.

1 Introduction

The monitoring of earth, society and personal life through various sensors has led to a ubiquity of large-scale collections of time series. Correlation analysis for all pairs of time series is often the first step of analysis of such data. In the past decade, many works have used estimates of the full pairwise correlation matrix among time series, e.g., to infer functional brain networks [17], for portfolio selection in empirical finance [9], to detect periods of financial crisis [19] and to better understand the climate system [20]. Since the time and space complexity for computing the full pairwise correlation matrix is quadratic in the number of time series, analyses that rely on exact computation of the full matrix do not scale with the increasing size of time series collections. For this reason, there is

© Springer International Publishing AG, part of Springer Nature 2018
D. Phung et al. (Eds.): PAKDD 2018, LNAI 10938, pp. 458–470, 2018.
https://doi.org/10.1007/978-3-319-93037-4_36

a need for approaches that estimate all pairwise correlations without computing and storing the entire matrix.

We introduce the novel problem of low redundancy estimation for correlation matrices. A low redundancy estimate describes the *complete correlation matrix R* of a time series collection using a *smaller representation* $\mathfrak{R}$ and *without computing all pairwise correlations*. Our estimation approach COREQ (CORrelation EQuivalence) is driven by the observation that many time series collections show inherent group structure that leads to blocks of redundant entries in the correlation matrix. We exploit this structure by computing equivalence classes of highly correlated time series and pooling the redundant correlation estimates into a single class estimate. The resulting estimate is visualized in Fig. 1. We describe an algorithm to obtain the estimate $\mathfrak{R}$ on the right directly from the data after computing only a small fraction of the actual correlations in R. The computational problem lies in finding—with as few correlation computations as possible—a suitable partition of the time series collection into equivalence classes that allows correlation estimation with bounded loss.

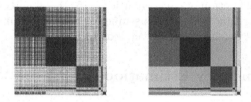

Fig. 1. Example correlation matrix R (left) and low redundancy estimate $\mathfrak{R}$ (right).

Our contributions are as follows. We formalize low redundancy estimation as an approximation problem and formally derive low redundancy estimates with error guarantees. Furthermore, we propose a greedy approximation algorithm and two powerful heuristics to obtain high-quality estimates with few correlation computations. We carefully evaluate our algorithm on 85 time series collections from the UCR Time Series Classification Archive [1] and a large satellite image time series dataset from the geoscientific domain as a real-life use case.

2   Related Work

There are two challenges for efficient correlation estimation for large time series collections. The first challenge is the increasing *number* of time series that are jointly analyzed, while the second challenge is the increasing *velocity* of newly arriving observations in streaming time series.

COREQ addresses the first challenge. Most work in the field has been done on rapidly retrieving all pairs of highly correlated time series [16,23,25] and avoiding the computation of weak correlations. Conceptually, all these approaches discard information about weak correlations. In contrast, our COREQ algorithm provides estimates for the *complete correlation structure*, including weak correlations. Low-rank approximations to a correlation matrix remove redundancies for a more space efficient representation of the full correlation structure, but existing methods [7,24] take fully estimated correlation matrices as inputs for their approximations. In contrast, we aim at low redundancy estimates *without computing all pairwise correlations* first. Mueen et al. [11] propose two algorithms to approximate all entries in the correlation matrix that are larger than some threshold τ. By design, they lose information about correlations below the hard threshold τ, while we provide accurate estimates for all correlations. We briefly describe their algorithms in Sect. 5 and evaluate COREQ against them.

Methodologically, COREQ exploits structure in time series collections by computing equivalence classes of time series that behave similarly under correlation. There is extensive literature on clustering time series with similar behavior for generic subsequent processing [10,14,18]. In contrast to these works, COREQ has *theoretical quality guarantees* for the resulting correlation estimates.

Orthogonal to our approach, works on streaming time series have focused on efficient updating schemes for correlation monitoring [4,12,25], robust correlation tracking [13], detection of lag correlations [15,21,22] and correlated windows [2,5,6] in streaming time series. We assume for now that our time series collections are static and defer streaming versions to future work.

3   Low Redundancy Estimation

3.1   Preliminaries

Let $\mathcal{X} = \{X_1, ..., X_N\}$ be a collection of N univariate time series of length T with $X_i = (X_{i1}, ..., X_{iT})$. We assume that the time series are equi-length and temporally aligned as in many use cases from the geosciences, neuroimaging, finance and other domains. The Pearson correlation coefficient between time series X_i and X_j (at lag 0) is given by $\rho_{ij} = \frac{1}{T} \sum_{t=1}^{T} \frac{X_{it} - \mu_i}{\sigma_i} \cdot \frac{X_{jt} - \mu_j}{\sigma_j}$, where μ_i and σ_i denote the mean and standard deviation of time series X_i, respectively. The correlation coefficient captures linear relationships and ranges from 1 (strong positive correlation) to -1 (strong negative correlation). A value of 0 means that time series are uncorrelated. The matrix $R \in [-1,1]^{N \times N}$ denotes the symmetric correlation matrix that contains all pairwise correlations between the input time series, i.e. $R = (\rho_{ij})_{i,j \in \{1,...,N\}}$. A useful property of Pearson's correlation coefficient is that it comes with triangular bounds similar to the triangle inequality in metric spaces [8]. These bounds allow estimating the correlation between two time series X_i and X_j via their correlations with a third time series X_k:

Theorem 1 (Triangular bounds). *For time series X_i, X_j and X_k it holds that* $\rho_{ik}\rho_{kj} - \sqrt{(1 - \rho_{ik}^2)(1 - \rho_{kj}^2)} \leq \rho_{ij} \leq \rho_{ik}\rho_{kj} + \sqrt{(1 - \rho_{ik}^2)(1 - \rho_{kj}^2)}$.

3.2   Problem Statement

Our goal is to obtain a small estimate $\mathfrak{R}$ that well approximates the full correlation matrix R without computing all pairwise correlations. Intuitively, the size of an estimate is the number of model parameters that need to be stored, and the quality is measured by the absolute deviation from the true correlation. Formally, let $\hat{\rho}(i, j \mid \mathfrak{R}) : \{1, ..., N\}^2 \longrightarrow [-1, 1]$ be an *estimator* for the correlation ρ_{ij} based on the representation $\mathfrak{R}$. The *loss* of an estimator is given by the absolute deviation from the true correlation $\ell_{ij} = |\hat{\rho}(i, j \mid \mathfrak{R}) - \rho_{ij}|$. The traditional brute force estimator is the special case $\mathfrak{R} = R$ and $\hat{\rho}(i, j \mid \mathfrak{R}) = \rho_{ij}$. The brute force approach has $\frac{1}{2}N(N+1)$ model parameters and incurs a loss of zero. The other extreme is the special case $\mathfrak{R} = c \in [-1, 1]$ and $\hat{\rho}(i, j \mid \mathfrak{R}) = c$, which has only a single parameter to store, but potentially high loss. We aim at trade-offs between these two extremes. The general problem is thus to find a *low redundancy representation* $\mathfrak{R}$ with a small number of parameters and an estimator $\hat{\rho}(i, j \mid \mathfrak{R})$ that incurs a small loss.

P_k $P_{k'}$

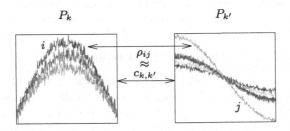

Fig. 2. Estimating pairwise time series correlations by inter-class correlations

We restrict ourselves to representations based on partitions of the dataset into classes of similar time series. The idea is illustrated in Fig. 2 for time series from two equivalence classes P_k and $P_{k'}$. All pairwise correlations between members of the two classes are redundant and can be collapsed to a single estimate for the inter-class correlation $c_{k,k'}$ with minor loss. Formally, we aim at representations of the form $\mathfrak{R} = (\mathcal{P}, C)$, where $\mathcal{P}$ is a partition of $\mathcal{X}$ into $K = |\mathcal{P}|$ equivalence classes and $C = \{c_{k,k'} \in [-1, 1] \mid 1 \leq k \leq k' \leq K\}$ is a set of inter-class correlations. The respective estimator is $\hat{\rho}(i, j \mid \mathcal{P}, C) = c_{k,k'}$ for $i \in P_k$ and j in $P_{k'}$. Such representations have $N + \frac{1}{2}K(K+1)$ parameters. The fewer classes K are necessary to capture all pairwise correlations with small loss, the lower the redundancy in the final estimate. We formalize our problem as the following approximation problem:

Problem 1. Given a collection of time series $\mathcal{X}$ and an error bound $\epsilon \geq 0$, find a partition $\mathcal{P}$ of $\mathcal{X}$ and a set of inter-class correlations C, such that the estimate $\mathfrak{R} = (\mathcal{P}, C)$ has a loss $\ell_{ij} = |c_{k,k'} - \rho_{ij}| \leq \epsilon$ for all $i \in P_k$ and $j \in P_{k'}$.

The challenge is to obtain such estimates with as few correlation computations as possible. In particular, without computing the full matrix R. A trivial solution for Problem 1 is the partition into N singleton classes $\mathcal{P} = \{\{X_1\}, ..., \{X_N\}\}$ such that the inter-class correlations are exactly the pairwise time series correlations. This solution collapses to the full correlation matrix R with zero loss but without reduction of redundancy or any computational efficiency improvements. In the following, we formally derive non-trivial approximations that guarantee a loss of at most ϵ with lower redundancy than R, and can be computed way more efficiently than the full matrix.

4   COREQ

The intuition behind our construction is that homogeneous equivalence classes with high *intra*-class correlations lend themselves to high-quality estimates for the *inter*-class correlations. Based on our formal analysis we propose the efficient greedy partitioning algorithm COREQ (CORrelation EQuivalence) and three estimators to obtain pairwise class correlations from the resulting partitions: an estimator with approximation guarantees and two powerful heuristics.

4.1   Approximations with Quality Guarantees

We start with the formal construction of a solution to Problem 1 with quality guarantees. The idea is to build homogeneous equivalence classes by a pivoting approach. Each class is identified with a unique pivot time series, and all other time series are assigned to classes such that the correlations to their respective pivot time series are at least $\alpha \in (0, 1]$. The parameter α controls the class homogeneity: the closer α to 1, the more homogeneous the equivalence classes, and the lower the estimation loss. Since we do not specify the number of classes K in advance, such partitions exist for any choice of α. The following theorem establishes how large α needs to be chosen to guarantee a loss of at most ϵ:

Theorem 2. *Let $\alpha \in (0, 1]$ and $\epsilon \geq 0$. Let $\mathcal{P} = \{P_k \mid k = 1, ..., K\}$ be a partition of $\mathcal{X}$ with associated pivot time series $X_{i_k} \in P_k$ such that $\forall X_i \in P_k : \rho_{i,i_k} \geq \alpha$. Furthermore, let the inter-class correlations C be the correlations between these pivot time series scaled by a correction factor that depends on α:*

$$c_{k,k'} = \frac{1}{2}\left(1 + \alpha^2\right)\rho_{i_k,i'_k}. \tag{1}$$

It holds that $\ell_{ij} \leq \epsilon$ for all $X_i, X_j \in \mathcal{X}$, if $\alpha \geq \sqrt{1 - \left(\frac{2\epsilon}{\sqrt{5}+2}\right)^2}$.

A proof based on the triangular bounds from Theorem 1 can be found in the Supplementary Material.[1] Sect. 4.2 provides an efficient greedy algorithm to compute such partitions. The scaling factor $\frac{1}{2}(1 + \alpha^2)$ in Eq. 1 can be interpreted as the

---

[1] Available on the project website https://hpi.de/mueller/coreq.html.

uncertainty about the representativeness of pivot correlations: the smaller α, the more heterogeneous the equivalence classes, and the less representative the pivots for their classes. Consequently, it is safer—in the general case—to estimate correlations close to zero instead of extremal values. Theorem 2 states that for any desired error bound ϵ we can find a (possibly) non-trivial solution $\mathfrak{R} = (\mathcal{P}, C)$ to Problem 1 that guarantees $\ell_{ij} < \epsilon$ for all pairs of time series. However, the quality guarantee is based on the worst-case bounds from Theorem 1 which do not make any assumptions on the distribution of correlations within a dataset. In particular, we do not assume that the time series cluster into homogeneous groups as motivated in Fig. 2 for many real-life time series collections. For any realistic choice of ϵ the theorem thus requires a threshold α very close to 1 to guarantee the quality on any possible input dataset. For example, a loss $\ell_{ij} \leq 0.1$ can only be guaranteed for all pairs of time series on any input dataset if we set $\alpha \geq 0.9989$. The downside of choosing a value of α close to 1 is that we will most likely obtain the trivial solution with high redundancy and no computational efficiency improvements. As we see in Sect. 5, we can efficiently obtain estimates with low redundancy *and* low losses on *many real-life datasets* for much lower values of α.

4.2   A Greedy Estimation Algorithm

We compute the pivot-based partitions formally defined in Theorem 2 as follows. We start by picking an arbitrary time series X_i from $\mathcal{X}$ as a pivot series and compute the correlations between X_i and all remaining time series. All time series with a correlation to X_i not smaller than α are stored in a new equivalence class P. The class P always contains X_i itself. All elements from P are removed from the original time series collection $\mathcal{X}$, and the procedure is repeated with a newly picked pivot series until all time series are processed. This procedure terminates with a partition as of Theorem 2 for any $\alpha \in (0, 1]$ with at most $\frac{1}{2}N(N + 1)$ correlation computations. In the best case, if all correlations are larger than α, it terminates with only N correlation computations. Given such a partition, the question is how to best estimate the inter-class correlations C. We propose three alternatives to obtain a complete correlation estimate:

(i) **COREQ-P1**: scaled pivot correlations from Eq. 1 in Theorem 2 which theoretically guarantee low errors on all datasets for $\alpha \longrightarrow 1$ but have a bias towards zero for smaller choices of α.

(ii) **COREQ-P2**: simplified estimate that uses unscaled pivot correlations $c_{k,k'} = \rho_{i_k, i'_k}$ to remove the bias for smaller choices of α.

(iii) **COREQ-A**: average estimate that samples a logarithmic number of correlations between pivot X_{i_k} and the class $P_{k'}$

$$c_{k,k'} = \frac{1}{\max\left(1, \lceil \log_2 N_{k'} \rceil\right)} \sum_{j'=1}^{\max(1, \lceil \log_2 N_{k'} \rceil)} \rho_{i_k, \text{rand}(P_{k'})},$$

where $N_{k'} = |P_{k'}|$ and $\text{rand}(P_{k'})$ returns a random time series from $P_{k'}$.

All of these estimates can be obtained from the correlations computed during class construction and do not require additional correlation computations. In COREQ-A we sample a logarithmic number of correlations to account for the heterogeneity in large equivalence classes. All three estimates converge to the pivot correlations for $\alpha \longrightarrow 1$ and differ only for $\alpha \ll 1$.

4.3 Formal Relation to Clustering Algorithms

There is a clear relationship between our equivalence class-based correlation matrix approximations and the well-known optimization problem of time series clustering. We could relax the goal of strict approximation guarantees for all pairs of time series towards estimation with minimal *aggregated* loss. Let $X \in \mathbb{R}^{N \times T}$ be a matrix representation of $\mathcal{X}$ where all time series are standardized to have zero mean and unit variance over time. Furthermore, let $R = \frac{1}{T} X X^\top$ be the true correlation matrix, $Z = \{0, 1\}^{N \times K}$ be an indicator matrix that encodes class memberships of a partition $\mathcal{P} = \{P_1, ..., P_K\}$, and $C \in [-1, 1]^{K \times K}$ be a matrix of inter-class correlations. The error function $E = \|R - ZCZ^\top\|^2$ encodes the goal of finding an estimate $\mathfrak{R} = (\mathcal{P}, C)$ that well represents all correlations within R. We observe that this error function is a quadratic form of the sum of squared errors (SSE) that is used extensively for clustering, most prominently in K-Means. To see this relation, let $M \in \mathbb{R}^{K \times T}$ be the matrix of cluster centroids in K-Means. The sum of squared errors is defined as SSE $= \|X - ZM\|^2$. Using the pairwise centroid correlations as estimates for the inter-class correlations $C = \frac{1}{T} M M^\top$, we obtain $E = \|\frac{1}{T} X X^\top - Z \frac{1}{T} M M^\top Z\|^2$. Due to the structural similarity of E and SSE, we use K-Means clustering as a baseline in our experiments. However, to the best of our knowledge, there is no clustering algorithm that allows approximating correlations up to an error bound ϵ.

5   Empirical Evaluation

Our empirical evaluation consists of two parts. In the first part, we extensively analyze the quality of the estimates obtained by COREQ in terms of average loss and model size on a large variety of datasets. In the second part, we compare the performance of COREQ against two state-of-the-art competitors and the K-Means baseline on a real-life dataset from the geoscientific domain. We implemented COREQ as a Python C module. All source codes necessary to reproduce our results are available on GitHub.[2] Additional information is provided on our project website.[3]

5.1   Experimental Setup

Performance Measures. The *average loss* for an estimate $\mathfrak{R}$ is given by $\bar{\ell} = \frac{1}{Z} \sum_{i=1}^{N} \sum_{j=i}^{N} \ell_{ij}$ with $Z = \frac{1}{2} N(N + 1)$. The closer to 0, the better. The

---

[2] https://github.com/KDD-OpenSource/coreq.
[3] https://hpi.de/mueller/coreq.html.

model size is given by the total number of model parameters that need to be stored by an algorithm, divided by the number of entries in the true correlation matrix. Model sizes close to 0 indicate a low redundancy, whereas values close to 1 indicate high redundancy. We also count the number of *correlation computations* necessary to obtain an estimate. All performance measures are averaged over ten independent runs to obtain stable results for each algorithm and dataset.

Data. To analyze the performance of COREQ over a large variety of time series collections, we run experiments on all 85 time series collections from the well-known publicly available *UCR Time Series Classification Archive* [1]. For a real-life comparison with state-of-the-art algorithms, we use satellite image time series obtained from the NASA Terra MODIS satellite mission [3]. The dataset contains 236,197 *EVI time series* (Enhanced Vegetation Index) for South America, captured with a temporal resolution of 16 days between 2000 and 2015 (length 368). The EVI is computed from multi-spectral satellite images and captures the level of greenness at a given point in time as a proxy for vegetation cover.

Competitors. As a baseline, we perform one iteration of K-Means clustering with a fixed K to obtain a partition of the dataset and use the pairwise centroid correlations as class correlations. Using more iterations is infeasible since it drastically increases the number of correlation computations. We also compare against two state-of-the-art algorithms proposed by Mueen et al. [11] to compute an Approximate Threshold Correlation Matrix (APPROXTHRESH) and a Threshold Boolean Correlation Matrix (THRESHBOOLEAN). APPROXTHRESH approximates (up to an error ϵ) all correlations larger than a threshold τ by exploiting a Discrete Fourier Transform-based early-abortion criterion for individual correlation computations; all correlations below τ are set to 0 without error guarantee. APPROXTHRESH is designed to reduce the number of operations for individual correlation computations. To compare the total costs of correlation estimation with our approach, we scale the number of correlation computations with the speedup factor per correlation computation. THRESHBOOLEAN uses a dynamic programming-based pruning strategy to reduce the number of pairwise comparisons. It estimates all (absolute) correlations above τ as ± 1 and all other correlations as 0, without any quality guarantees.

5.2 Quality of Estimates

We first analyze the performance of COREQ in terms of average loss and resulting model size on all 85 UCR datasets for various values of α. Figure 3 visualizes the distribution of average loss over all UCR datasets as boxplots along with the mean model size. We provide separate boxplots for COREQ-P1/P2 and COREQ-A; mean model sizes are identical. As expected, increasing α pushes the average loss on all datasets towards zero since equivalence classes become

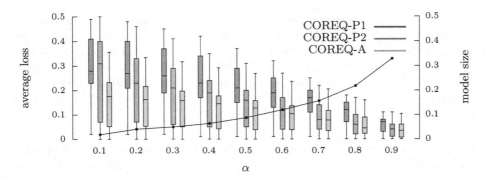

Fig. 3. Distribution of average loss (boxplots) and mean model size (line) across all UCR datasets for $\alpha \in [0, 1]$.

more homogeneous. At the same time, it increases the model size. COREQ-A outperforms COREQ-P1/P2 over the full parameter space, with the margin of improvement largest for low values of α. Lower values of α typically come with larger and more heterogeneous equivalence classes, such that the pivot correlations are not representative. The scaled pivot correlations from COREQ-P1 perform worse than the unscaled variant COREQ-P2 on many datasets. The datasets where COREQ-P2 outperforms COREQ-P1 contain time series that are all very strongly correlated. In these cases, the theoretically justified bias towards zero correlations is harmful. With $\alpha = 0.9$, all three estimation variants achieve high-quality estimates with average losses below 0.1 and a mean model size below 0.35.

Detailed scatter plots of the results of COREQ-A can be found in Fig. 4. Each point in a plot shows the model size and average loss achieved on a single dataset. The histograms below show the corresponding distributions of model sizes. We observe that even for $\alpha = 0.9$ the large majority of datasets can well

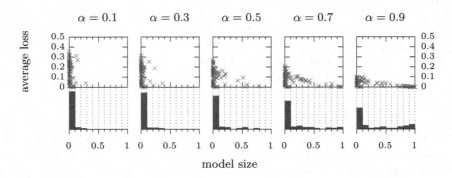

Fig. 4. Average loss against model size achieved by COREQ-A on all UCR datasets for $\alpha \in [0, 1]$, along with histograms over model size.

be estimated with model sizes below 0.1. Only a few datasets appear on the far right with model sizes close to 1. Manual inspection of these datasets revealed that they contain purely uncorrelated time series or ambiguous group structures. These instances cannot be estimated more efficiently with our approach. COREQ provides low redundancy estimates with low average losses on all datasets with strong group structures.

5.3   Comparison with Existing Methods

We now compare COREQ-A with the state-of-the-art algorithms introduced by Mueen et al. [11] and our K-Means baseline. We address two questions in our analysis: (1) How much loss does an algorithm incur at a given model size? (2) How many correlation computations are necessary to obtain an estimate with that model size? All algorithms in our evaluation depend on different input parameters that affect the estimation performance. These input parameters directly control the model size: the larger α in COREQ and K in K-Means, the more pairwise class correlations have to be estimated and stored, while a smaller threshold τ in APPROXTHRESH and THRESHBOOLEAN means that more pairwise time series correlations have to be stored. To compare these approaches in a meaningful and fair way, we run all algorithms over a wide range of parameterizations ($\alpha \in \{0.1, 0.2, ..., 0.9\}$, $K \in \{1, 2, 4, ..., 8192\}$, $\tau \in \{0.9, 0.8, ..., 0.1\}$) and use the resulting model size as the unified scale. The error bound for APPROXTHRESH is set to $\epsilon = 0.05$. We use the EVI dataset[4] as a real-life example from the geoscientific domain.[5]

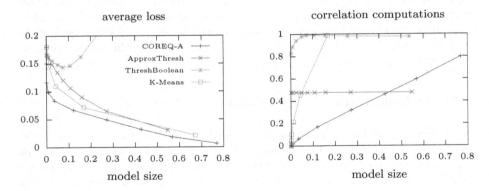

Fig. 5. Performance on EVI data over the full parameter space of each algorithm.

---

[4] Subsamples of 10,000 time series for COREQ/K-Means/APPROXTHRESH and 1,000 time series for THRESHBOOLEAN due to performance reasons.

[5] We also ran experiments on the chlorine concentration data used in the original publication by Mueen et al. [11]; the results are consistent with the results presented in this paper and reported for completeness in the Supplementary Material.

To answer the first question, Fig. 5 (left) shows the average loss of the resulting correlation estimates against the model size. If a curve is close to the origin, it means that small estimates obtained with that algorithm capture most of the information from the correlation matrix. COREQ-A clearly outperforms K-Means, APPROXTHRESH and THRESHBOOLEAN over the full parameter space: our algorithm has lower losses at the same model sizes. The improvement is largest for very small estimates. The THRESHBOOLEAN approach behaves unusually: since it can only estimate correlations as either 0 or ± 1, lowering the threshold τ means that more and more weak correlations are stored and estimated as ± 1. The algorithm is not designed to capture weak correlations accurately. Overall, COREQ-A provides the highest quality estimates for the full correlation structure, with improvements being largest for very small estimates.

For the second question, Fig. 5 (right) shows the number of correlation computations required to obtain the final estimates (normalized by the total number of pairs) against model size. Our approach scales linearly with the model size: the number of correlations that we compute is roughly the same as the number of model parameters we output. The K-Means baseline performs worst, even though we run only one iteration. More iterations or more sophisticated clustering algorithms could improve the quality of the estimates, but come with an even higher computational cost. APPROXTHRESH requires a constant number of correlation computations for all threshold values τ. The early abortion criterion yields an average speed-up of only 2 per correlation computation, meaning that the EVI time series are uncooperative [2]. APPROXTHRESH outperforms our approach in terms of correlation computations only in the large model size region on the right. The pruning strategy employed in THRESHBOOLEAN is effective at the far left of the plot, where the threshold τ is close to 1. For lower threshold values almost all pairwise correlations are computed. COREQ is the fastest algorithm in terms of correlation computations in the small model size region of the parameter space—with a large margin to all competitors. In the same region, we obtain the lowest average loss values.

6   Conclusion and Future Work

We provide a novel way to estimate correlation matrices for large time series collections that exploits redundancies in the input data to drastically reduce the number of parameters to estimate. We show that the partitions we obtain for estimation have theoretical approximation guarantees, allow for very small high-quality estimates on a large variety of real-life datasets, and outperform state-of-the-art approaches. There is still need for a robust way to select the parameter α optimally for any input dataset as to obtain the best trade-off between model size and average loss. Algorithmically, dynamically adapting α during estimation to process datasets with weak and strong group structures could be beneficial. We defer this challenge to future work. Furthermore, combining our estimation approach with a probabilistic model for time series collections would allow us to devise more concise probabilistic error guarantees on top of the worst-case bounds we used in Theorem 2. At last, an extension of COREQ for streaming time series would allow efficient monitoring of correlations for anomaly detection.

References

1. Chen, Y., Keogh, E., Hu, B., Begum, N., Bagnall, A., Mueen, A., Batista, G.: The UCR Time Series Classification Archive, July 2015. http://www.cs.ucr.edu/~eamonn/time_series_data/
2. Cole, R., Shasha, D., Zhao, X.: Fast window correlations over uncooperative time series. In: KDD (2005)
3. Didan, K.: MOD13C1 MODIS/Terra Vegetation Indices 16-Day L3 Global 0.05Deg CMG V006 (2015). https://doi.org/10.5067/modis/mod13c1.006
4. Guha, S., Gunopulos, D., Koudas, N.: Corrrelating synchronous and asynchronous data streams. In: KDD (2003)
5. Guo, T., Sathe, S., Aberer, K.: Fast Distributed correlation discovery over streaming time-series data. In: CIKM (2015)
6. Keller, F., Müller, E., Böhm, K.: Estimating mutual information on data streams. In: SSDBM (2015)
7. Kulis, B., Sustik, M.A., Dhillon, I.S.: Low-rank kernel learning with bregman matrix divergences. J. Mach. Learn. Res. **10**, 341–376 (2009)
8. Langford, E., Schwertman, N., Owens, M.: Is the property of being positively correlated transitive? Am. Stat. **55**(4), 322–325 (2001)
9. Ledoit, O., Wolf, M.: Improved estimation of the covariance matrix of stock returns with an application to portfolio selection. J. Empir. Financ. **10**(5), 603–621 (2003)
10. Liao, T.W.: Clustering of time series data: a survey. Pattern Recogn. **38**(11), 1857–1874 (2005)
11. Mueen, A., Nath, S., Liu, J.: Fast approximate correlation for massive time-series data. In: SIGMOD (2010)
12. Papadimitriou, S., Sun, J., Faloutsos, C.: Streaming pattern discovery in multiple time-series. In: VLDB (2005)
13. Papadimitriou, S., Sun, J., Yu, P.S.: Local correlation tracking in time series. In: ICDM (2006)
14. Paparrizos, J., Gravano, L.: k-Shape: efficient and accurate clustering of time series. In: SIGMOD (2015)
15. Sakurai, Y., Papadimitriou, S., Faloutsos, C.: BRAID: stream mining through group lag correlations. In: SIGMOD (2005)
16. Sathe, S., Aberer, K.: AFFINITY: efficiently querying statistical measures on time-series data. In: ICDE (2013)
17. Smith, S.M., Miller, K.L., Salimi-Khorshidi, G., Webster, M., Beckmann, C.F., Nichols, T.E., Ramsey, J.D., Woolrich, M.W.: Network modelling methods for FMRI. NeuroImage **54**(2), 875–891 (2011)
18. Ulanova, L., Begum, N., Keogh, E.: Scalable clustering of time series with U-shapelets. In: SIAM SDM (2015)
19. Wied, D., Galeano, P.: Monitoring correlation change in a sequence of random variables. J. Stat. Plan. Infer. **143**(1), 186–196 (2013)
20. Wiedermann, M., Radebach, A., Donges, J.F., Kurths, J., Donner, R.V.: A climate network-based index to discriminate different types of El Niño and La Niña. Geophys. Res. Lett. **43**(13), 7176–7185 (2016)

21. Wu, D., Ke, Y., Yu, J.X., Yu, P.S., Chen, L.: Detecting leaders from correlated time series. In: Kitagawa, H., Ishikawa, Y., Li, Q., Watanabe, C. (eds.) DASFAA 2010. LNCS, vol. 5981, pp. 352–367. Springer, Heidelberg (2010). https://doi.org/10.1007/978-3-642-12026-8_28
22. Xie, Q., Shang, S., Yuan, B., Pang, C., Zhang, X.: Local correlation detection with linearity enhancement in streaming data. In: CIKM (2013)
23. Xiong, H., Shekhar, S., Tan, P.N., Kumar, V.: TAPER: a two-step approach for all-strong-pairs correlation query in large databases. TKDE 18(4), 493–508 (2006)
24. Zhang, Z., Wu, L.: Optimal low-rank approximation to a correlation matrix. Linear Algebra Appl. 364, 161–187 (2003)
25. Zhu, Y., Shasha, D.: StatStream: statistical monitoring of thousands of data streams in real time. In: VLDB (2002)

Traffic Accident Detection with Spatiotemporal Impact Measurement

Mingxuan Yue[1]([⊠]), Liyue Fan[2], and Cyrus Shahabi[1]

[1] University of Southern California, Los Angeles, CA, USA
{mingxuay,shahabi}@usc.edu
[2] University at Albany SUNY, Albany, NY, USA
liyuefan@albany.edu

Abstract. Traffic incidents continue to cause a significant loss in deaths, injuries, and property damages. Reported traffic accident data contains a considerable amount of human errors, hindering the studies on traffic accidents. Several approaches have been developed to detect accidents using traffic data in real time. However, those approaches do not consider the spatiotemporal patterns inherent in traffic data, resulting in high false alarm rates. In this paper, we study the problem of traffic accident detection by considering multiple traffic speed time series collected from road network sensors. To capture the spatiotemporal impact of traffic accidents to upstream locations, we adopt Impact Interval Grouping (IIG), which compares real-time traffic speed with historical data, and generates *impact intervals* to determine the presence of accidents. Furthermore, we take a multivariate time series classification approach and extract three novel features to measure the *severity* of traffic accidents. We use real-world traffic speed and accident datasets in our empirical evaluation, and our solutions outperform state-of-the-art approaches in multivariate time series classification.

Keywords: Traffic accident · Multivariate time series classification

1 Introduction

Traffic accidents have been an essential concern in our society. In 2015 the total number of motor-vehicle deaths was 38,300, and the estimated cost of deaths, injuries, and property damage reached $412.1 billion [1]. Studying traffic accidents would help us understand the causes and potentially reduce the damage of such events. However, accident data is usually collected from various state and local agencies: these reports often contain duplicates, missing data, and/or inaccurate information as they are based on victim/witness estimates [2]. Therefore, we are in need of accurate characterization and detection of traffic accidents, which can be achieved utilizing the data from ubiquitous traffic sensors.

© Springer International Publishing AG, part of Springer Nature 2018
D. Phung et al. (Eds.): PAKDD 2018, LNAI 10938, pp. 471–482, 2018.
https://doi.org/10.1007/978-3-319-93037-4_37

In the past decades, a plethora of research studied automatic incident detection (AID) [3–8]. However, these approaches are prone to false alarms [9] due to improper calibration on insufficient and noisy data. On the other hand, video recordings by traffic cameras [8] and traffic flow data collected from probe cars [6] have also been used. But such data is expensive to acquire and does not provide a thorough coverage of the entire road networks. Traffic data, e.g., speed, has specific spatial and temporal patterns, which are unfortunately neglected in AID research. Such spatial and temporal pattern could be helpful in differentiating accidents from fluctuations that are often observed in ordinary traffic and attribute to false alarms. However, it is challenging to model and recognize such spatial and temporal patterns. Furthermore, real-world data could be noisy and has missing data or outliers. Hence it is also important to make the detection robust and not sensitive to such noise.

In this paper, we address the aforementioned challenges by considering traffic speed time series data collected from multiple sensors close to the accident location. We first adopt the Impact Interval Grouping (IIG) algorithm [2] to detect traffic accidents by recognizing the spatiotemporal patterns in traffic speed data. In addition, we propose a multivariate time series (MTS) classification technique, and define three novel features that measure the severity of traffic accidents. We propose two versions, i.e., Severity-I and Severity-T, to balance the algorithm's sensitivity to signals and noise in the traffic data. For evaluation, we utilize real-world traffic speed and accident dataset collected from Los Angeles and our proposed approaches are shown to outperform the state-of-the-art MTS classification methods. We conclude that our proposed severity features can sufficiently characterize the spatiotemporal impact of traffic accidents, and can be used for efficient and accurate traffic accident detection.

2   Related Work

Automatic Incident Detection. In the field of transportation, automatic incident detection (AID) has been an active area of research since the end of the last century. The classical and widely used approach is California Algorithms [3,4]. The basic idea is to use three levels of rules to raise an alarm, based on the difference between the upstream and downstream sensors. Other researchers also tried recognizing vehicles' behaviors from the camera videos. For example, Sadek et al. [8] estimated optical flow from the videos and constructed histograms of flow gradient to build a classifier. However, classical algorithms are not satisfactory due to high missing rate and false alarm rate. Moreover, the video data does not offer a sufficient coverage of most road networks. Yuan et al. [7] utilized SVM to classify accidents by traffic metrics but they do not consider the temporal and spatial correlations. We believe by using time series data from multiple sensors and utilizing multivariate time series classification approaches, the accidents could be modeled and recognized more accurately.

Multi-dimensional Dynamic Time Warping. To the best of our knowledge, most of the current study on Multivariate Time Series Classification (MTSC) are extended from the Univariate Time Series Classification (TSC) methods. For TSC, the kNN approach is typically used with certain distance measures, such as the Dynamic Time Warping (DTW) distance. kNN-DTW is proven to be a reliable approach in TSC [10]. Moreover, for DTW distance in multivariate case (MD-DTW), researchers usually calculate DTW distance between two MTS by summing over individual DTW distances of each variate [11], or formulating values in multiple variates at the same timestamp as a vector and computing p-norm of vectors [12]. A kNN classifier based on 2-norm MD-DTW is used as a baseline in our empirical evaluation in Sect. 5.

Feature Based MTS Classification. The feature-based representation of time series can be efficient in classification and can discover more latent patterns if the right set of features is extracted. Therefore, researchers developed various feature extraction approaches for time series. Yang et al. [13] proposed a feature selection technique for MTS based on Common Principal Component Analysis. Wang et al. [14] extracted statistical features such as trend and periodicity from MTS for clustering. Fulcher et al. [15] developed a comprehensive feature extraction approach which constructs features including statistics, correlations, and non-linear model fits, etc. Over 7000 features are extracted from the time series and forward selection is utilized to generate feature vectors for classification. This Feature Selection (FS) approach is also used as a baseline in Sect. 5.

3 Preliminaries

In this paper, the time series represents the sequence of traffic speed detected by a sensor every minute. Here, a sensor s is a loop detector located on major roads recording traffic metrics.

Since time series collected from a single sensor is noisy and sen-

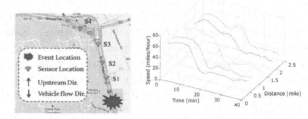

(a) Sample Accident [16]   (b) Traffic Speed MTS

Fig. 1. Multivariate Time Series

sitive to abnormal behaviors of vehicles, it is important to study multiple sensors at different locations. Thus, we represent the time series collected from multiple sensors close to an accident as a single MTS and label it as "positive". Accordingly, we also generate MTS of regular traffic from random locations when there is no accidents as the negative class. Specifically, the MTS is generated from time series of each *upstream sensor*, as defined below.

Definition 1 (Multivariate Time Series (MTS)). *A Multivariate Time Series is a set of time series: $X = \{x_1, x_2, ..., x_K\}$. In this paper, a time series*

x_k is a sequence of speed, indexed by time: $x_k = \{x_k(1), x_k(2), ..., x_k(n)\}$. The k_{th} time series x_k is collected by the k_{th} sensor s_k.

In the definition, n denotes the window size of the MTS, and K is the number of upstream sensors. The upstream sensors of a location l on a road are defined as a sequence of nearest sensors $S = \{s_1, s_2, ..., s_K\}$ ordered by the distance to l with the same direction as l, as shown in Fig. 1a. Figure 1b presents the MTS of upstream sensors relative to the accident location/time. The time series are generated at different distances in different colors. Hence we define the problem of traffic incident detection based on MTS classification as follows.

Problem 1 (MTS based Traffic Accident Classification Problem (MTS-TACP)). Given a multivariate time series $X = \{x_1, x_2, ..., x_K\}$, in which each time series x_k is collected from the k_{th} upstream sensor s_k of location l, during the same time window $[t_1, t_n]$, the goal is to identify whether an accident happened at location l from t_1 to t_n, classify the MTS to be accident or normal instance.

4 Traffic Accident Classification with Traffic MTS

4.1 A Discrete Unsupervised Solution: IIG

In our previous study [2], we proposed an approach to realign the start of existing accidents based on the upstream traffic speed around the reported time. Though it has different assumptions and problems, the idea of identifying the propagation pattern in MTS can be extended to detect an accident.

First, *impact interval* in Definition 2 is defined to reduce the dimensionality while keeping the accident patterns. Basically, a time series will be compared to its historical average and the intervals below a given ratio are extracted. Such definition is derived from real-world intuitions. Essentially, accidents usually result in visible and consecutive speed drops at upstream sensors. However, rush hours and narrow roads could also have regular speed drops. We assume the accidents cause unusual speed drops, i.e. the speed should be observed at a lower value than regular speed if an accident happened. Therefore the real-time speed should be compared to the historical speed to quantify the impact [16]. Here historical speed is calculated using the average of speed at the same location and same time [16]. So the unusual speed drop is modeled in a discrete way by extracting impact intervals. Via such discretization, we can convert the complex time series into a concise formulation which is easier to model as in Fig. 2a. The following paragraphs will describe the procedure of detecting accidents by the modified IIG method. The existence of accident in an MTS will be decided by the identification of propagated impact intervals via following steps.

Definition 2 (Impact Interval). *An impact interval is a tuple* (t_s, t_e), *s.t.* $\forall t, t_s \leq t \leq t_e, \frac{|x(t) - \bar{x}(t)|}{\bar{x}(t)} \geq \theta$. *Here* $x(t)$ *denotes the real-time speed at time* t, *and* $\bar{x}(t)$ *denotes the historical average speed of the same sensor, at time* t. θ *is a tuning parameter determining how strict the impact is measured.*

Discretization. First, for each upstream sensor s_k, we calculate a set of impact intervals $I_k = \{(t_{1s}, t_{1e}), (t_{2s}, t_{2e}), ...\}$ as described above. For example, in Fig. 2a, the time series (red line) will be compared to the historical time series (solid blue line). Then the impact intervals (solid bold segments) are generated by applying a threshold θ to the historical time series (dashed blue line). In this way, the MTS will be discretized into K sets of impact intervals $\{I_1, I_2, ...I_K\}$.

Smoothing. Since fluctuations and outliers could generate useless and noisy impact intervals, smoothing and cleansing should be applied before and after discretization. We smooth the time series before creating impact intervals by moving average with windows size 5 min for both real-time and historical data. After discretization, we concatenate adjacent impact intervals with distance less than 2 min, similar to one-dimensional clustering. Because close impact intervals may result from the same event. Then isolated intervals not concatenated to any other intervals with a small length, i.e., 2 min will be eliminated. Those impact intervals are actually noises, i.e. speed fluctuations. After that, we have a clean formation of impact intervals from all sensors.

Grouping. Intuitively, upstream sensors are affected by accident in the spatial order based on the traffic flow: a further upstream sensor is affected only after other upstream sensors closer to the accident. Temporally, the impacts of the same accident, observed

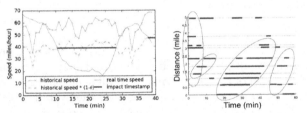

(a) Generate Impact Intervals (b) Impact Interval Groups

Fig. 2. Procedure of IIG (Color figure online)

at different upstream sensors, should be relatively close in time. So impact intervals can be grouped by their spatial and temporal distances, using heuristics such as maximum overlap and nearest center as in Fig. 2b. Specifically, each impact interval in the closest sensor will form a group. Then other impact intervals in the adjacent sensors will be iteratively added to the groups they belong to using maximum overlap or nearest center. The grouping procedure finishes until all impact intervals are visited. Groups with too few intervals will be filtered out. Then if a group of impact intervals exists after this IIG procedure, we classify the MTS as an accident. Otherwise, we label the MTS as a normal instance.

4.2 Severity Features Based Solution (Severity-I)

Admittedly, the IIG approach models the spatiotemporal propagation pattern intuitively using impact interval groups. However, it could be too strict to identify an accident. A slight turbulence or malfunction could disqualify the entire impact interval group. Thus, we need more robust characteristics of accidents. In this section, three types of features are defined and extracted from a traffic speed MTS based on empirical observations. Then the features can be adopted by various classifiers to detect accidents.

Dropping Severity λ. We first consider drops in traffic speed. Different from the IIG approach, the extent of speed drop should be essential in detecting accidents as opposed to the binary comparison to the threshold θ. Hence, we have the following observation: *An accident will cause a speed drop in upstream sensors to a certain extent.* A larger extent of speed drop can provide more confidence of an accident. For example, in Fig. 2a, the red line depicts the time series of speed from the nearest sensor to an accident. As shown in the figure, the speed drops from 60 miles/hour to around 20 miles/hour, which can indicate a severe accident. The dropping severity can be estimated by the ratio of speed change from historical data to real-time data. As described in IIG approach, the comparison to historical speed is necessary to eliminate the rush hour or other periodical effects. Given an MTS of traffic $X = \{x_1, x_2...x_k\}$, the historical average of speed is denoted by $\overline{X} = \{\overline{x}_1, \overline{x}_2...\overline{x}_k\}$. We define the measurement of dropping severity in the following equation.

$$\lambda_{max} = \max_{i,k}(1 - x_k(i)/\overline{x}_k(i)); \lambda_{avg} = \underset{i,k}{avg}(1 - x_k(i)/\overline{x}_k(i)) \qquad (1)$$

In the equation, we propose two options for this measurement for comprehensiveness. Dropping severity is first measured as the maximum speed drop ratio λ_{max}, which reflects the worst impact to all sensors of the accident. The other option is the average ratio λ_{avg}, aggregating the overall speed change in all sensors.

Figure 3a depicts the intermediate step in extracting dropping severity. The red line depicts the real-time speed reported by a certain sensor near an accident. The solid blue line represents the historical speed. Denoted by the red dot line, the maximum distance between real-time speed and historical speed is used for calculating λ_{max}, and such computation is applied to all upstream sensors.

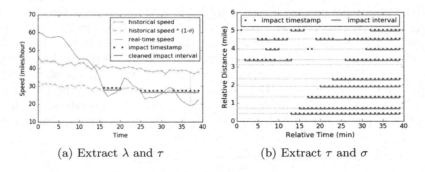

(a) Extract λ and τ (b) Extract τ and σ

Fig. 3. Severity features (Color figure online)

Lasting Severity τ. Not all accidents have grave speed drops and moreover normal traffic also has occasional drops, i.e., traffic MTS with moderate drops can be either normal traffic or accidents. To better differentiate these two conditions, we introduce the temporal criteria, lasting severity. For example, in Fig. 2a, the speed drops at 8 min and resumes at 28 min, staying at a low value for 20 min. Then we can conclude such speed drops are not caused by normal traffic fluctuations. *After an accident happens, the drop of speed will last for a certain time.* Significant dropping time can be the evidence of an accident. Impact interval is used to measure lasting severity because the discretization provides an easy extraction of temporal patterns. A list of impact intervals I_k (horizontal cyan segments in Fig. 3a) is generated. $|x_k|$ denotes the length of time series x_k. Then lasting severity is measured by the following formulation.

$$\tau_{max} = \max_{i,k}(I_k(i)[1] - I_k(i)[0])/|x_k|$$
$$\tau_{avg} = avg_{k}(\max_{i}(I_k(i)[1] - I_k(i)[0]))/|x_k| \tag{2}$$

The lasting severity is provided with two options as well. We assume each impact interval after smoothing and concatenation is individually impacted by a single event, as supposed in IIG. So the maximum length of impact intervals should be the upper bound of all the events(the longer cyan segment in Fig. 3a). Thus τ_{max} is the maximum of these maximum lengths in all sensors which indicates the longest impact. τ_{avg} is the average of the maximum lengths indicating the overall affected time in all sensors.

In addition, We also define a relaxed definition of τ, to overcome fluctuations in traffic speed which may prevent the formulation of impact intervals. Rather than extracting impact intervals, we compute impact timestamps(dark blue dots in Fig. 3a), which are the set of time index at which the relative speed drop is below θ. Then the sizes of the impact timestamps sets are used instead of the lengths of impact intervals in calculating τ'_{max} and τ'_{avg}.

Distant Severity σ. We also believe an accident usually influences a succession of cars rather than a single one, i.e. *an accident will affect a certain distance in the upstream traffic.* The longer stream of cars are affected, the traffic MTS is more likely to reflect an accident. To be consistent with lasting severity, the distant severity is measured based on the existence of impact intervals, as described in the following formulation. Here d_k denotes the distance of sensor s_k.

$$\sigma_{cons} = d_k/d_K, k = arg\max_{k}\{I_1 \text{ to } I_k \neq \emptyset\}$$
$$\sigma_{disc} = d_k/d_K, k = arg\max_{k}\{I_k \neq \emptyset\} \tag{3}$$

We provide two options for distant severity measurement to overcome the inconsistent behaviors among individual sensors. First we assume the impact at different locations be consecutive since the first sensor cannot affect the third sensor without influencing the second one. So σ_{cons} computes the furthest consecutively impacted distance. Optionally, σ_{disc} lifts such restriction to the furthest discrete impact location in case of noise or missing data.

Figure 3b shows the extraction of distant severity. The solid red lines depict impact intervals. The y-axis represents the distance to the detecting location l.

The distant severity is measured by the distance of furthest impacted sensor. The difference between σ_{cons} and σ_{disc} is the absence of impact interval in the 6th sensor s_6 (the bare-fine dotted blue line). σ_{cons} requires the impacts be consecutive so that it can separate events at different locations. σ_{disc} allows skipping some sensors to give more tolerance to malfunction of sensors. Similar to lasting severity, we also defined a relaxed definition of σ, to capture a loose distant severity. The relaxed impacted distant severity σ'_{cons} and σ'_{disc} will be computed based on the existence of the impact timestamps at different locations instead of the impact intervals. By replacing τ and σ with their relaxed version τ' and σ', we can derive a relaxed version of Severity-I, namely *Severity-T*.

Severity Based Classification. The advantage of measuring different types of severity (extent, time and space) is the capability of capturing enough variety of accidents. For example, an accident with low λ and large τ and σ could be caused by an emergency vehicle parked on the shoulder which does not reduce the traffic speed significantly but may last for 30 min and affects many upstream vehicles. An accident with low τ and large λ and σ may be a minor accident with no injury. So upstream drivers all brake and the speed drops a lot, but traffic is resumed soon after the accident is clear. Therefore, the combination of the three severity could describe accidents in a wide variety. After generating the three severity features from a traffic speed MTS, we can utilize various classifiers, e.g. Logistic Regression, SVM, Gradient Boosting Decision Tree (GBDT), etc. for classification. GBDT is adopted as the default classifier in Severity-I.

The complexity of extracting a severity-I feature is quite low. With n as the window length and K as the number of upstream sensors, it takes $O(Kn)$ time to traverse all historical and real-time speed to get Dropping Severity λ. Moreover, the extraction of impact intervals of one time series takes $O(n)$ time. So Lasting Severity τ and Distant Severity σ extractions also cost $O(Kn)$ time.

5   Empirical Evaluation

5.1   Experiments Settings

System Environment. For all experiments, we implemented the algorithms in Python to ensure a fair comparison. All the experiments were conducted on a 64-bit Windows machine with a 2.60 GHz CPU and 8.00 GB memory.

Data Set. In this paper, we use the accident reports in June 2012 as ground truths, which were reported by California Highway Patrol. Traffic metrics including speed, volume and occupancy, collected from more than 4,000 sensors during the same period are retrieved as real-time data. The historical average of traffic is generated from the sensor dataset during March to May 2012. 70% of real-time data is sampled as training data, and the rest 30% is testing data.

Baseline Approaches. As baseline approaches, we implemented kNN-MD-DTW and the FS approach [10,15]. In the implementation of kNN-MD-DTW, we set $k = 3$ and set the size of warping window in MD-DTW to 5 min. As for the extended FS approach using HTCSA [15], 4 features are selected from more than 7000 features. We generated 7000 features only using a 200 size sample because the extraction takes significant computation. The 4 feature extraction methods are SY_KPSStest_0_10.lagminstat (KPSS stationarity test), SY_LocalDistributions_5_each.meandiv (minimum divergence between two segments' distribution), MF_CompareAR_1_10_all.whereen4 (order of AR with little error), and CO_tc3_1.num (numerator of Normalized nonlinear autocorrelation function). We also tested the CA Algorithm [3] on real-world data, but the recall is lower than 5%. Therefore, it is not included in the following experiments.

5.2   Parameter Effects in Impact Based Approaches

Effect of θ. The impact threshold θ is a hyperparameter which determines how strict we are evaluating an impact. As shown in Fig. 4, we can make the following observations as θ increases. (1) Severity-I and Severity-T have sim-

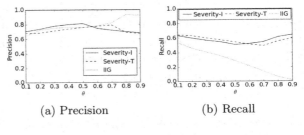

(a) Precision   (b) Recall

Fig. 4. Vary θ

ilar trends but Severity-T is less sensitive to large θ, because Severity-T uses impact timestamps which may still exist while impact intervals disappear when θ is large. (2) For severity based approaches, the precision increases first then drops. Since a higher threshold θ is more strict to impact generation, most detected accidents should have severe enough impacts and are always real accidents. However, as we use an overly strict threshold, only a small portion of impacts are captured in accidents, which will be mixed with the normal instances with drastic fluctuations. (3) On the other hand, the recall decreases first and then rise in severity approaches. (4) Though IIG can reach a very high precision as θ reaches 0.9, it fails to detect most real accidents. However, for the severity based approaches, the precision and recall are well-balanced, and $\theta = 0.3$ is chosen to strike a balance between the two metrics. Note that we can tune θ, e.g., choosing a high value for θ for detecting very severe accidents. In the following experiments, θ will be set to 0.3 as the impact threshold.

Effect of Severity Options. While controlling all other parameters, we vary the selection of different severity options and compare their effects. As listed in Table 1, λ_{avg} has a lower recall than λ_{max} since the average option can smooth a significant speed drop, the algorithm may miss those accidents. The effect and trend of τ are very similar to λ. In addition, the use of consecutive option in σ will exclude fluctuating normal traffic speeds. So σ_{cons} has a better precision than σ_{disc}. In the end, we choose $\lambda_{max}, \tau_{max}, \sigma_{cons}$ as the default severity options.

Effect of Classifier. After extracting severity features of Severity-I, we applied different approaches to make the classification. In Table 2, we can observe that

Table 1. Vary severity options

λ	prec.	rec.	f-score	τ	prec.	rec.	f-score	σ	prec.	rec.	f-score
max	0.78	0.57	0.66	max	0.78	0.57	0.66	avg	0.79	0.55	0.65
avg	0.79	0.52	0.63	cons	0.78	0.57	0.66	disc	0.75	0.58	0.65

Table 2. Vary classifiers

Method	prec.	rec.	f-score	acc.
Logistic Regression	0.78	0.57	0.66	0.72
SVM	0.78	0.57	0.66	0.73
DecisionTree	0.66	0.58	0.62	0.67
GBDT	0.79	0.58	0.67	0.73
Neural Network	0.78	0.58	0.66	0.73
AdaBoost	0.78	0.59	0.67	0.73

Table 3. Comparison of approaches

Metric	prec.	rec.	f-score	acc.
Severity-I	**0.79**	0.58	**0.67**	**0.73**
FS	0.72	**0.61**	0.66	0.71
DTW	0.7	0.6	0.64	0.69
IIG	0.75	0.41	0.53	0.66

decision tree is not so good as others because the overall severity is the combination of the three severity features, which is not easily separable by decisions. GBDT, Logistic Regression, SVM and AdaBoost have similar performance.

5.3 Comparison of Different Approaches

To evaluate the reliability of the proposed approaches, we compare them using 4 metrics: precision, recall, f-score and accuracy. As shown in Table 3, we can make following observations: (1) IIG has a high precision but low recall, f-score and accuracy. The reason is that IIG is the most strict among all approaches since it identifies accidents based on the existence of a propagation behavior across all upstream sensors; (2) kNN-MD-DTW is neither too bad nor too good in all metrics. Due to the fluctuations often observed in traffic data, the optimal alignment may not be accurate; (3) Severity-I has better precision, f-score and accuracy than any other approaches. The FS approach has the highest recall, but requires exhaustive computation to select from more than 7000 features. Our Severity-I method only relies on three features and achieves the highest precision, f-score and accuracy for accident detection.

Understand Misclassification Cases. Real-world data always does not follow theoretical assumptions since some system noises and outliers are inevitable. Figure 5a shows a false negative case by Severity-I. Different colors of lines depict the speed at different sensors. We can observe that speed does not change much in all sensors, except a small drop at the 4th sensor. However this case is an accident in our record. Such case is difficult to be detected by any approach without any observable impact. It is possible that some reported accidents may not show much impact to the traffic because of a wrongly reported time or location. Figure 5b shows a false positive case of Severity-I. It is observable that there are many nontrivial propagated speed drops, which does not like a normal instance. For these instances, none of the proposed approaches can successfully differentiate accidents with normal cases.

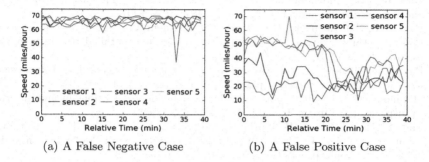

(a) A False Negative Case   (b) A False Positive Case

Fig. 5. Severity-I misclassification cases (Color figure online)

6   Conclusions

We presented IIG and Severity-I, two techniques that detect traffic accidents from the traffic speed data by exploiting the spatiotemporal impact of traffic accidents. IIG detects the presence of propagation behavior by impact interval groups thus has a high precision with proper impact threshold. Severity-I takes an MTS classification approach and extracts three features i.e., drop ratio, lasting time, and impact distance, to measure the severity of an accident. As the three severity features capture different aspects of the accident impact, we are able to classify traffic speed MTS generated from nearby sensors into accidents and normal instances. Our proposed methods are evaluated and compared to state-of-the-art MTS classification methods with real-world accident and traffic data: Severity-I is shown to be superior to other methods. Future work includes integrating our algorithms into real-time traffic streaming systems. Our work can also be applied to accident severity evaluation for routing problems.

Acknowledgement. This research has been funded in part by NSF grants CNS-1461963, Caltrans-65A0533, the USC Integrated Media Systems Center (IMSC), the USC METRANS Transportation Center, and a UAlbany FRAP-A award. Any opinions, findings, and conclusions or recommendations expressed in this material are those of the authors and do not necessarily reflect the views of any of the sponsors such as NSF.

References

1. National Safety Council: NSC motor vehicle fatality estimates (2016). http://www. nsc.org/NewsDocuments/2016/mv-fatality-report-1215.pdf. Accessed 13 Feb 2017
2. Yue, M., Fan, L., Shahabi, C.: Inferring traffic incident start time with loop sensor data. In: Proceedings of the 25th ACM International Conference on Information and Knowledge Management, pp. 2481–2484. ACM (2016)
3. Payne, H., Helfenbein, E., Knobel, H.: Development and testing of incident detection algorithms, volume 2: research methodology and detailed results. Technical report, FHWA (1976)
4. Payne, H., Tignor, S.: Freeway incident-detection algorithms based on decision trees with states. Technical report, National Research Council (1978)
5. Stephanedes, Y.J., Chassiakos, A.P.: Freeway incident detection through filtering. Transp. Res. C: Emerg. Technol. **1**(3), 219–233 (1993)
6. Zhu, T., Wang, J., Lv, W.: Outlier mining based automatic incident detection on urban arterial road. In: Proceedings of the 6th International Conference on Mobile Technology, Application & Systems (29), September 2009
7. Yuan, F., Cheu, R.: Incident detection using support vector machines. Transp. Res. Part C: Emerg. Technol. **11**(3–4), 309–328 (2003)
8. Sadek, S., Al-Hamadi, A., Michaelis, B., Sayed, U.: A statistical framework for real-time traffic accident recognition. J. Sig. Inf. Process. **1**(1), 77–81 (2010)
9. Parkany, E., Xie, C.: A complete review of incident detection algorithm & their deployment: what works and what doesn't. Technical report, New England Transportation Consortium (2005)
10. Ding, H., Trajcevski, G., Scheuermann, P., Wang, X., Keogh, E.: Querying and mining of time series data: experimental comparison of representations and distance measures. Proc. VLDB Endowment **1**(2), 1542–1552 (2008)
11. Bashir, M., Kempf, J.: Reduced dynamic time warping for handwriting recognition based on multidimensional time series of a novel pen device. Int. J. Intell. Syst. Technol. WASET **3**(4), 194 (2008)
12. Ten Holt, G.A., Reinders, M.J., Hendriks, E.: Multi-dimensional dynamic time warping for gesture recognition. In: Thirteenth Annual Conference of the Advanced School for Computing and Imaging, vol. 300 (2007)
13. Yoon, H., Yang, K., Shahabi, C.: Feature subset selection and feature ranking for multivariate time series. IEEE Trans. Knowl. Data Eng. **17**(9), 1186–1198 (2005)
14. Wang, X., Wirth, A., Wang, L.: Structure-based statistical features and multivariate time series clustering. In: Seventh IEEE International Conference on Data Mining, ICDM 2007, pp. 351–360. IEEE (2007)
15. Fulcher, B.D., Jones, N.S.: Highly comparative feature-based time-series classification. IEEE Trans. Knowl. Data Eng. **26**(12), 3026–3037 (2014)
16. Pan, B., Demiryurek, U., Shahabi, C., Gupta, C.: Forecasting spatiotemporal impact of traffic incidents on road networks. In: 2013 IEEE 13th International Conference on Data Mining (ICDM), pp. 587–596. IEEE (2013)

MicroGRID: An Accurate and Efficient Real-Time Stream Data Clustering with *Noise*

Z. Tari[1(⊠)], A. Thompson[1], N. Almusalam[1], P. Bertok[1], and A. Mahmood[2]

[1] School of Science, RMIT University, Melbourne, Australia
zahir.tari@rmit.edu.au
[2] School of Engineering and Mathematical Science,
La Trobe University, Melbourne, Australia

Abstract. Data stream clustering aims to produce clusters from a data-stream in a real-time. Many of existing algorithms focus however on solving a single problem, leaving anomalous *noise* in data streams at the wayside. This paper describes the MicroGRID approach to cluster data from single data-streams to handle noisy data streams, accurately identifying and separating noise-affected data points from outlier points. In particular, MicroGRID utilises a combination of micro-cluster and grid-based prospectives, an approach that has not been attempted when clustering data-streams. The experimental results clearly show that MicroGRID significantly outperforms the baseline methods: MicroGRID is up 87% faster and up to 80% more accurate clustering outputs.

1 Introduction

Data stream clustering has been used for many different real world applications, such as analyzing stock market data and network intrusion detection systems. A number of different innovations have been made recently into improving different aspects of real-time data stream clustering, such as adapting to multiple data streams [8,15] and data streams with a high number of dimensions [2,14]. However, many of these innovations only focus on improving one such a problem, leaving other weaknesses not handled, such as *noise* [10].

Noise is data points transmitted by the data streams, which have had their contained information modified into content which is not related/relevant to the data values that should be transmitted by the stream. This can be as a result of sensor failure [7] or interference in data transmission [13], for example. Noise-affected points can still contain relevant information to the clustering algorithm, and as such should still be expected to be clustered appropriately by a clustering algorithm. The problem arises in correctly identifying these noisy points as valid data points, and not incorrectly identifying these points as possible outliers, or attributing these points to the incorrect cluster. This reduces the accuracy of the clusters generated, and this is caused by modified information within these points affecting their identification by a clustering algorithm.

© Springer International Publishing AG, part of Springer Nature 2018
D. Phung et al. (Eds.): PAKDD 2018, LNAI 10938, pp. 483–494, 2018.
https://doi.org/10.1007/978-3-319-93037-4_38

As with noisy data points, outliers can also be the result of noise. However, noise-created outliers [5] means that these outlier points will not form any cluster, containing dimension values that are not uniform with the rest of the data set, and are not indicative of the system operation. This is a problem, as if these points are placed in the same clusters as valid, non-outlier data points, the resulting clusters will be inaccurate to a degree [7]. The aim of this paper to provide both more efficient and accurate clustering results than similar, existing, algorithms in the field of real-time data streams. This paper introduces the MicroGRID approach that makes use of a joint micro-clusters array and grid-based clustering algorithm. The MicroGRID approach adapts micro-clusters to represent information received from data streams allowing many data points to be summarised into a single element, thereby reducing the amount of points within a clustering grid to improve both the accuracy and efficiency of cluster generation. By limiting the amount of micro-clusters that are stored and processed by the proposed approach at any given time via a micro-cluster array, this is expected to reduce the execution time of cluster generation, while also allowing outlier data to be removed from clustering equations at a faster rate. MicroGRID was experimentally compared with two baseline methods including, namely CluStream [3] and DenStream [7], on real data sets (i.e. Forest Cover-Type, Sensor Readings, Electricity Market and Spam) and the results show the followings: up to 87% faster average execution times than the comparison algorithms, and up to 96% more accurate clustering outputs.

Next section provides overviews existing work, and Sect. 3 provides details of the MicroGRID approach. In Sect. 4, the experimental results are discussed, and Sect. 5 concludes the paper.

2   Related Work

There are several problems to address with when dealing with data streams. One of the first ones is handling data streams which evolve over time [1]. CluStream uses of micro-clusters for summarising data, and the online-offline structure for processing data streams. Data points are summarised by creating a vector for each data point. These vectors include the time the point reached the system, a weight value of the point, and the weighted sum of all the data values within a point. The other problem relates to the variance of data when this is unknown [11]. A similarity between existing algorithms is that most of them are based on k-means. SUBCFM [16] is an example that extends k-means to create clusters with "fuzzy" boundaries. These clusters have set radius boundaries, but they also include a second boundary that surround the cluster.

While k-mean [1] is a popular method of implementing data-stream clustering, it is not the only proposed method. Another such method is the density-based algorithm, called D-Stream. First proposed by Chen et al. [9], this method still uses an online-offline approach (as in CluStream), but maps the summarised data points to a grid within the online-component. As updating a grid cell only takes a single update of the cell's summary vector for each new data point added, processing each data point in the online component is a constant time complexity, allowing D-Stream to be a suitable real-time clustering algorithm. To keep storage costs of the grid lower, each grid cell contains a summary vector of all data points contained within the cell, rather than the points themselves. Like CluStream, *D-Stream* has been expanded upon to solve evolving data streams with noise (unwanted data) [7], as well as a number of different possible improvements to time and space complexity [5, 6, 12].

Another type of algorithms is the grid-based clustering algorithms, which have been designed for data streams. *MG-Join* [4] is a grid-based clustering that takes a user defined window of each stream, and summarises a function of each streams dimension values into a set number of coefficients. The values of a streams coefficients dictate its position on a grid, where each coefficient is a different axis of the grid. Streams are mapped onto the grid, with each stream being represented by a single point on the grid. Once all streams are mapped onto the grid according to their coefficient values, streams are then clustered together with other streams in the same grid cell, as well as streams located within neighbouring grid cells. *MG-Join* chooses the grid cell containing the largest number of stream points as the central clustering point. All non-empty neighbouring cells are then attached to this cluster, with each of these cells having their own neighbouring cells checked for non-empty cells, in a recursive fashion.

3   The MicroGRID Approach

As shown in Algorithm 1, MicroGRID can efficiently and accurately clusters and summarizes the incoming data. Firstly, each data point received from the data stream is converted into a micro-cluster. MicroGRID keeps an array of a user-defined length of micro-clusters, in which all incoming micro-clusters are added into, removed from or merged into. The purpose of this array is to capture a summary of the recent history and trends of the data stream. This is possible as each micro-cluster generated must be inserted into the array, so older micro-clusters are removed or merged, keeping the information within the array recent. The second stage of MicroGRID is the mapping of each micro-cluster within the array to a 2D Grid. The grid is updated each time the array is updated, to constantly reflect the status and trends of the stream in real-time. The purpose of mapping these micro-clusters to a grid is to organise these micro-clusters by their content, representing the current data trends of the stream, as well as isolating any micro-clusters representing outliers to different grid areas. Finally, clusters are generated at regular intervals based on the locations of micro-clusters on this grid. As this clustering is performed based on the grid, the time taken to produced clusters can be greatly reduced, as this clustering is generated based on the grid cells, rather than the individual micro-clusters themselves. These cluster results will also highlight the outlier patterns of the data stream, as grid cells containing outlier micro-clusters will not be contained in the clusters.

The advantage of MicroGRID over other clustering algorithms is twofold. **Firstly**, the use of a micro-cluster array of limited size allows a summary of

Algorithm 1. MicroGRID

1: **Begin**
2:     St = Data Stream
3:     Ar = Micro-Cluster Array
4:     G = Clustering Grid
5:     H = Horizon time
6:     m = Current size of Ar
7:     q = Maximum size of Ar
8: **Repeat**
9:        Receive next data stream point X_i
10:       Convert X_i into a new micro-cluster M_i
11:    **If** $m < q$
12:          Add M_i to Ar
13:    **Else If** an underweight micro-cluster M_u exists in Ar
14:          Add M_i to Ar and remove M_u
15:    **Else** find two closest micro-clusters M_a and M_b in Ar
16:          Merge M_a and M_b into M_{ab}
17:          Add M_i to Ar
18:          Update clustering grid G with changes to Ar
19:    **If** it has been H time units since last clustering
20:          Generate Clusters from clustering grid G
21:    **Until** Data Stream St ends
22: **End**

---

the streams current pattern and trends to be used for clustering, rather than the entire history of the stream. Assuming an array of appropriate size is used, this allows the clustering results of MicroGRID to be produced more efficiently, as less micro-clusters are used for clustering. This also reduces the impact that micro-clusters from further in the past (which are less relevant) have on the most recent clustering results, as these are naturally merged or removed within the array as more data points are processed. This produces more accurate clustering results than other algorithms, which can include irrelevant or outdated data points in their most recent clustering results. **Secondly**, a clustering grid is used to generate clusters to also improve efficiency. Clusters are generated based on grid cells. As many micro-clusters can be contained within a single grid cell, this greatly increases the efficiency of generating clusters, as opposed to algorithms where each individual micro-cluster must be considered when generating clusters.

3.1   The Micro-cluster Structure

The micro-clusters are used to store the information of a point in a summarized form [1]. A point is defined as a set of d-dimensions, where a point X can be represented as $X = (x_1, x_2, x_3, \ldots, x_d)$. In the MicroGRID algorithm described in this paper, each micro cluster stores a set of information used to calculate its position and validity within the central grid. A micro-cluster of $O(2d + 5)$ size, of a fixed radius r and containing a set of d-dimensional points, stores the following information:

- The sum of values per each dimension, across all points stored within a micro-cluster. This results in a vector $\mathbf{S}$, containing d values. Of a micro-cluster storing n points, the p-th entry of $\mathbf{S}$ is described as $\sum_{k=1}^{n} x_k^p$.
- The squared sum of values per each dimension, across all points stored within a micro-cluster. This results in a vector $\mathbf{SS}$, containing d values. Of a micro-cluster storing n points, the p-th entry of $\mathbf{SS}$ is described as $\sum_{k=1}^{n} (x_k^p)^2$.
- The combined sum of values of all dimensions from all points stored within a micro-cluster. This results in a single value $SumTotal$, which can be described as $\sum_{k=1}^{n} \sum_{l=1}^{p} x_k^l$.
- The combined squared sum of values of all dimensions from all points stored within a micro-cluster. This results in a single value $SqSumTotal$, which can be described as $\sum_{k=1}^{n} \sum_{l=1}^{p} x_k^l$.
- The number of points stored within the micro-cluster, represented by an value N.
- The time stamp at which the micro-cluster was created, $CreateTime$.
- The time stamp at which the last point was added to the micro-cluster, $LastTime$.

$CreateTime$ and $LastTime$ are used to track the weight of a micro-cluster, and are used in calculating the minimum weight-limit and current weight of a micro-cluster defined by Cao et al. [7]. However, some modifications to the algorithm (as shown in Sect. 2) for calculating the current weight have been made. The modified weight calculation is as follows:

$$W(t) = e^{-\lambda t}, \ \lambda \in [0, 1]$$

where t is $LastTime$, which is either (a) a new data point was added to the micro-cluster, or (b) the last time another micro-cluster was merged into this micro-cluster. $W(t)$ is used during the maintenance of the micro-cluster array, to determine if the current micro-cluster is underweight and should be removed. The use of this weight calculation is shown Line 11 of Algorithm 2.

As micro-clusters now only keep the latest time of modification within $LastTime$, instead of time stamps for all included data points, the space requirements of each micro-cluster is reduced, especially as the running time of the algorithm increases and more possible additions to micro-clusters can be made. In addition, the computational complexity of calculating a micro-cluster's weight is reduced, as only a single weight must be calculated, rather than the weight of each data point within a micro-cluster. This changes the current weight algorithm's time complexity from $O(N)$ to $O(1)$. The space requirements of micro-clusters is reduced to improve MicroGRID's scalability. As the data-stream is presumed infinite, the space requirements of keeping time-stamps of each data point within each micro-cluster increases and can exceed any set space limits for the algorithm. Therefore only requiring a single time stamp value, regardless of data stream length, removes this potential space limitation.

Once the micro-cluster array has no remaining empty spaces, at worst $O(q)$ weight calculations will be made for every incoming data point. Therefore, reducing the weight calculation to a constant time complexity greatly increases the speed at which every new data point can be processed, in order for the micro-cluster array to be maintained in real time. **S**, **SS** and N are used in calculating the centre of the micro-cluster, needed for clustering evaluation, and the distance of the cluster from other clusters or incoming points. *SumTotal* and *SqSumTotal* are used in calculating the mapping of the micro-cluster onto the central grid.

The use of micro-clusters to represent data points is beneficial, as storing this information in micro-clusters means that these attributes only need to be calculated once per each micro-cluster creation and addition. This reduces execution times, as these values (cluster centre, distance and mapping) are used frequently in the clustering process. These micro-clusters are different from micro-clusters used in other algorithms in a number of ways. One major difference is that the radius of the micro-clusters is fixed. This eliminates the chance that a single micro-cluster will expand to too large of a size from having many data points added to it. The larger a micro-cluster becomes, the higher a chance that outlier data points, or data points of differing classes (in particular noise-affected points), being included into these large micro-clusters. In addition, the single values of *SumTotal* and *SqSumTotal* are kept within each micro-cluster, rather than just the vector summations (**S** and **SS**) that are stored in other micro-clusters. These values are used for faster calculations when mapping the micro-clusters to the grid (as explained in Sect. 3.2). The computation and maintenance of both of these values can also be performed at the same time as those for both vector values, thus incurring no further calculation costs to produce these values.

3.2   The Proposed Grid Clustering

The grid structure is used for distributing all the current micro-clusters over a fixed plane, based on the values of dimensions of all points represented by the cluster. A grid structure is used for generating clusters, as it provides a plane of fixed size that all micro-clusters are mapped to. This allows for more dense clusters to be generated, as all micro-clusters must fit inside this grid of fixed size, rather than the size of the clustering plane being unbounded. Each grid cell can store multiple micro-clusters within it at any given time, which means that when clusters are generated, they are generated based on the grid cells, rather than the individual micro-clusters.

Clustering based on grid cells eliminates the need to perform distance calculations between micro-clusters when detecting clusters. The use of grid structure of fixed size also eliminates the need for assumptions about the number of clusters that will be generated by the clustering algorithm. This is because the clustering grid is already a fixed size, and only non-empty cells are used in cluster calculation, so no limits on the number of clusters made within this grid need to be defined. This improves the accuracy of the clustering results obtained, as the number of clusters generated can change to reflect the current status of the data stream much better. This lowers the time taken to generated these clusters, while keeping the clustering content accurate. The method chosen is useful as it results in both identifying accurate clusters of current data stream trends, as

Algorithm 2. Micro-cluster Array Maintenance

1: **Begin**
2:     M_n = New Micro-Cluster to insert
3:     Ar = Micro-Cluster Array
4:     m = Current size of Ar
5:     q = Maximum size of Ar
6:     **If** $m < q$
7:         Add M_n to Ar
8:         Increment m
9:     **End**
10:     **For each** M_i in Ar
11:         Calculate current weight of $M_i \rightarrow W_c$
12:         Calculate current weight threshold of $M_i \rightarrow W_t$
13:         **If** $W_c < W_t$
14:             Replace M_i with M_n
15:         **End**
16:     $M_1 \rightarrow Close_a$
17:     $M_2 \rightarrow Close_b$
18:     Distance from $Close_a$ to $Close_b \rightarrow Close_{curr}$
19:     **For each** M_i in Ar
20:         **For each** M_j in Ar
21:             Distance from M_i to $M_j \rightarrow Close_{next}$
22:             **If** $Close_{next} < Close_{curr}$
23:                 $Close_{next} \rightarrow Close_{curr}$
24:                 $M_i \rightarrow Close_a$
25:                 $M_j \rightarrow Close_b$
26:             Merge $Close_b$ into $Close_a$
27:             Replace $Close_b$ with M_n
28: **End**

---

well identifying outlier micro-clusters from those non-empty grid cells which are not attributed to a cluster.

The first stage of the proposed grid clustering is maintaining the content of the grid each time the micro-cluster array, in order for the grid to reflect the most up-to-date content available from the data stream. Maintenance of this grid involves adding, removing and remapping of micro-clusters within the grid, in order to always reflect the current micro-clusters contained within the micro-cluster array. The second stage of the grid clustering is the generation of the clusters, which is undertaken at regular intervals set by the horizon. This clustering algorithm is accesses each grid cell that currently contains micro-clusters, and attempts to generate a cluster based on non-empty neighbouring grid cells.

This proposed clustering method is used as it will naturally cluster the most densely packed grid regions first, quickly reducing the number of cells to be checked for clusters, as many cells will be assigned to these first, larger clusters in most scenarios. The results of this clustering algorithm both show the current trends of a data stream, as well as the size of these trends (in comparison to the rest of the current clusters). This clustering algorithm also identifies any recent outlier data points, which are presented as non-empty grid cells that are not attributed to a generated cluster.

Grid Maintenance. A single square grid of size $s \times s$ cells is stored, which is used for determining the resulting clusters of the data stream. Multiple micro-clusters can be stored within a grid cell, but the location of a micro-cluster within the grid does not modify any information within the micro-cluster. When a change is made to the micro-cluster array, this change is communicated to the grid, to reflect the current array content. The grid changes as follows:

1. If a micro-cluster M_i has been added to the array with no further modifications, M_i is mapped to a grid cell.
2. If M_i has been added to the array and an underweight micro-cluster (as identified in Step 2 of the Micro-Cluster Array Maintenance process), M_j, is removed, M_i is mapped to a grid cell and M_j is removed from the grid.
3. If M_i has been added to the array and micro-cluster M_b has been merged into M_a, M_i is mapped to a grid cell, M_b and M_a are removed from the grid and M_{ab} is mapped to the grid.

One of these three actions are performed every time a new data point is received from the data stream. This allows each stored micro-cluster's grid location to always accurate, while only requiring a recalculation of a micro-cluster on the grid when it is either new, being removed or modified.

Micro-cluster Mapping Function. To map a micro-cluster to the grid, the appropriate cell reflecting the micro-cluster's dimensional values must be calculated. This is done by generating two coefficient values from a micro-cluster, xd and xy, each reflecting a micro-cluster's position along the x and y axis of the grid respectively. Initially, these two coefficients were to be generated using a Discrete Fourier Transformation (DFT) function, as explained by Aghbari et al. [4]. The use of a DFT as a mapping function is beneficial as it reduces the number of dimensions of a micro-cluster (the number of components within **S** and **SS**), down to a user specified number of coefficients, without sacrificing accuracy. The number of coefficients can be set as the number of axis within a clustering grid, allowing the DFT coefficients to be used as the location of a micro-cluster within the grid. To calculate this function, **S** and **SS** are used to determine the coefficients. However, the time complexity of using this function would be $O(d^2 + d)$ to calculate the cell of a single micro-cluster. Therefore, when adding a micro-cluster, or adding/deleting a micro-cluster, the mapping function would take place in $O(d^2 + d)$ time. However, when adding a micro-cluster, and adding a merged micro-cluster, the mapping of this action would take $O(2d^2 + 2d)$ time, as the mapping of a micro-cluster must be performed twice.

4   Experiments and Analysis

This section describes the experiment environment, variable values used for the MicroGRID algorithm, and the process undertaken to insert random noise into the experiment data sets. In particular, the experiments were conducted to measure this algorithm using four (4) different data sets to measure the algorithms time-based performance and clustering accuracy. The MicroGRID algorithm is implemented in Java, with version 1.8 of the Java Runtime Environment. The *DenStream* and *CluStream* algorithms were evaluated using their implementations provided by the *MOA* (Massive Online Analysis) software, produced by the University of Waikato (http://moa.cms.waikato.ac.nz/).

The following algorithms are used as baseline methods in these experiments: *CluStream* [1] and *DenStream* [7]. The former is a popular clustering algorithm which has been used for comparison in data stream clustering [3,7,17]. The data sets used in our experiments are: *Forest Covertype, Electricity Market, Sensor Readings*, and *Spam*. Finally, two metrics are used to evaluate the accuracy (Rand Index) and efficiency (Execution Time) of the MicroGRID approach, and the baseline methods: *Rand Index* and *Execution Time*.

Several results were obtained to show the advantage of MicroGRID over the baseline methods in efficient and effective clustering, when tested upon the four real-world data sets explained in the previous section. As mentioned earlier, because of the space limitation, we will only present the experimental results for *Rand Index*. Figures 1, 2, 3 and 4 show the average Rand Index results of the MicroGRID, CluStream and DenStream algorithms when tested with the *Forest Covertype, Electricity Market, Sensor Readings* and *Spam* data sets.

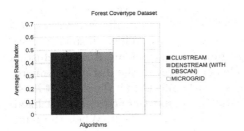
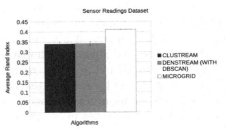

Fig. 1. Average rand index with the Forest Covertype Dataset

Fig. 2. Average rand index with the Sensor Readings Dataset

From the results presented in Figs. 1, 2, 3 and 4, it is clear that under most cases, MicroGRID provided significantly higher Rand Index values than both *CluStream* and *DenStream*. This is true when using the *Forest Covertype, Sensor Readings* and *Electricity Market* data sets, where MicroGRID is more accurate than both baseline methods. When using the *Spam* data set, the average results of MicroGRID are more accurate than the average result of *DenStream*, however

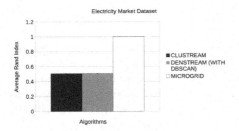
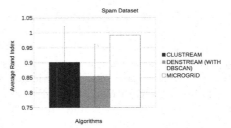

Fig. 3. Average rand index with the Electricity Market Dataset

Fig. 4. Average rand index with the Spam Dataset

a statistically significant conclusion cannot be drawn. When using the three previous data sets, it is evident that *CluStream* and *DenStream* both give extremely similar Rand Index values, while MicroGRID outperforms these results by 22%, 20% and 96% respectively. In the case of the *Covertype* and *Sensor Readings* data sets, both of these data sets contain 7 and 4 possible class values, used for calculating the rand index of generated clusters. These two data sets also produced the lowest overall Rand Index values for all algorithms. This is indicative of the difficulty faced by clustering algorithms as the number of different classes increases. However, MicroGRID still performs better than both baseline methods. The *Electricity Market* data set contains 2 possible class values, and MicroGRID has shown the highest advantage in accuracy.

These results indicate that having only 2 distinct classes allows for the greatest amount of separation between micro-clusters of each class upon the grid, greatly reducing the chances of false positives or false negatives when clustering. As the number of classes increases, the separation between each group of micro-clusters (of different classes) on the grid will naturally decrease, leading to a higher chance of false positive and negatives. When clustering, both *CluStream* and *DenStream* use a distance function to calculate if micro-clusters should be included in the same created cluster. However, the distance of micro-clusters that should be grouped together is greatly affected by the data set itself. Secondly, the radius of micro-clusters in *CluStream* and *DenStream* are not fixed, and can expand as other micro-clusters are merged into them. When the number of classes increases, this can lead to a number of very large micro-clusters, increasing the possibility that micro-classes attributed to different classes are assigned to the same cluster. Micro-clusters in MicroGRID are of a fixed radius, making it more difficult for micro-clusters of different classes to be assigned to the same cluster, as their distance together depends purely on dimension values, instead of cluster radius also. MicroGRID presents a greater advantage in accuracy in the *Electricity Market* data set, as even if many of the resulting micro-clusters are mapped very close to each other upon the grid, these micro-clusters do not expand, keeping the overall area covered by these micro-clusters smaller. However, even under 2 possible class values, many micro-clusters may be merged together, if their resulting distances are very close. This once again can lead to larger micro-clusters, increasing the chance of these large micro-clusters being clustered with other micro-clusters of different class values.

In summary, MicroGRID outperforms both CluStream and DenStream in terms of both clustering accuracy and efficiency across the majority of data sets. MicroGRID outperforms both CluStream and DenStream in average execution time across all four data sets, and also outperforms these algorithms in average Rand Index across the *Forest Covertype, Sensor Readings* and *Electricity Market* data sets.

5 Conclusion

This paper proposed a new approach for clustering data-streams in real time, combining a micro-cluster array of limited size, a decaying weight function and a central clustering grid to cluster both single data streams. In addition, the use of a decaying weight function allows the proposed MicroGRID approach to still produce accurate clustering results when faced with both noisy and outlier data points. We have shown that the use of a limited-size micro-cluster array, and clustering based on grid cells rather than individual micro-clusters, produces more efficient clustering results than other common clustering algorithms that also make use of micro-clusters. We have shown that these efficiency results are consistent across multiple real-world data sets, which illustrates that these measures, along with grid mapping and weight calculation functions being of a constant time complexity, make a significant efficiency difference when clustering a single data stream. We also show that the use of a decaying weight function to remove outdated and outlier micro-clusters from a micro-cluster array, and the use of mapping micro-clusters to a grid of fixed size for clustering, results in more accurate clustering results across all data sets, which also indicates an improved ability to handle noise when clustering.

References

1. Aggarwal, C.C., Han, J., Wang, J., Yu, P.S.: A framework for clustering evolving data streams. In: Proceedings of the 29th International Conference on Very Large Data Bases, pp. 81–92 (2003)
2. Aggarwal, C.C., Han, J., Wang, J., Yu, P.S.: A framework for projected clustering of high dimensional data streams. In: Proceedings of the 13th International Conference on Very Large Data Bases, pp. 852–863 (2004)
3. Aggarwal, C.C., Yu, P.S.: A framework for clustering uncertain data streams. In: 24th Proceedings of the IEEE International Conference on Data Engineering, pp. 150–159 (2008)
4. Al Aghbari, Z., Kamel, I., Awad, T.: On clustering large number of data streams. Intell. Data Anal. **16**(1), 69–91 (2012)
5. Amini, A., Wah, T.Y., Saybani, M.R., Yazdi, S.R.: A study of density-grid based clustering algorithms on data streams. In: Proceedings of the 8th IEEE International Conference on Fuzzy Systems and Knowledge Discovery, pp. 1652–1656 (2011)

6. Amini, A., Saboohi, H., Herawan, T., Wah, T.Y.: Mudi-stream: s multi density clustering algorithm for evolving data stream. J. Netw. Comput. Appl. **59**, 370–385 (2016)
7. Cao, F., Ester, M., Qian, W., Zhou, A.: Density-based clustering over an evolving data stream with noise. In: Proceedings of the SIAM International Conference on Data Mining, vol. 6, pp. 328–339 (2006)
8. Chen, L., Zou, L.J., Tu, L.: A clustering algorithm for multiple data streams based on spectral component similarity. Inf. Sci. **183**(1), 35–47 (2012)
9. Chen, Y., Tu, L.: Density-based clustering for real-time stream data. In: Proceedings of the 13th ACM SIGKDD International Conference on Knowledge Discovery and Data Mining, pp. 133–142. ACM (2007)
10. Ciampi, A., Appice, A., Malerba, D.: Summarization for geographically distributed data streams. In: Proceedings of Knowledge-Based and Intelligent Information and Engineering Systems, pp. 339–348 (2010)
11. de Andrade Silva, J., Hruschka, E.R.: Extending k-means-based algorithms for evolving data streams with variable number of clusters. In: Proceedings of the 10th International Conference on Machine Learning and Applications, pp. 14–19 (2011)
12. Hahsler, M., Bolaos, M.: Clustering data streams based on shared density between micro-clusters. IEEE Trans. Knowl. Data Eng. **28**, 1449–1461 (2016)
13. Huang, G., Zhang, Y., Cao, J., Steyn, M., Taraporewalla, K.: Online mining abnormal period patterns from multiple medical sensor data streams. World Wide Web **17**(4), 569–587 (2014)
14. Liu, W., and J. OuYang. Clustering algorithm for high dimensional data stream over sliding windows. In: Proceedings of 10th IEEE International Conference on Trust, Security and Privacy in Computing and Communications, pp. 1537–1542 (2011)
15. Qi, Z., Jinze, L., Wei, W.: Approximate clustering on distributed data streams, pp. 1131–1139 (2008)
16. Sabit, H., Al-Anbuky, A., Gholam-Hosseini, H.: Distributed WSN data stream mining based on fuzzy clustering. In: Proceedings of Symposia on Ubiquitous, Autonomic and Trusted Computing, pp. 395–400 (2009)
17. Wang, C.D., Lai, J.H., Huang, D., Zheng, W.S.: SVStream: a support vector-based algorithm for clustering data streams. IEEE Trans. Knowl. Data Eng. **25**(6), 1410–1424 (2013)

UFSSF - An Efficient Unsupervised Feature Selection for Streaming Features

Naif Almusallam[1,2]($\boxtimes$), Zahir Tari[1]($\boxtimes$), Jeffrey Chan[1]($\boxtimes$),
and Adil AlHarthi[3]($\boxtimes$)

[1] Royal Melbourne Institute (RMIT), Melbourne, Australia
{naif.almusallam,zahir.tari,jeffrey.chan}@rmit.edu.au
[2] Al Imam Muhammad Bin Saud Islamic University (IMSIU),
Riyadh, Kingdom of Saudi Arabia
[3] Albaha University, Albaha, Kingdom of Saudi Arabia
alharthi.adil@gmail.com

Abstract. Streaming features applications pose challenges for feature selection. For such dynamic features applications: (a) features are sequentially generated and are processed one by one upon their arrival while the number of instances/points remains fixed; and (b) the complete feature space is not known in advance. Existing approaches require class labels as a guide to select the representative features. However, in real-world applications most data are not labeled and, moreover, manual labeling is costly. A new algorithm, called Unsupervised Feature Selection for Streaming Features (UFSSF), is proposed in this paper to select representative features in streaming features applications without the need to know the features or class labels in advance. UFSSF extends the k-mean clustering algorithm to include linearly dependent similarity measures so as to incrementally decide whether to add the newly arrived feature to the existing set of representative features. Those features that are not representative are discarded. Experimental results indicates that UFSSF significantly has a better prediction accuracy and running time compared to the baseline approaches.

1 Introduction

High dimensionality is a major challenge for machine learning algorithms in data stream environments. Irrelevant features decrease the prediction accuracy and the running time of such algorithms. Feature selection has been widely used as a pre-processing technique to select representative features from streams in order to tackle the dimensionality issue. However, existing feature selection approaches assume that features are *static* because they need to be known in advance so as to accurately select a set of representative features. Therefore these approaches are not appropriate for data stream applications, where features are not static and instead arrive one by one.

Data streams can be broadly classified into *streaming data* and *streaming features* [6]. In streaming data, the number of features is fixed while the instances

© Springer International Publishing AG, part of Springer Nature 2018
D. Phung et al. (Eds.): PAKDD 2018, LNAI 10938, pp. 495–507, 2018.
https://doi.org/10.1007/978-3-319-93037-4_39

arrive sequentially. Regarding streaming features, however, which is the focus of this paper, the number of instances remains fixed while the features arrive sequentially and are processed one by one. In real-world applications such as Twitter, features such as slang words are dynamically created and therefore need to be processed upon their creation instead of waiting for all features to arrive, as required by traditional feature selection approaches. Actually waiting for all features to arrive before starting the selection process is impractical, as the number of streaming features is unknown in advance and new features appear over time. The process of feature selection in streaming features applications consists of two main parts [7]: (1) the evaluation of the new feature to check whether this is a representative one based on a specific criterion (e.g. dependency of the features), and (2) the evaluation of the selected set of features to check whether they remain representative. The non-representative features are discarded. By following this process, we ensure that only the representative features are added. Additionally, we ensure that features that tend to be no longer representative over time are removed from the selected set of features, as new more representative features will be added.

Only a few previous works [10,13,15] have addressed feature selection for streaming features. These, however, require class labels so to guide the selection of representative features. To the best of our knowledge, the only unsupervised feature selection approach for streaming features is the one proposed in [7]. Although this approach does not require class labels, their model is limited to the scenarios where link information could be established (i.e. a friendship relationship between Twitter users). Although a trick can be used to replace the link information by computing the similarity of the data, their model is no different from traditional feature selection approaches because it is limited to using the link information to evaluate the relevance of the features.

The proposed UFSSF approach extends the k-mean algorithm to cluster a stream of features that are not known in advance. It integrates three linearly dependent similarity measures, namely PCC (Pearson Correlation Coefficient), LSRE (Least Square Regression Error) and MICI (Maximal Information Compression Index), to incrementally measure the dependency of the newly arrived streaming features to decide whether or not to add them to the existing set of representative features. The features arrive sequentially and they are processed upon their arrivals one by one in a real-time manner. Linearly dependent measures are used because they are not sensitive to the order and the scatter of the distribution of the features. Additionally, UFSSF incrementally updates the centroids to cope with concept drift in streaming features, as one feature is relevant at a given time. After assigning a feature to its relevant cluster, the mean is updated and we compare the similarity of the arrived feature with the existing representative features.

Extensive experiments have been carried out to benchmark the proposed UFSSF against two well-known unsupervised approaches, namely SPEC [14] and the one proposed in [8]. These approaches are evaluated in terms of the prediction accuracy and the running time. The evaluation work is carried out in two parts. In the first part, we simulated the streaming features environment such that: (a) features are not completely known in advance; and (b) they are processed in real time. In the second part of the evaluation, we assume the existence of the entire stream in order to test the stability of the results. We therefore vary the number of features selected from the whole stream (i.e. select 10, 15, 30, etc. from the entire stream). In both experiments, UFSSF outperforms these two selected approaches in both prediction accuracy and running time.

This paper is organized as follows. The next section discusses existing solutions, and a description of the problem is followed. Section 3 provides full details of the proposed approach and Sect. 4 presents the experimental results in details. We conclude with our findings in Sect. 5.

2   Related Work

To the best of our knowledge, there are few studies that have been conducted on feature selection in streaming features applications. Perkins et al.'s work [10] proposed an approach, called *grafting*, which selects a subset of streaming features that have arrived so far as an integral part of a regularized learning process. It incrementally and gradually builds the selected subset of features in addition to training the predictive model using gradient descent. Because it works in an incremental way, this approach can efficiently cope with the dynamic nature of the streams. However, in order to specify a good regulariser parameter value, this approach requires an insight into the complete feature space in advance. Therefore, it cannot process streaming features of an unknown size. Alpha-investing [15] evaluates the relevance of the arrived feature based on a dynamic threshold of error reduction (called p-value). In particular, the p-value is introduced to determine whether or not to add a feature to the selected set of features. Although Alpha-investing can process the unknown size of streaming features, no selected features can be removed. Finally, Online Streaming Feature Selection (OSFS) was proposed in [13] to select in real time relevant features and remove redundant ones. Whenever a feature arrives, OSFS measures its dependency on the available class labels and then adds the feature to the best candidate feature if this meets a specific criterion. OSFS can dynamically remove redundant features using the Markov Blanket.

The approaches discussed above require the class label as a guide to select representative features. However, in real-world applications most of the data is un-labeled and, moreover, labeling is a time consuming. To the best of our knowledge, the only approach that is unsupervised (i.e. no labels) and is applicable for streaming features applications is proposed in [7]. Although this approach has good performance, it is limited to scenarios where link information must be established (i.e. a friendship relationship between Twitter users). Also, the authors assume that the link information is stable, which obviously is not true as this could dynamically change.

Preliminaries and Problem Statement. We formally introduce the notation used in this paper and describe the problem of *unsupervised feature selection for streaming features* in this section. We assume a stream of feature vectors, $F = \{f_1, f_2, \ldots\}$ (possibly infinite in their number), where each f_i is a vector of the feature values for n instances. Let F_t be the features observed up to time t. E.g., if F represents a stream of tweets from Twitter, then the features are individual words, and each post is an instance, and a feature vector would represent the frequency with which that word (feature) appears in each of the tweets. F_t is the feature/word vectors observed up to time t. Each feature vector in F arrives one by one, there are no restrictions on the order in which they arrive, and they do not have class labels.

We wish to maintain a representative set of features that approximates the feature stream seen so far. As the feature stream is potentially infinite in length and the relevant set of features could change with time due to concept drift, it is not efficient to wait for all the features to be collected. Let $R_t = \{f_1^R, f_2^R, \ldots, f_k^R\}$, $f_i^R \subset F_t$, $1 \leq i \leq k$, denote the set of k representative features at time t. k can range from 1 to k_{max}, the maximum number of representative features.

As features arrive one by one, the problem of unsupervised feature selection for streaming features is to maintain a set of representative features R_t, such that R_t approximates the features F_t observed up to time t. Each representative feature f_j^R of R_t represents a subset/cluster of features in F_t.

For each incoming feature f_i, the problem we are addressing in this paper considers the following two issues:

1. How to determine which existing representative feature and associated feature cluster f_i must be assigned to?
2. How to update the feature cluster and representative feature?

For both (1) and (2) above, the following three similarity measures are selected: Pearson Correlation Coefficient (PCC) [9], Least Square Regression Error (LSRE) [12], and Maximal Information Compression Index (MICI) [8]. We have chosen these linearly dependent measures because they are known not to be sensitive to the order and scatter of the features [8]. These similarity measures will measure the dependency of streaming features in order to (1) allocate a feature to a relevant cluster; (2) decide whether to add a feature to a set representative features; and (3) dynamically update a set of selected features by removing those that are no longer representative.

3   The UFSSF Approach

This section provides details of the proposed approach. Let's first define the concepts of *cluster centroid* and *representative feature*.

Definition 1 (Cluster Centroid). We represent each feature cluster by a centroid, which is a weighted mean of all the features assigned to it. The weights are largest for recently arrived features, and smallest for features that arrived in the distant past.

Definition 2 (Representative Feature). A feature assigned to a cluster is considered to be *representative* if it has the maximum similarity to the cluster's centroid amongst all other stream features assigned to the same cluster. Given a centroid c_r, $f_r \in F_t$ is a *representative feature* in c_r, namely $f_r \in R_t$, if and only if we have one of the following properties:

$$\text{PCC}(f_r, c_r) > \text{PCC}(f_j, c_r)$$
$$\text{LSRE}(f_r, c_r) < \text{LSRE}(f_j, c_r)$$
$$\text{MICI}(f_r, c_r) < \text{MICI}(f_j, c_r)$$

where R_t is the set of current representative features and f_j is any feature of c_r.

Any feature that is not representative is therefore discarded. This will lead to less usage of space and will also allow UFSSF to rapidly filter out non-representative features in dynamic feature space.

3.1 The UFSSF Model

This section explains how the proposed UFSSF approach computes the set of representative features. The model consists of two parts: (1) adding features to the set of representative features; and (2) updating the set of representative features by removing the ones that are no longer representative. To do so, we employ the similarity measures provided above. We rely on clustering approaches that are capable to select the representative features without the requirement of class labels. The k-mean algorithm [5] works well with multi-dimensional datasets, and is therefore well suited for streaming features.

Linearly dependent measures are more efficient for the purpose of feature selection as they are not sensitive to the order and the scatter of the distribution of the features. Three well-known linearly dependent measures (i.e. PCC, LSRE, MICI) are used for the following reasons. Firstly, a single similarity measure might produce bias towards a specific model, and therefore produces better selection of representative features for that model over other models. Secondly, the three measures proved their effectiveness for feature selection as experimentally shown in [8]. Therefore, PCC, LSRE and MICI are used in the k-mean algorithm to compute the dependency between features and cluster centroids.

The following steps show how UFSSF selects a set of representative features from a stream of features. Features are processed one by one upon their arrivals in a first-in-first-out strategy as they are not known in advance. The first step is the initialisation of the clusters and the representative features:

- UFSSF assigns the first arrived k features from a stream as centroids of k number of clusters. For example, if $k = 10$ then the first ten features collected from a stream are the initial centroids of 10 clusters.

- UFSSF sets the initial centroid of every cluster as the initial representative feature of that cluster.

Whenever a Feature f_j Arrives
For every similarity measure PCC/LSRE/ MICI, the following steps are carried out to update the representative feature set:

- The similarity between f_j and the centroid of every cluster is computed. f_j is assigned to a cluster C_{lus} if f_j has the maximum similarity to C_{lus}'s centroid amongst all other clusters centroids. The mean of C_{lus} is then incrementally updated and f_j is assigned to C_{lus}.
- In C_{lus}, we compare the similarity (say S) between f_j and the representative feature (i.e. f_r) with C_{lus}'s centroid c_r. If $S(f_j, c_r) > S(f_r, c_r)$, f_j is set as the representative feature and f_r is removed.
- The representative feature from every cluster comprises the set of representative features.

UFSSF has one-pass over data, as it reads the stream of the data only once. Additionally, UFSSF incrementally updates the mean of the clusters: (i) to accurately measure the representativeness of the features (as a feature f_r might be representative at time t but not in $t + 1$); and (ii) to tackle the concept drift in clusters that could result from the dynamic nature of the stream. This helps to improve the prediction accuracy of the classifiers.

Finally, UFSSF requires only a reasonable storage capacity as it stores only the representative feature and the centroid of every cluster. Because UFSSF is able to meet the requirements of major streaming applications, we believe that it is capable to efficiently work in streaming features applications as shown by the experimental results. The pseudo code of UFSSF is given in Algorithm 1.

4   Experimental Evaluation

This section describes the experimental setup of the proposed UFSSF approach. This approach is compared against two well-known traditional unsupervised feature selection approaches, namely the one proposed in [8] and SPEC [14]. To the best of our knowledge, no other unsupervised feature selection approach has been developed for streaming features applications without the requirement of link information. Therefore, these two approaches have been selected as they are the most common traditional unsupervised feature selection approaches (e.g. batch applications).

The respective computed representative feature sets of these three approaches are evaluated by taking the average of three well-known classifiers, namely Naive Bayes [4], J48 Decision Tree [11] and Lazy Nearest Neighbour [1] (also called IB1). In addition to these classifiers, the k-fold-cross validation is applied on all selected features to produce better results by avoiding the problem of over fitting data. The selected feature set is first divided into subsets of equal size depending on the selected k folds. Then, only one k is used as a testing subset and the rest

---

Algorithm 1. UFSSF

Input: $D = \{f_1, f_2, ..., f_n\}$, a stream of features vectors
Input: $j = \{1 = PCC, 2 = LSRE, 3 = MICI\}$, similarity measure
Input: n, number of clusters centroids
Output: representative_ features
// Initialization of centroids matrix and representative features matrix
1 cluster_ centroids=NaN(size(D,1),n);
2 representative_ features(:,q)=D(:,q);
   // Assigning the first n features to be both the centroids and the representative
      features of the first n clusters
3 **for** $q=1{:}n$ **do**
4 $\quad$ cluster_ centroids(:,q)=D(:,q);
5 $\quad$ representative_ features(:,q)=D(:,q);
6 **end**
7 feature_ indexes=zeros(1,n);
8 **for** $u=1{:}n$ **do**
9 $\quad$ feature_ indexes(1,u)=u;
10 **end**
   // Looping over the remaining stream of features
11 **for** $w=n+1{:}size(D,2)$ **do**
12 $\quad$ **for** $r=1{:}n$ **do**
   $\quad\quad$ // Compute the similarities between the arriving feature and every centroid
13 $\quad\quad$ similarity(r,1)=calcDistance(D(:,r),D(:,w),j);
14 $\quad$ **end**
   $\quad$ // Finding the most similar cluster centroid to the arriving feature
15 $\quad$ **if** $j==1$ **then**
16 $\quad\quad$ cluster_most_similar=find(similarity==max(similarity));
17 $\quad$ **else**
18 $\quad\quad$ cluster_most_similar=find(similarity==min(similarity));
19 $\quad$ **end**
20 $\quad$ **end**
   $\quad$ // Incremental mean computation
21 $\quad$ cluster_ centroids(:,cluster_ most_ similar)=mean([cluster_ centroids(:,cluster_ most_
      similar),D(:,w)],2);
   $\quad$ // Computing the similarity of arriving feature f_j and representative feature f_r to
      the cluster centroid
22 $\quad$ f_j=calcDistance(cluster_centroids(:,cluster_ most_ similar),D(:,w),j);
23 $\quad$ f_x=calcDistance(cluster_ centroids(:,cluster_ most_ similar),representative_
      features(:,cluster_ most_ similar),j) // checking the representativeness of feature
24 $\quad$ **if** $j==1$ **then**
25 $\quad\quad$ **if** $f_j > f_r$ **then**
26 $\quad\quad\quad$ representative_ features(:,cluster_ most_ similar)=D(:,w);
27 $\quad\quad\quad$ feature_ indexes(1,cluster_ most_ similar)=w;;
28 $\quad\quad$ **end**
29 $\quad\quad$ **else if** $f_j < f_r$ **then**
30 $\quad\quad\quad$ representative_ features(:,cluster_ most_ similar)=D(:,w);
31 $\quad\quad\quad$ feature_ indexes(1,cluster_ most_ similar)=w;;
32 $\quad\quad$ **end**
33 $\quad$ **end**
34 **end**
35 **Return** representative_ features;

---

are used as training subsets. Finally, the average value of all folds is set to be the average result. In the evaluation, k is set to 10 to demonstrate the efficiency of our proposed algorithm, as suggested in [3]. Following the experimental settings given in [7,13], the experiments conducted in two phases:

- In the first part of the evaluation, we simulated the streaming features environment such that: (a) features are not completely known in advance; and (b) they are processed in real time [13]. The feature space is split into five subsets:

20%, 40%, 60%, 80% and 100%. First, we pick the 20% subset of streaming features and then 40% and so on to sequentially simulate the arrival of the features. In each subset of streaming features, we apply UFSSF, SPEC [8,14] to select representative features. We ensure that all the approaches select the same number of features for a fair comparison.

- In the second part of the evaluation, we vary the number of features selected from the full feature stream, i.e., 100% of features. In this case, we assume the existence of the entire space. The reason is to test the stability of the results and to avoid the randomness.

Although the benchmarked approaches are not designed for streaming features applications, the way we conducted the experiments ensures the fairness of the comparison. We apply the UFSSF, SPEC [8,14] to every subset individually and select the same number of features. Also, the entire dataset is tested with different numbers of selected representative features for every approach in order to compare the non-streaming features benchmark approaches.

Two datasets, namely *Waveform (5000 × 40)*[1] and *Ionosphere (351 × 34)*[2] are used to evaluate the performance of the proposed UFSSF approach. They are commonly used for data mining algorithms and they are from diverse domains. They are used for classification and clustering purposes.

The evaluation of UFSSF will provide answers to the following questions:

- How accurate is UFSSF in selecting a set of representative features?
- How efficient is UFSSF in terms of running time?

Two evaluation metrics are used to answer the above questions, namely F-measure and the running time in seconds. F-measure is the harmonic mean of precision and recall, which precisely demonstrates the accuracy of the classification task [2]. First, we present the results related to the stream of features, where features are not known in advance. Then, we present the results relating to the one that considers the existence of the entire stream, where we investigate the selection of different numbers of features to show the stability of the provided results. Finally, we present the results relating to the efficiency of UFSSF along with those for the two other approaches in terms of the running time. For each dataset, every approach runs its own similarity measure/s to investigate its prediction accuracy. The approach proposed in [8] already includes these three similarity measures (i.e. PCC, LSRE and MICI), while SPEC works with the RBF Kernel similarity measure.

Figure 1 shows the experiment results of the streaming features when features are not completely known in advance but arrive sequentially. For the Waveform dataset shown Fig. 1(a), UFSSF outperforms [8] and SPEC [14] for *every* percentage of streaming features as well as for *every* similarity measure. The selected representative features by UFSSF has the highest prediction accuracy

---

[1] https://archive.ics.uci.edu/ml/datasets/Waveform+Database+Generator+(Version +2).

[2] https://archive.ics.uci.edu/ml/datasets/Ionosphere.

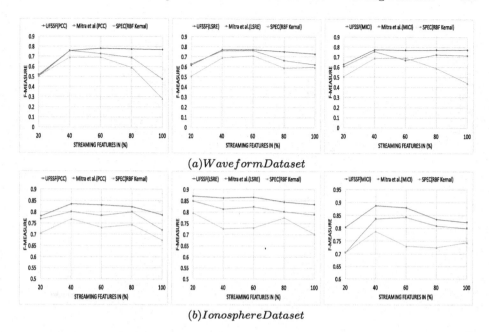

Fig. 1. Comparison of prediction accuracy of the proposed UFSSF scheme along with the two baseline approaches on two datasets. The columns in the figure show the results produced by using different similarity measures. The x-axis denotes the percentage of streaming features while the y-axis denotes the corresponding F-measure

(F-measure) compared to the baseline approaches. Although UFSSF and [8] have a similar prediction accuracy at the early stage of arrival of streaming features (i.e. 20% and 40%), the accuracy of UFSSF distinctly increases for all other percentages of arriving streaming features (i.e. 60%–100%). UFSSF waits for the arrival of more features from these this dataset to significantly perform well. Indeed, UFSSF gradually builds the model due to the incremental updating of the clusters, which affects the selection of representative features. UFSSF processes a stream of features one by one and incrementally selects the representative feature seen so far from a cluster. Therefore, in a few scenarios where we do not really have good representative features, UFSSF is forced to select the maximum representative features that have just arrived. Therefore, the accuracy gradually improves with the arrival of more features. Conversely, the other two approaches statistically search the complete subset of the streaming features that have arrived and select the best of them.

For the Ionosphere dataset, as shown in Fig. 1(b), UFSSF has significantly the highest prediction accuracy for all different percentages of streaming features compared to the two other approaches. This is valid when using either PCC, LSRE or MICI as the similarity measure for UFSSF and [8].

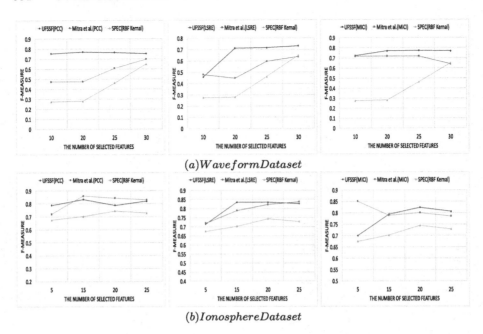

(a)WaveformDataset

(b)IonosphereDataset

Fig. 2. Comparison of prediction accuracy of the proposed UFSSF scheme along with the two baseline approaches on two datasets. The columns in the figure show the results produced by using different similarity measures. The x-axis denotes the selected number of features while the y-axis denotes the corresponding F-measure

SPEC [8,14] do not incrementally update their models to cope with the dynamic nature of the streams. A feature arriving from the stream can be representative at only a specific time due to dynamic nature of the stream. By contrast, UFSSF incrementally updates its clusters to check whether the selected representative features are still representative for every arriving new feature. Therefore, UFSSF outperforms the other two approaches in terms of prediction accuracy. Figure 2 demonstrates the accuracy of UFSSF along with baseline approaches when selecting different numbers of features by considering the entire feature space as a stream (i.e. 100%). For the Waveform dataset, as depicted in Fig. 2(a), UFSSF significantly outperforms [8] and SPEC [14] in terms of prediction accuracy. This holds for all the different numbers of selected features and for all the different similarity measures. However, when only 10 features are selected, UFSSF and [8] are quite similar in terms of prediction accuracy when LSRE and MICI are used as similarity measures.

For the Ionosphere datasets, as illustrated in Fig. 2(b), UFSSF has either a slightly better or a competitive prediction accuracy compared to [8] and SPEC [14]. This is valid for all different numbers of selected features and for all different similarity measures. Baseline approaches are indeed designed to work with statistical datasets. In contrast, UFSSF is designed to work in a

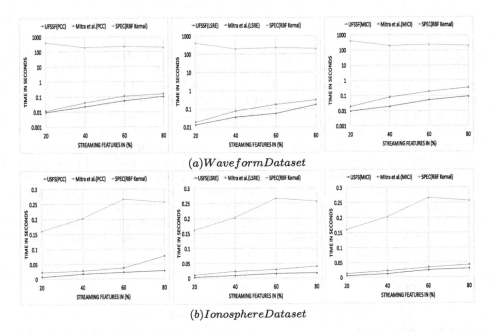

(a) *Waveform Dataset*

(b) *Ionosphere Dataset*

Fig. 3. Comparison of running time of the proposed UFSSF approach along with the two baseline approaches on two datasets

stream environment where features are not completely known in advance but arrive sequentially. Although UFSSF has a lower accuracy than the baseline approaches for a few of selected features, the accuracy difference is negligible.

It worth pointing out that the UFSSF is not sensitive to the order of the features. This is due to the methodology of selecting the representative features. Indeed, every cluster retains only the feature that has the maximum similarity to the cluster centroid. Therefore, it does not matter which feature arrives first as the similarity is computed based on its values.

The running time for the three approaches is depicted in Fig. 3. UFSSF has the lowest running time for all different similarity measures on the provided datasets. It consistently outperforms the baseline approaches on all the different percentages of the streaming features. The approach in [8] is competitive with UFSSF while SPEC [14] has a higher running time. [8] relies on K-Nearest Neighbour (K-NN) search to partition the subset of arrived features. As a result, it has a higher running time due to the computation of the similarity between features. The performance of SPEC [14] is the worst in terms of running time due to the time required to build the Laplacian matrix, which is computationally expensive. The reason for UFSSF having the best running time is that it does not have to search the entire subset of newly arrived features as the other two approaches do to select features. Instead, UFSSF processes the arriving features one by one by computing their dependency on only the cluster's centroids, which are very few compared to the number of streaming features.

5   Conclusion

This paper proposed an unsupervised feature selection approach to reduce the dimensionality of a stream in streaming features applications. Unlike existing streaming features approaches that require class labels, UFSSF can efficiently select a set of representative features without requiring class labels or information such as the link between users. In UFSSF, a k-mean clustering algorithm is extended to work in streaming features applications. It clusters a stream of features that are not known in advance. It uses three similarity measures namely, PCC, LSRE and MICI, in order to: (a) allocate a feature to a relevant cluster, (b) decide whether to add the arrived feature to the set of representative features and (c) decide whether to dynamically update a set of selected features by removing those that are no longer representative. Experimental results show that UFSSF generates a representative feature set with the lowest running time. The selected set of representative features has mostly achieved the best prediction accuracy according to F-measure evaluation metric.

References

1. Aha, D.W., Kibler, D., Albert, M.K.: Instance-based learning algorithms. Mach. Learn. **6**(1), 37–66 (1991)
2. Almalawi, A.: Designing unsupervised intrusion detection for SCADA systems (2014)
3. Chen, H.-L., Yang, B., Liu, J., Liu, D.-Y.: A support vector machine classifier with rough set-based feature selection for breast cancer diagnosis. Expert Syst. Appl. **38**(7), 9014–9022 (2011)
4. John, G.H., Langley, P.: Estimating continuous distributions in Bayesian classifiers. In: Proceedings of the Eleventh Conference on Uncertainty in Artificial Intelligence, pp. 338–345. Morgan Kaufmann Publishers Inc. (1995)
5. Kanungo, T., Mount, D.M., Netanyahu, N.S., Piatko, C.D., Silverman, R., Wu, A.Y.: An efficient k-means clustering algorithm: analysis and implementation. IEEE Trans. Pattern Anal. Mach. Intell. **24**(7), 881–892 (2002)
6. Li, J., Cheng, K., Wang, S., Morstatter, F., Trevino, R.P., Tang, J., Liu, H.: Feature selection: a data perspective. arXiv preprint arXiv:1601.07996 (2016)
7. Li, J., Hu, X., Tang, J., Liu, H.: Unsupervised streaming feature selection in social media. In: Proceedings of the 24th ACM International on Conference on Information and Knowledge Management, pp. 1041–1050. ACM (2015)
8. Mitra, P., Murthy, C.A., Pal, S.K.: Unsupervised feature selection using feature similarity. IEEE Trans. Pattern Anal. Mach. Intell. **24**(3), 301–312 (2002)
9. Onwuegbuzie, A.J., Daniel, L., Leech, N.L.: Pearson product-moment correlation coefficient. In: Encyclopedia of Measurement and Statistics, pp. 751–756 (2007)
10. Perkins, S., Lacker, K., Theiler, J.: Grafting: fast, incremental feature selection by gradient descent in function space. J. Mach. Learn. Res. **3**, 1333–1356 (2003)
11. Quinlan, J.R.: C4.5: Programs for Machine Learning. Elsevier, Amsterdam (2014)
12. Radhakrishna Rao, C.: Linear Statistical Inference and Its Applications, vol. 22. Wiley, New York (2009)
13. Xindong, W., Kui, Y., Ding, W., Wang, H., Zhu, X.: Online feature selection with streaming features. IEEE Trans. Pattern Anal. Mach. Intell. **35**(5), 1178–1192 (2013)

14. Zhao, Z., Liu, H.: Spectral feature selection for supervised and unsupervised learning. In: Proceedings of the 24th International Conference on Machine Learning, pp. 1151–1157. ACM (2007)
15. Zhou, J., Foster, D., Stine, R., Ungar, L.: Streaming feature selection using alpha-investing. In: Proceedings of the Eleventh ACM SIGKDD International Conference on Knowledge Discovery in Data Mining, pp. 384–393. ACM (2005)

Online Clustering for Evolving Data Streams with Online Anomaly Detection

Milad Chenaghlou[1(✉)], Masud Moshtaghi[1], Christopher Leckie[1], and Mahsa Salehi[2]

[1] Department of Computing and Information Systems,
The University of Melbourne, Melbourne, VIC 3010, Australia
`mchenaghlou@student.unimelb.edu.au`,
`{masud.moshtaghi,caleckie}@unimelb.edu.au`
[2] Faculty of Information Technology, Monash University,
Melbourne, VIC 3168, Australia
`mahsa.salehi@monash.edu`

Abstract. Clustering data streams is an emerging challenge with a wide range of applications in areas including Wireless Sensor Networks, the Internet of Things, finance and social media. In an evolving data stream, a clustering algorithm is desired to both (a) assign observations to clusters and (b) identify anomalies in real-time. Current state-of-the-art algorithms in the literature do not address feature (b) as they only consider the *spatial* proximity of data, which results in (1) poor clustering and (2) poor demonstration of the temporal evolution of data in noisy environments. In this paper, we propose an online clustering algorithm that considers the *temporal* proximity of observations as well as their spatial proximity to identify anomalies in real-time. It identifies the evolution of clusters in noisy streams, incrementally updates the model and calculates the minimum window length over the evolving data stream without jeopardizing performance. To the best of our knowledge, this is the first online clustering algorithm that identifies anomalies in real-time and discovers the temporal evolution of clusters. Our contributions are supported by synthetic as well as real-world data experiments.

1 Introduction

Data stream clustering [1] has become a fundamental part of data analysis and data mining in online digital applications. In data streaming environments the volume of data is unbounded, while memory is limited, in contrast to the assumptions made by traditional data clustering that all data can be stored indefinitely. Stream clustering has a wide range of applications, such as Wireless Sensor Networks (WSNs) [2], meteorological analysis [3] and the Internet of Things (IoT). While traditional clustering has been widely studied, it differs from stream clustering in several important ways. First, traditional algorithms have a *global* view of the data, i.e., all observations are randomly accessible whereas in the streaming environment a window is defined that slides over the stream, creating a

© Springer International Publishing AG, part of Springer Nature 2018
D. Phung et al. (Eds.): PAKDD 2018, LNAI 10938, pp. 508–521, 2018.
https://doi.org/10.1007/978-3-319-93037-4_40

partial view of the data. Second, in streaming environments, clusters of data emerge, evolve and change over time but static datasets lack this kind of temporal evolution within the current data window and therefore an effective approach to stream clustering should discover the temporal evolution of clusters as well.

There are three main approaches for data stream clustering: (1) data summarization clustering [4,5], (2) online (real-time) clustering [6,7], and (3) time-series clustering [8,9]. The primary focus of this paper is online clustering where the main challenges are (1) the stream *evolves* gradually and identifying the evolution of patterns in noisy environments is not trivial, and (2) algorithms do not have a global view of the data but instead a partial view, i.e., a window that slides over the stream. Current state-of-the-art algorithms in this category are sequential K-means and competitive neural network based algorithms such as *Adaptive Resonance Theory* (ART-2) [6] and *Self-Organizing Map* (SOM) [10]. A limitation of the current algorithms is that they do not identify the evolution of clusters in streams effectively, mainly because they only consider the *spatial proximity* of data in the stream. In this paper, we propose an algorithm that considers the *temporal proximity* of observations as well as their spatial proximity to identify anomalies in real-time. Our algorithm incrementally updates the model and identifies the evolution of clusters. It takes a single input parameter to calculate the minimum window length without jeopardizing performance, and detects the number of clusters automatically. Accordingly, the contributions of this paper are two-fold: (1) we provide an online clustering algorithm with online anomaly detection that identifies the evolution of clusters, and (2) we only require a single parameter to be specified to calculate the minimum window length without sacrificing performance. To the best of our knowledge, this is the first online clustering algorithm with these capabilities.

This paper is structured as follows. In Sect. 2, we provide a brief review of data stream clustering. Section 3 formally defines the problem of online clustering in evolving data streams. In Sect. 4, we describe our proposed algorithm in detail. Then we analyze the performance of our algorithm by modeling its worst-case time complexity. In Sect. 5, we present the results of our experiments on both synthetic and real datasets, and in Sect. 6, we propose several possible future directions for research.

2   Related Work

Data stream clustering algorithms can be categorized into three groups : (1) data stream summarization, (2) online clustering and (3) time-series clustering. The algorithms in the first group, such as ClusTree [4] and DenStream [5] mainly comprise two phases: (online) data abstraction phase and (offline) clustering phase. In the online phase, they maintain statistical information of data locality with the help of particular data structures to deal with memory constraints, and in the offline phase a batch-mode clustering algorithm is applied to the summary to find clusters. The main drawback of these algorithms is that cluster labels for observations are not identified in real-time.

Online clustering algorithms (the second group) produce cluster labels on the fly and can be subdivided into two major subgroups: (1) sequential K-means and (2) competitive neural networks. K-means [7] is a well-known batch clustering algorithm that has been extensively studied in the literature. The authors of [7] introduced sequential (online) K-means that updates the cluster centers with the arrival of each observation. This algorithm has been improved in [11] by setting an upper bound on the performance of the resulting clusters. However, in noisy environments these algorithms tend to partition the data into Voronoi cells which restricts the shape of clusters found, and cannot determine the temporal evolution of the data.

Online K-means can be extended to a competitive neural network, where the output layer neurons are the K centres of the clusters, and with the arrival of a new observation, the nearest neuron (winner) is identified and is updated. One type of network structure for choosing the winner is through lateral inhibition [12], where each neuron reinforces itself and undermines others. Another competitive network is Hebbian learning [8], where correlated neurons are updated to reinforce the correlation. In these networks only the winner neuron is updated, which usually results in having dead neurons in the network. To avoid this, one way is to update a group of neurons in the neighborhood of the winner neuron, i.e., using self-organizing maps (SOM) [10].

The abovementioned algorithms all require the number of clusters in advance. A neural network that automatically identifies the number of clusters is presented in [6] named *adaptive resonance theory* (ART-2). In this algorithm, clusters are defined with hyper-spheres where the radius is calculated from an input parameter named *vigilance*. The algorithm starts with a single cluster (neuron) and if a new observation does not belong to any clusters, a new cluster is defined. In this network, setting a suitable value for the vigilance parameter is difficult because it defines the radius of the clusters and if it is not set correctly, the algorithm will have a poor performance.

The third group of algorithms is for time-series data [9,13] where data points are indexed in time order. These algorithms exploit the principle of locality, limiting their practicality in various domains of application and we do not consider them in this paper. Table 1 briefly compares current data stream clustering algorithms in the literature where none of the current algorithms are capable of detecting anomalies in real-time.

3 Problem Definition

In this section, the problem of evolving data stream clustering is defined. Suppose a stream of observations $X = (x_1, x_2, x_3, \ldots, x_n, \ldots)$ are generated from a set of γ unknown distributions denoted by $H = \{h_1, h_2, \ldots, h_\gamma\}$ where each $h_i \in H$ is a mixture of K_i components. Let $\Theta = \{\theta_{i,j}, j = 1, \ldots, K_i\}$ denote the parameters corresponding to the components of h_i, e.g., if h_i is a multivariate Gaussian distribution, $\theta_{i,j} = \{\mu_{i,j}, \Sigma_{i,j}\}$ determines the mean and the covariance matrix of the jth component and the mixture weights for $h_i \in H$ is denoted by $\Phi = \{\phi_{i,j}, j = 1, \ldots, K_i\}$.

Table 1. Related Work Summary - OL: Online Labelling, EVS: Evolving Data Streams, OAD: Online Anomaly Detection

Alg. Name	ClusTree [4]	DenStream [5]	On. KMeans [7]	SOM [10]	ART-2 [6]	OnCAD
OL	✗	✗	✓	✓	✓	✓
EVS	✓	✓	✗	✗	✓	✓
OAD	✗	✗	✗	✗	✗	✓

Each observation $x_i \in X$ (drawn from a random $h_i \in H$) is a feature vector $(x_1, x_2, \ldots, x_d) \in \mathbb{R}^d$ where each $x_i, 1 \leqslant i \leqslant d$ is an arbitrary type of variable such as temperature in x_i. A small subset of observations are i.i.d random anomalies, i.e., do not come from any distributions in H. X is potentially unbounded and there is no control over the order of which observations arrive. At any time interval, one of the distributions in H is active (contributing to the stream) and active distributions change over time. When a distribution becomes active for the first time it is called an *emerging* distribution We aim to cluster X into several dense and separated partitions and identify the temporal evolution of data. A good clustering algorithm should reflect the structure imposed by $\theta_{i,j}$ in the input space, and filter out anomalies.

Data Presentation: Observations become available to the algorithm one at a time and data is presented in a window of length L. When a new observation arrives, all observations in the window move one cell to the left (the observation at the leftmost cell is discarded) and the new observation is put in the rightmost cell.

4   Methodology

In this section we propose OnCAD (Online Clustering and Anomaly Detection). In a non-stationary environment where clusters of data emerge, evolve and change over time, a new observation may: (1) *belong to an existing cluster in the model*, (2) *be an anomaly*, or (3) *be part of an emerging cluster*. Our goal is to determine the fate of an observation, update the model, and represent each underlying component with a cluster. We name the leftmost cell in the window as the *check cell* and the rightmost cell as the *update cell* because of their importance.

When a new observation x_{n+1} arrives, the algorithm performs two main steps:

Step 1: It checks if x_{n+1} belongs to case 1 by determining its membership to the clusters in the model and updating them.

Step 2: If x_{n+1} does not belong to any clusters in the model, it should be assigned to either case 2 or 3. Distinguishing between these two cases is impossible at this stage, therefore we postpone the assignment of x_{n+1} until it is in the check cell where either a new cluster is formed or the observation is labelled as anomalous. At this stage one iteration of the algorithm terminates and the algorithm proceeds to the next iteration. However, to perform these steps the main questions that we need to answer are:

1. How to define the membership of observations to clusters?
2. How to distinguish between anomalies and emerging clusters?
3. How to calculate the minimum window length (L)?

Answering these questions is not trivial because the algorithm has only a partial view of the data that evolves over time.

4.1   Step 1 (Cluster Update Rule):

In this section, we answer question 1 by building a cluster model for the underlying components in the stream using Gaussian clusters $C = \{C_1, C_2, \ldots, C_v\}$ where the cluster prototypes are hyper-ellipsoidal and each cluster is represented by its mean and inverse of its covariance matrix. To determine the membership of x_{n+1} w.r.t the clusters in the model, we employ the boundary definition in [14] for hyper-ellipsoids:

$$(x - m)^T S^{-1}(x - m) \le t^2. \tag{1}$$

In this formula, x is an observation, m is the center of the hyper-ellipsoid (the mean of the cluster), S^{-1} is the inverse of covariance matrix of the cluster and t^2 is a constant that marks a specific level set of the cluster. When t^2 is chosen from the cumulative inverted chi-squared distribution with d degrees of freedom for a probability value P, $((\chi_d^2)_P^{-1})$ sets the boundary of the cluster such that the probability of an observation falling outside the boundary is $1 - P$. In our experiments, we set the value of P to 0.99.

If x_{n+1} falls inside the boundary of a cluster, we update it by the weighted incremental update formulas in [15] (Eqs. 20 and 23). These formulas are used to update $m_{n,i}$ (the mean of the cluster at time n) and $S_{n,i}^{-1}$ (the inverse of covariance matrix of the cluster at time n) with x_{n+1}. By determining the membership and updating the selected clusters step 1 terminates.

4.2   Step 2 (Detecting Emerging Clusters)

The key to identifying emerging clusters from anomalies is the window length L that defines the temporal locality of observations. If L is large enough to fit the whole dataset, it is identical to batch-mode clustering, and if L is too small the algorithm would not be able to capture any statistically significant patterns. Hence, our aim is to calculate the smallest L that enables emerging clusters to be distinguished from anomalies.

The prominence of the K_i components of a distribution in the window can be described by the *mixture-weight* and *sample size*. Therefore, we set a threshold p (parameter set by the user) on mixture-weights to cap the number of emerging components and calculate the minimum number of observations to model each component with a cluster (sample size). As a result, the value of L is calculated such that the algorithm is capable of detecting an emerging mixture distribution $h_i \in H$ with K_i components where their mixture-weights are greater than p. For instance, if p is set to 0.2, the algorithm can detect any emerging $h_i \in H$ where $\phi_{i,j} \geq 0.2$ for all $j = 1, \ldots, K_i$, i.e., it can detect at most 5 simultaneous emerging components.

Assume $X_s = \{x_1, x_2, \ldots, x_q\}$ is a sample of size q from a single random component with mean μ and covariance matrix Σ. Also, let m and S be the maximum likelihood estimates (unbiased) of μ and Σ based on X_s, respectively. We compute q such that the distance between m and μ is less than a threshold by calculating an elliptical confidence region (ECR) for μ with a high confidence level.

A confidence region $A \subset \mathbb{R}^d$ for $\mu \in \mathbb{R}^d$ with confidence level $1 - \alpha \in (0, 1)$ is a region where $P(\mu \in A) \geq 1 - \alpha$ and it is defined as [16]:

$$A = \left\{ \mu \in \mathbb{R}^d : (m - \mu)S^{-1}(m - \mu)' < \frac{d}{q - d}F_{1-\alpha;d,q-d} \right\} \tag{2}$$

where d is the dimensionality of the data, q is the sample size, F is the F-inverse cumulative distribution and α is the confidence level. Here, the term $(m - \mu)S^{-1}(m - \mu)'$ is the Mahalanobis distance between m and μ based on S^{-1} and the term $\frac{d}{q-d}F_{1-\alpha;d,q-d}$ sets the boundary of the confidence region where the user usually sets α to a fixed value as the confidence level. Therefore, it is the value of q that determines the boundary of the ECR and as q grows, the area (volume) of the ECR shrinks and the cluster becomes a more accurate representation of the component.

To model a component with a cluster, Eq. 2 gives us an effective tool for achieving any desired precision. Our aim is to minimize q but the question is how small q should be. The criterion that we define for minimizing q is that q should be set to a value such that only one cluster is formed for each component. Having X_s in memory and creating the cluster C_s based on it, we need the majority of the subsequent observations of the component to fall inside C_s. Otherwise, the algorithm may form another cluster for the same underlying component. Therefore, the ECR should be small enough (or q should be large enough) such that C_s covers at least half of the component. To do so, we combine the boundary definitions in Eqs. 1 and 2 to extract the maximum q. Therefore, we find the maximum q such that

$$\frac{d}{q - d}F_{1-\alpha;d,q-d} \leqslant P((\chi_d^2)_P^{-1}). \tag{3}$$

This equation means that the true mean μ is within the boundary of the cluster C_s that covers at least $P\%$ of the observations of C_s. Accordingly, by having P set to 0.5, μ would be within the boundary of C_s such that it covers at least half of the observations in the component. In practice, the value of q with a 95% confidence level in a two dimensional environment is 11 and it grows to 69, in a 20 dimensional environment. Finally, the value of L is calculated as $L = \lfloor 1/p \rfloor \times q$. **Detecting Emerging Clusters:** Detecting emerging clusters is achieved by performing a batch clustering algorithm on the window. In the second step, if the observation at the check cell is not assigned to case 1, we extract all such observations from the window and apply DBScan [17] to them. DBScan is a well-known clustering algorithm on static data that requires two parameters $MinPts$ and ϵ that we extract in the initialization phase. Accordingly, any calculated cluster by DBScan is added to the model. Although using DBScan limits the practicality of the algorithm to data streams where clusters have similar densities, it can be replaced by any other batch clustering algorithm. This concludes one iteration of the algorithm for updating the current model and identifying anomalies from emerging clusters.

Initialization: We have demonstrated that q observations are sufficient to represent a cluster in this methodology, so we set $MinPts$ to q and we assume that the first q observations in the stream are coming from a single component without anomalies. Hence, with the first q observations we form the first cluster by calculating the mean and the inverse of the covariance matrix. Then, we calculate ϵ as the distance between the farthest and the closest observation from the mean.

Time Complexity: In the initialization phase, the mean and the inverse of the covariance matrix of the first q observations are calculated with complexity $O(qd^2)$. The first step of the algorithm includes calculating the Mahalanobis distance of the new observation from all the clusters in the model in $O(|C|d^2)$ and updating the mean and the covariance matrix of the selected clusters (we assume all of the clusters) in $O(|C|d^2)$. This step applies to any observation that belongs to at least one cluster. Therefore, for a stream of size N, the total time complexity is $O(N|C|d^2)$.

In the second step, the time complexity of calling DBScan on the window (we assume the whole window) is $O(L^2)$. At this point, since DBScan is called for each underlying component (in the worst-case), we can rewrite the time complexity as $O(|C|L^2)$. Calculating the mean and covariance matrix for each new cluster is $O(d^2)$, and we can rewrite it as $O(|C|d^2)$ for all components. As a result, the total time complexity of OnCAD is $O(qd^2 + N|C|d^2 + |C|L^2)$ in the worst-case. When the algorithm models all the underlying components, the time complexity for the rest of the stream is $O(N|C|d^2)$, which is linear in terms of the number of observations and underlying components.

5 Evaluation

We have conducted an extensive experimental evaluation on both synthetic and real-world datasets. We generated streams of data in various numbers of dimensions in the range $[2, 20]$, in both clean and noisy environments (with 5% noise) and various minimum mixture-weights $(0.1, 0.2, 0.5, 1)$ for distributions. The size of the streams is at least 500,000 with a random number of underlying components in the range $[100, 110]$.

For comparison, we used two state-of-the-art online clustering algorithms, i.e., online KMeans and the ART-2 network. We implemented online KMeans in MatLab and employed the ART-2 network implementation in [18]. We first present the parameter settings for each algorithm, then we present the measures that are employed for comparison, and finally, the results are presented.

Parameter Setting: Online KMeans takes a single parameter that determines the number of clusters in the data stream. While setting this number in static datasets is not trivial, in evolving streams it is even more challenging. Consequently, we set the KMeans parameter to the true number of clusters for each stream. When the vigilance parameter of ART-2 is set close to 1, ART-2 creates many small clusters (more vigilant), and when it is close to 0, it creates a few big clusters (less vigilant). In order to get the best performance of this algorithm, the first 500 observations of each data stream are from a randomly selected single component without anomalies. We start this algorithm with the value 0.999995 and reduce this parameter by the value 0.000003 until it forms a single cluster for the first 500 observations to calculate the optimal vigilance value.

Finally, OnCAD takes a parameter $p \in (0, 1]$ that determines the minimum weight of a component in a mixture distribution that the algorithm is able to detect. Having p set to a value close to 1 makes the algorithm a time-series clustering algorithm where there is one emerging cluster at any time. In contrast, setting p to a value in the range $(0, 0.5]$ causes the algorithm to detect mixture distributions with various numbers of components. In our experiments, we set this parameter to 0.1 and recommend a value in the range $[0.1, 0.15]$ to be chosen for this parameter.

Performance Measure: To compare the performances we used the Normalized Mutual Information (NMI_{max}) [19], which is a well-known information theoretic cluster validity index (higher values determine better clusterings), and computation time.

Results on Synthetic Datasets: The results on synthetic clean data streams over different data dimensions are presented in Fig. 1. The results on noisy datasets are omitted because of space limitations, as they are similar to clean datasets. Figure 1a illustrates the NMI_{max} value of the online KMeans in clean data streams, which is modest in all dimensions. The reason for this is that it generally converges to a local minimum by partitioning the input space into Voronoi cells. We observed that the noise in the stream does not affect the results greatly, but surprisingly the NMI_{max} value improves slightly by a few percent. The reason for this behavior is that the noise in the stream causes the cluster centers to be distributed better in the input space.

The results of the ART-2 network on clean data are illustrated in Fig. 1b. This network normalizes the input points to unit length and employs hyper-spherical cluster prototypes. When the number of dimensions is 3, an abrupt drop of NMI_{max} value is observed. A close investigation revealed that when the number of dimensions is 3, the components are densely packed and it is difficult for ART-2 to characterize all of these densely packed components. In contrast, in other number of dimensions the individual component distributions are more clearly separated and it is easier for them to be covered by ART-2. The behavior of this algorithm is similar in noisy environments. Finally, Fig. 1c illustrates

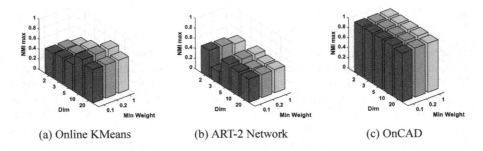

(a) Online KMeans   (b) ART-2 Network   (c) OnCAD

Fig. 1. NMI_{max} values on clean synthetic data streams.

the results of OnCAD, where it outperforms the existing state-of-the-art algorithms. The main reason for this is that OnCAD identifies emerging clusters from anomalies in real-time and filters them out of subsequent clusterings.

In order to demonstrate how real-time anomaly detection contributes to the results of OnCAD, in Table 2 we briefly compare this capability of OnCAD with an ensemble model anomaly detection algorithm [20] and an efficient anomaly detection algorithm [21] on a randomly generated noisy stream. Online KMeans and ART-2 label all observations normal, achieving an accuracy of 95%. The methods in [20,21] achieve the best accuracy of 98% (rounded to two decimal places) and OnCAD has the best anomaly detection rate of 93% with a slightly lower accuracy.

Table 2. Anomaly detection table.

Algorithms	Online KMeans [7]	ART-2 net. [6]	Eff. method [21]	Ens. model [20]	OnCAD
Sensitivity	0%	0%	86%	82%	**93%**
Specificity	**100%**	**100%**	99%	99%	98%
Accuracy	95%	95%	**98%**	**98%**	97%

To elaborate the results in Table 2, the clustering results and the cluster membership figures of the algorithms are presented in Fig. 2. The cluster membership figures (second row of Fig. 2) show the evolution of clusters over time, where the x-axis is the timestamp of an observation and the y-axis is its cluster label. In these figures, we present the first 25,000 observations of the stream for better visualization.

The clustering of online KMeans (Fig. 2a) shows the Voronoi cells created by this algorithm. Moreover, Fig. 2d shows that this algorithm exhibits hardly any knowledge about the evolution of clusters. The ART-2 network performs poorly as well, as illustrated in Fig. 2b since it cannot identify anomalies in the stream. Besides, the similarity measure that this algorithm employs is the dot product between the network pattern and the input point, therefore, maximizing this measure is equivalent to minimizing the angle between them. The cluster membership (Fig. 2e) is not informative as this algorithm does not identify anomalies in real-time.

The results of OnCAD are presented in Fig. 2c where anomalies are represented by blue points and clusters are represented by various colors. Since OnCAD identifies anomalies in real-time, it detects emerging clusters, and from the cluster membership of OnCAD in Fig. 2f, the time of emergence of each cluster is clearly apparent. Moreover, the times when a cluster becomes active and inactive can be determined.

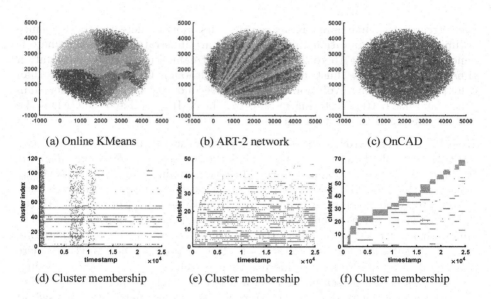

(a) Online KMeans   (b) ART-2 network   (c) OnCAD

(d) Cluster membership   (e) Cluster membership   (f) Cluster membership

Fig. 2. The clustering results and the evolution of clusters in synthetic data. (Color figure online)

The second row of Fig. 2 provides useful insights into how these algorithms work. For instance, the first few hundred observations in online KMeans (Fig. 2d) are assigned to all cluster centers, but then a few clusters dominate the input space and the majority of observations are assigned to them. Figure 2e shows that the ART-2 network is unable to identify emerging clusters because of the noise in the stream. However, the results of OnCAD in Fig. 2f demonstrate that OnCAD identifies emerging clusters by filtering out anomalies (anomalies are assigned to cluster 0).

The computation times of the algorithms on clean synthetic datasets are illustrated in Fig. 3. Since the computation times on noisy datasets are similar to these figures, they are not included here. Figure 3a shows the computation times of the online KMeans, which is the fastest algorithm and takes about 30 s for

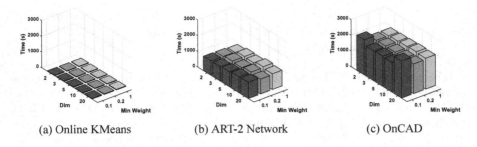

(a) Online KMeans   (b) ART-2 Network   (c) OnCAD

Fig. 3. Computation times on synthetic data streams.

each stream. Although online KMeans is fast, its anomaly detection performance is limited. The computation times of ART-2 are illustrated in Fig. 3b where it generally takes about 15 to 20 min for each stream. Finally, the computation times of OnCAD are illustrated in Fig. 3c. Although OnCAD is slower than others, it models the evolution of clusters (second row of Fig. 2) better than others and has better clustering results, i.e., the NMI_{max} value of OnCAD is the highest compared to other algorithms.

Results on real-world datasets: For the real-world datasets, we used two publicly available datasets. The first one is the *gas sensor array under dynamic gas mixtures*[1] where we compare the clustering qualities in terms of NMI_{max} values. In this labeled dataset gas mixtures of varying concentration levels were exposed to 16 chemical sensors. We chose 12 of these 16 sensors along with the data acquired from Ethylene and CO mixtures. After exposing the gas to the environment, it takes about 4 s for the sensors to capture the change. Therefore, we chose a subset of the data after about 4 s of exposing a new mixture where it reaches an equilibrium, considering them as clusters. The NMI_{max} values on this dataset are presented in Table 3 where OnCAD outperforms the other algorithms. Note that since the clusters are well-separated in this real-world dataset, ART-2 outperforms online KMeans.

Table 3. NMI_{max} values on the gas sensor array dataset.

Algorithm name	Online KMeans	ART-2	OnCAD
NMI_{max}	0.47	0.79	**0.92**

For the second real-world dataset we evaluate the capabilities of the algorithms to identify the temporal evolution of clusters on the *global terrorist attack* dataset[2] which contains the coordinates of the terror attacks around the

---

[1] https://goo.gl/zcAijP.
[2] https://www.kaggle.com/START-UMD/gtd.

world from 1970 through 2016. In this context, detecting emerging clusters from anomalies is crucial as an anomaly represents a single attack but an emerging cluster is a new group of attacks.

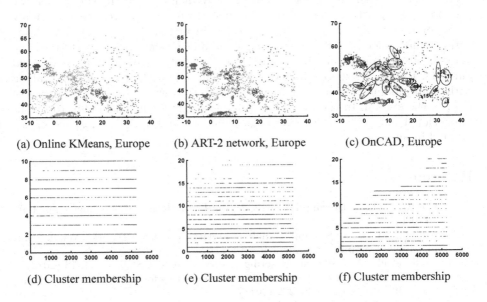

(a) Online KMeans, Europe   (b) ART-2 network, Europe   (c) OnCAD, Europe

(d) Cluster membership   (e) Cluster membership   (f) Cluster membership

Fig. 4. The evolution of clusters in Europe on global terrorist attack dataset.

We chose Europe as the area of interest (latitude in range (35,70) and longitude in range (−10,35)) and the cluster membership of online KMeans are illustrated in Figs. 4a and d. We set the parameter of this algorithm to 10 using the elbow method to determine the optimal number of clusters. This algorithm fails to show the evolution of the clusters in Fig. 4d because it only considers the spatial proximity of observations.

The results of ART-2 are presented in Fig. 4e. Since it does not identify anomalies in real-time, it is not clear when a new cluster emerges. As an example, clusters like 2, 9, 11, 15, 18, 19 are formed with a single observation (an anomaly), which is not effective. However, with OnCAD (Fig. 4f) the emergence of the clusters can be determined as it filters out anomalies in real-time. For instance, clusters 11 (Hungary, Serbia, and Croatia), 12 (East Germany, west Poland) are formed later than clusters 1 (Ireland), 2 (north of Spain), 3 (West Germany and Netherlands), and clusters 16 (North Africa), 17 (Ukraine) and 20 (Denmark) are formed towards the end of the stream. Meanwhile, some clusters become inactive towards the end of the stream such as cluster 2 (north of Spain) as there have not been any terror attacks in that region.

6   Conclusion and Future Directions

In evolving data streams where patterns of data emerge, mature and evolve over time, identifying the temporal evolution of data is crucial. In this paper, we proposed an algorithm that performs online clustering and real-time anomaly detection on evolving data streams. It scales linearly with the number of observations and underlying clusters. In our experiments OnCAD outperformed the state-of-the-art algorithms in the literature on both synthetic and real-world datasets. It employs the flexibility of hyper-ellipsoidal cluster prototypes along with online anomaly detection to better explain the distribution and evolution of data in the stream. In the future, we intend to extend this algorithm to higher dimensional data where clusters of data tend to appear in a subspace of the original input space.

References

1. Guha, S., et al.: Clustering data streams. In: Data Stream Management, pp. 169–187 (2000)
2. Moshtaghi, M., et al.: Streaming analysis in wireless sensor networks. Wirel. Commun. Mobile Comput. **14**(9), 905–921 (2014)
3. Silva, J., et al.: Data stream clustering: a survey. ACM CSUR **46**(1), 13 (2013)
4. Kranen, P., et al.: The clustree: indexing micro-clusters for anytime stream mining. Knowl. Inf. Syst. **29**(2), 249–272 (2011)
5. Cao, F., et al.: Density-based clustering over an evolving data stream with noise. In: SIAM International Conference on Data Mining, pp. 328–339 (2006)
6. Carpenter, G.A., et al.: Art 2-a: an adaptive resonance algorithm for rapid category learning and recognition. In: IEEE International Joint Conference on Neural Networks, pp. 151–156 (1991)
7. MacQueen, J., et al.: Some methods for classification and analysis of multivariate observations. In: Berkeley Symposium on Mathematical Statistics and Probability, Oakland, CA, USA, vol. 1(14), pp. 281–297 (1967)
8. Krogh, A., Hertz, J.A.: A simple weight decay can improve generalization. In: Advances in Neural Information Processing Systems, pp. 950–957 (1992)
9. Angelov, P.: Evolving takagi-sugeno fuzzy systems from streaming data. In: Evolving Intelligent Systems: Methodology and Applications, vol. 12, p. 21 (2010)
10. Kohonen, T.: The self-organizing map. Neurocomputing **21**(1), 1–6 (1998)
11. Charikar, M., et al.: Incremental clustering and dynamic information retrieval. SIAM J. Comput. **33**(6), 1417–1440 (2004)
12. Feldman, J.A., Ballard, D.H.: Connectionist models and their properties. Cogn. Sci. **6**(3), 205–254 (1982)
13. Moshtaghi, M., et al.: Online clustering of multivariate time-series. In: SIAM International Conference on Data Mining, pp. 360–368 (2016)
14. Rajasegarar, S., et al.: Elliptical anomalies in wireless sensor networks. ACM Trans. Sensor Netw. **6**(1), 7 (2009)
15. Moshtaghi, M., et al.: Evolving fuzzy rules for anomaly detection in data streams. IEEE Trans. Fuzzy Syst. **23**(3), 688–700 (2015)
16. Härdle, W., Simar, L.: Applied Multivariate Statistical Analysis. Springer, Heidelberg (2007)

17. Ester, M., et al.: A density-based algorithm for discovering clusters in large spatial databases with noise. In: KDD 1996, vol. 34, pp. 226–231 (1996)
18. Bielecki, A., Wójcik, M.: Hybrid system of ART and RBF neural networks for online clustering. Appl. Soft Comput. **58**, 1–10 (2017)
19. Lei, Y., et al.: Generalized information theoretic cluster validity indices for soft clusterings. In: IEEE Symposium on CIDM, pp. 24–31 (2014)
20. Salehi, M., Leckie, C.A., Moshtaghi, M., Vaithianathan, T.: A relevance weighted ensemble model for anomaly detection in switching data streams. In: Tseng, V.S., Ho, T.B., Zhou, Z.-H., Chen, A.L.P., Kao, H.-Y. (eds.) PAKDD 2014. LNCS (LNAI), vol. 8444, pp. 461–473. Springer, Cham (2014). https://doi.org/10.1007/978-3-319-06605-9_38
21. Chenaghlou, M., et al.: An efficient method for anomaly detection in nonstationary environments. In: IEEE Globecom (2017)

An Incremental Dual nu-Support Vector Regression Algorithm

Hang Yu, Jie Lu$^{(\boxtimes)}$ (iD), and Guangquan Zhang

Decision System and e-Service Intelligence Laboratory,
Centre for Artificial Intelligence, Faculty of Engineering
and Information Technology, University of Technology Sydney,
Sydney, Australia
Hang.Yu@student.uts.edu.au,
{Jie.Lu,Guangquan.Zhang}@uts.edu.au

Abstract. Support vector regression (SVR) has been a hot research topic for several years as it is an effective regression learning algorithm. Early studies on SVR mostly focus on solving large-scale problems. Nowadays, an increasing number of researchers are focusing on incremental SVR algorithms. However, these incremental SVR algorithms cannot handle uncertain data, which are very common in real life because the data in the training example must be precise. Therefore, to handle the incremental regression problem with uncertain data, an incremental dual nu-support vector regression algorithm (dual-v-SVR) is proposed. In the algorithm, a dual-v-SVR formulation is designed to handle the uncertain data at first, then we design two special adjustments to enable the dual-v-SVR model to learn incrementally: incremental adjustment and decremental adjustment. Finally, the experiment results demonstrate that the incremental dual-v-SVR algorithm is an efficient incremental algorithm which is not only capable of solving the incremental regression problem with uncertain data, it is also faster than batch or other incremental SVR algorithms.

Keywords: Support vector regression · Regression learning algorithm
Incremental regression problem · Uncertain data

1 First Section

1.1 A Subsection Sample

Support vector regression (SVR) has been a hot research topic for several years because it is an effective regression learning algorithm [1–4]. It aims to minimize a combination of the empirical risk and a regularization term [5]. Early studies on SVR mostly focus on solving large-scale problems [6–8]. Nowadays, an increasing number of researchers are focusing on incremental SVR algorithms [9–13]. Junshui et al. introduced ε-SVR and developed an accurate online support vector regression (AOSVR) [9]. Omitaomu et al. propose AOSVR with varying parameters that uses varying SVR parameters rather than fixed SVR parameters [11]. Later, Gu et al. proposed an exact incremental v-SVR algorithm (INSVR) [13].

D. Phung et al. (Eds.): PAKDD 2018, LNAI 10938, pp. 522–533, 2018.
https://doi.org/10.1007/978-3-319-93037-4_41

These incremental SVR algorithms require precise data for the training examples. However, the data in many practical applications is not precise yet represented by an uncertain data. For example, the height of a man is between 180 cm and 185 cm. Therefore, some researchers have proposed many improved SVR algorithms [14–16] which explicitly handle uncertain data and perform better than traditional SVRs. Hao et al. incorporate the concept of fuzzy set theory into the SVM regression model [15]. Peng proposed an interval twin support vector regression algorithm for interval input-output data [16]. Several SVR algorithms treat uncertain data as random noise [17–19]. By replacing the constraints in the standard ε-SVR with probability constraints, chance-constrained, robust regression formulations can be obtained. For example, an robust SVR algorithm which is robust to bounded noise was proposed in [19]. However, the quadratic programming problems (QPPs) of these algorithms is too complex to translate these algorithms into incremental algorithms directly.

Hence, to handle the incremental regression problem with uncertain data, the incremental dual nu-support vector regression (dual-v-SVR) algorithm is proposed. In the algorithm, a dual-v-SVR formulation be designed to handle the uncertain data at first, then we design two special adjustments to enable the dual-v-SVR model to learn incrementally: incremental adjustment and decremental adjustment. Finally, the experiment results demonstrate that the incremental dual-v-SVR algorithm is an efficient incremental algorithm which is not only capable of solving the incremental regression problem with uncertain data, it is also faster than batch or other incremental SVR algorithms.

The rest of this paper is organized as follows. In Sect. 2, we describe the formulation, KKT conditions and two adjustments of the incremental dual-v-SVR algorithm. The experimental setup, results and discussions are presented in Sect. 3. Section 4 provides the concluding remarks.

2   An Incremental Dual-v-SVR

As previously mentioned, the QPPs of many SVR algorithms are too complex to translate into online algorithms directly. Hence we propose a dual-v-SVR algorithm estimates the upper bound functions $f_1(x) = \langle w_1 \cdot x \rangle + b_1$ and lower bound functions $f_2(x) = \langle w_2 \cdot x \rangle + b_2$ at same time, and the final regression function is constructed as follows: $f(x) = \frac{1}{2}[f_1(x) + f_2(x)]$.

2.1   The Formulation

For cases with data uncertainties, we suppose the independent variables are perturbed by noise: $x_i = \widetilde{x}_i + \delta_i$, such that $\|\delta_i\| \leq \tau$, where δ_i represents a bounded perturbation with $\tau > 0$ and x_i constructs a nominal vector $X = (x_1, x_2, \ldots, x_N)$. The dependent variable Y is also perturbed by noise: $Y = \widetilde{y} + \sigma = [u, l]$, such that $\|\sigma\| \leq \widehat{\tau}$, where σ represents a bounded perturbation with $\widehat{\tau} > 0$. Thus, we can get:

$$\langle w \cdot \Phi(\widetilde{x}_i + \delta_i) \rangle = \langle w \cdot \Phi(\widetilde{x}_i) \rangle + \langle w \cdot \Phi(\delta_i) \rangle \tag{1}$$

By the Cauchy-Schwarz inequality, we have:

$$|\langle w \cdot \Phi(\delta_i) \rangle| \leq \|w\| \cdot \|\delta_i\| \leq \tau \|w\| \tag{2}$$

Hence a formulation of dual-v-SVR is:

$$\min_{w_1, b_1, \xi_{1i}} \frac{N}{2} \|w_1\|^2 + C_1 \left(v_1 b_1 N + \sum_{i=1}^{N} \xi_{1i} \right) \tag{3}$$

$$\text{s.t. } \langle w_1 \cdot \Phi(\widetilde{x}_i) \rangle + \tau \|w_1\| + b_1 \geq u - \xi_{1i}, \xi_{1i} \geq 0, i = 1, \dots, N$$

and

$$\min_{w_2, b_2, \xi_{2i}} \frac{N}{2} \|w_2\|^2 + C_2 \left(v_2 b_2 N + \sum_{i=1}^{N} \xi_{2i} \right) \tag{4}$$

$$\text{s.t. } \langle w_2 \cdot \Phi(\widetilde{x}_i) \rangle + \tau \|w_2\| + b_2 \geq l + \xi_{2i}, \xi_{2i} \geq 0, i = 1, \dots, N$$

where Φ is a nonlinear transform: $R^N \to F$ to map the data points into a higher dimensional feature space F, $\|w_{1,2}\|^2$ is the regularization term, C_1, $C_2 \geq 0$ are the regularization parameters and ξ_{1i}, ξ_{2i} are the slack variables. Parameter $v_1 \in (0, 1)$ controls the tradeoff between the minimization of $b_{1,2}$ and the minimization of errors (Fig. 1).

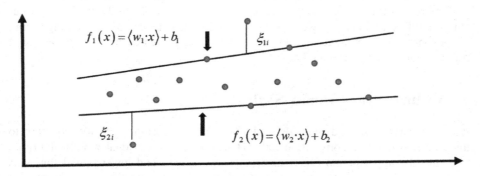

Fig. 1. Incremental dual-v-SVR

2.2   KKT Conditions

Let $Q_{i,j} = \frac{1}{N} k \langle x_i, x_j \rangle = \frac{1}{N} \langle (\Phi(x_i) + \tau) \cdot (\Phi(x_j) + \tau) \rangle$, the dual problem of (3) can be written as:

$$\min_{\alpha} \frac{1}{2} \sum_{i,j=1}^{N} \alpha_{1i} \alpha_{1j} Q_{ij} - \sum_{i=1}^{N} y_{1i} \alpha_{1i} \tag{5}$$

$$\text{s.t. } \sum_{i=1}^{N} \alpha_{1i} = C_1 v_1 N, 0 \leq \alpha_{1i} \leq C_1, i = 1, \dots, N$$

From Eq. (5), we can see the box constraints of α_{1i} are independent of the size of the training sample set.

Then, we introduce the extended training set S, which is defined as $S = S^- \cup S^+$, where $S^- = \{(x_{1i}, y_{1i}, z_{1i} = -1)\}_{i=1}^N$, $S^+ = \{(x_{1i}, y_{1i}, z_{1i} = +1)\}_{i=1}^N$ and z_i is the label of the training sample (x_{1i}, y_{1i}). Thus, the minimization problem (5) can be further rewritten as:

$$\min_{\alpha} \frac{1}{2} \sum_{i,j=1}^{2N} \alpha_{1i}\alpha_{1j}Q_{ij}$$

$$\text{s.t.} \sum_{i=1}^{2N} z_{1i}\alpha_{1i} = 0, \sum_{i=1}^{2N} \alpha_{1i} = 2C_1\nu_1 N, 0 \leq \alpha_{1i} \leq C_1, i = 1, \ldots, 2N \tag{6}$$

The solution of the minimization problem (6) can also be obtained by minimizing the following convex quadratic objective function under constraints:

$$\min_{0 \leq \alpha_{1i} \leq C_1} W = \frac{1}{2} \sum_{i,j=1}^{2N} \alpha_{1i}\alpha_{1j}Q_{ij} + \mu \left(\sum_{i=1}^{2N} z_{1i}\alpha_{1i} \right) + \varepsilon \left(\sum_{i=1}^{2N} \alpha_{1i} - 2C_1\nu_1 N \right) \tag{7}$$

Then by the KKT theorem, the first-order derivative of W leads to the following KKT conditions:

$$\frac{\partial W}{\partial \mu} = \sum_{i=1}^{2N} z_{1i}\alpha_{1i} = 0 \tag{8}$$

$$\frac{\partial W}{\partial \varepsilon} = \sum_{i=1}^{2N} \alpha_{1i} = 2C_1\nu_1 N \tag{9}$$

$$\forall i \in S : g_{1i} = \frac{\partial W}{\partial \alpha_{1i}} = \sum_{j=1}^{2N} \left(Q_{ij}\alpha_{1i} + z_{1i}\mu + \varepsilon \right)$$

$$\begin{cases} \geq 0 & \text{for } \alpha_{1i} = 0 \\ = 0 & \text{for } 0 < \alpha_{1i} < C_1 \\ \leq 0 & \text{for } \alpha_{1i} = C_1 \end{cases} \tag{10}$$

According to the value of the function g_{1i}, the extended training set S is partitioned into three independent sets (see Fig. 2):

$$\begin{array}{ll} \text{Support Set} & S_S = \{i | g_{1i}(x_{1i}) = 0, 0 < \alpha_{1i} < C_1\} \\ \text{Error Set} & S_E = \{i | g_{1i}(x_{1i}) \leq 0, \alpha_{1i} = C_1\} \\ \text{Remaining Set} & S_R = \{i | g_{1i}(x_{1i}) \geq 0, \alpha_{1i} = 0\} \end{array} \tag{11}$$

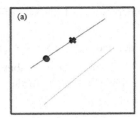 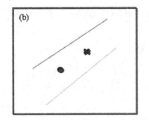 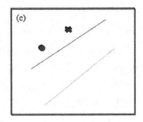

Fig. 2. The partitioning of the training samples S into three independent sets by KKT conditions. (a) S_S. (b) S_E. (c) S_R.

The function $f_2(x)$ can use the same procedure for the analyses and the KKT condition is same as Eq. (9). Therefore, the two functions build an insensitive zone. Furthermore, only errors outside the insensitive zone contribute to the cost function, and only those points (i.e. SVs [2]) determine the final regression model. Hence, the obtained regression model of dual-v-SVR is sparse.

2.3 Incremental and Decremental Adjustment

From Sect. 2.2, we know when a new sample arrives, the weights of the new sample are set to 0 initially, and then it needs to be assigned into a set to satisfy the KKT conditions. Hence, if the assignment violates the KKT conditions, the weights of the new sample will be adjusted. Furthermore, due to a conflict between Eqs. (8) and (9), the adjustment of dual-v-SVR involves two steps: incremental and decremental.

Incremental Adjustment
In the incremental adjustment step, we need to ensure all the samples satisfy the KKT conditions, but the restriction $\sum_{i=1}^{N} \alpha_{1i} = 2C_1 vN$ does not need to hold for all the weights, so we have the following linear system:

$$\Delta g_{1i} = \sum_{j \in S_S} \Delta\alpha_{1j}Q_{ij} + z_{1i}\Delta\mu + \Delta\varepsilon - \Delta\alpha_c Q_{ic} = 0 \tag{12}$$

$$\sum_{j \in S_S} z_{1j}\Delta\alpha_{1j} + z_{1c}\Delta\alpha_{1c} = 0 \tag{13}$$

where $\Delta\alpha_{1j}$, z_{1j}, $\Delta\varepsilon$ and Δg_{1i} denote the corresponding variations. Then we define $e_{S_S} = [1, \ldots, 1]^T$ as the $|S_s|$ dimensional column vector with all ones, let $z_{S_S} = [z_1, \ldots z_{|S_S|}]^T$, and let $Q_{S_S S_S}$ denote the matrix, the above liner system can be further rewritten as:

$$\begin{bmatrix} 0 & z_{S_S}^T \\ z_{S_S} & Q_{S_S S_S} \end{bmatrix} \cdot \begin{bmatrix} \Delta\mu \\ \Delta\alpha_{S_S} \end{bmatrix} = -\begin{bmatrix} z_{1c} & 0 \\ Q_{S_S c} & e_{S_S} \end{bmatrix} \cdot \begin{bmatrix} \Delta\alpha_{1c} \\ \Delta\varepsilon \end{bmatrix} \tag{14}$$

where the underbrace of the left matrix is $\widetilde{Q}_{\backslash\varepsilon\varepsilon}$ and the underbrace of $\begin{bmatrix}\Delta\mu\\\Delta\alpha_{S_S}\end{bmatrix}$ is $\Delta h_{\backslash\varepsilon}$.

where $\widetilde{Q}_{\backslash\varepsilon\varepsilon}$ is the abbreviation of the matrix above the under brace, $\Delta h_{\backslash\varepsilon}$ is also the abbreviation which is defined in the same way.

Let $R = \widetilde{Q}_{\backslash\varepsilon\varepsilon}^{-1}$ and $\Delta\varepsilon = 0$, then the linear relationship between $\Delta h_{\backslash\varepsilon}$ and $\Delta\alpha_{1c}$ can be easily solved as follows:

$$\Delta h_{\backslash\varepsilon} = \begin{bmatrix} \Delta\mu \\ \Delta\alpha_{S_S} \end{bmatrix} = -R \cdot \begin{bmatrix} z_{1c} \\ Q_{S_S c} \end{bmatrix} \equiv \underbrace{\begin{bmatrix} \beta_\mu^c \\ \beta_{S_S}^c \end{bmatrix}}_{\beta_{\backslash\varepsilon}^c} \Delta\alpha_{1c} \tag{15}$$

where $\beta_{\backslash\varepsilon}^c$ stands for the dimension corresponding to μ in the vector $\beta_{\backslash\varepsilon}^c$. $\beta_{S_S}^c$ is the vector with the same meaning of $\beta_{\backslash\varepsilon}^c$. Accordingly, let $\beta_{\backslash\varepsilon}^c = 0$, then the relationship between $\Delta h_{\backslash\varepsilon}$ and $\Delta\alpha_c$ can also be defined as:

$$\Delta h_\varepsilon = \Delta\varepsilon = \beta_\varepsilon^c \Delta\alpha_c \tag{16}$$

By substituting (15) and (16) into (12), we can get the linear relationship between Δg_{1i} and $\Delta\alpha_c$ as follows:

$$\Delta g_i = \left(\sum_{j\in S_S} \beta_j^c Q_{ij} + z_{1i}\beta_\mu^c + \beta_\varepsilon^c + Q_{ic} \right) \Delta\alpha_{1c} \equiv \gamma_{1i}^c \Delta\alpha_{1c}, \forall i \in S \tag{17}$$

Obviously $\forall i \in S_S$, so we have $\gamma_{1i}^c = 0$. Thus, for each incremental adjustment, we can compute the maximal increment of $\Delta\alpha_{1c}$ (here denoted as $\Delta\alpha_{1c}^{\max}$), update α, g, S_S, S_E, S_R and the inverse matrix R, similar to the approaches in [20].

Decremental Adjustment

In the decremental adjustment step, we gradually adjust $\sum_{i\in S} \alpha_{1i}$ to restore the equality $\sum_{i\in S} \alpha_{1i} = 2Cv(N+1)$, so that the KKT conditions are satisfied by all the samples. For each adjustment of $\sum_{i\in S} \alpha_{1i}$, in order to ensure all the samples satisfy the KKT conditions, the weights of the samples in S_S, the Lagrange multipliers μ and ε should also be adjusted accordingly, and these have the following linear system:

$$\Delta g_{1i} = \sum_{j\in S_S} \Delta\alpha_{1j}Q_{ij} + z_{1i}\Delta\mu + \Delta\varepsilon = 0, \forall i \in S_S \tag{18}$$

$$\sum_{j\in S_S} z_{1i}\Delta\alpha_{1j} = 0 \tag{19}$$

$$\sum_{j \in S_S} \Delta \alpha_{1j} + \eta \Delta \varepsilon + \Delta \omega = 0 \tag{20}$$

where $\Delta \omega$ is the introduced variable of adjusting $\sum_{i \in S} \alpha_{1i}$, η is any negative number, and $\eta \Delta \varepsilon$ is incorporated in (20) as an extra term. Using this extra term can prevent the recurrence of conflicts between Eqs. (8) and (9). The above linear system can also be further rewritten as:

$$\underbrace{\begin{bmatrix} 0 & 0 & z_{S_S}^T \\ 0 & \eta & e_{S_S}^T \\ z_{S_S} & e_{S_S} & Q_{S_S S_S} \end{bmatrix}}_{\widehat{Q}} \underbrace{\begin{bmatrix} \Delta \mu \\ \Delta \varepsilon \\ \Delta \alpha_{S_S} \end{bmatrix}}_{\Delta h} = - \begin{bmatrix} 0 \\ 1 \\ 0 \end{bmatrix} \Delta \omega \tag{21}$$

Let $\widehat{R} = \widehat{Q^{-1}}$, then the linear relationship between Δh and $\Delta \omega$ can be obtained as follows:

$$\Delta h = \begin{bmatrix} \Delta \mu \\ \Delta \varepsilon \\ \Delta \alpha_{S_S} \end{bmatrix} = -\widehat{R} \begin{bmatrix} 0 \\ 1 \\ 0 \end{bmatrix} \Delta \omega \equiv \underbrace{\begin{bmatrix} \widehat{\beta_\mu} \\ \widehat{\beta_\varepsilon} \\ \widehat{\beta_{S_S}} \end{bmatrix}}_{\widehat{\beta}} \Delta \omega \tag{22}$$

From Eq. (20), we have $\sum_{i \in S} \Delta \alpha_{1i} = -\left(1 + \eta \widehat{\beta_\varepsilon}\right) \Delta \omega$, which implies that the control of the adjustment of $\sum_{i \in S} \Delta \alpha_{1i}$ is achieved by $\Delta \omega$.

Finally, substituting Eq. (20) into Eq. (16), we can also get the linear relationship between Δg_{1i} and $\Delta \omega$ as follows:

$$\Delta g_i = \left(\sum_{j \in S_S} \widehat{\beta_j} Q_{ij} + z_{1i} \widehat{\beta_\mu} + \widehat{\beta_\varepsilon} \right) \Delta \omega \equiv \gamma_{1i}^c \Delta \omega, \forall i \in S \tag{23}$$

Obviously, $\forall i \in S_S$, and we also have $\widehat{\gamma_{1i}} = 0$. However, the decremental adjustments cannot be used directly to obtain the optimal solution to the minimization problem (6). Hence, to solve this problem, we need to compute the adjustment quantity $\Delta \omega^*$ for each decremental adjustment such that a certain sample migrates among the sets S_S, S_R and S_E. If $\sum_{i \in S} \alpha_{1i} > 2Cv(N+1)$, we will compute the maximal adjustment quantity $\Delta \omega^{\max}$ and let $\Delta \omega^* = \Delta \omega^{\max}$; otherwise, we will compute the minimal adjustment quantity $\Delta \omega^{\min}$ and let $\Delta \omega^* = \Delta \omega^{\min}$.

3   Experiment Result

3.1   Experiment Setup

In this section, we validate the performance of the proposed incremental dual-v-SVR algorithm (IDVSVR) through several experiments which compare our algorithm with standard TSVR [22], AOSVR [9], and INSVR [12]. All regression models are implemented in MATLAB 2016a version on Windows 7 running on a PC with system configuration Intel Core i5 processor (2.40 GHz) with 8-GB RAM. We also use *cadata*, and *Friedman* data sets. Cadata [12] is a real data set, *Friedman* is an artificial data set [21]. The details of the three data sets are shown in Table 1.

Table 1. Data sets used in the experiments

Data set	# training set	# attributes
cadata	20000	8
Friedman	40000	10

For simplicity, the Gaussian radial basis function kernel is adopted for all examples. We set the model parameters $C_1 = C_2 = C$ and $v_1 = v_2 = v$. The values of parameter C, q, ε, v are, respectively, selected from the sets $\{10^i \mid i = 0, 1, ..., 6\}$, $\{2^i \mid i = -9, -8, ..., 2\}$, $\{0.01, 0.02, ..., 0.4\}$, and $\{0.01, 0.02, ..., 0.6\}$. As for uncertainty, as the aforementioned data sets are not noisy, we artificial introduce a noisy e_i into predictor variable X and dependent variable Y, and e_i is drawn from a uniform distribution on U $(-k, k)$. Here, U$(-k, k)$ represents the uniformly random variable in $[-k, k]$.

3.2   Performance Evaluation

In the first experiment, we compare their trends in relation to regression risk on noisy data when the data size N is increased. We use RMSE [22] to represent the accuracy where a smaller RMSE represents a lower risk. Figure 3 shows the comparison result for different data sets, different data size N and different k.

From Fig. 3, we can see that when the data size N is increased, no matter how much k is, the RMSE of any regression is gradually decreasing. When $k = 0$, IDVSVR and INSVR have the better performance. These results identify that both IDVSVR and INSVR have the advantage of using parameter v to control the bounds on the fraction of SVs and errors. However, when k increases, or in other words, when the data is perturbed by noise, the IDVSVR still has the better performance, as the performance of INSVR is very poor. Furthermore, Fig. 3 also shows that the accuracy of IDVSVR remains relatively stable even when k = 1. Hence, a major advantage of the proposed IDVSVR over the other algorithms is its effectiveness in handling uncertain data.

In the second experiment, we compare the training speed on noisy data when the data size N is increased. In the previous experiment, we know TSVR has the worst generalization ability. Hence, we only test the training time of AOSVR, INSVR, and IDVSVR. Figure 4 shows the comparison results in terms of time for the different data sets, different data size N and different k.

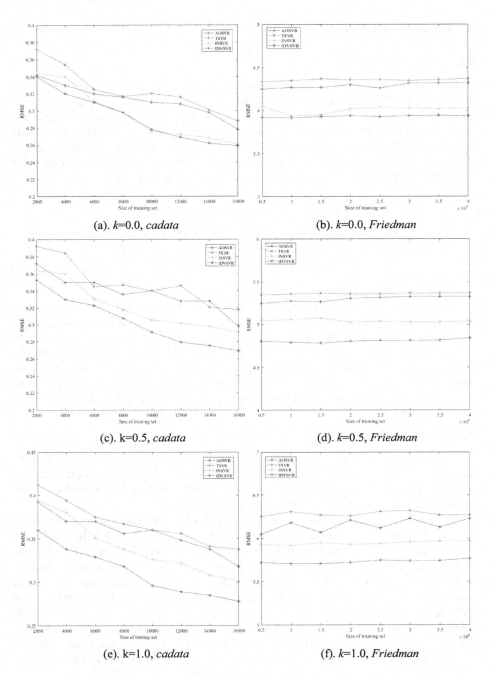

(a). *k*=0.0, *cadata*

(b). *k*=0.0, *Friedman*

(c). *k*=0.5, *cadata*

(d). *k*=0.5, *Friedman*

(e). k=1.0, *cadata*

(f). k=1.0, *Friedman*

Fig. 3. Comparison result of RMSE

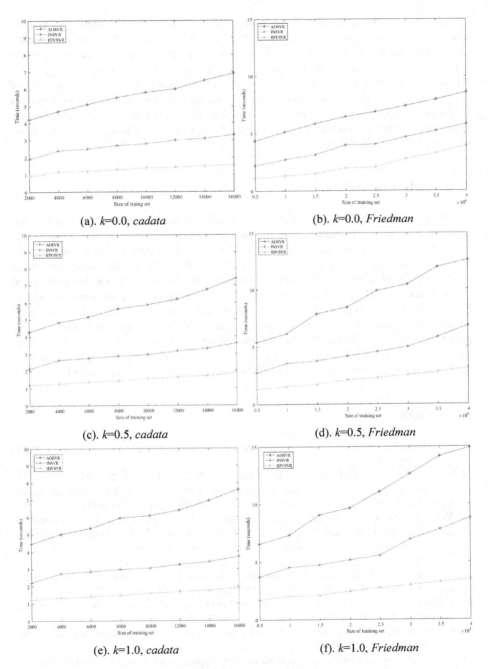

(a). k=0.0, cadata

(b). k=0.0, Friedman

(c). k=0.5, cadata

(d). k=0.5, Friedman

(e). k=1.0, cadata

(f). k=1.0, Friedman

Fig. 4. Comparison result of training speed.

Figure 4 demonstrates that the learning speed of the IDVSVR algorithm is generally much faster than the other online SVR algorithms when data size N increases. One reason for this is that in IDVSVR, the two nonparallel functions are estimated by solving two SVR-type QPPs of smaller size at the same time, so the learning speed of IDVSVR is faster. The second reason is using an extra term to prevent the recurrence of conflicts between the Eqs. (8) and (9) and reduce the number of adjustment. Figure 4 furthermore suggests that the prediction speed of IDVSVR model remained relatively stable in different k. But the training speed of other online SVR algorithms becomes slower with the k increased.

4 Concluding Remarks

As there is no effective SVR algorithm which can handle incremental regression problem with uncertain data, we design an incremental dual-v-SVR algorithm in this paper. Our proposed incremental dual-v-SVR has good robustness against uncertainty and can handle the incremental regression problem efficiently. Furthermore, there are a total of five advantages of our proposed incremental algorithm: (1) the learning speed of incremental dual-v-SVR algorithm is fast; (2) the sparsity of incremental dual-v-SVR algorithm is improved; (3) the incremental dual-v-SVR algorithm has good generalization performance; (4) the incremental dual-v-SVR algorithm also can use parameter v to control the bounds on the fractions of SVs and errors; and (5) in incremental dual-v-SVR algorithm, the box constraints are independent of the size of the training sample set. The experimental results also prove our conclusion.

References

1. Scholkopf, B., Smola, A.J.: New support vector algorithms. Neural Comput. Learn. II **1245**, 1207–1245 (1998)
2. Chang, C.-C., Lin, C.-J.: Libsvm. ACM Trans. Intell. Syst. Technol. **2**, 1–27 (2011)
3. Gu, B., Sheng, V.S.: A robust regularization path algorithm for v-support vector classification. IEEE Trans. Neural Netw. Learn. Syst. **28**, 1241–1248 (2017)
4. Wang, G., Zhang, G., Choi, K., Lu, J.: Deep additive least squares support vector machines for classification with model transfer. IEEE Trans. Syst. Man Cybern. Syst. 1–14 (2017)
5. Smola, A.J., Schölkopf, B.: A tutorial on support vector regression. Stat. Comput. **14**, 199–222 (2001)
6. Shevade, S.K., Keerthi, S.S., Bhattacharyya, C., Murthy, K.R.K.: Improvements to the SMO algorithm for SVM regression. IEEE Trans. Neural Netw. **11**, 1188–1193 (2000)
7. Takahashi, N., Guo, J., Nishi, T.: Global convergence of SMO algorithm for support vector regression. IEEE Trans. Neural Netw. **19**, 971–982 (2008)
8. Collobert, R., Williamson, R.C.: SVMTorch: support vector machines for large-scale regression problems. J. Mach. Learn. Res. **1**, 143–160 (2001)
9. Ma, J., Theiler, J., Perkins, S.: Accurate on-line support vector regression. Neural Comput. **15**, 2683–2703 (2003)

10. Omitaomu, O.A., Jeong, M.K., Badiru, A.B., Hines, J.W.: Online support vector regression approach for the monitoring of motor shaft misalignment and feedwater flow rate. IEEE Trans. Syst. Man Cybern. Part C Appl. Rev. **37**, 962–970 (2007)
11. Omitaomu, O.A., Jeong, M.K., Badiru, A.B.: Online support vector regression with varying parameters for time-dependent data. IEEE Trans. Syst. Man Cybern. Part A Syst. Hum. **41**, 191–197 (2011)
12. Gu, B., Sheng, V.S., Tay, K.Y., Romano, W., Li, S.: Incremental support vector learning for ordinal regression. IEEE Trans. Neural Netw. Learn. Syst. **26**, 1403–1416 (2015)
13. Gu, B., Sheng, V.S., Wang, Z., Ho, D., Osman, S., Li, S.: Incremental learning for ν -support vector regression. Neural Netw. **67**, 140–150 (2015)
14. Hong, D.H., Hwang, C.H.: Interval regression analysis using quadratic loss support vector machine. IEEE Trans. Fuzzy Syst. **13**, 229–237 (2005)
15. Yang, X., Zhang, G., Lu, J., Ma, J.: A kernel fuzzy c-Means clustering-based fuzzy support vector machine algorithm for classification problems with outliers or noises. IEEE Trans. Fuzzy Syst. **19**, 105–115 (2011)
16. Peng, X., Chen, D., Kong, L., Xu, D.: Interval twin support vector regression algorithm for interval input-output data. Int. J. Mach. Learn. Cybern. **6**, 719–732 (2015)
17. Chen, G., Zhang, X., Wang, Z.J., Li, F.: Robust support vector data description for outlier detection with noise or uncertain data. Knowl.-Based Syst. **90**, 129–137 (2015)
18. Yang, X., Tan, L., He, L.: A robust least squares support vector machine for regression and classification with noise. Neurocomputing **140**, 41–52 (2014)
19. Huang, G., Member, S., Song, S., Wu, C., You, K.: Robust support vector regression for uncertain input and output data. IEEE Trans. Neural Netw. Learn. Syst. **23**, 1690–1700 (2012)
20. Gu, B., Wang, J., Yu, Y., Zheng, G., Huang, Y., Xu, T.: Accurate on-line ν-support vector learning. Neural Netw. **27**, 51–59 (2012)
21. Meyer, D., Leisch, F., Hornik, K.: The support vector machine under test. Neurocomputing **55**, 169–186 (2003)
22. Peng, X.: TSVR: an efficient twin support vector machine for regression. Neural Netw. **23**, 365–372 (2010)

Text Stream to Temporal Network - A Dynamic Heartbeat Graph to Detect Emerging Events on Twitter

Zafar Saeed[1,2], Rabeeh Ayaz Abbasi[1], Abida Sadaf[3],
Muhammad Imran Razzak[2], and Guandong Xu[2(✉)]

[1] Department of Computer Science, Quaid-i-Azam University,
Islamabad, Pakistan
zsaeed@cs.qau.edu.pk, rabbasi@qau.edu.pk
[2] Advanced Analytics Institute, University of Technology Sydney,
Sydney, Australia
mirpak@gmail.com, guandong.xu@uts.edu.au
[3] Institute of Information Technology,
Quaid-i-Azam University, Islamabad, Pakistan
abida.sadaf@qau.edu.pk

Abstract. Huge mounds of data are generated every second on the Internet. People around the globe publish and share information related to real-world events they experience every day. This provides a valuable opportunity to analyze the content of this information to detect real-world happenings, however, it is quite challenging task. In this work, we propose a novel graph-based approach named the Dynamic Heartbeat Graph (DHG) that not only detects the events at an early stage, but also suppresses them in the upcoming adjacent data stream in order to highlight new emerging events. This characteristic makes the proposed method interesting and efficient in finding emerging events and related topics. The experiment results on real-world datasets (i.e. FA Cup Final and Super Tuesday 2012) show a considerable improvement in most cases, while time complexity remains very attractive.

Keywords: Dynamic graph · Time series analysis · Event detection
Text stream · Big data · Emerging trend

1 Introduction

In recent years, with the unprecedented growth of social media and blog networks, huge amounts of diverse types of data are being generated every day. The information that is collectively generated on such platforms is of great value. In addition to its huge volume and diversity, much of the data is inter-dependent in nature. The analysis of such data is quite important and helps to successfully

© Springer International Publishing AG, part of Springer Nature 2018
D. Phung et al. (Eds.): PAKDD 2018, LNAI 10938, pp. 534–545, 2018.
https://doi.org/10.1007/978-3-319-93037-4_42

detect meaningful information that could be used for searching, discovering patterns and sensing trends. The detection of emerging trends from social media text streams has recently become a research area of great interest. However, real-time streaming data is quite complicated to analyze. Recent work mainly focuses on event detection using bursty features or graph similarity patterns using subgraph matching [6,7], however, there is a need for a more scalable and localized pattern analysis approach to detect emerging events in text streams.

Analyzing large, diverse and noisy data, especially social media, requires addressing scalability, accuracy as well as complexity challenges. Documents describing the same event and story have a similar set of collocated keywords that could be used to identify time and its description. In order to identify significant and unusual patterns, recently graph-based methods have been extensively applied to deal real-world data efficiently [9,11–13].

Graph mining has received considerable attention in the data analytic community [4,5,9,10,12]. Most of the time, data is gathered as a stream of time, thus traditional graph-based algorithms are not efficient to process data of such complex nature (i.e. dynamic and non-stationary). Most of the existing graph-based methods focus on frequent, co-occurrent, and highly weighted patterns to highlight the significance of data entities, but ignores the fact that the burstiness often dominates the other related details that exist in the data which, sometimes, can be very important. In this work, we present a novel graph-based approach named Dynamic Heartbeat Graph (DHG) based on the differences between temporal graphs. The proposed DHG approach not only detects events at an early stage but also suppresses the burstiness of event related topics in the upcoming data stream for a certain time interval in order to highlight new emerging events. This characteristic makes the DHG approach unique and efficient in finding new emerging events and related topics.

We formulate the text stream as a series of disjoint temporal graphs. These disjoint graphs are further processed to generate heartbeats within each time window of fixed temporal length. We design three features *growth factor*, *trend probability*, and *topic centrality*. Based on these features, we use a binary classifier to detect emerging events in data stream.

The **goal** of this paper is to address the key aforementioned problems. By employing the proposed DHG approach which analyzes the patterns in adjacent time windows, we can overcome the limitations of the state-of-the-art work by identifying key occurrences efficiently. We describe the theoretical and empirical **key contributions** of this work as follows:

- A novel graph-based approach named Dynamic Heartbeat Graph (DHG) which is efficient in the detection of events.
- Low computational complexity of proposed method, which generates a series of DHGs in $O(K(|V|)^2)$, where K is total number of DHGs within a sliding window, which is a considerably small in value. Event candidate DHGs are identified in $O(K(|V| + N))$.

- The latter method is evaluated empirically on the FA Cup and Super Tuesday datasets. The experiment results on data show that the DHG outperforms state-of-the-art methods.

2   Preliminaries

A **Micro-document** is short textual content consisting of words that are published online through some micro-blog. It is defined as 3-tuple $d_i = (t, u, W)$, where u is a user who publishes a micro-document d_i with some set of words W at a specific time instance t. A **Text Stream** is a set of micro-documents $\mathcal{D} = \{d_1, d_2, d_3, ..., d_n\}$, where d_i and $d_{(i-1)}$ are the ith and $(i-1)^{th}$ micro-documents published at time $\pi_1(d_i)$ and $\pi_1(d_{i-1})$ respectively, such that $\pi_1(d_i) \geq \pi_1(d_{(i-1)})$. The lengths of micro-documents are usually short hence, the measures based on burstiness, similarity as well as distance may not yield good results, however this issue could be resolved by creating a super-document. Let $\mathcal{D}$ be the set of all micro-documents available in a text stream, then a **Super-document** d_i^ρ is a continuous temporal accumulation of each $d_i \in \mathcal{D}$ separated at t_a and $t_{(a+b)}$ time intervals (we refer as t_i later in the paper). To create a super-document, instead of merging the micro-documents into one core document, we create k partitions in text stream $\mathcal{D}^\rho = \{\{d_1, d_2, ..., d_p\}, \{d_{p+1}, ..., d_{p+q}\}, ..., \{..., d_n\}\}$. By doing so, we are able to retain the identity of each micro-document that we use later to generate a network series (See Sect. 3.1 for details) which increases the cohesiveness among the topics and keywords. Thus, this super-document can be defined as k number of mutually exclusive partitions i.e. $\bigcap_{i=1}^{|\mathcal{D}^\rho|} d_i^\rho = \varnothing$. A **Sliding Window** is a set of super-documents (chunk of data) whose temporal length Δt. Each sliding window is processed independently in each sliding window to detect event related information. A set of word(s) in a text stream may refer to a *topic*. When more people are using specific topic in their micro-document, it becomes a trend, often called a trending topic. Similar to other research studies, we use the terms "trend" and "event" interchangeably and also the terms "word(s)" and "topic" [1–3,8,14,15].

3   Dynamic Heartbeat Graph (DHG)

Using text stream, we devise a technique that creates a series of dynamic disjoint graphs and then maps each adjacent pair of graphs in a network series on to another DHG series. In order to classify DHGs as candidates for events, we design and use trend probability, change in burstiness, and normalized degree centrality in the DHGs as key features. This section defines all the components involved in the transformation of text stream into series of temporal networks.

3.1   Network Series

Against each super-document $d_i^\rho \in D^\rho$, a network G_i is created in such a way that each node is a "word" and an edge between two nodes represents co-occurrence relationship. A network series is a set of disjoint graphs

$\mathcal{G} = \{G_1, G_2, G_3, ..., G_{|\mathcal{D}^\rho|}\}$, where each $G_i \in \mathcal{G}$ is built against $d_i^\rho \in \mathcal{D}^\rho$ such that G_i is a labeled graph, i.e. $G_i = (V, E, \mathcal{W}, \mathcal{S})$, where V is a set of nodes such that $\forall v_i \in V$ are the unique words which appear in d_i^ρ, and $E \subseteq V \times V$ is a set of edges such that $e_k = (v_k, v_k^{'}) \wedge v_k \neq v_k^{'}$. $\mathcal{W} : V \to \mathbb{R}$ and $\mathcal{S} : E \to \mathbb{R}$ are the functions that assign weights to each node and edge in the graph G_i as shown in Eqs. 1 and 2, where $|d_i^\rho(v_k)|$ is the term-frequency of v_k and $|d_i^\rho(v_k, v_k^{'})|$ is the frequency of co-occurrence of nodes v_k and $v_k^{'}$ in super-document d_i^ρ.

$$\mathcal{W}(v_k) = |d_i^\rho(v_k)| \tag{1}$$

$$\mathcal{S}(e_k) = |d_i^\rho(v_k, v_k^{'})| \tag{2}$$

The network is created in such way that it retains the coherence among the words of each micro-document d_i participating in the building of the network. The coherence is enhanced by creating a clique among the words of each d_i. *Clique* — each node $v_k \in d_i$ is connected to every other node $v_k^{'} \in d_i$. This results in an increase in the central tendency of topics within the large network of diverse words.

3.2 DHG Series

To create a DHG series, Algorithm 1 linearly combines and maps every pair of adjacent graphs G_i and G_{i-1} on to a new DHG G_i^h which is used further for emerging trend detection. The goal of generating a set of DHG $\mathcal{G}^h$ is to discriminate among topics and the drift in their popularity within each subsequent graph.

Algorithm 1. Generate Set of Dynamic Heartbeat Graphs

---

    input : $\mathcal{G} = \{G_1, G_2, G_3, ..., G_{|P|}\}$ set of a graph series
             where $\exists G_i \in \mathcal{G}$ is generated against $\exists d_i^\rho \in D^\rho$
    output: $\mathcal{G}^h = \{G_1^h, G_2^h, ..., G_{|\mathcal{G}|-1}^h\}$
           $\varepsilon = \{\epsilon_1, \epsilon_2, \epsilon_3, ..., \epsilon_{|\mathcal{G}|-1}\}$

1   **for** $i \leftarrow 1$ **to** $|\mathcal{G}| - 1$ **do**
2 $U \leftarrow \text{Join}(V^{G_i}, V^{G_{i+1}})$
3 $A \leftarrow \text{RegenerateMatrix}(G_i , U)$
4 $B \leftarrow \text{RegenerateMatrix}(G_{i+1} , U)$
5 $\varepsilon[i] \leftarrow \text{EstimateHeartbeat } A, B, V^{G_i}, V^{G_{i+1}}$

6   **end**

The DHG algorithm takes network series G as input and generates another series of networks which we call Dynamic Heartbeat Graph (DHG) series. For every adjacent pair of graphs G_{i-1} and G_i, the algorithm aligns the dimensions of the adjacency matrices by taking union of the vertices in both graphs and then reorders them canonically. In later step (at Line 5), the algorithm estimates the change in the node and edge weights (see Algorithm 2) and stores it in an indexed vector $\varepsilon[i] \in \mathbb{R}^{n \times 3}$. The step-by-step implementation detail is given in Algorithm 1. An example in Fig. 1 shows how the DHG between two networks is calculated, where node weights (given by "()"), and edge weights can be seen in the graphs as well as in the adjacency matrices. Reordering each graph G_i canonically and transforming the DHG into vector-space reduce the computational complexity significantly from $O(K|V|^4)$ to $O(K(|V|)^2)$, where K is considerably a small value, i.e. $K = |\mathcal{G}^h|$ and $V = Max(|V^{G_i^h}|)$.

Algorithm 2. Estimate change in burstiness

---

    input : A, B are adjacency matrices that represent G_{i-1} and G_i respectively.
               V^A and V^B are lists of vertices in G_{i-1}, G_i respectively

    output: e vector that represents a DHG against G_{i-1} and G_i
               V^H list of vertices in DHG

  1 **for** $k \leftarrow 1$ **to** $|V^B|$ **do**
  2 | $V^H[k] \leftarrow V^B[k] - V^A[k]$
  3 **end**
  4 **for** $x \leftarrow 1$ **to** $|V^B|$ **do**
  5   **for** $y \leftarrow 1$ **to** x **do**
  6      $edgeWt \leftarrow B[x,y] - A[x,y]$
  7   **if** $edgeWt! = 0$ **then**
  8   e.**Add**($x, y, edgeWt$)
  9   **end**
10   **end**
11 **end**

The DHG approach implicitly suppresses and handles the dominance of bursty topics by calculating the change in the weights of each node and edge between each pair of adjacent graphs $G_{(i-1)}$ and G_i in order to highlight other details which are less frequent. Algorithm 2 estimates and labels all the corresponding nodes and edges with new weights in DHG G_i^h. The DHG series is a set of disjoint graphs, generated in a streaming fashion; therefore, it is temporally well aligned with the text stream. Furthermore, these DHGs are used to detect emerging events. The detail of the detection method is given in next section.

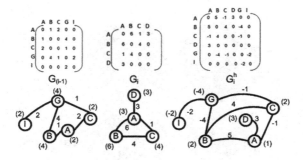

Fig. 1. Example for the creation of DHG from two subsequent time intervals

3.3   Event Detection Method

In the following section, we present our event detection method using DHGs. The event detection method works on the following assumptions:

- The text stream has diverse contents, but an emerging event may only occur in a text stream whenever there is a significant change in either burstiness displacement of existing topics or the appearance of new topics in the text stream between two adjacent time intervals at $t_{(i-1)}$ and t_i.
- The significant change is not only dependent upon the burstiness of topics, but also the change in their probability distribution and central tendencies within the network.

In all of the following equations, for the simplification of the notations, let $\psi = G_i^h$ where G_i^h is i^{th} heartbeat graph. The detection method uses the fusion of three key features *growth factor, trend probability*, and *topic centrality* to calculate *HeartbeatScore* $\mathscr{H}(\psi)$. *growth factor* $Gr_{fact}(\psi)$, *trend probability* $Tr_{Prob}(\psi)$, and *topic centrality* $\Sigma\mathscr{C}(v^\psi)$ represent the significance of the change in the burstiness of topics, possibility of an emerging event at time interval t_i, and central tendency and coherence among different topics in DHG ψ, respectively. The *growth factor* of DHG ψ is calculated as shown in Eq. 3 where $\mathscr{W}(v_k^\psi)$ is the kth node weight that represents a change in burstiness of a topic between G_i and G_{i-1} (see Algorithm 2). A higher score of *growth factor* shows that the topics are appearing with high frequency in sliding window $k\Delta t$.

$$Gr_{fact}(\psi) = \sum_{k=0}^{|V^\psi|} \mathscr{W}(v_k^\psi) \tag{3}$$

A node in DHG ψ can have negative and positive weights. To calculate *trend probability* $Tr_{Prob}(\psi)$, the probability distribution against positive $\mathscr{W}(v_k^{\psi+})$ and negative $\mathscr{W}(v_k^{\psi-})$ weights of each word are calculated within the DHG ψ. The probability distribution over positive and negative weights are then linearly combined, as shown in Eq. 4, which shows the convergence of DHG ψ towards trending topics, where β_1 and β_2 are 1 and -1 respectively. $Tr_{Prob}(\psi) > 0$ indicates

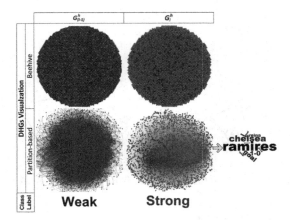

Fig. 2. Graph visualization (Beehive and Partition-based) of two subsequent DHGs $G_{(i-1)}^h$ and G_i^h at time $t_{(i-1)}$ and t_i respectively, from the FA Cup 2012 dataset. At t_i a significant event "Goal" occurs. Red and green are the nodes with positive and negative weights respectively. The DHG approach shows hyper-sensitivity to burstiness and as well as newly emerging topics. (Color figure online)

that topics are gaining popularity, thus, denoting the possibility of emerging event(s) in sliding window $k\Delta t$.

$$Tr_{Prob}(\psi) = \beta_1 \sum_{k=0}^{|V^{\psi+}|} \frac{\mathscr{W}(v_k^{\psi+})}{\sum_{l=0}^{|V^{\psi}|} |\mathscr{W}(v_l)|} + \beta_2 \sum_{k=0}^{|V^{\psi-}|} \frac{|\mathscr{W}(v_k^{\psi-})|}{\sum_{l=0}^{|V^{\psi}|} |\mathscr{W}(v_l)|} \tag{4}$$

topic centrality $\mathscr{C}(v_k^{\psi})$ is then calculated to highlight the central tendency of topics in each DHG ψ, as shown in Eq. 6 where v_k^{ψ}, $\epsilon_i \in \varepsilon$, $\pi_3(e_i)$, and $|V^{\psi}|$ represent the topic, indexed edge vector, weight of edge e_i connected to v_k^{ψ}, and the total number of topics in DHG ψ, respectively. In the calculation of *topic centrality*, all the edges with negative weights are dropped because of the initial assumption (see Sect. 3), which positively influences the centrality of newly emerging topics with respect to the existing ones. A higher aggregated centrality score shows that the emerging topics are coherent and concurrently appearing in text stream in sliding window $k\Delta t$. The detection method comprises two steps:

1. If $Tr_{Prob}(\psi) \le 0$ then DHG is assigned to the *"Weak"* class. Once the highly frequent topics reach their peak, they start to lose their importance because of the decay in their burstiness. If the weights of certain topics are reduced at time t_i compared to t_{i-1} and there is no significant increase in the weights of the other topics, then the *trend probability* score is always negative, therefore indicating the fact that the heartbeat between G_{i-1} and G_i is not significant.
2. Otherwise calculate heartbeat score $\mathscr{H}(\psi)$ which is the product of *growth factor* $Gr_{fact}(\psi)$, *trend probability* $Tr_{Prob}(\psi)$, and aggregated *topic centrality* $\Sigma\mathscr{C}(v^{\psi})$ in DHG ψ (as shown in Eq. 7).

To assign a binary class membership (i.e. [$Strong, Weak$]) to each DHG ψ, where "$Strong$" means DHG ψ contains emerging trends and "$Weak$" means an insignificant heartbeat, a classification function $Est(\psi)$ (as shown in Eq. 8) estimates and assigns two-class labels to each DHG $\psi \in \mathcal{G}^h$. Here, θ is an adoptive measure that finds the local optimum value in each sliding window $k\Delta t$ to set a threshold for classification function $Est(\psi)$ as shown in Eq. 5, where Δt, τ are the temporal length of each sliding window and super-document d_i^{ρ} respectively such that $\Delta t(\text{mod } \tau) = 0$, i is the index of the first DHG in the sliding window under consideration, $Gr_{fact}(\psi)$ is the $growth\ factor$ (as shown in Eq. 3) of each DHG ψ, and ω is the adjustment parameter. We set ω as 1 and 0.6 for the FA Cup and Super Tuesday datasets, respectively.

$$\theta_{(k\Delta t)} = \frac{\tau \sum_{i=k}^{k+\frac{\Delta t}{\tau}} (\mathcal{H}(\psi))}{\Delta t} + \omega \sqrt{\frac{\tau \sum_{i=k}^{k+\frac{\Delta t}{\tau}} (\mathcal{H}(\psi) - \frac{\tau \sum_{i=k}^{k+\frac{\Delta t}{\tau}} (\mathcal{H}(\psi))}{\Delta t})^2}{\Delta t}} \tag{5}$$

$$\mathcal{C}(v_k^{\psi}) = \frac{\sum_{i=1}^{|\varepsilon^{\psi}|} [(\pi_1(\varepsilon_i^{\psi}) = k \vee \pi_2(\varepsilon_i^{\psi}) = k) \wedge \pi_3(\epsilon_i^{\psi}) > 0]}{|V^{\psi}|} \tag{6}$$

$$\mathcal{H}(\psi) = Gr_{fact}(\psi) \times Tr_{Prob}(\psi) \times \sum_{k=0}^{|V^{\psi}|} \mathcal{C}(v_k^{\psi}) \tag{7}$$

$$Est(\psi) = \begin{cases} \text{'Strong'}, & \text{if } \mathcal{H}(\psi) \geq \theta_{(k\Delta t)} \\ \text{'Weak'}, & \text{otherwise} \end{cases} \tag{8}$$

The transformation of the DHG series into vector-space ε (see Sect. 3.2) results in reducing the computational complexity of binary classification from $O(|V| + |E|)$ to $O(|V| + N))$, where $V = Max(|V^{\psi}|)$ and $N = Max(|\epsilon_i|)$. Here, the value of $N \ll |E|$. In the worst case scenario, $O(|V| + |E|) = O(|V| + N))$ if and only if there is a continuous increase in temporal frequency of all the words in the text stream, however, the occurrence of such scenarios is nearly impossible due to the fact that real-world events have an evolutionary pattern (i.e. build-up, stable/peak, decay). Later, in each sliding window $k\Delta t$ a ranked topic list in the candidate DHGs that are classified as "$Strong$" is generated by calculating the score of each topic, as shown in Eq. 9. Figure 2 shows the heartbeats of two subsequent DHGs and their class labels using classification function $Est(\psi)$ with top ten trending topics.

$$Rank(v_k^{\psi}) = \mathcal{C}(v_k^{\psi}) \times \mathcal{W}(v_k^{\psi}) \tag{9}$$

4   Experiment and Results

In this section, we evaluated the performance of proposed dynamic heartbeat graph as discussed in Table 1.

4.1   Evaluation

To evaluate the performance of proposed DHG method, two benchmark datasets "FA Cup" and "Super Tuesday", and the framework introduced by [1] are used for the comparison. We create partitions for the super-document accumulation as one minute and five minutes for the "FA Cup" and "Super Tuesday" datasets, respectively. The DHG method takes input data in a streaming fashion and create time series disjoint network and then DHG series based on the temporal length of accumulation.

Two evaluation measures: *Topic-Recall@K* (T-Rec), which is the percentage of ground truth topics detected correctly at top K retrieved topics; and *Keyword-Precision@K* (K-Pre), which is the percentage of keywords detected correctly out of the top K number of keywords are used. *T-Rec* and *K-Prec* are calculated by micro-averaging the individual *T-Rec* and *K-Prec* scores. Similar to our baseline framework [1], we did not use Topic-Precision@K because the ground truth is formulated through main stream media that contains only those topics which are reported by the main stream media. The data may also have other newsworthy topics that took place at the same time. Moreover, it is impractical to list all possible topics directly from the huge data. Therefore, the Topic-Precision@K cannot be measured in this context. We did not use Topic-Precision@K because the ground truth is formulated through main stream media that contains only those topics which are reported by the main stream media. The data may also have other newsworthy topics that took place at the same time, and it is impractical to analyze all possible topics directly from the huge data. Therefore, the Topic-Precision@K cannot be measured in this context.

4.2   Dataset

We conducted our experiments on a well-known benchmark datasets ("FA Cup final" and "Super Tuesday" [1]). The "FA Cup" is one of the oldest knock-out football competition and very popular among the fans around the world. The dataset consists of a text stream of the final match held on May 5th, 2012, between the Chelsea and Liverpool teams. The ground truth comprised 13 topics, including goals, bookings and fouls, kick-off, half-time and match ending. The "Super Tuesday" dataset is the US presidential primary election held on Tuesday 6 March 2012, the key moment when it is likely that the party nominee is elected as presidential candidate. The ground truth comprised 22 topics covering stories related to the projection and success of nominees in particular states and their speeches. For evaluation purposes, the temporal length Δt of each sliding window is set to one minute and one hour for the FA Cup and Super Tuesday datasets, respectively. The ground truth contains topics with respect to each sliding window. To reduce noise, the datasets are pre-processed. To improve data quality, retweets, tweets containing URLs or those containing less than three (3) words are removed. Furthermore, common words, stop words, the words which have less than three letters, and punctuation are removed.

4.3 Results

We present the results for *Topic-Recall* (T-Rec) at $K = 2, 4, 6, ..., 20$ in Fig. 3, and *Keyword-Precision* (K-Prec) in Table 1 to compare the six different event detection methods including DHG. Our method gives best results on FA Cup, because of the users those publishing contents on micro-blogs, are very focused, consistent, and to the point due to the popularity and limited time of the underlying event. Therefore, the topics reported in the text stream are less diverse, making them easier to detect compared to the Super Tuesday dataset. The topics which are reported by the ground truth are taken from the mainstream media and cover a broader semantic prospective. For instance at time window 17:56 in the FA Cup, ground truth marked "Andy, Carroll" as topic, whereas "header, cech, over, claim, equalize" are among the other keywords therefore, it is more likely that the topics are among the top trends but they do not necessarily appear in the top most position every time. The DHG method has comparable T-Rec at $K = 2, 4, 6$, and 8. It eventually achieves the maximum possible T-Rec at $K \geq 10$ for the FA Cup dataset. Similarly, the DHG method outperforms the other detection methods after $K > 30$ for the Super Tuesday dataset. The results for T-Rec are shown in Fig. 3.

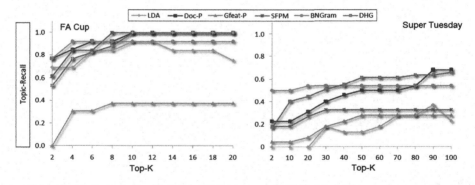

Fig. 3. Topic-Recall@K for six (6) different well-known methods for the FA Cup (left) and Super Tuesday (right) datasets.

Conversely, the DHG method combines the scores of change in topics' burstiness and central tendency in the graph, therefore it is able to detect relevant keywords with high precision compared to the other methods for both datasets, as shown in Table 1. Hence, DHG exhibits one of the effective detection method in terms of performance and accuracy.

Figure 4 shows user participation, network size, and the heartbeat score signals using the DHG series across the temporal data of the FA Cup from 17:21 to 18:11. User participation is the total number of unique users who published at least one micro-document, and network size is the total number of unique words in the DHG ψ at time t_i. It is observed that the DHG method detects emerging

Table 1. DHG outperforms all other detection methods for K-Prec@2 for both the FA Cup and Super Tuesday datasets

Method	FA Cup	Super Tuesday
LDA	0.164	0.000
Doc-P	0.337	0.511
Gfeat-P	0.000	0.375
SFPM	0.233	0.471
BNGram	0.299	0.628
DHG	**0.682**	**0.875**

events at an early stage. Whenever an event occurs on a particular time interval, the proposed method detects the event-related topics before the diversity in the text stream increases. On the other hand, we also observe that user participation also increases whenever an event occurs.

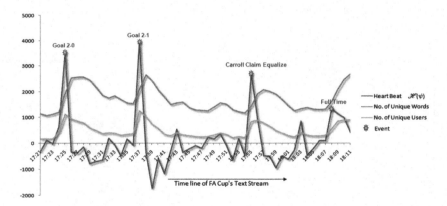

Fig. 4. Detected events with respect to the heartbeat pattern. The figure also shows the variations in the number of unique words and user participation across different time intervals

5   Conclusion

In this paper, a novel Dynamic Heartbeat Graph (DHG)-based method is developed that is efficient for text streams such as Twitter. We formulated the text stream as a series of disjoint temporal graphs that are further processed to generate heartbeats within each time interval of fixed temporal length. Furthermore, we have designed three unique features growth factor, trend probability and topic centrality to identify the emerging events using DHG. In order to evaluate the performance of DHG, we have used two publicly available benchmark datasets (the FA Cup Final 2012 and Super Tuesday 2012). The quantitative evaluation shows that the DHG method is sensitive to the dynamic nature of text streams and detected emerging events with high precision compared to the state-of-the-art methods. Empirical evaluation showed that DHG method is robust in terms of computational complexity and scalability thus, it could be used for live streaming as well.

References

1. Aiello, L.M., Petkos, G., Martin, C., Corney, D., Papadopoulos, S., Skraba, R., Göker, A., Kompatsiaris, I., Jaimes, A.: Sensing trending topics in twitter. IEEE Trans. Multimed. **15**(6), 1268–1282 (2013)
2. Benhardus, J., Kalita, J.: Streaming trend detection in twitter. Int. J. Web Based Communities **9**(1), 122–139 (2013)
3. Buntain, C.: Discovering credible events in near real time from social media streams. In: Proceedings of the 24th International Conference on World Wide Web, pp. 481–485. ACM, New York (2015)
4. Garimella, K., Morales, G.D.F., Gionis, A., Mathioudakis, M.: Quantifying controversy on social media. ACM Trans. Soc. Comput. **1**(1), 3 (2018)
5. Giannis Nikolentzos, P.M.a.V.: Matching node embedding for graph similarity. In: Proceedings of the Thirty-First AAAI Conference on Artificial Intelligence (AAAI 2017), pp. 2429–2435 (2017)
6. Johansson, F., Jethava, V., Dubhashi, D., Bhattacharyya, C.: Global graph kernels using geometric embeddings. In: Proceedings of the 31st International Conference on Machine Learning, ICML 2014, Beijing, China, 21–26 June 2014 (2014)
7. Johansson, F.D., Dubhashi, D.: Learning with similarity functions on graphs using matchings of geometric embeddings. In: Proceedings of the 21st ACM SIGKDD International Conference on Knowledge Discovery and Data Mining, pp. 467–476. ACM (2015)
8. Nguyen, D.T., Jung, J.E.: Real-time event detection for online behavioral analysis of big social data. Future Gener. Comput. Syst. **66**, 137–145 (2017)
9. Shabunina, E., Pasi, G.: A graph-based approach to ememes identification and tracking in social media streams. Knowl. Based Syst. **139**, 108–118 (2018)
10. Velampalli, S., Eberle, W.: Novel graph based anomaly detection using background knowledge. In: Proceedings of the Thirtieth International Florida Artificial Intelligence Research Society Conference, pp. 538–543 (2017)
11. Yanardag, P., Vishwanathan, S.: Deep graph kernels. In: Proceedings of the 21th ACM SIGKDD International Conference on Knowledge Discovery and Data Mining, pp. 1365–1374. ACM (2015)
12. Yanardag, P., Vishwanathan, S.: A structural smoothing framework for robust graph comparison. In: Advances in Neural Information Processing Systems, pp. 2134–2142 (2015)
13. Yao, Y., Holder, L.B.: Detecting concept drift in classification over streaming graphs. In: KDD Workshop on Mining and Learning with Graphs (MLG), San Francisco, CA, 14 August 2016, pp. 2134–2142 (2016)
14. Zhou, D., Chen, L., He, Y.: An unsupervised framework of exploring events on twitter: filtering, extraction and categorization. In: Proceedings of 29th AAAI Conference on Artificial Intelligence, pp. 2468–2475. AAAI, USA (2015)
15. Zhou, X., Chen, L.: Event detection over twitter social media streams. VLDB J. Int. J. Very Larg. Data Bases **23**(3), 381–400 (2014)

Model the Dynamic Evolution of Facial Expression from Image Sequences

Zhaoxin Huan and Lin Shang[✉]

Department of Computer Science and Technology,
State Key Laboratory for Novel Software Technology,
Nanjing University, Nanjing, Jiangsu, China
huanzhx@smail.nju.edu.cn, shanglin@nju.edu.cn

Abstract. Facial Expression Recognition (FER) has been a challenging task for decades. In this paper, we model the dynamic evolution of facial expression by extracting both temporal appearance features and temporal geometry features from image sequences. To extract the pairwise feature evolution, our approach consists of two different models. The first model combines convolutional layers and temporal recursion to extract dynamic appearance features from raw images. While the other model focuses on the geometrical variations based on facial landmarks, in which a novel 2-distance representation and resample technique are also proposed. These two models are combined by weighted method in order to boost the performance of recognition. We test our approach on three widely used databases: CK+, Oulu-CASIA and MMI. The experimental results show that we achieve state-of-the-art accuracy. Moreover, our models have minor-setup parameters and can work for the variable-length frame sequences input, which is flexible in practical applications.

1 Introduction

Facial Expression Recognition (FER) gives machines the ability to perceive and understand human emotions. Recognizing human emotions has been the focus of attention in computer vision and pattern recognition [6, 12, 15].

With the rapid development of deep learning, Convolutional Neural Networks (CNNs) are used in facial expression recognition recently [15]. However, traditional CNNs can not model the sequential data such as the consecutive frames. While another deep learning model Recurrent Neural Networks (RNNs) are good at learning relations from sequence inputs [2, 7], it is encouraging to combine CNNs and RNNs for extracting the spatial information from image frames as well as the temporal features between consecutive frames. Such jointly trained convolutional and recurrent networks are promising for many temporal classification tasks such as video description and activity recognition [4].

Noticing that facial expressions are actually tiny activities which perform on face, in this work, we propose a new neural networks called Facial Appearance Convolutional Recurrent Network (FACRN) which combines convolutional layers

© Springer International Publishing AG, part of Springer Nature 2018
D. Phung et al. (Eds.): PAKDD 2018, LNAI 10938, pp. 546–557, 2018.
https://doi.org/10.1007/978-3-319-93037-4_43

and temporal recursion to extract temporal appearance features and model the dynamic evolution from consecutive frames. Specifically, in our FACRN, a simple CNN is used to extract appearance features from each frame among the whole sequence. Then RNN takes the features extracted by CNN as input and learns the temporal information between different frames. Finally, the output of RNN passes through a fine-tuning average pooling layer to give the prediction of facial expression.

Moreover, facial expression recognition is not totally same as other image classification problems. In particular, the dynamic geometrical variations of several key areas such as eyes, mouth and eyebrows play important roles in facial expression recognition. Therefore, capturing the geometry features of different key areas becomes the main challenge [1,3].

In this work, we propose another deep neural network – Facial Geometry Recurrent Network (FGRN) which focuses on the dynamic geometrical variations of facial key areas based on facial landmarks. First, we use a new 2-distance representation method to normalize the facial landmarks, which can eliminate the effect of out-of-plane rotations. Then the facial landmarks are resampled into two groups based on the facial physical structure. In our FGRN, we use two independent RNN subnets to take the resampled two parts of landmarks as input and extract the low-level geometry features of facial landmarks. These features concatenate in the upper layers to form the global high-level features.

We use weighted summation to combine FACRN and FGRN in order to boost the performance of recognition. The main contributions of our work can be listed as follows:

1. We model the dynamic evolution of facial expression from consecutive frames by proposing two deep network models: Facial Appearance Convolutional Recurrent Network (FACRN) and Facial Geometry Recurrent Network (FGRN).
2. In order to normalize the facial landmarks and extract more powerful local features, we propose a new 2-distance representation and resample technique.
3. Our model has moderate parameters and can handle the variable-length frame sequences input, which is flexible in practical applications.

The remainder of this paper is structured as follows. We present some related work about Facial Expression Recognition in Sect. 2. Section 3 explains our FACRN and FGRN models in detail and Sect. 4 gives an introduction of the databases and results in our experiments. Section 5 concludes the paper.

2   Related Work

Recent CNN-based approaches have been studied in the field of FER. These methods only use single CNN to capture the appearance features without modeling the temporal information of image sequences. 3DCNN-DAP [11] utilized a deformable parts learning component in 3DCNN framework to capture the expression features from video sequences. Jung et al. [8] proposed 3D filters in a small CNN to capture the dynamical variations of appearance. These 3D-based methods take temporal information into account, but they are confined to fixed and preset inputs which is not flexible in practical application. Moreover, 3D-based methods consume a lot of computing resources, which spend much time during training. Therefore, to address these problems, we propose FACRN method which combines CNNs and RNNs in order to extract the appearance features from still frames as well as learning the temporal features between consecutive frames. Besides, the parameters in our FACRN are moderate, and it is an end-to-end trainable deep neural network without being constrained to fixed length inputs.

Facial landmark, which is another important factor in facial expression recognition, has also been studied in recent years. Daniel et al. [1] proposed the geometric-based descriptor to model the triangle formed by facial landmarks. Paul et al. [3] analysed the trajectories of facial landmarks from different expression frames. Jung et al. [8] utilized a small DNN to capture the dynamical variations of facial landmarks. Inspired by these method, the FGRN we proposed also focuses on facial landmarks which reflect the dynamic variation of several key facial parts. Moreover, in order to eliminate the effect of out-of-plane rotation in two-dimensional images, we preprocess facial landmarks by using a new 2-distance representation method. We also resample the facial landmarks into two groups based on facial physical structure, which can learn more powerful local features.

3   Our Approach

Our approach consists of two independent deep neural networks: Facial Appearance Convolutional Recurrent Network (FACRN) and Facial Geometry Recurrent Network (FGRN). Specifically, FACRN combines convolutional layers and temporal recursion to extract temporal appearance features from frame sequences. FGRN focuses on geometrical variations of facial key areas based on facial landmarks. The outputs of the FACRN and FGRN are integrated to boost the performance of expression recognition. Our model is shown in Fig. 1.

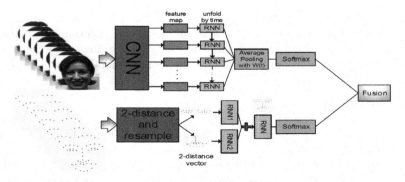

Fig. 1. The architecture of our model

3.1   Facial Appearance Convolutional Recurrent Network

CNN for Extracting Appearance Features. Some powerful CNN structures which are mainly used for image classification have been proposed in [17]. Unlike these methods, the classes and frame sequences in facial expression databases are insufficient. If directly using some well-known CNN models such as GoogleNet [17] for facial expression recognition, it can easily fall into overfitting when training.

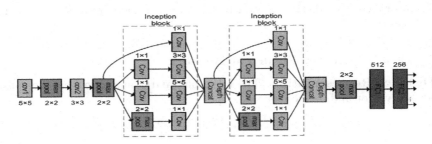

Fig. 2. CNN for extracting appearance features

To address this problem, in this work, we propose a CNN model with a moderate depth and a moderate number of parameters which is shown in Fig. 2. Specifically, first the network contains two convolution layers and pooling layers. Then we apply two Inception layers proposed by [17]. The benefit of Inception layer in this work is that the smaller convolution filters can improve the recognition of local features such as eyes and mouth. Noticing that the CNN we used is to extract the appearance features of each frame rather than classification, we remove the classification layer after the last full-connected layer.

RNN for Extracting Temporal Features. Seeing that we can view the variation of facial expression as an image sequence from neutral expression to peak expression, RNN can be modeled in accordance with people's understanding behavior of a facial expression. Therefore, after CNN extracts appearance features from each frame, instead of using 3D filters to extract temporal information like [8], we use RNN to model the temporal relations between expression frames. Moreover, in [6] a CRF module was added to extract the temporal information. But the model is a two-step network. On contrast, by applying the RNN for modeling consecutive frames in vision problems, our approach can jointly train convolutional and recurrent networks in order to make the FACRN end-to-end trainable. Another advantage of RNN in our FACRN is that it is not confined to fixed length inputs allowing simple modeling for sequential data of varying lengths such as the expression frames in videos.

The RNN structure is shown in Fig. 1. The inputs of RNN are appearance features extracted by CNN of each frame. In order to average the predictions of each time step for final classification, we propose an average pooling layer with a weight function $W(t)$:

$$z_i = \frac{1}{N} * \sum_{t=1}^{N} h_{ti} * W(t), \tag{1}$$

where $H_i = (h_{t1}, h_{t2}, ..., h_{tn})$ is the output of RNN at time t, $Z = (z_1, z_2, ..., z_n)$ is the result of average pooling. $W(t)$ is a weight function which linearly increases from 0...1 over frames $t = 0...T$. By applying $W(t)$ in pooling layer, we emphasize the importance of prediction at later frames in which hidden units capture more information.

3.2   Facial Geometry Recurrent Network

2-Distance Representation. The facial landmarks at frame t can be considered as a vector and defined as:

$$L^{(t)} = [x_1^{(t)}, y_1^{(t)}, x_2^{(t)}, y_2^{(t)}, ..., x_n^{(t)}, y_n^{(t)}], \tag{2}$$

where n is the total number of landmark points in a facial expression frame, $x_i^{(t)}$ and $y_i^{(t)}$ are coordinates of the i-th facial landmark points at frame t.

Directly using these coordinates to extract the dynamical variations of facial landmarks is inappropriate and difficult to process. When there is an out-of-plane rotation in two-dimensional images, the absolute xy-coordinates are no longer accurate. To address this problem, we propose a **2-distance** representation method for facial landmarks.

First, we define a reference point of all facial landmarks, which is the xy-coordinates between the eyebrows. The **1-distance** refers to the Euclidean Distance between each facial landmark and reference point as following formulas:

$$d1_i^{(t)} = \sqrt{(x_i^{(t)} - x_0^{(t)})^2 + (y_i^{(t)} - y_0^{(t)})^2} \tag{3}$$

$$D1^{(t)} = [d1_1^{(t)}, d1_2^{(t)}, ..., d1_n^{(t)}], \tag{4}$$

where $d1_i^{(t)}$ is the Euclidean Distance between i-th facial landmark point and the reference point x_0 at frame t.

The **1-distance** eliminate the effect of out-of-plane rotations in two-dimensional images. But in practical, the position of the five facial organs varies from person to person. Therefore, the **1-distance** can not regulate these physical variations. Based on **1-distance**, we proposed the **2-distance** representation of facial landmarks. Specifically, after getting the **1-distance** vectors $D1^{(t)}$ at t frame, we calculate the difference between $D1^{(t)}$ and $D1^{(t-1)}$:

$$D2^{(t)} = D1^{(t)} - D1^{(t-1)} \quad t = 2, 3, ..., n \tag{5}$$

By using **2-distance** representation method, no matter how the position of facial organs varies from person to person, we only capture the dynamical variations of facial landmarks between frames.

Resample Technique and RNN Subnets. We propose a resample technique and two RNN subnets for partly extracting the dynamic geometrical variations. According to the physical structure of faces, we find that the eyes and eyebrows usually move together when perform an expression, and so are the mouths and noses. Therefore, we resample the facial landmarks into two groups: eyebrows and eyes, nose and mouth. Our FGRN is shown in the lower part of Fig. 1. The two resampled parts of landmarks are input into two independent RNN subnets in order to extract local low-level features. The whole facial high-level features are formed in the upper layers.

3.3 Integration Method

The outputs of FACRN and FGRN are integrated as follows:

$$o_i = \varepsilon a_i + (1 - \varepsilon)g_i \quad 0 \leq \varepsilon \leq 1, \tag{6}$$

where o_i is the output of the entire network, a_i is the output of FACRN, and g_i is the output of FGRN. The ε is a coefficient used to balance the performance of each network. For all the experiments, we set ε to 0.5. $i = 1, 2, ..., 7$ is the total number of emotion class.

4 Experiments

In this section, we evaluate our model on three widely used databases: CK+ [13], MMI [19] and Oulu-CASIA [18]. For all experiments, we use the 10-fold validation protocol.

4.1 Databases

CK+. This database has 327 image sequences with seven emotion labels: happiness, anger, surprise, fear, disgust, sadness and contempt.

Oulu-CASIA. The database includes 480 image sequences with 80 subjects taken under normal illumination conditions.

MMI. A database with 205 facial image sequences and only about 30 subjects. The size of MMI database is much smaller than CK+ and Oulu-CASIA, which is challenging to use the deep learning method.

4.2 Evaluation of the FACRN

In order to determine the optimal RNN structure in our FACRN, we compare three different models in our experiment, including standard RNN model, LSTM [7] and GRU [2]. Intuitively, the curves in Fig. 3 show the test accuracies of three

RNN models during one training process. From Fig. 3 we find that the convergence rate of LSTM is slower than the standard RNN and GRU, in large part because of the more network parameters in LSTM training. Unlike other video classification tasks [4], the number of frames in facial expression databases are normally less than 20. Therefore, the weakness of standard RNN for capturing long-term dependencies does not come out in this work. But we do not know the exact number of frames extracted from facial expression videos in practice and the number of frames will be more than 20 compared with the databases. Therefore, we can use LSTM or GRU in the FACRN in order to avoid gradients vanish (Fig. 3).

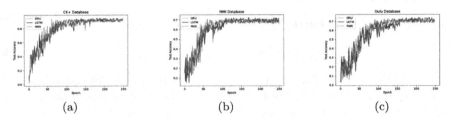

Fig. 3. Test accuracies with different RNN models. (a) CK+. (b) MMI. (c) Oulu-CASIA

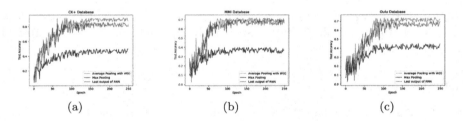

Fig. 4. Test accuracies with different pooling. (a) CK+. (b) MMI. (c) Oulu-CASIA

We also conduct experiments to evaluate our pooling methods, including average pooling with a weight function $W(t)$, max pooling and directly using the last output of RNN without pooling. The result is shown in Fig. 4. From the curves we can clearly see that the average pooling method with $W(t)$ in FACRN performs better.

4.3 Evaluation of the FGRN

To show the 2-distance representation and resample technique have a promoting effect on capturing dynamic geometrical variations, we design an experiment to compare FGRN with other two models. The first model inputs the absolute coordinates of facial landmarks to a simple DNN. The second model transforms the facial landmarks into 2-distance vectors but use a single RNN without resampling the landmarks into two groups.

Figure 5 shows the results of these models on three databases. Since our FGRN and the second model preprocess the facial landmarks by using 2-distance vectors, these two models can obtain higher accuracies than the first model, which shows the effectiveness of 2-distance representation for normalizing the facial landmarks. Moreover, the performance of FGRN is better than the second model, which indicates that, by resampling the facial landmarks into two groups and using two RNN subnets to extract the local low-level features, it can achieve significant performance for modeling dynamic geometrical variations. Because the parameters in our FGRN are much more than the other two models. From Fig. 5 we also notice that FGRN has relatively slow convergence rate.

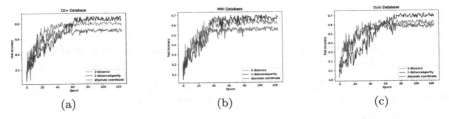

Fig. 5. (a) (b) (c) show the test accuracies with different methods of facial landmarks

4.4   Classification Results

Accuracies and Error Analysis. Tables 1, 2 and 3 show the recognition accuracies of our model achieved on each database and comparison with the state-of-the-art methods. Specifically, we mainly compare our model with the hand-crafted methods and 3D-based models. For the CK+ and Oulu-CASIA databases, our proposed model outperform the previous methods which can be seen from Tables 1 and 2. The parameters in our model are moderate compared with 3D-based methods. Besides, our model is not confined to fixed length inputs as described in Sect. 3.1, which is more flexible in practical applications.

However, for MMI database, the number of subjects and frame sequences is much smaller than that in the other two databases, which means that it is especially not competent for deep learning methods. Moreover, the frame sequences in MMI database are also different from the other two databases. Each subject starts and ends with a neutral facial expression. The peak expressions of each emotion are in the middle of the frame sequence. This characteristic will limit the performance of our model, because the RNN and pooling methods we proposed can perform better in capturing the unidirectional variations rather than modeling bidirectional evolution.

It is worth mentioning that after integrating our FACRN and FGRN, the performance reaches the highest. In other words, FACRN and FGRN are com-

Table 1. CK+ database   **Table 2.** MMI database   **Table 3.** Oulu database

Method	Accuracy
HOG-3D [9]	91.44
SPTS+CAPP [13]	83.33
ITBN [20]	86.31
CRF [6]	93.04
3DCNN [9]	85.9
Cov3D [15]	92.3
STM-ExpLet [12]	94.19
3DCNN-DAP [9]	92.4
FACRN	**91.78**
FGRN	**92.15**
Integration	**95.63**

Method	Accuracy
3D SIFT [16]	64.39
HOG-3D [9]	60.89
ITBN [20]	59.7
3DCNN [9]	53.2
3DCNN-DAP [9]	63.4
AUDN [10]	74.76
STM-ExpLet [12]	**75.12**
PCA-SR [14]	**78.51**
FACRN	71.54
FGRN	72.69
Integration	75.0

Method	Accuracy
LBP-TOP [21]	68.13
HOG-3D [9]	70.63
AdaLBP [21]	73.54
Atlases [5]	75.52
STM-ExpLet [12]	74.59
3D SIFT [16]	55.83
FACRN	**74.18**
FGRN	**73.24**
Integration	**76.50**

Table 4. Oulu database

	An	Di	Ha	Sa	Fe	Su
An	**70.5**	15.4	2.5	2.5	9.1	0
Di	19.3	**69.2**	0	3.8	7.7	0
Ha	0	0	**89**	0	0	11
Sa	12.8	0	0	**81.2**	6	0
Fe	0	0	5.8	8.6	**73.6**	12
Su	0	6.3	0	0	18.1	**75.6**

Table 5. MMI database

	An	Di	Ha	Sa	Fe	Su
An	**58.3**	32.5	0	9.2	0	0
Di	23	**65.1**	0	11.9	0	0
Ha	0	0	**95.4**	0	0	4.6
Sa	13.2	2.1	0	**69**	15.7	0
Fe	0	0	10.7	10.7	**73.2**	5.4
Su	0	0	20	0	5.7	**74.3**

plementary to each other. FACRN has the ability to extract the temporal appearance features from raw frame sequences, while FGRN can model the dynamic geometrical variations based on facial landmarks. As a result, integrating FACRN and FGRN can boost the performance of facial expression recognition.

Confusion Matrix. Tables 4, 5 and 6 show the resulting confusion matrices of our model on three databases. It can be seen that our model achieved high recognition accuracies of each emotion. In particular, our model performs well on happiness (Ha), surprise (Su), sad (Sa) and fear (Fe). The high confusion in contempt (Co) expression can be caused by the few number of contempt sequences in CK+ database. Another reason is that the geometrical variations of contempt expressions are slight which creates more confusions for FGRN.

Intuitively, we visualize the 2-distance vectors between the first and the last frames which are shown in Fig. 67. The 2-distance vectors of six basic expression are obviously different from each other except the disgust (Di) and angry (An), which leads to the high confusion in these two emotions. This is because that the motions of facial key areas of these two expressions are similar based on 2-distance representation.

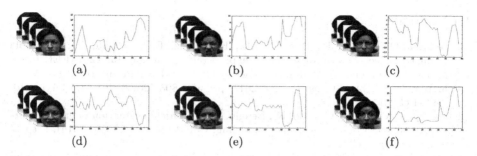

(a)   (b)   (c)

(d)   (e)   (f)

Fig. 6. 2-distance vector of (a) angry (b) disgust (c) sad (d) fear (e) happy (f) surprise

Table 6. CK+ database

	An	Co	Di	Ha	Sa	Fe	Su
An	**94.1**	0	4.2	0	1.7	0	0
Co	8.4	**89.4**	0.7	0	1.5	0	0
Di	5.1	0	**93.8**	0	0	0	1.1
Ha	0	0	0	**97.3**	0	0	2.7
Sa	0	0	1.7	0	**95.6**	0	2.7
Fe	2	0.7	0	0	3.1	**94.2**	0
Su	0	0	0	0.9	0	3.6	**95.5**

5   Conclusion

In this paper, two deep networks are presented in order to model the dynamic evolution of facial expression from image sequences. FACRN combines convolutional layers and temporal recursion to extract temporal appearance features from raw frame sequences. FGRN focuses on dynamic geometrical variations of facial expressions based on facial landmarks. We also proposed a new 2-distance representation and resample technique for normalizing the facial landmarks and extracting more powerful local features. We evaluated two deep network models on CK+, MMI, and Oulu-CASIA databases respectively. Meanwhile, we showed the test accuracies and the recognition accuracy of each emotion based on confusion matrix. Furthermore, we analysed that our model has much fewer network parameters than some 3D-based methods and can handle the variable-length frame sequence input, which is more flexible in practical applications.

Acknowledgements. This work is supported by the National Natural Science Foundation of China (No. 61672276) and Natural Science Foundation of Jiangsu, China (BK20161406).

References

1. Acevedo, D., Negri, P., Buemi, M.E., Fernández, F.G., Mejail, M.: A simple geometric-based descriptor for facial expression recognition. In: 12th IEEE International Conference on Automatic Face & Gesture Recognition, FG 2017, Washington, DC, USA, 30 May–3 June 2017, pp. 802–808 (2017). https://doi.org/10.1109/FG.2017.101
2. Chung, J., Gülçehre, Ç., Cho, K., Bengio, Y.: Empirical evaluation of gated recurrent neural networks on sequence modeling. CoRR abs/1412.3555 (2014). http://arxiv.org/abs/1412.3555
3. Desrosiers, P.A., Daoudi, M., Devanne, M.: Novel generative model for facial expressions based on statistical shape analysis of landmarks trajectories. In: 23rd International Conference on Pattern Recognition, ICPR 2016, Cancún, Mexico, 4–8 December 2016, pp. 961–966 (2016). https://doi.org/10.1109/ICPR.2016.7899760
4. Donahue, J., Hendricks, L.A., Guadarrama, S., Rohrbach, M., Venugopalan, S., Darrell, T., Saenko, K.: Long-term recurrent convolutional networks for visual recognition and description. In: IEEE Conference on Computer Vision and Pattern Recognition, CVPR 2015, Boston, MA, USA, 7–12 June 2015, pp. 2625–2634 (2015). https://doi.org/10.1109/CVPR.2015.7298878
5. Guo, Y., Zhao, G., Pietikäinen, M.: Dynamic facial expression recognition using longitudinal facial expression atlases. In: Computer Vision - ECCV 2012–12th European Conference on Computer Vision, Florence, Italy, 7–13 October 2012, Proceedings, Part II, pp. 631–644 (2012). https://doi.org/10.1007/978-3-642-33709-3_45
6. Hassani, B., Mahoor, M.H.: Spatio-temporal facial expression recognition using convolutional neural networks and conditional random fields. In: 12th IEEE International Conference on Automatic Face & Gesture Recognition, FG 2017, Washington, DC, USA, 30 May–3 June 2017, pp. 790–795 (2017). https://doi.org/10.1109/FG.2017.99
7. Hochreiter, S., Schmidhuber, J.: Long short-term memory. Neural Comput. $9(8)$, 1735–1780 (1997). https://doi.org/10.1162/neco.1997.9.8.1735
8. Jung, H., Lee, S., Yim, J., Park, S., Kim, J.: Joint fine-tuning in deep neural networks for facial expression recognition. In: 2015 IEEE International Conference on Computer Vision, ICCV 2015, Santiago, Chile, 7–13 December 2015, pp. 2983–2991 (2015). https://doi.org/10.1109/ICCV.2015.341
9. Kläser, A., Marszalek, M., Schmid, C.: A spatio-temporal descriptor based on 3D-gradients. In: Proceedings of the British Machine Vision Conference 2008, Leeds, September 2008, pp. 1–10 (2008). https://doi.org/10.5244/C.22.99
10. Liu, M., Li, S., Shan, S., Chen, X.: Au-aware deep networks for facial expression recognition. In: 10th IEEE International Conference and Workshops on Automatic Face and Gesture Recognition, FG 2013, Shanghai, China, 22–26 April 2013, pp. 1–6 (2013). https://doi.org/10.1109/FG.2013.6553734
11. Liu, M., Li, S., Shan, S., Wang, R., Chen, X.: Deeply learning deformable facial action parts model for dynamic expression analysis. In: Computer Vision - ACCV 2014 - 12th Asian Conference on Computer Vision, Singapore, Singapore, 1–5 November 2014, Revised Selected Papers, Part IV, pp. 143–157 (2014). https://doi.org/10.1007/978-3-319-16817-3_10
12. Liu, M., Shan, S., Wang, R., Chen, X.: Learning expressionlets on spatiotemporal manifold for dynamic facial expression recognition. In: 2014 IEEE Conference on Computer Vision and Pattern Recognition, CVPR 2014, Columbus, OH, USA, 23–28 June 2014, pp. 1749–1756 (2014). https://doi.org/10.1109/CVPR.2014.226

13. Lucey, P., Cohn, J.F., Kanade, T., Saragih, J.M., Ambadar, Z., Matthews, I.A.: The extended cohn-kanade dataset (CK+): a complete dataset for action unit and emotion-specified expression. In: IEEE Conference on Computer Vision and Pattern Recognition, CVPR Workshops 2010, San Francisco, CA, USA, 13–18 June 2010, pp. 94–101 (2010). https://doi.org/10.1109/CVPRW.2010.5543262

14. Mohammadi, M.R., Fatemizadeh, E., Mahoor, M.H.: Pca-based dictionary building for accurate facial expression recognition via sparse representation. J. Vis. Commun. Image Represent. **25**(5), 1082–1092 (2014). https://doi.org/10.1016/j.jvcir.2014.03.006

15. Sanin, A., Sanderson, C., Harandi, M.T., Lovell, B.C.: Spatio-temporal covariance descriptors for action and gesture recognition. In: 2013 IEEE Workshop on Applications of Computer Vision, WACV 2013, Clearwater Beach, FL, USA, 15–17 January 2013, pp. 103–110 (2013). https://doi.org/10.1109/WACV.2013.6475006

16. Scovanner, P., Ali, S., Shah, M.: A 3-dimensional sift descriptor and its application to action recognition. In: Proceedings of the 15th International Conference on Multimedia 2007, Augsburg, Germany, 24–29 September 2007, pp. 357–360 (2007). https://doi.org/10.1145/1291233.1291311

17. Szegedy, C., Liu, W., Jia, Y., Sermanet, P., Reed, S.E., Anguelov, D., Erhan, D., Vanhoucke, V., Rabinovich, A.: Going deeper with convolutions. In: IEEE Conference on Computer Vision and Pattern Recognition, CVPR 2015, Boston, MA, USA, 7–12 June 2015, pp. 1–9 (2015). https://doi.org/10.1109/CVPR.2015.7298594

18. Taini, M., Zhao, G., Li, S.Z., Pietikäinen, M.: Facial expression recognition from near-infrared video sequences. In: 19th International Conference on Pattern Recognition (ICPR 2008), Tampa, Florida, USA, 8–11 December 2008, pp. 1–4 (2008). https://doi.org/10.1109/ICPR.2008.4761697

19. Valstar, M., Pantic, M.: Induced disgust, happiness and surprise: an addition to the MMI facial expression database. In: Proceedings International Workshop on Emotion Corpora for Research on Emotion & Affect, pp. 65–70 (2010)

20. Wang, Z., Wang, S., Ji, Q.: Capturing complex spatio-temporal relations among facial muscles for facial expression recognition. In: 2013 IEEE Conference on Computer Vision and Pattern Recognition, Portland, OR, USA, 23–28 June 2013, pp. 3422–3429 (2013). https://doi.org/10.1109/CVPR.2013.439

21. Zhao, G., Huang, X., Taini, M., Li, S.Z., Pietikäinen, M.: Facial expression recognition from near-infrared videos. Image Vis. Comput. **29**(9), 607–619 (2011). https://doi.org/10.1016/j.imavis.2011.07.002

Unsupervised Disaggregation of Low Granularity Resource Consumption Time Series

Pantelis Chronis[1,2]([✉]), Giorgos Giannopoulos[2], Spiros Athanasiou[2], and Spiros Skiadopoulos[1]

[1] Department of Informatics and Telecommunications, University of Peloponnese, Tripoli, Greece
chronis@uop.gr
[2] Athena Research Center, Athens, Greece

Abstract. Resource consumption is typically monitored at a single point that aggregates all activities of the household in one time series. A key task in resource demand management is disaggregation; an operation that decomposes such a composite time series in the consumption parts that comprise it, thus, extracting detailed information about how and when resources were consumed. Current state-of-the-art disaggregation methods have two drawbacks: (a) they mostly work for frequently sampled time series and (b) they require supervision (that comes in terms of labelled data). In practice, though, sampling is not frequent and labelled data are often not available. With this problem in mind, in this paper, we present a method designed for unsupervised disaggregation of consumption time series of low granularity. Our method utilizes a stochastic model of resource consumption along with empirical findings on consumption types (e.g., average volume) to perform disaggregation. Experiments with real world resource consumption data demonstrate up to 85% Recall in identifying different consumption types.

1 Introduction

Resource conservation, concerning for instance water, energy or fuel, is an important challenge for modern societies. Monitoring and analysing the consumption of resources is a valuable tool in developing resource conservation policies. Analysis of consumption time series includes several tasks, one of which, disaggregation, is the focus of this paper.

Disaggregation is the process of analysing a composite time series into the individual components that it consists of. In the case of resource consumption, the composite time series consists of several discrete *consumption types*. As an example of resource disaggregation we consider a household's water consumption: A household's water consumption is measured at the main supply where the consumption of all the various consumption types (e.g., clothes-washing, showering) is aggregated. The goal of a disaggregation algorithm would be to identify when a shower was taken, when the washing-machine was being used, etc.

© Springer International Publishing AG, part of Springer Nature 2018
D. Phung et al. (Eds.): PAKDD 2018, LNAI 10938, pp. 558–570, 2018.
https://doi.org/10.1007/978-3-319-93037-4_44

An important property that affects disaggregation performance is the relation between consumption type duration and measurement interval length. When the measurement interval is smaller than the expected consumption type duration, disaggregation can be effectively approached as a pattern recognition problem, since there are enough measurements for the pattern of each consumption type to be identified. However, if the measurement interval is equal or larger than the duration of a consumption type, an occurrence of a consumption type can start and finish inside the interval of a single measurement. This means that the pattern of the consumption type is essentially lost. This makes the disaggregation problem much more challenging. An example of this is presented in Fig. 1. Generally, in residential resource consumption, major consumption types have durations ranging from several minutes to 2–3 h. Thus, in this setting, time series with measurement interval of 15 min or larger (e.g., 30 min, 1 h) can be considered of *low granularity*. In practice, resources, especially water, are measured at low granularity. The reasons for this are limitations of the sensors, usually due to battery life, and increased infrastructure costs required for the transmission of high frequency measurements from a very large number of sensors. However, very few works have handled the problem of resource disaggregation in low granularity data.

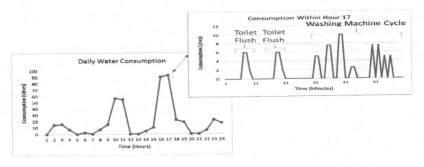

Fig. 1. An example of a group of patterns aggregated within an one hour measurement. In this example two toilet flushes and a washing machine cycle are aggregated within the measurement of 17:00 h.

Another shortcoming of most existing disaggregation algorithms is that they need to be trained on a labelled dataset [1,6,16]. This means they require a dataset with time series of each consumption type measured *separately* and labelled, so that the algorithm can learn to identify its pattern. However, gathering such datasets requires costly and intrusive measurement trials.

All existing algorithms have one or both of the above requirements, which makes them unsuitable for many real world applications. Motivated by this, we present a disaggregation method for low granularity time series that does not require a labelled dataset. Our method is based on a stochastic model of resource consumption that we have developed. Instead of labelled data, it requires approximate assumptions concerning the volume, frequency and usual time of occurrence of each consumption type. These assumptions are simple and intuitive and can be retrieved from the literature [12–15], provided by experts or be requested from the users. Utilizing those assumptions and the stochastic model, our method calculates the probability that each consumption type has occurred at each time.

In order to thoroughly evaluate our algorithm, we test it in two datasets: one consisting of residential water consumption data, and one consisting of residential energy consumption data, both of hourly granularity. Our algorithm achieves good performance on both datasets.

The rest of the paper is organized as follows: in Sect. 2 we present related work, in Sect. 3 we describe and formulate our method and, finally, in Sect. 4 we present the experimental evaluation.

2   Related Work

Reviewing the literature on time series disaggregation, we can discriminate existing work into two categories. The high granularity algorithms, that are designed for data with measurements intervals ranging from milliseconds up to one minute and the low granularity algorithms, that can be applied to data with measurement intervals from several minutes to several hours. Our distinction between low and high granularity is based on the relation of the measurement interval to the average consumption type duration, as we described in Sect. 1.

In the high granularity setting [2,3,7–11], existing works usually scan the time series to identify significant *step changes* in consumption that indicate the start or the end of a specific consumption event. Then, using a Machine Learning model and labelled data, they identify the consumption types each consumption event. Existing methods mainly vary in the adopted Machine Learning model. In [9], the authors use a Hidden Markov Model (HMM) to classify the events. [8] use a convex optimization approach, similar to Support Vector Machines (SVM), and [11] use a Neural Network. There also exist a few unsupervised approaches in the setting of high granularity data. In [3], the authors model consumption using an extension of the HMM. Instead of labelled data, they use detailed assumptions about each device's consumption pattern and usage. They evaluate their algorithm on data with granularity of 3 s. In the same line of work, [2] use a version of HMM to perform disaggregation on time series with 1 min granularity. The algorithm does not require labelled data, however, it requires information concerning each device's exact consumption pattern. We note that HMM models are successful in high granularity time series, where the transitions between different operating stages of an appliance are detectable in the time series pattern. On the contrary, those transitions are, generally, not detectable in low granularity data.

In the setting of low granularity data [1,6], the general approach is to model the aggregate time series as a sum of separate components, that represent the various consumption types, and apply an optimization algorithm in order to decompose the time series into those components. In [1], the authors apply Sparse Coding, a model that allows the combination of a large number of basis functions by imposing sparsity constraints, on low granularity (15 min) time series. [6] use the same idea as [1], but modify the algorithm to work iteratively, disaggregating only one consumption type in each iteration. These approaches also require a set of labelled time series for the various consumption types, which they use as basis functions.

3   Model Formulation and Disaggregation Algorithms

In this section, we present our method. We start by providing an intuitive overview and, then, we describe the details of the method.

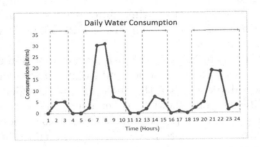

Fig. 2. A time series of the water consumption of a day, for a single household, divided in its major consumption events.

We can obtain an intuitive understanding of the challenge by looking at an example. Given a consumption time series like the one depicted in Fig. 2, we aim at identifying the occurrences of each consumption type in time. There are two sources of information about the occurrence of a consumption type: (i) The *footprint* that the occurrence leaves on the aggregate time series. For example, in a water consumption time series, if a shower is taken at some time, we would expect to observe consumption of around 50 L or more at that time; (ii) The external information we have about each consumption type: how frequently it occurs and at which hours within a day.

Our disaggregation approach is based on developing a model for the consumption behaviour of the household that incorporates the aforementioned sources of information and which we can use to infer the occurrence of the various consumption types. In order for disaggregation to be performed effectively, the model needs to capture the structure of the problem, use only available information and handle the variability of human behaviour. The model is based on the assumption that there exists a set of major *consumption types*, each of which is represented by an amount of consumption, an expected time and a frequency of occurrence. The model is stochastic in nature since: (i) These quantities may randomly vary between different occurrences of the consumption type and between different households; (ii) There is inherent uncertainty in our estimates of these quantities for each household. The model and the disaggregation process are described in detail in the next subsection.

3.1 Method Description

Events Identification. The first issue that arises is that an occurrence of a consumption type may be divided in more than one measurements. For example, if an activity started at 09:50 and finished at 10:10, its consumption would be distributed in two consecutive measurements. However, if we select a part of the time series that starts and ends with (near) zero consumption, given the assumption that no consumption type can have a pause of one hour or greater, we can be certain that all consumption types that started inside this interval have also ended, i.e., this interval comprises only complete consumption types. We refer to those parts of the time series, that start and end at *near zero* consumption, as *consumption events* or just *events*. Figure 2 shows an example of a day's consumption events.

Thus, our first step is to identify the distinct consumption events. To achieve this, we sequentially scan the time series and isolate the sequences of all *consecutive* points whose value exceeds a threshold θ_e, above which the consumption is considered significant. Threshold θ_e is set using the assumptions about the volume of each consumption type, so that it is only exceeded if some consumption type is occurring. We denote as $t_{j_{start}}, t_{j_{end}}$ the times of start and end of consumption event j.

Model Description. We consider a set of n major consumption types. For each consumption type $i, 1 \leq i \leq n$, we assume that its total consumption c_i is distributed according to a normal distribution with mean μ_i and standard deviation σ_i. We treat all minor consumption types as *background* noise and model them using variable b with mean μ_b and standard deviation σ_b. We denote the probability of consumption type i occurring at time h as τ_{ih}. In order to limit the computational complexity of the model, we make the assumption that each occurrence of a consumption type is only affected by other occurrences of the same consumption type within a specified *time period*. For most cases, this period would be a *day* or *a week* (e.g., people tend to shower once a day or use the washing machine two or three times a week). We denote as v_{ik} the probability that consumption type i occurs k times in the duration of the predefined time period. We also define K as the maximum value of k. For simplicity, we do not include in the model any dependencies between the total number of occurrences of different consumption types, i.e., v_{ik} is independent of $v_{lk}, l \neq i$.

Given a set of m consumption events, each event $j, 1 \leq j \leq m$ is represented by the following: the time it started $t_{j_{start}}$, the time it ended $t_{j_{end}}$ and its total consumption s_j. We denote as It_j a vector containing both $t_{j_{start}}$ and $t_{j_{end}}$, which defines the time interval of event j. The total consumption of event j is the result of the aggregation of each major consumption type i occurring x_{ij} times, plus the background noise:

$$s_j = \sum_{i=1}^{n} c_i x_{ij} + b \tag{1}$$

with $x_{ij} \in \mathbb{N}$. As x_{*j} we denote a vector $[x_{1j}, \cdots, x_{mj}]$, that contains the number of occurrences of every consumption types in the interval of event j. All notation is gathered in Table 1 for convenience.

The probability of occurrence of consumption types x_{*j} in an event j depends on the total consumption of the event s_j, the time of the event It_j and all other occurrences of consumptions of the same type x_{*l}, inside the time period. For further analysis, it is convenient to formulate these dependencies into a Bayesian Network (BN). For example, the BN of Fig. 3 illustrates a period containing three events. More events may be handled in a similar fashion. We note that the directions of the arrows show the order of decomposition of the joint probability, however the dependencies between the variables are bidirectional [17].

Table 1. Notation

$i, 1 \le i \le n$	index of consumption type	μ_i, σ_i	mean and stand. dev. of c_i
$j, 1 \le m \le m$	index of consumption event	b	volume of background noise
s_j	total consumption of event j	b_μ, b_σ	mean and stand. dev. of b
x_{ij}	occurrences of type i in event j	It_j	interval $(t_{j_{start}}, t_{j_{end}})$ of event j
x_{*j}	all occurrences within event j	τ_{ih}	prob. of cons. type i at hour h
c_i	consumption volume of type i	v_{ik}	prob. of type i occurring k times

The purpose of the disaggregation algorithm is to infer the values of $x_{*j}, 1 \le j \le m$. Since we consider each period to be independent from the others, we can treat each period separately. We assume that period d has m_d consumption events $j, 1 \le j \le m_d$. We denote as $\cap_j x_{*j}$ the joint possibility of all occurrences $x_{*j}, 1 \le j \le m_d$, i.e. the possibility of occurrences $x_{*1}, x_{*2}, \cdots, x_{*m_d}$ happening in the same period. $\cap_j x_{*j}$ contains the occurrences of all consumption types in all the events of a period. $\cap_j s_j$ and $\cap_j It_j$ are defined in the same way.

The probability of $\cap_j x_{*j}$ can be written as:

$$p(\cap_j x_{*j} \mid \cap_j It_j, \cap_j s_j) = \frac{p(\bigcap_j It_j, \bigcap_j s_j \mid \bigcap_j x_{*j}) p(\bigcap_j x_{*j})}{p(\bigcap_j It_j, \bigcap_j s_j)} \tag{2}$$

Given the dependencies modelled by the BN, Eq. (2) is transformed to:

$$p(\cap_j x_{*j} \mid \cap_j It_j, \cap_j s_j) = \frac{p(\bigcap_j x_{*j} \mid \bigcap_j It_j) \prod_j p(s_j \mid x_{*j})}{p(\bigcap_j s_j \mid \bigcap_j It_j)} \tag{3}$$

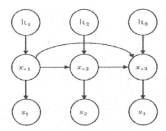

Fig. 3. The Bayesian Network that describes the dependencies within a period comprised of three consumption events. Each consumption type occurrence depends on the observed consumption, the time of day and the previous occurrences in the given period.

From Eq. (1) we have:

$$p(s_j \mid x_{*j}) = Normal(\sum_{i=1}^{n} \mu_i * x_{ij} + \mu_b, \sum_{i=1}^{n} \sigma_i^2 * x_{ij} + \sigma_b^2) \qquad (4)$$

as s_j is a sum of normally distributed variables. The term $p(s_j|x_{*j})$ models the probability of observing consumption s_j, given that consumption types x_{*j} have occurred.

The term $p(\cap_j x_{*j}| \cap_j It_j)$ corresponds to the prior probability of the consumption types occurring at the specific times $\cap_j It_j$, irrespective of the observed $\cap_j s_j$. It depends on the probability that the activities defined by $\cap_j x_{*j}$ occur all in a single period and that the occurrences are distributed accordingly in the time of the observed consumption events. We model it using the assumptions about frequency and time of occurrence of each consumption type:

$$p(\cap_j x_{*j} \mid \cap_j It_j) = \prod_{i=1}^{n} v_{ik_i} * Multinomial(x_{ij}\ \forall j,\ \pi_{ij}\ \forall j) \qquad (5)$$

As defined, the term v_{ik_i} is the probability of consumption type i occurring k times overall. The multinomial distribution models the probability for the consumption types to occur at the specific intervals of the consumption events j. We can break the joint probability into a product because we have assumed that the occurrence of the different consumption types are independent. In Eq. (5), k_i is the total number of occurrences of consumption type i and π_{ij} is the probability of consumption type i occurring in the interval defined by It_j :

$$k_i = \sum_{j=1}^{m_d} x_{ij},\ \pi_{ij} = \sum_{h=t_{j_{start}}}^{t_{j_{end}}} \tau_{ih} \qquad (6)$$

Finally, the denominator of Eq. (3) is the sum of the probability of all possible joint events $\cap_j x_{*j}$ and is constant for all x_{*j}:

$$p(\cap_j s_j \mid \cap_j It_j) = \sum_{\cap_j x_{*j}} p(\cap_j x_{*j} \mid \cap_j It_j) \prod_j p(s_j \mid x_{*j}) \qquad (7)$$

In order to perform the disaggregation, we need to find $\cap_j x_{*j}$ that maximizes the probability of Equation (3):

$$argmax_{\cap_j x_{*j}} p(\cap_j x_{*j} \mid \cap_j It_j, \cap_j s_j) \tag{8}$$

The above maximization problem is too complex to solve exhaustively, since the number of possible combinations grows exponentially. For example, in a case with five events, and five activities that can occur up to five times, there are more than 10^{17} combinations. In order to find a solution for Eq. (8), we implement two different algorithms: a greedy approximation and a Markov Chain Monte Carlo simulation. Next we describe each method.

Algorithm: GREEDYAPPROXIMATION

Input   : $It_j, s_j, 1 \leq j \leq m$
Output   : $\cap_j x_{*j}$ of maximum probability
1 **for** $j = 1$ to m_d **do**
2 $\quad\lfloor \quad x_{*j} = $ EVENTOFMAXPROBABILITY$(It_j, s_j, x_{*1}, \ldots, x_{*j-1})$
3 **return** $\cap_j x_{*j}$

---

1. Greedy Approximation. We start by calculating the most probable occurrences for the first event, ignoring all next events. Then we incrementally calculate the most probable occurrences for following events, given the occurrences of all previously calculated ones. This process is described in Algorithm GREEDYAPPROXIMATION.

The probability of occurrences given the previous events are:

$$p(x_{*j} \mid It_j, \; s_j, \; \cap_{l=1}^{j-1} x_{*l}) = \frac{p(s_j \mid x_{*j}) \cdot p(x_{*j} \mid It_j, \; \cap_{l=1}^{j-1} x_{*l})}{p(s_j \mid It_j, \; \cap_{l=1}^{j-1} x_{*l})} \tag{9}$$

where $p(s_j | x_{*j})$ is calculated as in Eq. (4). The term $p(x_{*j}|It_j, \cap_{l=1}^{j-1} x_{*l})$, is only conditioned on the occurrences of the previous events $\cap_{l=1}^{j-1} x_{*l}$. Since only one event is examined at each step, the binomial distribution is used instead of the multinomial, to calculate the probability of occurrence of each consumption type in the specific time of the event:

$$p(x_{*j} \mid It_j, \; \cap_{l=1}^{j-1} x_{*l}) = \prod_{i=1}^{n} \sum_{k=k_i}^{K} v_{ik} * Binomial(x_{ij}, \; k, \; \pi_{ij}), \quad k_i = \sum_{l=1}^{j} x_{il} \tag{10}$$

For each x_{*j}, we exhaustively search all its possible values and select the one with the maximum probability (Algorithm EVENTOFMAXPROBABILLITY).

Algorithm: EVENTOFMAXPROBABILLITY

Input   : $It_j, s_j, x_{*1}, \ldots, x_{*j-1}$
Output   : x_{*j} of maximum probability

1 $x_{max} = \emptyset$, $p_{max} = 0$
2 **for** *every possible value of* x_{*j} **do**
3 $\quad$ $p = p(x_{*j}|It_j, s_j, \cap_{l=1}^{j-1} x_{*l})$
4 $\quad$ **if** $p > p_{max}$ **then**
5 $\quad$ $\quad$ $p_{max} = p$, $x_{max} = x_{*j}$

6 **return** x_{max}

---

2. Markov Chain Monte Carlo (MCMC).

MCMC is a set of algorithms used to sample from the joint probability distribution described by a Bayesian Network. Given a set of observed variables ($\cap_j s_j, \cap_j It_j$ in our case), we want to sample from the distribution of the unobserved variables ($\cap_j x_{*j}$), in order to find their most probable values. To achieve this we apply the Gibbs sampling algorithm, which sets the observed variables to their observed values, randomly initialises the unobserved variables and sequentially updates the value of each unobserved variable with a value sampled from its conditional distribution, conditioned on all other variables. The conditional distribution of x_{*j} is:

$$p(x_{*j}| \cap_j It_j, \cap_j s_j, \cap_{l=1}^{m_d, l \neq j} x_{*l}) = \frac{p(s_j|x_{*j})p(x_{*j}| \cap_j It_j, \cap_{l=1}^{m_d, l \neq j} x_{*l})}{p(\cap_j s_j| \cap_j It_j, \cap_{l=1}^{m_d, l \neq j} x_{*l})} \quad (11)$$

The terms of Eq. (11) can be derived in a straightforward way from Eqs. (4) and (5). Due to lack of space we skip those derivations. After many iterations of this process, the sampled values follow the joint probability of the Bayesian Network. Based on the obtained samples we find the most probable value for $\cap_j x_{*j}$.

4 Evaluation

4.1 Baseline

To the best of our knowledge our work is the first to handle the problem of unsupervised disaggregation in time series of low granularity (>1 min). To obtain some comparative results, we device a baseline method that uses clustering similarly to [7]. Each consumption event is represented as a triple, containing the starting time, the total consumption and the total length of the event. The k-means algorithm is used to cluster the consumption events and the known instances of all consumption types are assigned to the clusters accordingly. Then, in order to perform disaggregation of an event, we find its closest cluster and take the most probable consumption types of the cluster. For the water consumption dataset, where, as we explain in Sect. 4.2, the negative events are not known with certainty, we select as representing of showering behaviour a set of clusters that has a total number of events close to the known total number of showers.

4.2 Water Consumption Dataset

The water consumption dataset was gathered from a real world trial performed in the context of DAIAD[1] project, that addresses the issue of water sustainability through the use of Information Technology. The dataset consists of water consumption data for 17 households, measured hourly. Also, for each household, there are measurements that contain starting, ending time and total consumption of numerous shower occurrences. Due to the real-world conditions of the experiment, the time of occurrence of a significant portion of the showers is actually unknown. Thus, we cannot state with certainty that at a given day and time a shower was not taken. This means that, while we can directly measure the recall of identifying the showers, we are unable to directly measure the accuracy of the algorithm. However, we have knowledge of the total number of showers, which we use to compensate for the latter. We achieve that by using appropriate evaluation metrics, which we describe next.

The first metric we use is *Recall (RC)*, which measures how many of the known occurrences are retrieved by the algorithm. In order to evaluate if the algorithm is overly biased towards positive classification, we use the following two metrics: *Total Positive Ratio (TPR)* and *Positive Rate (PR)*. Let A be the total number of showers that the algorithm predicts, B the total number of showers that have actually occurred and C the total number of consumption events. Then TPR and PR are defined as:

$$TPR = \frac{A}{B}, \quad PR = \frac{A}{C} \tag{12}$$

Finally, we use the *Average Length of an Event (AEL)*, in hours, to evaluate the precision of the disaggregation in time. The results are presented next.

As we can see in Fig. 4, our proposed methods achieve very good RC, with acceptable values of TPR and PR. We note that the optimal value for TPR is 1.0. For PR, values significantly different than 0 and 1 indicate a non-trivial behaviour. The greedy approximation algorithm achieves the highest RC (0.85). From the PR metric, we can see that such high RC is achieved without classifying excessively many instances as positive. The TPR metric shows that the greedy approximation overestimates the total number of showers by an acceptable factor of 20%. The MCMC algorithm is balanced in both metrics (0.84 RC, 1.08 TPR). The baseline method does not perform well in terms of RC, while it also severely underestimates the total number of showers. The fact that the baseline's PR is similar to that of the other methods, while its TPR is much lower, is because the algorithm does not predict multiple showers in the same event. Due to that, we experimented with a modified version of the baseline, which predicted two

---

[1] http://daiad.eu/.

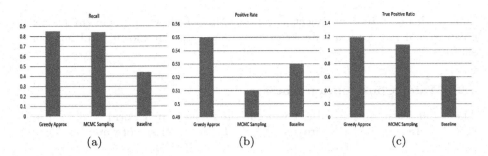

Fig. 4. The performance of all algorithms in terms of (a) Recall, (b) Positive Rate and (c) True Positive Ratio, on the water dataset.

showers in each event that it classifies as positive. This did not improve RC while it severely increased TPR. Finally, the AEL value for all algorithms is 2.71 h, which means that each occurrence of the consumption type is specified within a window of 2.71 h, by average. It is the same for all algorithms because they share a common consumption events identification step.

4.3 Energy Consumption Dataset

For energy consumption, we use the Reference Energy Disaggregation Dataset (REDD) [18]. REDD contains separate consumption time series for several appliances, as well as the aggregate power consumption, for 6 households. The interval of measurement is 1 s. Since we are intereset in lower granularity datasets, we downsample the data to 1 h. In this dataset, both positive and negative labels are available, thus we can use the Accuracy (ACC) measure. We calculate the AEL measure for this dataset as well.

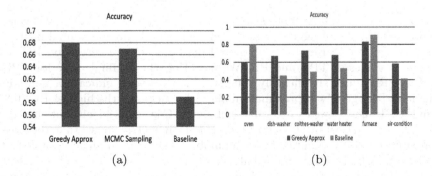

Fig. 5. The performance of each algorithm on the energy consumption dataset: (a) The average Accuracy (b) The average Accuracy per consumption type.

In Fig. 5a, we see the performance of the algorithms on the energy consumption dataset. We can see that our algorithm achieves relatively high ACC (0.68) and outperforms the baseline (0.59). Figure 5b presents the performance on each consumption type separately. We can see that our algorithm achieves high ACC and outperforms the baseline in four consumption types (clothes-washer, dish-washer and air-conditioner). On the other hand, the baseline performs better in two consumption types (oven and furnace). The most likely explanation for this is that our assumptions were not sufficiently accurate for those particular consumption types, thus comprising a subject for further investigation, in future work.

5   Conclusion

In this paper, we presented a novel method for resource consumption disaggregation, that works effectively on low granularity data (e.g., 1 h). Our method does not have the demanding requirement for labelled observations, which are hard to obtain. To our knowledge, our algorithm is the first that addresses the disaggregation problem under those constraints. This is particularly important in real world settings, especially for water consumption, where high frequency and labelled data are rarely available. We evaluated our algorithm in two residential consumption datasets and showed that it achieves high performance (up to 85% Recall) in identifying consumption types. Thus, our algorithm constitutes an effective solution for analysing resource consumption in real world settings.

References

1. Kolter, J.Z., Batra, S., Ng, A.Y: Energy disaggregation via discriminative sparse coding. In: NIPS (2010)
2. Parson, O., Ghosh, S., Weal, M., Rogers, A.: Non-intrusive load monitoring using prior models of general appliance types. In: AAAI (2012)
3. Kim, H., Marwah, M., Arlitt, M., Lyon, G., Han, J.: Unsupervised disaggregation of low frequency power measurements. In: SDM (2011)
4. Ellert, B., Makonin, S., Popowich, F.: Appliance water disaggregation via Non-Intrusive Load Monitoring (NILM). In: Leon-Garcia, A., et al. (eds.) SmartCity 360 2015-2016. LNICST, vol. 166, pp. 455–467. Springer, Cham (2016). https://doi.org/10.1007/978-3-319-33681-7_38
5. Cole, G., Stewart, R.A.: Smart meter enabled disaggregation of urban peak water demand: precursor to effective urban water planning. UWJ **10**(3), 174–194 (2013)
6. Dong, H., Wang, B., Lu, C.-T.: Deep sparse coding based recursive disaggregation model for water conservation. In: JCAI (2013)
7. Kozlovskiy, I.: Non-intrusive disaggregation of water consumption data in a residential household. In: IIWHEC Workshop (2016)
8. Vasak, M., Banjac, G., Novak, H.: Water use disaggregation based on classification of feature vectors extracted from smart meter data. In: CCWI (2015)
9. Nguyen, K.A., Stewart, R.A., Zhang, H.: An intelligent pattern recognition model to automate the categorisation of residential water end-use events. EMS **47**, 108–127 (2013)
10. DeOreo, W.B., Heaney, J.P., Mayer, P.W.: Flow trace analysis to assess water use. AWWA **88**(1), 79–90 (1996)

11. Kelly, J., Knottenbelt, W.: Neural NILM: deep neural networks applied to energy disaggregation. In: ICESEBE (2015)
12. Hunter Water Utility, Australia. Water Calculator. http://www.hunterwater.com. au/Save-Water/Water-Usage-Calculator.aspx
13. Mayer, P., et al.: Residential End Uses of Water. AWWA (1999)
14. Zimmermann, J.-P., et al.: Household Electricity Survey A study of domestic electrical product usage
15. Energy.gov. Energy Calculator. https://www.energy.gov/energysaver/estimating-appliance-and-home-electronic-energy-use
16. Froehlich, J., et al.: Disaggregated end-use energy sensing for the smart grid. IEEE Pervasive Comput. **10**(1), 28–39 (2011)
17. Bishop, C.M.: Graphical models. In: Pattern Recognition and Machine Learning. Springer, New York (2006). Chap. 8
18. Kolter, J.Z., Johnson, M.J.: REDD: a public data set for energy disaggregation research. In: SustKDD Workshop (2011)

STARS: Soft Multi-Task Learning for Activity Recognition from Multi-Modal Sensor Data

Xi Liu[1]([⊠]), Pang-Ning Tan[1], and Lei Liu[2]

[1] Michigan State University, Lansing, MI, USA
{liuxi4,ptan}@cse.msu.edu
[2] Huawei Technologies, Santa Clara, USA
leiliu@ieee.org

Abstract. Human activity recognition from ubiquitous sensor data is an important but challenging classification problem for applications such as assisted living, energy management, and security monitoring of smart homes. In this paper, we present a soft probabilistic classification model for human activity recognition from multi-modal sensors in a smart home environment. The model employs a softmax multi-task learning approach to fit a joint model for all the rooms in the smart home, taking into account the diverse types of sensors available in different rooms. The model also learns the transitional dependencies between activities to improve its prediction accuracy. Experimental results on a real-world dataset showed that the proposed approach outperforms several baseline methods, including k-nearest neighbors, conditional random field, and standard multinomial logistic regression.

1 Introduction

Rapid advances in the development of inexpensive, low-power, wireless sensing technology have enabled the deployment of sensors ubiquitously in a smart home environment to support various applications, from personal safety and security to water conservation and energy management. Real-time data generated from the myriad of sensors in the smart home provide a unique opportunity for monitoring daily living activities, alerting the residents or the authorities if any unusual activities are detected. The ability to accurately recognize human activities from the multi-modal sensor data is essential to support such applications.

However, classifying human activities from smart home sensor data is not a trivial task for several reasons. First, the sensor data are often noisy, and thus, require substantial preprocessing to extract discriminative features for the classification task. Second, the data are heterogeneous and may vary depending on the type of sensors deployed for monitoring the user activities. For example, wearable sensors such as accelerometers would generate data continuously at all times unlike other sensors such as motion detectors and surveillance cameras, which may only be available in certain rooms. For example, Fig. 1 shows the

© Springer International Publishing AG, part of Springer Nature 2018
D. Phung et al. (Eds.): PAKDD 2018, LNAI 10938, pp. 571–583, 2018.
https://doi.org/10.1007/978-3-319-93037-4_45

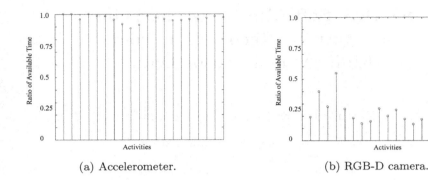

(a) Accelerometer.   (b) RGB-D camera.

Fig. 1. Percentage of time data from an accelerometer and RGB-D camera are available for each human activity. The list of activities are shown in Table 1.

percentage of time in which data from two sensors—accelerometer and surveillance camera—are available for each human activity in the smart home dataset investigated in this study. The results suggest that the accelerometer data is available at all times for most of the classes (human activities) whereas the surveillance camera data has a more imbalanced and irregular distribution as they are affected by the user's location as well as the rooms where the cameras are deployed. Thus, one of our key challenges is to develop a modeling approach that can handle the multi-modal sensor data, whose availability varies from one location to another depending on the sensor placement. Furthermore, the modeling approach must consider the imbalanced class distribution in different rooms since some activities could be restricted to certain locations only (e.g., one will more likely lie down in a bedroom or living room than in a kitchen).

The activities performed by each user can be represented by a sequence of actions, where the transition from one action to another proceeds in a continuous fashion. Since the data are collected and annotated at discrete time periods, some activities could be interleaved together in the same time period (e.g., walking and turning at the same time or going from a standing posture to a bending and eventually kneeling position). For example, Fig. 2 shows a 30-s segment of user activity from the labeled data used in this study. Since there could be more than one activity performed in each second, each class label (human activity) is associated with a confidence score, represented by its gray scale color. One of our goals of this study is to develop a modeling approach that can leverage the soft labels to determine the probability an activity is performed at a given time period. The temporal dependency between activities is another factor that must be taken into consideration. For example, the lie-to-sit transition activity typically occurs between the lie and sit postures. However, we do not expect the sequences to contain transitions from lie to jump activities. How to effectively incorporate such temporal dependencies into the modeling framework is another challenge that needs to be addressed. Although such constraints can be pre-defined from domain knowledge, they may vary depending on the dataset used. Instead of encoding them

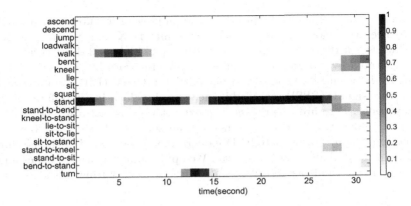

Fig. 2. A segment of ground truth activities.

as hard constraints, our goal is to infer the temporal dependencies automatically from the data.

To address these challenges, this paper presents a soft classification approach for activity recognition in a smart home environment. The approach employs a softmax classifier to predict user activities in a sequence based on the multimodal sensor data available. Training a global softmax classifier is not effective since some features (e.g., surveillance camera data) are only available in certain rooms. Imputing their missing values may introduce errors into the model while discarding the data with incomplete features may lead to suboptimal models. Conversely, training a local model for each room is also not the answer due to the limited training data available for some rooms and the large number of classes that need to be predicted. To overcome this limitation, we propose a multi-task learning framework that allows the local models for all the rooms to be jointly trained, taking into account the relationship between the tasks and the varying types of features available. Specifically, the framework enables the model for predicting, say, the `walk` activity in one room, to be related to the same activity in another room even if their features are not identical. This is accomplished by decomposing the weight matrix associated with the prediction of each class into a set of low rank latent factors, where the decomposition is performed only on the common features for the rooms. Using a real-world multi-modal sensor dataset [16] as our case study, we showed that the proposed framework is more effective than other sequential and non-sequential classification algorithms, including multinomial logistic regression and conditional random fields.

2   Preliminaries

Consider a multi-modal sensor dataset, $\mathcal{D} = \{\mathcal{D}_1, \mathcal{D}_2, \cdots, \mathcal{D}_R\}$, where each $\mathcal{D}_r = (\mathbf{X}^r, \mathbf{Y}^r)$ is the training set for room r. Furthermore, each $\mathbf{X}^r = \mathbb{R}^{N_r \times d_r}$ corresponds to the data matrix derived for room r, where N_r is the number of training examples available and d_r is the number of features. For notational convenience, we denote $\mathbf{X}_{i:}$ as the i-th row of matrix $\mathbf{X}$ and $\mathbf{X}_{j:}$ as its j-th column. The sensor data considered in this study [16] include 3-d acceleration features generated by a portable triaxial accelerometer worn by the subject, RGB-D camera data, and location data from passive infrared (PIR) and strength of acceleration signals (RSSI) recorded by access points location in different rooms. The raw sensor data are preprocessed to extract various features (e.g., kurtosis, frequency, and entropy of accelerometer time series and bounding box information about subjects from RGB-D camera data) associated with the human activities measured at every 1 s interval. We apply similar feature extraction and preprocessing methods as described in [12] for the sensor time series data.

Table 1. List of human activity classes from the Sphere challenge data [16].

Ascend	Bent	Stand	Sit-to-stand
Descend	Kneel	Stand-to-bend	Stand-to-kneel
Jump	Lie	Kneel-to-stand	Stand-to-sit
Loadwalk	Sit	Lie-to-sit	Bend-to-stand
Walk	Squat	Sit-to-lie	Turn

Let $\mathbf{Y}^r \in [0,1]^{N_r \times K}$ be the class membership matrix for all N_r observations in room r, where $\mathbf{Y}_{ik}^r \in [0,1]$ denotes the confidence score for the i-th training instance in room r belonging to the k-th class. There are altogether 20 classes in our dataset, which are divided into 3 groups: (1) Active motion (a), which include activities such as ascending or descending stairs, jumping, and walking, (2) Stationary postures (p), which include bending, sitting down, and standing, and (3) Transition movements (t), which include stand-to-bend, lie-to-sit, sit-to-lie, and stand-to-kneel. The complete list of classes is shown in Table 1.

3 Methodology

3.1 Multi-class Learning with Softmax Regression

Softmax regression can be used to compute the posterior probability that the i-th instance in room r belongs to class k as follows [3]:

$$P(Y_i^r = k | \mathbf{X}_{i:}^r) = \frac{\exp(\mathbf{X}_{i:}^r \mathbf{W}_{k:}^r)}{\sum_{s=1}^{K} \exp(\mathbf{X}_{i:}^r \mathbf{W}_{s:}^r)} \equiv \mathbf{P}_{ik}^r, \tag{1}$$

where $\mathbf{W}^r \in \Re^{K \times d_r}$ is the model parameter matrix for room r. The parameters can be estimated by minimizing the following cross entropy loss function:

$$\mathbf{W}^r = arg \min_{\mathbf{W}^r} \sum_{i}^{N_r} \sum_{k}^{K} -\mathbf{Y}_{ik}^r \log \mathbf{P}_{ik}^r \tag{2}$$

Intuitively, the loss function measures the discrepancy between the estimated posterior probability and annotated confidence score of each class for the training examples. The loss function is well-suited for handling soft labels in the human activity recognition problem shown in Fig. 2, in which multiple activities may occur in the same time period.

3.2   Proposed Method: STARS

Although the softmax regression approach can be applied to the smart home data, it has several limitations. First, it does not account for the temporal dependencies of activities in the sequence data. Second, it is designed for learning models independently for each room. Since the amount of training examples available in each room may vary, this may lead to suboptimal local models. Furthermore, the features available to classify the human activities can be different from one room to another. It would be useful to develop a multi-task learning approach that can jointly train the models for all the rooms, taking into account the relationships among the prediction tasks and variable features of the rooms.

To overcome these limitations, we propose the following soft multi-task learning framework called STARS, which is designed to optimize the following objective function:

$$\min_{\Theta} \quad \mathscr{L}_1 + \mathscr{L}_2 + \mathscr{L}_3 \tag{3}$$

$$\text{s.t.} \quad \mathbf{W}_{k:}^r = [\mathcal{W}_{rk:}^{com}, \mathbf{W}_{k:}^{dif,r}]_{1 \times d_r}, \quad \mathcal{W}_{rk:}^{com} = \mathcal{U}_{kr:}\mathcal{V}_{k::} \tag{4}$$

$$\mathbf{P}_{ik}^r = \frac{\exp(\mathbf{X}_{i:}^r \mathbf{W}_{k:}^r{}^T + \mathbf{Z}_{i-1:}\mathcal{F}_{rk:}^1{}^T + \mathbf{Z}_{i+1:}\mathcal{F}_{rk:}^2{}^T)}{\sum_s^K \exp(\mathbf{X}_{i:}^r \mathbf{W}_{s:}^r{}^T + \mathbf{Z}_{i-1:}\mathcal{F}_{rs:}^1{}^T + \mathbf{Z}_{i+1:}\mathcal{F}_{rs:}^2{}^T)} \tag{5}$$

$$\mathbf{Z}_{i:}^r = \mathbf{X}_{i:}^r \mathbf{W}^{rT} \tag{6}$$

$$\text{where} \quad \mathscr{L}_1 = \sum_r^R \sum_i^{N_r} \sum_k^K -\mathbf{Y}_{ik}^r \log \mathbf{P}_{ik}^r$$

$$\mathscr{L}_2 = \sum_r^R \beta \|\mathbf{P}^r - \mathbf{G}^r \mathbf{P}^r\|_F^2$$

$$\mathscr{L}_3 = \sum_k^K (\lambda_U \|\mathcal{U}_{k,:,:}\|_1 + \lambda_V \|\mathcal{V}_{k,:,:}\|_1)$$

$$+ \sum_r^R (\lambda_W \|\mathbf{W}^{dif,r}\|_1 + \lambda_{F1} \|\mathcal{F}_{r::}^1\|_1 + \lambda_{F2} \|\mathcal{F}_{r::}^2\|_1)$$

$$\mathbf{G}^r = \begin{bmatrix} 1 & 0 & 0 & \dots & 0 & 0 \\ 1 & 0 & 0 & \dots & 0 & 0 \\ 0 & 1 & 0 & \dots & 0 & 0 \\ \dots & & & & \dots \\ 0 & 0 & 0 & \dots & 1 & 0 \end{bmatrix}_{N_r \times N_r}$$

where $\Theta = \{\mathcal{U}, \mathcal{V}, \mathbf{W}^r, \mathcal{F}^1, \mathcal{F}^2\}$ corresponds to the set of model parameters for all the rooms $r = 1, 2, \cdots, R$. The framework assumes a linear model for each room r, parameterized by the matrix $\mathbf{W}^r = [\mathbf{W}^{com,r}, \mathbf{W}^{dif,r}]$, where $\mathbf{W}^{com,r}$ is represented by the r-th slice of tensor $\mathcal{W}^{com}$, and denotes the weight matrix associated with the common features for all the rooms. And $\mathbf{W}^{dif,r}$ denotes the weight matrix associated with the unique features of the room. For example, the common features may include those derived from accelerometer sensors worn by the users whereas the unique features may correspond to those derived from surveillance cameras located only in certain rooms.

Note that the objective function consists of three parts: (1) $\mathscr{L}_1$, which is the cross entropy loss function associated with the classification error, (2) $\mathscr{L}_2$, which captures the temporal persistence of the classes (to be explained below), and (3) sparsity constraints $\mathscr{L}_3$. The posterior probability $\mathbf{P}^r_{ik}$ in the proposed formulation depends not only on the features $\mathbf{X}^r_{i:}$ at time i, but also on the temporal features $\mathbf{Z}_{i-1:}$ and $\mathbf{Z}_{i+1:}$ at time $i-1$ and $i+1$, respectively. We consider $\mathbf{Z}^r_{i-1:} = \mathbf{X}^r_{i-1:}\mathbf{W}^{rT}$ and $\mathbf{Z}^r_{i+1} = \mathbf{X}^r_{i+1:}\mathbf{W}^{rT}$ as temporal features because they are related to the predicted probabilities in the previous and next timesteps. Our model also encapsulates information about the class transitions by using the transition tensors $\mathcal{F}^1$ and $\mathcal{F}^2$. Specifically, $\mathcal{F}^1_{r::}$ encodes the relationship between the activity at previous timestep $i-1$ to the activity at current timestep i in room r. Conversely, $\mathcal{F}^2_{r::}$ encodes the relationship between the activity at the next timestep $i+1$ and the activity at current timestep i in room r. These transition tensors are estimated automatically in the STARS framework. The second term of the objective function, $\mathscr{L}_2$, is a regularization term to ensure the temporal persistence of the classes. As illustrated in Fig. 2, most activities tend to last for more than several seconds. This suggests a trivial approach to predict user activity in the next timestep is by using the predicted activity for the current timestep. The temporal persistence of an activity between two adjacent time steps is reflected by the soft constraint $||\mathbf{P}^r_{i:} - \mathbf{P}^r_{i-1:}||^2_F$, where $\mathbf{P}^r_{i-1:} = (\mathbf{G}^r\mathbf{P}^r)_{i::}$. Finally, the third term in the objective function, $\mathscr{L}_3$, is used to ensure sparsity of the model parameters and to avoid model overfitting.

One unique feature of the proposed STARS framework is that it uses a multi-task learning approach to train the models for all rooms simultaneously. Furthermore, instead of treating classification task for different rooms as independent learning problems, it assumes the tasks are related via the common features shared by all the rooms. Specifically, although the weight matrix $\mathbf{W}^{com,r}$ for all the rooms can be different, they share a pair of common low-rank factors, $\mathcal{U}$ and $\mathcal{V}$. Here, we use the notation $\mathcal{W}^{com}$ to represent a 3-dimensional tensor, where the r-th slice of the tensor corresponds to the weight matrix $\mathbf{W}^{com,r}$ for room r.

3.3 Optimization

Observe that the $\mathscr{L}_1$ and $\mathscr{L}_2$ terms of the objective function given in Eq. 3 are differentiable functions unlike $\mathscr{L}_3$, which is non-differentiable due to ℓ_1 regularization. The accelerated proximal gradient descent (PGD) method [4,13] can

---

Algorithm 1. Training Phase for STARS Framework

Input: training set $\{\mathbf{X}^r, \mathbf{Y}^r\}$ and set of regularizers $\{\beta, \lambda_U, \lambda_V, \lambda_W, \lambda_{F1}, \lambda_{F2}\}$.

Output: $\Theta^{(t)} = \{\mathcal{U}^{(t)}, \mathcal{V}^{(t)}, \mathbf{W}^{diff,r(t)}, \mathcal{F}^{1(t)}, \mathcal{F}^{2(t)}\}$

Set $t = 0$ and initialize $\Theta^{(0)} = \{\mathcal{U}^{(0)}, \mathcal{V}^{(0)}, \mathbf{W}^{diff,r(0)}, \mathcal{F}^{1(0)}, \mathcal{F}^{2(0)}\}$.

repeat

    $t = t + 1$

    $\forall r, q : \mathcal{U}_{qr:} \leftarrow S_{\lambda_U t}(\mathcal{U}_{qr:} - \alpha^{(t)}(\frac{\partial \mathscr{L}_1}{\partial \mathcal{U}_{qr:}} + \frac{\partial \mathscr{L}_2}{\partial \mathcal{U}_{qr:}}))$

    $\forall q : \mathcal{V}_{q::} \leftarrow S_{\lambda_V t}(\mathcal{V}_{q::} - \alpha^{(t)}(\frac{\partial \mathscr{L}_1}{\partial \mathcal{V}_{q::}} + \frac{\partial \mathscr{L}_2}{\partial \mathcal{V}_{q::}}))$

    $\forall r, q : \mathbf{W}_{q:}^{dif,r} \leftarrow S_{\lambda_W t}(\mathbf{W}_{q:}^{dif,r} - \alpha^{(t)}(\frac{\partial \mathscr{L}_1}{\partial \mathbf{W}_{q:}^{dif,r}} + \frac{\partial \mathscr{L}_2}{\partial \mathbf{W}_{q:}^{dif,r}}))$

    $\forall r, q : \mathcal{F}_{rq:}^1 \leftarrow S_{\lambda_{F1} t}(\mathcal{F}_{rq:}^1 - \alpha^{(t)}(\frac{\partial \mathscr{L}_1}{\partial \mathcal{F}_{rq:}^1} + \frac{\partial \mathscr{L}_2}{\partial \mathcal{F}_{rq:}^1}))$

    $\forall r, q : \mathcal{F}_{rq:}^2 \leftarrow S_{\lambda_{F2} t}(\mathcal{F}_{rq:}^2 - \alpha^{(t)}(\frac{\partial \mathscr{L}_1}{\partial \mathcal{F}_{rq:}^2} + \frac{\partial \mathscr{L}_2}{\partial \mathcal{F}_{rq:}^2}))$

until convergence

Algorithm 2. Prediction Phase for STARS Framework

Input: test example, $\mathbf{X}_{i:}^r$, its adjacent predictors, $\mathbf{X}_{i-1:}^r$ and $\mathbf{X}_{i+1:}^r$, and estimated model parameters, $\Theta = \{\mathcal{U}, \mathcal{V}, \mathbf{W}^{dif,r}, \mathcal{F}^1, \mathcal{F}^2\}$, $r = 1, 2, ..., R$

Output: predicted probability

$P(y = k | \mathbf{X}_{i:}^r, \mathbf{X}_{i-1:}^r, \mathbf{X}_{i+1:}^r, \Theta) = \mathbf{P}_{ik}^r$, k = 1,2,...K, with formula 5 and 6

---

be applied to learn the model parameters. A pseudo-code of the algorithm for inferring the model parameters is provided in Algorithm 1 while the pseudo-code for predicting the next activity of the sequence is shown in Algorithm 2.

Specifically, the soft thresholding proximal mapping operator is used to update the model parameters as follows:

$$\Theta^{(t)} = S_{\lambda t}\left[\Theta^{(t-1)} - \alpha^{(t)}\left(\frac{\partial \mathscr{L}_1}{\partial \Theta^{(t-1)}} + \frac{\partial \mathscr{L}_2}{\partial \Theta^{(t-1)}}\right)\right] \tag{7}$$

where $S_\delta(x) = (x - \delta)_+ - (-x - \delta)_-$, and $\alpha^{(t)}$ is the gradient descent step size during iteration t. Since there are multiple model parameters, $\Theta = \{\mathcal{U}, \mathcal{V}, \mathbf{W}^{dif,r}, \mathcal{F}^1, \mathcal{F}^2\}$, the parameters are each updated in an alternating fashion. The hyper-parameter of the soft thresholding operator λ depends on the Lasso regularizer for each model parameter, e.g., λ is set to λ_U when updating $\mathcal{U}$, λ_V when updating $\mathcal{V}$, and so on. A backtracking line search is also implemented to adaptively choose the step size of the gradient descent [4] and ensure faster convergence.

In the remainder of this section, we show the gradient computation of the differentiable part of the loss function with respect to each model parameter.

Gradient Computation for $\mathcal{U}$. Taking the partial derivative of $\mathcal{L}_1$ w.r.t. $\mathcal{U}_{qr:}$, where $k = 1, 2, ..., K$, and $q = 1, 2, ..., K$, yields the following:

$$\frac{\partial \mathcal{L}_1}{\partial \mathcal{U}_{qr:}} = \sum_i^{N_r} (\mathbf{P}_{iq}^r - \mathbf{Y}_{iq}^r) \mathcal{X}_{ri:}^{com} \mathcal{V}_{q::}^T$$

$$+ \sum_i^{N_r} \sum_k^K (\mathbf{P}_{ik}^r - \mathbf{Y}_{ik}^r)(\mathcal{F}_{rkq}^1 \mathcal{X}_{r(i-1):}^{com} \mathcal{V}_{q::}^T + \mathcal{F}_{rkq}^2 \mathcal{X}_{r(i+1):}^{com} \mathcal{V}_{q::}^T) \quad (8)$$

where $\mathcal{X}_{ri:}^{com} \mathcal{V}_{q::}^T$ denote the latent representation of the common features $\mathcal{X}_{ri:}^{com}$ for room r. The preceding equation suggests that the update formula for $\mathcal{U}_{qr:}$ depends on two terms. The first term on the right hand side of Eq. (8), $\sum_i^{N_r} (\mathbf{P}_{iq}^r - \mathbf{Y}_{iq}^r) \mathcal{X}_{ri:}^{com} \mathcal{V}_{q::}^T$, measures the difference between the predicted and true class in terms of the latent, common feature vectors. The second term, $\sum_i^{N_r} \sum_k^K (\mathbf{P}_{ik}^r - \mathbf{Y}_{ik}^r)(\mathcal{F}_{rkq}^1 \mathcal{X}_{r(i-1):}^{com} \mathcal{V}_{q::}^T + \mathcal{F}_{rkq}^2 \mathcal{X}_{r(i+1):}^{com} \mathcal{V}_{q::}^T)$ measures the difference in terms of the latent feature vectors for adjacent time periods, taking into account the temporal dependencies between activities, $\mathcal{F}_{r::}^1$ and $\mathcal{F}_{r::}^2$.

Furthermore, the gradient of $\mathcal{L}_2$ w.r.t. $\mathcal{U}_{qr:}$ is given by

$$\frac{\partial \mathcal{L}_2}{\partial \mathcal{U}_{qr:}} = \sum_i^N \sum_k^K 2\beta(\mathbf{P}_{ik}^r - (\mathbf{G}^r \mathbf{P}^r)_{ik}) \times (\frac{\partial \mathbf{P}_{ik}^r}{\partial \mathcal{U}_{qr:}} - \sum_j^{N_r} \mathbf{G}_{ij}^r \frac{\partial \mathbf{P}_{jk}^r}{\partial \mathcal{U}_{qr:}})$$

where,

$$\frac{\partial \mathbf{P}_{ik}^r}{\partial \mathcal{U}_{qr:}} = \mathbf{P}_{ik}^r \Bigg((1\{q = k\} - \mathbf{P}_{iq}^r) \mathcal{X}_{ri:}^{com} \mathcal{V}_{q::}^T + (\mathcal{F}_{rkq}^1 \mathcal{X}_{r(i-1):}^{com} \mathcal{V}_{q::}^T + \mathcal{F}_{rkq}^2 \mathcal{X}_{r(i+1):}^{com} \mathcal{V}_{q::}^T)$$

$$- \sum_s^K (\mathcal{F}_{rsq}^1 \mathcal{X}_{r(i-1):}^{com} \mathcal{V}_{q::}^T + \mathcal{F}_{rsq}^2 \mathcal{X}_{r(i+1):}^{com} \mathcal{V}_{q::}^T) \mathbf{P}_{is}^r \Bigg)$$

Gradients Computation for $\mathcal{V}$. Similarly, the gradients w.r.t. $\mathcal{V}$ are:

$$\frac{\partial \mathcal{L}_1}{\partial \mathcal{V}_{q::}} = \sum_r^R \sum_i^{N_r} (\mathbf{P}_{iq}^r - \mathbf{Y}_{iq}^r) \mathcal{U}_{qr:} \mathcal{X}_{ri:}^{com}$$

$$+ \sum_r^R \sum_i^{N_r} \sum_k^K (\mathbf{P}_{ik}^r - \mathbf{Y}_{ik}^r)(\mathcal{F}_{rkq}^1 \mathcal{U}_{qr:} \mathcal{X}_{r(i-1):}^{com} + \mathcal{F}_{rkq}^2 \mathcal{U}_{qr:} \mathcal{X}_{r(i+1):}^{com})$$

$$\frac{\partial \mathcal{L}_2}{\partial \mathcal{V}_{q::}} = \sum_r^R \sum_i^{N_r} \sum_k^K 2\beta(\mathbf{P}_{ik}^r - (\mathbf{G}^r \mathbf{P}^r)_{ik}) \times (\frac{\partial \mathbf{P}_{ik}^r}{\partial \mathcal{V}_{q::}} - \sum_j^{N_r} \mathbf{G}_{ij}^r \frac{\partial \mathbf{P}_{jk}^r}{\partial \mathcal{V}_{q::}})$$

Gradients Computation for $\mathbf{W}^{dif,r}$.

$$\frac{\partial \mathcal{L}_1}{\partial \mathbf{W}_{q:}^{dif,r}} = \sum_i^{N_r} (\mathbf{P}_{iq}^r - \mathbf{Y}_{iq}^r)\mathbf{X}_{i:}^{dif,r} + \sum_i^{N_r}\sum_k^K (\mathbf{P}_{ik}^r - \mathbf{Y}_{ik}^r)(\mathcal{F}_{rkq}^1 \mathbf{X}_{i-1:}^{dif,r} + \mathcal{F}_{rkq}^2 \mathbf{X}_{i+1:}^{dif,r})$$

$$\frac{\partial \mathcal{L}_2}{\partial \mathbf{W}_{q:}^{dif,r}} = \sum_i^{N_r}\sum_k^K 2\beta(\mathbf{P}_{ik}^r - (\mathbf{G}^r\mathbf{P}^r)_{ik}) \times (\frac{\partial \mathbf{P}_{ik}^r}{\partial \mathbf{W}_{q:}^{dif,r}} - \sum_j^{N_r} \mathbf{G}_{ij}^r \frac{\partial \mathbf{P}_{jk}^r}{\partial \mathbf{W}_{q:}^{dif,r}})$$

Gradients Computation for $\mathcal{F}^1$ (or $\mathcal{F}^2$).

$$\frac{\partial \mathcal{L}_1}{\partial \mathcal{F}_{rq:}^1} = \sum_i^{N_r} (\mathbf{P}_{iq}^r - \mathbf{Y}_{iq}^r)\mathbf{Z}_{i-1:}^r$$

$$\frac{\partial \mathcal{L}_2}{\partial \mathcal{F}_{rq:}^1} = 2\beta \sum_i^{N_r}\sum_k^K (\mathbf{P}_{ik}^r - (\mathbf{G}^r\mathbf{P}^r)_{ik}) \times (\frac{\partial \mathbf{P}_{ik}^r}{\partial \mathcal{F}_{rq:}^1} - \sum_j^{N_r} \mathbf{G}_{ij}^r \frac{\partial \mathbf{P}_{jk}^r}{\partial \mathcal{F}_{rq:}^1})$$

The gradients $\frac{\partial \mathcal{L}_1}{\partial \mathcal{F}_{rq:}^2}$ and $\frac{\partial \mathcal{L}_2}{\partial \mathcal{F}_{rq:}^2}$ can be obtained in a similar way.

4 Experimental Evaluation

We performed our experiments on a real-world data set from the *SPHERE Challenge* competition [16]. The dataset contains classes of human activities recorded in a house containing 9 different rooms. The raw data contains 10 sequences, where each sequence corresponds to a series of activities performed by a subject for a time period lasting between 1,392 to 1,825 s. Using the sensor observations at each second as a data instance, we ended up with a total 16,124 instances. Each instance was labeled by a team of 12 annotators [16], whose results are aggregated to obtain a confidence score for each class label. For evaluation purposes, we apply 5-fold cross-validation and report the mean and standard deviation of their prediction accuracies.

Following the approach described in [16], we employ the weighted Brier score to evaluate the classification performance. The metric is defined as follows:

$$\text{BS} = \frac{1}{N} \sum_{i=1}^N \sum_{k=1}^K l_k(\mathbf{Y}_{ik} - \mathbf{P}_{ik})^2 \tag{9}$$

where $\mathbf{Y}_{ik}$ is the confidence score for the i-th instance and k-th class, computed based on the labels provided by a team of annotators, whereas $\mathbf{P}_{ik}$ is the predicted posterior class. The weight for each class, l_k, is defined in [1], which is negatively correlated with the class size.

4.1 Baseline Algorithms

We compare the performance of our proposed framework, **STARS**, against the following baseline algorithms:

- **SR:** Softmax regression, which trains a local softmax regression model for each room with Eqs. (1) and (2). Unlike our proposed framework, it is a single-task learning model and does not incorporate temporal dependencies.
- **KNN:** A k-nearest neighbor classifier, which is another baseline used in [1] for the SPHERE competition data. The parameter k is tuned on the validation data in order to provide the best result. We sums up the weights associated with each neighboring instance $\mathbf{Y}_{ik}$ for the test instance i and normalize the weighted sum to obtain the predicted posterior probabilities.
- **CRF:** Conditional Random Field [15], which is a widely used model for sequence classification problems [8, 10] and has been applied to activity recognition problems [9, 17].

4.2 Experimental Results

The results comparing the weighted brier score of the proposed framework STARS, against the baseline methods (SR, KNN and CRF) are shown in Table 2. We reported the weighted Brier score for all rooms (denoted as Overall) as well as for individual rooms. The results suggest that the overall performance of STARS is significantly better than the baseline methods. In terms of performances for individual rooms, STARS achieves the best (i.e., lowest score) in 7 out of the 9 rooms. The performance of STARS is slightly worse than SR for `bedroom2` and `hallway` due to the lack of transitional activities, making it harder to learn the temporal dependency accurately based on their limited training data.

In addition to its lower Brier score, another advantage of using the STARS framework is that the model can be used to learn the transition between activities

Table 2. Weighted Brier Scores for various competing algorithms.

Room	SR	KNN	CRF	STARS
Bathroom	0.1334 ± 0.0149	0.1315 ± 0.0103	0.1479 ± 0.0152	$\mathbf{0.1269 \pm 0.0173}$
Bedroom1	$\mathbf{0.0853 \pm 0.0129}$	0.1026 ± 0.0085	0.0935 ± 0.0118	0.0920 ± 0.0156
Bedroom2	0.2817 ± 0.0148	0.2862 ± 0.0178	0.2886 ± 0.0223	$\mathbf{0.2675 \pm 0.0172}$
Hallway	$\mathbf{0.1926 \pm 0.0953}$	0.2323 ± 0.0675	0.2115 ± 0.0816	0.1953 ± 0.0849
Kitchen	0.0842 ± 0.0122	0.0915 ± 0.0099	0.0917 ± 0.0133	$\mathbf{0.0820 \pm 0.0106}$
Living room	0.1594 ± 0.0181	0.1774 ± 0.0171	0.1710 ± 0.0201	$\mathbf{0.1468 \pm 0.0142}$
Stairs	0.3827 ± 0.0705	0.4366 ± 0.0552	0.3834 ± 0.0288	$\mathbf{0.3373 \pm 0.0505}$
Study room	0.0441 ± 0.0407	0.0649 ± 0.0240	0.0506 ± 0.0556	$\mathbf{0.0381 \pm 0.0365}$
Toilet	0.1440 ± 0.0380	0.1368 ± 0.0304	0.1442 ± 0.0358	$\mathbf{0.1360 \pm 0.0417}$
Overall	0.1700 ± 0.0095	0.1815 ± 0.0089	0.1794 ± 0.0121	$\mathbf{0.1598 \pm 0.0087}$

via $\mathcal{F}^1$ and $\mathcal{F}^2$. Figure 3a and b depict a heat map of the two tensor slices $\mathcal{F}^1_{r::}$ and $\mathcal{F}^2_{r::}$ for the living room. The results shown in these figures are mostly consistent with our common sense knowledge. For example, Fig. 3a shows that the bent posture is mostly followed by the activity bend-to-stand whereas stand-to-bend often leads to the bent posture. Similarly, Fig. 3b shows that the stand-to-kneel activity would lead to the kneel posture in the next time step, while lie-to-sit begins with lie posture and ends with the sit posture.

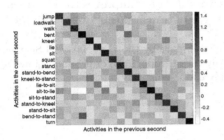

(a) Transition matrix from $(i - 1)$ to i timestep.

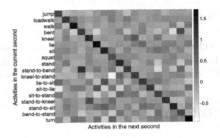

(b) Transition matrix from i to $(i + 1)$ timestep.

Fig. 3. The estimated transition matrices $\mathcal{F}^1_{r::}$ (left) and $\mathcal{F}^2_{r::}$ (right) for living room. The ordering of the classes on the horizontal and vertical axes are the same.

5 Related Work

Numerous approaches have been developed for human activity recognition. Classic approaches include decision tree [5], support vector machine [18], logistic regression, and Bayesian networks [7]. For example, a comparison between logistic regression and non-linear SVM on human activity recognition is given in [12]. These approaches do not consider the temporal/sequential dependencies between activities. In contrast, methods such as Hidden Markov Model (HMM) and Conditional Random Fields (CRF) are more well-suited for handling sequential data [8,21], and thus, have been widely utilized for activity recognition tasks [11,17]. However, these approaches are primarily designed for single-task learning, unlike the multi-task approach proposed in this study. The success of multi-task learning for activity recognition has been well-documented [2,6,14,19,20]. In [14], a structured multi-task classification method was proposed, where each task corresponds to the classification of a specific person. [19] presented a multi-task clustering framework for analyzing daily living activities from visual data collected by wearable cameras. In addition, [20] focused on multi-task feature selection whereas [2] focused on online matrix regularization. Unlike other existing works, our proposed framework considers the classification in different rooms as separate tasks, with possibly different types of features.

6 Conclusion

In this paper, we present a soft multi-task learning technique for human activity recognition from multi-modal sensor data in a smart home. Our technique incorporates the temporal dependencies between classes in a multi-task learning setting. Experimental results using a public human activity recognition dataset showed that the proposed technique outperforms baseline methods including K-Nearest Neighbor, Conditional Random Field, and single-task learning with multinomial softmax regression. The framework not only improves the classification performance, it also reveals the typical type of transitions between activities.

Acknowledgments. This research is supported in part by the U.S. National Science Foundation through grant NSF III-1615612. Any use of trade, firm, or product names is for descriptive purposes only and does not imply endorsement by the U.S. Government.

References

1. The sphere challenge: activity recognition with multimodal sensor data. http://blog.drivendata.org/2016/06/06/sphere-benchmark/
2. Agarwal, A., Rakhlin, A., Bartlett, P.: Matrix regularization techniques for online multitask learning. Technical report UCB/EECS-2008-138, EECS Department, University of California, Berkeley (2008)
3. Bishop, C.M.: Pattern Recognition and Machine Learning. Springer, New York (2006)
4. Boyd, S., Vandenberghe, L.: Convex Optimization. Cambridge University Press, New York (2004)
5. Carós, J.S., Chételat, O., Celka, P., Dasen, S., CmÃral, J.: Very low complexity algorithm for ambulatory activity classification. In: Proceedings of the 3rd European Medical and Biological Conference, EMBEC, pp. 16–20 (2005)
6. Cavallanti, G., Cesa-Bianchi, N., Gentile, C.: Linear algorithms for online multitask classification. J. Mach. Learn. Res. **11**, 2901–2934 (2010)
7. Dougherty, J., Kohavi, R., Sahami, M., et al.: Supervised and unsupervised discretization of continuous features. In: Proceedings of the 12th International Conference of Machine Learning, vol. 12, pp. 194–202 (1995)
8. Gao, Q., Doering, M., Yang, S., Chai, J.: Physical causality of action verbs in grounded language understanding. In: Proceedings of the 54th Annual Meeting of the Association for Computational Linguistics, vol. 1, pp. 1814–1824 (2016)
9. Kim, E., Helal, S., Cook, D.: Human activity recognition and pattern discovery. IEEE Pervasive Comput. **9**(1) (2010)
10. Lafferty, J., McCallum, A., Pereira, F.C.: Conditional random fields: probabilistic models for segmenting and labeling sequence data (2001)
11. Lester, J., Choudhury, T., Borriello, G.: A practical approach to recognizing physical activities. In: Fishkin, K.P., Schiele, B., Nixon, P., Quigley, A. (eds.) Pervasive 2006. LNCS, vol. 3968, pp. 1–16. Springer, Heidelberg (2006). https://doi.org/10.1007/11748625_1
12. Liu, X., Liu, L., Simske, S.J., Liu, J.: Human daily activity recognition for healthcare using wearable and visual sensing data. In: 2016 IEEE International Conference on Healthcare Informatics (ICHI), pp. 24–31. IEEE (2016)

13. Nesterov, Y.: A method of solving a convex programming problem with convergence rate o(1/k2). Sov. Math. Dokl. **27**, 372–376 (1983)
14. Sun, X., Kashima, H., Ueda, N.: Large-scale personalized human activity recognition using online multitask learning. IEEE Trans. Knowl. Data Eng. **25**(11), 2551–2563 (2013)
15. Sutton, C., McCallum, A.: An introduction to conditional random fields for relational learning. In: Sutton, C., McCallum, A. (eds.) Introduction to Statistical Relational Learning, pp. 93–128. MIT Press, Cambridge (2006)
16. Twomey, N., Diethe, T., Kull, M., Song, H., Camplani, M., Hannuna, S., Fafoutis, X., Zhu, N., Woznowski, P., Flach, P., Craddock, I.: The SPHERE challenge: Activity recognition with multimodal sensor data. arXiv:1603.00797 (2016)
17. Vail, D.L., Veloso, M.M., Lafferty, J.D.: Conditional random fields for activity recognition. In: Proceedings of the 6th International Joint Conference on Autonomous Agents and Multiagent Systems, p. 235. ACM (2007)
18. Wang, S., Yang, J., Chen, N., Chen, X., Zhang, Q.: Human activity recognition with user-free accelerometers in the sensor networks. In: Proceedings of the International Conference on Neural Networks and Brain, vol. 2, pp. 1212–1217. IEEE (2005)
19. Yan, Y., Ricci, E., Liu, G., Sebe, N.: Egocentric daily activity recognition via multitask clustering. IEEE Trans. Image Process. **24**(10), 2984–2995 (2015)
20. Yang, H., King, I., Lyu, M.R.: Online learning for multi-task feature selection. In: Proceedings of the 19th ACM International Conference on Information and Knowledge Management, pp. 1693–1696. ACM (2010)
21. Yang, S., Gao, Q., Liu, C., Xiong, C., Zhu, S.C., Chai, J.Y.: Grounded semantic role labeling. In: Proceedings of the North American Chapter of the Association for Computational Linguistics: Human Language Technologies, pp. 149–159 (2016)

A Refined MISD Algorithm Based on Gaussian Process Regression

Feng Zhou[1,2]([⊠]), Zhidong Li[2], Xuhui Fan[1], Yang Wang[2],
Arcot Sowmya[1], and Fang Chen[2]

[1] University of New South Wales, Sydney, NSW 2052, Australia
[2] Data61, CSIRO, 13 Garden St., Eveleigh, NSW, Australia
feng.zhou@data61.csiro.au

Abstract. Time series data is a common data type in real life, and modelling of time series data along with its underlying temporal dynamics is always a challenging job. Temporal point process is an outstanding method to model time series data in domains that require temporal continuity, and includes homogeneous Poisson process, inhomogeneous Poisson process and Hawkes process. We focus on Hawkes process which can explain self-exciting phenomena in many real applications. In classical Hawkes process, the triggering kernel is always assumed to be an exponential decay function, which is inappropriate for some scenarios, so nonparametric methods have been used to deal with this problem, such as model independent stochastic de-clustering (MISD) algorithm. However, MISD algorithm has a strong dependence on the number of bins, which leads to underfitting for some bins and overfitting for others, so the choice of bin number is a critical step. In this paper, we innovatively embed a Gaussian process regression into the iterations of MISD to make this algorithm less sensitive to the choice of bin number.

Keywords: Hawkes process · MISD · Gaussian process
Nonparametric

1 Introduction

In a real application, data is always collected in sequential mode. How to model time series data to discover the underlying temporal dynamics is a challenging problem in this domain. To solve it, different models have been proposed in the past such as recurrent neural network (RNN) [1] and temporal point process [2]. There are many variants of the latter, such as homogeneous Poisson process [3], inhomogeneous Poisson process [4] and Hawkes process [5].

Hawkes process is a self-exciting temporal point process which can explain the self-exciting phenomenon in time series data. In real applications, the occurrence of events in the past will usually have a triggering influence on the future which leads to a clustering effect, for example, in the earthquake domain [6], the crime

© Springer International Publishing AG, part of Springer Nature 2018
D. Phung et al. (Eds.): PAKDD 2018, LNAI 10938, pp. 584–596, 2018.
https://doi.org/10.1007/978-3-319-93037-4_46

domain [7] and the social network domain [8]. In classical Hawkes process, the conditional intensity function can be expressed as:

$$\lambda(t) = \mu + \sum_{t_i < t} \gamma(t - t_i) \tag{1}$$

where $\mu > 0$ is the baseline intensity which is a constant, $\{t_i\}$ are the timestamps of observed events before time t indexed by i, and $\gamma(\cdot)$ is the triggering kernel representing the influence from t_i to t. Generally, the triggering kernel $\gamma(t - t_i)$ is always assumed to be an exponential decay function: $\alpha \cdot \exp(-\beta(t - t_i))$, which is inadequate to represent the actual influence in scenarios where it is not like that. Furthermore, in some new fields, there could be lack of prior knowledge about the form of $\gamma(t - t_i)$ or there is no analytic form to describe it [9,10]. In this case, nonparametric methods can be used to estimate the general form of the triggering kernel and the baseline intensity.

An expectation-maximization (EM) algorithm called model independent stochastic de-clustering were proposed to perform nonparametric estimation of the triggering kernel and baseline intensity [11]. Essentially MISD is a histogram density estimator, so there are problems with it: the triggering kernel obtained from MISD is a discrete function and the number of bins used in the model has a vital impact on learning results. It can be seen from the experiments in this paper that the learned triggering kernel is underfitting when fewer bins are used and overfitting when using more. How to determine the optimal number of bins? We can compute the log-likelihood conditioned on bin number M: $\log \mathcal{L}(\{t_i\}|M)$ and compute $\hat{M}$ from maximum likelihood estimation (MLE), or from an un-normalized posterior distribution by multiplying the likelihood with a prior distribution on M such as Poisson distribution[1]. But both these methods will lead to extra computation which is undesirable. Can we propose a refined MISD algorithm which does not depend on the choice of bin number severely? In this paper we innovatively embed a Gaussian process (GP) regression into the iterations of MISD to design a refined algorithm which is less sensitive to the choice of bin number; we call it GP-MISD. In this new method, M can be set to a large number to use over-segmented bins since it can prevent the learning result from overfitting to some extent.

The remainder of the paper is organized as follows: In Sect. 2, we summarize the related work in Hawkes process and its nonparametric estimations. In Sect. 3, we describe the background knowledge about Hawkes process, MISD algorithm and Gaussian process regression and propose our new algorithm GP-MISD. Synthetic data and real data experiments and the detailed discussion are provided in Sects. 4, and 5 concludes this paper.

2   Related Work

Temporal point process has been used as a continuous mathematical model to reflect temporal dynamics and to predict the arrived time of the next event in

---

[1] We assume all the bins are equally wide.

many domains such as seismology [12], financial engineering [13], and stock market [14]. Recently, the self-exciting process has become a hot topic for explaining the clustering phenomenon in social networks [15] and crime. The classical self-exciting processes, such as Hawkes process, have a limitation that the latent triggering effect is always assumed to be parametric, which introduces computational convenience but limits the expressive ability of the model. To conquer this problem, various nonparametric methods have been proposed, such as considering the triggering kernel as a linear combination of some kernels [16,17], approximating the triggering kernel by an RNN [18] and empirically estimating the triggering kernel using a histogram density estimator (MISD) where the resolution can be adapted by setting different number of bins for the histogram [19]. Although maximum penalized likelihood estimation (MPLE) has been proposed [19], which is a regularized MISD with an l_2 norm on the gradient to avoid overfitting, the gradient information can only regularize the local variance which limits the use of this method. Based on MISD, the GP-MISD algorithm we propose can produce a continuous triggering kernel function which introduces dependence on all the locations on the triggering kernel. As a result, the method is less likely to be overfitting when the bin number is chosen improperly.

3   Proposed Model

The GP-MISD algorithm is closely related to Hawkes process, MISD and Gaussian process regression, so in Sects. 3.1 and 3.2 the preliminary knowledge about these is provided. Most of the details about MISD are draw from [19]. GP-MISD is formally described in Sect. 3.3.

3.1   Hawkes Process

Temporal point process is a stochastic process, whose realization is a sequence of timestamps $\{t_i\}_{i=1}^N$ in $[0, T]$ where t_i is the occurrence time of i-th event and T is the observation time for the process. In temporal point process, an important characterization is the conditional intensity function $\lambda(t)$ which is defined as:

$$\lambda(t) = \lim_{\delta t \to 0} \frac{P(event\ occurring\ in\ [t, t + \delta t] | \mathcal{H}_t)}{\delta t} \tag{2}$$

where $\mathcal{H}_t = \{t_i | t_i < t\}$ is the history before time t. Different temporal point processes will have different conditional intensity functions to distinguish them. For example, $\lambda(t)$ is a constant for homogeneous Poisson process, a function of time $f(t)$ for inhomogeneous Poisson process, and a function of time and history for Hawkes process. The specific intensity form of Hawkes process is already given in (1). The summation of triggering kernels explains the nature of self-excitation, which is the occurrence of events in the past will intensify events occurring in the future. Given a sequence of observed data $\{t_i\}_{i=1}^n$ in time interval $[0, T]$, the log-likelihood of this list of event times can be expressed as:

$$\log \mathcal{L} = \sum_{i=1}^n \log \lambda(t_i) - \int_0^T \lambda(t)dt \qquad (3)$$

which can be used in MLE to perform inference for the parameters in the model.

3.2   MISD

Lewis and Mohler [19] provide details on how to use MISD algorithm in one dimension Hawkes, which is an EM-based nonparametric algorithm to ease MLE. Firstly, when the branching structure of a Hawkes process is observable, we can define the following matrix:

$$\mathcal{X}_{ij} = \begin{cases} 1 & \text{if event } i \text{ is caused by event } j \\ 0 & \text{otherwise} \end{cases}$$
$$\mathcal{X}_{ii} = \begin{cases} 1 & \text{if event } i \text{ is a baseline event} \\ 0 & \text{otherwise.} \end{cases} \qquad (4)$$

Let us assume baseline intensity μ is a constant and there is no prior knowledge about the form of $\gamma(\cdot)$, so given the branching matrix, the log-likelihood (3) could be decoupled into two independent parts: part μ and part $\gamma(\cdot)$,

$$\log \mathcal{L}(\{t_i\}|\mu,\gamma) = \left[\sum_{i=1}^n \mathcal{X}_{ii} \log(\mu)\right] - \mu T$$
$$+ \sum_{i=2}^n \left[\sum_{j=1}^{i-1} \mathcal{X}_{ij} \log\left(\gamma(t_i - t_j)\right)\right] - \sum_{i=1}^n \int_{t_i}^T \gamma(t - t_i)dt. \qquad (5)$$

It is straightforward to rewrite this problem into an EM framework, which is the MISD algorithm. When the branching structure is unobservable, the MISD algorithm works by maximizing the expectation of the log-likelihood. Thus $\mathcal{X}_{ij}$ is replaced by p_{ij}, which is the probability of event i caused by event j. The matrix p_{ij} is a lower triangular matrix

$$\begin{bmatrix} p_{11} & & & & \\ p_{21} & p_{22} & & & \\ p_{31} & p_{32} & p_{33} & & \\ & \vdots & & \ddots & \\ p_{n1} & p_{n2} & p_{n3} & \cdots & p_{nn} \end{bmatrix} \qquad (6)$$

where $\sum_{j=1}^i p_{ij} = 1$, because event i must be caused by previous events or the baseline event.

Then the EM iteration is:

(1) E-step: The update for the matrix P:

$$p_{ij}^s = \frac{\gamma^s(t_i - t_j)}{\mu^s + \sum_{j=1}^{i-1} \gamma^s(t_i - t_j)}$$

$$p_{ii}^s = \frac{\mu^s}{\mu^s + \sum_{j=1}^{i-1} \gamma^s(t_i - t_j)} \tag{7}$$

where s is the iteration step.

(2) M-step: The update for baseline intensity:

$$\mu^{s+1} = \frac{1}{T} \sum_{i=1}^{n} p_{ii}^s \tag{8}$$

where T is the observation duration.

Assuming the duration of $\gamma(\Delta t)$ is limited: $[0, M\delta t]$ where M is the number of bins, δt is the bin width, the update for rates is given by:

$$\gamma_m^{s+1} = \frac{1}{N_m \delta t} \sum_{i,j \in A_m} p_{ij}^s \tag{9}$$

where A_m is the set of pairs of events s.t. $m\delta t \leqslant |t_i - t_j| \leqslant (m+1)\delta t$, $\gamma_m = \gamma(m\delta t)$ where $0 \leqslant m \leqslant M-1$, and N_m is the corresponding normalizing parameter with respect to m-th bin. Equations (8) and (9) are derived from $\frac{\partial}{\partial \mu}\mathbb{E}[\log \mathcal{L}] = 0$ and $\frac{\partial}{\partial \gamma_m}\mathbb{E}[\log \mathcal{L}] = 0$.

3.3   GP-MISD

The key idea in GP-MISD is to embed a Gaussian process regression into the EM iterations, which makes use of those rates learned in each iteration step to perform a regression and get a smooth mean triggering kernel. This smooth mean triggering kernel will be used in the next iteration step, so the iteration goes on.

Gaussian process is an infinite dimensional extension of multivariate normal distribution. In GP, every finite set of points has a multivariate normal distribution, so it can be expressed as a distribution over functions in a continuous domain. GP is specified by the mean function $m(x)$ and covariance kernel $k(x, x')$:

$$f(x) \sim \mathcal{GP}(m(x), k(x, x')) \tag{10}$$

where $f(x)$ is a sample function drawn from GP. Without loss of generality, the prior mean function can be assumed to be zero: $m(x) = 0$, and the only work left is to define the covariance kernel $k(x, x')$. A widely used kernel is squared exponential kernel:

$$k(x_i, x_j) = \theta_0 \exp\left(-\frac{\theta_1}{2}\|x_i - x_j\|^2\right) \tag{11}$$

where θ_0, θ_1 are the hyperparameters.

After getting the observation points $(\gamma_1^s, \gamma_2^s, \cdots, \gamma_M^s)$ in iteration step s in MISD, the GP regression is used to evaluate the posterior mean function $m(x|(\gamma_1^s, \cdots, \gamma_M^s))$ which will be used as the $\gamma(\Delta t)$ in the next iteration step. Specifically, the new algorithm can be divided into three steps:

(1) E-step: The update for the matrix P:

$$p_{ij}^s = \frac{\bar{\gamma}^s(t_i - t_j)}{\mu^s + \sum_{j=1}^{i-1} \bar{\gamma}^s(t_i - t_j)}$$
$$p_{ii}^s = \frac{\mu^s}{\mu^s + \sum_{j=1}^{i-1} \bar{\gamma}^s(t_i - t_j)} \tag{12}$$

(2) M-step: The update for baseline intensity and rates is same as before.
(3) GP-step: The update for Gaussian process predictive distribution:

$$\bar{\gamma}^{s+1}(\Delta t) = \boldsymbol{k}^T \boldsymbol{C}_M^{-1} \boldsymbol{\gamma}^{s+1} \tag{13}$$

where $\boldsymbol{C}_M$ is the matrix of $C(\Delta t_n, \Delta t_m) = k(\Delta t_n, \Delta t_m) + \sigma_{noise}^2 \delta_{nm}$, $\{\Delta t_i\}_{i=1}^M$ are the x-values of M rate points, $k(\cdot)$ is the covariance kernel, and σ_{noise}^2 is the variance of observation points' noise, $\boldsymbol{k} = (k(\Delta t_1, \Delta t), k(\Delta t_2, \Delta t), \cdots, k(\Delta t_M, \Delta t))^T$, $\boldsymbol{\gamma}^{s+1} = (\gamma_1^{s+1}, \gamma_2^{s+1}, \cdots, \gamma_M^{s+1})^T$ are the y-values of M rate points on step $s + 1$. The final triggering kernel we obtain from this algorithm is $\bar{\gamma}(\Delta t)$. Equation (13) is derived from the standard Gaussian process regression [20].

4   Experiment

4.1   Synthetic Data

For simplicity, we assume the true triggering kernel is an exponential decay function: $\mu = 1$, $\gamma(t - t_i) = 1 \cdot \exp(-2 \cdot (t - t_i))$. Two sets of synthetic data are generated from the Hawkes process specified above using the thinning algorithm [12]. For each set, the observation duration T is set to 400, resulting in a realization of about 850 events. The first set is used as the training data, and the second one is the test data.

For the inference, it is assumed that the baseline intensity is a constant and the form of the triggering kernel is unknown, so the goal is to infer μ and $\gamma(\Delta t)$. For MISD algorithm, we apply the training data for different bin numbers ranging from 3 to 100. $\gamma(\Delta t)$ is assumed to be zero outside the interval $[0, 3]$ and the number of iterations is set to 100. In the evaluation, the training error is defined as $-\log \mathcal{L}$ of the training data. Then the model learned is applied to the test data to get the test error which is defined as $-\log\mathcal{L}$ of the test data. The same process is also applied to the GP-MISD algorithm. The hyperparameters θ_0, θ_1, σ^2_{noise} are set to 2.3, 2.3 and 0.01 in the GP step.

The training error and test error for both algorithms appear in Fig. 1. It can be seen that as the number of bins increases from 3 to 100, the training error of MISD will decrease monotonically, while the test error will increase after #bin = 8. But when we look at GP-MISD, the training error will not decrease rapidly after #bin = 8 and the test error is almost constant after #bin = 8. These results show that GP-MISD is less sensitive to the choice of bin number than MISD which is very likely to be overfitting when too many bins are used. More importantly, from test error we can see that GP-MISD is always superior to MISD no matter how many bins are used, and this can also be found from the fitting result of $\gamma(\Delta t)$ in Fig. 2 which is based on #bin = 10, 40 and 100. It is clear that the $\gamma(\Delta t)$ learned from GP-MISD is closer to the ground truth and more stable, which shows the superiority of GP-MISD.

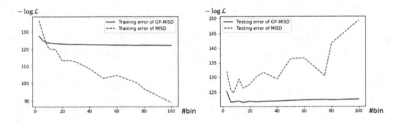

Fig. 1. The training error and test error of MISD and GP-MISD.

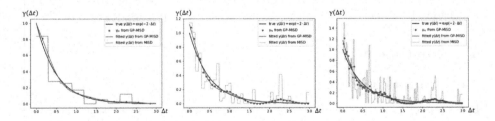

Fig. 2. The fitting result of $\gamma(\Delta t)$ from MISD and GP-MISD based on 10 bins (left), 40 bins (middle) and 100 bins (right).

4.2 Real Data

We evaluate the performance of GP-MISD and MISD on real world datasets from two different domains.

Motor Vehicle Collisions in New York City: This motor vehicle collision dataset[2] is provided by the New York City Police Department (NYPD). It contains about 1.05 million vehicle collision records in New York City from July, 2012 to September, 2017. The dataset includes the collision date, time, borough, location, contributing factor and so on. For our model, the most valuable information is the date and time. We filter out the collision records in Manhattan, Queens and Bronx caused by 'Alcohol Involvement'. For each borough, half of the records are used as the training data and the other half as the test data. Just as the synthetic data, we define the test error as $-\log \mathcal{L}$ of the test data. There are some collisions happening at the same time, as the resolution is at minute level, which violates the definition of the temporal point process. To avoid this, we add a small time interval to all the simultaneous records to separate them. The hyperparameters θ_0, θ_1, σ_{noise}^2 are set to 3.5, 3.5, 0.01 for Manhattan, 4.5, 4.5, 0.01 for Queens and 3.9, 3.9, 0.01 for Bronx. 100 iterations are performed in both algorithms. The duration of $\gamma(\Delta t)$ is set to 3.0 and the time unit is 1.16 day.

NYPD Complaint Data 2017: This dataset[3] includes all valid felony, misdemeanour and violation crimes reported to the NYPD for all complete quarters so far in 2017. It includes 228 thousand complaint records in New York City. The columns are complaint number, date, time, offense description, Borough etc. We filter out the complaints in Manhattan, Queens and Brooklyn, and the offense description is 'THEFT-FRAUD'. Again, for each borough, half the records are used as training data and the others as test data. Add a small time interval to separate all the simultaneous records. The hyperparameters θ_0, θ_1, σ_{noise}^2 are set to 6.45, 6.45, 0.01 for all boroughs. 100 iterations are performed in both algorithms. The duration of $\gamma(\Delta t)$ is set to 3.0 and the time unit is 11.6 days.

Experiment Results: For Motor Vehicle Collisions in New York City, the learned μ, $\gamma(\Delta t)$ and the test errors of both algorithms for #bin = 20, 50, 80, 100 are shown in Table 1 and Fig. 3.

Table 1. Motor Vehicle Collisions in New York City: the learned baseline intensity μ from MISD and GP-MISD based on #bin = 20, 50, 80, 100.

#bin / borough	20		50		80		100	
	μ_{MISD}	$\mu_{GP-MISD}$	μ_{MISD}	$\mu_{GP-MISD}$	μ_{MISD}	$\mu_{GP-MISD}$	μ_{MISD}	$\mu_{GP-MISD}$
Manhattan	0.408	0.384	0.391	0.393	0.375	0.399	0.363	0.398
Queens	0.496	0.462	0.488	0.477	0.465	0.482	0.448	0.481
Bronx	0.445	0.456	0.420	0.441	0.400	0.438	0.391	0.437

[2] https://data.cityofnewyork.us/Public-Safety/NYPD-Motor-Vehicle-Collisions/h9gi-nx95.

[3] https://data.cityofnewyork.us/Public-Safety/NYPD-Complaint-Data-Current-YTD/5uac-w243.

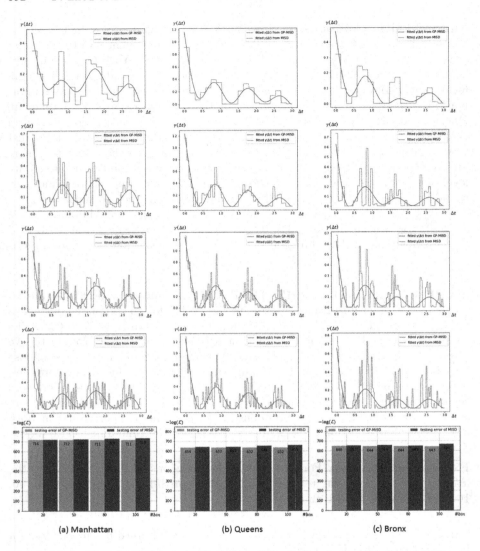

Fig. 3. Motor Vehicle Collisions in New York City: the learned $\gamma(\Delta t)$ from MISD and GP-MISD based on #bin = 20, 50, 80, 100 (upper, time unit is 1.16 day), and test errors of both algorithms for #bin = 20, 50, 80, 100 (lower).

For NYPD Complaint Data 2017, the learned μ, $\gamma(\Delta t)$ and the test errors of both algorithms for #bin = 30, 50, 75, 100 are shown in Table 2 and Fig. 4.

Table 2. NYPD Complaint Data 2017: the learned baseline intensity μ from MISD and GP-MISD based on #bin = 30, 50, 75, 100.

#bin / borough	30		50		75		100	
	μ_{MISD}	$\mu_{GP-MISD}$	μ_{MISD}	$\mu_{GP-MISD}$	μ_{MISD}	$\mu_{GP-MISD}$	μ_{MISD}	$\mu_{GP-MISD}$
Manhattan	0.084	0.102	0.084	0.102	0.077	0.103	0.075	0.102
Queens	0.039	0.041	0.039	0.041	0.038	0.041	0.038	0.041
Brooklyn	0.044	0.047	0.043	0.046	0.043	0.047	0.042	0.046

From both experimental results, we can see that $\gamma(\Delta t)$ from GP-MISD is smoother and more stable than that from MISD and the test error of GP-MISD is always lower than MISD, which is consistent with the synthetic data result: the former effectively avoids the overfitting phenomenon and makes this algorithm less sensitive to the choice of #bin. For vehicle collision, the triggering patterns in different boroughs are similar and the triggering effect lasts for about 4.5 days; for crime complaint, the triggering patterns in different boroughs are similar and the triggering effect lasts for almost one month, but significant in the first 10 days. Moreover, we can see that the trend of triggering kernel is quite dynamic, especially in the short period after the source event happened, e.g., within about 0.5 day after the initial collision in Fig. 3, or about 5 days after the initial complaint in Fig. 4. To capture the trend, the #bin must be set to be large enough so that the resolution is high, however, too large a #bin will cause overfitting, such as spikes in the triggering kernel. This is the advantage of GP-MISD to represent the triggering kernel with continuity, capturing any dynamic trends without overfitting.

Setting hyperparameters θ_0 and θ_1 is also a key step in all GP-based methods. The hyperparameters used to determine the GP kernel values implicitly encode the information on how flexible the GP could be. The optimization of hyperparameters in GP has been proved to be a non-convex problem [20], which may introduce some difficulty in learning hyperparameters. In our experiments, we use grid search to find the optimal hyperparameters and find that setting the hyperparameters in a reasonable range does not severely affect the final result.

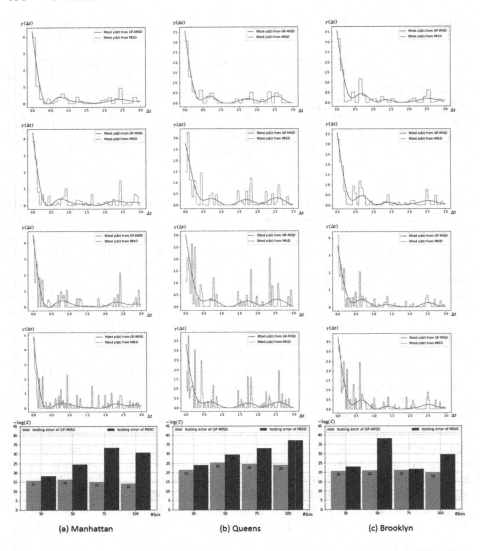

Fig. 4. NYPD Complaint Data 2017: the learned $\gamma(\Delta t)$ from MISD and GP-MISD based on #bin = 30, 50, 75, 100 (upper, time unit is 11.6 days), and test errors of both algorithms for #bin = 30, 50, 75, 100 (lower).

5   Conclusion

To conclude, in this paper we propose a refined MISD algorithm for Hawkes process: GP-MISD algorithm which can effectively avoid overfitting when more bins are used. The key thought of embedding a Gaussian process regression into the EM iterations actually can be applied to most algorithms based on bins, resulting in a smooth effect to avoid overfitting. GP-MISD inherits the advantage from MISD to predict the baseline intensity and triggering kernel without any prior knowledge of the function form of latent triggering kernel. We have performed experiments on both synthetic and real datasets demonstrating that GP-MISD is less sensitive to the choice of #bin and has consistent superiority to MISD.

References

1. Mikolov, T., Karafit, M., Burget, L., Cernock, J., Khudanpur, S.: Recurrent neural network based language model. In: Interspeech, vol. 2, p. 3 (2010)
2. Schoenberg, F.P., Brillinger, D.R., Guttorp, P.: Point processes, spatialtemporal. In: Encyclopedia of Environmetrics (2002)
3. Thompson Jr., W.A.: Homogeneous Poisson processes. In: Point Process Models with Applications to Safety and Reliability, pp. 21–31. Springer, Boston (1988). https://doi.org/10.1007/978-1-4613-1067-9_3
4. Weinberg, J., Brown, L.D., Stroud, J.R.: Bayesian forecasting of an inhomogeneous Poisson process with applications to call center data. J. Am. Stat. Assoc. **102**(480), 1185–1198 (2007)
5. Hawkes, A.G.: Spectra of some self-exciting and mutually exciting point processes. Biometrika **58**(1), 83–90 (1971)
6. Vere-Jones, D.: Stochastic models for earthquake occurrence. J. Roy. Stat. Soc. Ser. B (Methodological) **32**, 1–62 (1970)
7. Short, M.B., Mohler, G.O., Brantingham, P.J., Tita, G.E.: Gang rivalry dynamics via coupled point process networks. Discret. Contin. Dyn. Syst. Ser. B **19**(5), 1459–1477 (2014)
8. Mitchell, L., Cates, M.E.: Hawkes process as a model of social interactions: a view on video dynamics. J. Phys. Math. Theor. **43**(4), 045101 (2009)
9. Mohler, G.O., Short, M.B., Brantingham, P.J., Schoenberg, F.P., Tita, G.E.: Self-exciting point process modeling of crime. J. Am. Stat. Assoc. **106**(493), 100–108 (2011)
10. Lewis, E., Mohler, G., Brantingham, P.J., Bertozzi, A.L.: Self-exciting point process models of civilian deaths in Iraq. Secur. J. **25**(3), 244–264 (2012)
11. Marsan, D., Lengline, O.: Extending earthquakes reach through cascading. Science **319**(5866), 1076–1079 (2008)
12. Ogata, Y.: Space-time point-process models for earthquake occurrences. Ann. Inst. Stat. Math. **50**(2), 379–402 (1998)
13. Bacry, E., Jaisson, T., Muzy, J.F.: Estimation of slowly decreasing Hawkes kernels: application to high-frequency order book dynamics. Quant. Financ. **16**(8), 1179–1201 (2016)
14. Hardiman, S., Bercot, N., Bouchaud, J.P.: Critical reflexivity in financial markets: a Hawkes process analysis. Eur. Phys. J. B **86**, 442 (2013)
15. Zhou, K., Zha, H., Song, L.: Learning social infectivity in sparse low-rank networks using multi-dimensional Hawkes processes. In: Artificial Intelligence and Statistics, pp. 641–649 (2013)
16. Zhou, K., Zha, H., Song, L.: Learning triggering kernels for multi-dimensional Hawkes processes. In: Proceedings of the 30th International Conference on Machine Learning, pp. 1301–1309 (2013)
17. Du, N., Song, L., Yuan, M., Smola, A.J.: Learning networks of heterogeneous influence. In: Advances in Neural Information Processing Systems, pp. 2780–2788 (2012)

18. Du, N., Dai, H., Trivedi, R., Upadhyay, U., Gomez-Rodriguez, M., Song, L.: Recurrent marked temporal point processes: embedding event history to vector. In: Proceedings of the 22nd ACM SIGKDD International Conference on Knowledge Discovery and Data Mining, pp. 1555–1564 (2016)
19. Lewis, E., Mohler, G.: A nonparametric EM algorithm for multiscale Hawkes processes. J. Nonparametric Stat. 1(1), 1–20 (2011)
20. Bishop, C.M.: Pattern Recognition and Machine Learning. Springer, New York (2006)

Author Index

Printed in the United States
By Bookmasters